Handbook of Research on Nano-Strategies for Combatting Antimicrobial Resistance and Cancer

Muthupandian Saravanan
Mekelle University, Ethiopia & Saveetha Dental College, Saveetha Institute of Medical and Technical Sciences (SIMATS), India

Venkatraman Gopinath
University of Malaya, Malaysia

Karthik Deekonda
Monash University, Malaysia

A volume in the Advances in Medical Technologies and Clinical Practice (AMTCP) Book Series

IGI Global
PUBLISHER of TIMELY KNOWLEDGE

Published in the United States of America by
IGI Global
Medical Information Science Reference (an imprint of IGI Global)
701 E. Chocolate Avenue
Hershey PA, USA 17033
Tel: 717-533-8845
Fax: 717-533-8661
E-mail: cust@igi-global.com
Web site: http://www.igi-global.com

Library of Congress Cataloging-in-Publication Data

Names: Saravanan, Muthupandian, 1977- editor. | Gopinath, Venkatraman, 1987- editor. | Deekonda, Karthik, 1987- editor.
Title: Handbook of research on nano-strategies for combatting antimicrobial resistance and cancer / Muthupandian Saravanan, Venkatraman Gopinath, Karthik Deekonda, editors.

Description: Hershey, PA : Medical Information Science Reference, [2020] | Includes bibliographical references and index. | Summary: "This book examines (1) how to overcome the existing, traditional approaches to combat antimicrobial resistance and cancer; (2) how to apply multiple mechanisms to target the cancer cells and microbes; and (3) how the nanomaterials can be used as carriers"-- Provided by publisher.
Identifiers: LCCN 2020002056 (print) | LCCN 2020002057 (ebook) | ISBN 9781799850496 (hardcover) | ISBN 9781799850502 (ebook)
Subjects: MESH: Nanostructures--therapeutic use | Nanomedicine--methods | Drug Resistance, Microbial--drug effects | Neoplasms--drug therapy
Classification: LCC QR46 (print) | LCC QR46 (ebook) | NLM QT 36.5 | DDC 616.9/041--dc23
LC record available at https://lccn.loc.gov/2020002056
LC ebook record available at https://lccn.loc.gov/2020002057

This book is published in the IGI Global book series Advances in Medical Technologies and Clinical Practice (AMTCP) (ISSN: 2327-9354; eISSN: 2327-9370)

British Cataloguing in Publication Data
A Cataloguing in Publication record for this book is available from the British Library.

All work contributed to this book is new, previously-unpublished material. The views expressed in this book are those of the authors, but not necessarily of the publisher.

For electronic access to this publication, please contact: eresources@igi-global.com.

Advances in Medical Technologies and Clinical Practice (AMTCP) Book Series

Srikanta Patnaik
SOA University, India
Priti Das
S.C.B. Medical College, India

ISSN:2327-9354
EISSN:2327-9370

Mission

Medical technological innovation continues to provide avenues of research for faster and safer diagnosis and treatments for patients. Practitioners must stay up to date with these latest advancements to provide the best care for nursing and clinical practices.

The **Advances in Medical Technologies and Clinical Practice (AMTCP) Book Series** brings together the most recent research on the latest technology used in areas of nursing informatics, clinical technology, biomedicine, diagnostic technologies, and more. Researchers, students, and practitioners in this field will benefit from this fundamental coverage on the use of technology in clinical practices.

Coverage

- Patient-Centered Care
- E-Health
- Telemedicine
- Nutrition
- Medical Imaging
- Diagnostic Technologies
- Biomedical Applications
- Medical Informatics
- Clinical Data Mining
- Neural Engineering

IGI Global is currently accepting manuscripts for publication within this series. To submit a proposal for a volume in this series, please contact our Acquisition Editors at Acquisitions@igi-global.com or visit: http://www.igi-global.com/publish/.

The Advances in Medical Technologies and Clinical Practice (AMTCP) Book Series (ISSN 2327-9354) is published by IGI Global, 701 E. Chocolate Avenue, Hershey, PA 17033-1240, USA, www.igi-global.com. This series is composed of titles available for purchase individually; each title is edited to be contextually exclusive from any other title within the series. For pricing and ordering information please visit http://www.igi-global.com/book-series/advances-medical-technologies-clinical-practice/73682. Postmaster: Send all address changes to above address.

Titles in this Series

For a list of additional titles in this series, please visit: www.igi-global.com/book-series

Handbook of Research on Solving Modern Healthcare Challenges With Gamification
Ricardo Alexandre Peixoto de Queirós (uniMAD, Escola Superior de Media Artes e Design, Portugal) and António José Marques (LabRP, School of Health, Polytechnic of Porto, Portugal)
Medical Information Science Reference • © 2021 • 382pp • H/C (ISBN: 9781799874720) • US $425.00

Diagnostic Applications of Health Intelligence and Surveillance Systems
Divakar Yadav (National Institute of Technology, Hamirpur, India) Abhay Bansal (Amity University, India) Madhulika Bhatia (Amity University, India) Madhurima Hooda (Amity University, India) and Jorge Morato (Universidad Carlos III de Madrid, Spain)
Medical Information Science Reference • © 2021 • 332pp • H/C (ISBN: 9781799865278) • US $285.00

AI Innovation in Medical Imaging Diagnostics
Kalaivani Anbarasan (Department of Computer Science and Engineering, Saveetha School of Engineering, India & Saveetha Institute of Medical and Technical Sciences, Chennai, India)
Medical Information Science Reference • © 2021 • 248pp • H/C (ISBN: 9781799830924) • US $345.00

Optimizing Health Monitoring Systems With Wireless Technology
Nilmini Wickramasinghe (Swinburne University of Technology, Australia & Epworth HealthCare, Australia)
Medical Information Science Reference • © 2021 • 338pp • H/C (ISBN: 9781522560678) • US $265.00

Deep Learning Applications in Medical Imaging
Sanjay Saxena (International Institute of Information Technology, India) and Sudip Paul (North-Eastern Hill University, India)
Medical Information Science Reference • © 2021 • 274pp • H/C (ISBN: 9781799850717) • US $245.00

Design and Quality Considerations for Developing Mobile Apps for Medication Management Emerging Research and Opportunities
Kevin Yap (La Trobe University, Australia) Eskinder Eshetu Ali (Addis Ababa University, Ethiopia) and Lita Chew (National University of Singapore, Singapore)
Medical Information Science Reference • © 2021 • 256pp • H/C (ISBN: 9781799838326) • US $225.00

Expert Approaches to Health IT Tools in Clinical Practice
Ramgopal Kashyap (Amity University, Raipur, India)
Medical Information Science Reference • © 2021 • 300pp • H/C (ISBN: 9781799840510) • US $245.00

IGI Global
PUBLISHER of TIMELY KNOWLEDGE

701 East Chocolate Avenue, Hershey, PA 17033, USA
Tel: 717-533-8845 x100 • Fax: 717-533-8661
E-Mail: cust@igi-global.com • www.igi-global.com

List of Contributors

Table of Contents

Detailed Table of Contents

prime focus in the generation of tuneable nanostructures with versatile properties. This chapter highlights the importance, mode of synthesis, applications, and limitations of biopolymer nanomaterials from natural origin.

In recent years, the irrational use of antibiotics has escalated the evolution of multidrug-resistant (MDR) bacterial strains. The infectious diseases caused by these MDR bacterial strains remain a major threat to human health and have emerged as the leading cause of morbidity and mortality. The WHO and CDC have expressed serious concern regarding the continued increase in the development of multidrug resistance among bacteria. The antimicrobial resistance (AMR) poses a severe global threat of growing concern to human health and economic burden. Bacteria have developed the ability to resist antimicrobials by altering target site/enzyme, inactivation of the enzyme, decreasing cell permeability, increasing efflux due to over-expression of efflux pumps, target protection, target overproduction, and many other ways. The shortage of new antimicrobials and rapid rise in antibiotic resistance demands pressing need to develop alternate antibacterial agents.

The worldwide misuse of antimicrobials and subsequent rise of the multi-drug resistant (MDR) opportunistic intracellular pathogens have led to a paradigm shift in the established perspective of antimicrobials and bacterial-human relations. Antibiotic resistant strains of mycobacterium tuberculosis have threatened the progress in management and control of worldwide epidemic of tuberculosis. Mycobacterium tuberculosis intrinsically resists many antimicrobials, limiting the number of compounds available for treatment. Typically, conventional treatments for these diseases consist of long-term therapy with a combination of bioactive(s) that can cause side effects and contribute to low patient compliance. The escalation of antimicrobial resistance poses compelling demand for new ways of treating bacterial infections. The main objective of this chapter is to review the potential applications of nanotechnology-based drug delivery systems in combating drug resistance associated with opportunistic intracellular pathogens with special reference to tuberculosis.

The antimicrobial resistance has significantly risen in last few decades as a result of too much usage of antibiotics. In the case of failure of drug therapy, the multidrug resistance has become a serious issue in a clinical and food sectors. The rapid emergence of antibiotic resistance has become a great concern for public health worldwide. Therefore, the novel approaches of nano-biomaterials (NBM) and bacteriophage (phage) have received a new attention as an alternative treatment for antibiotic-resistant bacterial infections. Moreover, the advanced molecular biology and genetic engineering have been successfully applied for the rapid detection and control of multidrug resistant pathogenic strains. This chapter specifically summarizes the prevalence of common as well as under-researched antimicrobial resistance of bacterial pathogens and the application of nano-biomaterials and bacteriophage to fight against antimicrobial resistance. Ultimately, the synergistic approach will be a highly promising prospect to threat of antimicrobial resistance.

Antibiotic therapy for life-threatening bacterial infections has greatly enhanced the lifespan and health of human beings. But continued increase in the improper usage of antibiotics and emergence of multidrug resistant bacteria (MDR) remains a major reason for the morbidity and mortality worldwide. Delay in the detection of pathogens, poor penetration power of the antimicrobial agent, and susceptibility for mutational resistance also resulted in clinical failure of conventional antibiotics. Innovation of nanoparticles (NPs) serves as a promising strategy to overcome MDR bacteria. The site-specific action of NPs, high penetration power, minimal dosage, and multidimensional mechanism makes NPs highly efficient to tackle MDR bacteria than conventional drugs. This chapter presents a brief overview on various types of NPs, strategies to combat drug resistance, mode of action of NPs as antibacterial agents, applicability in various medical fields, clinical manifestation, challenges in clinical translation, and future prospects.

The technological advancements have not only made humans more civilized but have also caused the micro-organisms to develop several survival strategies via antimicrobial resistance to keep pace. Such

Nanotechnology has explored tremendous development during the last decade because of its specific properties. Nanotechnology offers generous prospects in the improvement of agriculture, water treatment, and food industry. In this study, modern extraction techniques of agricultural and plant samples with advanced characterization methods were discussed. Additionally, various factors affecting during synthesis of NPs are also briefly described. The assessments of NPs in these samples are very complex and various techniques are to be used to get essential data. The outcomes estimated by these various techniques and methodologies are not constantly identical because of different samples different standardization methodology. A new challenge emerges when testing samples with low concentration. For this situation, expository techniques with high affectability are wanted to gauge low convergences of NPs. A perfect analytical technique should be able to detect plant-NP association, for example, structure, morphology, natural speciation, size, mass concentration, etc.

Nanomotors are the new generation drug delivery devices that find their major application in the pharmaceutical sector. This chapter highlights the basic aspects, mechanisms, and applications of nanomotors in antibacterial therapy, cancer therapy, nano-surgery, and imaging. Because of their higher penetration ability, rapid transportation, and controlled mobility, micro and nanomotors are referred as new generation targeted drug delivery devices. The detecting and sensing ability of nanomotors find potential applications in diagnostics and therapeutics. Bio-hybrid nanomotors are most attracting candidates for effective drug delivery. Bio-functionalized nanomotor can be used for detection and identification of tumor cells. Thus, nanomotors and micromotors can pave way for development of diagnostic and therapeutic tools for the future. Further development in these areas facilitate the discovery of lab-on-chip devices that can be used for super-fast screening and clinical diagnostic applications.

Cancer is one of the prime rationales for mortality in humanity and remains a difficult disease to treat. Contemporary problems allied with conventional cancer chemotherapies embrace the insolubility of drugs in an aqueous medium, delivery of sub-therapeutic doses to target cells, lack of bioavailability, and most importantly, non-specific toxicity to normal tissues. Recent advances in nanotechnology investigation tackle potential solutions to these riddles. However, there are challenges regarding targeting specific sites, tracking the delivery system and control over the release of the drug to the target site. The nanodevices are 100 to 1000 times smaller than cells in humans; their size is comparable to the enzymes, the receptors. This enables them to have a large surface area and ability to interact with biomolecules on both the surface and inside cells. Nanomedicines between 8-100 nm have an enhanced permeability and retention (EPR) effect, which make these medicines to target passively the solid tumours.

Cancer remains the important cause for the mortality worldwide. In recent days, different biomedical strategies are in the limelight for the control of mortality due to cancer. As a promising strategy in cancer treatment, nano-formulations with advantages in drug delivery, health, and pharmacy are booming. The conventional nanocarrier continues, however, to suffer from low drug load efficiency, possible toxicity, unknown metabolism, and other uncertainties. To overcome these problems, carrier-free nanodrugs with desirable bioactivity were developed quickly and attracted considerable attention. In the meantime, the nanoarchitecture developed by a simple "natural" method with a multifunctional self-life has major advantages in multi-drug resistant synergistic cancer therapies and inhibition. Until now, the carrier-free nanoparticles for tumour therapy, phototherapy, chemotherapy, diagnostics, and synergistic therapy have made significant progress. In this review, the authors provide an integrated and detailed review of recent literature on nanodrug delivery systems composed of several active agents.

Immunotherapy has become a preferable candidate for many diseases in recent days. The infusion or administration of immune complexes or components to elicit the own immune response against a particular disease by attracting the antigen presenting cells against the disease causing organism and eliciting the T-cell mediated killing and further activating cell mediated immunity based on the processed surface antigens underlies the basic concept behind the immunotherapy. Immunotherapy can be applied for all course of diseases even in the treatment of cancer. The limitation in using immunotherapy is that it needs a proper delivery vehicle to reach the diseased spot to shows its pharmacokinetic property. In case of cancer, the immune components administered itself are not able to pertain and penetrate the solid tumor mass. Nanoparticles are small-sized particles which are generally specific in action used in the field of medicine. Nanoparticles aid in targeted drug delivery to the specific spots and immune targeting of nanoparticles is due to its enhanced permeability and retention (EPR).

Due to drug resistance, lack of cancer cell selectivity, and solubility, conventional cancer treatments lose their therapeutic uses, and as such, new therapeutic agents need to be developed. Nanomaterials and peptides are increasingly being used in the fields of cancer diagnosis, biomarker discovery due to their therapeutic values and novel way of targeting and curing the disease. Synergism among the peptide-conjugated nanoparticles is an exhilarating group of materials, not only sharing the benefits of conventional nanomedicine, but also possessing the unique properties of excellent biocompatibility, biodegradability, versatile sensitivity, specific biological purpose, and synthetic feasibility. These virtues inspired by the scientists and have taken advantage in the peptide-conjugated nano drugs for the accurate delivery of drugs reliably to the site of the lesion. This chapter offers a summary of emerging technologies that have recently been developed in the broad field of peptide-conjugated nanoparticles and offers guidance for targeted drug delivery and cancer therapy.

Metals such as silver, gold, and copper were used in ancient times for their medicinal properties. When these metals are converted to nanoparticles, they show unique and advanced physicochemical and biological properties due to their enhanced surface to volume ratio. Hence, these properties are utilized by researchers to develop highly specific diagnostic tools as well as a therapeutic agent against cancer. Cancer is a complex disease-causing desolation and death. Early detection and treatment is the only way to evade mortality. This chapter focuses on metal nanoparticles used as a theranostic agent against cancer. It summarizes the synthesis methodology along with their advantages, drawbacks and characterizations. Their recent application in diagnosing and treating cancer has also been highlighted.

The magnetic nanoparticles are said to be a class of nanoparticles or nanomaterials that can be manipulated by the help of externally applied magnetic field. These magnetic nanoparticles constitute materials such as nickel, cobalt, iron, and their derivatives. These are normally smaller than 1 μm in diameter possess wide range of properties and attractive characteristics suitable for biomedical such as used as hyperthermia, enhancing magnetic resonance imaging (MRI) data, supplementing tissue engineering efforts, and improving the target-based drug delivery and many other technological applications. In the field of cancer research, the role of nanoparticles and nanotechnology-based methods and novel strategies have been increasing swiftly for cancer identification and cancer therapy. The iron oxide (Fe3O4, γ-Fe2O3) nanoparticles (NPs) are widely used for the drug delivery, magnetic nanoparticle-enhanced hyperthermia, and also as MRI contrast agents due to its biocompatibility, low toxicity, etc. lead to the growth of novel biopharmaceutical technologies.

Nanoparticles are also proving a great alternative to antibiotics in combating the deadly bacterial infections. Copper nanoparticles, in particular, are being utilized as antimicrobial agents as well as their interaction with other particles helps to improve the antimicrobial efficacy. Compared to other metallic nanoparticles, copper nanoparticles are highly reactive, and their small size allows them to be used in different areas. When exposed to air, copper nanoparticles form copper oxide, and to prevent this oxidation, silica and carbon are utilized to coat copper nanoparticles. Metallic copper nanoparticles have been used in place of silver nanoparticles and other noble metals as anti-infective agents and also utilized in in water purification.

Preface

Multidrug-resistant bacteria play a significant role in public health by destroying the potency of existing antibiotics. Meanwhile, cancer remains one of the most common health problems that impact society, resulting in many deaths worldwide. Novel strategies are required to combat antimicrobial resistance and create efficient anticancer drugs that could revolutionize treatment.

Nano-Biotechnology is a promising field of science that encompasses biology and nanotechnology to benefit the humankind. The current edited book deals with synthesis methods, and various applications of nanomaterials in the field of antimicrobial and anticancer. Featuring a range of topics such as bacteriophage, nanomedicine, and oncology, this book is ideal for molecular biologists, microbiologists, nanotechnologists, academicians, chemists, pharmacists, oncologists, researchers, healthcare professionals, and students.

INSIDE THIS BOOK

The emergence of antimicrobial resistance (AMR) is the major threat to the public, In brief, AMR is the ability of a microorganisms to resist the effects of antibiotics or the other existing antimicrobials that were applied to treat the microbial infections. The term antibiotic resistance (AR) is a kind of AMR, as it applies only to bacteria that become resistant to antibiotics. In recent years the resistant microbial infections are more difficult to treat, thus requiring alternative medications over the current antibiotics and other antimicrobials. Nanomaterial-based therapeutical approach are promising tools to combat bacterial infections that are challenging to treat, advancing the capability to elude existing mechanisms correlated with drug resistance mechanisms. Lastly, we discuss about the current progress and futuristic challenges for the extensive clinical utilization of nanomaterials as antimicrobial and anticancer therapeutics. The above-mentioned reason, this edited book is very exciting to the researchers who are involving in the antimicrobial and anticancer therapeutics.

IMPORTANCE OF EACH CHAPTER

Chapter 1: Antimicrobial Resistance and Antimicrobial Nanomaterials – An Overview

The emergence of antibacterial, antifungal, antiviral resistance and the treatment complications in modern medicine. The growing rise of antimicrobial resistance inflicts a remarkable economic and social burden on the health care system globally. The replacement of conventional antimicrobials by new technology to counteract and lessen antimicrobial resistance is currently ongoing. Nanotechnology is an advanced approach to overcome challenges of such resisted conventional drug delivery systems mainly based on the development and fabrication of nanoparticulate structures. Numerous forms of nanoparticulate systems have been discovered and tried as prospective drug delivery systems, comprising organic and inorganic nanoparticles. The chapter also covers the area of advancements in nanomedicine to combat antimicrobial resistance and nanoparticles microbial interaction mechanism. Despite, nanoparticles pose few drawbacks, they provide hope for the development of effective antimicrobial agents for the future were extensively studied.

Chapter 2: Biopolymer-Based Nanomaterials for Biomedical Applications

Biopolymers derived from natural materials are widely utilized for biomedical application for its biocompatibility, biodegradability, lesser toxicity, low cost and abundant nature compared to the synthetic counterparts. In recent years, the emergence of nanoscience has evidenced notable preference in the development of safe and potent nano-biomaterials. Among the various methods reported for the fabrication of nanomaterials, electrospinning has received prime focus in the generation of tunable nanostructures with versatile properties. This chapter highlights the importance, mode of synthesis, applications and limitations of biopolymer nanomaterials from natural origin.

Chapter 3: Addressing Antimicrobial Resistance Through Nano-Antibiotics – Challenges and Novel Strategies

The current work significantly analyzes the recent advances on nano-antibiotics systems developed to enhance therapeutic efficiency of antimicrobial resistance. In recent years, the irrational use of antibiotics has escalated the evolution of multidrug-resistant (MDR) bacterial strains. The Infectious diseases caused by these MDR bacterial strains remain a major threat to human health and have emerged as the leading cause of morbidity and mortality. The WHO and CDC have expressed serious concern regarding the continued increase in the development of multidrug resistance among bacteria. The antimicrobial resistance (AMR) poses a severe global threat of growing concern to human health and economic burden. Bacteria have developed the ability to resist antimicrobials by altering target site/enzyme, inactivation of the enzyme, decreasing cell permeability, increasing efflux due to over-expression of efflux pumps, target protection, target overproduction and many other ways. The shortage of new antimicrobials and rapid rise in antibiotic resistance demands pressing need to develop alternate antibacterial agents. Over the past decade, the global use of nanomedicines as innovative tools to combat the high rates of antimicrobial resistance has seen a substantial increase. There has also been a paradigm shift in the management of biofilms and MDR bacteria with polymeric nanocomposites and antibiotic-loaded polymeric

NPs. Improved therapeutic efficacy with a concomitant decline in side effects of antimicrobial drugs has also been achieved by surface modification of metallic NPs with ligands or antibodies for targeted delivery. This chapter will summarize antibiotic resistance strategies employed by bacteria and novel nonmaterial based antimicrobials under development which could be used to treat infections caused by multidrug-resistant bacteria. Further, the translation of nanocomposites, nanomaterials, surface-modified and antibiotic - functionalized NPs to clinics as next-generation nanotherapeutics is also highlighted.

Chapter 4: Nanomedicine and Its Applications in Combating Resistance Relating to Opportunistic Pathogens With Special Reference to Tuberculosis

The misuse of antibiotics leads to the subsequent rise of the multi-drug resistant (MDR) pathogens have led to a paradigm shift in the established perspective of antimicrobials and bacterial-human relations. These MDR bacterial pathogens cause diseases such as tuberculosis, hepatitis, and HIV/AIDS and are collectively responsible for global menace. Nanotechnology offers an innovative platform to overcome these challenges, with their potential to effectively treat and manage even infections involving antibiotic resistance. Specifically, nanomaterials may provide additional advantages owing to their size in the nanometer scale, potential to overcome physiological barriers, and their ability to access the intended targets. Additionally, the high surface-area-to-volume ratio may facilitate the interaction between pathogenic membranes and cell walls. The main objective of this chapter is to review the potential applications of nanotechnology-based drug delivery systems in combating drug resistance associated with opportunistic intracellular pathogens with special reference to tuberculosis.

Chapter 5: Novel Synergistic Approach of Nano-Biomaterials and Bacteriophage for Combating Antimicrobial Resistance

The antimicrobial resistance has significantly risen last few decades as a result of too much usage of antibiotics. In the case of failure of drug therapy, the multidrug resistance has become a serious issue in a clinical and food sectors. The rapid emergence of antibiotic resistance has become a great concern for public health worldwide. Therefore, the novel approaches of nano-biomaterials and bacteriophage have received a new attention as an alternative treatment for antibiotic-resistant bacterial infections. Moreover, the advanced molecular biology and genetic engineering have been successfully applied for the rapid detection and control of multidrug resistant pathogenic strains. This chapter specifically summarize the prevalence of common as well as under-researched antimicrobial resistance of bacterial pathogens and the application of nano-biomaterials and bacteriophage to fight against antimicrobial resistance. Ultimately, the synergistic approach will be a highly promising prospect to threat of antimicrobial resistance.

Chapter 6: Medicine at Nanoscale – A New Paradigm for Emerging Trend of Multi-Drug Resistance in Bacteria

Nanoparticles have already proven its promising role in various medical fields and treatment of diseases. However, in depth knowledge on antimicrobial and anticancer mechanism is still in developmental stage. Even though nano-antibiotics exhibited several advantages over conventional antibiotics in treating multiple drug resistant pathogens, more interdisciplinary understanding from various disciplines such as microbiology, pathology, immunology, biomaterials, pharmacology, toxicology, and nanotechnology are needed. Metallic NP provides a promising approach when utilized as such or in combination with other antimicrobial agents to combat bacterial drug resistance. In addition, targeted or site-specific action of NPs displays its potential in controlling and mitigating various MDR strains. But a systematic investigation becomes imperative before the clinical translation of NPs against MDR organisms. Pharmacology aspects of nano-antibiotics should be addressed as the dosage level to be administered is essential for clinical translation of NPs.

Chapter 7: Encountering the Survival Strategies Using Various Nano Assemblages – A Mechanistic Approach

The technological advancements have not only made humans more civilized but also the micro-organisms to develop several survival strategies via antimicrobial resistance to keep in pace. Such highly developed microbial systems have been classified as superbugs, exhibiting Trojan-horse mechanism. This uncertain behavior in microbes has challenged humans to scour around novel moiety to shield themselves from the detrimental effects. One such natural phenomenon that has drawn the attention of researchers is the metal-microbe interaction where microbes were found to be controlled during their interaction with metals. Fine tuning could bestow them with enhanced physico-chemical properties capable of controlling life threatening micro-organisms. Nano forms of metals (nanoparticles, quantum dots, polymeric nanostructures) exhibiting medicinal properties have been implied toward biomedical theranostics. This chapter highlights on the mechanistic antimicrobial resistance and the containment strategy using various nano assemblage highlighting its fabrication and bio-molecular interaction.

Chapter 8: Nanotechnology-Based Emerging Approaches to Combat Malaria and Dengue Fever

Malaria and dengue fever are devastating diseases causing serious illnesses in Asia. Nanotechnology has potential role to combat both malaria and dengue fever by controlling larvae of mosquitoes or to stop the breeding of the mosquitoes. Nanomaterials have shown potential as larvicides especially food-grade nanoparticles which are economical, ecofriendly and no toxic to human health. In this chapter, types, sources and composition of NPs, how they control the vectors, data on nanomaterials used for the control of vectors, toxicity and ecotoxicity of NPs and comparison of NPs with antiparasitic drugs have been discussed in detail.

Chapter 9: Current Strategies in Peptide Conjugated Nanoparticles – A Novel Approach in the Field of Biomedicine

In the arena of biomedicine, a Nano-peptide conjugates (NPCs) has been newly arising as a multifunctional agent. The synergistic effect of the two groups directly heightened the various biological properties. There are thousands of NPCs has been developed for both the narrow and broad bioactivities. In this review, we highlighted the novel technologies like single and multi-walled carbon nanotubes conjugates, trivalent and multivalent conjugates and their importance. This chapter provides a clear picture about the emerging technologies in the arena of NPCs and also bids a standard path for budding researchers, particularly for the beginners.

Chapter 10: Modern Nanomaterials Extraction and Characterization Techniques in Plant Samples and Their Biomedical Potential

In this chapter, modern extraction techniques along with advanced characterization procedure of agricultural and plant samples were highlighted. Additionally, various factors inter linking green synthesis of nanoparticles were also briefly described.

Chapter 11: Role of Micro and Nano Motors in Therapeutics and Diagnostics

Nanomotors emerge as a novel new generation drug delivery device with versatile applications. In the medical field, non-invasive and minimally invasive surgery techniques are very arduous and is highly preferred. The recent focus is on nanojets and microdrillers which can be used to perform minimally invasive and precise surgery. These techniques are still under development as the materials currently used for the propulsion of the motors are toxic to biological system. Magnetically controlled nanojets are now under research to develop a fuel free motor that is also non-toxic. Nanomotors find their applications in environmental remediation also. Studying immune response of these devices are also the promising area of research. The robust design, suitable material selection and bio-compatible nature outweigh the drawback of using nanomotors in physiological environment, hence these areas will drive the future of the nanomotor research.

Chapter 12: Overview of Nano-Strategies for Combating Cancer – Introduction, Nanostructure-Mediated Drug Delivery, Therapeutics, Diagnosis, and Imaging

Cancer is one of the prime rationales for mortality in humanity and remains complicated disease to treat. Applications of Nanotechnology in cancer diagnosis and treatment are indeed a fact that contains a wide variety of new technologies and opportunities, from extremely early diagnosis by imaging and treatment by the targeted area. Nanodevices could prevent or regulate cancer from recurrence be-sides destroying any cancer cells. Diagnosis of cancer by Circulating tumor cells (CTCs) is non-invasive, predictable, and also provide novel insights about the disease pathogenesis of lethal neoplasms. Cancer diagnostic systems embedded in nanomaterials, for example, NanoVelcro CTC Assays are unique techniques which work by cell sorting tactic, assessing CTCs in peripheral blood and providing an opportunity for noninvasive surveillance of progression of various cancers like solid cancer, prostate cancer, pancreatic cancer, lung cancer, and melanoma.

Chapter 13: Emerging Nano-Based Drug Delivery Approach for Cancer Therapeutics

Cancer is one of the foremost causes of global mortality. In recent days different biomedical strategies are in lime light to control the cancer mortality, however the outcome is therapeutically not convincing. The existing traditional anticancer drug has demonstrated undesired side effects more than therapeutic benefits. In recent decades, nano-based materials were extensively applied to treat cancer and the results are encouraging. The current survey discussed the advantages and clinical concern facing the cancer diagnostic therapy by applying nano-drug delivery system. The conventional nanocarrier continues, however, to suffer from low drug load efficiency, possible toxicity, unknown metabolism and other uncertainties. To overcome these problems, carrier-free nanodrugs with desirable bioactivity were developed quickly and attracted considerable attention. In the meantime, the nanoarchitecture developed by a simple "natural" method with a multifunctional self-life has major advantages in multi-drug resistant synergistic cancer therapies and inhibition. Until now, the carrier-free nanoparticles for tumour therapy, phototherapy, chemotherapy, diagnostics and synergistic therapy have made significant progress. In this review, we provide an integrated and detailed review of recent literature on nanodrug delivery systems composed of several active agents. We summarize briefly the design and synthesis, the biological properties and its broad variety of cancer therapy applications.

Chapter 14: Immune-Targeted Nanomedicine

Immunotherapy becomes a preferable candidate for the treatment of cancer therapy. However, it needs a proper delivery vehicle to reach the targeting area to show their pharmacokinetic property. In case of cancer, the immune components administered itself cannot able to pertain and penetrate the solid tumor mass. Thus the clinical appliances of nanoparticles aid in targeted drug delivery to the specific spots and immune targeting of nanoparticles is highly emerging due to its Enhanced Permeability and Retention (EPR). The specificity of the nanoparticles are based on the porosity, cell membrane permeability and its surface markers. The chapter focuses on the implications of nano-based biopolymeric targeted drug delivery system. The study also emphasis on actively modulating host anticancer immunity using nanomaterials provides new avenues for developing novel therapeutics.

Chapter 15: Current Therapies and Future Prospects of Peptide-Conjugated Nanoparticle for Targeted Drug Delivery to Cancer

Due to drug resistance, loss of specificity and solubility of cancer cells, traditional cancer drugs are losing their therapeutic applications, and new therapeutic agents need to be created. Fabrication of nanomaterials with bioactive peptides are widely used in the area of cancer research, biomarker exploration due to their medicinal utility and innovative approach to disease management. The combination of peptide-conjugated nanoparticles is an exhilarating community of materials, not only do they share the benefits of traditional nanotechnology, but they also have the special properties of outstanding biocompatibility, biodegradability, flexible resilience, precise biological activities and artificial feasibility. Such principles, inspired by scientists and used in peptide-conjugated nano drugs for the precise delivery of medications efficiently to the site of the lesion. This chapter offers a detailed review of emerging technologies that have recently been introduced in the broad area of peptide-conjugated nanoparticles and provides guidelines for selective drug delivery and cancer treatment.

Chapter 16: Smart Theranostic Applications of Metal Nanoparticles Against Cancer

This chapter described various procedures that were employed for the synthesis and characterization of metal nanoparticles. Further, the biomedical importance of silver, gold and copper nanoparticles in the field of cancer theranostics have also been highlighted. These nanoparticles overcome the limitation of the existing technique thereby providing a step ahead in cancer therapy.

Chapter 17: Magnetic Nanoparticles – Role in Next Generation Nanomedicine

The magnetic nanoparticles can be manipulated by the help of externally applied magnetic field. Due to their unique biophysical properties and attractive characteristics magnetic nanoparticles are highly suitable for biomedical such as used as hyperthermia, enhancing magnetic resonance imaging (MRI) data, supplementing tissue engineering efforts, and improving the target-based drug delivery and many other technological applications. In the field of cancer research, the role of nanoparticles and nanotechnology-based methods and novel strategies have been increasing swiftly for cancer identification and cancer therapy. The iron oxide (Fe_3O_4, γ-Fe_2O_3) nanoparticles (NPs) are widely used for the drug delivery, magnetic nanoparticle-enhanced hyperthermia and also as MRI contrast agents due to its biocompatibility, low toxicity that leads to the growth of novel biopharmaceutical technologies. The present study elaborates the clinical significance of magnetic nanoparticles in the field of nanomedicine.

Chapter 18: Copper Nanoparticles – Easily Accessible Nano-Weapons Against the Escalating Antibiotic Resistance

Metal nanoparticles have shown promising effects that mimics for therapeutic antibiotics and other antimicrobial materials; however, the microbial interaction of metal states and related antibacterial mechanisms are mostly unknown. Here, the bacterial drug resistance, pathogenicity and the multifunctional behavior of copper nanoparticles were extensively studied including drug delivery, antibacterial, biosensors, catalyst and its pharmaceutics potential. Our report provides a deeper understanding of copper nanoparticles antibacterial behaviors and emphasize a new approach for fabricating selective antibacterial treatments based on copper nanoparticles.

Chapter 19: Toxicity and Risk Assessment of Nanomaterials

The development and advancement of nanotechnology has led to widespread use of nanomaterials (NMs) in multidisciplinary sectors, including the environment, agriculture, biomedicine, industry and human health. Nanomaterials may interfere with biological systems due to their smaller size and increased surface area ratios in terms of volume which may raise unanticipated toxicological apprehensions. The ultrasmall nature of the particles may cross biological margins and induce toxicity in vital organs like the human brain and placenta. Moreover, environmental exposures to NMs are inevitable, which have become crucial parts of our daily routine lives and consequently, search for the nanotoxicity is gaining attention to lessen or get rid of the toxic impacts of the NMs. The safety concerns and risk assessment of NMs have been raised significantly in research community circles, industries and regulatory bodies. The information in the concerned section, obtained from current literature and toxicological effects of NMs research are assorted. In this chapter, we focused on the hazardous impact of NMs on human health, environment, and soil due to the exposure of various sorts of NMs and their risk assessment with control measures. In summary, the information discussed in this chapter will provide the requisite foundation for the regulating agencies/society, and other organizations to assess the toxic effects of these NMs.

TARGET AUDIENCE

This book chapter we have focused to address the major ongoing challenges of antimicrobial and anti-cancer therapeutics by applying nano-materials based therapeutic approach. The chapter also discussed with the conclusive remarks in which research challenges and future trends are discussed. Moreover, the topic is very useful to the academician, researchers and graduate students in the field of applied and Medical science, nanotechnology and nanomedicine.

CONCLUSION

Nanomedicine is one such innovation that plays a significant role in developing alternative and more effective treatment strategies for antimicrobial resistance and cancer theranostics. Nano-Strategies for Combatting Antimicrobial Resistance and Cancer is an essential scholarly resource that examine:

1. How to overcome the existing, traditional approaches to combat antimicrobial resistance and cancer.
2. How to apply multiple mechanisms to target the cancer cells and microbes.
3. How the nanomaterials can be used as carriers for delivering the drug into target site.

Chapter 1
Antimicrobial Resistance and Antimicrobial Nanomaterials:
An Overview

Muthupandian Saravanan
https://orcid.org/0000-0002-1480-3555
Mekelle University, Ethiopia & Saveetha Dental College, Saveetha Institute of Medical and Technical Sciences (SIMATS), India

Melaku Ashagrie Belete
https://orcid.org/0000-0002-0706-9903
Wollo University, Ethiopia

Selam Niguse
Mekelle University, Ethiopia

Ephrem Tsegay
Mekelle University, Ethiopia

Tadele Araya
Mekelle University, Ethiopia

Birhanu Hadush
Mekelle University, Ethiopia

Kebede Nigussie
Mekelle University, Ethiopia

P. Prakash
Thiagarajar College, India

ABSTRACT

Back in the mid-nineties, the discovery of antimicrobials denoted a profound and remarkable achievement in medicine which was capable of saving lives. However, recently, antimicrobial resistance became a major global issue facing modern medicine and significantly increased among bacteria, fungi, and viruses which results in reduced efficacy of many clinically important and lifesaving antimicrobials. The growing rise of antimicrobial resistance inflicts a remarkable economic and social burden on the health care system globally. The replacement of conventional antimicrobials by new technology to counteract and lessen antimicrobial resistance is currently ongoing. Nanotechnology is an advanced approach to overcome challenges of such resisted conventional drug delivery systems mainly based on the development and fabrication of nanoparticulate structures. Numerous forms of nanoparticulate systems have been discovered and tried as prospective drug delivery systems, comprising organic and inorganic nanoparticles.

DOI: 10.4018/978-1-7998-5049-6.ch001

INTRODUCTION

Back in the mid-nineties, the discovery of antimicrobials denotes a profound and remarkable achievement in medicine which was capable of saving lives of millions of populations [1]. Antimicrobial agents, with their static or cidal ability for numerous microorganisms such as bacteria (antibacterials), fungi (antifungals), and viruses (antivirals). Commonly used antimicrobials may be synthetic, may be natural compounds which are modified chemically or may be of animal or plant origin, (Von Nussbaum et al., 2006). Having such variant forms, antimicrobials have a substantial effect on the disease outcome of an infected individual mainly responsible for recovery if they are chosen and used appropriately. Their use ranges from chemotherapy (treatment) to prophylaxis (prevention) of various infections. The history of antimicrobials officially started in 1928, when penicillin, which is the first antibiotics, was discovered. The discovery of penicillin put a major landmark in medicine and started, what is called "the antibiotic revolution" (Davey et al., 2017).

Antibiotics have different mode of actions including the inhibition of cell wall synthesis, inhibition of DNA replication, inhibition of protein synthesis, and alteration or inhibition of metabolism (Awad et al., 2012). In order to behave such mode of actions and access targets located inside the bacterial cell wall, antimicrobials necessarily need to penetrate into the cell. Hence, antimicrobial agents must be capable of penetrating to the site of action which is usually attained by diffusion or by active transport mechanisms. However, antimicrobial agent's penetration into the microbial cell and reaching the sensitive intracellular targets is largely influenced by the presence of lipopolysaccharide-lipoprotein complexes which are mainly found in the cell wall of Gram-negative microorganisms. Despite such impeding mechanism, some antibacterial agents deploy aqueous transmembrane channels called porins in the bacterial outer membrane to gain entry into Gram negative organisms. Moreover, both Gram positive and Gram-negative bacteria possess an outer membrane structure known as Peptidoglycan, which forms a rigid layer. However, Gram positive organisms have a very thick peptidoglycan layer (cross-linked with interpeptide bridges) and Gram-negative organisms have a very thin peptidoglycan layer. Numerous antibiotics, including penicillins, fosfomycin, cycloserine, bacitracin, cephalosporins, teicoplanin and vancomycin selectively inhibit peptidoglycan layer synthesis at different stages (Awad et al., 2012). On the other hand, antimicrobials such as ionophores affect the transport of cations through the cell membrane. The intracellular targets for antimicrobials include DNA replication and protein synthesis. Antibiotic agents such as chloramphenicol, puromycin, aminoglycosides, tetracyclines, fusidic acid, macrolides, lincosamides, mupirocin, streptogramins, and oxazolidinones interfere with the process of protein synthesis. Some antibiotics such as novobiocin, quinolones, diaminopyrimidines, nitroimidazoles, rifampicin and sulfonamides, inhibit DNA synthesis by targeting topoisomerases which are vital components for modulation of DNA supercoiling, an essential step in DNA replication (Drlica et al., 2008). However, microorganisms revealed an astonishing capability to adapt, evolve, and survive by developing resistance mechanisms to antimicrobial compounds; and antimicrobial resistance became a major global issue and significantly increased among bacteria, fungi and viruses which results in reduced efficacy of many clinically important and lifesaving antimicrobials (Seiffert et al., 2013) The development of antimicrobial resistance, which may be intrinsic or acquired occurring through mutation (Martinez & Baquero, 2018) or gene transfer from other species or strains(Hegstad et al., 2010; Palmer et al., 2010) is principally caused by alteration in the binding sites; modification of the metabolic pathways; alteration or inactivation of the drug; or decreased permeability (increased flux) of antimicrobial agents (Schmieder

& Edwards, 2012). The growing rise of antimicrobial resistance inflicts a remarkable economic and social burden on the health care system globally.

The discovery of nanotechnology is an advanced approach to overcome challenges of such resisted conventional drug delivery systems mainly based on the development and fabrication of nanoparticulate structures. Nanotechnology, due to enhanced durability, performance, strength, flexibility, and the inimitable physicochemical characteristics of nanomaterials, currently become the groundbreaking technology; and have a thoughtful impact on improving human health. Nanomaterials can mainly be used in a number of treatment modalities including targeted drug delivery, prognostic monitoring of therapy, and also in the detection of tumors (Jena et al., 2013). The antimicrobial actions of nanoparticles include blockage of microbial enzymatic pathways, cidal destruction of cell membranes, microbial cell wall alterations, and DNA or RNA pathways (Clarence & Geoffrey, 2015). Numerous forms of nanoparticulate systems have been discovered and tried as prospective drug delivery systems, comprising Organic Nanoparticles (such as Polymeric Antimicrobials, Chitosan, Organometallic Polymers, Peptides Nanomaterials, Quaternary Ammonium Compounds, Benzoic Acid, Phenol, Metallic and p-Hydroxy Benzoate Esters nano polymers etc...) and Inorganic Nanomaterials (including Silver, Gold, Zinc Oxide, Titanium Oxide, Iron Oxide, Magnesium Oxide Nanomaterials etc...). Nowadays, encapsulation of antimicrobial drugs in nanoparticle systems has begun being a ground-breaking and promising therapeutic options which heightens treatment effectiveness and lessens the undesirable side effects of antimicrobial drugs.

This chapter focus on the common drug resistance mechanisms in microbes and the role of emerging therapeutic alternatives particularly organic and inorganic nanoparticles pursuing to address prevailing issue of antimicrobial resistance; provide explicit information on the antimicrobial activity of different types of nanomaterials.

Figure 1. The main mechanisms of action and cellular targets of antibiotics within the bacterial cell (Mariana et al., 2016).

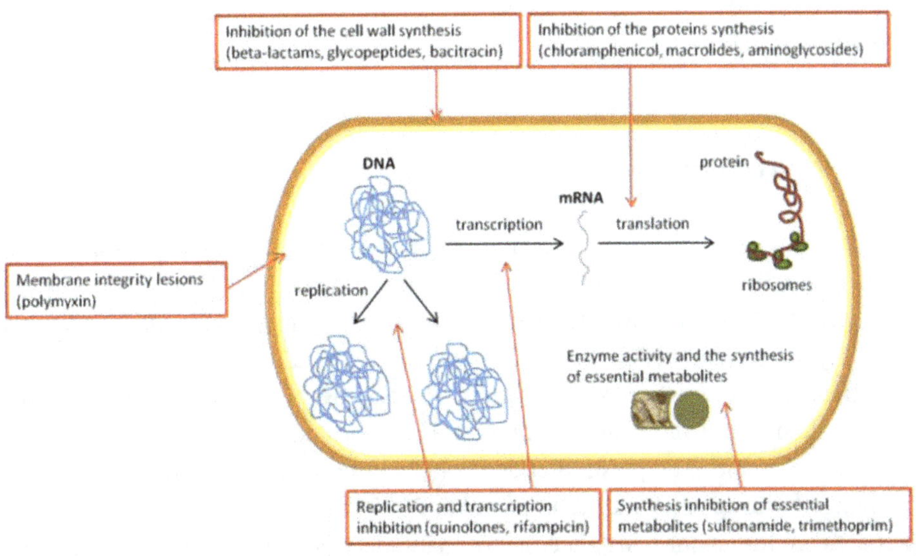

Antibacterial Resistance

Antimicrobial resistance is one of the main challenges facing modern medicine. The emergence and hasty spreading of resistant bacteria is currently decreasing the efficiency of antibiotic agents in the community; and recent assessments report that more than 700,000 individuals die each year mainly as a result of antimicrobial related problems (O'Neill et al., 2016). This leads to the growing concern regarding antimicrobial resistance worldwide, particularly to antibacterial agents intended to combat gram positive and negative bacterial agents causing disease to human. Bacteria deploy different resistance mechanisms against commonly used drugs including beta lactamase production, carbapenemase production, methicillin resistance and vancomycin resistance.

Figure 2. History of antimicrobial agent development Vs subsequent acquaintance of resistance by microorganisms.

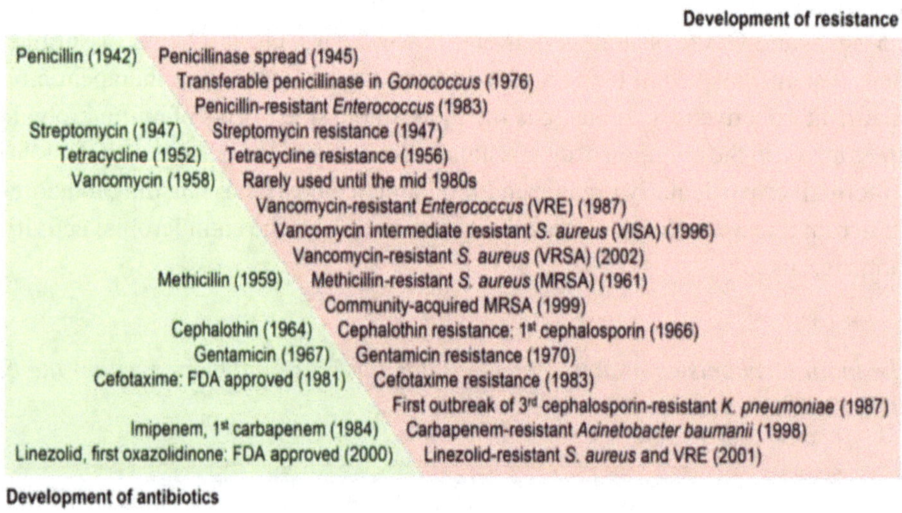

Beta Lactamase Production

Bacteria have a complex cell wall structure comprises an outer membrane imbedded with protein channels called porins. The β- lactam antibiotics have three important characteristics (mechanisms) to pass through this porin channels: sufficiently small in size, positive in charge, and being hydrophilic. However, bacteria particularly gram negative bacilli develop resistance to these drugs by altering the permeability of the channels. Beta lactamase are enzymes produced by bacteria and have ability to hydrolyze all types of beta lactam ring found in beta lactam antibiotics. The genes code for these enzymes may be coded in chromosome or in plasmid. The plasmid coded beta lactamase are transferable to other bacteria and becomes a challenge for drug resistance (Ruppé et al., 2015).

Beta lactamases are the major resistance determinant enzymes for beta lactam antibiotics in bacteria. The resistance mechanism pertained by beta lactamase enzyme producing bacteria is hydrolyzing the beta lactam ring of the antibiotics. *Enterobacteriaceae, P. aeruginosa* and *A. baumannii* are gram negative bacilli (GNB) and infections they cause are primarily treated with beta-lactam (β-lactam) class of antibiotics. Beta-lactams are the group of antibiotics which contain a common beta-lactam ring and comprise pencillines, cephalosporins, monbactams and carbapenems. Beta-lactams act on the newly forming bacterial cell wall peptidoglycan by binding and inactivating the penicillin binding protein. Resistance of GNB for β-lactams can be intrinsic (chromosomal) or acquired (plasmid), both includes modifying the porin channel, efflux out the drug and/or importantly production of hydrolyzing enzymes like extended spectrum beta-lactamase (ESBL), Amp C and carbapenemase (O'Neill et al., 2016; Santajit & Indrawattana, 2016).

Extended spectrum β-lactamases can hydrolyze penicillin, first-, second-, and third- generation cephalosporins and monbactam, while carbapenemase have a wide hydrolyzing activity that can degrade all β-lactams including carbapenems and β-lactams inhibitor like clavulinic acid. ESBLs are inhibited by -lactam inhibitors and do not hydrolyze cephamycins. The genes for ESBL that found on mobile genetic element are important for the spread of drug resistance. Moreover, it allows GNB to acquire resistance to other antibiotics classes like aminoglycosides, trimethoprim, and quinolones (Lovleva & Bonomo, 2017).

Carbapenemase Production

Carbapenemase are enzymes that are capable of hydrolyzing one or more carbapenem antibiotics such as doripenem, ertapenem, impenem, meropenem. Carbapenemase enzymes belong to class A, B or D of Ambler classification. The bacteria that produce carbapenemase have a limited treatment choice; for instance, Carbapenemase (AmpC) enzymes, unlike ESBL enzymes, are not inhibited by classic β-lactam inhibitors and confer resistance to cephamycins but do not efficiently hydrolyze cefepime (Mehrad et al., 2015). Carbapenemase genes mediate a stable and transferable form of resistance, which is able to spread through clonal expansion or by horizontal transfer of genes to another bacterium. Nowadays, infection due to carbapenem producing bacteria is increasing and become associated with mortality. It is predominantly seen among gram negative bacteria particularly among *Enterobacteriaceae*. Routine identification carbapenemase production knowing its type is significant for the follow up of the spread of carbapenemase producers, the early identification or recognition of outbreaks and the choice of treatment with novel drugs such as ceftazidime/avibactam combinations or using nanoparticulate antibacterial agents (Nordmann & Poirel, 2014).

Methicillin Resistance

Methicillin resistant, predominantly studied on *Staphylococcus aureus* (MRSA) has been identified as one of the major threat associated with the development of antimicrobial resistance (AMR). MRSA strains pose multidrug resistance patterns mainly achieved due to the acquisition of extra resistance determinants such as conjugative plasmid carrying gentamicin resistance (Ohlsen et al., 2013). Methicillin-resistant Staphylococcus aureus (MRSA) was first observed in 1960, less than one year after the introduction of this second generation beta-lactam antibiotic into clinical practice. It is a general fact that infections due to methicillin resistant strains of *S. aureus* are inflict higher mortality rates than infections caused by methicillin susceptible strains. The genetic basis of methicillin resistance in *S. aureus* is associated with carriage of a mobile cassette of genes known as the staphylococcal cassette chromosome mec which contain mecA gene responsible for resistance to beta lactam antibiotics including methicillin (Katayama et al., 2010). Most importantly epidemiological evidences regarding MRSA suggest that

MRSA resistance commonly occur when *mecA* gene which is capable of encoding methicillin resistance was able to horizontally transferred to another methicillin sensitive strain of *S. aureus*. Poor infection control measures along with continued indiscriminate exposure of humans to antibiotics have resulted in this enormous problem of attainment and spreading of MRSA, which in turn restrict the choice of treatment for MRSA infections. Several studies have reported that MRSA have developed resistance to most common antibiotics used for the treatment of its infections.

Vancomycin Resistance

Vancomycin most importantly used for the treatment of methicillin resistant bacteria. It has been an effective drug of choice potent against methicillin resistant strains such as MRSA. However, recently bacterial isolates with reduced susceptibility or complete resistance to vancomycin have emerged and are still spreading. Resistance to vancomycin was discovered among enterococci bacteria in 1980, and this discovery prompted a remarkable concern regarding the forthcoming usage of vancomycin as a potent treatment for MRSA. Vancomycin resistance is predominantly seen among *S. aureus* and *enterococcus* species; and currently becoming a challenging problem worldwide (Hidayat et al., 2006) Some researchers confirmed the fact that *S. aureus* can acquire enterococcal plasmids during distinct conjugation processes and exhibit vancomycin resistance ability(Zhu et al., 2010).

ANTIFUNGAL RESISTANCE

In spite of the fact that therapeutic drugs are predominantly used to control infections caused by prokaryotic microorganisms, microbial disease due to eukaryotic microorganisms has also been controlled using therapeutic drugs intended controlling or managing and preventing disease caused by fungi pathogens. Numerous types of antifungals, pertaining diverse drug resistance profiles, are used for treating disease caused by fungi (Table 1). The earliest and oldest antifungal is amphotericin B, a broad spectrum agent, which is categorized under polyene macrolide antibiotics and remains a forefront choice for fungal infections having unknown etiology. Similarly, as with prokaryotic infectious agents, resistance against antifungal drugs is becoming an increasingly significant clinical problem particularly in tropical countries and among immune compromised population (Denning & Bromley, 2015). Some known cases exist showing inherent resistance of fungi to particular antifungal agents, however antifungal resistance is most commonly attributable to and mutations and induced changes.

Likewise, the azoles, including imidazoles and triazoles found to be the leading class of antifungal agents commonly prescribed for the management of fungal infections. Several modifications of the triazole groups have generated an advanced forms of antifungals including ravuconazole, itraconazole, posaconazole, voriconazole, and the recently licensed isavuconazole, which have enhanced action against *Aspergillus* and filamentous fungal species. Such advancements also lead to emergence of fluconazole which are used mainly in the treatment of *Candida* infections. These antifungals indeed have significant variations in potencies, bioavailability, spectrum of activities, potential toxicity and drug interactions (Epaulard et al., 2013). Moreover, sterol inhibitors including the allylamines squalene epoxidase inhibitors, isomerase inhibitors and phenylmorpholine reductase are also as topical antifungal against dermatophyte fungal infections.

Table 1. Disease caused by fungi and conventional treatment options. Data from the World Health Organization (Leroux & Ullmann, 2013).

Disease	*Pathogen*	First line (second line) treatment of choice
Invasive candidiasis	*C. albicans, C. glabrata, C. parapsilosis*	Echinocandins; fluconazole; liposomal amphotericin B.
Aspergillosis	*A. fumigatus*	Voriconazole (amphotericin B formulations; caspofungin; micafungin; posaconazole; itraconazole).
Pneumocystis pneumonia	*Pneumocystis carinii*	Sulfamethoxazole–trimethoprim (clindamycin–primaquine).
Cryptococcal meningitis	*Cryptococcus neoformans*	Amphotericin B plus flucytosine; amphotericin B plus fluconazole.

Fungal resistance against antifungal drugs pertain different mechanisms. For instance, some fungi species such as *Candida krusei* are inherently resistant to azole groups (Shor & Perlin, 2015). Similarly, increased triazole resistant strains are emerging nowadays (Kidd et al., 2015). Moreover, fungal strains showing cross resistance to azoles and echinocandins, signifying the emergence of multidrug resistant species phenotypes among medically important fungi pathogens (Van der Linden et al., 2015). A recent study depicts that threat from multi-azole-resistant strains of *A. fumigatus* might be occurring as a result of the selective pressure of agricultural azole fungicides and subsequent transmission of such azole resistant fungal strains to the health facility via spore dispersal mechanisms (Alan et al., 2017).

The most lately developed class of antifungals are the echinocandin antibiotics which comprises of clinically important micafungin, caspofungin and anidulafungin. These antifungal agents have relatable pharmaco-kinetic characteristics; and are fungi static or fungicidal against *Aspergillus* species and fungicidal against *Candida* species, resulting bud tip or hyphal lysis. However, echinocardin antifungals are found to be ineffective against *Pneumocystis jiroveci* and other species (Hegstad et al., 2010).

In recent times, multidrug azole resistance specifically multidrug echinocandin resistance has been recognized among fungi particularly in *C. glabrata* strain, which is frequent in patients with solid tumors and hematological malignancies (Alexander et al., 2013). Such multidrug resistant strains of *C. glabrata* can only be managed using intravenous injections of amphotericin; however, due to poor penetration of amphotericin into urine, such infections are non-respondent to treatment and are principally untreatable (Pfaller & Diekema, 2012).

Antiviral Drug Resistance

Discovery of antiviral drugs lack behind antibacterial drugs. This is mainly due to the fact that treating viral disease is difficult due to their replication mechanisms. Viruses mostly exist as pathogen intracellularly, use host cell machinery for their replication and some have ability to conjugate into host genetic material (DNA or RNA) and undergo rapid replication, which conclusively make antiviral treatments difficult. However, antiviral drug therapy is being used for common viral disease. Antiviral drugs can be categorized into the inhibitors of fusion, uncoating, nucleic acid synthesis, integration, protease, and release (Biswas et al., 2019). Conventional antiviral drugs are limited by the inadequate response, the rapid development of resistance, and adverse effects. It is well established that using even standard and the same doses of antiviral drugs may result in different inter-individual serum concentrations and clinical outcomes, the reasons for which may be multifactorial including differences in concomitant medications, underlying diseases, treatment compliance, genetic factors, and gender-related metabolism. (Antonelli & Turriziani, 2012; Pommier et al., 2005). Altogether, the best antiviral agents would include those more effective, having less resistance and which pose low toxicity. The general principles or mechanisms of antiviral agents are based on the mutations in the enzyme, receptor modification, and decreased efflux of antiviral agents (table 3). The concept of drug resistance has been well recognized and has become a serious concern in clinical virology. For example, HIV/AIDS and Hepatitis infections are among the most prevalent viral diseases worldwide. Nowadays viral pathogens are mediating resistance against antiviral agents. Among the antiviral agents, drugs against Hepatitis B and HIV/AIDS are currently pertaining resistance and treatment failures (Pommier et al., 2005).

Table 2. Mechanism and mode of actions of drug resistance among fungi pathogens (Clarence & Geoffrey, 2015).

Drug	Drug class	Mode of action	Resistance mechanism
Amphotericin B, amphotericin deoxycholate	Polyene macrolide antibiotics	Binds ergosterol more avidly than human cholesterol, disrupting the semi-permeable membrane and causing leakage of essential metabolites and the collapse of electrochemical gradients. Binding of low-density lipoprotein receptors and amphotericin mediated oxidative damage may also contribute.	Laboratory mutants with lower ergosterol content are less sensitive to amphotericin B, but are rare clinically. Aspergillus terreus is intrinsically less amphotericin sensitive but resistant strains have a normal ergosterol content, suggesting that membrane permeability may not be the only mechanism of amphotericin action.
Fluconazole, itraconazole, voriconazole, posaconazole, ravuconazole, isavuconazole	Azoles	Bind haem groups and inhibit the P450-mediated 14α-emethylation (Erg11p or Cyp51p) of lanosterol in the ergosterol biosynthetic pathway. Leads to impaired membrane permeability, membrane protein action and cell wall synthesis	Resistance involves the overexpression of drug efflux pumps and point mutations in the target ERG11/ CYP51A. Azole resistance are likely to have arisen from environmentally generated mutations.
Caspofungin, micafungin, anidulafungin, CD101 (formerly biofungin)	Echinocandins	Cyclic hexapeptides with an antifungal bioactive lipid side chain that binds the fungal-specific β-1,3-glucan synthase Fks cell membrane proteins, disrupting cell wall integrity.	Resistance through point mutations in two major hotspots in the β-1,3 glucan synthase genes FKS1, and, in C. glabrata, FKS2 these reduce drug binding Upregulation of cell wall chitin can protect cell wall damage.
Flucytosine (5-fluorocytosine)	Fluoropyrimdines	Converted to 5- fluorouracil by cytosine deaminase, which becomes incorporated into RNA, resulting in inhibition of DNA synthesis.	Resistance results from mutations in the genes encoding cytosine permease transporter, cytosine deaminase, which converts 5- fluorocytosine to 5-fluorouracil, or the uracil phosphoribosyl transferase required to convert 5- fluorocytosine into a substrate for nucleic acid synthesis.

Table 3. Antiviral agents, and its resistance mechanisms in different emerging and reemerging viruses

Drug	Examples of targeted Viruses	Mode of action	Resistance Mechanism
Polysulphatcs, polysulphonates, other enveloped viruses polycarboxylates, polyoxometalates, chicoric acid, zintevir, cosalane derivatives, negatively charged albumins	HIV, HSV, CMV, RSV	Virus adsorption inhibitors	Viral thymidine kinase or DNA polymerase mutation and Cross resistance among different antiviral drugs (Tozzi, 2010)
AMD3100, TAK779 and T20 paramyxoviruses derivatives	HIV, RSV and paramyxoviruses	Virus–cell fusion inhibitors	Clustering resistant genes in in between catalytic regions and Modifying the receptor binding sites in case in paramyxoviruses(Strasfeld & Chou, 2010)
Acyclovir, valaciclovir, ganciclovir, valganciclovir, penciclovir, famciclovir, brivudin, foscarnet, Bicyclic furopyrimidine nucleoside analogues, A5021, cyclohexenylguanine	Herpesviruses (HSV-1, -2,VZV, CMV, EBV, HHV-6, -7, -8)	Viral DNA polymerase inhibitor	Mutation in the DNA polymerase, Resistant mutations cluster at particular VZV thymidine kinase gene loci, and Phosphorylation of ganciclovir without impairing the important functions of enzymes used in replication (Fearns & Plemper, 2017)
Cidofovir, adefovir, tenofovir	DNA viruses (polyoma-,papilloma-, herpes-, adeno-and poxviruses), HIV, HBV	Acyclic nucleoside phosphonates	Mutations in the reverse transcriptase domians of antiviral polymerase gene, with most substitutions occurring in the domians B, C and D, Conformational changes in the domians, Mutations in the thyrosine-methionine-aspartate-aspartate motif, and N236T and A181V/T mutations reducing viral replicative capacity and conferring cross resistance 39,40]
saquinavir, ritonavir, indinavir, nelfinavir, amprenavir, lopinavir, atazanavir, mozenavir, tipranavir, AG7088	HIV, herpesviruses, rhinoviruses, HCV	Viral protease inhibitors	Mutations/changes in the viral protease enzyme reducing its efficiency to bind with its substrate polyproteins-Gag and GagPol, and Substrate envelope selects active site resistant mutations (Balzarini et al., n.d.)
Zanamivir, oseltamivir, RWJ270201	Influenza A and B virus	Viral neuraminidase and hemaaglutinin inhibitors	Mutations in the H274Y confer resistance Mutations at E119, D198, I222, R292 and N294 confer nuraaminidase sensitivity(Maarseveen & Boucher, 2006)
Ribavirinll, Mycophenolic acid, EICAR, VX497	HCV, RSV	IMP dehydrogenase inhibitors	Amplifying the IMP dehydrogenase gene, Mutations in the domain of the enzyme, and Cross resistance with other bacterial and viral receptor modifications (Oakley et al., 2010)

HIV Drug Resistance

The global practice of antiretroviral therapy (ART) has led to vivid reductions in HIV-1 incidence, mother-to-child transmission as well as mortality (Tanser et al., 2013). These extraordinary public health achievements were made possible by the accessibility and widespread administration of combinations of two nucleoside reverse transcriptase inhibitors (NRTIs) and a non-nucleoside reverse transcriptase inhibitor (NNRTI) (Rhee et al., 2015). However, HIV drug resistance is posing a potential threat to the long-standing success of ART and a threat to the eradication of AIDS as a public health problem especially in developing countries. Mainly due to various factors such as patient antiviral drug usage habit (incomplete adherence) and viral mutation, nowadays, there is increase in the prevalence of acquired drug resistance (ADR) in treated individuals and transmitted drug resistance (TDR) in newly infected individuals, particularly in low and middle income countries (Phillips et al., 2014). Even though both acquired and transmitted HIV-1 drug resistance are public health concerns,

TDR has the potential to more rapidly reverse the effectiveness of first-line antiretroviral (ARV) therapy in the community. HIV patients with TDR who begin ARV therapy with a lower genetic barrier to resistance have a higher risk of virological failure (Chung et al., 2014). Drug resistant HIV Viruses can occur naturally every day in untreated HIV/AIDS patients. Moreover, clinically significant HIV drug resistance occurs mainly due to drug resistance mutation resulting from incompletely suppressive therapy and as a result of HIV genetic variability from the high rate of HIV reverse transcriptase processing errors and viral recombination with nearby viruses (Perelson & Ribeiro, 2013).

Hepatitis B Drug Resistance

Worldwide, more than 240 million people are chronically infected with HBV and more than 686,000 people die every year due to liver disease, liver cancer and hepatocellular carcinoma (World Health Organization, n.d.). However, quality of life and disease progression is significantly improved after patients were treated with interferon and nucleoside/ nucleotide analogs (NAs) (Ma et al., 2013). Most importantly NAs are currently serving as the primary therapeutic options for chronic hepatitis B patients. However, emergence of drug resistance, predominantly attributable to mutations, is currently becoming a major limitation of long term NA treatment (Kwon & Lok, 2011).

Rational treatment of HBV infected cases is important to suppress HBV DNA replication, reduce inflammatory reactions of liver and prevent progression to cirrhosis and hapatic carcinoma. Currently, different therapeutic agents are approved by the United States Food and Drug administration (FDA) like interferon (IFN), pegylated-interferon (peg-IFN), Lamivudine, Emtricitabine, Adefovir dipivoxil, Entecavir, Telbivudine and Tenofovir disoproxil fumarate (Lok & McMahon, 2009). Isothiafludine efficiently inhibited the replication of HBV mutants that were resistant to nucleoside and nucleotide analogues. Myrcludex B, a novel peptide therapeutic that blocks viral entry in vitro and in animal models, has advanced to the clinical validation stage (Petersen et al., 2008).

Among five NAs, lamivudine was the first NA to be approved for the treatment of HBV patients (Kwon & Lok, 2011). Hepatitis B patients receiving lamivudine treatment develop resistance mainly as a result of the emergence of M204I and M204V primary mutations, which can significantly affect viral fitness and later accompanied by secondary or compensatory mutations (V173L, L180M/I, and L80V/I), which most commonly can either boost the level of antiviral resistance or restore the fitness loss (Santajit & Indrawattana, 2016).

ANTIMICROBIAL NANOMATERIALS

Antimicrobial nanomaterials become the preferable choice of antimicrobial agent owing to the global emergence of multidrug resistant microorganisms (bacteria, viruses and fungi) which has made conventional treatment of infectious diseases difficult. Therefore, the advent of alternative new classes of antibiotics, antiviral and antifungal agents that can treat resistant strains is paramount achievement of modern medicine (Clarence & Geoffrey, 2015).

Figure 3. Various functions and broad spectrum activity of nanoparticulate antimicrobial agents.

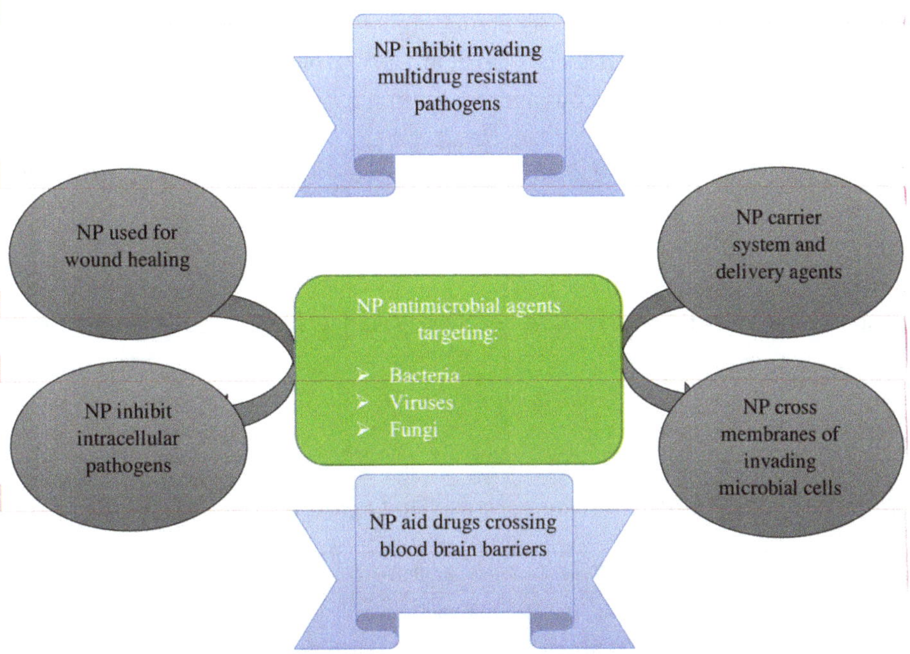

Antimicrobial nanoparticles are pose a new and novel nano-sized platform for efficient antibiotics delivery in which microbial pathogens cannot be able to develop resistance. For instance, recent study reports have revealed that some metal nano-constructs possess antimicrobial activities, which is used to combat infectious agents (Rai et al., 2009). Antimicrobial nanoparticles (NPs) provide several typical benefits in overcoming resistance, plummeting toxicity and dropping expense (cost), when compared to conventional antimicrobial agents(Weir et al., 2008). Many nano-sized drug carriers (nanoparticulate drug delivery systems) such as Polymeric drug carriers, Liposomes, Niosomes, Solid Lipid Nanoparticles (SLNs), Fullerenes and Dendrimers are also offered to effectively deliver antibiotics by improving pharmacokinetics and accumulation, while reducing the adverse side effects of the antibiotics. Nanoparticless are also known to be retained for longer duration in the body than conventional antibiotics, which could be useful for attaining persistent therapeutic effects [Table.4 and Table 5].

Nanomedicine has potential impact on the control of infectious microorganisms mainly through use of nanotechnology in vaccination, design and delivery of antimicrobial drugs, and diagnosis and control of cross infections, particularly in overcoming drug resistant pathogens which has been given attention as a promising alternative to the current conventional antibiotics based approaches (Allaker & Ren, 2008). Metal and metal oxide nanoparticles are known for their increasing use in antimicrobial formulations through production of reactive oxygen species. More specifically, nano sized zinc, silver, titanium and zinc dioxide have been extensively used and proven to be effective in inactivating and killing various pathogenic microorganisms (Mühling et al., 2019). Moreover, recent studies reported that antibiotics formulated in polymeric nanoparticles have shown a significant and potent antimicrobial activities including anti-MRSA activities when compared with non-polymerized forms of conventional antibiotics such as penicillin and beta lactams (Turos et al., 2017). Similarly, Vancomycin-capped gold

(Au) NPs has also exhibited enhanced antimicrobial activities against vancomycin resistant bacterial strains (Gu et al., 2013).

Antibacterial NPs consist of naturally occurring organic antibacterial substances, inorganic metals, metal oxides and carbon based nanomaterials (Li et al., 2018). Nanomaterials deploy the following antimicrobial mechanisms to combat infectious agents: Antimicrobial mechanisms of nanomaterials include:

- photocatalytic production of reactive oxygen species (ROS) that damage cellular and viral components,
- damaging the bacterial cell membrane/wall,
- interruption of energy transduction, and
- inhibition of enzyme activity and DNA synthesis (Huh & Kwon, 2011) (Figure 3).

Table 4. Antibacterial and Antifungal nanomaterials and their outcome

Type of Nanoparticle	Type of bacteria inhibited	Outcome
Fe-Oxide NP & AgNP	Bacillus, E. coli and Staphylococcus species	Fe-Oxide NPs were sensitive against Bacillus, E. coli and Staphylococcus species.
Ag NPs.	Escherichia coli, Klebsiella pneumoniae, Pseudomonas spp. Bacillus species, Staphylococcus species	Inhibited the growth and multiplication of E. coli, Pseudomonas species, Bacillus spp. And Staphylococcus species,
Polyvinyl-N-carbazole (PVK) and single-walled carbon nanotubes (SWNTs) (PVK:SWNT)	E. coli and B. subtilis	The nano-composite showed anti-microbial activity against both Gram-positive and negative bacterial isolates.
Multi walled coated nanotubes (MWCNT)-lysine functionalized	S. aureus, Streptococcus agalactiae, S. dysgalactiae, E. coli, K. pneumonia, S. typhimurium	The functionalized MWCNT with lysine expressed high antimicrobial effect against all bacterial cells
MWCNT-AgNPs	Escherichia coli	MWCNT-AgNPs exhibited strong antimicrobial activities and reduce biofilm formation.
Silicon nanotubes (SNTs), silicon nanoparticles (SNPs)	Multidrug-resistant Staphylococcus aureus	SCSNTs were effective in limiting the growth of multidrug-resistant S aureus
Ag–Fe/SWCNTs	Escherichia coli.	Purified Ag–Fe/SWCNT hybrid nanoparticles were effective against E. coli.
SWCNTs combine with H2O2 or NaOCl	Bacillus anthracis Spores	The combined effect of SWCNTs and H2O2 or NaOCl exhibited sporicidal effect on B. anthracis spores
Zirconia (ZrO2) nanoparticles	Staphylococcus aureus, Escherichia coli,	Zirconia (ZrO2) nanoparticles exhibited antibacterial against the test organisms.
Au/CuS core/shell nanoparticles (NPs)	B. anthracis spores and cells	The Au/CuS NPs were highly efficient in inactivating B. anthracis cells, but not effective to the spores.
Ag NPs.	Aspergillus niger, Aspergillus flavus, Penicillium, Candidia species	Inhibited the growth and multiplication of A. niger, A. flavus, Penicillium spp and candidia spp. Inactivating their infectiousness
Zirconia (ZrO2) nanoparticles	Candida albicans, Aspergillus niger	Zirconia (ZrO2) nanoparticles exhibited antifungal activity

Table 5. Antiviral nanoparticulate agents and their outcome (Clarence & Geoffrey, 2015).

Type of Nanoparticle	Type of viral microorganisms inhibited	Outcome
Polyvinylpyrrolidone (PVP)-coated silver nanoparticles	HIV-1	PVP-coated AgNP exhibit potent cyto-protective and post-infected anti-HIV-1 activities toward CCR5 cells.
PVP-coated silver nanoparticles	HIV-1	PVP-coated AgNPs inhibited cell associated HIV-1 and cell-free HIV-1 transmission.
mercaptoethane sulfonate (MES)-coated silver and gold nanoparticles	Herpes simplex virus type 1 (HSV-1)	The MES-coated silver and gold nanoparticles inhibited HSV-1 infection in cell culture
PVP-coated silver nanoparticles	Respiratory syncytial virus (RSV)	Inhibited RSV infection
AgNP and polysaccharidecoated AgNP	Monkey pox virus (MPV)	The AgNPs of approximately 10 nm inhibit MPV infection in vitro, as an anti-viral
AgNPs	Hepatitis B virus (HBV)	AgNPs inhibited in vitro HBV RNA and extracellular virions
Mycosynthesized silver nanoparticles	HSV 1 and 2 and with human parainfluenza virus type 3.	Smaller-sized AgNPs were able to inhibit the infectivity of the viruses
Sialic-acid functionalized gold nanoparticles	Influenza virus	The NPs inhibition influenza virus infection

In recent times, intracellular or extracellular synthesis of inorganic (metallic) nanoparticles including Silver (Ag), Gold (Au), Zinc Oxide, Titanium Oxide, Iron Oxide, Magnesium Oxide nanomaterials acting against pathogenic microbial cells have been explored as novel nanoparticle formulations.

Figure 4. Antimicrobial mechanisms of nanomaterials (Perelson & Ribeiro, 2013).

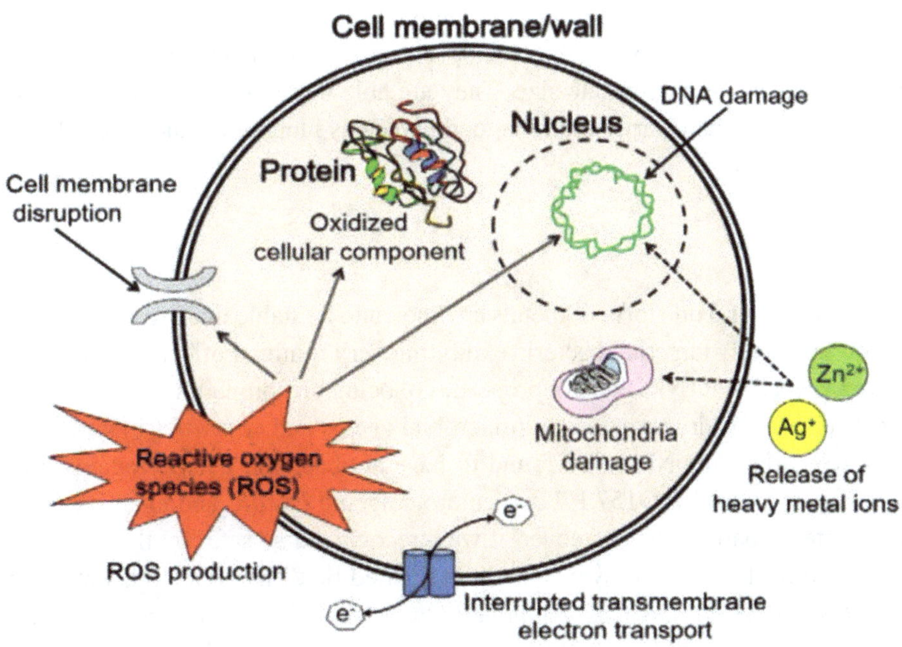

Inorganic Nanomaterials

Numerous forms of inorganic metallic elements and their oxide forms, such as Silver, Gold, Zinc Oxide, Titanium Oxide, Iron Oxide, Magnesium Oxide Nanomaterials have been demonstrated for their antimicrobial activities against bacterial, viral and fungal pathogenic microorganisms.

Silver (Ag) NPs

Amongst the diverse forms of metallic and metal oxide nanoparticles, Silver (Ag) NPs have been demonstrated to be the most effective against bacteria, viruses, and fungi microorganisms (Chamundeeswari et al., 2010). Ag NPs mainly synthesized through 3 methods, namely chemical (involving use of toxic and hazardous substances), physical (by grinding Silver metal) and biological (using fungi, bacteria and plant extracts). The most prominent outcome of silver nanoparticles (Ag NPs) is on the cellular metabolic activity, arresting the respiratory chain of microbes and on the cell division through membrane inflicting damage to the microbial cells, occasioning in numerous subordinate effects, including generation of reactive oxygen species (ROS) and DNA damage which finally lead to bacterial and fungi cell death(Park et al., 2011). Moreover, Ag NPs have the ability to attach to bacterial cell wall and form 'pits' on the microbial surface increasing its accumulation. Microbial cell death subsequently results from structural changes in the cell membrane and formation of free silver radicals, which damage cell membrane and make it porous (Prabhu & Poulose, 2012).

Generally, Ag NPs interfere with the electron transport chain and transfer of energy through the membrane, Inhibit DNA replication and respiratory chain in bacteria and fungi being effective against Methicillin-resistant *Staphylococcus aureus*, *Staphylococcus epidermidis*, Vancomycin-resistant *Enterococcus faecium*, *Klebsiella pneumoniae* and *Aspergillus* species. The potency (antimicrobial efficacy) of Ag NPs is dependent on the cell type; as well as size and shape of Ag NPs. Improved and synergistic antimicrobial effect can be obtained by combining Ag NPs with antibiotics, such as penicillin, erythromycin, amoxicillin, and vancomycin, most commonly against Gram-positive and Gram-negative bacteria (Fayaz et al., 2010). Moreover, silver nanoparticles have well developed surface chemistry and chemical stability, and are of appropriate size. They are able to maintain a constant shape and size in solution. As a result, silver nanoparticles are a good choice as inorganic nanomaterials for use against pathogenic microorganisms.

Zinc oxide (ZnO) NPs

Nanoparticles composed of oxide form of metals are known to be stable under tough processing settings and have toxicity selectively targeting bacteria exhibiting very minimal effect on human cells (Brayner et al., 2016). For instance, ZnO NPs, behave properties of being biocompatible and non-toxic; moreover, ZnO NPs have been used as drug carriers, antimicrobial agents and cosmetics ingredients (Perelson & Ribeiro, 2013). Recently, ZnO NPs were found to have antibacterial activity against important food borne pathogens, such as *E. coli* O157:H7 and enterotoxigenic *E. coli*; and against gram positive and gram-negative bacterial pathogens; moreover, it was 'recognized as safe' by the U.S. Food and Drug Administration (Xie et al., 2011). Previous studies revealed that ZnONPs show antibacterial activity against *E. coli*, *Listeria monocytogenes*, *Salmonella*, and *Staphylococcus aureus* (Tayel et al., 2011). The antimicrobial mechanism of ZnO NPs comprises of the release of Zn2+ ions and the production of highly reactive oxygen species including hydrogen peroxide (H2O2) and OH´ which subsequently penetrate the bacterial cells inhibit microbial cell growth (Yamamoto, 2011). In addition, ZnO NPs are known to destruct proteins and lipids of the bacterial cell membrane, causing leakage of intracellular contents and eventually the death of bacterial cells (Perelson & Ribeiro, 2013). Simply due to their low production cost and UV-blocking properties, ZnO NPs are considered advantageous over Ag NPs (Uğur et al., 2010). As a result, ZnO NPs are widely used in the current advent nanoparticulate technology.

Iron Oxide (Fe_3O_4) NPs

Attributable to their intrinsic characteristics, iron containing nanoparticles have valuable scientific use. Iron oxide (Fe_3O_4) NPs are used in food preservation agents, biosensors, antimicrobial agents, storage media, anti-cancer agents, magnetic resonance imaging (MRI), and targeted drug delivery. Fe_3O_4 NPs have high magnetic property and enhanced biological compatibility. This property makes them preferable for biomedical applications (Chen et al., 2008) . Iron oxide nanoparticles can be synthesized and characterized using different approaches including sol-gel and forced hydrolysis, hydrothermal processing, surfactant-mediated synthesis, co-precipitation, laser pyrolysis, electrochemical processing, and micro-emulsion processing through application of techniques such as absorption spectrophotometry, scanning electron microscopy (SEM) and X-ray diffraction. Regarding biological actions, Fe_3O_4 NPs pertain similar antimicrobial mechanisms as ZnO, including formation of ROS, oxidative stress, superoxide radicals (O_2), hydroxyl radicals (OH´), and hydrogen peroxide (H_2O_2) (Behera et al., 2012).

According to recent reports, Fe_3O_4 NPs has been demonstrated to impose antimicrobial activity against various bacteria including *S. aureus*, *S. epidermidis*, *E. coli*, *Xanthomonas*, *Staphylococcus saprophyticus*, *S. pyogenes*, Methicillin resistant S. aureus (MRSA) and *P. vulgaris* employing the above mentioned mechanisms (Prabhu et al., 2015). Moreover, a recent study (Arokiyaraj et al., 2013) assessed the antimicrobial efficiency of Fe_3O_4 NPs by comparing the antimicrobial action of plant leaf extract called *Argemone Mexicana* alone and in combination with Fe_3O_4 NPs, against bacterial pathogens. Remarkably, they witnessed a significant inhibition of *P. mirabilis* and *E. coli* bacterial species by Fe_3O_4 NPs treated plant leaf extract. Similarly, another study (Anghel et al., 2012) have also reported the high level inhibition of biofilm formation by *Candidia albicans* when using Fe_3O_4 NP-coated textile dressings more than their counterparts (Fe_3O_4 uncoated textile dressing). It is a general fact that Fe_3O_4 can be coated with various structures. For instance, Fe_3O_4 NPs can be coated with an essential oil called *Rosmarinus officinalis* and impose effective inhibitory action against biofilm forming fungi pathogens such as *Candidia albicans* and *Candidia tropicalis*(Chifiriuc et al., 2012).

Gold (Au) NPs

Even before its use as nanoparticulate agents, Gold (Au) has been assumed to possess healing abilities and was being consumed orally in ancient times. This profound property enable Au NPs to be the earliest recognized forms of nanoparticulate materials. Gold nanoparticles also have distinct optical properties and are naturally inert and compatible. Such properties make Au NPs suitable nanomaterials in modern medicine. The most commonly used Au NPs have size range from 0.8 to 250 nm (Li, Zhao, & Astruc, 2014). Synthesis of Au NPs encompasses both biological and chemical methods. Despite the involvement of toxic chemicals and being expensive, chemical synthesizing method used techniques including tetrachloroauric acid reduction to prepare colloidal AuNPs. Currently, nanomedicine technology is mainly using biological techniques for the synthesis of Au NPs yielding less expensive and non-toxic metallic nanomaterials (Alex & Tiwari, 2015). Numerous researchers have used the biological technique to synthesize Au NPs that have different form, size and antimicrobial activity. For instance, researchers were able to synthesize Au NPs using plant leaf extracts (Badri Narayanan & Sakthivel, 2018), or using microorganisms such the fungi *Candidia albicans* (Chauhan et al., 2011) or the bacterium *Rhodopseudomonas capsulate*(He et al., 2017) . The Au NPs antimicrobial and anticancer activity is further enhanced though photo thermal heating which is possible due to their property of being inert naturally (Méndez-Vilas, Ravishankar, & Jamuna, 2011). Moreover, Au NPs can be conjugated with several forms to boost their antimicrobial potency; for instance, recent research findings reported that Au NPs conjugated with a photosensitizer (toluidine blue) and a specific antibody have been reported to exhibit an enhanced antimicrobial activity against methicillin resistant *S. aureus* (MRSA) (Gil-Tomas et al., 2007).

The antimicrobial mode of action of Au NPs is through formation of holes in the cell wall of bacteria, which results in cell death due to leakage of cellular content. Furthermore, Au NPs behave a character of binding to DNA, preventing its uncoiling, which inhibit transcription pathway (Rai et al., 2010). Recent researchers reported that Au NPs used for preventing biofilm formation by the nosocomial fungi, *Candidia albicans*; and also have effective antimicrobial activity against multidrug resistant uropathogens including *Pseudomonas aeruginosa, Escherichia coli*, and methicillin resistant *Staphylococcus aureus* (Li, Robinson, Gupta et al, 2014). Studies also demonstrated that Au NPs conjugated with conventional antibiotics such as ciprofloxacin, gentamycin, rifampicin and vancomycin were effective in killing *S. haemolyticus* and *S. epidermidis* compared with using the antibiotics alone (Roshmi et al., 2015). Therefore, Au NPs are favorable adjuvants for antibiotics treatment in combating bacterial infections at a reduced antibiotics dosage with minimal adverse effects.

Copper Oxide (CuO) NPs

Largely due to its numerous physical properties such as super conductivity, copper compounds have vital application in current medicine. Several techniques such as laser radiation are used for the synthesis of CuO NPs (Sampath et al., 2014). Recent researchers have reported that CuO NPs, which are most commonly synthesized by the biological technique, showed antibacterial activity against both Gram positive bacteria such as *Staphylococcus aureus* and Gram negative bacteria including *Escherichia coli* (Chatterjee et al., 2012). CuO NPs commonly pose their antibacterial activity through adhesion to bacterial cell wall, creating negative opposite electric charges, resulting in the bacterial cell wall damage. Moreover, copper ions produce ROS which results in oxidative stress DNA or membrane damage while resulting cytotoxicity by interacting either directly with the cellular membrane or intracellularly (Naika et al., 2015; Weitz et al., 2015). It is also revealed that various gram positive bacterial pathogens such as S. aureus and gram negative microbes including such as *Pseudomonas desmolyticum, Klebsiella aerogenes* and *E. coli* were found to be successfully inhibited using CuO NPs synthesized from Gloriosa superba L. plant extract (Weitz et al., 2015). Furthermore, recent scientific work reported the improved antibacterial and antifungal activity of CuO NPs against *Candidia albicans, Escherichia coli, Pseudomonas aeruginosa, Staphylococcus aureus* and *Proteus vulgaris* when combined with fluconazole (Khashan et al., 2016; Weitz et al., 2015).

Aluminum (Al) NPs

Aluminum and Aluminum oxide nanoparticles (Al_2O_3 NPs) have inhibitory and cidal effects against microorganisms mainly through disruption of cell wall. A number of methods are used for the synthesis of Al_2O_3 NPs including hydrothermal processing, laser ablation and sputtering, in which laser ablation method the most commonly used technique due to its quick and high purity process(Ghorbani, 2014). Al_2O_3 and Al NPs have a numerous application in therapeutic nanomedicine. Plenty of studies have elicited the antimicrobial activities of Al_2O_3 NPs against microbial pathogens. For instance, Al NPs and Al_2O_3 NPs showed greater toxicity against *E. coli*, *B. subtilis* and *Pseudomonas species*(Mukherjee et al., 2011). Likewise, Al_2O_3 NPs have been found to show greater potency against *P. fluorescens* (Balasubramanyam et al., 2010), while disrupting bacterial cell wall through ROS leading to cell death is the predominant antimicrobial mechanism (Ansari et al., 2014).

Titanium Oxide (TiO_2) NPs

Titanium, which is one of the strongest metal, forms titanium dioxide (TiO_2) which is nontoxic molecule having stability and optical properties. TiO_2 is currently providing plenty of advantages in medicine, particularly the nanoparticulate form (TiO_2 NPs) is known for its vital antimicrobial activities against bacteria, fungi and viruses (Pushpanathan et al., 2013). Currently, electrochemical and sol-gel techniques are commonly used to synthesize TiO_2 NPs antimicrobial agents. The killing and inhibitory activity of TiO_2 NPs is mainly due to photocatalytic production of vigorous oxidizing power when illumined with ultraviolet light at a specific wavelength, which yields active free hydroxyl (OH-) radicals which subsequently form superoxide radicals. Moreover, same as other metal and metal oxide nanomaterials, TiO_2 NPs harm microbes through generating ROS which acts on bacterial membrane and cause DNA damage (Besinis et al., 2014). Several recent researches have reported the antimicrobial activity of TiO_2 NPs against bacteria such as *E. coli* biofilms, *S. aureus* and fungi pathogens (Méndez-Vilas, Ravishankar Rai, & Jamuna Bai, 2011).

Magnesium Oxide (MgO) NPs

Similarly, like other inorganic metal oxides such as ZnO, magnesium oxide (MgO) is stable and considered safe for use in humans. Unlike titanium dioxide nanomaterial MgO do not require photo activation to attain its antimicrobial activity (Tang & Feng, 2014). MgO NPs have a remarkable property of ability to adsorb and carry plenty of active halogens for longer duration. The process of being halogenated helped MgO NPs become stronger and have faster effect on the killing action of microorganisms. The MgO NPs exhibited killing activity against gram positive bacteria, gram negative bacteria and even some form of bacterial spores mainly due to various mechanisms including the formation of ROS, alkaline effects, electrostatic interactions, and lipid peroxidation. For instance, the surface structure of MgO NPs has high pH; and when it comes in contact with bacterial cell, the high pH (alkaline effect) damage the cell membrane of the bacteria, subsequently leading to death(Jin & He, 2011). Different scientific reports have showed that MgO NPs have antibacterial activity against ram positive and gram negative bacteria including *E. coli*, *S. aureus*, *P. aeruginosa* and *Bacillus subtilis*(Krishnamoorthy et al., 2012).

Organic Nanomaterials

Organic nanoparticles including Polymeric Antimicrobials, Chitosan, Organometallic Polymers, Peptides Nanomaterials, Quaternary Ammonium Compounds, Benzoic Acid, Phenol, Metallic and p-Hydroxy Benzoate Esters nano polymers etc... are used as antimicrobial agents in the current advent of nanomedicine. However, they are naturally less stable when compared with the inorganic NPs.

Polymeric NPs

Polymeric NPs are the most commonly available organic nanoparticles. Polymeric NPs are known to exhibit antimicrobial activity against pathogenic microbes. Polymeric NPs most commonly deploy contact killing or damage microorganisms through release of potent antimicrobial agents. They are nonvolatile, chemically stable and capable of binding to the surface of microbial cells. However, they have poor ability to influx through microbial cell membrane(Kenawy et al., 2017) .

Organometallic Polymers

Organometallic polymers are a result of a combination of metal elements combined with the polymer nanomaterials. Numerous organometallic polymers comprising metal elements in the external backbone chain (pendant groups) have been manufactured by varying the size of the alkyl groups (ethyl, methyl, octyl and butyl). These organometallic polymers have showed effective inhibition against DNA and RNA viruses viral pathogens (Beyth et al., 2015).

Peptides

Numerous peptides have been known to be synthesized using different techniques from hydrophilic amino acids such as lysine and hydrophobic amino acids including alanine, leucine and phenylalanine). Peptides synthesized using these amino acids particularly lysine and phenylalanine exhibit potent antimicrobial activity against *E. coli* and *P. aeruginosa.* Moreover, study reports showed that Peptides have high potenct and broad spectrum antimicrobial activity against *Serratia marcescens*, and *C. albicans*(Zhou et al., 2010).

Chitosan

Chitin is a long polymer of N-acetyl glucosamine chain and when deacetylated partially it forms chitosan. Chitosan has a broad spectrum antimicrobial activity. Furthermore, chitosan nanoparticles are effective antimicrobials against bacteria, fungi and viruses (Cheung et al., 2015). Chitosan is reported to have more effect against fungal and viral infections than disease caused by bacteria. The antimicrobial effect of chitosan is strongly reliant on the molecular weight of chitosan and inherent variances in the target bacterial cell wall structure. Most commonly chitosan poses its antimicrobial action by binding with the bacterial surface structure and cause agglutination, in turn enhancing leakage of intracellular components. Moreover, it penetrates the nucleus of fungi and inhibits RNA and protein syntheses (Qi et al., 2014). Chitosan predominantly showed basic characteristics including non-toxicity, biocompatibility, ability to enhance absorption and low level of immunogenicity. Chitosan NPs exhibit effective antifungal action against *Candidia albicans*. On the other hand, *A. niger* is comparatively resistant to chitosan NPs (Yien et al., 2012). Using chitosan NPs as antimicrobial agent is a promising, cost-effective, and technologically affordable antimicrobial killing method having tow toxicity and enhanced broad spectrum activity.

Quaternary Ammonium Compounds

Quaternary ammonium compounds most commonly form familiar disinfectants including cetrimonium chloride and benzalkonium chloride. Besides, these quaternary ammonium compounds also have antimicrobial activity particularly against bacteria and fungi. Their mechanism of action is through binding with negatively charged surface membrane which most commonly results in denaturation of enzymes and structural proteins of the bacterial surface membrane, subsequently causing cell membrane rupture in gram positive and gram negative bacteria; as well as *Candidia albicans* (Xue et al., 2015). Quaternary ammonium compounds exhibit better antimicrobial agents when used in combination with and Au NPs and TiO_2 NPs. For instance, quaternary ammonium compounds functionalized with SiNPs have been reported to have enhanced bactericidal activity against *E. coli* and *S. aureus*(Song et al., 2011).

FUTURE PERSPECTIVES

Nanoparticles are currently providing good hope for the treatment of many diseases that are challenging to manage and treat by conventional treatment techniques. Despite this fact, certain concerns need to be addressed regarding detailed understanding of mechanism of action, desirable synthesis methods, toxicity and over-all impact on the environment and population.

CONCLUSION

The emergence and spread of drug resistant microorganisms nowadays pose huge challenge for clinicians and new form of drugs are required for the treatment of many diseases. Microbial pathogens including bacteria, fungi and bacteria develop drug resistance through various mechanisms. Currently, advancements in nanomedicine open a new era of synthesizing nanoparticles which have numerous functions including antimicrobial activity against common disease causing microorganisms. Nanoparticles synthesized by biological methods have lessened environmental concerns than chemically synthesized nanoparticles. The mechanism of action of nanoparticles differs with the type and composition of the nanoparticles. Despite nanoparticles currently have substantial applications in biomedicine, the associated health risks and the limitations of nanomaterials need to be given due attention. More studies need to be conducted and tangible guidelines need to be developed by the experts from industries, governments, and the scientific community to reduce the toxicity and other unpredictable effects of nanoparticles on human health. Even if nanoparticles pose few drawbacks, they provide hope for the development of effective antimicrobial agents for the future.

ACKNOWLEDGMENT

This research received no specific grant from any funding agency in the public, commercial, or not-for-profit sectors.

REFERENCES

Alan, H. F., Neil, A., Keith, R. M., & Andrew, P. W. (2017). Drug resistance in eukaryotic microorganisms. *Nature Microbiology*, *1*(7), 16092. doi:10.1038/nmicrobiol.2016.92 PMID:27572976

Alex, S., & Tiwari, A. (2015). Functionalized gold nanoparticles: Synthesis, properties and applications—A review. *Journal of Nanoscience and Nanotechnology*, *15*(3), 1869–1894. doi:10.1166/jnn.2015.9718 PMID:26413604

Alexander, B. D., Johnson, M. D., Pfeiffer, C. D., Jiménez-Ortigosa, C., Catania, J., Booker, R., Castanheira, M., Messer, S. A., Perlin, D. S., & Pfaller, M. A. (2013). Increasing echinocandin resistance in Candida glabrata: Clinical failure correlates with presence of FKS mutations and elevated minimum inhibitory concentrations. *Clinical Infectious Diseases*, *56*(12), 1724–1732. doi:10.1093/cid/cit136 PMID:23487382

Allaker, R. P., & Ren, G. (2008). Potential impact of nanotechnology on the control of infectious disease. *Transactions of the Royal Society of Tropical Medicine and Hygiene*, *102*(1), 1–2. doi:10.1016/j.trstmh.2007.07.003 PMID:17706258

Anghel, I., Grumezescu, A. M., Andronescu, E., Anghel, A. G., Ficai, A., Saviuc, C., Grumezescu, V., Vasile, B. S., & Chifiriuc, M. C. (2012). Magnetite nanoparticles for functionalized textile dressing to prevent fungal biofilms development. *Nanoscale Research Letters*, *7*(1), 7. doi:10.1186/1556-276X-7-501 PMID:22950367

Ansari, M. A., Khan, H. M., Khan, A. A., Cameotra, S. S., Saquib, Q., & Musarrat, J. (2014). Interaction of Al2O3 nanoparticles with Escherichia coli and their cell envelope biomolecules. *Journal of Applied Microbiology*, *116*(4), 772–783. doi:10.1111/jam.12423 PMID:24354999

Antonelli, G., & Turriziani, O. (2012). Antiviral therapy: Old and current issues. *International Journal of Antimicrobial Agents*, *40*(2), 95–102. doi:10.1016/j.ijantimicag.2012.04.005 PMID:22727532

Arokiyaraj, S., Saravanan, M., Prakash, N. U., Arasu, M. V., Vijayakumar, B., & Vincent, S. (2013). Enhanced antibacterial activity of iron oxide magnetic nanoparticles treated with Argemone mexicana L. leaf extract: An in vitro study. *Materials Research Bulletin*, *8*(9), 3323–3327. doi:10.1016/j.materresbull.2013.05.059

Awad, H. M., Kamal, Y. S., Aziz, R., Sarmidi, M. R., & El-Enshasy, H. A. (2012). Antibiotics as microbial secondary metabolites: Production and application. *Jurnal Teknologi*, *59*, 101–111. doi:10.11113/jt.v59.1593

Badri Narayanan, K., & Sakthivel, N. (2018). Coriander leaf mediated biosynthesis of gold nanoparticles. *Materials Letters*, *62*(30), 4588–4590. doi:10.1016/j.matlet.2008.08.044

Balasubramanyam, A., Sailaja, N., Mahboob, M., Rahman, M. F., Hussain, S. M., & Grover, P. (2010). In vitro mutagenicity assessment ofaluminium oxide nanomaterials using the Salmonella/microsome assay. *Toxicology In Vitro, 24*(6), 1871–1876. doi:10.1016/j.tiv.2010.07.004 PMID:20624453

Balzarini, J., Ford, A., Maguire, N. M., John, J., Das, K., & Arnold, E. (n.d.). Alpha-carboxynucleoside phosphonates: direct-acting inhibitors of viral DNA polymerases. *Future Medicinal Chemistry, 11*(2). doi:10.4155/fmc-2018-0324

Behera, S. S., Patra, J. K., Pramanik, K., Panda, N., & Thatoi, H. (2012). Characterization and Evaluation of antibacterial activities of chemically synthesized Iron Oxide nanoparticles. *World J. Nano Sci. Eng.*, *2*(04), 196–200. doi:10.4236/wjnse.2012.24026

Besinis, A., de Peralta, T., & Handy, R. D. (2014). The antibacterial effects of silver, titanium dioxide and silica dioxide nanoparticles compared to the dental disinfectant chlorhexidine on Streptococcus mutans using a suite of bioassays. *Nanotoxicology*, 8. PMID:23092443

Beyth, N., Houri-Haddad, Y., Domb, A., Khan, W., & Hazan, R. (2015). Alternative antimicrobial approach: Nano-antimicrobial materials. *Evidence-Based Complementary and Alternative Medicine*, *20*, 151–169. PMID:25861355

Biswas, A., Haldane, A., Arnold, E., & Levy, R. (2019). Epistasis and entrenchment of drug resistance in HIV-1 subtype B. *eLife, 50*(24), 1-30.

Brayner, R., Ferarri-Iliou, R., Brivois, N., Djediat, S., Benedetti, M. F., & Fiévet, F. (2016). Toxicological impact studies based on Escherichia coli bacteria in ultrafine ZnO nanoparticles colloidal medium. *Nano Letters*, *6*(4), 866–870. doi:10.1021/nl052326h PMID:16608300

Chamundeeswari, M., Sobhana, S. S., Jacob, J. P., Kumar, M. G., Devi, M. P., Sastry, T. P., & Mandal, A. B. (2010). Preparation, characterization and evaluation of a biopolymeric gold nanocomposite with antimicrobial activity. *Biotechnology and Applied Biochemistry*, *55*(1), 29–35. doi:10.1042/BA20090198 PMID:19929854

Chatterjee, A. K., Sarkar, R. K., Chattopadhyay, A. P., Aich, P., Chakraborty, R., & Basu, T. (2012). A simple robust method for synthesis of metallic copper nanoparticles of high antibacterial potency against E. coli. *Nanotechnology*, *23*(8), 85–103. doi:10.1088/0957-4484/23/8/085103 PMID:22293320

Chauhan, A., Zubair, S., Tufail, S., Sherwani, A., Sajid, M., Raman, S. C., Azam, A., & Owais, M. (2011). Fungus-mediated biological synthesis of gold nanoparticles: Potential in detection of liver cancer. *International Journal of Nanomedicine*, *6*, 2305–2319. PMID:22072868

Chen, W. J., Tsai, P. J., & Chen, Y. C. (2008). Functional Fe3O4/TiO2 core/shell magnetic nanoparticles as photokilling agents for pathogenic bacteria. *Small*, *4*(4), 485–491. doi:10.1002mll.200701164 PMID:18348230

Cheung, R. C. F., Ng, T. B., Wong, J. H., & Chan, W. Y. (2015). Chitosan: An update on potential biomedical and pharmaceutical applications. *Marine Drugs*, *13*(8), 5156–5186. doi:10.3390/md13085156 PMID:26287217

Chifiriuc, C., Grumezescu, V., Grumezescu, A. M., Saviuc, C., Lazˇar, V., & Andronescu, E. (2012). Hybrid magnetite nanoparticles/Rosmarinus officinalis essential oil nanobiosystem with antibiofilm activity. *Nanoscale Research Letters*, *7*(1), 14. doi:10.1186/1556-276X-7-209 PMID:22490675

Chung, M. H., Beck, I. A., Dross, S., Tapia, K., Kiarie, J. N., Richardson, B. A., Overbaugh, J., Sakr, S. R., John-Stewart, G. C., & Frenkel, L. M. (2014). Oligonucleotide ligation assay detects HIV drug resistance associated with virologic failure among antiretroviral-naive adults in Kenya. *Journal of Acquired Immune Deficiency Syndromes*, *67*(3), 246–253. doi:10.1097/QAI.0000000000000312 PMID:25140907

Clarence, Y., & Geoffrey, S. (2015). Nanoparticles as potential new generation broad spectrum antimicrobial agents. *Daru: Journal of Faculty of Pharmacy, Tehran University of Medical Sciences*, *23*(1), 43. doi:10.118640199-015-0125-6 PMID:26329777

Davey, P., Sneddon, J., & Nathwani, D. (2017). Overview of strategies for overcoming the challenge of antimicrobial resistance. *Expert Review of Clinical Pharmacology*, *3*(5), 667–686. doi:10.1586/ecp.10.46 PMID:22111749

Denning, D. W., & Bromley, M. J. (2015). Infectious disease. How to bolster the antifungal pipeline. *Science*, *347*(6229), 1414–1416. doi:10.1126cience.aaa6097 PMID:25814567

Drlica, K., Malik, M., Kernsm, R. J., & Zhaom, X. (2008). Quinolone-mediated bacterial death. *Antimicrobial Agents and Chemotherapy*, *52*(2), 385–392. doi:10.1128/AAC.01617-06 PMID:17724149

Duraffour, S., Andrei, G., Topalis, D., Krečmerová, M., Crance, J. M., Garin, D., & Snoeck, R. (2012, July 01). Mutations Conferring Resistance to Viral DNA Polymerase Inhibitors in Camelpox Virus Give Different Drug-Susceptibility Profiles in Vaccinia Virus. *Journal of Virology*, *86*(13), 7310–7325. Advance online publication. doi:10.1128/JVI.00355-12 PMID:22532673

Epaulard, O., Villier, C., Ravaud, P., Chosidow, O., Blanche, S., Mamzer-Bruneel, M.-F., Thiebaut, A., Leccia, M.-T., & Lortholary, O. (2013). A multistep voriconazole-related phototoxic pathway may lead to skin carcinoma: Results from a French nationwide study. *Clinical Infectious Diseases*, *57*(12), e182–e188. doi:10.1093/cid/cit600 PMID:24046296

Fayaz, A. M., Balaji, K., Girilal, M., Yadav, R., Tech, M., Kalaichelvan, P. T., & Venketesan, R. (2010). Biogenic synthesis of silver nanoparticles and their synergistic effect with antibiotics: A study against gram-positive and gram-negative bacteria. *Nanomedicine (London)*, *6*(1), 103–109. doi:10.1016/j.nano.2009.04.006 PMID:19447203

Fearns, R., & Plemper, R. K. (2017, April 15). Polymerases Of Paramyxoviruses And Pneumoviruses. *Virus Research*, *234*, 87–102. doi:10.1016/j.virusres.2017.01.008 PMID:28104450

Ghorbani, H. R. (2014). A review of methods for synthesis of Al nanoparticles. *Oriental Journal of Chemistry*, *30*(4), 1941–1949. doi:10.13005/ojc/300456

Gil-Tomas, J., Tubby, S., Parkin, I. P., Narband, N., Dekker, L., Nair, S. P., Wilson, M., & Street, C. (2007). Lethal photosensitisation of Staphylococcus aureus using a toluidine blue O-tiopronin-gold nanoparticles conjugate. *Journal of Materials Chemistry*, *17*(35), 3739–3746. doi:10.1039/b706615e

Gu, H., Ho, P. L., Tong, E., Wang, L., & Xu, B. (2013). Presenting vancomycin on nanoparticles to enhance antimicrobial activities. *Nano Letters*, *3*(9), 1261–1263. doi:10.1021/nl034396z

He, S., Guo, Z., Zhang, Y., Zhang, S., Wang, J., & Gu, N. (2017). Biosynthesis of gold nanoparticles using the bacteria Rhodopseudomonas capsulata. *Materials Letters*, *61*(18), 3984–3987. doi:10.1016/j.matlet.2007.01.018

Hegstad, K., Mikalsen, T., Coque, T. M., Werner, G., & Sundsfjord, A. (2010). Mobile genetic elements and their contribution to the emergence of antimicrobial resistant Enterococcus faecalis and Enterococcus faecium. *Clinical Microbiology and Infection*, *16*(6), 541–554. doi:10.1111/j.1469-0691.2010.03226.x PMID:20569265

Hidayat, L. K., Hsu, D. I., Quist, R., Shriner, K. A., & Wong-Beringer, A. (2006). A. High-dose vancomycin therapy for methicillin-resistant Staphylococcus aureus infections: Efficacy and toxicity. *Archives of Internal Medicine*, *166*(19), 2138–2144. doi:10.1001/archinte.166.19.2138 PMID:17060545

Huh, A. J., & Kwon, Y. J. (2011). "Nanoantibiotics": A new paradigm for treating infectious diseases using nanomaterials in the antibiotics resistant era. *Journal of Controlled Release*, *156*(2), 128–145. doi:10.1016/j.jconrel.2011.07.002 PMID:21763369

Jena, M., Mishra, S., Jena, S., & Mishra, S. S. (2013). Nanotechnology-future prospect in recent medicine: A review. *International Journal of Basic and Clinical Pharmacology*, *2*, 353–359. doi:10.5455/2319-2003.ijbcp20130802

Jin, T., & He, Y. (2011). Antibacterial activities of magnesium oxide (MgO) nanoparticles against food-borne pathogens. *Journal of Nanoparticle Research*, *13*(12), 6877–6885. doi:10.100711051-011-0595-5

Katayama, Y., Ito, T., & Hiramatsu, K. (2010). A new class of genetic element, staphylococcus cassette chromosome mec, encodes methicillin resistance in Staphylococcus aureus. *Antimicrobial Agents and Chemotherapy*, *44*(6), 1549–1555. doi:10.1128/AAC.44.6.1549-1555.2000 PMID:10817707

Kenawy, E. R., Worley, S. D., & Broughton, R. (2017). The chemistry and applications of antimicrobial polymers: A state-of-the-art review. *Biomacromolecules*, *8*(5), 1359–1384. doi:10.1021/bm061150q PMID:17425365

Khashan, K. S., Sulaiman, G. M., & Abdulameer, F. A. (2016). Synthesis and Antibacterial Activity of CuO Nanoparticles Suspension Induced by Laser Ablation in Liquid. *Arabian Journal for Science and Engineering*, *41*(1), 301–310. doi:10.100713369-015-1733-7

Kidd, S. E., Goeman, E., Meis, J. F., Slavin, M. A., & Verweij, P. E. (2015). Multi-triazole-resistant Aspergillus fumigatus infections in Australia. *Mycoses*, *58*(6), 350–355. doi:10.1111/myc.12324 PMID:25885568

Krishnamoorthy, K., Manivannan, G., Kim, S. J., Jeyasubramanian, K., & Premanathan, M. (2012). Antibacterial activity of MgO nanoparticles based on lipid peroxidation by oxygen vacancy. *Journal of Nano Research*, 14.

Kwon, H., & Lok, A. S. (2011). Hepatitis B therapy. *Nature Reviews. Gastroenterology & Hepatology*, *8*(5), 275–284. doi:10.1038/nrgastro.2011.33 PMID:21423260

Leroux, S., & Ullmann, A. J. (2013). Management and diagnostic guidelines for fungal diseases in infectious diseases and clinical microbiology: Critical appraisal. *Clinical Microbiology and Infection, 19*(12), 1115–1121. doi:10.1111/1469-0691.12426 PMID:24118188

Li, N., Zhao, P., & Astruc, D. (2014). Anisotropic Gold Nanoparticles: Synthesis, Properties, Applications, and Toxicity. *Angewandte Chemie International Edition, 53*(7), 1756–1789. doi:10.1002/anie.201300441 PMID:24421264

Li, Q., Mahendra, S., Lyon, D. Y., Brunet, L., Liga, M. V., Li, D., & Alvarez, P. J. (2018). Antimicrobial nanomaterials for water disinfection and microbial control: Potential applications and implications. *Water Research, 42*(18), 4591–4602. doi:10.1016/j.watres.2008.08.015 PMID:18804836

Li, X., Robinson, S. M., Gupta, A., Saha, K., Jiang, Z., Moyano, D. F., Sahar, A., Riley, M. A., & Rotello, V. M. (2014). Functional gold nanoparticles as potent antimicrobial agents against multi-drug-resistant bacteria. *ACS Nano, 8*(10), 10682–10686. doi:10.1021/nn5042625 PMID:25232643

Lok, A. S., & McMahon, B. J. (2009). Chronic hepatitis B: Update. *Hepatology (Baltimore, Md.), 50*(3), 661–662. doi:10.1002/hep.23190 PMID:19714720

Lovleva, A., & Bonomo, R. A. (2017). The Ecology of Extended-Spectrum β-lactamases (ESBLs) in the Developed World. *Journal of Travel Medicine, 24*(suppl_1), S44–S51. doi:10.1093/jtm/taw102 PMID:28521000

Ma, J., Zhang, Y., Chen, X., Jin, Y., Chen, D., Wu, Y., Cui, J., Wang, H., Liu, J., Li, N., & Gao, F. (2013). Association of Preexisting Drug-Resistance Mutations and Treatment Failure in Hepatitis B Patients. *PLoS One, 8*(7), e67606. doi:10.1371/journal.pone.0067606 PMID:23935839

Maarseveen, N., & Boucher, C. (2006). *Resistance to protease inhibitors.* Antiretroviral Resistance in Clinical Practice.

Mariana, C., Alina, M. H., Carmen, C., Grigore, M., & Veronica, L. (2016). *Antibiotic Drug Delivery Systems for the Intracellular Targeting of Bacterial Pathogens.* Smart Drug Delivery System INTECH. Retrieved from https://www.intechopen.com/books/smart-drug-delivery-system

Martinez, J. L., & Baquero, F. (2018). Mutation frequencies and antibiotic resistance. *Antimicrobial Agents and Chemotherapy, 44*(7), 1771–1777. doi:10.1128/AAC.44.7.1771-1777.2000 PMID:10858329

Mehrad, B., Clark, N. M., Zhanel, G. G., & Lynch, J. P. III. (2015). Antimicrobial Resistance in Hospital-Acquired Gram-Negative Bacterial Infections. *Chest, 147*(5), 1413–1421. doi:10.1378/chest.14-2171 PMID:25940252

Mühling, M., Bradford, A., Readman, J. W., Somerfield, P. J., & Handy, R. D. (2019). An investigation into the effects of silver nanoparticles on antibiotic resistance of naturally occurring bacteria. *Marine Environmental Research, 68*(5), 278–283. doi:10.1016/j.marenvres.2009.07.001 PMID:19665221

Mukherjee, A., Mohammed Sadiq, I., Prathna, T. C., & Chandrasekaran, N. (2011). Antimicrobial activity of aluminium oxide nanoparticles for potential clinical applications. *Sci. Microb. Pathog. Commun. Curr. Res. Technol. Adv., 1*, 245–251.

Naika, H. R., Lingaraju, K., Manjunath, K., Kumar, D., Nagaraju, G., Suresh, D., & Nagabhushana, H. (2015). Green synthesis of CuO nanoparticles using Gloriosa superba L. extract and their antibacterial activity. *Journal of Taibah University for Science: JTUSCI*, *9*(1), 7–12. doi:10.1016/j.jtusci.2014.04.006

Nordmann, P., & Poirel, L. (2014). The difficult-to-control spread of carbapenemase producers among Enterobacteriaceae worldwide. *Clinical Microbiology and Infection*, *20*(9), 821–830. doi:10.1111/1469-0691.12719 PMID:24930781

O'Neill, J., Davies, S., Rex, J., White, L. J., & Murray, R. (2016). *Tackling Drug-Resistant Infections Globally: Final Report and Recommendations*. Wellcome Trust and UK Government.

Oakley, A. J., Barrett, S., Peat, T. S., Newman, J., Streltsov, V. A., Waddington, L., Saito, T., Tashiro, M., & McKimm-Breschkin, J. L. (2010). Structural and Functional Basis of Resistance to Neuraminidase Inhibitors of Influenza B Viruses. *Journal of Medicinal Chemistry*, *53*(17), 6421–6431. doi:10.1021/jm100621s PMID:20695427

Ohlsen, K., Ternes, T., Werner, G., Wallner, U., Löffler, D., Ziebuhr, W., Witte, W., & Hacker, J. (2013). Impact of antibiotics on conjugational resistance gene transfer in Staphylococcus aureus in sewage. *Environmental Microbiology*, *5*(8), 711–716. doi:10.1046/j.1462-2920.2003.00459.x PMID:12871238

Palmer, K. L., Kos, V. N., & Gilmore, M. S. (2010). Horizontal gene transfers and the genomics of enterococcal antibiotic resistance. *Current Opinion in Microbiology*, *13*(5), 632–639. doi:10.1016/j.mib.2010.08.004 PMID:20837397

Park, M. Z., Neigh, A. M., Vermeulen, J. P., de la Fonteyne, L. J., Verharen, H. W., Briedé, J. J., van Loveren, H., & de Jong, W. H. (2011). The effect of particle size on the cytotoxicity, inflammation, developmental toxicity and genotoxicity of silver nanoparticles. *Biomaterials*, *32*(36), 9810–9817. doi:10.1016/j.biomaterials.2011.08.085 PMID:21944826

Perelson, A. S., & Ribeiro, R. M. (2013). Modeling the within-host dynamics of HIV infection. *BMC Biology*, *11*(1), 96. doi:10.1186/1741-7007-11-96 PMID:24020860

Petersen, J., Dandri, M., Mier, W., Lutgehetmann, M., Volz, T., von Weizsacker, F., Haberkorn, U., Fischer, L., Pollok, J.-M., Erbes, B., Seitz, S., & Urban, S. (2008). Prevention of hepatitis B virus infection in vivo by entry inhibitors derived from the large envelope protein. *Nature Biotechnology*, *26*(3), 335–341. doi:10.1038/nbt1389 PMID:18297057

Pfaller, M. A., & Diekema, D. J. (2012). Progress in antifungal susceptibility testing of Candida spp. by use of Clinical and Laboratory Standards Institute broth microdilution methods, 2010 to 2012. *Journal of Clinical Microbiology*, *50*(9), 2846–2856. doi:10.1128/JCM.00937-12 PMID:22740712

Phillips, A. N., Cambiano, V., Miners, A., Revill, P., Pillay, D., Lundgren, J. D., Bennett, D., Raizes, E., Nakagawa, F., De Luca, A., Vitoria, M., Barcarolo, J., Perriens, J., Jordan, M. R., & Bertagnolio, S. (2014). Effectiveness and cost-effectiveness of potential responses to future high levels of transmitted HIV drug resistance in antiretroviral drug-naive populations beginning treatment: Modelling study and economic analysis. *The Lancet. HIV*, *1*(2), 85–93. doi:10.1016/S2352-3018(14)70021-9 PMID:26423990

Pommier, Y., Johnson, A. A., & Marchand, C. (2005). Integrase inhibitors to treat HIV/AIDS. *Nature Reviews. Drug Discovery*, *4*(3), 236–248. doi:10.1038/nrd1660 PMID:15729361

Prabhu, S., & Poulose, E. K. (2012). Silver nanoparticles: Mechanism of antimicrobial action, synthesis, medical applications, and toxicity effects. *International Nano Letters*, *32*(1), 2–10. doi:10.1186/2228-5326-2-32

Prabhu, Y. T., Rao, K. V., Kumari, B. S., Kumar, V. S., & Pavani, T. (2015). Synthesis of Fe3O4 nanoparticles and its antibacterial application. *International Nano Letters*, *5*(2), 85–92. doi:10.100740089-015-0141-z

Pushpanathan, M., Gunasekaran, P., & Rajendhran, J. (2013). Antimicrobial Peptides: Versatile Biological Properties. *International Journal of Peptides*, *12*, 92–119. PMID:23935642

Qi, L., Xu, Z., Jiang, X., Hu, C., & Zou, X. (2014). Preparation and antibacterial activity of chitosan nanoparticles. *Carbohydrate Research*, *339*(16), 2693–2700. doi:10.1016/j.carres.2004.09.007 PMID:15519328

Rai, A., Prabhune, A., & Perry, C. C. (2010). Antibiotic mediated synthesis of gold nanoparticles with potent antimicrobial activity and their application in antimicrobial coatings. *Journal of Materials Chemistry*, *20*(32), 6789–6798. doi:10.1039/c0jm00817f

Rai, A., Yadav, A. G., & Gade, A. (2009). Silver nanoparticles as a new generation of antimicrobials. *Biotechnology Advances*, *27*(1), 76–83. doi:10.1016/j.biotechadv.2008.09.002 PMID:18854209

Ravishankar, R. V., & Jamuna, B. A. (2011). Nanoparticles and their potential application as antimicrobials. In A. Méndez-Vilas (Ed.), *Science against Microbial Pathogen: Communicating Current Research and Technological Advances* (Vol. 2, pp. 197–209). Formatex Research Center.

Ravishankar Rai, V., & Jamuna Bai, A. (2011). Nanoparticles and their potential application as antimicrobials. In A. Méndez-Vilas (Ed.), *Science against Microbial Pathogen: Communicating Current Research and Technological Advances* (Vol. 2, pp. 197–209). Formatex Research Center.

Rhee, S. Y., Blanco, J. L., Jordan, M. R., Taylor, J., Lemey, P., Varghese, V., Hamers, R. L., Bertagnolio, S., de Wit, T. F. R., Aghokeng, A. F., Albert, J., Avi, R., Avila-Rios, S., Bessong, P. O., Brooks, J. I., Boucher, C. A. B., Brumme, Z. L., Busch, M. P., Bussmann, H., ... Shafer, R. W. (2015). Geographic and Temporal Trends in the Molecular Epidemiology and Genetic Mechanisms of Transmitted HIV-1 Drug Resistance: An Individual-Patient- and SequenceLevel Meta-Analysis. *PLoS Medicine*, *12*(4), e1001810. doi:10.1371/journal.pmed.1001810 PMID:25849352

Roshmi, T., Soumya, K. R., Jyothis, M., & Radhakrishnan, E. K. (2015). Effect of biofabricated gold nanoparticle-based antibiotic conjugates on minimum inhibitory concentration of bacterial isolates of clinical origin. *Gold Bulletin*, *48*(1-2), 63–71. doi:10.100713404-015-0162-4

Ruppé, É., Woerther, P.-L., & Barbier, F. (2015). Mechanisms of Antimicrobial Resistance in Gram-Negative Bacilli. *Annals of Intensive Care*, *5*(1), 21. doi:10.118613613-015-0061-0 PMID:26261001

Sampath, M., Vijayan, R., Tamilarasu, E., Tamilselvan, A., & Sengottuvelan, B. (2014). Green synthesis of novel jasmine bud-shaped copper nanoparticles. *Journal of Nanotechnology*, *2014*, 1–7. Advance online publication. doi:10.1155/2014/626523

Santajit, S., & Indrawattana, N. (2016). Mechanisms of Antimicrobial Resistance in ESKAPE Pathogens. *BioMed Research International*, *2016*, 1–8. doi:10.1155/2016/2475067 PMID:27274985

Schmieder, R., & Edwards, R. (2012). Insights into antibiotic resistance through metagenomic approaches. *Future Microbiology*, *7*(1), 73–89. doi:10.2217/fmb.11.135 PMID:22191448

Seiffert, S. N., Hilty, M., Perreten, V., & Endimiani, A. (2013). Extended spectrum cephalosporin-resistant gram-negative organisms in livestock: An emerging problem for human health? *Drug Resistance Updates*, *16*(1), 22–45. doi:10.1016/j.drup.2012.12.001 PMID:23395305

Shor, E., & Perlin, D. S. (2015). Coping with stress and the emergence of multidrug resistance in fungi. *PLoS Pathogens*, *11*(3), e1004668. doi:10.1371/journal.ppat.1004668 PMID:25790300

Song, J., Kong, H., & Jang, J. (2011). Bacterial adhesion inhibition of the quaternary ammonium functionalized silica nanoparticles. *Coll. Surf.*, *82*(2), 651–656. doi:10.1016/j.colsurfb.2010.10.027 PMID:21115282

Strasfeld, L., & Chou, S. (2010, June). Antiviral Drug Resistance: Mechanisms and Clinical Implications. *Infectious Disease Clinics of North America*, *24*(2), 413–437. doi:10.1016/j.idc.2010.01.001 PMID:20466277

Tang, Z. X., & Feng, B. (2014). MgO nanoparticles as antibacterial agent: Preparation and activity. *Brazilian Journal of Chemical Engineering*, *31*(3), 591–601. doi:10.1590/0104-6632.20140313s00002813

Tanser, F., Bärnighausen, T., Grapsa, E., Zaidi, J., & Newell, M. L. (2013). High coverage of ART associated with decline in risk of HIV acquisition in rural KwaZulu-Natal, South Africa. *Science*, *339*(6122), 966–971. doi:10.1126cience.1228160 PMID:23430656

Tayel, A. A., El-Tras, W. F., Moussa, S., El-Baz, A. F., Mahrous, H., Salem, M. F., & Brimer, L. (2011). Antibacterial action of zinc oxide nanoparticles against foodborne pathogens. *Journal of Food Safety*, *31*(2), 211–218. doi:10.1111/j.1745-4565.2010.00287.x

Tozzi, V. (2010). Pharmacogenetics of antiretrovirals. *Antiviral Research*, *85*(1), 190–200. doi:10.1016/j.antiviral.2009.09.001 PMID:19744523

Turos, E., Reddy, G. S., Greenhalgh, K., Ramaraju, P., Abeylath, S. C., Jang, S., Dickey, S., & Lim, D. V. (2017). Penicillin-bound polyacrylate nanoparticles: Restoring the activity of β- lactam antibiotics against MRSA. *Bioorganic & Medicinal Chemistry Letters*, *17*(12), 3468–3472. doi:10.1016/j.bmcl.2007.03.077 PMID:17420125

Uğur, S. S., Sarıışık, M., Aktaş, A. H., Uçar, M. C., & Erden, E. (2010). Modifying of cotton fabric surface with nano-ZnO multilayer films by layer-by-layer deposition method. *Nanoscale Research Letters*, *5*(7), 1204–1210. doi:10.100711671-010-9627-9 PMID:20596450

Van der Linden, J. W., Arendrup, M. C., Warris, A., Lagrou, K., Pelloux, H., Hauser, P. M., Chryssanthou, E., Mellado, E., Kidd, S. E., Tortorano, A. M., Dannaoui, E., Gaustad, P., Baddley, J. W., Uekötter, A., Lass-Flörl, C., Klimko, N., Moore, C. B., Denning, D. W., Pasqualotto, A. C., ... Verweij, P. E. (2015). Prospective multicenter international surveillance of azole resistance in Aspergillus fumigatus. [*Emerging Infectious Diseases*, *21*(6), 1041–1044. doi:10.3201/eid2106.140717 PMID:25988348

Van Hoek, A., Mevius, D., Guerra, B., Mullany, P., Roberts, A., & Aarts, H. (2011). Acquired antibiotic resistance genes: An overview. *Frontiers in Microbiology*, *2*, 203. doi:10.3389/fmicb.2011.00203 PMID:22046172

Von Nussbaum, F., Brands, M., Hinzen, B., Weigand, S., & Habich, D. (2006). Antibacterial natural products in medicinal chemistry—Exodus or revival? *Angewandte Chemie International Edition*, *45*(31), 5072–5129. doi:10.1002/anie.200600350 PMID:16881035

Weir, E., Lawlor, A., Whelan, A., & Regan, F. (2008). The use of nanoparticles in anti-microbial materials and their characterization. *Analyst (London)*, *133*(7), 835–845. doi:10.1039/b715532h PMID:18575632

Weitz, I. S., Maoz, M., Panitz, D., Eichler, S., & Segal, E. (2015). Combination of CuO nanoparticles and fluconazole: Preparation, characterization, and antifungal activity against Candida albicans. *Journal of Nanoparticle Research*, *17*(8), 342. doi:10.100711051-015-3149-4

World Health Organization. (n.d.). *Hepatitis B, Fact sheet N° 204* [updated July 2016]. Available from: http: // www. who. int/mediacentre /factsheets/fs204/en/

Xie, Y., He, Y., Irwin, P. L., Jin, T., & Shi, X. (2011). Antibacterial activity and mechanism of action of zinc oxide nanoparticles against Campylobacter jejuni. *Applied and Environmental Microbiology*, *77*(7), 2325–2331. doi:10.1128/AEM.02149-10 PMID:21296935

Xue, Y., Xiao, H., & Zhang, Y. (2015). Antimicrobial Polymeric Materials with Quaternary Ammonium and Phosphonium Salts. *International Journal of Molecular Sciences*, *16*(2), 3626–3655. doi:10.3390/ijms16023626 PMID:25667977

Yamamoto, O. (2011). Influence of particle size on the antibacterial activity of zinc oxide. *International Journal of Inorganic Materials*, *3*(7), 643–646. doi:10.1016/S1466-6049(01)00197-0

Yien, L., Zin, N.M., Sarwar, A., & Katas, H. (2012). Antifungal activity of chitosan nanoparticles and correlation with their physical properties. *Int. J. Biomater.*. doi:10.1155/2012/632698

Zhou, C., Qi, X., Li, P., Chen, W. N., Mouad, L., Chang, M. W., Leong, S. S. J., & Chan-Park, M. B. (2010). High potency and broad-spectrum antimicrobial peptides synthesized via ring-opening polymerization of α-aminoacid-N-carboxyanhydrides. *Biomacromolecules*, *11*(1), 60–67. doi:10.1021/bm900896h PMID:19957992

Zhu, W., Murray, P. R., Huskins, W. C., Jernigan, J. A., McDonald, L. C., Clark, N. C., Anderson, K. F., McDougal, L. K., Hageman, J. C., Olsen-Rasmussen, M., Frace, M., Alangaden, G. J., Chenoweth, C., Zervos, M. J., Robinson-Dunn, B., Schreckenberger, P. C., Reller, L. B., Rudrik, J. T., & Patel, J. B. (2010). Dissemination of an Enterococcus Inc18-Like vanA plasmid associated with vancomycin-resistant Staphylococcus aureus. *Antimicrobial Agents and Chemotherapy*, *54*(10), 4314–4320. doi:10.1128/AAC.00185-10 PMID:20660665

Chapter 2
Biopolymer-Based Nanomaterials for Biomedical Applications:
Biomedical Applications of Electrospun Nanofibers

Karthikeyan Mahendhran
Thiagarajar College, India

Murugappan Ramanathan
Thiagarajar College, India

ABSTRACT

Biopolymers derived from natural sources like plants, microorganisms, and animals have attracted significant attention towards biomedical applications due to their biodegradability, biocompatibility, zero toxicity, and abundant nature compared to the synthetic counterparts. In recent years, the emergence of nanoscience has evidenced notable preference in the development of safe and potent nano-biomaterials. Among the various methods reported for the fabrication of nanomaterials, electrospinning has received prime focus in the generation of tuneable nanostructures with versatile properties. This chapter highlights the importance, mode of synthesis, applications, and limitations of biopolymer nanomaterials from natural origin.

INTRODUCTION

Biopolymers include a wide range of materials from plants, animals, insects, microorganisms and synthetic materials which are biodegradable in nature. The term "biopolymers" encompass not only polymeric materials that occur naturally but also natural substances that have been polymerized into high molecular weight materials by biological systems (Lau *et al.*, 2011). Compared to the synthetic polymers,

DOI: 10.4018/978-1-7998-5049-6.ch002

biopolymers possess diverse and versatile biomedical applications. Natural polymers such as cellulose, chitosan, alginate, pectin, starch etc., exhibit remarkable properties such as purity, crystallinity, tensile solidity, improved elasticity, and extensive surface area than the synthetic polymers. Biodegradable polymers are broadly classified into three types. 1. Chemically synthesized biopolymers: It comprises of biodegradable polymers synthesized by chemical methods with biodegradable properties. It includes diverse polymers such as polycaprolactone, polyethylene oxide and polylactic acid etc., They are amenable to enzymatic modification or microbial degradation. 2. Biobased biodegradable polymers: These groups of polymers are biodegradable and compostable ex: cellulose, starch etc., these polymers are partially degradable and their degraded fragments are obstinate and prevail in nature for a long period of time. 3. Microbial biopolymers: Microbially derived polymers are synthesized by microbes under stress conditions both intra and extra cellular to cope up the harsh environmental conditions. Based on the structure and chemical composition, microbial biopolymers are grouped as bacterial cellulose, poly (γ- glutamic acid), hyaluronic acid, dextran and polyhydroxyalkanoates. Microbially derived polymers are completely safe, biodegradable, and biocompatible and zero toxic in nature. In recent years, studies on microbially derived biopolymers pretend to possess tunable properties and applications compared to other biodegradable polymers from natural sources. General outline classifications of biopolymers are summarized in Figure 1.

Nanotechnology is garnering significant attention in scientific research due to its diverse and commercial applications in diverse fields. Nanomaterials are given prime importance in the development of smart materials with versatile properties. The unique physicochemical properties of nanomaterials such as size, high surface to volume ratio and quantum nature enable them for the application in multiple fields like medicine, biology and chemistry. The field of bionanotechnology ought to have a special focus due to its fascinating interdisciplinary approach of research that brings together biology, materials science, and nanotechnology for better clinical endings. Application of biopolymer based nanomaterials is gathering interest in diverse biomedical applications such as drug and gene delivery, tissue engineering, wound healing, cancer therapy etc., Medical applications of biopolymers are depicted in Figure 2. Different nanomaterials such as nanofibers, nanocomposites and nanowires fabricated by nanotechniques such as electrospinning, sol-gel method, chemical vapour deposition and self assembly techniques have received importance in the field of bionanotechnology. Therefore, in this chapter it is intended discuss about the features, applications and limitations of biopolymer nanomaterials and their role in biomedical applications.

Nanofibers

In recent decades, a steady progress has been witnessed in the development of novel and functional polymer fibers. Natural fibers, due to their versatile material properties and their abundant nature have become inevitable in the day to day life applications (Kotek, 2008). Nanofibres have emerged as an important group of nanostructures with multiple functionalities compared to the other nanomaterials. Nanofibers are nonwoven, three dimensional, porous and exhibit increased surface to volume ratio with controlled porosity makes them suitable for various applications including tissue engineering (Chae et al., 2011) drug delivery (Kenawy et al., 2009) wound dressing (Khil et al., 2003) filtration, energy storage, defense, and security (Ramakrishna et al., 2006; Thavasi et al., 2008). Nanofibers fibers have received more attention in material and biological sciences due to their ease of fabrication and the ability to control their compositional, structural and functional properties. A wide range of polymers can be used to produce polymer non woven mats, consisting of nanofibers.

Figure 1.Outline classification of biopolymers

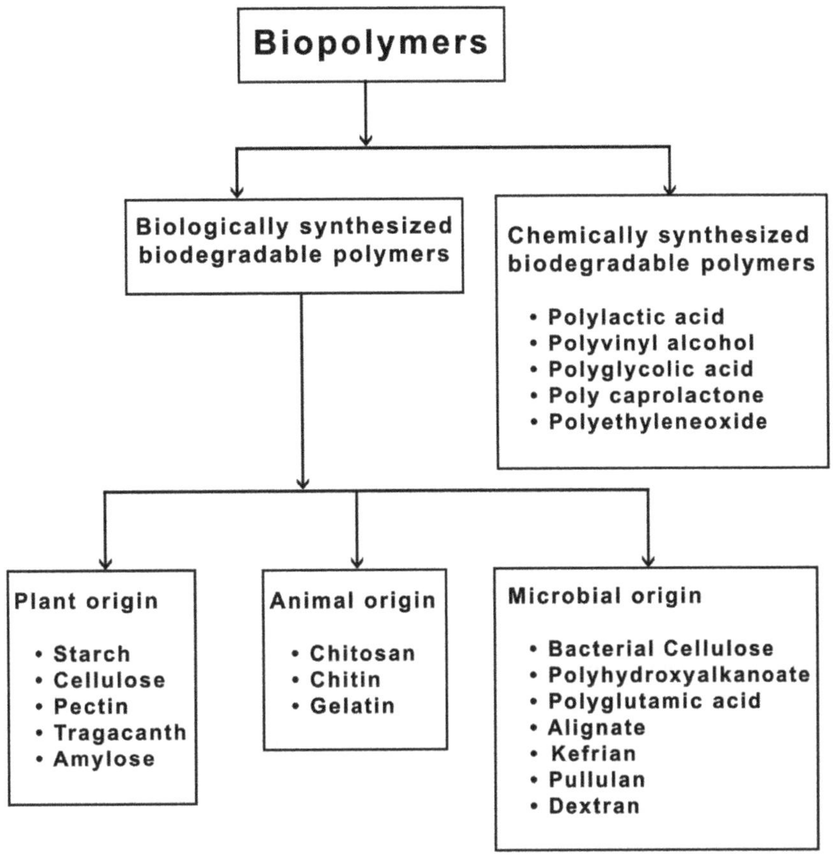

Nanofibers synthesized from natural biodegradable polymers are proven to be a good matrix for the generation biocompatible materials with wide scope in medical applications (Vlierberghe et al., 2011; Raveendran et al., 2013). In recent years, different modes of nanofiber fabrication has extensively explored in scientific literatures. Successive modification in the equipment set up and geometrical post-processing treatments favors the inter-fiber positional ordering and intra-fiber molecular alignment of nanofibers. Further, wide distinct polymer materials in combination with different solvent systems played a vital role in the formation of electrospun nanofiber with tailor specific properties and functionalities (Quiros et al., 2016). Nanofiber fabrication techniques can be broadly divided in to two types namely top-down and bottom-up approaches that includes physical, chemical and biological fabrication methods. Top-down approaches includes ball milling, sequential cutting, refining and griding (Barhoum et al., 2019). Bottom-up methods include template synthesis (Martin 1996; Feng et al., 2002) phase separation (Ma and Zhang, A) self-assembly (Xu and Liu 1997), drawing techniques (Ondarçuhu and Joachim, 1998), centrifugal spinning (Weitz etal., 2008), solution blowing (Medeiros et al., 2009) and electrospinning (Ray et al., 2016). Among the different techniques carried out for developing nanofiber mats, electrospinning was found to be the efficient and affordable fabrication technique for the development of nanofibers.

Figure 2 .Biomedical applications of biopolymers

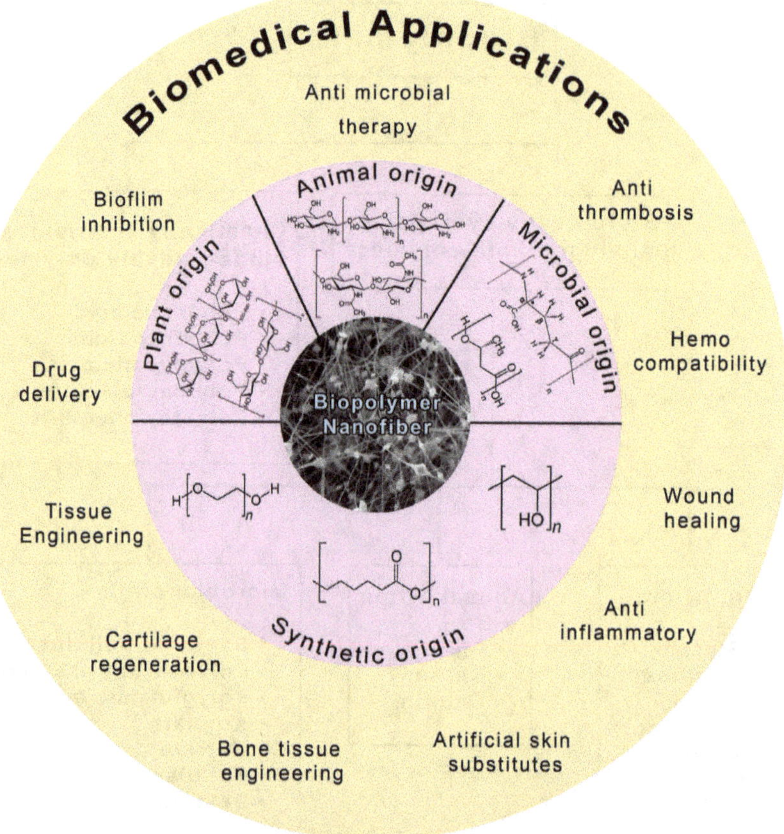

Electrospinning: Fabrication of Biopolymer Nanofiber

Electrospinning is an egressing nanotechnology technique that parallels the current scientific research towards multidisciplinary approach. For more than a decade nanofibers fabricated by electrospinning process is garnering considerable attention in material research. (Aravindhan et al., 2015; Han et al., 2008). Electrospinning is the process in which nanofibers are produced from a solution of polymer or melt in presence of electric field. The process is based on Rayleigh's studies on electric fields reported in nineteenth century and was patented in 19th century (Cooley, 1900). Unlike conventional spinning process, electrospinning is intelligible, elementary and economical. Further, electrospinning apply electrostatic force for the fabrication of electrospun nanofibers (Feng et al., 2010; Lu and Ding, 2008). Electrospinning allows the rapid evaporation of solvent and results in solidification of charged droplets to form fibers. Electrospinning setup basically consists of syringe and a needle, in which the tip of the needle is connected to a high voltage supply. Supplementation of high voltage to the polymer solution results in the formation of high surface to volume ratio nanofibrous scaffold in presence of electric field. Size controlled nanofibers can be fabricated by optimizing the processing conditions. Till date different morphologies of nanofiber scaffolds were fabricated by optimizing the parameters such as voltage, flow rate, tip to collector distance, solvent and polymer concentration etc., relatively low cost equipment, simple as well as basic operating procedures and possibility of large-scale nanofiber production has resulted in a rapid development of electrospinning technique towards laboratory scale applications. Multiple electrospinning techniques have been emerged in the last two decades, such as needless electrospinning, bubble electrospinning, magnetic field assisted electrospinning, electroblowing, coaxial electrospinning, charge injection and electrospinning (Liu and He, 2010; Weng et al., 2005; Nayak et al., 2012). Coaxial electrospinning was developed using special bilayer spinneret with two different types of polymer solutions for the formation of core-shell filaments especially for slow release of drugs and other compounds (Jiang et al., 2014; Moulton and Wallace, 2014). In magnetic field assisted electrospinning, uniform nanofibers are fabricated in presence of magnetic field for modulating the nanofiber deposition and alignment (Liu et al., 2010). Solvent evaporation during spinning process is the major limiting factor that hinders mass production of specific nanofibers and confers negative effect to the environment. In order overcome these obstacles solvent free electrospinning methods such as super critical CO_2 facilitated electrospinning, melt electrospinning and UV curing electrospinning was developed in recent days (Barhoum et al., 2019). Solvent free electrospinning is considered as an ecofriendly method for the fabrication of nanofibers with large diameter. However, different factors such as higher precursor viscosity and specific set up requirement limits their usage priority. Supercritical CO_2 facilitated electrospinning employ supercritical CO_2 instead of liquid solvents for spinning un dissolved bulk polymer samples (Zhang et al., 2017). In UV curing electrospinning nanofibers are electrospun in presence of UV light under N_2 atmopshere (Zhang et al., 2016). Basic schematic experimental set up of electrospinning units are illustrated in Figure 3.

Figure 3. Schematic Ilustration of Electrospinning process

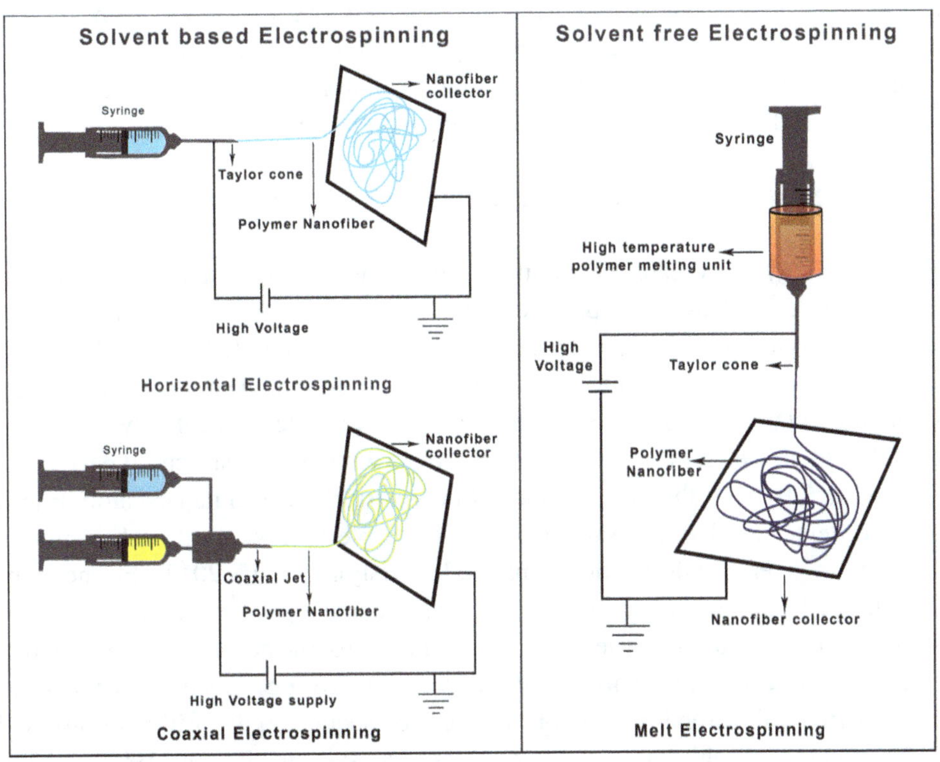

Various biomimetic nanostructures resembling honey comb (Si et al., 2015; Si et al., 2017) spider web (Ding et al., 2011; Zainab et al., 2017), tubular (Zhao et al., 2007; Liang et al., 2017), bamboo stem (Li et al., 2012), bird feather (Liu et al., 2008; Wang et al., 2016), lotus leaf (Koch et al., 2009) are synthesized by electrospinning process. Electrospun polymeric nanofibers mimics the structure and function of the natural extracellular matrix (ECM) and enhances the function of human tissues. The small pore size with high porous nature of the nanofibers enhances the gas exchange and filtration property. Further, electrospun nanomaterials represent an efficient formulation of natural and synthetic drugs in high concentration with controlled and sustained release properties.

Polymer Nanocomposites

Composites are generally considered as combination of two molecules, usually consisting of a polymer and organic /inorganic materials. Generally composites consist of simple fillers dispersed in to polymers. The properties of the filler influence the composite behavior depending upon the shape, size, physio-chemical dimensions and other inherent properties. Nanocomposites are defined as composite materials in which at least a part of a component has nanoscale dimensions. (Qi et al., 2018). Nanocomposites have engrossed significant attention in therapeutic and industrial application due to its multifunctional materialistic properties. Recently, polymer nanocomposites have escalated as superior nanomaterials with exceptional physical and chemical properties (Jeevanandam et al., 2018). Nowadays, tailor made polymer nanocomposites with unique functional groups confer peculiar properties for multiple applications in modern science. (Li et al., 2015). Annexation of nanofiller within the polymer host and their composition is considered as the key factor that influences the physical and chemical properties of the nanocomposites. Especially, bionanocomposites are considered as a boon for scientific research that encircles the innovative approach of nanotechnology with biomedical sciences (Blaiszik et al., 2010; Gaharwar et al., 2014). The molecular interactions between inorganic/organic fillers and polymer matrix controls the properties of the bionanocomposites (Hule and Pochan, 2007).

Methods Involved in Fabrication of Biopolymer Nanocomposites

The past few years of polymer research have authenticated a number of methods for the synthesis of biopolymer nanocomposites such as *in-situ* polymerization, melt interaction, solution interaction, polymer grafting surface modifying process, layer by layer assembly technique (LBL) etc., In all the methods fabrication and properties of the nanocomposite depends upon different factors such as type of polymer, solvent system, molecular weight and size of the nanofiller. Controlled radical polymerization is one of the popular methods followed for the synthesis of polymers with explicit structural design and molecular weight (Qi et al., 2018). Of the different radical polymerizations techniques, atom transfer radical polymerization technique (ATRP) is documented as a favorable method, for the fabrication of polymer nanocomposites with diverse multifunctional materials. (Matyjaszewski, 2012). Further, surface initiated atom transfer radical polymerization (SIATRP) has become a powerful tool for the fabrication of various polymer bionanocomposites. The technique is suitable for the development of different polymer composite materials at various reactions conditions with diverse functionalities (Kumar et al., 2016; Rajender et al., 2016). LBL assembly technique depends on sequential deposition of inversely charged species, and this method is highly suitable for polyelectrolyte systems (Decher, 1997). In situ polymerization method favours the dispersion of nanofiller in to the momomer with the suitable polymer matrix.

This process facilitates the effective distribution of nanofiller in to the polymer matrix compared to the other methods (Okamoto et al., 2001). Melt intercalation is a solvent free method where the dispersion of the composites happens in molten state of the polymer (Chen et al., 2003; Kim et al., 2009). Here, the nanofiller polymer matrix forms a network with the polymeric materials to form nanocomposite (Weng et al., 2005). Solution intercalation is preferred as most favorable and simple method for the fabrication of biopolymer nanocomposites such as polyvinyl alchol, polyvinyl chloride and polylactic acid etc.,. It involves simple mixing of colloidal suspension of nanofiller in to the polymer matrix with suitable solvent system followed by solvent evaporation (Wanga et al., 2004; Kalaitizidou et al., 2007).

BIOMEDICAL APPLICATIONS OF BIOPOLYMERS BASED NANOMATERIALS

Cellulose

Cellulose is considered as the most copious renewable biopolymer in the biosphere. Plants are considered as authentic and prime source of cellulose. Apart from plant source, cellulose is also produced by certain bacteria such as *Gluconacetobacter*, *Acetobacter* and *Rhizobium* in extracellular mode (Picheth et al., 2017). Cellulose and its derivative forms such as cellulose acetate, carboxymethylethyl cellulose cellulose have emerged as an important materials in biomedical, pharmaceutical and industrial applications, Nanoscale applications of cellulose nanomaterials are highly documented in scientific literatures due to their biocompatibility and nontoxic behavior towards biological systems (Menon et al., 2017). Some common nanofabrication methods and their corresponding nanofiber characteristic features are mentioned in Table 1.

Table 1.List of literature reports on nanofabrication techniques and solvent systems of cellulose nanomaterials

Biopolymer	*Nanomaterial type*	*Fabrication Technique*	*Solvent*	*Characteristic features*	*Reference*
Starch	*Nanofiber*	*Electrospinning*	*Water/formic acid*	*Highly stable, homogenous, cylindrical, randomly oriented nanofibers*	*Adeli et al. (2019)*
Starch/PVA/Chitosan	*Nanofiber*	*Electrospinning*	*Water/Acetic acid*	*Uniform, bead free, porous nanofibrous mats*	*Fonseca et al. (2019)*
Starch/PCL	*Nanofiber*	*Melt electrospining*	-	*Multilayered mesh like nanofibers*	*Martins et al. (2009)*
Starch/PVA/PEO	*Composite polymer nanofiber*	*Electrospinning*	*Water*	*Randomly oriented smooth imporous nanofibers*	*Tang et al.(2016)*
Starch/Silkfibrion	*Polymer nanocomposite*	*Electrospinning and freeze drying*	*Methanol/ water*	*Uniform, smooth homogenous nanofibers*	*Hadisit et al. (2015)*

Cellulose is composed of (1-4)-D-glucose units and Bacterial Cellulose (BC) is made up of (1-4)-D-anhydroglucopyranose chains connected through β-glycosidic linkages. Production of cellulose nanofibers from bacteria is relatively cheap, efficient and eco-friendly compared to the plant sources. Chawla et al. (2009) suggested that, by adopting different fermentation strategies, tailor made BC can be synthesized in different size, thickness and porosity for diverse applications. BC nanocomposite displays high tissue regeneration efficiency and wound healing capacity in chronic wounds (Ullah et al., 2016). Picheth et al. (2017) highlighted BC as an outstanding, indispensible and versatile biomaterial for tissue engineering and wound repair. However, BC in its native form does not possess antimicrobial activity. In order to inherit antimicrobial properties BC have been amalgamated with metals to form BC nanocomposites. Wahab et al. (2019) demonstrated the thermal synthesis of AgNPs blended cellulose nanofibers with antimicrobial property. Further, Ullah etal.(2016) fabricated antibacterial bacterial cellulose blended silver nanoparticle membrane by TEMPO (2,2,6,6-tetramethylpiperidine-1-oxyl) mediated oxidation reaction. Pinto et al.(2013) employed synthesized stable copper nanofiller and nanowires in cellulose matrices with excellent antimicrobial activity. Further, cellulose acetate nanofiber embedded, TiO2 coated silver nanofibers found to exhibit tremendous antimicrobial activity towards multiple pathogens with 36-72 hours of incubation (Jotai et al., 2019). In addition to antibacterial effect, Yadav et al. (2010) investigated the antiviral property of the NAcG blended BC *exvivo.*

Cellulase integrated Bioresorbable BC composites were developed by Hu and Catchmark (2011). The results of the study indicates loss of 90% enzyme activity in a suboptimal pH conditions. To overcome the loss of enzyme activity the authors had optimized the buffer components of BC composite solutions in optimal pH microenvironment. Amin et al. (2012) designed pH sensitive BC-PAA nanocomposite hydrogels for controlled delivery of bovine serum albumin. Similar assay was carried out with BC-PAA by Ahmad et al. (2014) for oral delivery of proteins. Nanocellulose is reported to be the most explored nanomaterial in recent decades. It includes two main nanoforms namely cellulose nanofibers (CNF) and cellulose nanocrystals (CNC). Nanocellulose materials are tested in a number of biomedical and biotechnological applications as composites in drug delivery, bone replacements, anticancer drug delivery, artificial skin scaffolds and as wound dressing materials (Jorfi& Foster, 2015; Smyth et al., 2017). Electrospun cellulose nanofiber exhibits inherent properties of cellulose and characteristic features such as porosity, nanoscale dimension, breathability, high surface area and mechanical stability enable them to transform as an ideal material for healthcare and therapeutic sectors (Missoum et al., 2014; Liu et al., 2017). CNCs and CNF possess mechanically stable self-assembled structural alignment due to their specific higher surface area and uniform particle size distribution (Gu et al., 2018). Assessment of mechanical strength and cytotoxicity of CNC incorporated gelatin nanofibers at different concentrations was reported by Hivechi et al.(2018). Their findings highlighted the increased biodegradability and zero toxicity of the composite nanomaterials. Souza et al. (2018) investigated the role of physical properties of nanocellulose and its biocompatibility nature against vero cell lines. Khoshnevisan et al. (2018) explored the possible application of cellulose acetate nanofibers as drug delivery agents for the treatment of transdermal patches with anti-inflammatory potential. Golizadeh et al. (2019) evaluated the improved cellular attachment and proliferation of electrospun surface charged cellulose nanofibers.

Starch

Starch is the second important renewable, inexpensive, and abundant polymer material found next to cellulose. Starch is semi-crystalline in nature and composed of two types of carbohydrate molecules

Table 2. Literature data on fabrication mode and characteristic feature of starch nanomaterials

Biopolymer	Nanomaterial type	Fabrication Technique	Solvent	Characteristic features	Reference
Starch	Nanofiber	Electrospinning	Water/formic acid	Highly stable, homogenous, cylindrical, randomly oriented nanofibers	Adeli et al. (2019)
Starch/PVA/Chitosan	Nanofiber	Electrospinning	Water/Acetic acid	Uniform, bead free, porous nanofibrous mats	Fonseca et al. (2019)
Starch/PCL	Nanofiber	Melt electrospining	-	Multilayered mesh like nanofibers	Martins et al. (2009)
Starch/PVA/PEO	Composite polymer nanofiber	Electrospinning	Water	Randomly oriented smooth imporous nanofibers	Tang et al.(2016)
Starch/Silkfibrion	Polymer nanocomposite	Electrospinning and freeze drying	Methanol/ water	Uniform, smooth homogenous nanofibers	Hadisit et al. (2015)

namely amylose and amylopectin. Amylose and amylopectin are branched polymers made up of α (1, 4) glucopyranose linked by α (1, 6) bonds. The composition of amylopectin and amylose differs with that of the plant source and genetic makeup of the plant species. Research outputs of the recent studies have declared starch as a promising biomaterial for successful application in tissue engineering, wound healing, drug delivery and antimicrobial therapy. Initial attempts on the fabrication of starch nanofibers were carried out by extruding amylose from starch (Muetgreet& Hiemstra, 1959). But the process was failed due to the inexpensive methodology and lower occurrence of amylose in starch composition. Transforming starch in to fiber form is complex due to the heat instability, water resistibility, low strength and poor process ability. Bailey and Mackey (2012) insisted that high amount of amylopectin affects the formation of starch nanofibers through electrospinning.

Kong and Ziegler (2002) produced starch fibers in dual solvent medium composed of DMSO and water. The results highlighted the role of solvents in the fiber formation with respect to molecular orientation. Compilation of literature reports on the effect of different solvents on starch nanofibers are summarized in Table 2. Li et al. (2016) applied centrifugal electrospinning technique for the synthesis of starch microfibers from various sources. Further, studies on the development of starch nanofibers highlighted the potential of starch acetate (SA) obtained by esterification favors nanofiber formation (Luncuski et al., 2017). Luncuski et al.(2015) reported the electrospinability of starch was enhanced in presence of 90% formic acid. The results further confirmed the development of bead free fibers with improved mechanical strength. Similarly, Oktay et al. (2013) prepared highly stable electrospun starch nanoweb composite in presence of 10% (poly (ethylene-alt-maleic anhydride) by thermal modification.

Encapsulation potential of *lactobacillus paracasei* with starch nanofibers was reported by (Luncuski et al., 2017). The report of the study confirms the increased survival rate of the bacteria suitable for probiotic applications. Prashanthini and Kalyani (2015) designed a hybrid starch nanofibers with PVA and rice husk for bone tissue engineering. Electrospun starch/polycaprolactone composite nanofiber was fabricated by coaxial electrospinning method with effective blood clotting property (Giridev and Hemamalini, 2018). Hadisi et al.(2018) investigated the antibacterial and anti-inflammatory activity of gelatin blended starch nanofiber on burn wound in mice model. The results indicated the clear observation of re epithelialization and angiogenesis with promising wound healing property.

Polyhydroxybutyrate

Polyhydroxyalkanotes (PHAs) comprises a large biopolymer family produced by bacteria as intracellular carbon reserve material under stress conditions. Polyhydroxybutyrate (PHB) is the predominant homopolymer of 3-hydroxybutyrate and significant member in PHA family. More than 100 types of PHAs and their co-monomers have been considered for commercial applications in multiple disciplines (Lu et al., 2013). PHB along with their corresponding copolymers and composites are recently receiving considerable attention in biomedical sectors. Especially, PHB nanofiber composites with large surface area, high porosity, cytocompatibility and biodegradability are considered as appreciable materials in anticancer drug delivery and tissue engineering (Kuppan et al., 2011; Bhattacharjee et al., 2016). Recent studies has suggested, PHB as a preferable candidate for bone tissue engineering due to the favorable mechanical property and zero chronic inflammatory response even after years of implantation (Thadavirul et al., 2017; Zarei et al., 2018). Gunes et al. (2018) fabricated 3D silk impregnated HAp/PHBV nanofibrous scaffolds for bone regeneration by wet electrospinning process. Amini et al.(2018) developed gentamycin loaded PVDF/PHB/chitosan nanofiber for the treatment of postsurgical ulcers. However, brittleness and high degree of crystallinity of PHB is assumed to be the most important factor limiting its role in tissue engineering (Toloue et al., 2019). Further, low hydrophilicity and degradation rate of PHB undermines its application in bone tissue engineering (Zhijiang et al., 2016). Xu et al.(2017) substantiated the importance of solvent system and their effects on morphological features of PHB nanofibers . The results of their study insisted the use of binary solvent for the effective fabrication of PHB and their corresponding copolymer nanofibers. The effect of different solvents, fabrication modes and their impact on PHB nanofiber structures are represented in Table 3. Therefore different attempts have been made to improve the physical, mechanical and biological properties by blending PHAs with biomaterials such as chitosan, alginate, cellulose etc., Mohammadalizadeh et al. (2019) suggested that low hydrophilicity and degradability can be overcomed by formulating chitosan/PHB nanocomposites. Followed by chitosan, hydroxyapatite (HA)/ PHBV nanofiber was designed with better osetogenic potential with increased surface roughness and cell adhesion (Noorani et al., 2018). Fernandes et al. (2014) highlighted the application of antibacterial effect of chlorhexidine loaded PHB/PEO nanofiber membranes.

Chitin and Chitosan

Chitin is considered as an important degradable polymer of biological origin. Chitin is composed of poly (β-(1–4)-N-acetyl-D-glucosamine) moieties. Chitin microfibrils occurs as structural entity of arthropods and cell wall exoskeleton of unicellular and multi cellular eukaryotes such as yeast and fungi (Rinaudo, 2006). Chitosan, the deacytlated form of chitin is a polysaccharide containing one amino and two hydroxyl group with repeated glycosidic residues. Both chitin and chitosan nanomaterials has drawn significant consideration in nanoscience due to their renewable property, biodegradability, biocompatibility, non-toxic nature, porous nature and intrinsic antibacterial behavior. (Zamora-Mora et al., 2017; Ho et al., 2014). However, chitosan is the preferred choice of polymer for in biomedical applications due to the poor soluble behavior of chitin. Therefore, different approaches were followed by modifying the solvent type to enhance the application of the chitin nanoformulations. Different fabrication methods for the development of chitin and chitosan nanomaterials are depicted in Table 4. A novel water dispensable graphene coated chitin nanofibers was synthesized by ball milling process (Wang et al., 2018). The hybrid nanomaterial was highly stable and biocompatible in nature. PCL incorporated chitin derived glycosamine sulfate nanofibers was fabricared by coaxial electrospinning for cartilage regeneration (Chen et al., 2019b). Chen et al. (2019a) designed chitin nanofiber based hydrogels with improved mechanical strength via surface deactylation. Madhumathi et al. (2010) prepareda α-chitin/ silver nanocomposite scaffolds with antibacterial and blood clotting properties. Kong et al.(2010) reported that high degree deacylation leads to the generation of increased positive charge density and confers potent antibacterial activity.

Table 3. Literature reports on fabrication type, solvent and morphological features of PHB Nanomaterials

Biopolymer	Nanomaterial type	Fabrication Technique	Solvent	Characteristic features	Reference
PHB	Nanofiber	Electrospinning	Chloroform	Bead free, loopy nanofibers	Acevedo et al. (2018)
PHB	Nanofiber	Electrospinning	Chloroform/ NaCl	Uniform, continuous, small diameter nanofibers	Kuntzler et al. (2018)
PHB/Multiwalled carbon nanotube	Polymer nanocomposite fiber	Electrospinning	Chloroform/DMF	Ultrafine, circular, random, highly porous nanofibers with large surface area	Zhijiang et al. (2018)
PHB/Gelatin	Nanofiber	Co-axial Electrospinning	1, 1, 1,3,3,3 hexafluoro-2-propanaol/ 2,2,2 trifluroethanol	Highly stable porous nanofibers with high surface area	Nagiah et al. (2013)
PHB/Chitosan/Alumina	Nanowires	Electrospinning	Tri fluroaceticacid	Uniform, fully porous, bead free nanofibers	Toloueet al.(2019)

Chitosan have been emerged as predominant biopolymer in the fabrication of nanocomposite polymer scaffolds for tissue engineering (Elie-Ali-Komi and Hamblin, 2016). Electrospinning of chitosan is quite complex due to their polycationic specification and high molecular weight distribution. Homayoni et al. (2009) explored that several parameters such as electric field, tip to collector space and choice solvents strongly influences the fiber morphology and properties. Solubility of gluteraldehyde cross linked chitosan nanofibers was assessed by Schiffman and Schauer (2007). Considerable improvement in structural and chemical properties was observed. Electrospun chitosan nanofiber matrix along with nanoscale properties and high surface to volume ratio was believed to mimic the extracellular matrix of the mammalian cells. Feng et al.(2009) fabricated a novel galactosylated chitosan nanofiber in presence of formic acid for tissue engineering application using electrospinning. Time dependent degradation behavior of chitosan nanofiber materials have demonstrated to provide an excellent space for osteoblast adhesion. Improved effects of chitosan nanofiber scaffolds on osteoblast proliferation and maturation was highlighted by Ho et al.(2014). Similarly, chitosan/ polycaprolactone nanofibrous scaffolds have reported to promote bipolar elongation of neurons and stimulation of cell proliferation (Wang et al.,

2009). A novel chitosan/PVA blend nanofibers was fabricated in presence of quartenery ammonium salts by Ignatova et al. (2006) for wound-healing applications. The average diameter of the fiber was between the range of 60-200nm. Jiang et al. (2004) highlighted the controlled drug delivery potential of ibuprofen loaded electrospun PEG-g-chitosan with PLGA. Sasmal and Datta (2019) investigated the controlled release of tranexamic acid loaded chitosan nanofiber with antibiofilm activity and heamocompatibility.

Tragacanth Gum

Tragacanth Gum (TG) is a natural polysaccharide obtained from dried exudates of stem and branches of Asiatic species (Zare et al., 2019). TG is a made of complex mixture of D-galacturonic acids. Compositional studies on the different species indicate the presence of rhamnose, xylose, glucose, galactose, arabinose and fucose in their structure (Mayes, 2011; Balaghi et al., 2011). TG hydrocolloid has been declared as safe material for food safety and biomedical applications (Anderson and Bridgemen , 1985). In recent years immense efforts have been riveted on therapeutic applications including wound repair, artificial skin substitutes and antibacterial activity (Zarekhalili et al., 2016; Morad et al., 2019). Scaffolds fabricated with *Aloe vera* loaded PCL/TG nanofibers demonstrated the anti-inflammatory and wound healing properties in fibroblast cells (Ranjbar-Mohammadi, 2018). However, electrospinning of pure tragacanth poses many challenges due to the hindrance in fiber formation. High viscosity and polyanoionic properties of TG favors chain entanglement by repulsive interaction between the polyanioni chains, prevents fiber formation (Zarekhalili et al., 2016). The significance of single and dual solvent systems and their impact on nanofiber morphology reported in literature studies are represented in Table 5.Ranjbar-Mohammadi et al. (2013) investigated the spinnability of PVA blended TG nanofiber matrix. Their result suggested the possible fabrication of smooth PVA/TG nanofibers with antimicrobial property and biocompaitability in human fibroblast cells. Further, controlled release of curcumin by PVA/TG nanofibers under optimized conditions was investigated by Ranjbar-MohammadiandBharami (2016).Ghayempour and Montazer (2019) explored the possible production of peppermint loaded TG nanofibers by sonochemical/microemulsion method.

Table 4. Nanofabrication techniques of chitin and chitosan nanomaterials reported in literature

Biopolymer	Nanomaterial type	Fabrication Technique	Solvent	Characteristic features	Reference
Chitin	Nanofiber hydrogel	Ballmilling/ ultrahomogenization	Water	Monodispresed swirled nanofiber networks	Song et al. (2019)
Chitin/PCL	Nanofiber	Co-axial Electrospinning	Water/ 1,1,1,3,3,3-Hexafluoro-2-propanol	Highly Porous nanofibers	Chen et al.(2020)
Chitin/PVA/ Graphene	Composite hybrid polymer nanofibers	Ballmilling/TEMPO mediated oxidation	Water/ethanol	Stable nanocomposites	Wang et al.(2018)
Chitosan	Nanofiber	Electrospinning	Acetic acid/water	Slightly beaded, uniform nanofbers	Sasmal and Datta (2019)
Chitosan/Gelatin/ Hydroxyappatite	Composite nanofiber	Co-axial Electrospinning	Acetic acid/water	Uniform, smooth, stable nanofibers	Chen et al.(2019)
Chitosan/PEO	Nanofiber	Electrospinning/ Photo cross linking	Acetic acid/water	Uniform, defect free nanofibers	Kianfar et al. (2019)

Table 5. Compilation of nanomaterial type, solvent system and fabrication method of Tragacanth Gum based nanofibers from literature

Biopolymer	Nanomaterial type	Fabrication Technique	Solvent	Characteristic features	Reference
Tragacanth Gum	Nanofiber	Sonochemical/micro emulsion method	Triton-X/Acetone	Uniform nanofibers with smooth surface and large surface to volume ratio	Ghayempour and Montazer(2019)
Tragacanth Gum	Nanofiber	Electrospinning	Acetic acid	Smooth, bead free narrow nanofibers	Mohammadi and Bharami (2016)
Tragacanth Gum/Zein	Core-shell nanofiber	Co-axial electrospinning	Water/Ethanol	Ultrathin nanofiber	Dehcheshmeh and Fathi (2019)
Tragacanth Gum/PVA/ PCL	Hybrid Nanofiber	Co-axial electrospinning	Chloroform/DMF	Uniform, thick nanofibers	Zarekhaliliet al.(2016)
Tragacanth Gum/PVA/ Nanoclay	Polymer nanocomposite fiber	Electrospinning	Water	Porous nanofibers with large surface area to volume ratio	Heydaryet al.(2015)

Alginate

Alginate is a natural, linear, anionic un branched copolymer made up of (1–4)-linked α -L-guluronic acid (G) and β-D-mannuronic acid (M) monomers synthesized by wide range of brown algae and bacteria. Structure, stability and properties of the alginate vary with composition and sequence pattern of the G and M residues (Lee and Moneey, 2012). Alginate is considered as a safe biological material for biomedical applications due its non-toxic, biodegradable, biocompatible and non-immunogenic by FDA (Paques et al., 2014; Daemi et al., 2017; Ching et al., 2017). Electrospun alginate nanofiber materials, have found to possess broad applications in biomedical divisions such as, artificial skin tissue engineering and wound dressing systems, and drug delivery systems (Dami et al., 2018). However, fabrication of electrospun alginate nanofiber is still a challenge due to the stable complex hydrogen bonding and rigid nature of alginate molecular chains (Fang et al., 2011).Lim et al. (2017) reported that electrospinnability of alginate can be improved by blending PVA and PEO. Alginate nanofibers blended with other biodedegradabe polymers and their characteristic features are listed in Table.6. A similar report on improvement of mechanical strength of the alginate nanofibers was achieved by blending with PEO (Bhattari et al., 2006). Kim et al. (2018) fabricated nanofibrous dressings composed of PCL/alginate/spirulina by coaxial electrospinning. Tang et al. (2019) fabricated honey loaded alginate/PVA nanofinrous wound dressing material in aqueous conditions. Majidi et al (2018) developed 3D Alginate/gelatin nanofibrous hydrogel with improved greater cell adhesion, motility, proliferation and maturation in human dermal fibroblast cells (HDFs) by wet electrospinning. Further, modifications in alginate structure by sulfation have been reported in recent years to enhance the both stability and compatibility (Arlov et al., 2014).

Table 6. Alginate nanomaterials and their characteristics features reported in literature

Biopolymer	Nanomaterial type	Fabrication Technique	Solvent	Characteristic features	Reference
Alginate	Nanofiber	Electrospinning	Formamide	Uniform nano fibers with limited beads	Dameiet al.(2018)
Alginate/PEO	Nanofiber	Electrospinning	TritionX-100	Smooth, uniform, bead free nanofibers	Kyziolet al.(2017)
Alginate/PVA	Nanofiber	Electrospinning	Water/Tween-20	Uniform, continuous, miscible nanofibers	Aadilet al.(2018)
Alginate/Chitosan/ PEO	Polycomplex nanofiber	Coaxial electrospinning	Acetic acid/water/ ethanol	Small beaded, irregular nanofibers	Nistaet al.(2015)
Alginate/ Hydroxyappatite	Nano composite fibers	Electrospinning	Water/DMSO	Random, continuous, thick nanofibers	Chaeet al.(2013)

CONCLUSION

Continuous efforts on the exploitation of natural polymers in the development of advanced functional materials have led to the sustainable fabrication of biomimetic nanomaterials. Electrospinning has evolved as an inevitable nanomaterial fabrication technology for the synthesis of nanoscale materials such as nanofibers, nanorods and nanowires. This chapter summarized the methods, applications and limitations of biopolymer nanomaterials with respect to the electrospinning technique. Development of novel biomaterial based on electrospinning by natural polymers will lead to significant progress in biomedical science. Further, research on the usage of cheap solvent and solvent free nanofiber fabrication methods will transform electrospinning as an outstanding nanofabrication technique in the economic point of view.However, large scale production of biopolymer nanomaterials is still scarcely explored. More research effort needs to be investigated in the mass production of biopolymer nanomaterials. In depth analysis on the development of controllable polymer nanomaterials with specific medical applications will overcome the hurdles in the commercialization of bionanomaterials. Further, Integrated approach on the simultaneous synthesis and modification of nanoscale materials with desired applications will replace the role of synthetic polymers in health sector. Further studies on the cheap production and purification strategies will improve the adoption of biopolymers in biomedical field.

REFERENCES

Aadil, K. R., Nathani, A., Sharma, C. S., Lenka, N., & Gupta, P. (2018). Fabrication of biocompatible alginate-polyvinyl (alcohol) nanofibers scaffold. *Materials Technology*, *33*(8), 507–512. doi:10.1080/10667857.2018.1473234

Abral, H., Kadriadi, Mahardika, M., Handayani, D., Sugiarti, E., & Muslimin, A. N. (2019). Characterization of disintegrated bacterial cellulose nanofibers/PVA bionanocomposites prepared via ultrasonication. *International Journal of Biological Macromolecules*, *135*, 591–599. doi:10.1016/j.ijbiomac.2019.05.178 PMID:31132440

Acevedo, F., Villegas, P., Urtuvia, V., Hermosilia, J., Navia, R., & Seeger, M. (2018). Bacterial polyhydroxybutyrate for electrospun fiber production. *International Journal of Biological Macromolecules*, *106*, 692–697. doi:10.1016/j.ijbiomac.2017.08.066 PMID:28823514

Adeli, H., Khorasani, T., & Parvazinia, M. (2019). Development of antimicrobial and antioxidant electrospun soluble potato starch nanofibers loaded with carvacrol. *International Journal of Biological Macromolecules*, *1*(122), 238–254. doi:10.1016/j.ijbiomac.2018.10.115 PMID:30342125

Ahmad, N., Amin, M. C. I. M., Mahali, S. M., Ismail, I., & Chuang, V. T. G. (2014). Biocompatible and mucoadhesive bacterial cellulose-g-poly (acrylic acid) hydrogels fororal protein delivery. *Molecular Pharmaceutics*, *11*(11), 4130–4142. doi:10.1021/mp5003015 PMID:25252107

Amin, M. C. I. M., Ahmad, N., Halib, N., & Ahmad, I. (2012). Synthesis and characterization of thermo- and pH-responsive bacterial cellulose/acrylic acid hydrogels for drug delivery. *Carbohydrate Polymers*, *88*(2), 465–473. doi:10.1016/j.carbpol.2011.12.022

Amini, F., Semnani, D., Karbasi, S., & Banitaba, S. N. (2018). A novel bilayer drug-loaded wound dressing of PVDF and PHB/Chitosan nanofibers applicable for post-surgical ulcers. *International Journal of Polymeric Materials and Polymeric Biomaterials*, *68*(13), 772–777. doi:10.1080/00914037.2018.1506982

Anderson, D. M. W., & Bridgeman, M. M. E. (1985). The composition of the proteinaceous polysaccharides exuded by astragalus microcephalus, A. Gummifer and A. Kurdicus - The sources of turkish gum tragacanth. *Phytochemistry*, *24*(10), 2301–2304. doi:10.1016/S0031-9422(00)83031-9

Arlov, O., Aachmann, F. L., Sundan, A., Espevik, T., & Skjåk-Bræk, G. (2014). Heparin-like properties of sulfated alginates with defined sequences and sulfation degrees. *Biomacromolecules*, *15*(7), 2744–2750. doi:10.1021/bm500602w PMID:24844124

Aravindan, V., Sundaramurthy, J., Suresh Kumar, P., Lee, Y. S., Ramakrishna, S., & Madhavi, E. (2015). Electrospun nanofibers:a prospective electroactive material for constructing high performance Li-ion batteries. *Chemical Communications*, *51*(12), 2225–2234. doi:10.1039/C4CC07824A PMID:25493289

Azarniya, M., Tamjid, E., Eslahi, N., & Simchi, A. (2019). Modification of bacterial cellulose/keratin nanofibrous mats by a tragacanth gum-conjugated hydrogel for wound healing. *International Journal of Biological Macromolecules*, *134*, 280–289. doi:10.1016/j.ijbiomac.2019.05.023 PMID:31071405

Balaghi, S., Mohammadifar, M. A., Zargaraan, A., Gavlighi, H. A., & Mohammadi, M. (2011). Compositional analysis and rheological characterization of gum tragacanth exudates from six species of Iranian Astragalus. *Food Hydrocolloids*, *25*(7), 1775–1784. doi:10.1016/j.foodhyd.2011.04.003

Barhoum, A., Pal, K., Rahierd, H., Uludage, H., Kim, I. S., & Bechelanya, M. (2019). Nanofibers as new-generation materials: From spinning and nano-spinning fabrication techniques to emerging applications. *Applied Materials Today*, *17*, 1–35. doi:10.1016/j.apmt.2019.06.015

Bailey, V. A., Mackey, L. N., & Trokhan, P. D. (2010). *Starch fiber.* US Patent, 7,704328.

Bhattarai, N., Li, Z., Edmondson, D., & Zhang, M. (2006). Alginate- based Nanofibrous Scaffolds: Structural, mechanical, and Biological Properties. *Advanced Materials*, *18*(11), 1463–1467. doi:10.1002/adma.200502537

Bhattacharjee, A., Kumar, K., Arora, A., & Katti, D. S. (2016). Fabrication and characterization of Pluronic modified poly(hydroxybutyrate) fibers for potential wound dressing applications. *Materials Science and Engineering C*, *63*, 266–273. doi:10.1016/j.msec.2016.02.074 PMID:27040219

Blaiszik, B. J., Kramer, S. L. B., Olugebefola, S. C., Moore, J. S., Sottos, N. R., & White, S. R. (2010). Self-healing polymers and composites. *Annual Review of Materials Research*, *40*(1), 179–211. doi:10.1146/annurev-matsci-070909-104532

Chae, W. P., Xing, Z. C., Kim, Y. J., Sang, H. S., Huh, M. W., & Kang, I. K. (2011). Fabrication and biocompatibility of rutin-containing PHBV nanofibrous scaffolds. *Polymer Korea*, *35*(3), 210–215. doi:10.7317/pk.2011.35.3.210

Chae, T., Yang, H., Leung, V., Ko, F., & Troczynski, T. (2013). Novel biomimetic hydroxyapatite/alginate nanocomposite fibrous scaffolds. *Journal of Materials Science. Materials in Medicine*, *24*(8), 1885–1894. doi:10.100710856-013-4957-7 PMID:23695359

Chen, G., Wu, C., Weng, W., Wu, D., & Yan, W. (2003). Preparation of polystyrene graphite nanosheet composites. *Polymer*, *44*(6), 1781–1784. doi:10.1016/S0032-3861(03)00050-8

Chen, P., Liu, L., Pan, J., Mei, J., Li, C., & Zheng, Y. (2019). Biomimetic composite scaffold of hydroxyapatite/gelatinchitosan core-shell nanofibers for bone tissue engineering. *Materials Science and Engineering C*, *97*, 325–335. doi:10.1016/j.msec.2018.12.027 PMID:30678918

Chen, C., Li, D., Yano, H., & Abe, K. (2019a). Bioinspired hydrogels: Quinone crosslinking reaction for chitin nanofibers with enhanced mechanical strength via surface deacetylation. *Carbohydrate Polymers*, *207*, 411–417. doi:10.1016/j.carbpol.2018.12.007 PMID:30600023

Chen, W., Wang, C., Gao, Y., Wu, Y., Wu, G., Shi, X., Du, Y., & Deng, H. (2019b). Incorporating Chitin Derived Glucosamine Sulfate into Nanofibers via Coaxial electrospinning for Cartilage Regeneration. *Carbohydrate Polymers*, *229*, 115544. doi:10.1016/j.carbpol.2019.115544 PMID:31826435

Chawla, P. R., Bajaj, I. B., Survase, S. A., & Singhal, R. S. (2009). Microbial cellulose: Fermentative production and applications. *Food Technology and Biotechnology*, *47*(2), 107–124.

Ching, S. H., Bansal, N., & Bhandari, B. (2017). Alginate gel particles–A review of production techniques and physical properties. *Critical Reviews in Food Science and Nutrition*, *57*(6), 1133–1152. doi:10.1080/10408398.2014.965773 PMID:25976619

Cooley, J.F. (1900). *Improved methods of apparatus for electrically separating the relatively volatile liquid component from the component of relatively fixed substances of composite fluids*. United Kindom patent, 6385.

Daemi, H., Barikani, M., & Sardon, H. (2017). Transition-metal-free synthesis of supramolecular ionic alginate-based polyurethanes. *Carbohydrate Polymers, 157*, 1949–1954. doi:10.1016/j.carbpol.2016.11.086 PMID:27987915

Daemi, H., Mashayekhi, M., & Modaress, M. P. (2018). Facile fabrication of sulfated alginate electrospun-nanofibers. *Carbohydrate Polymers, 198*, 481–485. doi:10.1016/j.carbpol.2018.06.105 PMID:30093025

Decher, G. (1997). Fuzzy nanoassemblies: Toward layered polymeric multicomposites. *Science, 277*(5330), 1232–1237. doi:10.1126cience.277.5330.1232

Dehcheshmeh, M. A., & Fathi, M. (2019). Production of core-shell nanofibers from zein and tragacanth for encapsulation of saffron extract. *International Journal of Biological Macromolecules, 1*(22), 272–279. doi:10.1016/j.ijbiomac.2018.10.176 PMID:30416096

Ding, B., Wang, X., Yu, J., & Wang, M. (2011). Polyamide 6 composite nano-fiber/net func- tionalized by polyethyleneimine on quartz crystal microbalance for highly sensitive formaldehyde sensors. *Journal of Materials Chemistry, 21*(34), 12784–12792. doi:10.1039/c1jm11847a

Elie-Ali-Komi, D., & Hamblin, M. R. (2016). Chitin and Chitosan: Production and Application of Versatile Biomedical Nanomaterials. *International Journal of Advanced Research, 4*(3), 411–427. PMID:27819009

Fang, D., Liu, Y., Jiang, S., Nie, J., & Ma, G. (2011). Effect of intermolecular interaction on electrospinning of sodium alginate. *Carbohydrate Polymers, 85*(1), 276–279. doi:10.1016/j.carbpol.2011.01.054

Feng, L., Li, S., Li, H., Zhai, J., Song, Y., Jiang, L., & Zhu, D. (2002). Super- Hydrophobic Surface of Aligned Polyacrylonitrile Nanofibers. *Angewandte Chemie International Edition, 41*(7), 1221–1223. doi:10.1002/1521-3773(20020402)41:7<1221::AID-ANIE1221>3.0.CO;2-G PMID:12491265

Feng, Z. Q., Chu, X., Huang, N. P., Wang, T., Wang, Y., Shi, X., Ding, Y., & Gu, Z. Z. (2009). The effect of nanofibrous galactosylated chitosan scaffolds on the formation of rat primary hepatocyte aggregates and the maintenance of liver function. *Biomaterials, 30*(14), 2753–2763. doi:10.1016/j.biomaterials.2009.01.053 PMID:19232710

Feng, C., Khulbe, K. C., & Matsuura, T. (2010). Recent progress in the preparation, characterization, and applications of nanofibers and nanofiber membranes via electrospinning/interfacial polymerization. *Journal of Applied Polymer Science, 115*(2), 756–776. doi:10.1002/app.31059

Fernandes, J. G., Correia, D. M., Botelho, G., Padrão, J., Dourado, F., Ribeiro, C., Lanceros-Méndez, S., & Sencadas, V. (2014). PHB- PEO electrospun fiber membranes containing chlorhexidine for drug delivery applications. *Polymer Testing, 34*(1), 64–71. doi:10.1016/j.polymertesting.2013.12.007

Fonseca, L. M., Cruxen, C. E. D. S., Bruni, G. P., Fiorentini, A. M., Zayareze, E. D. R., & Lim, L. T. (2019). Development of antimicrobial and antioxidant electrospun soluble potato starch nanofibers loaded with carvacrol. *International Journal of Biological Macromolecules, 15*(139), 1182–1190. doi:10.1016/j.ijbiomac.2019.08.096 PMID:31415859

Gaharwar, A. K., Peppas, N. A., & Khademhosseini, A. (2014). Nanocomposite hydrogels for biomedical applications. *Biotechnology and Bioengineering, 111*(3), 441–453. doi:10.1002/bit.25160 PMID:24264728

Ghayempour, S., & Montazer, M. (2019). A novel controlled release system based on Tragacanth nanofibers loaded pepper mint oil. *Carbohydrate Polymers*, *205*, 589–595. doi:10.1016/j.carbpol.2018.10.078 PMID:30446145

Giri Dev, V. R., & Hemamalini, T. (2018). Porous electrospun starch rich polycaprolactone blend nanofibers for severe hemorrhage. *International Journal of Biological Macromolecules*, *118*, 1276–1283. doi:10.1016/j.ijbiomac.2018.06.163 PMID:29964114

Golizadeh, M., Karimi, A., Gandomi-Ravandi, S., Vossoughi, M., Khafaji, M., Johateai, M. T., & Faghihi, F. (2019). Evaluation of cellular attachment and proliferation on different surface charged functional cellulose electrospun nanofibers. *Carbohydrate Polymers*, *207*, 796–805. doi:10.1016/j.carbpol.2018.12.028 PMID:30600068

Gunes, O. C., Unalan, I., Cecen, B., Albayrak, A. Z., Havitcioglu, H., Ustun, O., & Ergur, B. U. (2018). Three-dimensional silk impregnated HAp/PHBV nanofibrous scaffolds for bone regeneration. *International Journal of Polymeric Materials and Polymeric Biomaterials*, *68*(5), 217–228. doi:10.1080/00914037.2018.1443928

Gu, F., Wang, W., Cai, Z., Xue, F., Jin, Y., & Zhu, J. Y. (2018). Water retention value for characterizing fibrillation degree of cellulosic fibers at micro and nanometer scales. *Cellulose (London, England)*, *25*(5), 2861–2871. doi:10.100710570-018-1765-8

Han, S. O., Son, W. K., Youk, J. H., & Park, W. H. (2008). Electrospinning of ultrafine cellulose fibers and fabrication of poly (butylene succinate) biocomposites reinforced by them. *Journal of Applied Polymer Science*, *107*(3), 1954–1959. doi:10.1002/app.26643

Hadisi, Z., Nourmohammadi, J., & Nassiri, S. M. (2018). The antibacterialand anti-inflammatory investigation of Lawsonia Inermis gelatin starch nanofibrous dressing in burn wound. *International Journal of Biological Macromolecules*, *107*, 2008–2019. doi:10.1016/j.ijbiomac.2017.10.061 PMID:29037870

Heydary, A. H., Karamian, E., Poorazizi, E., Khandan, A., & Heydaripour, J. (2015). A novel nanofiber of Iranian Gum Tragacanth-Polyvinyl Alcohol/ Nanoclay composite for wound healing applications. *Procedia Materials Science*, *11*, 176–182. doi:10.1016/j.mspro.2015.11.079

Hivechi, A., Bahrami, S. H., & Siegel, R. A. (2018). Investigation of morphological, mechanical and biological properties of cellulose nanocrystal reinforced electrospun gelatin nanofibers. *International Journal of Biological Macromolecules*, *124*, 411–417. doi:10.1016/j.ijbiomac.2018.11.214 PMID:30476513

Hule, R. A., & Pochan, D. J. (2007). Polymer nanocomposites for biomedical applications. *MRS Bulletin*, *32*(4), 354–358. doi:10.1557/mrs2007.235

Hu, Y., & Catchmark, J. M. (2011). In vitro biodegradability and mechanical properties of bioabsorbable bacterial cellulose incorporating cellulases. *Acta Biomaterialia*, *7*(7), 2835–2845. doi:10.1016/j.actbio.2011.03.028 PMID:21459165

Huang, C., Soenen, S. J., Gulck, E., Vanhamb, G., Rejman, J., Van Calenbergh, S., Vervaet, C., Coenye, T., Verstraelen, H., Temmerman, M., Demeester, J., & De Smwdt, C. (2012). Electrospun cellulose acetate phthalate fibers for semen induced anti-HIV vaginal drug delivery. *Biomaterials*, *3*(33), 962–969. doi:10.1016/j.biomaterials.2011.10.004 PMID:22018388

Ho, M. H., Liao, M. H., Lin, Y. L., Lai, C. H., Lin, P. I., & Chen, R. M. (2014). Improving effects of chitosan nanofiber scaffolds on osteoblast proliferation and maturation. *International Journal of Nanomedicine*, *9*, 4293–4304. PMID:25246786

Homayoni, H., Ravandi, S. A. H., & Valizadeh, M. (2009). Influence of the Molecular Weight of Chitosan on the spinnability of chitosan/poly(vinyl alcohol) blend nanofibers. *Journal of Applied Polymer Science*, *113*(4), 2507–2513. doi:10.1002/app.30148

Ignatova, M., Starbova, K., Markova, N., Manolova, N., & Rashkov, I. (2006). Electrospunnano-fibre mats with antibacterial properties from quaternized chitosan and poly(vinyl alcohol). *Carbohydrate Research*, *341*(12), 2098–2107. doi:10.1016/j.carres.2006.05.006 PMID:16750180

Jeevanandham, J., Barhoum, A., Chan, Y., Dufresne, A., & Danquah, M. (2018). Review on nanoparticles and nanostructured materials: History, sources, toxicity and regulations. *Beilstein Journal of Nanotechnology*, *9*, 1050–1074. doi:10.3762/bjnano.9.98 PMID:29719757

Jiang, H., Fang, D., Hsiao, B., & Chen, W. (2004). Preparation and characterization of ibuprofen loaded poly(lactide-co-glycolide)/poly(ethylene glycol)-g-chitosan electrospun memberanes. *Journal of Biomaterials Science. Polymer Edition*, *15*(3), 279–296. doi:10.1163/156856204322977184 PMID:15147162

Jiang, H. L., Wang, L. Q., & Zhu, K. J. (2014). Coaxial electrospinning for encapsulation and controlled release of fragile water-soluble bioactive agents. *Journal of Controlled Release*, *193*, 296–303. doi:10.1016/j.jconrel.2014.04.025 PMID:24780265

Jorfi, M., & Foster, E. J. (2015). Recent advances in nanocellulose for biomedical applications. *Journal of Applied Polymer Science*, *132*(14), 1–19. doi:10.1002/app.41719

Jotai, A. W., Kim, I. S., & Ni, Q. Q. (2019). Cellulose acetate nanofibers embedded with AgNPs anchored TiO2 nanoparticles for long term excellent antibacterial applications. *Carbohydrate Polymers*, *207*, 640–649. doi:10.1016/j.carbpol.2018.12.029 PMID:30600049

Kalaitzidou, K., Fukushima, H., & Drzal, L. T. (2007). A new compounding method for exfoliated graphite- polypropylene nanocomposites with enhanced flexural properties and lower percolation threshold. *Composites Science and Technology*, *67*(10), 2045–2051. doi:10.1016/j.compscitech.2006.11.014

Kenawy, E. R., Abdel-Hay, F. I., El-Newehy, M. H., & Wnek, G. E. (2009). Processing of polymer nanofibers through electrospinning as drug delivery systems. *Materials Chemistry and Physics*, *113*(1), 296–302. doi:10.1016/j.matchemphys.2008.07.081

Khil, M.S., Cha, D.I., Kim, H.Y., Kim, I.S., & Bhattarai, N. (2003). Electrospun nanofibrous polyurethane membrane as wound dressing. *Journal* of *Biomedical* Materials *Research Part B*: *Applied Biomaterial*, *67B*(2), 675–679.

Khoshnevisan, K., Maleki, H., Samadian, H., Shahsavari, S., Sarrafzadeh, M. H., Larijani, B., Dorkoosh, F. A., Haghpanah, V., & Khorramizadeh, M. R. (2018). Cellulose Acetate ElectrospunNanofibers for Drug Delivery Systems: Applications and Recent Advances. *Carbohydrate Polymers*, *198*, 131–141. doi:10.1016/j.carbpol.2018.06.072 PMID:30092983

Kianfara, P., Vitalea, A., Dalle Vacchea, S., & Bongiovanni, R. (2019). Photo-crosslinking of chitosan/poly(ethylene oxide) electrospun nanofibers. *Carbohydrate Polymers*, *217*, 144–151. doi:10.1016/j.carbpol.2019.04.062 PMID:31079670

Kim, S., Do, I., & Drzal, L. T. (2009). Thermal stability and dynamic mechanical behavior of exfoliated graphite nanoplatelets-LLDPE nanocomposites. *Polymer Composites*, *31*(5), 755–761. doi:10.1002/pc.20781

Kim, M. S., Kim, H. J., Jang, J. Y., & Shin, H. S. (2018). Development of Coaxial Alginate-PCL Nanofibrous Dressing for Controlled Release of Spirulina Extract. *Journal of Biomaterials Science*, *29*(12), 1–0. doi:10.1080/09205063.2018.1462931 PMID:29667511

Kotek, R. (2008). Recent Advances in Polymer Fibers. *Polymer Reviews (Philadelphia, Pa.)*, *48*(2), 221–229. doi:10.1080/15583720802020038

Koch, K., Bhushan, B., Jung, Y. C., & Barthlott, W. (2009). Fabrication of artificial lotus leaves and significance of hierarchical structure for super hydrophobicity and low adhesion. *Soft Matter*, *5*(7), 1386–1393. doi:10.1039/b818940d

Kong, M., Chen, X. G., Xing, K., & Park, H. J. (2010). Antimicrobial properties of chitosan and mode of action: A state of the art review. *International Journal of Food Microbiology*, *144*(1), 51–63. doi:10.1016/j.ijfoodmicro.2010.09.012 PMID:20951455

Kong, L., & Ziegler, G. R. (2012). Patents on fiber spinning from starches. *Recent Patents on Food, Nutrition & Agriculture*, *4*, 210–2019. doi:10.2174/2212798411204030210 PMID:23013410

Kuppan, P., Vasanthan, K. S., Sundaramurthi, D. M., Krishnan, U., & Sethuraman, S. (2011). Development of poly (3-hydroxybutyrate-co-3-hydroxyvalerate) fibers for skin tissue engineering: Effects of topography, mechanical, and chemical stimuli. *Biomacromolecules*, *12*(9), 3156–3165. doi:10.1021/bm200618w PMID:21800891

Kuntzler, S. G., Almedia, A. C. A., Costa, J. A. V., & Morais, M. G. (2018). Polyhydroxybutyrate and phenolic compounds microalgae electrospun nanofibers: A novel nanomaterial with antibacterial activity. *International Journal of Biological Macromolecules*, *1*(113), 1008–1014. doi:10.1016/j.ijbiomac.2018.03.002 PMID:29505877

Kumar, A., Bansal, A., Behera, B., Jain, S. L., & Ray, S. S. (2016). Ternary hybrid polymeric nanocomposites through grafting of polystyrene on graphene oxide-TiO2 by surface initiated atom transfer radical polymerization (SI-ATRP). *Materials Chemistry and Physics*, *172*, 189–196. doi:10.1016/j.matchemphys.2016.01.064

Kyziol, A., Michna, J., Morneo, I., Gamez, E., & Irusta, S. (2017). Preparation and characterization of electrospun alginate nanofibers loaded with ciprofloxacin hydrochloride. *European Polymer*, *96*, 350–360.

Lancuski, A., Vasilyev, G., Luc Putaux, J., & Zussman, E. (2015). Rheological properties and electrospinnability of high amylose starch in formic acid. *Biomacromolecules*, *16*(8), 2529–2536. doi:10.1021/acs.biomac.5b00817 PMID:26192477

Lancuski, A., Abu Ammar, A., Avrahami, R., Vilensky, R., Vasilyev, G., & Zussman, E. (2017). Design of starch formate compound fibers as encapsulation platform for biotherapeutics. *Carbohydrate Polymers, 20*(158), 68–76. doi:10.1016/j.carbpol.2016.12.003 PMID:28024544

Lau, N. S., Tsuge, T., & Sudesh, K. (2011). Formation of new polyhydroxyalkanoate containing 3-hydroxy-4-methylvalerate monomer in *Burkholderia* sp. *Applied Microbiology and Biotechnology, 89*(5), 1599–1609. doi:10.100700253-011-3097-6 PMID:21279348

Lee, K. Y., & Mooney, D. J. (2012). Alginate: Properties and biomedical applications. *Progress in Polymer Science, 37*(1), 106–126. doi:10.1016/j.progpolymsci.2011.06.003 PMID:22125349

Li, S., Lu, X., Li, X., Xue, Y., Zhang, C., Lei, J., & Wang, C. (2012). Preparation of bamboo-like ppy nanotubes and their application for removal of Cr(VI) ions in aqueous solution. *Journal of Colloid and Interface Science, 378*(1), 30–35. doi:10.1016/j.jcis.2012.03.065 PMID:22551477

Li, Q., Chen, L., Gadinski, M. R., Zhang, S., Zhang, G., Li, H. U., Iagodkine, E., Haque, A., Chen, L.-Q., Jackson, T. N., & Wang, Q. (2015). Flexible high-temperature dielectric materials from polymer nanocomposites. *Nature, 523*(7562), 576–579. doi:10.1038/nature14647 PMID:26223625

Li, X., Chen, H., & Yang, B. (2016). Centrifugally spun starch-based fibers from amylopectin rich starches. *Carbohydrate Polymers, 137*, 459–465. doi:10.1016/j.carbpol.2015.10.079 PMID:26686151

Liang, M., Chen, X., Xu, Y., Zhu, L., Jin, X., & Huang, C. (2017). Double-grooved nanofibre surfaces with enhanced anisotropic hydrophobicity. *Nanoscale, 9*(42), 16214–16222. doi:10.1039/C7NR05188C PMID:29043355

Liu, M., Duan, X. P., Li, Y. M., Yang, D. P., & Long, Y. Z. (2017). Electrospunnanofibers for wound healing. *Materials Science and Engineering C, 76*, 1413–1423. doi:10.1016/j.msec.2017.03.034 PMID:28482508

Liu, Y., Chen, X., & Xin, J. H. (2008). Hydrophobic duck feathers and their simulation on textile substrates for water repellent treatment. *Bioinspiration & Biomimetics, 3*(4), 046007. doi:10.1088/1748-3182/3/4/046007 PMID:18997276

Liu, Y., & He, J. H. (2010). Bubble electrospinning for mass production of nanofibers. *International Journal of Nonlinear Sciences and Numerical Simulation, 8*(3), 393–396.

Liu, Y., Zhang, X., Xia, Y., & Yang, H. (2010). Magnetic-Field-Assisted Electrospinning of Aligned Straight and Wavy Polymeric Nanofibers. *Advanced Materials, 22*(22), 2454–2457. doi:10.1002/adma.200903870 PMID:20376855

Lim, C. T. (2017). Nanofiber technology: Current status and emerging developments. *Progress in Polymer Science, 70*, 1–17. doi:10.1016/j.progpolymsci.2017.03.002

Lu, P., & Ding, B. (2008). Applications of electrospun fibers. *Recent Patents on Nanotechnology, 2*(3), 169–182. doi:10.2174/187221008786369688 PMID:19076051

Lu, X., Wang, L., Yang, Z., & Lu, H. (2013). Strategies of polyhydroxyalkanoates modification for the medical application in neural regeneration/nerve tissue engineering. *Advances in Bioscience and Biotechnology, 4*(6), 731–740. doi:10.4236/abb.2013.46097

Martin, C. (1996). Membrane-based of nanomaterials. *Chemistry of Materials, 8*(43), 1739–1746. doi:10.1021/cm960166s

Ma, P., & Zhang, R. (1999). Synthetic nano-scale fibrous extracellular matrix. *Journal of Biomedical Materials Research, 46*(1), 60–72. doi:10.1002/(SICI)1097-4636(199907)46:1<60::AID-JBM7>3.0.CO;2-H PMID:10357136

Martins, A., Chung, S., Pedro, A. J., Sousa, R. A., Marques, A. P., Reis, R. L., & Neves, N. M. (2009). Hierarchical starch-based fibrous scaffold for bone tissue engineering applications. *Journal of Tissue Engineering and Regenerative Medicine, 3*(1), 37–42. PMID:19021239

Madhumathi, K., Sudheesh Kumar, P. T., Abhilash, S., Sreeja, V., Tamura, H., Manzoor, K., Nair, S. V., & Jayakumar, R. (2010). Development of novel chitin/nanosilver composite scaffolds for wound dressing applications. *Journal of Materials Science. Materials in Medicine, 21*(2), 807–813. doi:10.100710856-009-3877-z PMID:19802687

Majidi, S. S., Slemming-Adamsen, P., Hanif, M., Zhang, Z., Wang, Z., & Chen, M. (2018). Wet electrospun alginate/gelatin hydrogel nanofibers for 3D cell culture. *International Journal of Biological Macromolecules, 118*, 1648–1654. doi:10.1016/j.ijbiomac.2018.07.005 PMID:29981331

Mayes, J. M. (2011). Gum tragacanth and karaya. In A. Imeson (Ed.), *Food stabilisers, thickeners and gelling agents Singapore* (pp. 167–179). John Wiley & Sons.

Matyjiaszewski, K. (2012). Atom transfer radial polymerization (ATRP): Current status and future perspectives. *Macromolecules, 45*(10), 4015–4039. doi:10.1021/ma3001719

Medeiros, E. S., Glenn, G. M., Klamczynski, A., Orts, W. J., & Mattoso, L. H. C. (2009). Solution blow spinning: A new method to produce micro-and nanofibers from polymer solutions. *Journal of Applied Polymer Science, 113*(4), 2322–2330. doi:10.1002/app.30275

Menon, M. P., & Selvakumar, R. (2017). Extraction and modification of cellulose nanofibers derived from biomass for environmental application. *RSC Advances, 7*(68), 42750–42773. doi:10.1039/C7RA06713E

Missoum, K., Sadocco, P., Causio, J., Belgacem, M. N., & Bras, J. (2014). Antibacterial activity and biodegradability assessment of chemically grafted nanofibrillated cellulose. *Materials Science and Engineering C, 45*, 477–483. doi:10.1016/j.msec.2014.09.037 PMID:25491853

Muetgreet, J. & Hiemstra. (1959). *Process for the production of amylose articles by extrusion of aqueous sodium hydroxide solution thereof into concentrated aqueous ammonium sulphate solution.* US Patent, 2,902,336.

Moulton, S. E., & Wallace, G. G. (2014). 3-dimensional (3D) fabricated polymer based drug delivery systems. *Journal of Controlled Release, 193*, 27–34. doi:10.1016/j.jconrel.2014.07.005 PMID:25020039

Mohammadalizadeh, Z., Karbasi, S., & Arasteh, S. (2019). (in press). Physical, mechanical and biological evaluation of poly (3-hydroxybutyrate)-chitosan/MWNTs as a novel electrospun scaffold for cartilage tissue engineering applications. *Polymer-Plastics Technology and Materials.*

Morad, F. N., Rashid, A., Khavi, R., Rahimi, M. K., & Bahador, A. (2019). Producing of loxacin Loaded Gum Tragacanth/Bassorin Electrospun Nano Fibers and Evaluation Its Antibacterial Activity. *Ekoloji, 28*(107), 391–400.

Nayak, R., Padhye, R., Kyratzis, I. L., Truong, Y. B., & Arnold, L. (2012). Recent advances in nanofibre fabrication techniques. *Textile Research Journal, 82*(2), 129–147. doi:10.1177/0040517511424524

Nagiah, N., Madhavi, L., Anitha, R., Anandan, C. Srinivasan, N.T., & Sivagnanam, U.T. (2013). Development and characterization of coaxially electrospun gelatin coated poly (3-hydroxybutyric acid) thin films as potential scaffolds for skin regeneration. *Materials Science and Engineering C*, 33-4444-4452.

Nista, S. V., Bettini, J., & Mei, L. H. (2015). Coaxial nanofibers of chitosan-alginate-PEO polycomplex obtained by electrospinning. *Carbohydrate Polymers, 127*, 222–228. doi:10.1016/j.carbpol.2015.03.063 PMID:25965478

Noorani, B., Yazdian, F., Khoshraftar, A., Ghaemi, R. V., Alihemmati, Z., Rashedi, H., & Shahmoradi, S. (2018). Fabrication and evaluation of nanofibrouspolyhydroxybutyratevalerate scaffolds containing hydroxyapatite particles for bone tissue engineering. *International Journal of Polymeric Materials and Polymeric Biomaterials, 67*(17), 987–995. doi:10.1080/00914037.2017.1417283

Okamoto, M., Morita, S., & Kotaka, T. (2001). Dispersed structure and ionic conductivity of smectic clay/polymer nanocomposites. *Polymer, 42*(6), 2685–2688. doi:10.1016/S0032-3861(00)00642-X

Oktay, B., Basturk, E., Kayaman Apohan, N., & Vezir Kahraman, M. (2013). Highly Porous starch/ Poly (ethylene-alt-maleic anhyride) composite nanofiber mesh. *Polymer Composites, 34*(8), 1321–1324. doi:10.1002/pc.22545

Ondarçuhu, T., & Joachim, C. (1998). Drawing a single nanofibre over hundreds of microns. *Europhysics Letters, 42*(2), 215–220. doi:10.1209/epl/i1998-00233-9

Paques, J. P., van der Linden, E., van Rijn, C. J. M., & Sagis, L. M. C. (2014). Preparation methods of alginate nanoparticles. *Advances in Colloid and Interface Science, 209*, 163–171. doi:10.1016/j.cis.2014.03.009 PMID:24745976

Picheth, G. F., Pirich, C. L., Sierakowski, M. R., Woehl, M. A., Sakakibara, C. N., de Souza, C. F., Martin, A. A., da Silva, R., & de Freitas, R. A. (2017). Bacterial cellulose in biomedical applications: A review. *International Journal of Biological Macromolecules, 104*, 97–106. doi:10.1016/j.ijbiomac.2017.05.171 PMID:28587970

Pinto, R. J. B., Diana, S., Sadocco, P., Pascol Neto, C., & Trindade, T. (2013). Antibacterial Activity of Nanocomposites of Copper and Cellulose. *BioMed Research International, 2013*, 1–6. doi:10.1155/2013/280512 PMID:24455681

Prashanthini, L. K., & Kalyani, D. (2015). Preparation of scaffold using rice husk and starch for bone regeneration. *International Journal for Research in Applied Science and Engineering Technology, 3*(6), 598–603.

Qi, W., Zhang, X., & Wang, H. (2018). Self-assembled polymer nanocomposites for biomedical application. *Current Opinion in Colloid & Interface Science, 35*, 36–41. doi:10.1016/j.cocis.2018.01.003

Quirós, J., Boltes, K., & Rosal, R. (2016). Bioactive Applications for Electrospun Fibers. *Polymer Reviews (Philadelphia, Pa.)*, *56*(4), 631–667. doi:10.1080/15583724.2015.1136641

Ramakrishna, S., Fujihara, K., Teo, W. E., Yong, T., Ma, Z., & Ramaseshan, R. (2006). Electrospun nanofibers: Solving global issues. *Materials Today*, *9*(3), 40–50. doi:10.1016/S1369-7021(06)71389-X

Ranjbar-Mohammadi, M., Bahrami, S. H., & Joghataei, M. T. (2013). Fabrication of novel nanofiber scaffolds from gum tragacanth/poly (vinyl alcohol) for wound dressing application: In vitro evaluation and antibacterial properties. *Materials Science and Engineering C*, *33*(8), 4935–4943. doi:10.1016/j.msec.2013.08.016 PMID:24094207

Ranjbar-Mohammadi, M., & Bahrami, S. H. (2016). Electrospun curcumin loaded poly (ε-caprolactone)/gum tragacanth nanofibers for biomedical application. *International Journal of Biological Macromolecules*, *84*, 448–456. doi:10.1016/j.ijbiomac.2015.12.024 PMID:26706845

Ranjbar-Mohammadi, M., (2018). Characteristics of aloe vera incorporated poly (ε-caprolactone)/gum tragacanthnanofibers as dressings for wound care. *Journal of Industrial Textiles*, *47*(7), 144–1477.

Rinaudo, M. (2006). Chitin and chitosan: Properties and applications. *Progress in Polymer Science*, *31*(7), 603–632. doi:10.1016/j.progpolymsci.2006.06.001

Rajender, N., & Suresh, K. I. (2016). Surface-initiated atomtransfer radical polymerization (SI-ATRP) from graphene oxide: Effect of functionalized graphene sheet (FGS) on the synthesis and material properties of PMMA nanocomposites. *Macromolecular Materials and Engineering*, *301*(1), 81–92. doi:10.1002/mame.201500256

Raveendran, S., Dhandayuthapani, B., Nagaoka, Y., Yoshida, Y., Maekawa, T., & Sakthi Kumar, D. (2013). Biocompatible nanofibers based on extremophilic bacterial polysaccharide, Mauran from Halomonas-Maura. *Carbohydrate Polymers*, *92*(2), 1225–1233. doi:10.1016/j.carbpol.2012.10.033 PMID:23399150

Ray, S. S., Chen, S. S., Li, C. W., Nguyen, N. C., & Nguyen, H. T. (2016). A comprehensive review: Electrospinning technique for fabrication and surface modification of membranes for water treatment application. *RSC Advances*, *6*(88), 85495–85514. doi:10.1039/C6RA14952A

Robles-Gracia, M. A., Del-Toro-Sanchez, C. L., Marquez-Rios, E., Barrera-Rodriguez, A., Aguilar, J., Aguilar, J. A., Reynoso-Marin, F. J., Ceja, I., Dorame-Miranda, R., & Rodriguez-Felix, F. (2018). Nanofibers of cellulose bagasse from Agave tequilana Weber var. azul by electrospinnimg: Preparation and Characterization. *Carbohydrate Polymers*, *15*(192), 69–74. doi:10.1016/j.carbpol.2018.03.058

Sasmal, P., & Datta, P. (2019). Tranexamic acid-loaded chitosan electrospunnanofibers as drug delivery system for hemorrhage control applications. *Journal of Drug Delivery Science and Technology*, *52*, 559–567. doi:10.1016/j.jddst.2019.05.018

Seoane, I. T., Cerrutti, P., Vazquez, A., Manfredi, B., & Cyras, V. P. (2017). Polyhydroxybutyrate-Based Nanocomposites with Cellulose Nanocrystals and Bacterial Cellulose. *Journal of Polymers and the Environment*, *25*(3), 586–598. doi:10.100710924-016-0838-8

Si, Y., Fu, Q., Wang, X., Zhu, J., Yu, J., Sun, G., & Ding, B. (2015). Superelastic and super- hydrophobic nanofiber-assembled cellular aerogels for effective separation of oil/ water emulsions. *ACS Nano*, *9*(4), 3791–3799. doi:10.1021/nn506633b PMID:25853279

Si, Y., Wang, L., Wang, X., Tang, N., Yu, J., & Ding, B. (2017). Ultrahigh-water-content, super- elastic, and shape-memory nanofiber-assembled hydrogels exhibiting pressure-re- sponsive conductivity. *Advanced Materials*, *29*(24), 1700339. doi:10.1002/adma.201700339 PMID:28417597

Smyth, M., Fournier, C., Driemeier, C., Picart, C., Foster, E. J., & Bras, J. (2017). Tunable Structural and Mechanical Properties of Cellulose Nanofiber Substrates in Aqueous Conditions for Stem Cell Culture. *Biomacromolecules*, *18*(7), 2034–2044. doi:10.1021/acs.biomac.7b00209 PMID:28485582

Souza, S. F., Mariano, M., Reis, D., Lombello, C. B., Ferreira, M., & Sain, M. (2018). Cell interactions and cytotoxic studies of cellulose nanofibers from curaua natural fibers. *Carbohydrate Polymers*, *201*, 87–95. doi:10.1016/j.carbpol.2018.08.056 PMID:30241866

Song, S., Zhao, Y., Yuan, X., & Zhang, J. (2019). β-Chitin nanofiber hydrogel as a scaffold to in situ fabricate monodispersed ultra-small silver nanoparticles. *Colloids and Surfaces. A, Physicochemical and Engineering Aspects*, *574*, 36–43. doi:10.1016/j.colsurfa.2019.04.047

Tang, S., Zhao, Z., Chen, G., Su, Y., Lu, L., Li, B., Liang, D., & Jin, R. (2016). Fabrication of ampicillin/starch/polymer composite nanofibers with controlled drug release properties by electrospinning. *Journal of Sol-Gel Science and Technology*, *77*(3), 594–603. doi:10.100710971-015-3887-x

Tang, Y., Lana, X., Lianga, C., Zhonga, Z., Xiea, R., Zhoua, Y., Miaoa, X., Wangc, H., & Wang, W. (2019). Honey loaded alginate/PVA nanofibrous membrane as potential bioactive wound dressing. *Carbohydrate Polymers*, *219*, 113–120. doi:10.1016/j.carbpol.2019.05.004 PMID:31151507

Thavasi, V., Singh, G., & Ramakrishna, S. (2008). Electrospun nanofibers in energy and environmental applications. *Energy & Environmental Science*, *1*(2), 205–221. doi:10.1039/b809074m

Thadavirul, N., Pavasant, P., & Supaphol, P. (2017). Fabrication and evaluation of poly- caprolactone–poly (hydroxybutyrate) or poly (3-hydroxybutyrate-co-3-hydro- xyvalerate) dual-leached porous scaffolds for bone tissue engineering applications. *Macromolecular Materials and Engineering*, *302*(3), 1600289. doi:10.1002/mame.201600289

Toloue, E. B., Karbasi, S., Salehi, H., & Rafienia, M. (2019). Potential of an electrospun composite scaffold of poly (3-hydroxybutyrate)- chitosan/alumina nanowires in bone tissue engineering applications. *Materials Science and Engineering C*, *99*, 1075–1091. doi:10.1016/j.msec.2019.02.062 PMID:30889640

Ullah, H., Wahid, F., Santos, H. A., & Khan, T. (2016). Advances in biomedical and pharmaceutical applications of functional bacterial cellulose-based nanocomposites. *Carbohydrate Polymers*, *150*, 330–352. doi:10.1016/j.carbpol.2016.05.029 PMID:27312644

Vlierberghe, S. V., Dubruel, P., & Schacht, E. (2011). Biopolymer-based hydrogels as scaffolds for tissue engineering applications: A review. *Biomacromolecules*, *12*(2), 1387–1408. doi:10.1021/bm200083n PMID:21388145

Wahaba, J. A., Kimc, I. S., & Ni, Q. Q. (2019). A comparative study on synthesis of AgNPs on cellulose nanofibers by thermal treatment and DMF for antibacterial activities. *Materials Science and Engineering C*, *98*, 1179–1195. doi:10.1016/j.msec.2019.01.017 PMID:30813001

Wanga, W. P., & Pana, C. Y. (2004). Preparation and characterization of polystyrene/graphite composite prepared by cationic grafting polymerization. *Polymer*, *45*(12), 3987–3995. doi:10.1016/j.polymer.2004.04.023

Wang, W., Itoh, S., Konno, K., Kikkawa, T., Ichinose, S., Sakai, K., Ohkuma, T., & Watabe, K. (2009). Effects of Schwann cell alignment along the oriented electrospun chitosan nanofibers on nerve regeneration. *Journal of Biomedical Materials Research. Part A*, *91*(4), 994–1005. doi:10.1002/jbm.a.32329 PMID:19097155

Wang, S., Yang, Z., Gong, G., Wang, J., Wu, J., Yang, S., & Jiang, L. (2016). Icephobicity of penguins spheniscushumboldti and an artificial replica of penguin feather with air infused hierarchical rough structures. *The Journal of Physical Chemistry C*, *120*(29), 15923–15929. doi:10.1021/acs.jpcc.5b12298

Wang, K., Ma, Q., Pang, K., Ding, B., Zhang, J., & Duan, Y. (2018). One-pot Synthesis of Graphene/Chitin Nanofibers Hybrids and Their Remarkable Reinforcement on Poly (vinyl alcohol). *Carbohydrate Polymers*, *194*, 146–153. doi:10.1016/j.carbpol.2018.04.036 PMID:29801822

Weng, W., Chen, G., & Wu, D. (2005). Transport properties of electrically conducting nylon 6/foliated graphite nanocomposites. *Polymer*, *46*(16), 6250–6257. doi:10.1016/j.polymer.2005.05.071

Weitz, R., Harnau, L., Rauschenbach, S., Burghard, M., & Kern, K. (2008). Polymer nanofibers via nozzle-free centrifugal spinning. *Nano Letters*, *8*(4), 1187–1191. doi:10.1021/nl080124q PMID:18307320

Xu, S., & Liu, G. (1997). Nanometer-scale fabrication by simultaneous nanoshaving and molecular self assembly. *Langmuir*, *13*(2), 127–129. doi:10.1021/la962029f

Xu, Y., Lu, H., & Kang, T. (2017). Effect of different solvent systems on PHBV/PEO electrospun fibers. *RSC Advances*, *7*(7), 4000–4010. doi:10.1039/C6RA26783A

Yadav, V., Paniliatis, B. J., Shi, H., Lee, K., Cebe, P., & Kaplan, D. L. (2010). Novel in vivo- degradable cellulose chitin copolymer from metabolically engineered *Gluconacetobacterxylinus*. *Applied and Environmental Microbiology*, *76*(18), 6257–6265. doi:10.1128/AEM.00698-10 PMID:20656868

Zainab, G., Iqbal, N., Babar, A., Huang, C., Wang, X., Yu, J., & Ding, B. (2017). Free-standing,spider-web-like polyamide/carbon nanotube composite nanofibrous membrane impregnated with polyethyleneimine for CO2 capture. *Composites Communications*, *6*, 41–47. doi:10.1016/j.coco.2017.09.001

Zarei, M., & Karbasi, S. (2018). Evaluation of the effects of multiwalled carbon nanotubes on electrospun poly (3-hydroxybutirate) scaffold for tissue engineering applications. *Journal of Porous Materials*, *25*(1), 259–272. doi:10.100710934-017-0439-5

Zamora-Mora, V., Fernández-Gutiérrez, M., González-Gómez, Á., Sanz, B., Román, J. S., Goya, G. F., & Mijangos, C. (2017). Chitosan nanoparticles for combined drug delivery and magnetic hyperthermia: From preparation to in vitro studies. *Carbohydrate Polymers*, *157*, 361–370. doi:10.1016/j.carbpol.2016.09.084 PMID:27987939

Zare, E. N., Makvandi, P., & Tay, F. R. (2019). Recent progress in the industrial and biomedical applications of tragacanth gum: A review. *Carbohydrate Polymers*, *212*, 450–467. doi:10.1016/j.carbpol.2019.02.076 PMID:30832879

Zarekhalili, S.Z., Bahrami, H.M., Mohammadi, R., & Milan, P.B. (2016).Fabrication and characterization of PVA/Gum tragacanth/PCL hybrid nanofibrous scaffolds for skin substitutes. *International Journal of Biological Macromolecules*, *94*(Pt A), 679-690.

Zhang, L. H., Duan, X. P., Yan, X., Yu, M., Ning, X., Zhao, Y., & Long, Y.-Z. (2016). Recent advances in melt electrospinning. *RSC Advances*, *6*(58), 53400–53414. doi:10.1039/C6RA09558E

Zhang, B., Yan, X., He, H.W., Yu, M., Ning, X., & Long, Y.Z. (2017). Solvent-free electrospinning: opportunities and challenges. *Polymer Chemistry, 8*(2) 333–352.

Zhao, Y., Cao, X., & Jiang, L. (2007). Bio-mimic multichannel microtubes by a facile method. *Journal of the American Chemical Society*, *129*(4), 764–765. doi:10.1021/ja068165g PMID:17243804

Zhijiang, C., Yi, X., Haizheng, Y., Jia, J., & Liu, Y. (2016). Poly(hydroxybutyrate)/cellulose Acetate Blend Nanofiber Scaffolds: Preparation, Characterization and Cytocompatibility. *Materials Science and Engineering C*, *58*, 757–767. doi:10.1016/j.msec.2015.09.048 PMID:26478369

Zhijiang, C., Cong, Z., Jie, G., Qing, Z., & Kongyin, Z. (2018). Electrospun carboxy multiwalled carbon nanotubes grafted polyhydroxybutyrate composite nanofibers membrane scaffolds: Preparation, characterization and cytocompatibility. *Materials Science and Engineering C*, *82*(1), 29–30. doi:10.1016/j.msec.2017.08.005 PMID:29025660

Chapter 3
Addressing Antimicrobial Resistance Through Nanoantibiotics:
Challenges and Novel Strategies

Krishnanand Nagarajan
University College of Engineering, Bharathidasan Institute of Technology Campus, Anna University, Tiruchirappalli, India

Sathish Kumar Perumal
Anna University, Tiruchirappalli, India

Sathish Kumar Marimuthu
Anna University, Tiruchirappalli, India

Selvamani Palanisamy
Anna University, Tiruchirappalli, India

Latha Subbiah
Anna University, Tiruchirappalli, India

ABSTRACT

In recent years, the irrational use of antibiotics has escalated the evolution of multidrug-resistant (MDR) bacterial strains. The infectious diseases caused by these MDR bacterial strains remain a major threat to human health and have emerged as the leading cause of morbidity and mortality. The WHO and CDC have expressed serious concern regarding the continued increase in the development of multidrug resistance among bacteria. The antimicrobial resistance (AMR) poses a severe global threat of growing concern to human health and economic burden. Bacteria have developed the ability to resist antimicrobials by altering target site/enzyme, inactivation of the enzyme, decreasing cell permeability, increasing efflux due to over-expression of efflux pumps, target protection, target overproduction, and many other ways. The shortage of new antimicrobials and rapid rise in antibiotic resistance demands pressing need to develop alternate antibacterial agents.

DOI: 10.4018/978-1-7998-5049-6.ch003

INTRODUCTION

Infectious diseases caused by pathogenic bacteria's are a leading cause of morbidity and mortality among all ages. The severity of the infection varies from mild inflammation to a catastrophic epidemic. Cholera, dysentery, gonorrhea, measles, meningitis, necrotizing fasciitis, pneumonia and tuberculosis are some of the most severe bacterial infections that have ravaged humanity and had changed the course of history. Many microbial pathogens have evolved diverse escape strategies to surpass the anatomical defenses of the immune system and develop intracellular infections. Through the circulatory system, contagious bacteria spreads inside the body, multiplies, colonizes, and sustains itself, making them difficult to eradicate. Thus, these bacterial infections turn into a life-threatening condition at this stage. Most often, macrophages and other immune cells present in the body's major immune organs, such as bone marrow, lymph nodes, spleen and thymus, involves in the elimination of pathogens from the bloodstream (Infection, n.d.). However, antibiotic treatment is the only viable approach to destroy and inhibit the further growth of deadly pathogenic bacteria. Antimicrobials typically kill bacteria by hampering cell wall synthesis or inhibit vital cellular / metabolic activities in the bacterial metabolism. For example, beta-lactams, glycopeptides and lipopeptides class of antibiotics inhibit bacterial cell wall synthesis; sulfonamides inhibit folic acid synthesis, aminoglycosides, chloramphenicol, macrolides, streptogramins, tetracyclines inhibit protein synthesis; ansamycins inhibit the RNA production, and quinolones interfere with DNA replication and transcription of bacteria cells (A Brief Overview of Classes of Antibiotics, n.d.). Although antimicrobial development has advanced considerably, many bacterial infections remain hard to treat. Some major reasons include poor absorption/diffusion, rapid renal clearance, poor solubility and toxicity to host cells (Jijie et al., 2017). Another most serious and critical issue in treating bacterial infections today is the rapid acquisition of resistance to the antibiotic by the infectious microbes. Unfortunately, antimicrobial resistance (AMR) is now a pandemic at an ever-increasing pace. Many bacteria are entirely resistant to commonly prescribed first-line broad-spectrum antibiotics such as amoxicillin, clarithromycin and erythromycin, resulting in relapse and treatment failure due to recurrence of infection. Recent reports on bacterial resistance to carbapenems and colistin, considered to be last-line antibiotics, are worrying.(). Many of the antibiotics prescribed for decades have been derived primarily from another microbial source which makes it easy for the microbes to acquire resistant mechanisms. Many of the antibiotics prescribed for decades have been derived primarily from another microbial source, making it easy for microbes to acquire resistant mechanisms. In addition, as a result of natural selection, virulence gene transfer through horizontal gene transfer mechanism, environmental stress, and artificial gene transformation practices, the microbes have quickly gained resistance and had now evolved into a multi-resistant superbug. Other factors contributing towards the rise and spread of antibiotic resistance include inappropriate use of antibiotics, failure to complete prescribed antibiotic courses, overuse of antibiotics in forestry, cattle, poultry and fish farms, lack of measures to control infections, poor environmental hygiene, and lack of discovery and development of new antibiotics (Nagarajan et al., 2018). In order to overcome the ill effects and spread of AMR, it is necessary to adopt revolutionary approaches and develop novel antimicrobial therapy and formulation with minimal side effects and improved pharmacokinetic profiles. Antibiotics loaded nanoparticles are one such proposed alternative antimicrobial drug system. Nanoantibiotics (nAbts) offers a unique way of circumventing the existing antibiotic discovery approach and provides a new attack mechanism which can contribute to a longer-term solution for the evolution of drug resistance. It offers environment-specific activation, thus improves the efficacy of the nAbts and also open up new design approaches for various other clinical applications.

These nAbts hold a promising future in combating multidrug resistance. However, there is still plenty of efforts required for their successful development into a therapeutic aid for addressing antimicrobial-resistant infections. These “Nanoantibiotics” facilitate to alter and enhance the pharmacokinetic and therapeutic profile of antibiotic drugs, in contrast to the drugs in its free form (). The multiple benefits upon employing nanocarriers include; targeted drug delivery, sustained drug release, enhanced solubility, improved bioavailability, delivery of multiple therapeutic agents for polytherapy and several others, which is also applicable to antimicrobial nanosystems (). Therefore, the nanoantibiotic constructs hold a huge promise to prioritize disease-specific bactericidal treatment at a lower dose to ensure safety with sustained immunity and may hopefully revolutionize the field of nanobiomedicine.

Figure 1. The estimated number of deaths at every continent in 2050 attributed to antimicrobial resistance (Refer, 119)

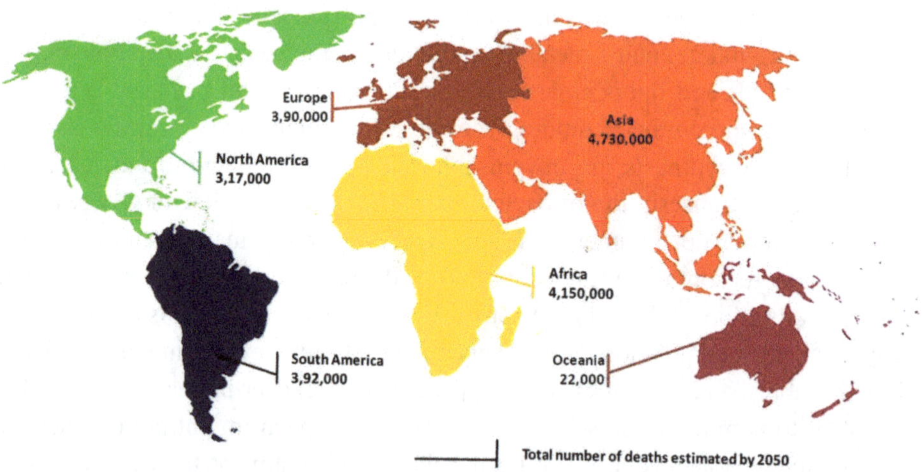

NANOPARTICLES TYPES FOR ANTIBACTERIAL APPLICATIONS

In the current nanotechnology scenario, there is a keen interest in the development of nano-drug carriers primarily for their use in cancer therapy. Several types of research have also extensively studied the antimicrobial potential of nanomaterials against a wide range of microbes. Advances in nanotechnology have opened up new ways to fight bacterial infections and resistance and have demonstrated groundbreaking prospects. The small size and high surface area strongly facilitate the antimicrobial activity of NPs. In some cases, metallic NPs, including AgNPs, AuNPs and metal oxides, such as ZnO and CuO, and other organic NPs, such as fullerene and carbon-CNTs, have inherent antibacterial activity (). In other cases, NPs do not have inherent activity but are used as carriers to supply antibacterial agents such as antibiotics, AMPs, small molecular drugs, proteins, nucleic acids, and enzymes. Besides, hybrid nanoantibiotic systems, together with bioinspired and biomimetic methods, provide additional functionality for the treatment of infectious diseases in clinical settings. Importantly, many of the NPs reported to date have shown intense antibacterial action against various pathogenic MDR bacteria, such as MRSA and Mycobacterium tuberculosis, and are even capable of combating biofilm infections ().

Besides, NPs offer a promising alternative forum for traditional methods for managing and reducing bacterial and biofilm infections. Nanoantimicrobials are considered as next-generation antibiotics because they have a remarkable potential to control and overcome the problems caused by multidrug-resistant pathogenic microbes. Depending on the bacterial and bacteriostatic potential of developed nanoantibiotics, it will be used in agriculture, biomedical, food processing and packaging, health and personal care products, livestock, textile and water treatment industries. Inorganic, metallic, organic and polymeric nanoparticles are often considered for nanoantibiotic development in the choice of materials in drug delivery (Álvarez-Paino et al., 2017). Just as the nanoparticle acts as an inert carrier, and the antibiotic agent expresses its antimicrobial activity. Antibacterial properties often originate from the antibiotic compound that is bound to or attached to the surface of the nanoparticles, which, in direct contact with the microbe, exerts antimicrobial effects on the microbe. However, some nanoparticles have inherent wide-spectrum inhibitory potential against Gram-positive and Gram-negative bacteria (Edson & Kwon, 2016). Pioneering nanoparticles for nanomedicine have also been extensively developed over the last two decades, such as albumin-drug conjugates, anti-arthritis gold nanoparticles, anti-body-drug conjugates, antimicrobial silver nanoparticles, block-copolymer mice, DNA-drug complexes, liposomes, polymeric nanocapsules, polymer-drug conjugates, and polymer-protein conjugates. The FDA has approved many nanotherapeutics, some have been in early and late clinical trials with a focus on the personalized treatment of infectious diseases ().

Figure 2. Antibiotic resistance strategies in bacteria

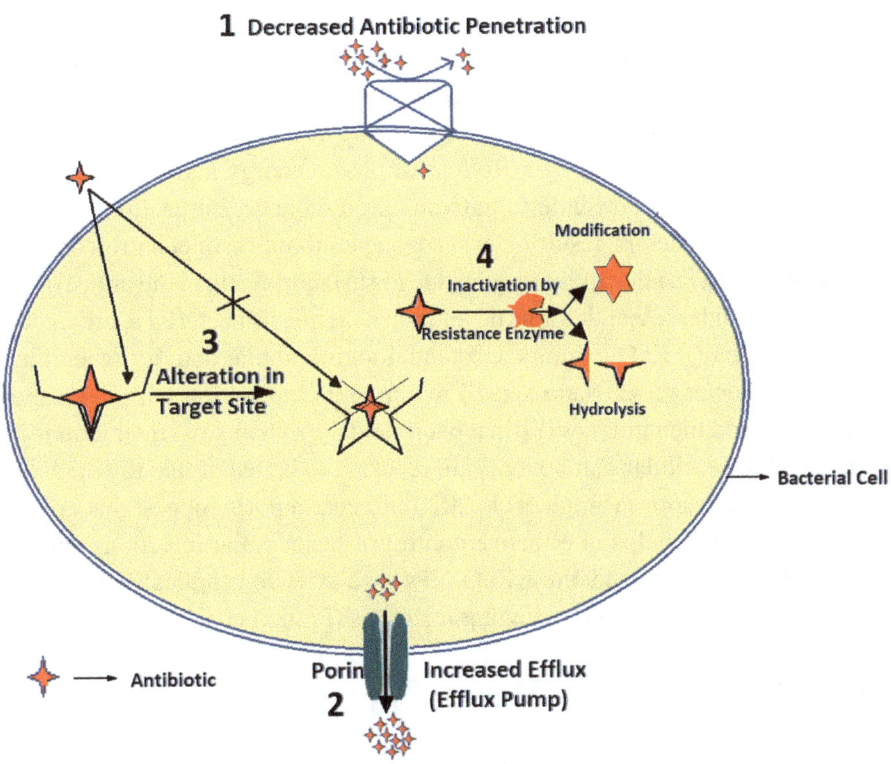

METALLIC NANOANTIBIOTICS

The metals have been used for thousands of years for their antimicrobial properties. Metal NPs are known for their extremely ionic characteristics and are synthesized with different morphologies that exhibit remarkable crystallinity and high surface area. The surfaces of these NPs are reactive due to plentiful corners and edges. The primary metal nanoparticles that were currently used as antibacterial agents include copper oxide (CuO), gold (Au), magnesium oxide (MgO), silver (Ag), titanium oxide (TiO_2), zinc oxide (ZnO) and few other metals (Al, Ga, Ge, As, Se, Sn, Sb, Te, Pb and Bi) in groups 13–16 of the periodic table. The metal nanoparticles help to combat AMR and mediate biocidal effect by inducing oxidative stress, interfering with bacterial protein synthesis, disruption of the bacterial cell membrane and hindering with biofilm formation ().

Zinc Oxide Nanoparticles

Zinc oxide nanoparticles (ZnO-NPs) antibacterial activity has received significant interest worldwide, particularly in the domain of antimicrobials. ZnO-NPs exhibit attractive antibacterial properties due to the reduced size and increased surface area, which enables enhanced particle surface reactivity. The ZnO has high optical absorption in UVA (315–400 nm) and UVB (280–315 nm) regions which greatly favours broad-spectrum antibacterial and antibiofilm activity against some ESKAPE group of pathogens. Many researchers across the globe had reported and recorded the strong antibacterial exhibited by ZnO-NPs of varying morphologies. ZnO is also a non-toxic, bio-safe substance and possess photo-oxidizing and photo-catalysis impacts on chemical and biological entities. Studies of genotoxicity had shown that exposure to ZnO causes severe DNA damage to bacterial cells.

Further studies have also confirmed that the ZnO-NPs are capable of inducing reactive oxygen species (ROS) generation, protein oxidation and DNA damage with concomitant thiol depletion (). The ROS generation is one of the critical mechanisms in several pathways, including cell wall damage, membrane permeability, internalization of ZnO-NPs due to loss of proton energy and absorption of harmful dissolved zinc ions. Which in turn, contribute to mitochondrial damage, intracellular outflow, activation and release of oxidative stress factors, resulting in subsequent inhibition of cell growth and cell death. In some cases, enhanced antibacterial activity may be due to surface defects in the abrasive surface texture of ZnO. ZnO is currently under research for its development as an antibacterial agent in both microscale and nanoscale formulations (). ZnO exhibits substantial antimicrobial activity when the particle size gets reduced to the nanometer scale. Nano-sized ZnO upon its interaction with the bacterial cell wall internalizes into the cell and then initiates distinct bactericidal mechanisms. Interactions between these ZnO-NPs and bacterial intracellular components are reported to be lethal due to which they had found profound antimicrobial applications (Singh et al., 2020). Recent reports suggest that coating of hospital implants with 4% of ZnO-NPs helps in effective control of nosocomial infections contracted through medical instruments. On the other hand, these ZnO-NPs have extended applications for use in sunscreen skin lotions for UV protection and food packaging materials (Pinnell et al., 2000).

Iron Oxide Nanoparticles

Iron oxides such as magnetite (Fe_3O_4), maghemite (γ-Fe_2O_3), hematite (α-Fe_2O_3) and goethite (FeO(OH)) come under magnetic nanoparticles (MNPs). The iron oxide nanoparticles (IONPs), magnetite (Fe_3O_4), maghemite (γ-Fe_2O_3), hold great promise for biomedical and biotechnological applications. Due to their biocompatibility, IONPs are used as active drug carriers to treat various forms of cancer. In recent days MNPs are profoundly used for the site-directed drug delivery applications using an external magnetic field, which in turn improves drug stability against enzymatic or metabolic degradation. MNPs may also be used in other fields, including analytical chemistry, pathogen identification, antigen diagnosis, tissue repair and hyperthermia, due to its excellent magnetic, physicochemical, thermal and mechanical properties.

The results of strenuous research to understand the mechanism behind IONP's/MNPs antimicrobial activity have shown that magnetite NPs exhibit antimicrobial activity only in their nanoparticle form, but not in their bulk form (Arakha et al., 2015). Also, intact IONPs in contact with air loses its magnetic property and tend to agglomerate. In order to avoid these problems, numerous research groups have made untiring efforts to alter the surface of IONPs with a wide range of chemical and biological agents to provide enhanced in vitro and in vivo biological applications. Surface modification using natural or synthetic polymers enables more robust, hydrophilic nanostructures and provides a relatively higher number of variable surface functional groups that help in the proper binding of interactomes to the nano-structure. It should also be noted that superparamagnetic iron oxide nanoparticles (SPIONs) with a core diameter ranging from 10 to 20 nm are typically considered biocompatible and biodegradable(Edson & Kwon, 2016) as free iron is incorporated into the body's iron reserves, used for metabolic processes, and gets easily eliminated out of the body(Philosof-Mazor et al., 2013). The newly developed IONP-based nanosystems have been designed to hold multiple drug moieties drugs simultaneously with adequate release control at the target site, which is very useful in many clinical conditions. Furthermore, bare IONPs may release free iron that is toxic to bacterial cells as it catalyzes ROS production in Fenton's reaction (equation 1 & 2) (Dinali et al., 2017).

$$Fe^{2+} + H_2O_2 \rightarrow Fe^{3+} + OH^{\cdot} + OH^{-} \text{-----} \quad (1)$$

$$Fe^{3+} + H_2O_2 \rightarrow Fe^{2+} + OOH^{\cdot} + H^{+} \text{-----} \quad (2)$$

Another mechanism by which IONPs induces toxicity is by the release of metal ion from the metalc-ore. Released ions can increase membrane permeability and cause cell material leakage. Besides, the absorption of these ions by the bacterial cells results in the disruption of DNA replication. The potential of surface-modified IONPs with antibiotics, targeting and permeation enhancers have demonstrated their efficacy in a non-toxic manner in biofilm dispersion while significantly improving the outcome of antibiotic therapy. Amoxicillin conjugated magnetite nanoparticles showed significant antimicrobial activity against *Staphylococcus aureus* and *Escherichia coli* along with 3-4 fold reduction in the dose of amoxicillin, which usually has serious side effects when administrated at high doses (Grumezescu et al., 2014).

In an attempt to treat infection caused by methicillin-resistant *Staphylococcus aureus*, streptomycin loaded chitosan-coated magnetic nanoparticles (Strep-CS-MNPs) were used to maximize antibiotic therapeutic outcomes. Strep-CS-MNP showed sustained drug release profiles displaying remarkable enhancement of up to 18 folds compared to freeform streptomycin, thus reducing drug toxicity and side effects concerns, which indicates the ability of iron oxide nanosystems to operate in controlled release systems (Hussein-Al-Ali et al., 2014).

As a recent advancement, several antibiotics, such as Cefazolin, Cefotaxime, Ceftriaxone, Ciprofloxacin, Doxorubicin, Doxycycline and Rifampicin, have been loaded in silver-IONP conjugates aiming to improve therapeutic activity and targeted antimicrobial therapy. Antibiotic molecules were bound to magnetite/silver nanocomposites through chemical bonding, electrostatic interaction, and hydrogen bonding. Besides, antimicrobial findings have also shown that magnetite - silver - antibiotic nanocomposites exhibit enhanced antimicrobial activity against a wide range of bacterial organisms, including drug-resistant and pathogenic organisms without altering magnetite magnetic properties.() Many related antibiotics, including lactose functionalized IONPs-ciprofloxacin nanosystem, have also been studied. Also, nanosystems consisting of Amikacin, Amoxicillin, Bacitracin, Cefotaxime, Erythromycin, Gentamicin, Kanamycin, Neomycin, Penicillin, Polymyxin, Streptomycin and Vancomycin were directly linked to IONPs were analyzed for their therapeutic efficacy enhancements. Chlorhexidine (CHX), an antimicrobial agent used to control oral biofilms, is more effective in reducing biofilm biomass of *Enterococcus faecalis*, *Candida albicans*, and *Staphylococcus aureus* when bound to IONPs (Arias et al., 2018). The association of IONPs with the target pathogen has been shown to induce microbial toxicity due to several interactions, including membrane depolarization with consequent impairment of cell integrity, development of reactive oxygen species (ROS) along with lipid peroxidation, DNA damage, protein miscoordination and release of metal ions affecting cell homeostasis (Al-Shabib et al., 2018).

Silver Nanoparticles

Silver (Ag) and its compounds were better known for their antiseptic and antimicrobial properties for many decades. Dr Carl Crede, a German obstetrician, used silver nitrates in the early 19th century to treat and prevent microbial infections. However, as the years go, the microbicidal use of silver has gradually decreased due to the discovery of antibiotics (Bottagisio et al., 2019). Owing to the recent evolution of antibiotic resistance, the reduced efficacy of conventional antibiotics, the non-applicability of antibiotic resistance mechanisms for nanoparticles and the intrinsic antimicrobial capacity of silver, silver nanoparticles (AgNPs) have regained Prominence in the treatment of infectious diseases. Rather than penetration or pore formation, AgNPs exert an antibacterial effect by direct contact with the bacterial cell (Neethu et al., 2018). Today, the biomedical applications, such as diagnostics, therapeutics, surgery, drug delivery and coating material for surgical instruments, personal hygiene and health care, uses AgNPs in the majority. At the same time, other products have significance in industrial and environmental applications. Dose and size have a significant role to play in the antibacterial activity of AgNPs. AgNPs between sizes 5 and 10 nm have been shown to have the highest bactericidal capacity. Many studies suggest that, in addition to size, factors such as morphology, synthesis method, precursors used and the parameters adapted, determine the magnitude of the antimicrobial activity. In recent days, due to simple, reliable, biocompatible, eco-friendly, tunable and cost-effective features, the green chemistry approach for AgNP synthesis has received considerable attention (Dakal et al., 2016; Ferraris & Padoan, 2018). The primary benefit of biological synthesis methods is the ability to control the size and shape of the particle, which are the key factors deciding it to be used in biomedical applications. Several studies have shown that the antimicrobial propensity of AgNPs is strongly influenced by physicochemical parameters such as size, shape, concentration and colloidal state. Among which, the colloidal stability of the AgNPs is considered to be an important attribute of their bactericidal activity The colloidal state of silver is considered to be very effective in the fight against bacterial infections due to its ability to catalyze and destabilize enzymes that are crucial for the oxygen utilization. In contrast, the liquid systems lack colloidal stability hence considered less effective. It has also been proposed that AgNPs, in its colloidal form, appear to modulate signal transduction pathways in bacteria, thereby altering the protein profile of phosphotyrosine, which leads to inhibition of the respiratory chain but also for the subsequent production of ROS within the cell leading to oxidative stress, cell damage and death (). Electrostatic guided adhesion of AgNPs (positively charged) over the bacterial cell membrane (negatively charged) results in the creation of surface pits in the lipid bilayer, which in turn increases membrane permeability and ultimately leads to death through cytoplasmic leakage. In addition, their internalization and interaction with the sulfur and phosphorus-containing biomolecules arrest bacterial growth and replication by interfering with the vital functions of regulatory enzymes and DNA, which are involved in replication and metabolic activities. Interestingly, AgNPs have been shown to have bactericidal effects regardless of the pathogen type, ranging from pathogenic bacteria to antibiotic-resistant bacteria (Dakal et al., 2016; Haider & Kang, 2015). Recently, in order to improve antimicrobial effectiveness and to combat antimicrobial resistance, AgNPs have been used in combination with antibiotics. Many researchers have investigated the antimicrobial activity of this combination therapy using AgNPs and antibiotics of different types. The synergetic effects of AgNPs and antibiotics such as Amoxicillin, Clindamycin, Cephradine, Erythromycin, Penicillin G, Vancomycin and Vildagliptin against Gram-positive and Gram-negative bacteria including *Bacillus cereus*, *Escherichia coli*, *Klebsiella pneumonia*, *Pseudomonas aeruginosa*, *Streptococcus pyogenes* and methicillin-resistant *Staphylococcus aureus* (*MRSA*) have been reported. In another study, the antibacterial activity of biosynthesized AgNPs in combination with 14 antibiotics, namely Amikacin, Gentamicin, Kanamycin, Amoxicillin, Ampicillin, Ceftriaxone, Vancomycin, Ciprofloxacin, Doxycycline, Tetracycline, Chloramphenicol, Trimethoprim, Ceftazidime and Penicillin, was evaluated using the disk diffusion method. A huge degree of variation was observed in the antibacterial activities of antibiotics in the presence and absence of AgNPs. The findings demonstrated the benefit of the synergy achieved by combining nanomaterials and antibiotics. The overall highest synergistic activity was observed for vancomycin, followed by penicillin, amoxicillin and antibiotic class Aminoglycosides, except for Gentamicin and Kanamycin, which had antibacterial effects similar to vancomycin and penicillin against Pseudomonas aeruginosa and Acinetobacter baumannii. Similarly, drugs Cephradine and Vildagliptine demonstrated improved bactericidal activity when conjugated with AgNPs. In particular, Cephradine, a first-generation cephalosporin antibiotic that is not effective against most bacteria today due to resistance mechanisms, demonstrated enhanced antibacterial activity against Escherichia coli, Streptococcus pyogenes, and six other bacterial pathogens when conjugated with AgNPs (Ceph-AgNPs). Besides, the Ceph-AgNPs greatly increased the bioavailability of Cephradine and selectively decreased the pathogen-mediated host cell cytotoxicity (Masri et al., 2018).

Gold Nanoparticles

Gold is a noble transition metal, traditionally considered to be chemically inert in its bulk form. While this metal, when reduced to nanosize or subnanometer dimensions, these ultra-small nanoclusters tend to express surprising and exciting physicochemical properties (Ershov et al., 2016; Nanoantibiotics, n.d.). The practice of using gold (Au) in Ayurveda and other Indian medicines for the treatment of a variety of chronic diseases originated in about 2500 BC. *'Swarna Bhasma'* is a red colloidal gold-based, ayurvedic formulation used to treat a variety of diseases such as anaemia, diabetes, skin diseases, fever, and also in skin rejuvenation and anti-ageing therapies (). The transformation of bulk gold into gold nanoparticles (AuNPs) or gold nanoclusters (AuNCs) (core sizes less than 2 nm) modulates their biophysical and biochemical characteristics, including cytotoxicity, cellular absorption, biodistribution, pharmacokinetics and efficiency targeting.

In the way of substantiating the fact that Au in its nanoclusters form (AUNCs) pose robust catalytic activity, the Cambridge research group had demonstrated the robust catalytic activity of Au55NCs (contain 55 Au atoms) for the selective oxidation of styrene by dioxygen which has arisen due to the alteration of intrinsic electronic structure in small gold nanoparticles (Johnston, 2012). The ampicillin, streptomycin and kanamycin antibiotics conjugated AuNPs showcased an enhanced bactericidal efficiency against the bacterial strains, *E.coli DH5α*, *M. luteus* and *S. aureus*. The antibiotic conjugated AuNPs also showed a significant decrement in MIC values compared to AuNPs and antibiotics in their free. Since AuNPs are less toxic and biocompatible, they have been used as drug carriers to improve the bioavailability of the drug at infected sites and reduce drug-related toxicity due to the delivery of drug at the site of infection. Hence has been widely used for the release of the antibiotics such as ciprofloxacin, gentamycin, and tuberculosis drugs over an extended period for successful antibiotic therapy. Being exceptionally small in size, AuNCs is capable of interacting effectively with bacterial cell and subcellular organelles. Due to their unique features such as quantum mechanical behaviour, superior stability, high fluorescence intensity and strong hydrophilicity, the AuNCs has gained significant attention over these last few years in biomedical applications. The AuNPs conjugated with β-lactam antibiotics stabilize the β-lactam ring by delocalizing the electron in the carbonyl group of the β-lactam ring. This phenomenon makes the drug moiety more stable than its free form. It helps to preserve bactericidal activity at high temperatures and their storage at room temperature (). AuNPs (5nm) dispersed onto zeolites, mordenite, faujasite and clinoptilolite, have been reported to exhibit enhanced bactericidal activity by eliminating 90-95% of *E. coli* and *S. typhi* colonies (Lima et al., 2013). As a novel therapeutic strategy, ampicillin functionalized lysozyme capped gold nanoclusters (AuNC-L-Amp) were fabricated to act against methicillin-resistant *Staphylococcus aureus* (MRSA) infections. The study outcome reveals the reversion of MRSA resistance and enhanced bactericidal effect of AuNC-L-Amp against systemic MRSA. Which is due to targeted drug delivery and improved bioavailability, multivalent presentation and enhanced drug permeation, lysozyme mediated disruption of the cell wall, dysfunctioning of bacterial efflux pump and last but not least is the Au mediated destabilization of replication and metabolic processes (Lima et al., 2013).

Similarly, AuNPs loaded with imipenem (Ipm), and meropenem (mem) showed significant increases in antibacterial activity against carbapenem-resistant Gram-negative bacteria, namely Klebsiella pneumoniae, Proteus mirabilis and Acinetobacter baumanii. Further, the Ipm and Mem loaded AuNPs demonstrated four and three-fold decrease in the MIC, respectively in comparison with the free drugs over the bacterial isolates. These results proved carbapenems-loaded AuNPs to be, efficient drug delivery vehicle for improving the therapeutic activity and destroying the carbapenem-resistant pathogens (Turner et al., 2008).

Figure 3. Nanoantibiotic approaches employed for combating antibacterial resistance.

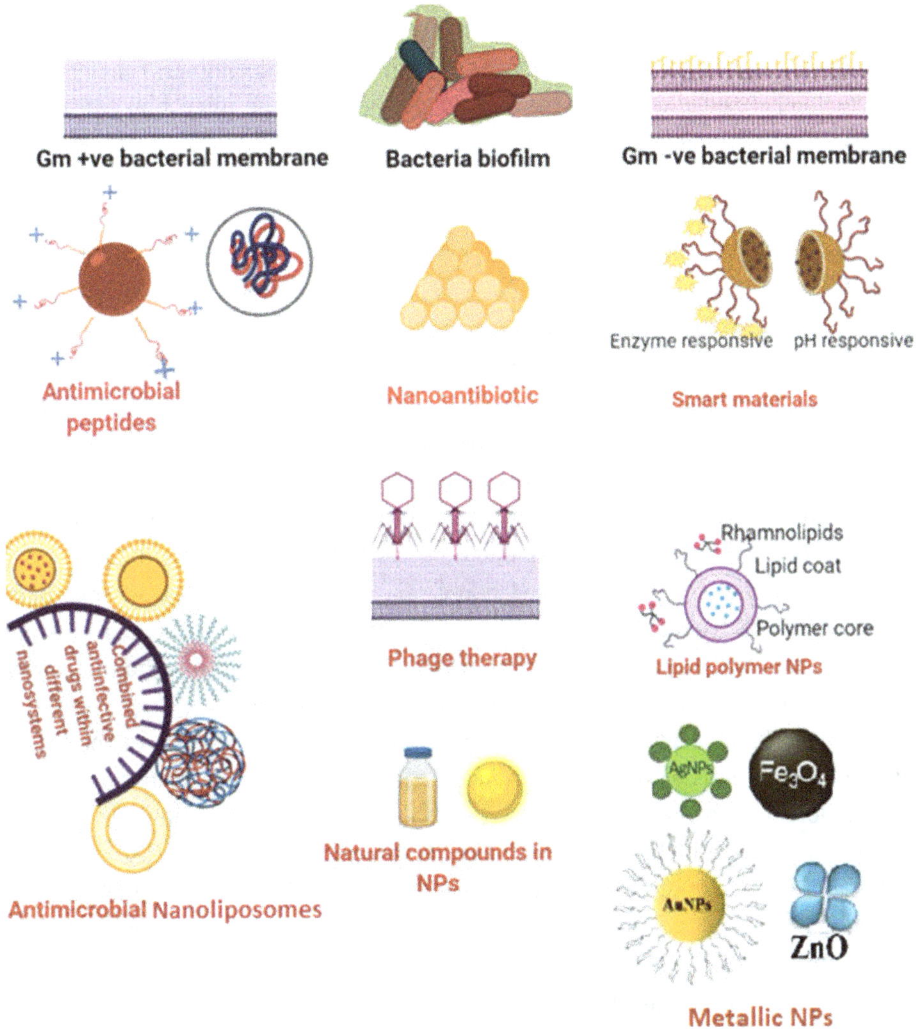

ANTIMICROBIAL PEPTIDES

Antimicrobial peptides (AMPs) are evolutionarily conserved therapeutic biomolecules. These molecules play a vital role in the first line of nonspecific immune defense. They are found in all classes of organisms ranging from prokaryotes to humans.() The AMPs are natural antagonistic agents, and their primary function is to protect the host from invading pathogens by exerting cell-mediated cytotoxicity over them. Apart from defending pathogens AMPs act as multifunctional effector molecules and mediate the release or activation of acute-phase proteins, complement cascades, cytokines, chemokines, mitogens, nitric oxides, prostaglandins and superoxides (). Since the discovery of Melittin in 1952 from bee venoms, more than 4000 natural AMPs have been identified in bacteria, fungi, plants, amphibians, insects and animals till date. These bioactive molecules exhibit selective antimicrobial activity against a wide range of pathogens (Zhang & Gallo, 2016). Several mechanisms have been proposed for the bactericidal activity mediated by AMPs, the most prominent of which is the membrane destabilization and permeability mechanism. To understand the action of AMPs, researchers have proposed many different pore formation models, such as the aggregated channel model, the barrel-stave model, the carpet model, and the toroidal-pore model. The first step in the process of membrane permeation is the development of the electrostatic interaction between the AMPs (positively charged) and the microbial cell wall (negatively charged). Subsequently, this interaction induces the formation of a cavity or the creation of temporary passage on the cell membrane, which disintegrates cell membrane leading to increased cell permeability which eventually causes the bacterial contents to ooze out, leading to cell lysis and death. Apart from the cationicity of the AMPs, the parameters such as peptide length, secondary structure and amino acid composition are also considered to be vital for determining the antibacterial activity (Jean-François et al., 2008). Natural antimicrobial peptides with a short half-life are usually not stable. It is, therefore, necessary to develop and synthesize long-acting peptide analogues for future clinical applications. Nanotechnology-based therapeutic approaches have attracted attention due to their ability to improve the biocompatibility, stability and therapeutic efficacy of drugs incorporated in the nanomaterials. The combination of AMPs with nanoparticles or nanocomposites ensures the use of synergistic therapeutic benefits of both the molecules, which improves the effectiveness of therapeutics against multidrug-resistant bacteria. Several studies are being conducted to assess the therapeutic benefits of combining nanoparticles with membrane-active peptides (Chen & Lu, 2020). Among the various nanomaterials reported lipids, metal and metal oxide-based nanomaterials are widely recommended for various biomedical applications, and their unique physicochemical properties enable us to use them as antimicrobials. Similar formulations have previously been used in traditional medicines.

The physicochemical factors that influence the pharmacokinetic parameters of the therapeutic formulation include solubility, particle size, surface charge and zeta potential of the nanosystem (Eleraky et al., 2020). Hence, before the development of AMP-based therapeutics, it is essential to optimize the vital parameters owing to its bactericidal activity. Stability and biocompatibility are essential considerations in the design of AMP. There are several examples where PBP10 peptide coated magnetic nanosystem expressed elevated antibacterial and antifungal activity against *Staphylococcus aureus* MRSA Xen 30, *Pseudomonas aeruginosa* Xen 5 and *Candida sp*. Similar studies using PBP10-containing nanosystems have exhibited potent anti-inflammatory and immunomodulatory activities.

Furthermore, this nanosystem was found to mediate antimicrobial activity through penetration and inactivation of efflux pump and metabolic enzymes (). In the same way human antibacterial peptide cathelicidin LL-37, synthetic ceragenins CSA-13 in combination with core-shell magnetic nanoparticles were tested against *P. aeruginosa* Xen 5 and *S. aureus* Xen 30 (a clinical MRSA isolate), which exhibited 64-fold and 32-fold decrease in the MIC value of LL-37 peptide (Niemirowicz et al., 2016). A synthetic rhodamine B-conjugated peptide (PBP10) which was found to be active against Gram-positive and Gram-negative bacteria that also limits bacterial pro-inflammatory factors upon binding with lipopolysaccharide (LPS) and lipoteichoic acid (LTA), when immobilized on the surface of IONPs limit the production of inflammatory mediators, such as reactive oxygen species (ROS), nitric oxide (NO), and interleukin-8 (IL-8). The PBP10 nanosystems also showed improved hemocompatibility and low microbial-induced inflammation when compared to free peptides (Piktel et al., 2019). Consistent with the results of previous studies, Cm-p5 conjugated citric acid-coated $MnFe_2O_4$-NPs demonstrated reduced toxicity and improved antimicrobial efficacy and biodistribution (Lopez-Abarrategui et al., 2016). For instance, the nisin loaded in the soy soluble polysaccharide nanoparticle (SSPS) exhibited remarkable stability. It sustained its antimicrobial activity against *Listeria monocytogenes*, *Bacillus subtilis* and

Staphylococcus aureus. The sustained release of Nisin from NSPNP significantly increased the shelf life of fresh tomato juice to 60 days. Thus, NSPNP holds a great promise for its use as a preservative in food industries (Luo et al., 2019). From the above studies, it is clear that the combination of nonmaterial's with membrane penetrating peptides and traditional antibiotics offers a revolutionary way to develop a therapy to treat and prevent infections caused by multidrug-resistant pathogens.

Figure 4. Biological function of Antimicrobial peptides. Adapted from (Zhang & Gallo, 2016)

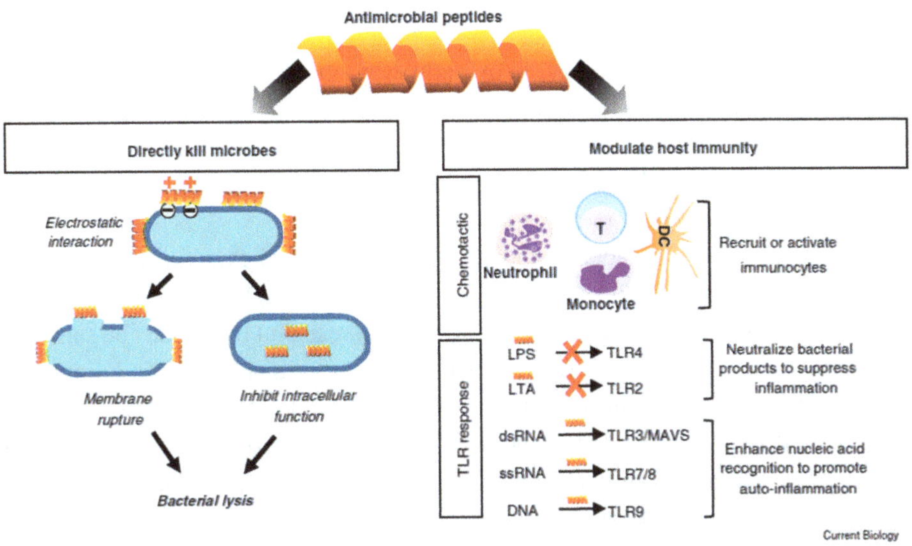

BACTERIOPHAGES

Bacteriophages or phages (20-200 nm in size) are viruses that can replicate, invade, and destroy bacteria, without damaging human or animal cells. They are one of the most prolific and ubiquitous species on Earth, playing an important role in microbial biology, population growth, ecology and medicine (Clokie et al., 2011). Currently, within a century of the discovery of antibiotics, health care is facing a serious challenge due to antimicrobial resistance. This situation demands rapid growth, benchmarking, and the adoption of innovative medical approaches to infectious diseases, thus, phage therapy is once again gaining prominence. A systematic phage therapy using viral bacteriophages to fight bacterial infections and diseases was first identified in 1917 by Felix de Herrell, where the phages kill the infected bacteria via the lytic phase. (). Currently, this technique is being reviewed for use alone or in combination with antibiotics and nanoparticles to treat infections caused by resistant bacteria. The use of naturally occurring phages to combat bacterial infection in Western medicine has a controversial past. However, the emerging phage-antimicrobial system has advanced beyond conventional approaches. Phage therapy has reiterated in recent years that drug-resistant bacterial infections are on the rise. In addition, phages were used to treat many bacterial infections, either alone or in combination with antibiotics, including bloodstream infections, lung infections, persistent otitis, skin burns, and lung infections (Domingo-Calap & Delgado-Martínez, 2018). Innovations such as bioengineered chimaeras of phage-derived lytic proteins have prompted the emergence of a new type of antibacterial drug in this rapidly growing field. Besides the prospect of replacing antibiotics, phage and its proteins have substantial uses involving bacterial biosensing, gene transfer, vaccine adjuvants, vaccine nanocarriers, medicines and therapeutic genes, cancer and tumour therapy, degenerative disease management, antibiofilm agents, multidrug-resistant (MDR) bacteria cure, glycobiology functions, surface disinfection, bacteriophage display and food biopreservation ().

Phage therapy was reported to be effective against pathogenic bacteria well before the discovery of antibiotics. The phage penetrates the bacterial cell by binding itself to the specific cell surface receptors that are present over the surface of the bacterium and kill the bacterial cells through activation of the lytic phase. The emergence of phage resistance is one of the most critical problems that phage therapy faces today. This phenomenon could easily be overcome by chemically altering of the phage surface with cell-penetrating peptides such as DGEA (Asp-Gly-Glu-Ala) or RGD (Arg-Gly-Asp). These surface modified phages may have improved therapeutic effects in a targeted delivery approach. Studies suggest that not alone the surface of phages could be modified. Still, their genetic material could also be manipulated to enhance their ability to deliver drugs and genes.

Apart from drugs and genes, therapeutic moieties such as antibodies, peptides, growth factors, and metabolites could also be delivered (Yoo et al., 2016). So far, several strategies have been proposed, one of which is the suppression of the SOS DNA repair system of the resistant bacteria. The same had also been experimented in-vitro against a wild-type *E. coli* EMG2 by co-administering *lexA3* repressor-producing phages with ofloxacin (a quinolone antibiotic). The results showed a 2-fold increase in the bactericidal activity against *E. coli* than ofloxacin alone. The engineered phage also enhanced the bactericidal activity of other antibiotics, such as gentamicin and ampicillin. The engineered phage was also found to be effective against persistent and biofilm cells and also reduced the development of antibiotic-resistant mutants. This strategy has also been found to be effective in *in-vivo* systems (). In another strategy, the genes encoding sensitivity to antibiotics were delivered into the bacterial cells to reduce the development of bacterial resistance to antibiotics. In regard, a study was conducted in which engineered temperate phages were made to deliver the dominant *rpsL* gene and *gyrA* gene encoding sensitivity to antibiotics streptomycin and nalidixic acid respectively into the bacteria (*E. coli* K-12). Upon evaluating the MICs after phage lysogenization, the bacterial susceptibility to both antibiotics found to be restored and the MICs of streptomycin and nalidixic acid had decreased by eight and two-fold, respectively ().

In addition to their therapeutic applications, bacteriophages were used in commercial food preservation technologies as food bio preservative. Hence to improve the stability of the phage particles during food processing, bacteriophages were encapsulated in nanovesicles such as niosomes, liposomes, and transfersomes. In an investigation, the stability and efficacy of phage phiIPLA-RODI encapsulated in different types of nanovesicles under different encapsulation conditions was studied. The results revealed that high encapsulation efficiency was observed with niosomes which also retained stability for an extended time than that of free or non-encapsulated phages (González-Menéndez et al., 2018).

NANOLIPOSOMES

Nanoliposomes are spherical bilayered nanoscale vesicles, widely used for encapsulation and delivery of bioactive compounds. The physical, chemical, structural and thermodynamic properties of nanoliposomes are similar to conventional liposomes and with a size less than 200 nanometers, which prevent the rapid degradation and clearance by the liver macrophages. Nanoliposomes offer excellent stability for bioactive compounds against several prominent physicochemical stress factors. Also, they can provide sustained drug release at the target site, resulting in improved bioavailability and efficacy. Overall, nanoliposomal delivery systems offer a larger surface area, enhanced drug stability, controlled release, solubility, bioavailability, and successful targeting properties compared to other liposomal encapsulations (Sarabandi et al., 2019). Due to enhanced biocompatibility and biodegradability; nanoliposomes finds various applications in nanotherapy, for example, in diagnosis, cancer treatment, tumour imaging, nutraceuticals, cosmetics, agriculture, food enrichment, and fortification. Specific nanoliposomal formulations were used in the delivery of genes or the production of vaccines. Besides, their applications have also expanded to include antibiotics, antiviral drugs, steroid and anti-inflammatory drugs. The cell-specific targeting seems to be a significant advantage of nanoliposomes, helping to achieve an optimum therapeutic efficacy at the target site alone while minimizing the adverse effects over surrounding or associated healthy cells and tissues ().

The physical parameters of the formulation such as vesicle size, lipid composition, surface charge and drug-to-lipid ratio determine the fate and the biological activity of liposomal drug preparations. In particular, varying one of the liposomal parameters, for example; the change in lipid composition (PC/cholesterol) leads to concomitant changes in other parameters such as liposome size, drug-to-lipid ratio and in vivo stability characteristics such as circulation longevity and toxicity. The critical ingredients of nanoliposomes are lipid and phospholipid molecules. Lipids are derivatives of fatty acids with various head group moieties. Phospholipids are amphiphilic molecules, containing a hydrophilic head group and the hydrophobic fatty acid tail (acyl chain). Apart from lipid and phospholipid molecules, nanoliposomes may contain other organic molecules such as sterols in their bilayer. Sterols are the core components of plasma membranes in most eukaryotic cells involved in maintaining the domain structure of the cell membrane. The most commonly used sterol in the preparation of the nanoliposome is cholesterol. Cholesterol is used in nanoliposome preparation to improve the stability and the fluidity of the lipid bilayer. In general, the cholesterol modulates membrane fluidity by preventing crystallization of phospholipid acyl chains and providing steric hindrance to their movement. This phenomenon contributes to the chemical stability of the nanoliposome formulation. It reduces the permeability of solute molecules across the lipid membrane.

The ratio of cholesterol to be used in the preparation of nanoliposomes varies according to its application. In general, 10 to 40% molar ratio of cholesterol is widely used in the preparation of nanoliposome vesicles (Mozafari, 2010; Reza Mozafari et al., 2008). Similar to cholesterol the stability of nanoliposomes can be improved by incorporating specific polymers and stabilizing agents, such as polyethylene glycol (PEG), chitosan, glycerol, phosphatidylglycerol, stearylamine and using other cationic or anionic ingredients. The formulation employing small unilamellar vesicle (SUV) with a particle size <100 nm was more efficacious than multilamellar vesicle (MLV) formulation due to extended circulation half-life and increased accessibility, which in turn enhances the therapeutic efficiency of this liposomal drug carrier system (Danaei et al., 2018). Another cofactor that is vital in determines the therapeutic efficiency of this liposomal formulation is the drug-to-lipid ratio which also influences the longevity of liposomes in the circulation and drug retention properties ().

Application of nanoliposomal antibiotics presents a promising strategy for improved drug delivery to bacterial cells and biofilms. Many gram-positive and gram-negative bacterial strains, which displayed resistance to antibiotics in its free state, have shown susceptibility while encapsulated in liposomal nanocarriers. Recently, many liposomal nanoparticles, aiming localized delivery of antibiotics were investigated with antibacterials such as Vancomycin Hydrochloride, Teicoplanin, Triclosan, Fusidic acid and Rifampin to eradicate MDR *Staphylococcus aureus*, *Staphylococcus epidermidis*, *Acinetobacter baumannii*, *Escherichia coli*, carcinogenic *Streptococcus* mutants and related pathogenic biofilms. The multimodal action of triclosan-loaded nanoliposomes such as biofilm penetration and accumulation, electrostatic targeting of the bacterial cell in a biofilm, and subsequent release of bacterial lipases from micelle core leads to degradation of the blood-accessible *staphylococcal* biofilms (Nicolosi et al., 2015; Scriboni et al., 2019; Su et al., 2018). Topical formulations containing drug-loaded nanoliposome holds excellent potential for localized antibiotic delivery in the treatment of skin infections caused by resistant bacterial strains and also dramatically reduces the drug dose and its related adverse side effects. Similarly, antibiotics in combination, for example, rifampin and N-acetylcysteine (NAC) loaded nanoliposomes (size less than 200 nanometers) had exerted mucolytic activity against *Staphylococcus epidermidis*, thus inhibiting biofilm formation (Hajiahmadi et al., 2019).

The vesicle type determines the efficacy of the antibiotic drug. In contrast, many studies have revealed that the multilamellar vesicles (MLVs) are best suited for encapsulating hydrophobic antibiotic drugs, and large unilamellar vesicles (LUVs) are preferred for loading hydrophilic antibiotics. Interestingly, cationic liposomes display better interaction with negatively charged surfaces of the bacterial cell. However, anionic liposomes composed of dipalmitoylphosphatidylcholine (DPPC), dioleoylphosphatidylglycerol (DOPC) are capable of encapsulating positively charged antimicrobials and was reported to have a loading capacity eight-fold higher than uncharged and natural-lipid based liposomes. Whereas, zwitterionic liposomes having both acidic and alkaline functional groups, holds flexibility for charge reversal below and above their iso-electric point. This feature helps the liposomes to circulate in blood without interacting with other negatively-charged components of the blood, and also acquires positive charge inside the acidic environment of a biofilm. They thus facilitate stronger interaction with negatively-charged bacteria present in the biofilm. Moreover, the zwitterionic liposomes were reported to be highly biocompatible and are capable of disrupting the periodontal biofilm (Eleraky et al., 2020).

POLYMERIC ANTIBIOTICS

Several new methods for controlling drug-resistant bacterial infections are currently under investigation to control or reverse the development of antibiotic resistance. Antimicrobial polymers and polymer-based antibacterials have now become one of the most promising solutions in recent years. These polymer-based antibacterials have the benefit of a broad-spectrum antibacterial activity by disrupting the bacterial membrane, with a lower risk of developing resistance in future. Most of the polymeric antibiotics investigated so far are capable of binding to the bacterial cell membrane, triggering membrane disruption and the cell content leakage. Interestingly, some antimicrobial polymers have also been reported to mimic and promote the activity of conventional antibiotics (Polymers to Combat Antibiotic-Resistant Bacterial Infections - Advanced Science News, n.d.). The nanotechnology-based polymers based drug delivery system with antibiotics emphasizes on the advantages that make nanoparticles as suitable carrier delivery systems for antibacterial drug delivery in the treatment of infections (Patel et al., 2017). The poly (D, L-lactic-co-glycolic acid)-b-poly (L-histidine)-b-poly (ethylene glycol) (PLGA-PLH-PEG) nanoparticles (NPs) encapsulated with vancomycin as core with pH-responsive, and surface charge switching behaviour was developed to enhance the antibacterial efficacy. These NPs are intended to protect non-target interactions and are intended to bind only to the bacterial surface in an acidic state and also to reduce the dose of the drug and thus to reduce drug-related side effects. The charge switching was achieved by selective protonation of the imidazole groups of PLH at reduced pH (Radovic-Moreno et al., 2012). The study reports of biodegradable poly lactide-co-glycolides (PLGA) NPs encapsulated over rifampicin (RIF) was developed adopting solvent diffusion method. The effect of nano encapsulated RIF on antibacterial activity were performed against gram-negative and positive bacterial strains. The results showed that the RIF NPs have considerably increased the antibacterial activity (Darvishi et al., 2015). Similarly, the formulation of poly (lactide-co-glycolide) (PLGA) nanoparticles loaded with 18-β-glycyrrhetinic acid (GLA) has been reported to show antibacterial activity. The antibacterial studies were performed through serial dilution technique to study the MIC of the nanoparticles; the initial characterization was performed reporting nanoscale size particles ranging from 175 to 212 nm were achieved. The highest encapsulation efficiency (53.2 ± 2.4%) was obtained when the ratio of drug to polymer was 1:4. Zeta potential of the developed nanoparticles was fairly negative (-11±1.5). The antimicrobial findings suggested that the nanoparticles against *P. aeuroginosa*, *S. aureus* and *S. epidermidis* were more successful than pure GLA. This increase the antibacterial behaviour of GLA loaded nanoparticles relative to pure GLA could be correlated with higher penetration of nanoparticles into infected cells and better distribution of GLA at its site of action (Esmaeili et al., 2007). The study aimed to develop clarithromycin (CLR) loaded PLGA nanoparticles, and the physicochemical properties of the NPs were studied. The antibacterial activity against *S.aureus* was determined by serial dilution technique and MIC was studied. The results revealed that NPs that encapsulate intact CLR against *S.aureus* showed 1/8 concentration of the intact drug with increased response. Further, they show potent response against *S.aureus* with increased MICs may be used for the treatment of susceptibility in inhibition of bacteria and could be appropriate for drug delivery systems through intravenous and other preparations in the treatment of infections (Mohammadi et al., 2011). To study reports that the effects of the antibiotic release profiles of levofloxacin-loaded polymeric nanoparticles on their antibacterial efficacy against E. coli biofilm cells. Biofilm cells that survive the initial antibiotic exposure have greater antibiotic resistance than new biofilm cells, where the lower the initial dose, the better the resistance of surviving biofilm cells. The lower antibiotic resistance of the remaining biofilm cells is passed to their planktonic cell progeny, which may eventually form new biofilm colonies with higher biofilm rates. The significance of high initial exposure to antibiotics makes the slow release profile inefficient, given the same amount of dosing (Cheow et al., 2010). Recently, to explore the relationship between polymer structure and antimicrobial activity, three variants of narrow-spectrum peptidomimetic antimicrobial polyurethanes varying in their structures was synthesized based on post-polymerization modification using decanoic acid and oleic acid. The results showed a remarkable transformation in the activity profile of these modified antimicrobial polyurethanes from narrow to broad-spectrum antibacterial activity. They exhibited broad-spectrum antimicrobial activity against Gram-positive bacteria such as *S. aureus* and *S. epidermidis* (Peng et al., 2019). Selected studies of antimicrobial formulations containing organometallic polymers such as silver and Cu(II) are extensively studied and were found to pose potent antimicrobial activity against a wide range of pathogenic bacteria species including the ESKAPE group of organisms. Further, their antimicrobial efficacy was found to vary with the state and type of formulation (solid/liquid). Findings have recorded potent antimicrobial activity in solid form than in the solution which has been attributed to the stability of the metal ion. Similarly, the AgNP/polystyrene polymers were found to be active against *E. coli*, *K. pneumoniae* and *S. aureus*. Numerous studies have been carried out using a different antibiotic - polymer nanosystems in order to improve the in-vivo biocompatibility, biodistribution and targeting. In a study incorporating ciprofloxacin in poly L-lactic acid (PLA) had greatly increased the solubility of the drug and had effectively arrested the growth of E. coli and S. aureus, thus enabling its use in the development of biodegradable non-woven nanofibers for biomedical applications ().

GREEN SYNTHESIZED NANOPARTICLES

Besides the antimicrobial potential of any nanoparticle/carrier, fundamental factors such as toxicity and bioavailability need to be considered before its development into clinical drugs. The use of toxic reagents, chemical-based solvents and stabilizers during the synthesize of metal nanoparticles (MNPs), and metal oxide nanoparticles (MONPs) often limit the therapeutic efficacy of metal oxide nanoparticles and therefore increases toxic side effects. Morteza Sagharjoghi Farahani *et al.* reported the in vitro toxicity impacts of metal nanoparticles such as gold, iron, silica, silver, zinc, and so many others in stemcells (Farahani et al., 2017). In order to develop nanotherapeutics with precise bioactivity, higher bioavailability and less toxicity it is essential to consider the greener synthesis of nanoparticles using phytochemicals (e.g., flavonoids, alkaloids, terpenoids, amides, and aldehydes), bacteria, fungi, yeasts and algae. Many research studies on the green synthesis of nanoparticles have reported that nanoparticles synthesized using plant compounds have higher stability, improved biocompatibility, low agglomeration and non-immunogenic than those prepared using conventional organic or inorganic chemical-based reducing agents such as citrate or sodium borohydride (El Shafey, 2020). In an experimental study, cytotoxic and neurotoxic effects of graphene oxide and chemical vs green synthesized zinc oxide nanoparticles was investigated against Drosophila melanogaster. The flies were exposed to varying concentrations (1- 300 µg/mL) of zinc nanoparticles for about 18 days. The findings of this experiment revealed that ZnO Nps synthesized chemically are more toxic (60% killing) than ZnO NPs (40% killing), with 100 µg / ml after 14 days of exposure (Sood et al., 2019).

Similarly, a previous study has reported that silver nanoparticles synthesized using aqueous extract of Desmodium gangeticum root exhibited relatively low oxidative stress, cortical and multifocal tubular damage than chemically synthesized AgNps. This nephroprotective effect attribute to the antioxidant potential of phytoconstituent, which has acted as a capping agent in the nanoparticle (Vasanth & Kurian, 2017). The Ag-NPs synthesized using aqueous cyanobacterial extracts by Hamouda et al., exhibited potent antibacterial activity against multidrug-resistant *B.cereus* and *E.coli*, further confirmed to be non- toxic for human erythrocytes (Hamouda et al., 2019). Some latest researches of green synthesis of MNPs and MONPs have also investigated the reduction of noble metals including gold, platinum and other metals such as copper. Furthermore, plant extract containing substances like antioxidants and sugars function as a reducing, capping, and dispersing agent during the synthesis of nanoparticles. For example, the caffeine/ polyphenols present in tea/coffee extracts and beet juice are used in controlled synthesis of MNPs such as Ag, Au, Pt, and Pd nanoparticles by the single-pot and microwave-supported method which uses no surfactant, capping agent, or template, and has yielded nanoparticles in the desired size range of 20–60 nm (). Armendariz *et al.* and also Bogireddy *et al.*, report that by controlling/varying the reaction conditions such as pH and temperature in a green synthesis procedure could influence the size, shape and crystallinity of the nanoparticles. These studies have also reported on synthesis AuNPs with different shapes such as decahedral, hexagonal, icosahedral, irregular and rod shape resulted upon varying the pH of the reaction medium (). Talamini et al. state that the size and shape of nanoparticles determine the kinetics of accumulation and excretion of AuNPs in kidneys, and it has been observed that only star-shaped AuNPs can accumulate in the lung. Further, it has been reported that the change in the geometry of the nanoparticle does not influence to cross the passage of the blood-brain barrier (Talamini et al., 2017) Likewise, Ubeyitogullari, A., and Ciftci, O. N had recently reported a novel green low-crystallinity curcumin nanoparticles which have enhanced the bioavailability of curcumin up to 173-fold when compared to the original curcumin (Ubeyitogullari & Ciftci, 2019)

Table 1. Nanoparticle type, size, synthesis and activity against different tested microorganisms.

Nanoparticle	Synthesized using	NP Size (nm)	Tested organism	Zone of inhibition (mm)	Reference
Ag NPs	Culture broth of Endophytic strain SYSU 333150	11 - 40	*S. warneri*	14	122
Ag NPs	Piper beetle leaves extract	10 - 30	*S. aureus* (ATCC 25923)	32.78±0.64	123
			S. typhi (ATCC 14028)	29.55±0.45	
			Escherichia coli (ATCC 25922)	27.12±0.38	
			P. aeruginosa (ATCC 27853)	21.95±0.55	
Ag NPs	Culture broth of Euphorbia hirta L	10 - 60	*B. subtilis* (MTCC 121)	13.00	124
			E. coli (MTCC 7410)	15.00	
			K. pneumoniae (MTCC 7407)	11.00	
			P. aeruginosa (MTCC 7903)	20.00	
			S. aureus (MTCC 7443)	14.00	
Chloromphenicol loaded Ag NPs	Culture broth of P. veronii AS 41G	5 - 50	*B. subtilis* (MTCC 121)	31.00	125
			E. coli (MTCC 7410)	32.00	
			K. pneumoniae (MTCC 7407)	29.00	
Glycolipoprotein-capped AgNPs	Exopolymer of Ochrobactrum rhizosphaerae	10	*V. cholerae* (N16961(	11.00 -15.00	126
CuNPs	Chemical precipitation method	13.13	*E. coli*	5.00	127
			K. pneumoniae	9.00	
			P. typhus	6.00	
NiNPs	Chemical precipitation method	24	*E. coli*	4.00	127
			K. pneumoniae	7.00	
			P. typhus	4.00	
Cu NPs	Extract of green alga Botryococcus braunii	40 –100	*P.aeruginosa* (MTCC 441)	17 ± 1.56	128
			E.coli (MTCC 442)	18 ± 0.1	
			K.pneumoniae (MTCC 109)	19 ± 0.0	
			S.aureus (MTCC 96)	22 ± 0.88	
Ag NPs	Extract of green alga Botryococcus braunii	10 – 70	*P.aeruginosa* (MTCC 441)	20 ± 1.78	128
			E.coli (MTCC 442)	20 ± 0.8	
			K.pneumoniae (MTCC 109)	21 ± 0.60	
			S.aureus (MTCC 96)	22 ± 0.75	

continues on following page

Table 1. Continued

Nanoparticle	Synthesized using	NP Size (nm)	Tested organism	Zone of inhibition (mm)	Reference
CuO NPs	Gum karaya natural biopolymer	4.8 ± 1.6 – 7.8 ± 2.3	*E. coli*	16.2 ± 0.8	129
			S. aureus	14.5 ± 0.6	
ZnO NPs	Sonochemical route	41	*S. aureus*	2.7	130
			K. pneumoniae	2.5	
			E. coli	2.1	
ZnO NPs	Chemical precipitation method	10	*P. gingivalis*	18.09	131
			A .naeslundii	12.05	
ZnO NPs	Extracts of chamomile flower, olive leaves, and red tomato fruit	40.5 - 124.0	*X. oryzae pv. oryzae* (GZ 0003)	22.00	132
Fe2O3-NPs	Chemical coprecipitation	10.4 ± 4.9 - 11.4 ± 5.2	*E. coli*	23.0	133
			B. subtilis	25.0	
Fe2O3-NPs	Skimmia laureola leaf extract	56 - 350	*R. solanacearum*	17.9 ± 0.5	134
Fe2O3-NPs	Commercially available	25 – 30	Methicillin resistant *S. aureus*	14 ± 10	135
			E.coli	12 ± 10	
CA-Fe2O3-NPs	Chemical co-precipitation method	113.3	*E. coli*	23.00	136
			S. aureus	22.00	

CONCLUSION

In conclusion, nanoantibiotics have a crucial role in the global effort to combat antimicrobial resistance. Over the past few years, many significant scientific and technological advancements have been made as a result of a major focus on understanding the dynamics and complexities involved between nanoparticles, bacteria, antimicrobials and biological resistance. Unlike conventional antibiotics, nanoantibiotics are multifunctional, dynamic, and biocompatible and are made of materials with intrinsic antibacterial properties and can be developed using bottom-up imaging methods. The killing of bacterial cells by nanoantibiotics is a multimodal phenomenon which includes, but not limited to impaired membrane function, generation of reactive oxygen species and antioxidant depletion, protein dysfunction, genotoxicity, interference with nutrient assimilation and many. Thus, nanoparticles - antibiotic conjugates are considered the most promising tool for addressing the increased threat and challenges in treating infections caused by antibiotic-resistant pathogens. With the knowledge that we have gained today about nanomaterials and antibiotics like never before, we can design a wide range of novel nano antimicrobials which are capable of use in medical and non-medical applications. However, a large amount of work remains to be done on unresolved challenges, such as long-term stability, half-life, functionality and bioavailability of nano antimicrobials. Further translational research is strongly needed in the future to make the possibility of transferring nanoantibiotics from the bench to the bedside.

CONFLICTS OF INTEREST

The authors declare no conflict of interest.

ACKNOWLEDGMENT

Krishnanand Nagarajan gratefully acknowledges the University Grants Commission, New Delhi, for providing financial support. Dr Selvamani Palanisamy acknowledges the Department of Science and Technology, New Delhi, India and Japan Society for the Promotion of Science, Japan for providing financial support under India Japan Cooperative Science Programme (DST Sanction Order No. DST / INT/ JSPS/ P-269/ 2018 Dt: 30.08.2018). Dr Sathish Kumar Perumal thank Science Engineering and Research Board, New Delhi, India for providing financial support for his research work. Mr Sathish Kumar Marimuthu thanks the Indian Council for Medical Research, New Delhi, India for financial assistance. Authors gratefully acknowledge the Department of Pharmaceutical Technology, University College of Engineering, Bharathidasan Institute of Technology Campus, Anna University, Tiruchirappalli, Tamil Nadu, India.

REFERENCES

A Brief Overview of Classes of Antibiotics – Compound Interest. (n.d.). Available at:https://www.compoundchem.com/2014/09/08/antibiotics/

Abedon, S. T., Thomas-Abedon, C., Thomas, A., & Mazure, H. (2011). Bacteriophage prehistory: Is or is not Hankin, 1896, a phage reference? *Bacteriophage*, *1*(3), 174–178. doi:10.4161/bact.1.3.16591 PMID:22164351

Abreu, A. S., Castanheira, E. M., Queiroz, M. J. R., Ferreira, P. M., Vale-Silva, L. A., & Pinto, E. (2011). Nanoliposomes for encapsulation and delivery of the potential antitumoral methyl 6-methoxy-3-(4-methoxyphenyl)-1H-indole-2-carboxylate. *Nanoscale Research Letters*, *6*(1), 482. doi:10.1186/1556-276X-6-482 PMID:21812989

Agarwala, M., Choudhury, B., & Yadav, R. N. S. (2014). Comparative study of antibiofilm activity of copper oxide and iron oxide nanoparticles against multidrug resistant biofilm forming uropathogens. *Indian Journal of Microbiology*, *54*(3), 365–368. doi:10.100712088-014-0462-z PMID:24891746

Al-Shabib, N. A., Husain, F. M., Ahmed, F., Khan, R. A., Khan, M. S., Ansari, F. A., ... Khan, J. M. (2018). Low temperature synthesis of superparamagnetic iron oxide (Fe3O4) nanoparticles and their ROS mediated inhibition of biofilm formed by food-associated bacteria. *Frontiers in Microbiology*, *9*, 2567. doi:10.3389/fmicb.2018.02567 PMID:30464757

Alam, T., Khan, R. A. A., Ali, A., Sher, H., Ullah, Z., & Ali, M. (2019). Biogenic synthesis of iron oxide nanoparticles via Skimmia laureola and their antibacterial efficacy against bacterial wilt pathogen Ralstonia solanacearum. *Materials Science and Engineering C*, *98*, 101–108. doi:10.1016/j.msec.2018.12.117 PMID:30812984

Álvarez-Paino, M., Muñoz-Bonilla, A., & Fernández-García, M. (2017). Antimicrobial polymers in the nano-world. *Nanomaterials (Basel, Switzerland)*, *7*(2), 48. doi:10.3390/nano7020048 PMID:28336882

Arakha, M., Pal, S., Samantarrai, D., Panigrahi, T. K., Mallick, B. C., Pramanik, K., Mallick, B., & Jha, S. (2015). Antimicrobial activity of iron oxide nanoparticle upon modulation of nanoparticle-bacteria interface. *Scientific Reports*, *5*(1), 14813. doi:10.1038rep14813 PMID:26437582

Arias, L. S., Pessan, J. P., Vieira, A. P. M., Lima, T. M. T. D., Delbem, A. C. B., & Monteiro, D. R. (2018). Iron oxide nanoparticles for biomedical applications: A perspective on synthesis, drugs, antimicrobial activity, and toxicity. *Antibiotics (Basel, Switzerland)*, *7*(2), 46. doi:10.3390/antibiotics7020046 PMID:29890753

Armendariz, V., Herrera, I., Jose-yacaman, M., Troiani, H., & Santiago, P. (2004). Size controlled gold nanoparticle formation by Avena sativa biomass: Use of plants in nanobiotechnology. *Journal of Nanoparticle Research*, *6*(4), 377–382. doi:10.100711051-004-0741-4

Arya, A., Gupta, K., Chundawat, T. S., & Vaya, D. (2018). Biogenic synthesis of copper and silver nanoparticles using green alga Botryococcus braunii and its antimicrobial activity. *Bioinorganic Chemistry and Applications*, *2018*, 2018. doi:10.1155/2018/7879403 PMID:30420873

Assa, F., Jafarizadeh-Malmiri, H., Ajamein, H., Anarjan, N., Vaghari, H., Sayyar, Z., & Berenjian, A. (2016). A biotechnological perspective on the application of iron oxide nanoparticles. *Nano Research*, *9*(8), 2203–2225. doi:10.100712274-016-1131-9

Baker, S., Pasha, A., & Satish, S. (2017). Biogenic nanoparticles bearing antibacterial activity and their synergistic effect with broad spectrum antibiotics: Emerging strategy to combat drug resistant pathogens. *Saudi Pharmaceutical Journal*, *25*(1), 44–51. doi:10.1016/j.jsps.2015.06.011 PMID:28223861

Baptista, P. V., McCusker, M. P., Carvalho, A., Ferreira, D. A., Mohan, N. M., Martins, M., & Fernandes, A. R. (2018). Nano-strategies to fight multidrug resistant bacteria—"A Battle of the Titans. *Frontiers in Microbiology*, *9*, 1441. doi:10.3389/fmicb.2018.01441 PMID:30013539

Bazrgari, F., Khameneh, B., Fazly Bazzaz, B. S., Mahmoudi, A., & Malaekeh-Nikouei, B. (2020). Effect of the nanoliposomal formulations of rifampin and N-acetyl cysteine on staphylococcus epidermidis biofilm. *Nanomedicine Journal*, *7*(2), 131–137.

Bogireddy, N. K. R., Pal, U., Gomez, L. M., & Agarwal, V. (2018). Size controlled green synthesis of gold nanoparticles using Coffea arabica seed extract and their catalytic performance in 4-nitrophenol reduction. *RSC Advances*, *8*(44), 24819–24826. doi:10.1039/C8RA04332A

Bottagisio, M., Lovati, A. B., Galbusera, F., Drago, L., & Banfi, G. (2019). A precautionary approach to guide the use of transition metal-based nanotechnology to prevent orthopedic infections. *Materials (Basel)*, *12*(2), 314. doi:10.3390/ma12020314 PMID:30669523

Brayner, R., Ferrari-Iliou, R., Brivois, N., Djediat, S., Benedetti, M. F., & Fiévet, F. (2006). Toxicological impact studies based on Escherichia coli bacteria in ultrafine ZnO nanoparticles colloidal medium. *Nano Letters*, *6*(4), 866–870. doi:10.1021/nl052326h PMID:16608300

Briuglia, M. L., Rotella, C., McFarlane, A., & Lamprou, D. A. (2015). Influence of cholesterol on liposome stability and on in vitro drug release. *Drug Delivery and Translational Research*, *5*(3), 231–242. doi:10.100713346-015-0220-8 PMID:25787731

Bucki, R., Niemirowicz-Laskowska, K., Deptuła, P., Wilczewska, A. Z., Misiak, P., Durnaś, B., Fiedoruk, K., Piktel, E., Mystkowska, J., & Janmey, P. A. (2019). Susceptibility of microbial cells to the modified PIP 2-binding sequence of gelsolin anchored on the surface of magnetic nanoparticles. *Journal of Nanobiotechnology*, *17*(1), 81. doi:10.118612951-019-0511-1 PMID:31286976

Bull, M. (2008). 基因的改变NIH Public Access. *Bone*, *23*, 1–7.

Chaudhary, J., Tailor, G., Yadav, B. L., & Michael, O. (2019). Synthesis and biological function of Nickel and Copper nanoparticles. *Heliyon*, *5*(6), e01878. doi:10.1016/j.heliyon.2019.e01878 PMID:31198877

Chen, C. H., & Lu, T. K. (2020). Development and challenges of antimicrobial peptides for therapeutic applications. *Antibiotics (Basel, Switzerland)*, *9*(1), 24. doi:10.3390/antibiotics9010024 PMID:31941022

Cheow, W. S., Chang, M. W., & Hadinoto, K. (2010). Antibacterial efficacy of inhalable levofloxacin-loaded polymeric nanoparticles against E. coli biofilm cells: The effect of antibiotic release profile. *Pharmaceutical Research*, *27*(8), 1597–1609. doi:10.100711095-010-0142-6 PMID:20407918

Clokie, M. R., Millard, A. D., Letarov, A. V., & Heaphy, S. (2011). Phages in nature. *Bacteriophage*, *1*(1), 31–45. doi:10.4161/bact.1.1.14942 PMID:21687533

Corrêa, A. C., Pereira, P. R., & Paschoalin, V. M. (2019). Preparation and Characterization of Nanoliposomes for the Entrapment of Bioactive Hydrophilic Globular Proteins. *JoVE*, (150), e59900. doi:10.3791/59900 PMID:31524878

Currie, C. J., Berni, E., Jenkins-Jones, S., Poole, C. D., Ouwens, M., Driessen, S., de Voogd, H., Butler, C. C., & Morgan, C. L. (2014). Antibiotic treatment failure in four common infections in UK primary care 1991-2012: Longitudinal analysis. *BMJ (Clinical Research Ed.)*, *349*(sep23 3), g5493. doi:10.1136/bmj.g5493 PMID:25249162

Dakal, T. C., Kumar, A., Majumdar, R. S., & Yadav, V. (2016). Mechanistic basis of antimicrobial actions of silver nanoparticles. *Frontiers in Microbiology*, *7*, 1831. doi:10.3389/fmicb.2016.01831 PMID:27899918

Danaei, M., Kalantari, M., Raji, M., Fekri, H. S., Saber, R., Asnani, G. P., ... Taheriazam, A. (2018). Probing nanoliposomes using single particle analytical techniques: Effect of excipients, solvents, phase transition and zeta potential. *Heliyon*, *4*(12), e01088. doi:10.1016/j.heliyon.2018.e01088 PMID:30603716

Darvishi, B., Manoochehri, S., Kamalinia, G., Samadi, N., Amini, M., Mostafavi, S. H., ... Dinarvand, R. (2015). Preparation and antibacterial activity evaluation of 18-β-glycyrrhetinic acid loaded PLGA nanoparticles. *IJPR*, *14*(2), 373. PMID:25901144

Dedrick, R. M., Guerrero-Bustamante, C. A., Garlena, R. A., Russell, D. A., Ford, K., Harris, K., ... Hatfull, G. F. (2019). Engineered bacteriophages for treatment of a patient with a disseminated drug-resistant Mycobacterium abscessus. *Nature Medicine*, *25*(5), 730–733. doi:10.103841591-019-0437-z PMID:31068712

Dinali, R., Ebrahiminezhad, A., Manley-Harris, M., Ghasemi, Y., & Berenjian, A. (2017). Iron oxide nanoparticles in modern microbiology and biotechnology. *Critical Reviews in Microbiology*, *43*(4), 493–507. doi:10.1080/1040841X.2016.1267708 PMID:28068855

Domingo-Calap, P., & Delgado-Martínez, J. (2018). Bacteriophages: Protagonists of a post-antibiotic era. *Antibiotics (Basel, Switzerland)*, *7*(3), 66. doi:10.3390/antibiotics7030066 PMID:30060506

Dong, Z. Y., Narsing Rao, M. P., Xiao, M., Wang, H. F., Hozzein, W. N., Chen, W., & Li, W. J. (2017). Antibacterial activity of silver nanoparticles against Staphylococcus warneri synthesized using endophytic bacteria by photo-irradiation. *Frontiers in Microbiology*, *8*, 1090. doi:10.3389/fmicb.2017.01090 PMID:28659903

Edgar, R., Friedman, N., Molshanski-Mor, S., & Qimron, U. (2012). Reversing bacterial resistance to antibiotics by phage-mediated delivery of dominant sensitive genes. *Applied and Environmental Microbiology*, *78*(3), 744–751. doi:10.1128/AEM.05741-11 PMID:22113912

Edson, J. A., & Kwon, Y. J. (2016). Design, challenge, and promise of stimuli-responsive nanoantibiotics. *Nano Convergence*, *3*(1), 26. doi:10.118640580-016-0085-7 PMID:28191436

El Haddad, L., Harb, C. P., Gebara, M. A., Stibich, M. A., & Chemaly, R. F. (2019). A systematic and critical review of bacteriophage therapy against multidrug-resistant ESKAPE organisms in humans. *Clinical Infectious Diseases*, *69*(1), 167–178. doi:10.1093/cid/ciy947 PMID:30395179

El Shafey, A. M. (2020). Green synthesis of metal and metal oxide nanoparticles from plant leaf extracts and their applications: A review. *Green Processing and Synthesis*, *9*(1), 304–339. doi:10.1515/gps-2020-0031

Eleraky, N. E., Allam, A., Hassan, S. B., & Omar, M. M. (2020). Nanomedicine fight against antibacterial resistance: An overview of the recent pharmaceutical innovations. *Pharmaceutics*, *12*(2), 142. doi:10.3390/pharmaceutics12020142 PMID:32046289

Ershov, B. G., Abkhalimov, E. V., Solovov, R. D., & Roldughin, V. I. (2016). Gold nanoparticles in aqueous solutions: Influence of size and pH on hydrogen dissociative adsorption and Au (III) ion reduction. *Physical Chemistry Chemical Physics*, *18*(19), 13459–13466. doi:10.1039/C6CP01996J PMID:27125624

Esmaeili, F., Hosseini-Nasr, M., Rad-Malekshahi, M., Samadi, N., Atyabi, F., & Dinarvand, R. (2007). Preparation and antibacterial activity evaluation of rifampicin-loaded poly lactide-co-glycolide nanoparticles. *Nanomedicine; Nanotechnology, Biology, and Medicine*, *3*(2), 161–167. doi:10.1016/j.nano.2007.03.003 PMID:17468055

Farahani, M. S., Falahati, M., & Askari, S. (2017). Nano-toxicology of metal nanoparticles in stem cells (in vitro). *Journal of Nanomedicine & Nanotechnology*, *8*(6). Advance online publication. doi:10.4172/2157-7439-C1-055

Ferraris, S., & Padoan, G. (2018). *Preparation and characterization of Ti-6Al-4V surfaces functionalized with silver for antibacterial purposes: study of ionic and nanoparticles release mechanism*. Academic Press.

Franci, G., Falanga, A., Galdiero, S., Palomba, L., Rai, M., Morelli, G., & Galdiero, M. (2015). Silver nanoparticles as potential antibacterial agents. *Molecules (Basel, Switzerland)*, *20*(5), 8856–8874. doi:10.3390/molecules20058856 PMID:25993417

Francolini, I., & Piozzi, A. (2019). *Polymeric Systems as Antimicrobial or Antifouling Agents*. Academic Press.

Gahlawat, G., Shikha, S., Chaddha, B. S., Chaudhuri, S. R., Mayilraj, S., & Choudhury, A. R. (2016). Microbial glycolipoprotein-capped silver nanoparticles as emerging antibacterial agents against cholera. *Microbial Cell Factories*, *15*(1), 25. doi:10.118612934-016-0422-x PMID:26829922

Ge, L., Li, Q., Wang, M., Ouyang, J., Li, X., & Xing, M. M. (2014). Nanosilver particles in medical applications: Synthesis, performance, and toxicity. *International Journal of Nanomedicine*, *9*, 2399. PMID:24876773

González-Menéndez, E., Fernández, L., Gutiérrez, D., Pando, D., Martínez, B., Rodríguez, A., & García, P. (2018). Strategies to encapsulate the Staphylococcus aureus bacteriophage philPLA-RODI. *Viruses*, *10*(9), 495. doi:10.3390/v10090495 PMID:30217072

Grumezescu, A. M., Gestal, M. C., Holban, A. M., Grumezescu, V., Vasile, B. Ş., Mogoantă, L., Iordache, F., Bleotu, C., & Mogoşanu, G. D. (2014). Biocompatible Fe3O4 increases the efficacy of amoxicillin delivery against Gram-positive and Gram-negative bacteria. *Molecules (Basel, Switzerland)*, *19*(4), 5013–5027. doi:10.3390/molecules19045013 PMID:24759068

Haider, A., & Kang, I. K. (2015). Preparation of silver nanoparticles and their industrial and biomedical applications: A comprehensive review. *Advances in Materials Science and Engineering*, *2015*, 2015. doi:10.1155/2015/165257

Hajiahmadi, F., Alikhani, M. Y., Shariatifar, H., Arabestani, M. R., & Ahmadvand, D. (2019). The bactericidal effect of lysostaphin coupled with liposomal vancomycin as a dual combating system applied directly on methicillin-resistant Staphylococcus aureus infected skin wounds in mice. *International Journal of Nanomedicine*, *14*, 5943–5955. doi:10.2147/IJN.S214521 PMID:31447553

Hamouda, R. A., Hussein, M. H., Abo-elmagd, R. A., & Bawazir, S. S. (2019). Synthesis and biological characterization of silver nanoparticles derived from the cyanobacterium Oscillatoria limnetica. *Scientific Reports*, *9*(1), 1–17. doi:10.103841598-019-49444-y PMID:31506473

Harada, L. K., Silva, E. C., Campos, W. F., Del Fiol, F. S., Vila, M., Dąbrowska, K., Krylov, V. N., & Balcão, V. M. (2018). Biotechnological applications of bacteriophages: State of the art. *Microbiological Research*, *212*, 38–58. doi:10.1016/j.micres.2018.04.007 PMID:29853167

Hozyen, H. F., Ibrahim, E. S., Khairy, E. A., & El-Dek, S. I. (2019). Enhanced antibacterial activity of capped zinc oxide nanoparticles: A step towards the control of clinical bovine mastitis. *Veterinary World*, *12*(8), 1225–1232. doi:10.14202/vetworld.2019.1225-1232 PMID:31641301

Hsieh, P. F., Lin, H. H., Lin, T. L., Chen, Y. Y., & Wang, J. T. (2017). Two T7-like bacteriophages, K5-2 and K5-4, each encodes two capsule depolymerases: Isolation and functional characterization. *Scientific Reports*, *7*(1), 1–13. doi:10.103841598-017-04644-2 PMID:28676686

Hussein-Al-Ali, S. H., El Zowalaty, M. E., Hussein, M. Z., Ismail, M., & Webster, T. J. (2014). Synthesis, characterization, controlled release, and antibacterial studies of a novel streptomycin chitosan magnetic nanoantibiotic. *International Journal of Nanomedicine*, *9*, 549. PMID:24549109

Infection: Types, causes, and differences. (n.d.). Available at: https://www.medicalnewstoday.com/articles/196271.php#types

Ivashchenko, O., Lewandowski, M., Peplińska, B., Jarek, M., Nowaczyk, G., Wiesner, M., Załęski, K., Babutina, T., Warowicka, A., & Jurga, S. (2015). Synthesis and characterization of magnetite/silver/antibiotic nanocomposites for targeted antimicrobial therapy. *Materials Science and Engineering C*, *55*, 343–359. doi:10.1016/j.msec.2015.05.023 PMID:26117765

Jean-François, F., Elezgaray, J., Berson, P., Vacher, P., & Dufourc, E. J. (2008). Pore formation induced by an antimicrobial peptide: Electrostatic effects. *Biophysical Journal*, *95*(12), 5748–5756. doi:10.1529/biophysj.108.136655 PMID:18820233

Jijie, R., Barras, A., Teodorescu, F., Boukherroub, R., & Szunerits, S. (2017). Advancements on the molecular design of nanoantibiotics: Current level of development and future challenges. *Molecular Systems Design & Engineering*, *2*(4), 349–369. doi:10.1039/C7ME00048K

Johnston, R. L. (2012). Metal nanoparticles and nanoalloys. *Frontiers of Nanoscience*, *3*, 1–42. doi:10.1016/B978-0-08-096357-0.00006-6

Kalita, S., Kandimalla, R., Bhowal, A. C., Kotoky, J., & Kundu, S. (2018). Functionalization of β-lactam antibiotic on lysozyme capped gold nanoclusters retrogress MRSA and its persisters following awakening. *Scientific Reports*, *8*(1), 1–13. doi:10.103841598-018-22736-5 PMID:29636496

Kamaruzzaman, N. F., Tan, L. P., Hamdan, R. H., Choong, S. S., Wong, W. K., Gibson, A. J., Chivu, A., & Pina, M. D. F. (2019). Antimicrobial polymers: The potential replacement of existing antibiotics? *International Journal of Molecular Sciences*, *20*(11), 2747. doi:10.3390/ijms20112747 PMID:31167476

Lasic, D. D. (1998). Novel applications of liposomes. *Trends in Biotechnology*, *16*(7), 307–321. doi:10.1016/S0167-7799(98)01220-7 PMID:9675915

Lei, J., Sun, L., Huang, S., Zhu, C., Li, P., He, J., ... He, Q. (2019). The antimicrobial peptides and their potential clinical applications. *American Journal of Translational Research*, *11*(7), 3919. PMID:31396309

Lima, E., Guerra, R., Lara, V., & Guzmán, A. (2013). Gold nanoparticles as efficient antimicrobial agents for Escherichia coli and Salmonella typhi. *Chemistry Central Journal*, *7*(1), 11. doi:10.1186/1752-153X-7-11 PMID:23331621

Liu, Y., Busscher, H. J., Zhao, B., Li, Y., Zhang, Z., van der Mei, H. C., Ren, Y., & Shi, L. (2016). Surface-adaptive, antimicrobially loaded, micellar nanocarriers with enhanced penetration and killing efficiency in staphylococcal biofilms. *ACS Nano*, *10*(4), 4779–4789. doi:10.1021/acsnano.6b01370 PMID:26998731

Liu, Y., Ren, Y., Li, Y., Su, L., Zhang, Y., Huang, F., ... Shi, L. (2018). Nanocarriers with conjugated antimicrobials to eradicate pathogenic biofilms evaluated in murine in vivo and human ex vivo infection models. *Acta Biomaterialia*, *79*, 331–343. doi:10.1016/j.actbio.2018.08.038 PMID:30172935

Lopez-Abarrategui, C., Figueroa-Espi, V., Lugo-Alvarez, M. B., Pereira, C. D., Garay, H., Barbosa, J. A., ... Franco, O. L. (2016). The intrinsic antimicrobial activity of citric acid-coated manganese ferrite nanoparticles is enhanced after conjugation with the antifungal peptide Cm-p5. *International Journal of Nanomedicine*, *11*, 3849–3857. doi:10.2147/IJN.S107561 PMID:27563243

Lu, T. K., & Collins, J. J. (2009). Engineered bacteriophage targeting gene networks as adjuvants for antibiotic therapy. *Proceedings of the National Academy of Sciences of the United States of America*, *106*(12), 4629–4634. doi:10.1073/pnas.0800442106 PMID:19255432

Luo, L., Wu, Y., Liu, C., Huang, L., Zou, Y., Shen, Y., & Lin, Q. (2019). Designing soluble soybean polysaccharides-based nanoparticles to improve sustained antimicrobial activity of nisin. *Carbohydrate Polymers*, *225*, 115251. doi:10.1016/j.carbpol.2019.115251 PMID:31521298

Mantravadi, P. K., Kalesh, K. A., Dobson, R. C., Hudson, A. O., & Parthasarathy, A. (2019). The quest for novel antimicrobial compounds: Emerging trends in research, development, and technologies. *Antibiotics (Basel, Switzerland)*, *8*(1), 8. doi:10.3390/antibiotics8010008 PMID:30682820

Masri, A., Anwar, A., Ahmed, D., Siddiqui, R. B., Raza Shah, M., & Khan, N. A. (2018). Silver nanoparticle conjugation-enhanced antibacterial efficacy of clinically approved drugs cephradine and vildagliptin. *Antibiotics (Basel, Switzerland)*, *7*(4), 100. doi:10.3390/antibiotics7040100 PMID:30445704

Mohammadi, G., Nokhodchi, A., Barzegar-Jalali, M., Lotfipour, F., Adibkia, K., Ehyaei, N., & Valizadeh, H. (2011). Physicochemical and anti-bacterial performance characterization of clarithromycin nanoparticles as colloidal drug delivery system. *Colloids and Surfaces. B, Biointerfaces*, *88*(1), 39–44. doi:10.1016/j.colsurfb.2011.05.050 PMID:21752610

Mozafari, M. R. (2010). Nanoliposomes: preparation and analysis. In Liposomes (pp. 29-50). Humana Press. doi:10.1007/978-1-60327-360-2_2

Mozafari, M. R., Khosravi-Darani, K., Borazan, G. G., Cui, J., Pardakhty, A., & Yurdugul, S. (2008). Encapsulation of food ingredients using nanoliposome technology. *International Journal of Food Properties*, *11*(4), 833–844. doi:10.1080/10942910701648115

Mudshinge, S. R., Deore, A. B., Patil, S., & Bhalgat, C. M. (2011). Nanoparticles: Emerging carriers for drug delivery. *Saudi Pharmaceutical Journal*, *19*(3), 129–141. doi:10.1016/j.jsps.2011.04.001 PMID:23960751

Muzammil, S., & Hayat, S. (2018). Nanoantibiotics: Future nanotechnologies to combat antibiotic resistance. *Frontiers in Bioscience (Elite Edition)*, *10*(2), 352–374. doi:10.2741/e827 PMID:29293463

Nagarajan, K., Marimuthu, S. K., Palanisamy, S., & Subbiah, L. (2018). Peptide therapeutics versus superbugs: Highlight on current research and advancements. *International Journal of Peptide Research and Therapeutics*, *24*(1), 19–33. doi:10.100710989-017-9650-0

Nano-Liposomes Formulation Production - Creative Biostructure. (n.d.). Available at: https://www.creative-biostructure.com/nano-liposomes-formulation-production-617.htm

Nanoantibiotics: Future nanotechnologies to combat antibiotic resistance. (n.d.). Available at: http://www.bioscience.org/2018/v10e/af/827/2.htm#a249

Ndieyira, J. W., Watari, M., Barrera, A. D., Zhou, D., Vögtli, M., Batchelor, M., ... Rayment, T. (2008). Nanomechanical detection of antibiotic–mucopeptide binding in a model for superbug drug resistance. *Nature Nanotechnology*, *3*(11), 691–696. doi:10.1038/nnano.2008.275 PMID:18989336

Neethu, S., Midhun, S. J., Radhakrishnan, E. K., & Jyothis, M. (2018). Green synthesized silver nanoparticles by marine endophytic fungus Penicillium polonicum and its antibacterial efficacy against biofilm forming, multidrug-resistant Acinetobacter baumanii. *Microbial Pathogenesis*, *116*, 263–272. doi:10.1016/j.micpath.2018.01.033 PMID:29366864

Nehra, P., Chauhan, R. P., Garg, N., & Verma, K. (2018). Antibacterial and antifungal activity of chitosan coated iron oxide nanoparticles. *British Journal of Biomedical Science*, *75*(1), 13–18. doi:10.1080/09674845.2017.1347362 PMID:28945174

Nicolosi, D., Cupri, S., Genovese, C., Tempera, G., Mattina, R., & Pignatello, R. (2015). Nanotechnology approaches for antibacterial drug delivery: Preparation and microbiological evaluation of fusogenic liposomes carrying fusidic acid. *International Journal of Antimicrobial Agents*, *45*(6), 622–626. doi:10.1016/j.ijantimicag.2015.01.016 PMID:25816979

Niemirowicz, K., Piktel, E., Wilczewska, A. Z., Markiewicz, K. H., Durnaś, B., Wątek, M., Puszkarz, I., Wróblewska, M., Niklińska, W., Savage, P., & Bucki, R. (2016). Core–shell magnetic nanoparticles display synergistic antibacterial effects against Pseudomonas aeruginosa and Staphylococcus aureus when combined with cathelicidin LL-37 or selected ceragenins. *International Journal of Nanomedicine*, *11*, 5443–5455. doi:10.2147/IJN.S113706 PMID:27799768

Nikolova, M. P., & Chavali, M. S. (2020). Metal Oxide Nanoparticles as Biomedical Materials. *Biomimetics*, *5*(2), 27. doi:10.3390/biomimetics5020027 PMID:32521669

Ogunyemi, S. O., Abdallah, Y., Zhang, M., Fouad, H., Hong, X., Ibrahim, E., Masum, M. M. I., Hossain, A., Mo, J., & Li, B. (2019). Green synthesis of zinc oxide nanoparticles using different plant extracts and their antibacterial activity against Xanthomonas oryzae pv. oryzae. *Artificial Cells, Nanomedicine, and Biotechnology*, *47*(1), 341–352. doi:10.1080/21691401.2018.1557671 PMID:30691311

Oppenheim, J. J. (2003). Roles of antimicrobial peptides such as defensins in innate and adaptive immunity. *Annals of the Rheumatic Diseases*, *62*(90002), 17–21. doi:10.1136/ard.62.suppl_2.ii17 PMID:14532141

Oppenheim, J. J., Biragyn, A., Kwak, L. W., & Yang, D. (2003). Roles of antimicrobial peptides such as defensins in innate and adaptive immunity. *Annals of the Rheumatic Diseases*, *62*(90002, suppl 2), ii17–ii21. doi:10.1136/ard.62.suppl_2.ii17 PMID:14532141

Padil, V. V. T., & Černík, M. (2013). Green synthesis of copper oxide nanoparticles using gum karaya as a biotemplate and their antibacterial application. *International Journal of Nanomedicine*, *8*, 889. PMID:23467397

Pal, D., Sahu, C. K., & Haldar, A. (2014). Bhasma: The ancient Indian nanomedicine. *Journal of Advanced Pharmaceutical Technology & Research*, *5*(1), 4. doi:10.4103/2231-4040.126980 PMID:24696811

Patel, S., Singh, D., Srivastava, S., Singh, M., Shah, K., Chauahn, D. N., & Chauhan, N. S. (2017). Nanoparticles as a platform for antimicrobial drug delivery. *Adv Pharma Pharmacy*, *5*(3), 31–43. doi:10.13189/app.2017.050301

Patra, J. K., Das, G., Fraceto, L. F., Campos, E. V. R., del Pilar Rodriguez-Torres, M., Acosta-Torres, L. S., ... Habtemariam, S. (2018). Nano based drug delivery systems: Recent developments and future prospects. *Journal of Nanobiotechnology*, *16*(1), 71. doi:10.118612951-018-0392-8 PMID:30231877

Peng, C., Zhang, T., Ortiz-Ortiz, D. N., Vishwakarma, A., Barton, H. A., & Joy, A. (2019). Modification of narrow-spectrum peptidomimetic polyurethanes with fatty acid chains confers broad-spectrum antibacterial activity. *Polymer International*, *68*(7), 1255–1262. doi:10.1002/pi.5773

Philosof-Mazor, L., Dakwar, G. R., Popov, M., Kolusheva, S., Shames, A., Linder, C., Greenberg, S., Heldman, E., Stepensky, D., & Jelinek, R. (2013). Bolaamphiphilic vesicles encapsulating iron oxide nanoparticles: New vehicles for magnetically targeted drug delivery. *International Journal of Pharmaceutics*, *450*(1-2), 241–249. doi:10.1016/j.ijpharm.2013.04.017 PMID:23623794

Piktel, E., Wnorowska, U., Cieśluk, M., Deptula, P., Pogoda, K., Misztalewska-Turkowicz, I., Paprocka, P., Niemirowicz-Laskowska, K., Wilczewska, A. Z., Janmey, P. A., & Bucki, R. (2019). Inhibition of inflammatory response in human keratinocytes by magnetic nanoparticles functionalized with PBP10 peptide derived from the PIP2-binding site of human plasma gelsolin. *Journal of Nanobiotechnology*, *17*(1), 22. doi:10.118612951-019-0455-5 PMID:30711007

Pinnell, S. R., Fairhurst, D., Gillies, R., Mitchnick, M. A., & Kollias, N. (2000). Microfine zinc oxide is a superior sunscreen ingredient to microfine titanium dioxide. *Dermatologic Surgery*, *26*(4), 309–314. doi:10.1046/j.1524-4725.2000.99237.x PMID:10759815

Pires, D. P., Cleto, S., Sillankorva, S., Azeredo, J., & Lu, T. K. (2016). Genetically engineered phages: A review of advances over the last decade. *Microbiology and Molecular Biology Reviews*, *80*(3), 523–543. doi:10.1128/MMBR.00069-15 PMID:27250768

Polymers to Combat Antibiotic-Resistant Bacterial Infections - Advanced Science News. (n.d.). Available at: https://www.advancedsciencenews.com/polymers-to-combat-antibiotic-resistant-bacterial-infections/

Pushpanathan, M., Gunasekaran, P., & Rajendhran, J. (2013). Antimicrobial peptides: Versatile biological properties. *International Journal of Peptides*. PMID:23935642

Radovic-Moreno, A. F., Lu, T. K., Puscasu, V. A., Yoon, C. J., Langer, R., & Farokhzad, O. C. (2012). Surface charge-switching polymeric nanoparticles for bacterial cell wall-targeted delivery of antibiotics. *ACS Nano*, *6*(5), 4279–4287. doi:10.1021/nn3008383 PMID:22471841

Raghupathi, K. R., Koodali, R. T., & Manna, A. C. (2011). Size-dependent bacterial growth inhibition and mechanism of antibacterial activity of zinc oxide nanoparticles. *Langmuir*, *27*(7), 4020–4028. doi:10.1021/la104825u PMID:21401066

Rashida, M. M. O., Islam, M. S., Haque, M. A., Rahman, M. A., Hossain, M. T., & Hamid, M. A. (2016). Antibacterial activity of polyaniline coated silver nanoparticles synthesized from Piper betle leaves extract. *IJPR*, *15*(2), 591. PMID:27642330

Reza Mozafari, M., Johnson, C., Hatziantoniou, S., & Demetzos, C. (2008). Nanoliposomes and their applications in food nanotechnology. *Journal of Liposome Research*, *18*(4), 309–327. doi:10.1080/08982100802465941 PMID:18951288

Saha, B., Bhattacharya, J., Mukherjee, A., Ghosh, A., Santra, C., Dasgupta, A. K., & Karmakar, P. (2007). In vitro structural and functional evaluation of gold nanoparticles conjugated antibiotics. *Nanoscale Research Letters, 2*(12), 614–622. doi:10.100711671-007-9104-2

Sarabandi, K., Rafiee, Z., Khodaei, D., & Jafari, S. M. (2019). Encapsulation of food ingredients by nanoliposomes. In *Lipid-Based Nanostructures for Food Encapsulation Purposes* (pp. 347–404). Academic Press. doi:10.1016/B978-0-12-815673-5.00009-X

Sayyed-Alangi, S. Z., & Nematzadeh, M. (2019). Formulation, development and evaluation of bifunctionalized nanoliposomes containing Trifolium resupinatum sprout methanolic extract: As effective natural antioxidants on the oxidative stability of soybean oil. *BMC Chemistry, 13*(1), 77. doi:10.118613065-019-0594-7 PMID:31384824

Scriboni, A. B., Couto, V. M., de Morais Ribeiro, L. N., Freires, I. A., Groppo, F. C., De Paula, E., ... Cogo-Müller, K. (2019). Fusogenic liposomes increase the antimicrobial activity of vancomycin against staphylococcus aureus biofilm. *Frontiers in Pharmacology, 10*, 10. doi:10.3389/fphar.2019.01401 PMID:31849660

Shaker, M. A., & Shaaban, M. I. (2017). Formulation of carbapenems loaded gold nanoparticles to combat multi-antibiotic bacterial resistance: In vitro antibacterial study. *International Journal of Pharmaceutics, 525*(1), 71–84. doi:10.1016/j.ijpharm.2017.04.019 PMID:28411141

Singh, R., Cheng, S., & Singh, S. (2020). Oxidative stress-mediated genotoxic effect of zinc oxide nanoparticles on Deinococcus radiodurans. *3 Biotech, 10*(2), 66.

Singh, R., Wagh, P., Wadhwani, S., Gaidhani, S., Kumbhar, A., Bellare, J., & Chopade, B. A. (2013). Synthesis, optimization, and characterization of silver nanoparticles from Acinetobacter calcoaceticus and their enhanced antibacterial activity when combined with antibiotics. *International Journal of Nanomedicine, 8*, 4277. PMID:24235826

Sirelkhatim, A., Mahmud, S., Seeni, A., Kaus, N. H. M., Ann, L. C., Bakhori, S. K. M., Hasan, H., & Mohamad, D. (2015). Review on zinc oxide nanoparticles: Antibacterial activity and toxicity mechanism. *Nano-Micro Letters, 7*(3), 219–242. doi:10.100740820-015-0040-x PMID:30464967

Sood, K., Kaur, J., Singh, H., Arya, S. K., & Khatri, M. (2019). Comparative toxicity evaluation of graphene oxide (GO) and zinc oxide (ZnO) nanoparticles on Drosophila melanogaster. *Toxicology Reports, 6*, 768–781. doi:10.1016/j.toxrep.2019.07.009 PMID:31428565

Stoimenov, P. K., Klinger, R. L., Marchin, G. L., & Klabunde, K. J. (2002). Metal oxide nanoparticles as bactericidal agents. *Langmuir, 18*(17), 6679–6686. doi:10.1021/la0202374

Su, Y., Zhao, L., Meng, F., Qiao, Z., Yao, Y., & Luo, J. (2018). Triclosan loaded polyurethane micelles with pH and lipase sensitive properties for antibacterial applications and treatment of biofilms. *Materials Science and Engineering C, 93*, 921–930. doi:10.1016/j.msec.2018.08.063 PMID:30274129

Talamini, L., Violatto, M. B., Cai, Q., Monopoli, M. P., Kantner, K., Krpetić, Z., ... Boselli, L. (2017). Influence of size and shape on the anatomical distribution of endotoxin-free gold nanoparticles. *ACS Nano, 11*(6), 5519–5529. doi:10.1021/acsnano.7b00497 PMID:28558193

Taviad, K. K., Vekariya, S., Bedarkar, P., Galib, R., & Patgiri, B. J. (2018). Process standardization of Swarna Makshika Shodhana (purification) in Triphala Kwatha (decoction). *Ayu*, *39*(3), 187. doi:10.4103/ayu.AYU_26_18 PMID:31000997

Thai, P. V. K., Ha, D. T. M., Hanh, N. T., Day, J., Dunstan, S., Nhu, N. T. Q., ... Thuong, N. T. (2018). Bacterial risk factors for treatment failure and relapse among patients with isoniazid resistant tuberculosis. *BMC Infectious Diseases*, *18*(1), 112. doi:10.118612879-018-3033-9 PMID:29510687

Turner, M., Golovko, V. B., Vaughan, O. P., Abdulkin, P., Berenguer-Murcia, A., Tikhov, M. S., Johnson, B. F. G., & Lambert, R. M. (2008). Selective oxidation with dioxygen by gold nanoparticle catalysts derived from 55-atom clusters. *Nature*, *454*(7207), 981–983. doi:10.1038/nature07194 PMID:18719586

Ubeyitogullari, A., & Ciftci, O. N. (2019). A novel and green nanoparticle formation approach to forming low-crystallinity curcumin nanoparticles to improve curcumin's bioaccessibility. *Scientific Reports*, *9*(1), 1–11. doi:10.103841598-019-55619-4 PMID:31836788

Van Giau, V., An, S. S. A., & Hulme, J. (2019). Recent advances in the treatment of pathogenic infections using antibiotics and nano-drug delivery vehicles. *Drug Design, Development and Therapy*, *13*, 327–343. doi:10.2147/DDDT.S190577 PMID:30705582

Vasanth, S. B., & Kurian, G. A. (2017). Toxicity evaluation of silver nanoparticles synthesized by chemical and green route in different experimental models. *Artificial Cells, Nanomedicine, and Biotechnology*, *45*(8), 1721–1727. doi:10.1080/21691401.2017.1282500 PMID:28278585

Viertel, T. M., Ritter, K., & Horz, H. P. (2014). Viruses versus bacteria—Novel approaches to phage therapy as a tool against multidrug-resistant pathogens. *The Journal of Antimicrobial Chemotherapy*, *69*(9), 2326–2336. doi:10.1093/jac/dku173 PMID:24872344

Wang, D. Y., Van Der Mei, H. C., Ren, Y., Busscher, H. J., & Shi, L. (2019). Lipid-based antimicrobial delivery-systems for the treatment of bacterial infections. *Frontiers in Chemistry*, 7. PMID:31998680

Wang, G., Li, X., & Wang, Z. (2015). The antimicrobial peptide database as a tool for research and education. *Nucleic Acids Research*, *44*(D1), 1087–1093. doi:10.1093/nar/gkv1278 PMID:26602694

Wang, G., Li, X., & Wang, Z. (2016). APD3: The antimicrobial peptide database as a tool for research and education. *Nucleic Acids Research*, *44*(D1), D1087–D1093. doi:10.1093/nar/gkv1278 PMID:26602694

Wang, J., Du, L., Fu, Y., Jiang, P., & Wang, X. (2019). ZnO nanoparticles inhibit the activity of Porphyromonas gingivalis and Actinomyces naeslundii and promote the mineralization of the cementum. *BMC Oral Health*, *19*(1), 1–11. doi:10.118612903-019-0780-y PMID:31088450

Wang, L., Hu, C., & Shao, L. (2017). The antimicrobial activity of nanoparticles: Present situation and prospects for the future. *International Journal of Nanomedicine*, *12*, 1227–1249. doi:10.2147/IJN.S121956 PMID:28243086

Yao, Q., Gao, Y., Gao, T., Zhang, Y., Harnoode, C., Dong, A., Liu, Y., & Xiao, L. (2016). Surface arming magnetic nanoparticles with amine N-halamines as recyclable antibacterial agents: Construction and evaluation. *Colloids and Surfaces. B, Biointerfaces*, *144*, 319–326. doi:10.1016/j.colsurfb.2016.04.024 PMID:27108209

Yashavantha Rao, H. C., Nagendra-Prasad, M. N., Prasad, A., Harini, B. P., Azmath, P., Rakshith, D., & Satish, S. (2016). Biomimetic synthesis of silver nanoparticles using endosymbiotic bacterium inhabiting Euphorbia hirta L. and their bactericidal potential. *Scientifica*. PMID:27403378

Yoo, S. Y., Jin, H. E., Choi, D. S., Kobayashi, M., Farouz, Y., Wang, S., & Lee, S. W. (2016). M13 Bacteriophage and Adeno-Associated Virus Hybrid for Novel Tissue Engineering Material with Gene Delivery Functions. *Advanced Healthcare Materials*, *5*(1), 88–93. doi:10.1002/adhm.201500179 PMID:26010471

Zhang, L., & Gallo, R. L. (2016). a Gallo, RL: Antimicrobial peptides. *Current Biology*, *26*(1), R14–R19. doi:10.1016/j.cub.2015.11.017 PMID:26766224

Zheng, K., Setyawati, M. I., Leong, D. T., & Xie, J. (2017). Antimicrobial gold nanoclusters. *ACS Nano*, *11*(7), 6904–6910. doi:10.1021/acsnano.7b02035 PMID:28595000

Chapter 4
Nanomedicine and Its Applications in Combating Resistance Relating to Opportunistic Pathogens With Special Reference to Tuberculosis

Priya Shrivastava
Dr. Harisingh Gour University, India

Laxmikant Gautam
Dr. Harisingh Gour University, India

Anamika Jain
Dr. Harisingh Gour University, India

Nikhar Vishwakarma
Dr. Harisingh Gour University, India

Sonal Vyas
Bundelkhand Medical College and Hospital, India

Suresh P. Vyas
Dr. Harisingh Gour University, India

ABSTRACT

The worldwide misuse of antimicrobials and subsequent rise of the multi-drug resistant (MDR) opportunistic intracellular pathogens have led to a paradigm shift in the established perspective of antimicrobials and bacterial-human relations. Antibiotic resistant strains of mycobacterium tuberculosis have threatened the progress in management and control of worldwide epidemic of tuberculosis. Mycobacterium tuberculosis intrinsically resists many antimicrobials, limiting the number of compounds available for treatment. Typically, conventional treatments for these diseases consist of long-term therapy with a combination of bioactive(s) that can cause side effects and contribute to low patient compliance. The escalation of antimicrobial resistance poses compelling demand for new ways of treating bacterial infections. The main objective of this chapter is to review the potential applications of nanotechnology-based drug delivery systems in combating drug resistance associated with opportunistic intracellular pathogens with special reference to tuberculosis.

DOI: 10.4018/978-1-7998-5049-6.ch004

INTRODUCTION

Tuberculosis (TB), an ancient human scourge, is an increasing health illness in both the developing and developed world. It is a dreadful health challenge that primarily affects the respiratory system. It is a communicable disease and one among the top 10 leading infectious killers in the world (MacLean et al., 2019). It is considered alongside HIV as the global menace. The pathogenic bacterial species responsible for TB is *Mycobacterium tuberculosis.* It spreads through the infected person who expels bacteria into the air, e.g. by coughing. The frequently affected organ is the lungs wherein it is manifested as pulmonary TB and it might spread to other organs causing extrapulmonary TB (Sharma & Mohan, 2019). A quarter of the world's population including both developing and developed countries are affected by *Mycobacterium tuberculosis* and therefore at greater risk of developing TB. As per the WHO report 2019, approximately 1.2 million deaths from TB among HIV-negative people and an additional 251000 deaths among HIV positive people were reported in 2018. People of both sexes are affected by TB among all age groups. However, the high TB burden is recorded in men (aged≥15 years) that represents 57% of all TB cases in 2018. In contrast to the fact, women only accounted for 32% and children (< 15 years of age) for 11%. 8.6% of all TB cases in HIV-positive people. Most of the TB cases were identified by WHO in the vicinity of South East Asia (44%), Africa (24%) and the western Pacific (18%), eastern Mediterranean (8%), America (3%), and Europe (3%). 2/3 rd of the global TB burden was identified in following countries: India (27%), China (9%), Indonesia (8%), Philippines (6%), Pakistan (6%), Nigeria (4%), Bangladesh (4%) and South Africa (3%). These and along with 22 other countries as per WHO's list of 30 high TB burden countries were found to have for 87% of total reported cases globally (World Health Organization, 2019).

Drug-resistant TB continues to pose a threat to human beings worldwide. The escalation in bioactive resistant strains involves multiple factors: poor public healthcare facilities (e.g., delay in the diagnosis of bioactive resistance, treatment interruptions, the inconsistent supply of anti-tubercular bioactive(s)), incomplete therapy (e.g., unmonitored therapy, non-adherence by patients, inappropriate or incorrect use of anti-tubercular bioactive(s)), and low immunity due to HIV infection. Several first and second-line anti-tubercular bioactive(s) have been developed against *Mycobacterium tuberculosis* (Jnawali & Ryoo, 2013). The mechanism of antimicrobial action of these bioactive(s) is used against causative pathogen i.e., *Mycobacterium tuberculosis.* Globally, a more alarming and recent public health threat has arisen after the identification of multidrug-resistant tuberculosis (MDR-TB) (Nyang'wa et al., 2019). In MDR-TB the mycobacterial strain is resistant to Isoniazid and Rifampicin, the two most powerful anti-tubercular bioactive(s), thus consequent upon a severe threat to conventional anti-tubercular treatment. The DOTs therapy (current conventional therapy) addresses this hurdle through a complex and prolongs treatment involving a combination of high dosage of bioactive(s) which are associated with side effects. Novel drug delivery modules (nanocarriers) are potentially seen as an option for the treatment of drug-resistant tuberculosis that can conquer the limitations of conventional anti-tubercular therapy. Anti-tubercular bioactive(s) entrapped or encapsulated in these nanocarrier systems leads to higher bioavailability. When these nanobiomaterials coupled to a targeting moiety, it may result in significantly boosted the drug efficacy with less or no drug-associated side effects (Mehta et al., 2019; Hamed et al., 2019). This chapter discusses the recent developments in the understanding of the mechanism of antibiotic resistance in tuberculosis with special reference to MDR-TB and also covers the applications and advantages of nanocarrier based anti-tubercular drug delivery systems to circumvent drug-resistant tuberculosis.

PATHOPHYSIOLOGY OF TUBERCULOSIS

Infection in a host is evoked by the inhalation of the droplets in the form of aerosols that contain *Mycobacterium tuberculosis***.** In the lungs (alveolar macrophages), pathogen internalized into the alveolar phagosomes through phagocytosis which is the resultant structure of membrane invagination (Pieters, 2008, pp. 399-407). Lungs macrophages i.e., alveolar macrophages that are activated by the appropriate stimulus will effectively shift the phagocytosed pathogen into the destructive environment of lysosomes through the formation of phagolysosome complex. Here, pathogens are exposed to the reactive oxygen species (ROS) generated via NADPH oxidase and lysosomal enzymes (Galagan, 2014, pp. 307-320). Within the cell, this degraded pathogenic material can further be carried through sequential vesicle trafficking events. Then this material is delivered to the antigen processing and presentation pathway where antigen after digestion as remnants are ligated to the major histocompatibility (MHC) class II molecules and resulting in clearance of potentially harmful pathogenic material (**Figure 1**). The peptide-loaded MHC class II molecules are shuttled to the plasma membrane. Here, they participate in the process of T cell activation to develop adaptive immune responses against the pathogen. But some bacteria like *Mycobacterium tuberculosis* inhibit the phagosome lysosome fusion and thus escape lysosomal digestion and continue to live within the hostile environment of the macrophages (Hunter & Actor, 2019). The pathogen-infected macrophages can either stay in the lung or spread to other organs in the body. Nevertheless, only ~10% of infected people develop tuberculosis. This is how the immune system keeps *Mycobacterium tuberculosis* in check so that the disease might not develop in most of the healthy individuals (Dicks & Stout, 2019).

Figure 1. Diagrammatic representation of pathophysiology of tuberculosis

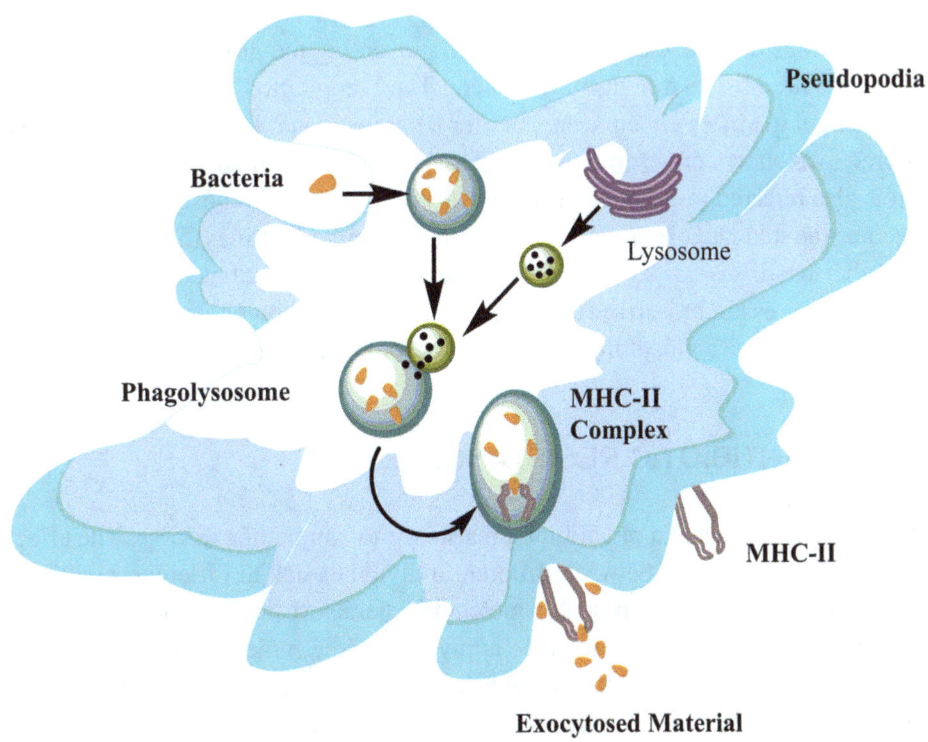

ANTIBIOTIC/DRUG RESISTANCE IN TUBERCULOSIS

Resistance to anti-tubercular bioactive(s) has been a concern since the beginning of chemotherapy. In the early 1990s, after a powerful surge of MDR-TB, resistance has become identified as a challenging problem worldwide. MDR-TB is now threatening the occupant of Africa, Asia, Europe, and the Americas. A thorough understanding of drug resistance at the molecular level is required for designing and development of new bioactive(s). As reported by WHO, drug-resistant TB may be classified into the following types:

- **Mono drug resistance:** Resistance towards one of the anti-tubercular bioactive(s).
- **Poly drug resistance:** Resistance towards more than one first-line anti-tubercular bioactive(s), other than Rifampicin and Isoniazid.
- **Multidrug resistance (MDR):** Resistance to both Isoniazid and Rifampicin.
- **Extensive drug resistance (XDR):** Resistance towards anyone fluoroquinolones (Ciprofloxacin, Ofloxacin, etc.) and at least one of the second-line injectable bioactive(s) (Amikacin, Capreomycin, Kanamycin).
- **Totally drug-resistant:** TB, which is known to be resistant to all first and second-line anti-tubercular bioactive(s) (World Health Organization, 2013; Pai & Memish, 2016).

From the time of use of first-line anti-tubercular bioactive i.e., Streptomycin, the bioactive resistance has been noticed in *Mycobacterium tuberculosis*. In the beginning, with the use of streptomycin, patients with TB could be refurbished to a healthy condition. Also, their sputum was found negative of the bacterial strain. Nevertheless, in 1947, Pyle noticed that despite consecutive therapy, bacilli were found in some patients sputum that were resistant to streptomycin (Pyle, 1947, pp. 465). In 1948, with the introduction of two new bioactive(s) i.e., Para aminosalicylic acid (PAS) and Thioacetazone and in 1952 the third one was Isoniazid, it has become clear that the key to preventing the drug resistance, the combination drug therapy was required for 18 months of treatment. The excogitation of Rifampicin in 1957 brought out the development of the more effective and shorter Rifampicin and Isoniazid based dosage regimens which are referred to as short-course drug therapy. Directly observed treatment short-course i.e., DOTS therapy which is also referred to as global TB control strategy involves those anti-tubercular regimens which have become the standard of care even under resource-limited settings (Seung, Keshavjee, & Rich, 2015). On the other hand, several surveillance measures have been initiated by the WHO to combat drug-resistant TB. Progress in bioactive susceptibility testing has shown that the prevalent behavior of drug resistance is growing as a global threat (Gelmanova et al., 2007).

MECHANISM OF ANTIBIOTIC RESISTANCE

It is needed to know the mechanism of antibiotic resistance to combat the disease. The effectiveness of chemotherapeutic regimens differs between different individuals due to differences in the mechanism of resistance. To overcome resistance based on molecular characteristics of antibiotic resistance, understanding the mechanism of antibiotic resistance is pivotal for the molecular targets identification and accordingly the individualized section of bioactive doses. This may potentially improve the outcome of the treatments (Vilchèze & Jacobs, 2019). Antibiotic/Drug resistance is of following types:

Natural/Intrinsic Antibiotic Resistance

The *Mycobacterium tuberculosis* species is intrinsically resistant to antimicrobials due to the presence of the cell wall. The cell wall contains chief waxy fatty acid components i.e., mycolic acid. The existence of the same makes these bacilli impermeable to conventional bioactive(s) and other antibiotics. This type of resistance towards aminoglycosides, tetracyclines, and fluoroquinolones has been observed in *Mycobacterium tuberculosis.* Such development of resistance not only restricts the applications of available anti-tubercular regimens for the treatment of TB but also impedes the development of new bioactive(s) (Rossi, Aínsa, & Riccardi, 2006).

Acquired Resistance

Acquired antibiotic/drug resistance is primarily provoked by spontaneous alterations in the chromosomal genes in *Mycobacterium tuberculosis.* Antibiotic resistance is not regulated by the mechanism of horizontal gene transfer of mobile genetic elements such as transposons and integrons in *Mycobacterium tuberculosis.* These mechanisms are usually seen in other bacilli. Among prokaryotes, these spontaneous alterations usually occur at a low rate of 0.0033 for each replication. Previously, literatures have reported that the mutation rate relies on the nature of bioactive selection. However, it occurs at a rate of 10^{-9} mutations for each cell division for most of the main anti-tubercular bioactive(s). This is the major reason why anti-tubercular bioactive(s) are prescribed in combination because the risk of mutant comprising of two resistance mutations is 10^{-18} (Almeida, Silva, & Palomino, 2011; Andersson & Hughes, 2010). Several major factors which are responsible for acquired antibiotic/anti-tubercular drug resistance; they are as follows:

Mutations in Antibiotic Targeting Genes

Numerous altered genes are involved in the development of the resistance to different anti-tubercular bioactive(s). An alteration in the KatG, KasA, inhA, ndh and ahpC genes leads to the development of resistance towards Isoniazid. Isoniazid, which is also a prodrug, is activated by the catalase-peroxidase enzyme encoded by the *KatG* gene. The product (activated Isoniazid) prevents the synthesis of mycolic acid which is the chief fatty acid component of the mycobacterial cell wall. The activated Isoniazid inhibits nicotinamide adenine dinucleotide hydrate (NADH)-dependent enoyl-ACP reductase. The NADH-dependent enoyl-ACP reductase is encoded by the *inhA* gene. An alteration in the inhA and KatG genes has been found as the primary cause of resistance to Isoniazid (Rawat, Whitty, & Tonge, 2003). Resistance towards Rifampicin is due to alterations in the *rpoB* gene. This gene encodes for the β-subunit of RNA polymerase. This is followed by conformational changes that cause a low bioactive affinity and subsequently the augmentation of the bioactive resistance (Almeida & Palomino, 2011). Similarly, resistance towards Pyrazinamide involves alteration in the *pncA* gene which encodes for the enzyme pyrazinamidase/nicotinamidase (PZase). Pyrazinamide is activated by this enzyme to its active form of Pyrazinoic acid. This activated form (Pyrazinoic acid) causes bacterial cell damage (Zhang & Mitchison, 2003). Streptomycin is another anti-tubercular bioactive which belongs to the aminoglycoside category. It kills actively growing bacilli on one hand and on other hand it inhibits initiation of the translation process of protein synthesis. At the molecular level, it binds to the 30S subunit of the ribosome at the ribosomal protein 16S rRNA and S1 encoded by the genes *rrs* and *rpsL* respectively. Alterations in the *rrs* and *rpsL* genes result in resistance to streptomycin (Palomino & Martin, 2014). Mutation in the embB gene is a major mechanism of resistance to Ethambutol. The mode of action of Ethambutol involves the inhibition of cell wall synthesis. It inhibits arabinotransferases enzymes involved in the arabinogalactan biosynthesis pathway of the mycobacterial cell wall. The mechanism of resistance to other second-line anti-tubercular bioactive(s) is described in **Table 1.** Fluoroquinolones, one of the second line anti-tubercular bioactive commonly used to treat drug-resistant TB. It inhibits DNA gyrase (topoisomerase II) encoded by *gyrA* and *gyrB* genes. Alterations in the *gyrA* and *gyrB* genes result in the resistance to fluoroquinolones (Sun et al., 2008). Other second line anti-tubercular bioactive(s) like Amikacin, Capreomycin, Kanamycin, and Viomycin interfere with the translation process of protein synthesis. Several alterations have been identified at their target site, which promotes drug resistance. The targets of Amikacin and Kanamycin are *rrs* genes encoding 16S rRNA. These genes are altered in drug-resistant strains. Likewise, an alteration in the *tlyA* gene which codes for rRNA methyltransferase is the major cause of resistance to Viomycin and Capreomycin (Johansen et al., 2006).

Phagosomal Association with TACO

Another factor responsible for the development of resistance is the deposition of tryptophan aspartate containing coat protein (TACO) on the phagosomal membrane (**Figure 2**). This protein is also referred to as P57 and coronin 1. It is recruited to phagosomes enclosing live bacteria. Previously, it is reported in the literature that phagosomes containing live *Mycobacterium tuberculosis* do not acquire Rab5 protein due to the deposition of TACO protein on the phagosomal membrane (Gatfield et al., 2005). Rab5 protein is required for phagosome maturation. The TACO protein is normally released from the phagosomes enclosing bacilli. It prevents phagosome-lysosome fusion by regulating calcium-dependent signaling processes. This action of TACO protein suggests that coronin 1 is a crucial factor in the development of drug resistance. Once macrophages are infected with *Mycobacterium tuberculosis*, they respond with a sustained calcium flux that largely depends on the existence of coronin 1 (Uetrecht & Bear, 2006; Jayachandran et al., 2008). This calcium influx stimulates calcium-dependent phosphatase calcineurin pathway. The coronin 1 dependent activation of calcineurin is required for blocking phagosome-lysosome fusion and facilitates the intracellular survival of *Mycobacterium tuberculosis* (Mueller et al., 2008).

CURRENTLY AVAILABLE TREATMENT REGIMENS AND THEIR LIMITATIONS IN CONQUERING ANTIBIOTIC RESISTANCE

In contempt of the augmentation of several first-line and second-line anti-tubercular bioactive(s), TB still continues to be the leading killer. Carelessness and mismanagement of anti-tubercular bioactive(s) result in the development of MDR-TB. Furthermore, for the treatment of MDR-TB; the second line anti-tubercular bioactive(s) used is more costly. However, they are associated with severe, harmful, and undesirable toxicological consequences. The prescribed dosage regimens in DOTS therapy for the treatment of TB require a multidrug combination of Isoniazid, Rifampicin, and Pyrazinamide. Initially these bioactive(s) are administered, along with Ethambutol for 2 months, followed by Isoniazid and Rifampicin for 4 months. The treatment of MDR-TB involves the use of Pyrazinamide in combination with second-line anti-tubercular bioactive(s) such as Capreomycin, Cycloserine, Para aminosalicylic acid, Fluoroquinolones, Ethionamide and Prothionamide (Onyebujoh et al., 2005). Although these conventional treatment regimens are associated with some limitations like all second-line anti-tubercular bioactive(s) possess high toxicity and are less active than first-line anti-tubercular bioactive(s), duration of therapy is prolonged (9-12 months) that leads to patient noncompliance, undesirable side effects (hepatotoxicity and neurotoxicity), poor permeability and stability, higher doses of anti-tubercular bioactive(s), limited bioavailability, etc. First-line and second-line anti-tubercular bioactive(s) along with their structure and mode of action are discussed in **Table 1**.

Figure 2. Phagosomal association with TACO (tryptophan aspartate containing coat protein): Major cause of multidrug resistant tuberculosis (MDR-TB)

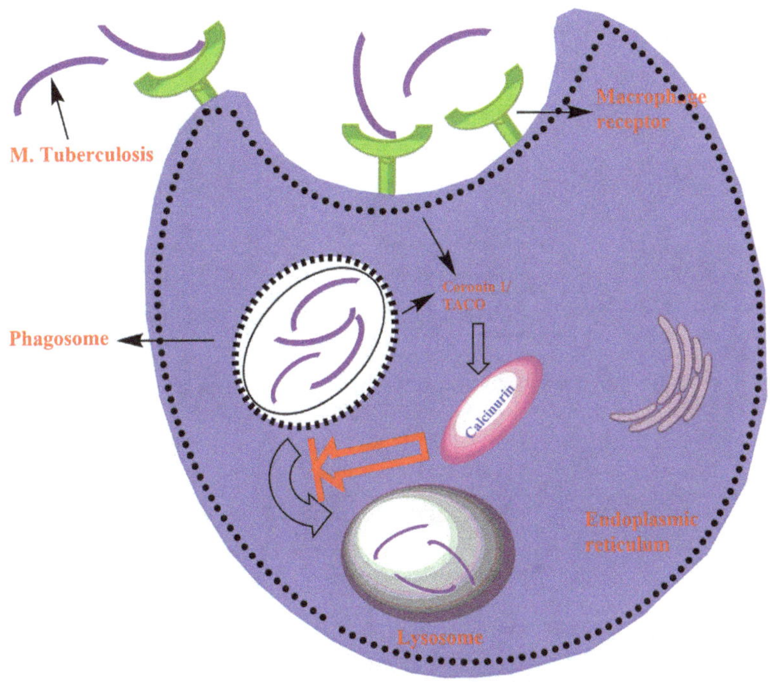

NANOMEDICINES: A RAY OF HOPE TO TREAT MULTI DRUG RESISTANT TUBERCULOSIS (MDR-TB)

The presence of the chief waxy layer of mycolic acid on the mycobacterial cell wall creates a problem in the delivery of the anti-tubercular bioactive(s). These bioactive(s) are also associated with the side effects on long term use. To overcome these limitations, a novel drug delivery module is needed that can increase the intracellular accumulation of the drug and as a consequence could increase anti-tubercular activity. This might result in improved efficacy of the bioactive(s) at lower doses with little side effects. The current need is to design and develop nanotechnology-based novel nanocarrier systems to improve the bioavailability and enhance site-specific targeting of anti-tubercular bioactive(s) in order to avoid the emergence of drug resistance (MDR) and drug-associated side effects (Saravanan et al., 2015). Some recent nanotechnological interventions to combat the MDR in tuberculosis are compiled in **Table 2.** Furthermore, at a given time in free plasma, such nanocarrier systems might result in reduced toxicity due to their low bioactive concentration. This became possible by entrapping anti-tubercular bioactive(s) in the nanocarrier systems, which will acts as a reservoir of bioactive, continuously delivering the required bioactive concentration at the target site (Vyas et al., 2004; Kaur & Singh, 2014). Some nanomaterial-based novel nanocarrier systems with huge potential for anti-tubercular drug delivery are discussed here **(Figure 3).**

Table 1. Currently available conventional treatment regimens for tuberculosis

Bioactive name	Structure of the bioactive	Mode of action	Gene involved in mechanism of resistance	Gene function	References
		First line anti-tubercular bioactive(s)			
Isoniazid	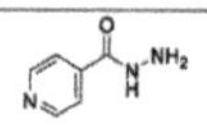	Inhibition of the mycolic acid synthesis (Chief fatty acid component of the mycobacterial cell wall)	Mutations in InhA and KatG gene	Catalase/peroxidase enoyl reductase	Hazbón et al., 2006)
Rifampicin	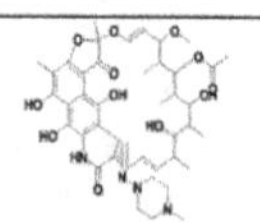	Inhibition of RNA biosynthesis	Mutations in rpoB gene	β subunit of RNA polymerase	Telenti et al., 1993)
Pyrazinamide	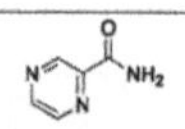	Interferes with the bacterium's ability to synthesize fatty acids required for growth, and replication, and depletion of membrane energy	Mutations in pncA gene	Nicotinamidase/ Pyrazinamidase	Juréen et al., 2008)
Ethambutol	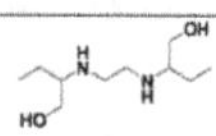	Inhibition of the arabinogalact--an synthesis	Mutations in the embB gene	Arabinosyl transferase	**(Palomino & Martin, 2014)**
Streptomycin	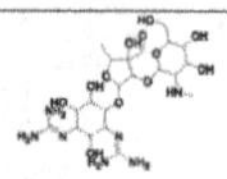	Protein synthesis inhibition	Mutations in rpsL, rrs and gidB genes	S12 ribosomal protein; 16S rRNA; rRNA methyl transferase	Gillespie, 2002; Okamoto et al., 2007)
Bioactive name	**Structure of the bioactive**	**Mode of action**	**Gene involved in mechanism of resistance**	**Gene function**	**References**
		Second line anti-tubercular bioactive(s)			
PAS (Para aminosalicylic acid)	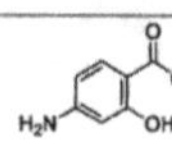	Folate synthesis inhibition	Mutations in thyA gene	Thymidylate synthetase	Rengarajan et al., 2004)
Ethionamide	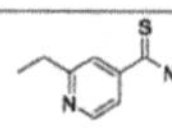	Mycolic acid synthesis inhibition	Mutations in etaA, ethA and inhA genes	Flavin monooxygenase	Brossier et al., 2011)
Kanamycin	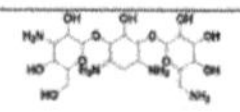	Binds to 30S subunit of the bacterial ribosome and inhibit protein synthesis	Mutations in rrs and eis genes	Aminoglycoside acetyl transferase	Zaunbrecher et al., 2009; Georghiou et al., 2012)
Amikacin	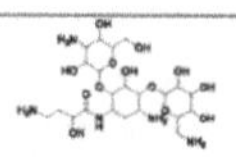	Alter 16S RNA level and inhibit protein synthesis	Mutations in rrs and eis genes	16S rRNA; aminoglycoside acetyl transferase	Zaunbrecher et al. (2009); Georghiou et al., (2012)
Viomycin	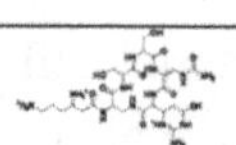	Inhibit protein synthesis by blocking elongation factor G (EF-G) catalyzed translocation of messenger RNA on ribosome	Mutations in thyA gene	2'O-methyltransferase (rRNA methyl transferase)	Georghiou et al., 2012)
Capreomycin	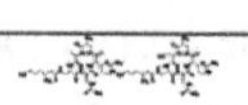	Protein synthesis inhibition	Mutations in thyA gene	2'O-methyltransferase (rRNA methyl transferase)	Georghiou et al., 2012)
Fluoroquinolon--es (Ciprofloxacin)	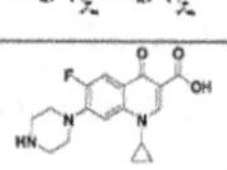	Inhibit DNA gyrase	Mutations in gyrA and gyrB genes	DNA gyrase subunit A; DNA gyrase subunit B	Sun et al., 2008)

Table 2. Recent nanotechnological interventions to combat the multidrug resistance (MDR) in tuberculosis

Type of nanoconstructs	Payload/ Bioactive	Route of administration	Type of ligand	Targeting mechanism	Comments/Benefits	References
Solid lipid nanoparticles (SLNs)	Isoniazid	-	Mannose	Active targeting	Mannose functionalized solid lipid nanoparticles reported to be effective to target alveolar macrophages.	(Costa, Sarmento & Seabra, 2018)
	Rifampicin	–	Methyl α-D mannopyranoside (Mannose residues)	Active targeting	Enhanced uptake of solid lipid nanoparticles by murine macrophage cell lines (J774)	(Maretti et al., 2017)
	Rifampicin	Oral	–	Passive targeting	The degradation of Rifampicin loaded SLNs was reduced in acidic medium when given in combination of Isoniazid SLNs.	(Singh, Bhandari & Kaur, 2013)
	Rifampicin	–	Chitosan	Active targeting	High binding efficiency to mucin. High permeability across the membrane of alveolar epithelial cells (A549).	(Vieira et al., 2018)
Nanostructured lipid carrier (NLCs)	Rifampicin	Intravenous	Mannan	Active targeting	Rifampicin loaded cationic mannosylated NLCs demonstrated tissue selectivity, lung accumulation, superior lung targeting ability and higher cellular uptake by NR8383 cells.	(Song et al., 2015)
Lipid nanoparticles	Bedaquiline	Intravenous	Chitosan	Active targeting	Minimum inhibitory concentrations (MIC) remained unchanged after bioactive encapsulation indicating intactness of activity after encapsulation.	(De Matteis et al., 2018)
Lipid-polymer hybrid nanoparticles (LPN)	Isoniazid and Ciprofloxacin hydrochloride	Pulmonary	-	Passive targeting	Maximum internalization efficiency and lungs deposition was recorded after nasopulmonary administration of LPNs	(Bhardwaj et al., 2016)
Nanoemulsion	Rifampicin	-	Chitosan	Active targeting	Ocular residence time was increased.	(Henostroza et al., 2020)
Liposomes	Levofloxacin	-	-	Passive targeting	Antimicrobial activity of levofloxacin loaded liposomes was more as compared to free drug against TB resistant strain.	(Gaidukevich et al., 2016)
Niosomes	Ethambutol hydrochloride	Subcutaneous	-	Passive targeting	Lower bacterial count and higher lung deposition for the ethambutol loaded niosomes than free drug.	(El-Ridy et al., 2015)

continues on following page

Table 2. Continued

Type of nanoconstructs	Payload/ Bioactive	Route of administration	Type of ligand	Targeting mechanism	Comments/Benefits	References
Lipospheres	Rifampicin	Intranasal	-	Passive targeting	Preferential pulmonary uptake of Rifampicin in case of Rifampicin loaded lipospheres and lower levels of drug in the systemic circulation and liver.	(Takenaga et al., 2008)
Microspheres	Rifampicin	-	-	Passive targeting	Rifampicin loaded microspheres exhibited sustained drug release of drug over a period of 5 days.	(Doan & Olivier, 2009)
Nanomicelles	Rifampicin and Isoniazid	-	-	Passive targeting	Rifampicin loaded Isoniazid conjugated copolymeric micelles demonstrated lower haemolytic toxicity than free drugs. These polymeric micelles were more effective against *Mycobacterium tuberculosis*.	(Upadhyay et al., 2017)
	Rifampicin	-	Hyaluronic acid	Active targeting	Hyaluronic acid-tocopherol succinate micelles increased intracellular concentration of Rifampicin in macrophages (MH-S cells) as compared to free counterparts.	(Gao et al., 2015)
Polymerosomes	Rifampicin	_	_	Passive targeting	Rifampicin loaded polymerosomes enhanced the intracellular concentration of bioactives in murine macrophage cells (Raw 264.7).	(Moretton et al., 2015)

Figure 3. Diagrammatic representations of different nanocarrier systems for anti-tubercular drug delivery

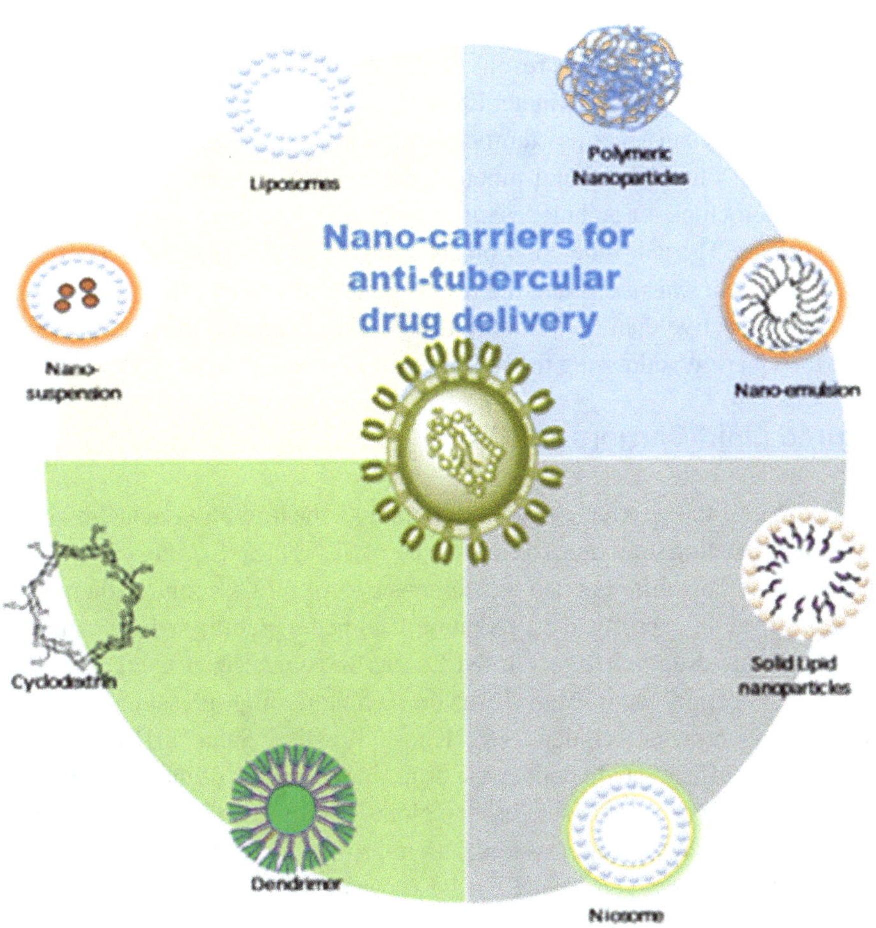

Solid Lipid Nanoparticles (SLNs)

The solid lipid nanoparticles are developed to address the problems associated with the conventional dosage forms. The SLNs offer various advantages over conventional dosage forms such as enhanced physical and chemical stability, higher drug loading, better biocompatibility, and extended drug release, etc. There are few applications of the SLNs for the effective delivery of the drugs for pulmonary drug delivery. Vieira et al., 2018 formulated Rifampicin (RIF) loaded mannosylated solid lipid nanoparticles for effective treatment of TB by enhancing their selective uptake by macrophages. The *in vitro* cytotoxicity study was conducted on human leukemia monocyte THP1 cell line. The study demonstrated a nearly comparable reduction in the parasite in both RIF loaded SLNs and free RIF solution. The uptake studies further revealed that the mannosylated SLNs had a 1.2 fold higher uptake by the THP1 differentiated macrophages as compared to the non-mannosylated ones. They concluded that the SLNs could provide a long term localized delivery with greater bioavailability of the drug when given through the pulmonary route (Vieira et al., 2018). Nemati et al., 2019 also developed Ethambutol (EMB) loaded solid lipid nanoparticles based on dry powder inhalable formulation. The SLNs were prepared by using hot homogenization and ultrasonication techniques and the encapsulation efficiency was found to be more than 98% and the particles were in the size range of 100 nm. The cytotoxicity studies were conducted on the A549 cell line using MTT assay. The results obtained suggested no significant cytotoxicity of EMB loaded SLNs when compared with pure intact EMB. A thorough evaluation of the formulation revealed that the EMB loaded SLNs can act as a potential system for the treatment of tuberculosis (Nemati et al., 2019). Khatak et al., 2018 formulated anti-tubercular drugs (Rifampicin, Isoniazid, and Pyrazinamide) loaded SLNs using microemulsion techniques for enhancing the anti-mycobacterial efficacy and combating multidrug resistance. The anti-tubercular bioactive(s) loaded SLNs were evaluated for antibacterial activity on *Mycobacterium smegmatis* and the results demonstrated that the formulation was comparatively 10 times more effective than the standard drugs. Thus it can be inferred that the system can be effectively used for squaring/addressing the multidrug resistance (Khatak et al., 2018).

Nanostructured Lipid Carriers (NLCs)

The NLCs are lipid-based nanocarriers developed to conquer the limitations and drawbacks of solid lipid nanoparticles. Instead of using solid lipids as in the case of SLNs the NLCs are comprised of liquid lipids blended with the solids. This difference in the composition of NLCs conjures them into a nanocarrier with improved drug loading capacity, drug retention, and better stability which can be utilized further to provide an efficacious drug delivery. The NLCs can be synthesized using emulsification solvent evaporation, solvent injection and solvent diffusion technique, high-pressure homogenization, ultrasonication, membrane contractor technique, etc. (Khosa, Reddi, & Saha, 2018). The NLCs due to their properties could be used effectively to deliver the drugs in the tropics of infection (macrophages). Such attempts have been made by researchers across the globe. Da Silva et al., 2018 formulated a ruthenium (II) complex loaded nanostructured lipid system with improved bioavailability for the treatment of infection caused by *Mycobacterium tuberculosis*. The NLCs were developed using the sonication technique. The average particle size was in the range of 10-200 nm. The complexes demonstrated promising anti-mycobacterial activity against *Mycobacterium tuberculosis* $H_{37}R_v$ strain. The ruthenium (II) complex loaded nanostructured lipid system showed intra-macrophagic accumulation and activity thereof even when evaluated at the minimum inhibitory concentration (> 50 µg/ml). The cytotoxicity assay was performed on J774A.1 macrophage cell lines. The formulations also exhibited reduced cytotoxicity to macrophage cell line whilst the bioavailability was enhanced (Da Silva et al., 2018). Vieira et al., 2017 developed Rifampicin loaded mannosylated NLCs using high shear homogenization and ultrasonication techniques followed by mannosylation. The prepared RIF loaded NLCs were evaluated for various parameters. The average size was 310 nm with an entrapment efficiency of 94-95%. The MTT assay revealed that the mannosylated RIF loaded NLCs are comparatively less toxic than the plain NLCs and free drug and can provide better cellular uptake by bone marrow-derived macrophages. The NLCs prepared thus can be used effectively for targeting the macrophages (Vieira et al., 2017). In another study Pinheiro et al., 2016 developed Rifabutin loaded mannosylated NLCs for the selective delivery of drugs to alveolar macrophages. The NLCs were prepared using high shear homogenization and ultrasonication techniques. The mannosylated NLCs showed an average size of around 198 nm with an entrapment efficiency of 90%. *In vitro* drug release study demonstrated that the release of the Rifabutin was faster at acidic pH than neutral pH. The cell viability studies were conducted on Calu-3, A549, and RAW 264.7

cells using MTT assay. This study suggested that the therapeutic drug concentration could be achieved on the administration of dose considerably lower than the conventional dose (Pinheiro et al., 2016).

Liposomes

Liposomes are vesicular structures composed of lipid bilayers formed on spontaneous hydration of phospholipids. Over the years the liposomes are proved to be an effective carrier for the bioactive(s) and antigens and thus can be used for targeting specific organs or tissue by active or passive methodologies. Liposomes can be formulated by lipid thin-film hydration, sonication, or extrusion techniques and are could be in the size range of 30 nm to several micrometers (Shrivastava et al., 2020). The characteristics of liposomes such as size, surface charge, and presence of ligands can be appropriately utilized for uptake by alveolar macrophages. Liposomes have been previously used as a nanocarrier for the delivery of anti-tubercular drugs (Pinheiro et al., 2011). Nkanga & Krause, 2019 formulated liposomes loaded with the Isoniazid-Hydrazone-Phthalocyanine-In-cyclodextrin complex. The author synthesized a complex system by complexing the Isoniazid-Hydrazone-Phthalocyanine conjugate with gamma-cyclodextrin and loaded into liposomes. The liposomal encapsulation was performed by the organic solvent-free method (heating method). The average size of the liposomes was in the range of 150-650 nm with an entrapment efficiency up to approximately 71%. The *in vitro* release studies indicate pH-dependent release behavior of the drug from formulation(s). Moreover these pH-dependent release characteristics of the liposome can be utilized for site-specific delivery (Nkanga & Krause, 2019). Ferraz-Carvalho et al., 2016 evaluated the effect of the encapsulation of Usnic acid (UA) into a liposome. They also studied the effect of interaction on combining Rifampicin (RIF) and Isoniazid (INH) (anti-tubercular drugs) with Usnic acid (UA) encapsulated in a liposome against the MDR-TB. The liposomes containing UA were prepared by lipid thin film hydration technique. The mean particle size of the liposomes was 146.46 nm. The anti-mycobacterial activity was assessed on *Mycobacterium tuberculosis* $H_{37}R_{v}$ ATCC 27294 strain and six MDR-TB clinical isolates (MDR-TB 1412, 1619, 0729, 1411, 1409, and 1484). The minimum inhibitory concentrations were found to be 31.25μg/mL and 0.98μg/mL respectively for Usnic acid and Usnic acid loaded liposomes. The *in vitro* interaction study of UA with anti-tubercular bioactive(s) was performed using the checkerboard method. The findings revealed a synergistic combination of Rifampicin and Usnic acid. The fractional inhibitory concentration index determined was to be 0.31 and 0.28 for free counterparts and UA loaded liposomes respectively. The improved activity of UA loaded liposomes could be attributed to the synergistic action of UA when given in combination with RIfampicin (Ferraz-Carvalho et al., 2016). Bhardwaj et al., 2013 developed ligand appended liposomes for the treatment of pulmonary tuberculosis using multiple drug therapy. The liposomes containing Rifampicin (RIF), Isoniazid (INH) and Pyrazinamide (PYZ) were prepared using the lipid thin-film hydration technique and mannan was used to coat the liposomes. The prepared liposomes were in a nanometric range with considerably good entrapment efficiency. The *in vivo* studies were performed on Wistar albino rats which demonstrated a higher lung uptake of mannan coated liposomes as compared to uncoated liposomes. Thus, the liposomal preparation seemingly offers exciting possibilities to deliver the bioactive(s) to the lungs for the treatment of pulmonary tuberculosis (Bhardwaj et al., 2013).

Niosomes

Niosomes are vesicular systems similar to liposomes in morphology. They are formed as a result of self-assembly of non-ionic surfactants in bilayers. Due to their inherent advantages over liposomes such as improved chemical stability, low production cost, etc., the niosomes are more preferred as a drug delivery system for the drugs due to their added ability to provide improved therapeutic efficacy with minimum side effects. Niosomes were explored extensively for the effective delivery of the bioactive(s) in the treatment of tuberculosis (Uchegbu & Vyas, 1998). Researchers have attempted successful delivery of the anti-tubercular drugs using vesicular system(s). Sadhu, Saisivam, & Debnath, 2019 designed and characterized Ethionamide loaded niosomes by thin-film hydration technique. The mean particle size of the Ethionamide loaded niosomes was reportedly 124.4 nm and entrapment efficiency up to 88.9%. The niosomes were found to be stable for 30 days and showed a prolonged/slow release of the drug. The prolonged-release from the niosome suggested that the formulation might lower the dosing frequency and could be used for the effective treatment of tuberculosis (Sadhu, Saisivam, & Debnath, 2019). Similarly, Kulkarni, Rawtani & Barot 2019 developed ethionamide and D-cycloserine loaded long-acting dual drug-loaded niosomes for simultaneous delivery of drug(s) in combination. The niosomes were optimized by Box Behnken experimental design and prepared using the ethanol injection method. The particle size of the niosomes was found to be 137.4 nm with % entrapment efficiencies (> 70%). The bacterial inhibition assay was conducted on *Mycobacterium smegmatis*. The minimum inhibitory concentrations (MIC) for ethionamide and D-cycloserine loaded niosomes and free drug(s) were reported to be 28.75μg/ml and 250μg/ml respectively. The co-loaded niosomal preparation showed a synergistic effect as compared to free drug counterparts and can successfully be used for anti-tubercular drug delivery (Kulkarni, Rawtani & Barot 2019). Singh et al., 2011 formulated Isoniazid loaded niosomes employing reverse-phase evaporation method. The zeta potential and polydispersity index (PDI) of the niosomal formulation were found to be 23 mV and 0.14 respectively. The release of the drug from the niosomes was slow at a steady rate protracted over 20-30 hours. The uptake studies carried out on J744A.1 (mouse macrophage) cells using a fluorescent marker demonstrated that the niosomal formulation was uptaken by the macrophages. They concluded that the Isoniazid loaded niosomes can be used for targeting macrophage infected with *Mycobacterium tuberculosis* (Singh et al., 2011). Mehta & Jindal, 2013 formulated anti-tubercular drug(s) loaded niosomes using tyloxapol a biocompatible surfactant. They incorporated Rifampicin (RIF), Isoniazid (INH) and Pyrazinamide (PZA) into the niosomes. The average size of niosomes was 150 nm while the loading capacity recorded for Rifampicin (RIF), Isoniazid (INH), and Pyrazinamide (PZA) was 97.95%, 98.89%, and 99.50% respectively. The stability analysis was carried out using absorption spectroscopy. The stability studies suggest that the niosomes remained fairly stable for a long period of time. The dissolution studies were performed using the dialysis bag method wherein the order of the drug(s) release rate was Rifampicin<Isoniazid <Pyrazinamide. The different kinetic models were applied to study the release rate of the bioactive(s). The release of isoniazid follows the Fickian diffusion model whilst the Rifampicin and Pyrazinamide release following a Non Fickian release model. The studies further suggested that the niosomes based on Tyloxapol can be used effectively for the long term delivery of anti-tubercular drugs in the treatment of tuberculosis (Mehta & Jindal, 2013).

Polymeric Nanoparticles

Polymeric nanoparticles are the biodegradable or biocompatible polymer-based nanoparticulate system with a size ranging from 10 to a few 100 nanometres. The drug can be easily incorporated, adsorbed, or chemically linked on to the surface of these particles. The polymeric nanoparticles offer several advantages including biodegradability, biocompatibility, stability, higher encapsulation efficiency, and prolonged drug release. Polymeric nanoparticles are capable of crossing the biological barriers and thus may target the cellular reservoirs containing *Mycobacterium tuberculosis.* These characteristics make them an ideal carrier for anti-tubercular drugs as well. The polymeric nanoparticles can be synthesized by utilizing techniques such as solvent evaporation, ionic gelation, interfacial polymerization, microemulsion, nano-precipitation, radical polymerization, etc. (Yadav et al., 2019). Earlier various researchers have attempted the incorporation and delivery of anti-tubercular drugs for the treatment of TB which are discussed in this chapter. Recently, Kalombo et al., 2019 developed PLGA based polymeric nanoparticles encapsulating first-line anti-tubercular drugs (Rifampicin, Pyrazinamide, and Isoniazid). The nanoparticles were prepared using a modified solvent evaporation spray drying technique. The average particle size of the nanoparticles was measured to be below 400 nm with a PDI between 0.2-0.4 and with drug entrapment efficiency for Rifampicin, Pyrazinamide, and Isoniazid was 62.4%, 75.2%, and 82.2% respectively. The bactericidal efficacies of the prepared nanoparticles were assessed on C57Bl/6 female mice against *Mycobacterium tuberculosis* $H_{37}R_{v}$ strain after they were given an aerosol challenge test. The data obtained from the study demonstrated that the polymeric nanoparticles were more effective in inhibiting the *Mycobacterium tuberculosis* bacterial replication as compared to that of the free drug forms administered orally. The nanoparticles were found to release the drug for weeks upon single administration. Thus, they concluded that the polymeric nanoparticles could provide cost-effectiveness, prolonged drug release with reduced dosing frequency for the treatment of tuberculosis (Kalombo et al., 2019). Costa-Gouveia et al., 2017 synthesized Ethionamide and booster molecule (BDM41906) co-loaded nanoparticles for combination therapy for the treatment of tuberculosis. Ethionamide loaded polylactic acid and poly (lactic-co-glycolic acid) nanoparticles were prepared using the nanoprecipitation method. The nanoparticles were in a size range from 254 to 277 nm with 76% drug entrapment efficiency for Ethionamide and 51% for BDM41906. The polylactic acid nanoparticles and Ethionamide booster containing pCD (cyclodextrin) nanoparticles were tested for their efficacy against *Mycobacterium tuberculosis.* The results indicated that the formulation containing Ethionamide as well as booster due to synergistic effect had two-fold higher inhibitory effects as compared to Ethionamide alone against *Mycobacterium tuberculosis* $H_{37}R_{v}$ strain. The *in vivo* studies were then performed on Balb/C female mice and on the basis of results it was concluded that this strategy could clinically be accepted as a better option for the treatment of pulmonary tuberculosis (Costa-Gouveia et al., 2017). Varma et al., 2015 prepared Pyrazinamide loaded polymeric nanoparticles (PNPs) for targeting alveolar macrophages to treat pulmonary tuberculosis. The PNPs were prepared using a double emulsion solvent evaporation/diffusion technique. The particle size of the optimized formulation was found to be below 136 nm with drug entrapment efficiency up to 73.3%. The nanoparticles were further evaluated for *in vitro* release of the drug in phosphate buffer (pH 6.8), the results showed the controlled drug release for more than 24 hours. The *in vivo* uptake studies were performed in Charles Foster strain rats while the *ex vivo* studies were performed on alveolar macrophages which were harvested on culture plates using fluorescein-labeled nanoparticles. The result demonstrated that the nanoparticles were taken up by the macrophages efficiently. The cellular uptake was attributed to the lipophilic nature of the polymer (Varma et al., 2015).

Nanoemulsion

Nanoemulsion contains two immiscible liquids as biphasic dispersion. It may be oil in water (o/w) or water in oil (w/o) droplets stabilized by using amphiphilic surfactant(s). The average mean droplet size of nanoemulsion is approx smaller than 500 nm (Rai et al., 2018). Clear or slightly hazy appearance is associated with small droplets size while milky white color is attained with coarse emulsion. Nanoemulsions are ultrafine dispersions that possess viscoelastic as well as visual properties with differential drug loading etc. It is composed of oil/lipid, surfactant, co-surfactant, preservatives, antioxidants, and chemoprotectants. Nanoemulsions having same droplets size as of microemulsions, while it differs in terms of long term thermodynamic stability and structural aspects (Singh et al., 2017). Ahmed et al., 2017 prepared a nanoemulsion based vaccine for tuberculosis. Potent mucosal IL-17 T-cell responses are induced when the vaccine is administered intranasally along with the *Mycobacterium tuberculosis* specific immunodominant antigens (NE-TB vaccine). Apart from this, vaccine exhibited/offered protection against *Mycobacterium tuberculosis* infection. When the NE-TB vaccine is administered along with the Bacillus Calmette-Guerin (BCG) vaccine, the severity of the disease is subsided significantly. Thus NE-TB vaccine is an effective, novel, and safe mucosal vaccine (Ahmed et al., 2017). Shah, Chan & Wong, 2017 reported Rifampicin-Oleic acid 1st generation nanoemulsion and chitosan-folate conjugate 2nd and 3rd generation nanoemulsions for pulmonary administration by nebulization. They synthesized the conjugate and formulated as a nanoemulsion by spontaneous emulsification method. The average droplet size of the nanoemulsion was 40-60 nm. All formulations exhibited more than 95% aerosol output and greater than 75% inhalational efficiency. The aerosolized and inhaled fine particles, aerosol output were governed by surface tension and the size of formulations following an inverse relationship. Thus, 3rd generation nanoemulsion shows higher cell internalization potential, greater lung drug content, and decreased plasma drug concentration (Shah, Chan & Wong, 2017).

Cyclodextrins

Cyclodextrins (CD) are the cyclic oligosaccharides which consisted of D-glucopyranose units linked via α-1,4 linkage. The common natural cyclodextrins are αCD, βCD, and γCD consisted of 6, 7, and 8 D-glucopyranose units respectively. Larger CDs are arduous to produce and not able to form an inclusion complex. Cyclodextrins have doughnet shape with a hydrophilic outer surface and a central cavity that is lipophilic in nature. In aqueous solutions, cyclodextrins form hydrophilic inclusion complex of the poorly soluble drugs by incorporating lipophilic moiety into their central cavity. In aqueous solution, drug and cyclodextrin are in the equilibrium of formation and dissociation. Thus, drug molecule is rapidly released in the body cavity such as the gastrointestinal tract. Cyclodextrins show similar biological and physiochemical properties as those of linear counterparts. They differ in terms of susceptibility towards enzymatic degradation and better solubilizers and complexing agents (Jansook, Ogawa & Loftsson, 2018). Machelart et al., 2019 have prepared nanoparticles of cross-linked poly-βCD for the pulmonary delivery of anti-tubercular drugs. Poly-βCD interferes with the lipid rafts of *Mycobacterium tuberculosis*, without showing any toxicity and thus hamper colonization of the macrophages. Poly-βCD aggravates macrophage apoptosis, which leads to the reduction of infected cells and making the lungs microenvironment unsuitable to *Mycobacterium tuberculosis*. Thus, these findings suggest that poly-βCD based nanoparticles loaded with or without antibody exhibits antibacterial action and could be an effective drug carrier for the treatment of tuberculosis (Machelart et al., 2019). Basha, Yunus, & Doble, 2019 have delivered a combination of drugs namely Rifampicin and Levofloxacin against *Mycobacterium tuberculosis.* Both the drugs exhibiting different mechanisms of action were complexed with cyclodextrin (CD) and then conjugated with curdlan nanoparticles. The formulation provides sustained release of both the bioactive for a prolonged period of time. The formulation is non-toxic to both L929 and RAW 264.7 cells. They are taken up through the dectin-1 receptor by the macrophages. Within the macrophages, more than 95% of *Mycobacterium smegmatis* is killed in 4 hours. These results suggested that curdlan-CD nanoparticles could be an attractive strategy for the encapsulation and intracellular release of water-insoluble bioactive(s) into macrophages (Basha, Yunus, & Doble, 2019).

Dendrimers

These nanocarrier systems are highly branched polymeric macromolecules with uniform and well-defined sizes and shapes. Dendrimers have three main components: a central core, branching units, and the terminal groups, which offer some unique modifiable surface functionalities. The generation number of dendrimers is defined by the repeated branching units and is responsible for the creation of the globular structure. Dendrimers are attractive drug and gene delivery carrier system(s) due to their molecular architecture. Dendrimers can encapsulate drugs and oligonucleotides into internal cavities or in the bounded form via electrostatic or hydrophobic interactions on their surface. Drugs may be attached via a covalent bond at their terminal end (Palmerston Mendes, Pan, & Torchilin, 2017). Haddaoui et al., 2019 have reported selective detection of mutated and wild rpoB genes in *Mycobacterium tuberculosis* with the use of an electrochemical DNA sensor, which is based on the nanocomposite of Fe3O4/polypyrrole containing redox naphthoquinone moiety on the poly amido amine or PAMAM. When given probe and complementary DNA target hybridized, the naphthoquinone redox signal is decreased and it is measured by square wave voltammetry. Apart from this, a single nucleotide mismatch DNA target does not show any cross-hybridization. Conducting property of the polypyrrole platform combined with the catalytic properties of iron oxide nanoparticles. The author(s) reported that detection of 1 fM of DNA target in the 50 μL drop which is equivalent to 3*104 copies of DNA. These sensors also detect the rpoB gene in the polymerase chain reaction amplified sample of genomic DNA and hence can distinguish single nucleotide mutated rpoB gene and wild type rpoB gene. Resistance towards Rifampicin is due to alterations in the *rpoB* gene. These findings suggest that the detection of mutated and wild DNA is possible with sensors without using PCR amplification (Haddaoui et al., 2019). Another study reported by Khoder & Korri-Youssoufi, 2020 for the amperometric recognition of DNA by the nanostructured polypyrrole. Nanowires of the polypyrrole are made up of electrochemical polymerization and oxidation of ethylenediamine or dendrimers PANAM, which provides an aminated surface. The ferrocenyl group, as redox reporter and DNA probe, were linked covalently to the surface of the nanowire of polypyrrole. Electrochemical properties of the nanostructured platform were characterized by cyclic voltammetry and electrochemical impedance spectroscopy (EIS). The chemical structures were characterized by Fourier transform infrared spectroscopy, X-ray photoelectron spectroscopy, and Scanning electron microscopy. The studies reported that greater surface area and hydrophilic character largely affect the electronic properties. Hence, electrochemical performance is enhanced as compared to other nanomaterials with the value of electron transfer of 18s-1. DNA can be detected with a detection limit 0.36 without PCR amplification. These findings suggest that the nanowire of polypyrrole modified with the redox marker is an attractive strategy for the diagnostic purpose (Khoder & Korri-Youssoufi, 2020).

Nanosuspension

Recently nanosuspensions demonstrated that they could be a promising approach for the new chemical entities, which are hydrophobic in nature and problematic to formulate using conventional techniques. Nanosuspensions are suitable for the delivery of water-insoluble drugs because it increases the dissolution velocity and saturation solubility of the bioactive. It also improves the biological performance of the bioactive(s). The formulation is stabilized by the surfactants wherein 100% pure drug is present without any vehicles or carriers. Thus nanosuspensions represent a different pharmaceutical approach as compared to other drug delivery systems like the polymeric matrix. High drug loading can be achieved via this approach. Thus greater therapeutic concentration can be achieved at the target site and maximizing the pharmacological effects (Wang et al., 2017). Mehanna, Mohyeldin, & Elgindy, 2019 have prepared a carbohydrate-based spray-dried nanocomposite of Rifampicin for the treatment of tuberculosis. This strategy provides a high distribution of drugs within the lungs, increased/enhanced bioavailability, showing localizing effects and thus decreased systemic exposure. Rifampicin nanosuspension was developed by spray-drying by using a combination of leucine, mannitol, and maltodextrin as matrix former of microparticles. *In vitro*, inhalational properties, and physiochemical properties of the nanocomposite were studied. Furthermore, the formulation was subjected to differential scanning calorimetry and Fourier transform infrared spectroscopy for compatibility study. *In vitro* cytotoxicity studies on alveolar basal epithelial cells was carried out and evaluated. The formulation consisted of particles with a size of 3.47-6.80 µm and a loading efficiency of 89.3% to 99.2% w/w. Inhalational efficiency of nanosuspension was enhanced after converting it into an inhalable carbohydrate composite. The IC_{50} value of the formulation was considerably high as compared to free drug. These results suggested that nanosuspension of Rifampicin by using carbohydrates as a matrix could significantly increase the inhalational performance with a decline in cellular toxicity. Thus anti-tubercular drugs can be delivered safely by pulmonary route (Mehanna, Mohyeldin, & Elgindy, 2019). Shiezadeh et al., 2019 have prepared Bovine Serum Albumin (BSA) loaded nanoparticles of streptomycin sulfate (STP) by high-pressure homogenization (HPH) method. Pulmonary dry powders of STP containing BSA microparticles, free STP and microparticles embedding STP loaded nanoparticles were formulated and characterized. The nanoparticles were of 200 nm in size. Spray-dried powder showed a size range of 3.3-3.8 µm and exhibited an excellent aerodynamic behavior. No interaction occurred between components which were confirmed by the Fourier transform infrared spectroscopy studies and differential scanning calorimetry. Dry powder was non-toxic, which was confirmed by the MTT assay and Trans-Epithelial Electrical Resistance (TEER). *In vitro* release occurred from all formulation within 4 hours. Biodistribution and pharmacokinetic studies in rats revealed low non-target organ bioavailability of dry powders as compared to intramuscular injection. Thus these findings suggest that Dry powder inhalers containing STP would be an attractive delivery option for the treatment of tuberculosis (Shiezadeh et al., 2019).

CONCLUSION AND FUTURE PROSPECTS

Nanotechnology-based drug delivery systems (nanocarriers) have proved to be potential modules owing to their utility for the advancement of existing conventional treatment of various diseases. Moreover, novel and innovative strategies as well as bioactive with rapid healing mechanisms must be searched and developed to stop or slow down the emergence of microbial resistance. Nanocarriers could reduce the duration of therapy, frequency of dosage regimens involving different intracellular targeting mechanisms. Furthermore nonocarriers possess some additional benefits such as high % drug entrapment, enhanced bioavailability, sustained, and prolonged release of bioactive(s), site-specific, and targeted delivery of the anti-tubercular bioactive(s). Using nanoparticles as carrier systems could render the possible treatment of MDR-TB with immense therapeutic efficacy and lesser side effects. Till now, there is no competitor for the first-line dosage regimens of MDR-TB. An effort is therefore required to design and develop an effective method to counter MDR-TB, improve patient compliance, and ensure that infection is effectively controlled globally. The scientists throughout the world have confirmed the advantageous nature of nanoparticles based anti-tubercular drug delivery, which will continue to expand to become an effective stratagem to be used and practiced clinically.

CONFLICT OF INTEREST

The authors declare no competing financial/personal interest whatsoever.

REFERENCES

Ahmed, M., Smith, D. M., Hamouda, T., Rangel-Moreno, J., Fattom, A., & Khader, S. A. (2017). A novel nanoemulsion vaccine induces mucosal Interleukin-17 responses and confers protection upon Mycobacterium tuberculosis challenge in mice. *Vaccine*, *35*(37), 4983–4989. doi:10.1016/j.vaccine.2017.07.073 PMID:28774560

Almeida Da Silva, P. E., & Palomino, J. C. (2011). Molecular basis and mechanisms of drug resistance in Mycobacterium tuberculosis: Classical and new drugs. *The Journal of Antimicrobial Chemotherapy*, *66*(7), 1417–1430. doi:10.1093/jac/dkr173 PMID:21558086

Andersson, D. I., & Hughes, D. (2010). Antibiotic resistance and its cost: Is it possible to reverse resistance? *Nature Reviews. Microbiology*, *8*(4), 260–271. doi:10.1038/nrmicro2319 PMID:20208551

Basha, R. Y., TS, S. K., & Doble, M. (. (2019). Dual delivery of tuberculosis drugs via cyclodextrin conjugated curdlan nanoparticles to infected macrophages. *Carbohydrate Polymers*, *218*, 53–62. doi:10.1016/j.carbpol.2019.04.056 PMID:31221343

Bhardwaj, A., Kumar, L., Narang, R. K., & Murthy, R. S. R. (2013). Development and characterization of ligand-appended liposomes for multiple drug therapy for pulmonary tuberculosis. *Artificial Cells, Nanomedicine, and Biotechnology*, *41*(1), 52–59. doi:10.3109/10731199.2012.702316 PMID:22889361

Bhardwaj, A., Mehta, S., Yadav, S., Singh, S. K., Grobler, A., Goyal, A. K., & Mehta, A. (2016). Pulmonary delivery of antitubercular drugs using spray-dried lipid–polymer hybrid nanoparticles. *Artificial Cells, Nanomedicine, and Biotechnology*, *44*(6), 1544–1555. doi:10.3109/21691401.2015.1062389 PMID:26178768

Brossier, F., Veziris, N., Truffot-Pernot, C., Jarlier, V., & Sougakoff, W. (2011). Molecular investigation of resistance to the antituberculous drug ethionamide in multidrug-resistant clinical isolates of Mycobacterium tuberculosis. *Antimicrobial Agents and Chemotherapy*, *55*(1), 355–360. doi:10.1128/AAC.01030-10 PMID:20974869

Costa, A., Sarmento, B., & Seabra, V. (2018). Mannose-functionalized solid lipid nanoparticles are effective in targeting alveolar macrophages. *European Journal of Pharmaceutical Sciences*, *114*, 103–113. doi:10.1016/j.ejps.2017.12.006 PMID:29229273

Costa-Gouveia, J., Pancani, E., Jouny, S., Machelart, A., Delorme, V., Salzano, G., ... Flipo, M. (2017). Combination therapy for tuberculosis treatment: Pulmonary administration of ethionamide and booster co-loaded nanoparticles. *Scientific Reports*, *7*(1), 1–14. doi:10.103841598-017-05453-3 PMID:28710351

da Silva, P. B., de Freitas, E. S., Solcia, M. C., de Souza, P. C., da Silva, M. M., Batista, A. A., ... Chorilli, M. (2018). A Nanostructured Lipid System to Improve the Oral Bioavailability of Ruthenium (II) Complexes for the Treatment of Infections Caused by Mycobacterium tuberculosis. *Frontiers in Microbiology*, *9*, 2930. doi:10.3389/fmicb.2018.02930 PMID:30574128

De Matteis, L., Jary, D., Lucía, A., García-Embid, S., Serrano-Sevilla, I., Pérez, D., ... De la Fuente, J. M. (2018). New active formulations against M. tuberculosis: Bedaquiline encapsulation in lipid nanoparticles and chitosan nanocapsules. *Chemical Engineering Journal*, *340*, 181–191. doi:10.1016/j.cej.2017.12.110

Dicks, K. V., & Stout, J. E. (2019). Molecular diagnostics for Mycobacterium tuberculosis infection. *Annual Review of Medicine*, *70*(1), 77–90. doi:10.1146/annurev-med-040717-051502 PMID:30125128

Doan, T. V. P., & Olivier, J. C. (2009). Preparation of rifampicin-loaded PLGA microspheres for lung delivery as aerosol by premix membrane homogenization. *International Journal of Pharmaceutics*, *382*(1-2), 61–66. doi:10.1016/j.ijpharm.2009.08.008 PMID:19682562

El-Ridy, M. S., Yehia, S. A., Kassem, M. A. E. M., Mostafa, D. M., Nasr, E. A., & Asfour, M. H. (2015). Niosomal encapsulation of ethambutol hydrochloride for increasing its efficacy and safety. *Drug Delivery*, *22*(1), 21–36. doi:10.3109/10717544.2013.868556 PMID:24359403

Ferraz-Carvalho, R. S., Pereira, M. A., Linhares, L. A., Lira-Nogueira, M. C., Cavalcanti, I. M., Santos-Magalhães, N. S., & Montenegro, L. M. (2016). Effects of the encapsulation of usnic acid into liposomes and interactions with antituberculous agents against multidrug-resistant tuberculosis clinical isolates. *Memorias do Instituto Oswaldo Cruz*, *111*(5), 330–334. doi:10.1590/0074-02760150454 PMID:27143488

Gaidukevich, S. K., Mikulovich, Y. L., Smirnova, T. G., Andreevskaya, S. N., Sorokoumova, G. M., Chernousova, L. N., Selishcheva, A. A., & Shvets, V. I. (2016). Antibacterial effects of liposomes containing phospholipid cardiolipin and fluoroquinolone levofloxacin on Mycobacterium tuberculosis with extensive drug resistance. *Bulletin of Experimental Biology and Medicine*, *160*(5), 675–678. doi:10.100710517-016-3247-z PMID:27021087

Galagan, J. E. (2014). Genomic insights into tuberculosis. *Nature Reviews. Genetics*, *15*(5), 307–320. doi:10.1038/nrg3664 PMID:24662221

Gao, Y., Sarfraz, M. K., Clas, S. D., Roa, W., & Löbenberg, R. (2015). Hyaluronic acid-tocopherol succinate-based self-assembling micelles for targeted delivery of rifampicin to alveolar macrophages. *Journal of Biomedical Nanotechnology*, *11*(8), 1312–1329. doi:10.1166/jbn.2015.2091 PMID:26295135

Gatfield, J., Albrecht, I., Zanolari, B., Steinmetz, M. O., & Pieters, J. (2005). Association of the leukocyte plasma membrane with the actin cytoskeleton through coiled coil-mediated trimeric coronin 1 molecules. *Molecular Biology of the Cell*, *16*(6), 2786–2798. doi:10.1091/mbc.e05-01-0042 PMID:15800061

Gelmanova, I. Y., Keshavjee, S., Golubchikova, V. T., Berezina, V. I., Strelis, A. K., Yanova, G. V., ... Murray, M. (2007). Barriers to successful tuberculosis treatment in Tomsk, Russian Federation: Non-adherence, default and the acquisition of multidrug resistance. *Bulletin of the World Health Organization*, *85*(09), 703–711. doi:10.2471/BLT.06.038331 PMID:18026627

Georghiou, S. B., Magana, M., Garfein, R. S., Catanzaro, D. G., Catanzaro, A., & Rodwell, T. C. (2012). Evaluation of genetic mutations associated with Mycobacterium tuberculosis resistance to amikacin, kanamycin and capreomycin: A systematic review. *PLoS One*, *7*(3), e33275. doi:10.1371/journal.pone.0033275 PMID:22479378

Gillespie, S. H. (2002). Evolution of drug resistance in Mycobacterium tuberculosis: Clinical and molecular perspective. *Antimicrobial Agents and Chemotherapy*, *46*(2), 267–274. doi:10.1128/AAC.46.2.267-274.2002 PMID:11796329

Haddaoui, M., Sola, C., Raouafi, N., & Korri-Youssoufi, H. (2019). E-DNA detection of rpoB gene resistance in Mycobacterium tuberculosis in real samples using Fe3O4/polypyrrole nanocomposite. *Biosensors & Bioelectronics*, *128*, 76–82. doi:10.1016/j.bios.2018.11.045 PMID:30640123

Hamed, A., Osman, R., Al-Jamal, K. T., Holayel, S. M., & Geneidi, A. S. (2019). Enhanced antitubercular activity, alveolar deposition and macrophages uptake of mannosylated stable nanoliposomes. *Journal of Drug Delivery Science and Technology*, *51*, 513–523. doi:10.1016/j.jddst.2019.03.032

Hazbón, M. H., Brimacombe, M., Del Valle, M. B., Cavatore, M., Guerrero, M. I., Varma-Basil, M., ... León, C. I. (2006). Population genetics study of isoniazid resistance mutations and evolution of multidrug-resistant Mycobacterium tuberculosis. *Antimicrobial Agents and Chemotherapy*, *50*(8), 2640–2649. doi:10.1128/AAC.00112-06 PMID:16870753

Henostroza, M. A. B., Melo, K. J. C., Yukuyama, M. N., Löbenberg, R., & Bou-Chacra, N. A. (2020). Cationic rifampicin nanoemulsion for the treatment of ocular tuberculosis. *Colloids and Surfaces. A, Physicochemical and Engineering Aspects*, *597*, 124755. doi:10.1016/j.colsurfa.2020.124755

Hunter, R., & Actor, J. (2019). The pathogenesis of post-primary tuberculosis. A game changer for vaccine development. *Tuberculosis (Edinburgh, Scotland)*, *116*, S114–S117. doi:10.1016/j.tube.2019.04.018 PMID:31076321

Jansook, P., Ogawa, N., & Loftsson, T. (2018). Cyclodextrins: Structure, physicochemical properties and pharmaceutical applications. *International Journal of Pharmaceutics*, *535*(1-2), 272–284. doi:10.1016/j.ijpharm.2017.11.018 PMID:29138045

Jayachandran, R., Gatfield, J., Massner, J., Albrecht, I., Zanolari, B., & Pieters, J. (2008). RNA interference in J774 macrophages reveals a role for coronin 1 in mycobacterial trafficking but not in actin-dependent processes. *Molecular Biology of the Cell, 19*(3), 1241–1251. doi:10.1091/mbc.e07-07-0640 PMID:18162581

Jnawali, H. N., & Ryoo, S. (2013). First-and second-line drugs and drug resistance. *Tuberculosis-Current Issues in Diagnosis and Management*, 163-180.

Johansen, S. K., Maus, C. E., Plikaytis, B. B., & Douthwaite, S. (2006). Capreomycin binds across the ribosomal subunit interface using tlyA-encoded 2′-O-methylations in 16S and 23S rRNAs. *Molecular Cell, 23*(2), 173–182. doi:10.1016/j.molcel.2006.05.044 PMID:16857584

Juréen, P., Werngren, J., Toro, J. C., & Hoffner, S. (2008). Pyrazinamide resistance and pncA gene mutations in Mycobacterium tuberculosis. *Antimicrobial Agents and Chemotherapy, 52*(5), 1852–1854. doi:10.1128/AAC.00110-08 PMID:18316515

Kalombo, L., Lemmer, Y., Semete-Makokotlela, B., Ramalapa, B., Nkuna, P., Booysen, L. L., Naidoo, S., Hayeshi, R., Verschoor, J. A., & Swai, H. S. (2019). Spray-Dried, Nanoencapsulated, Multi-Drug Anti-Tuberculosis Therapy Aimed at Once Weekly Administration for the Duration of Treatment. *Nanomaterials (Basel, Switzerland), 9*(8), 1167. doi:10.3390/nano9081167 PMID:31443150

Kaur, I. P., & Singh, H. (2014). Nanostructured drug delivery for better management of tuberculosis. *Journal of Controlled Release, 184*, 36–50. doi:10.1016/j.jconrel.2014.04.009 PMID:24732260

Khatak, S., Semwal, U. P., Pandey, M. K., & Dureja, H. (2018). *Investigation of Antimycobacterial potential of solid lipid nanoparticles against M. smegmatis*. Academic Press.

Khoder, R., & Korri-Youssoufi, H. (2020). E-DNA biosensors of M. tuberculosis based on nanostructured polypyrrole. *Materials Science and Engineering C, 108*, 110371. doi:10.1016/j.msec.2019.110371 PMID:31924004

Khosa, A., Reddi, S., & Saha, R. N. (2018). Nanostructured lipid carriers for site-specific drug delivery. *Biomedicine and Pharmacotherapy, 103*, 598–613. doi:10.1016/j.biopha.2018.04.055 PMID:29677547

Kulkarni, P., Rawtani, D., & Barot, T. (2019). Formulation and optimization of long acting dual niosomes using box-Behnken experimental design method for combinative delivery of ethionamide and D-cycloserine in tuberculosis treatment. *Colloids and Surfaces. A, Physicochemical and Engineering Aspects, 565*, 131–142. doi:10.1016/j.colsurfa.2019.01.004

Machelart, A., Salzano, G., Li, X., Demars, A., Debrie, A. S., Menendez-Miranda, M., ... Belhaouane, I. (2019). Intrinsic antibacterial activity of nanoparticles made of β-cyclodextrins potentiates their effect as drug nanocarriers against tuberculosis. *ACS Nano, 13*(4), 3992–4007. doi:10.1021/acsnano.8b07902 PMID:30822386

MacLean, E., Broger, T., Yerlikaya, S., Fernandez-Carballo, B. L., Pai, M., & Denkinger, C. M. (2019). A systematic review of biomarkers to detect active tuberculosis. *Nature Microbiology, 4*(5), 748–758. doi:10.103841564-019-0380-2 PMID:30804546

Maretti, E., Costantino, L., Rustichelli, C., Leo, E., Croce, M. A., Buttini, F., Truzzi, E., & Iannuccelli, V. (2017). Surface engineering of Solid Lipid Nanoparticle assemblies by methyl α-d-mannopyranoside for the active targeting to macrophages in anti-tuberculosis inhalation therapy. *International Journal of Pharmaceutics*, *528*(1-2), 440–451. doi:10.1016/j.ijpharm.2017.06.045 PMID:28624659

Mehanna, M. M., Mohyeldin, S. M., & Elgindy, N. A. (2019). Rifampicin-Carbohydrate Spray-Dried Nanocomposite: A Futuristic Multiparticulate Platform For Pulmonary Delivery. *International Journal of Nanomedicine*, *14*, 9089–9112. doi:10.2147/IJN.S211182 PMID:31819421

Mehta, M., Sharma, N., Vyas, M., Khurana, N., Maurya, P. K., Singh, H., ... Wadhwa, R. (2019). Interactions with the macrophages: An emerging targeted approach using novel drug delivery systems in respiratory diseases. *Chemico-Biological Interactions*, *304*, 10–19. doi:10.1016/j.cbi.2019.02.021 PMID:30849336

Mehta, S. K., & Jindal, N. (2013). Formulation of Tyloxapol niosomes for encapsulation, stabilization and dissolution of anti-tubercular drugs. *Colloids and Surfaces. B, Biointerfaces*, *101*, 434–441. doi:10.1016/j.colsurfb.2012.07.006 PMID:23010052

Moretton, M. A., Cagel, M., Bernabeu, E., Gonzalez, L., & Chiappetta, D. A. (2015). Nanopolymersomes as potential carriers for rifampicin pulmonary delivery. *Colloids and Surfaces. B, Biointerfaces*, *136*, 1017–1025. doi:10.1016/j.colsurfb.2015.10.049 PMID:26590894

Mueller, P., Massner, J., Jayachandran, R., Combaluzier, B., Albrecht, I., Gatfield, J., Blum, C., Ceredig, R., Rodewald, H.-R., Rolink, A. G., & Pieters, J. (2008). Regulation of T cell survival through coronin-1–mediated generation of inositol-1, 4, 5-trisphosphate and calcium mobilization after T cell receptor triggering. *Nature Immunology*, *9*(4), 424–431. doi:10.1038/ni1570 PMID:18345003

Nemati, E., Mokhtarzadeh, A., Panahi-Azar, V., Mohammadi, A., Hamishehkar, H., Mesgari-Abbasi, M., Ezzati Nazhad Dolatabadi, J., & de la Guardia, M. (2019). Ethambutol-loaded solid lipid nanoparticles as dry powder inhalable formulation for tuberculosis therapy. *AAPS PharmSciTech*, *20*(3), 120. doi:10.120812249-019-1334-y PMID:30796625

Nkanga, C. I., & Krause, R. W. M. (2019). encapsulation of isoniazid-conjugated phthalocyanine-in-cyclodextrin-in-Liposomes Using Heating Method. *Scientific Reports*, *9*(1), 1–16. doi:10.103841598-019-47991-y PMID:31391517

Nyang'wa, B. T., Berry, C., Fielding, K., & Nunn, A. J. (2019). Multidrug-resistant tuberculosis. *Lancet*, *394*(10195), 298–299. doi:10.1016/S0140-6736(19)30691-9 PMID:31354136

Okamoto, S., Tamaru, A., Nakajima, C., Nishimura, K., Tanaka, Y., Tokuyama, S., Suzuki, Y., & Ochi, K. (2007). Loss of a conserved 7-methylguanosine modification in 16S rRNA confers low-level streptomycin resistance in bacteria. *Molecular Microbiology*, *63*(4), 1096–1106. doi:10.1111/j.1365-2958.2006.05585.x PMID:17238915

Onyebujoh, P., Zumla, A., Ribeiro, I., Rustomjee, R., Mwaba, P., Gomes, M., & Grange, J. M. (2005). Treatment of tuberculosis: Present status and future prospects. *Bulletin of the World Health Organization*, *83*, 857–865. PMID:16302043

Pai, M., & Memish, Z. A. (2016). Antimicrobial resistance and the growing threat of drug-resistant tuberculosis. *Journal of Epidemiology and Global Health*, *6*(2), 45–47. doi:10.1016/j.jegh.2016.02.001 PMID:26896704

Palmerston Mendes, L., Pan, J., & Torchilin, V. P. (2017). Dendrimers as nanocarriers for nucleic acid and drug delivery in cancer therapy. *Molecules (Basel, Switzerland)*, *22*(9), 1401. doi:10.3390/molecules22091401 PMID:28832535

Palomino, J. C., & Martin, A. (2014). Drug resistance mechanisms in Mycobacterium tuberculosis. *Antibiotics (Basel, Switzerland)*, *3*(3), 317–340. doi:10.3390/antibiotics3030317 PMID:27025748

Pieters, J. (2008). Mycobacterium tuberculosis and the macrophage: Maintaining a balance. *Cell Host & Microbe*, *3*(6), 399–407. doi:10.1016/j.chom.2008.05.006 PMID:18541216

Pinheiro, M., Lúcio, M., Lima, J. L., & Reis, S. (2011). Liposomes as drug delivery systems for the treatment of TB. *Nanomedicine (London)*, *6*(8), 1413–1428. doi:10.2217/nnm.11.122 PMID:22026379

Pinheiro, M., Ribeiro, R., Vieira, A., Andrade, F., & Reis, S. (2016). Design of a nanostructured lipid carrier intended to improve the treatment of tuberculosis. *Drug Design, Development and Therapy*, *10*, 2467–2475. doi:10.2147/DDDT.S104395 PMID:27536067

Pyle, M. M. (1947, October). Relative numbers of resistant tubercle bacilli in sputa of patients before and during treatment with streptomycin. *Proceedings of the Staff Meetings. Mayo Clinic*, *22*(21), 465. PMID:20267460

Rai, V. K., Mishra, N., Yadav, K. S., & Yadav, N. P. (2018). Nanoemulsion as pharmaceutical carrier for dermal and transdermal drug delivery: Formulation development, stability issues, basic considerations and applications. *Journal of Controlled Release*, *270*, 203–225. doi:10.1016/j.jconrel.2017.11.049 PMID:29199062

Rawat, R., Whitty, A., & Tonge, P. J. (2003). The isoniazid-NAD adduct is a slow, tight-binding inhibitor of InhA, the Mycobacterium tuberculosis enoyl reductase: Adduct affinity and drug resistance. *Proceedings of the National Academy of Sciences of the United States of America*, *100*(24), 13881–13886. doi:10.1073/pnas.2235848100 PMID:14623976

Rengarajan, J., Sassetti, C. M., Naroditskaya, V., Sloutsky, A., Bloom, B. R., & Rubin, E. J. (2004). The folate pathway is a target for resistance to the drug para-aminosalicylic acid (PAS) in mycobacteria. *Molecular Microbiology*, *53*(1), 275–282. doi:10.1111/j.1365-2958.2004.04120.x PMID:15225321

Rossi, E. D., Aínsa, J. A., & Riccardi, G. (2006). Role of mycobacterial efflux transporters in drug resistance: An unresolved question. *FEMS Microbiology Reviews*, *30*(1), 36–52. doi:10.1111/j.1574-6976.2005.00002.x PMID:16438679

Sadhu, P. K., Saisivam, S., & Debnath, S. K. (2019). *Design and characterization of niosomes of ethionamide for multi drug resistance tuberculosis*. Academic Press.

Saravanan, M., Duche, K., Asmelash, T., Gebreyesus, A., Negash, L., Tesfay, A., Hailekiros, H., Niguse, S., Gopinath, V., & Barik, S. (2015). "Nano-Biomaterials"–A New Approach Concerning Multi-Drug Resistant Tuberculosis (MDR-TB). *Pharmaceutical Nanotechnology*, *3*(1), 5–18. doi:10.2174/221173 8503666150415230743

Seung, K. J., Keshavjee, S., & Rich, M. L. (2015). Multidrug-resistant tuberculosis and extensively drug-resistant tuberculosis. *Cold Spring Harbor Perspectives in Medicine*, *5*(9), a017863. doi:10.1101/cshperspect.a017863 PMID:25918181

Shah, K., Chan, L. W., & Wong, T. W. (2017). Critical physicochemical and biological attributes of nanoemulsions for pulmonary delivery of rifampicin by nebulization technique in tuberculosis treatment. *Drug Delivery*, *24*(1), 1631–1647. doi:10.1080/10717544.2017.1384298 PMID:29063794

Sharma, S. K., & Mohan, A. (2019). Extrapulmonary Tuberculosis. In *Mycobacterium Tuberculosis: Molecular Infection Biology, Pathogenesis, Diagnostics and New Interventions* (pp. 37–53). Springer. doi:10.1007/978-981-32-9413-4_4

Shiehzadeh, F., Hadizadeh, F., Mohammadpour, A., Aryan, E., Gholami, L., & Tafaghodi, M. (2019). Streptomycin sulfate dry powder inhalers for the new tuberculosis treatment schedule. *Journal of Drug Delivery Science and Technology*, *52*, 957–967. doi:10.1016/j.jddst.2019.05.052

Shrivastava, P., Vyas, S., Sharma, R., Mody, N., Gautam, L., Jain, A., Vishwakarma, N., & Vyas, S. P. (2020). Nanotechnology for oral drug delivery and targeting. In *Nanoengineered Biomaterials for Advanced Drug Delivery* (pp. 473–497). Elsevier., doi:10.1016/B978-0-08-102985-5.00020-6

Singh, G., Dwivedi, H., Saraf, S. K., & Saraf, S. A. (2011). Niosomal delivery of isoniazid-development and characterization. *Tropical Journal of Pharmaceutical Research*, *10*(2). Advance online publication. doi:10.4314/tjpr.v10i2.66564

Singh, H., Bhandari, R., & Kaur, I. P. (2013). Encapsulation of Rifampicin in a solid lipid nanoparticulate system to limit its degradation and interaction with Isoniazid at acidic pH. *International Journal of Pharmaceutics*, *446*(1-2), 106–111. doi:10.1016/j.ijpharm.2013.02.012 PMID:23410991

Singh, Y., Meher, J. G., Raval, K., Khan, F. A., Chaurasia, M., Jain, N. K., & Chourasia, M. K. (2017). Nanoemulsion: Concepts, development and applications in drug delivery. *Journal of Controlled Release*, *252*, 28–49. doi:10.1016/j.jconrel.2017.03.008 PMID:28279798

Song, X., Lin, Q., Guo, L., Fu, Y., Han, J., Ke, H., Sun, X., Gong, T., & Zhang, Z. (2015). Rifampicin loaded mannosylated cationic nanostructured lipid carriers for alveolar macrophage-specific delivery. *Pharmaceutical Research*, *32*(5), 1741–1751. doi:10.100711095-014-1572-3 PMID:25407545

Sun, Z., Zhang, J., Zhang, X., Wang, S., Zhang, Y., & Li, C. (2008). Comparison of gyrA gene mutations between laboratory-selected ofloxacin-resistant Mycobacterium tuberculosis strains and clinical isolates. *International Journal of Antimicrobial Agents*, *31*(2), 115–121. doi:10.1016/j.ijantimicag.2007.10.014 PMID:18164184

Takenaga, M., Ohta, Y., Tokura, Y., Hamaguchi, A., Igarashi, R., Disratthakit, A., & Doi, N. (2008). Lipid microsphere formulation containing rifampicin targets alveolar macrophages. *Drug Delivery*, *15*(3), 169–175. doi:10.1080/10717540801952530 PMID:18379929

Telenti, A., Imboden, P., Marchesi, F., Matter, L., Schopfer, K., Bodmer, T., Lowrie, D., Colston, M. J., & Cole, S. (1993). Detection of rifampicin-resistance mutations in Mycobacterium tuberculosis. *Lancet, 341*(8846), 647–651. doi:10.1016/0140-6736(93)90417-F PMID:8095569

Uchegbu, I. F., & Vyas, S. P. (1998). Non-ionic surfactant based vesicles (niosomes) in drug delivery. *International Journal of Pharmaceutics, 172*(1-2), 33–70. doi:10.1016/S0378-5173(98)00169-0

Uetrecht, A. C., & Bear, J. E. (2006). Coronins: The return of the crown. *Trends in Cell Biology, 16*(8), 421–426. doi:10.1016/j.tcb.2006.06.002 PMID:16806932

Upadhyay, S., Khan, I., Gothwal, A., Pachouri, P. K., Bhaskar, N., Gupta, U. D., Chauhan, D. S., & Gupta, U. (2017). Conjugated and entrapped HPMA-PLA Nano-polymeric micelles based dual delivery of first line anti TB drugs: Improved and safe drug delivery against sensitive and resistant Mycobacterium tuberculosis. *Pharmaceutical Research, 34*(9), 1944–1955. doi:10.100711095-017-2206-3 PMID:28685299

Varma, J. R., Kumar, T. S., Prasanthi, B., & Ratna, J. V. (2015). Formulation and characterization of pyrazinamide polymeric nanoparticles for pulmonary tuberculosis: Efficiency for alveolar macrophage targeting. *Indian Journal of Pharmaceutical Sciences, 77*(3), 258. doi:10.4103/0250-474X.159602 PMID:26180270

Vieira, A. C., Chaves, L. L., Pinheiro, M., Lima, S. A. C., Ferreira, D., Sarmento, B., & Reis, S. (2018). Mannosylated solid lipid nanoparticles for the selective delivery of rifampicin to macrophages. *Artificial cells, Nanomedicine, and Biotechnology, 46*(sup1), 653-663.

Vieira, A. C., Chaves, L. L., Pinheiro, S., Pinto, S., Pinheiro, M., Lima, S. C., Ferreira, D., Sarmento, B., & Reis, S. (2018). Mucoadhesive chitosan-coated solid lipid nanoparticles for better management of tuberculosis. *International Journal of Pharmaceutics, 536*(1), 478–485. doi:10.1016/j.ijpharm.2017.11.071 PMID:29203137

Vieira, A. C., Magalhães, J., Rocha, S., Cardoso, M. S., Santos, S. G., Borges, M., Pinheiro, M., & Reis, S. (2017). Targeted macrophages delivery of rifampicin-loaded lipid nanoparticles to improve tuberculosis treatment. *Nanomedicine (London), 12*(24), 2721–2736. doi:10.2217/nnm-2017-0248 PMID:29119867

Vilchèze, C., & Jacobs, W. R. Jr. (2019). The isoniazid paradigm of killing, resistance, and persistence in Mycobacterium tuberculosis. *Journal of Molecular Biology, 431*(18), 3450–3461. doi:10.1016/j.jmb.2019.02.016 PMID:30797860

Vyas, S. P., Kannan, M. E., Jain, S., Mishra, V., & Singh, P. (2004). Design of liposomal aerosols for improved delivery of rifampicin to alveolar macrophages. *International Journal of Pharmaceutics, 269*(1), 37–49. doi:10.1016/j.ijpharm.2003.08.017 PMID:14698575

Wang, L., Du, J., Zhou, Y., & Wang, Y. (2017). Safety of nanosuspensions in drug delivery. *Nanomedicine; Nanotechnology, Biology, and Medicine, 13*(2), 455–469. doi:10.1016/j.nano.2016.08.007 PMID:27558350

World Health Organization. (2013). *Global tuberculosis report 2013*. World Health Organization.

World Health Organization. (2019). The WHO global task force on TB impact measurement (No. WHO/CDS/TB/2019.2). World Health Organization.

Yadav, H. K., Almokdad, A. A., Sumia, I. M., & Debe, M. S. (2019). Polymer-based nanomaterials for drug-delivery carriers. In *Nanocarriers for Drug Delivery* (pp. 531–556). Elsevier. doi:10.1016/B978-0-12-814033-8.00017-5

Zaunbrecher, M. A., Sikes, R. D., Metchock, B., Shinnick, T. M., & Posey, J. E. (2009). Overexpression of the chromosomally encoded aminoglycoside acetyltransferase eis confers kanamycin resistance in Mycobacterium tuberculosis. *Proceedings of the National Academy of Sciences of the United States of America*, *106*(47), 20004–20009. doi:10.1073/pnas.0907925106 PMID:19906990

Zhang, Y., & Mitchison, D. (2003). The curious characteristics of pyrazinamide: A review. *The International Journal of Tuberculosis and Lung Disease*, *7*(1), 6–21. PMID:12701830

Chapter 5
Novel Synergistic Approaches of Nano-Biomaterials and Bacteriophage for Combating Antimicrobial Resistance

Maheswaran Easwaran
Sethu Institute of Technology, India

Nageshwari Raja
Sethu Institute of Technology, India

Juhee Ahn
Kangwon National University, South Korea

Hemalatha Karuppiah
Sethu Institute of Technology, India

Hyun Jin Shin
Chungnam National University, South Korea

ABSTRACT

The antimicrobial resistance has significantly risen in last few decades as a result of too much usage of antibiotics. In the case of failure of drug therapy, the multidrug resistance has become a serious issue in a clinical and food sectors. The rapid emergence of antibiotic resistance has become a great concern for public health worldwide. Therefore, the novel approaches of nano-biomaterials (NBM) and bacteriophage (phage) have received a new attention as an alternative treatment for antibiotic-resistant bacterial infections. Moreover, the advanced molecular biology and genetic engineering have been successfully applied for the rapid detection and control of multidrug resistant pathogenic strains. This chapter specifically summarizes the prevalence of common as well as under-researched antimicrobial resistance of bacterial pathogens and the application of nano-biomaterials and bacteriophage to fight against antimicrobial resistance. Ultimately, the synergistic approach will be a highly promising prospect to threat of antimicrobial resistance.

DOI: 10.4018/978-1-7998-5049-6.ch005

INTRODUCTION

Antimicrobial resistance (AMR) is the potentiality of bacteria for its growth and outspread in the state of existing of antimicrobial agents. Recent pathogens are getting virulent to recent antibiotics because of an unrestricted use of antibiotics (Kutateladze & Adamia, 2010), which are concern as an untreatable or uncontrollable bacterial infections. Antibiotics can act on bacteria by inhibiting or interfering cell-wall, protein, DNA and RNA synthesis. To overcome the influence of antibiotics, bacteria has the intrinsic potential to transfer the mutated genetic material between bacterial cells that leads to antibiotic resistance (Bengtsson-Palme, Kristiansson, & Larsson, 2018). Moreover, the characteristic features of resistance may gradually develop due to the modification of drug target-site, cell membrane permeability, enzymatic and metabolic activity (Miro et al., 2004). The World Health Organisation (WHO) has given prominence to the global threat of antibiotics resistance and alerted that the global death toll maybe ten million people per year due to antibiotic resistant infectious pathogens (de Kraker, Stewardson, & Harbarth, 2016).

Since the earliest antibiotic invention, penicillin have been applied as a commercially available antibiotics (McEwen, 2006). Moreover, antibiotics have been reported as a growth-inducing agent to improve the adequate quantity of animal feed (Silbergeld, Graham, & Price, 2008). As a result, resistance character have been enriched due to routine use of antibiotics in the domain of aquaculture, food animal and medicine to assist animal growth and to control bacterial infections (Cabello, 2006; Cantas et al., 2013). Those antibiotic-resistant strains and their genes can be transmitted from contaminated food animals to human and the other way round (Woolhouse & Ward, 2013). Precisely the food animals such as cattle, poultry and swine have been engaged in prominent public health crisis (Marshall & Levy, 2011).

Overall antibiotics are not beneficial approach for combating antimicrobial resistance and researchers also look in some novel synergistic approach. Simultaneously, a lot of new alternative approaches such as bacteriophage (Qadir & Chauhdary, 2018), antimicrobial peptides (Atterbury, 2009), combinations of antibiotics (Tyers & Wright, 2019) and nanomaterials (Gupta, Mumtaz, Li, Hussain, & Rotello, 2019) have been successfully examined to control multidrug resistant strains that are responsible for the prevalence of antibiotic resistance worldwide. This chapter mainly focuses on antibiotics and its impact on natural bacteria, synergistic approaches to combat antibiotic resistant pathogens and strategies in the development of new antibiotic therapies. In recent years, the combination of nanobiomaterials and bacteriophage has overcome the lab boundaries and barriers of pathogen detection and applied for combating antibiotic resistant strains. Here, we highlight the use of phage derived protein "endolysin" and Nano-biomaterials, specifically in synergistic approaches for effective modalities to fight against resistant pathogens.

URGENT TREATS AND ITS IMPACT

A bacterium has the ability to overcome the antibiotics which have been designed to control or kill. In most infectious pathogen, antibiotic resistant strain have been considered as one of the world's most serious public health threats now (Prestinaci, Pezzotti, & Pantosti, 2015).

Acinetobacter baumannii have considered as one of the notable pathogen in the hospital environment due to its acquired antimicrobial resistance (Peleg, Seifert, & Paterson, 2008). *A. baumannii* exhibited more than 90% of resistance for all the antibiotics such as cefepime, ceftazidime, ciprofloxacin, gentamycin, imipenem, levofloxacin and piperacillin-tazobactam (M. H. Kim et al., 2020). According to the recent report carbapenem-resistant *Acinetobacter* can cause bloodstream, pneumonia, wound, and urinary tract infections (Wong et al., 2017). Overall, the mortality rate is to be higher with single antibiotic therapy rather than combination therapy.

Clostridioides difficile is an intestinal spore-forming pathogen that causes of community-associated diarrhea that can provide a route to death (Sandhu & McBride, 2018). It has been recognized as a well-known resistant pathogen against novel antibiotics such as aminoglycosides, cephalosporins, clindamycin, fluoroquinolones, lincomycin, penicillins and tetracyclines in clinical sectors (Johanesen et al., 2015). Infectious *C. difficile* have considered as the leading causing agent of gastrointestinal infections acquired from hospital globally (Saleh & Petri, 2019). The Centers for Disease Control and Prevention reported that *C. difficile* infection caused approximately 29,000 deaths per annum in the United States (Balsells et al., 2019).

Carbapenem-resistant *Enterobacteriaceae* (CRE) have caused serious outbreaks that originating in hospitals (van Duin & Doi, 2017). Recently, Durante-Mangoni and his research team reported that CRE has been non-susceptible to multi-antimicrobials such as amikacin, fosfomycin, imipenem, meropenem, ceftazidime-avibactam, and piperacillin-tazobactam, an indicating resistance greater than 8 minimum inhibitory concentration (MIC) breakpoint (Durante-Mangoni, Andini, & Zampino, 2019). CRE colonization has been associated with maximum mortality rate in hospitalized and medical intensive care unit patients.

Neisseria gonorrhoeae is the major causative agent for gonorrhea infection that is the top most sexually transmitted infection worldwide. In 2018, Tshokey reported that the highly active antimicrobials such as ciprofloxacin, nalidixic acid, penicillin G and tetracycline were non-susceptible (>72%) with *N. gonorrhoeae (Tshokey et al., 2018)*. Average of 7.8×10^7 cases of gonorrhea infection have been reported each year (Newman et al., 2015). Complicated situation have been evidenced because of resistance mechanisms to multi- antimicrobials.

PREVALENCE OF ANTIBIOTIC RESISTANCE IN THE ENVIRONMENT

Previous studies reported that the avoparcin- resistant *Enterococcus faecium* has been raised in humans due to an animal feed (Marshall & Levy, 2011). In food animals, this characteristic feature can stimulate stress and weakens the immune system. As a consequence, the application of antibiotics in animal feed has been entirely prohibited in the United States of America (Cantas et al., 2013) and also the European countries (Casewell, Friis, Marco, McMullin, & Phillips, 2003).

Hospitalized patients with *Acinetobacter baumannii, Staphylococcus aureus, Streptococcus* and vancomycin-resistant enterococci, has been raised the infection even after complete sterilization process in the clinical regions (Carling & Bartley, 2010). Antibiotic polluted environments such as coastal area, hospital environment, clinical regions and natural water source have been reported by reason of improper use of antibiotics (V. K. Sharma, Johnson, Cizmas, McDonald, & Kim, 2016; Shimizu et al., 2013). The microbes from the contaminated environment may have high chance to get resistance against specific antibiotics (Gradmann, 2016).

A lot of health issues has been occurred by several bacterial pathogens such as *Escherichia coli, Campylobacter, Listeria, Shigella, Salmonella, Bacillus cereus, Vibrio* and *Staphylococcus* in human, goat, cattle, swine, sheep, and poultry (Hellberg & Chu, 2016; Strawn et al., 2013). *Salmonella en-*

terica is one of the major zoonotic bacterial pathogens cause gastrointestinal and food-borne diseases worldwide (Bloom, Merrett, Wilkinson, Lin, & Paulin, 2017). Particularly, *S.* Typhimurium has been acquired mobilized colistin resistance (*mcr*-1) gene as a result of the continuous use of polymyxin in animal production (Rau et al., 2018). Consumption of contaminated poultry and red meat can cause more than 70% of salmonellosis in human (Scallan et al., 2011) and also cause maximum economic damage in beef meat industries (R. Wang et al., 2017).

A further food-borne pathogen, *Campylobacter jejuni* provoked inflammation in digestive tract, gastroenteritis and diarrhea in human worldwide (Kaakoush, Castano-Rodriguez, Mitchell, & Man, 2015; Man, 2011; Vieira, Ramesh, Seddon, & Karlyshev, 2017). Similarly, shiga toxin-producing *E. coli*, enterohemorrhagic *Escherichia coli*, *E. coli* O25:H4 and *E. coli* 0157:H7 are highly resistant to multi-drugs such as cephalosporins, β- lactamases, fluroquinolones and tetracycline (Chopra & Roberts, 2001; Diarra et al., 2009; Hornish & Kotarski, 2002). Those multidrug resistant strains have been reported as major causing agents for uncontrolled diseases such as bloodborne and urinary tract infections (Johnson, Johnston, Clabots, Kuskowski, & Castanheira, 2010; Karmali, Gannon, & Sargeant, 2010). *Staphylococcus aureus* has been reported as a significant bacterial agent for fasciitis, pneumonia and urinary tract infection (Yonezawa, Kuwana, Kawamura, & Inamo, 2015).

NOTABLE SYNERGISTIC APPROACHES TO ENHANCE ANTIMICROBIAL ACTIVITY

Current antibiotics are becoming hopeless against emerging antibiotic-resistant bacteria due to the practical problem while treating with antibiotic resistance. An alternative antibacterial approach can be developed with the combination of two or more bactericidal agents for the control and treatment of infectious bacteria (Ma et al., 2020). Previous research revealed that the combination of antimicrobial compounds has potential to expand the antimicrobial activity, enhance the therapeutic effect, improve killing kinetics and reduce the emergence of resistance with broad inhibitory spectrum (Hu et al., 2015; Kalan & Wright, 2011; Soothill, Hu, & Coates, 2013).

Antimicrobial Activity in the Presence of Peptides and Antibiotics

Antimicrobial peptide (AMPs) has been reported as a traditional promising antimicrobial agent (Mahlapuu, Hakansson, Ringstad, & Bjorn, 2016). AMPs can generate the antibacterial characteristic feature by change in membrane permeability and denature of membrane structure (Teixeira, Feio, & Bastos, 2012). Such novel features also ineffective due to the development of resistance (Zharkova et al., 2019). The combination of peptides and antibiotics could be suggested for the effectiveness of antimicrobial agents against Gram-negative pathogens and multidrug-resistant bacteria (Baker et al., 2019; Shang et al., 2019; Wu et al., 2017). For example, tryptophan residues with antibiotics have been performed as an effective antimicrobial activity (Khandelia & Kaznessis, 2007).

Antimicrobial Activity in the Combination of Nanoparticles and Antibiotics

Nanoparticles (NPs) have been applied in case of fighting against antibiotic resistance and also for enhancing the antimicrobial activities (Singh, Garg, Pandit, Mokkapati, & Mijakovic, 2018). Characteristic features such as stability, electrostatic attraction and hydrophobic interactions can enhance antimicrobial activity of nanoparticles (Niño-Martínez, Salas Orozco, Martínez-Castañón, Torres Méndez, & Ruiz, 2019). Particularly, damage of cell membrane, proteins, internal components and DNA can generate nanoparticles as a suitable antimicrobial agent (Besinis, De Peralta, & Handy, 2014). Naturally, nanoparticle has the potential to bind with the bacterial surface and hence damaging cell wall thereby decelerating the biochemical pathways by the way of smashing cell organelles which leads to bacterial death (Arakha et al., 2015; Wang, Hu, & Shao, 2017). The prevalence of multidrug-resistant pathogens poses a serious problem to human health. These distinct features of nanoparticles have become purposeless because of the multidrug-resistant pathogens globally. Therefore the combined effect of nanoparticles with antibiotics could be the effective tool for handling the antibiotic resistant infectious pathogens. Latest report suggested that multidrug resistant pathogens could be controlled using a synergistic combination of antimicrobial drugs and silver nanoparticles against Gram-negative and Gram-positive pathogens (Lopez-Carrizales et al., 2018).

Synergistic Effect of Bacteriophages and Antibiotics

Bacteriophages (phages) are bacterial viruses that can infect and kill bacterial cells (Pires, Cleto, Sillankorva, Azeredo, & Lu, 2016). Bacteriophages were suggested as antibacterial agents once after discovered by Felix d'Herelle in 1919 (Fernández, Gutiérrez, García, & Rodríguez, 2019). Till today it has been successfully used in different fields such as agriculture, food safety, veterinary and clinical applications (Fernández, Gutiérrez, Rodríguez, & García, 2018). Even though bacteria can withstand during phage attack due to different molecular mechanisms such as gene mutations, restriction modification system and the CRISPR-Cas system (Labrie, Samson, & Moineau, 2010). Phage therapy also became ineffective in case of bacterial advantageous features due to the development of resistance. Phage can be functionally effective with the combination of antibiotics (Jo, Ding, & Ahn, 2016). A lot of researchers have revealed that combine activity of phages and antibiotics enhanced the antimicrobial activity and lytic activity of phages (Comeau, Tetart, Trojet, Prere, & Krisch, 2007; Kamal & Dennis, 2015).

NANO-BIOMATERIALS AGAINST ANTIBIOTIC RESISTANCE

Since the emergence of antibiotic resistance, the development of nano-biomaterial and its application has been repeatedly increased. Synthesis of nanoparticles through biological approaches has been encouraged for the control of multi-drug resistance. Since 2008, nano-materials obtained significant attention which are derived from natural sources such as bacteria, fungi, plant and yeast (Singh, Kim, Zhang, & Yang, 2016). In recent years, the applications of nanotechnology using Nano-biomaterials have been raised to detect, treat and prevent diseases due to its novel physicochemical and antimicrobial properties (Mohamed Hamouda, 2012). In nano-biomaterials, biomolecules such as proteins, DNA, enzyme, antigen and antibodies have been used for highly specific and recognition properties (Katz & Willner, 2004; Nagamune, 2017).

Functional Mechanism of Nano-Biomaterials

In recent years, nanotechnology has been improved their potential to cross the lab boundaries and its applications due to their different functional mechanism. Basically, NBM can inhibit the activity of enzyme, protein and gene in bacteria (L. Wang et al., 2017). Moreover, nano-biomaterial based diagnostic assay have used to diagnose infections pathogens. As an example, antibody-conjugated biomaterials have been successfully applied to quick spotting of a sole bacterium (Look, Bandyopadhyay, Blum, & Fahmy, 2010). As well, dextran-coated magnetic particles have been used for the identification of infections pathogen (Zhao et al., 2004). In particular, NBMs considered as the best antimicrobial activity against fungus Candida species (K. J. Kim et al., 2009). Specifically, it has been targeting cell membranes and creating pores on it and ultimately followed by discharge of intracellular substances and cell death (Gajbhiye, Kesharwani, Ingle, Gade, & Rai, 2009). Overall, the different functional mechanisms such as inhibition of protein synthesis, damage of genetic material, lysis of cell membrane and electron transport damage of nanobiomaterials have explained through the schematic diagram in Figure 1.

Some other nano-biomaterials such as magnesium oxide NBMs possess a notable amount of bromine and chlorine, which have been involved as an effective antimicrobial agent for pathogenic bacteria and its spores (Huang et al., 2005). Moreover, the positive charged magnesium oxide NBMs has been interacted with negatively charged bacterial cells or cell membrane for the enhancement of bactericidal effect (Koper et al., 2002). Correspondingly the stability and toxicity of zinc oxide NBM have been proved their potential antimicrobial activity (Reddy et al., 2007). ZnO NBMs have been suggested to apply for the food and agriculture field due to its selective and minimal toxicity towards bacteria and human cells, respectively (Brayner et al., 2006). ZnO NBMs can stop the growth of bacterium *E. coli* by breaking down the cell membrane, increasing cell membrane permeability and thereby inhibiting the bacterial growth by the generation of free radicals (Jiang, Mashayekhi, & Xing, 2009). Additionally, it also involved in breakage of the membrane components and interference in the function of protein (Stanković, Dimitrijević, & Uskoković, 2013).

Application of Nano-Biomaterials

In last few years, nanobiomaterials have been considered as valuable antimicrobial agent due to their selective antimicrobial efficacy against bacterial pathogen. The antimicrobial activity of DNA stabilized silver nanoparticles was able to be adapted by different oligonucleotide sequence (Javani et al., 2016). Moreover, nano-biomaterials like oligonucleotide sequence with maximum number of cytosine indicated the highest antimicrobial activity against both Gram strains. Additionally, nano-biomaterials also applied for cardiovascular diseases, cancer, diabetes, neurodegenerative and musculoskeletal treatment (Jiang, Rutherford, Vuong, & Liu, 2017). Specifically, it has been applied in the field of medicine and infectious diseases (Holban, Gestal, & Grumezescu, 2016). Nano-biomaterials are suitable agent for a variety of biological applications because of their electrical, mechanical and optical properties (Katz & Willner, 2004). In previous study, nano-biomaterials can be used not only as antimicrobial agents and also in water disinfection assay for the removal of pathogenic strains (Q. Li et al., 2008). Nino-Martinez et al. (2019) revealed that the halogens with metal oxide nanoparticles are extremely effective for the treatment of Gram-negative and Gram-positive strains. Researchers observed noticeable changes in the cell membranes and the cells due to abrasiveness, electrostatic attraction, and oxidizing properties of nano-biomaterials (Nino-Martinez, Salas Orozco, Martinez-Castanon, Torres Mendez, & Ruiz, 2019).

Protection of food from biological and chemical contaminants and improve the quality of food are played a vital role in food sectors (Fung, Wang, & Menon, 2018). For instance, silver nanoparticles based nano-composites possess magnificent antimicrobial activity in food package system (Honarvar, Hadian, & Mashayekh, 2016). Most of the biomaterials act as an effective antibacterial agent against *E. coli*. Similarly polymer-based packaging system utilizes nanoparticles as nanosensors for microbial contamination process (C. Sharma, Dhiman, Rokana, & Panwar, 2017). Biological nanomaterials like antibody, enzymes, peptides, proteins and viruses have been used in the field of biosensors designing mechanism. Those biological receptors have typically applied to recognize interested microbes with high-affinity binding character (Whaley, English, Hu, Barbara, & Belcher, 2000). Particularly, the nanosensors are applied for the detections of three major pathogenic bacteria such as *Bacillus cereus*, *Vibrio parahemolyticus*, and *Salmonella* species in food industries (Law, Ab Mutalib, Chan, & Lee, 2014).

BACTERIOPHAGE AS AN ANTIMICROBIAL AGENT

Naturally, bacteriophages have the potential to control and kill infectious pathogen in the way of host adsorption, insertion of genetic material, and break down of host cell (Bai, Kim, Ryu, & Lee, 2016). Bacteriophages act as an ideal agent for the pathogen detection, their growth control with its host specificity and eradication of pathogenic bacteria (Golkar, Bagasra, & Pace, 2014; Yosef, Manor, Kiro, & Qimron, 2015). For instance, phage has been examined as an antibacterial agent in case of food-borne and other bacterial diseases (Coffey, Mills, Coffey, McAuliffe, & Ross, 2010). In fact, the spot and the control of infectious bacterial pathogen are high-cost and prolonged process. Hence the phage-based assays such as biosensor, endolysin therapy and engineered phage have been successfully applied for the screening purpose of pathogens in a quick and inexpensive manner (O'Sullivan, Buttimer, McAuliffe, Bolton, & Coffey, 2016; Pires et al., 2016). For example, phage-based therapy have successfully identifies and eradicates the presence of antibiotic resistant strains in the surface of chicken and raw beef meats (Moye, Woolston, & Sulakvelidze, 2018).

Phage Derived Proteins and its Mechanism

Phage derived proteins are considered as antimicrobial agents using advanced molecular biology. Particularly, phage lysis proteins (holins and endolysins) are responsible for the denaturation of bacterial membrane and eradication of bacteria (Drulis-Kawa, Majkowska-Skrobek, & Maciejewska, 2015). Endolysins has been successfully examined for the treatment of bacterial infections (Love, Bhandari, Dobson, & Billington, 2018). Accessory protein - holin has been regulated for bacterial lysis and it supports to form a lesion in cytoplasmic membrane (Catalao, Gil, Moniz-Pereira, & Pimentel, 2011). Moreover, the signaled sequence of holin has been recognized as a responsible substance, which is involved for translocation of endolysin. Like this advantageous characteristic featured phage derived proteins can successfully applied for the control of multidrug resistant strains (Schmelcher, Donovan, & Loessner, 2012). Figure 2 showed the differentiation of phage therapy and endolysin therapy and its efficacy. Phage endolysin thrapy highlights better lytic activity (a few seconds) then phage-based therapy.

Significance of Engineered Phage

Engineered phage is a powerful tool to produce phages with extended host-range and those phages also involved to arrest bacterial resistance against phage (Yehl et al., 2019). Engineered phages have practically used in the field of healthcare centers, hospitals, food sectors for the purpose of rapid pathogen detection and inhibition (Velusamy, Arshak, Korostynska, Oliwa, & Adley, 2010). Specifically, it has the potential for the rapid spotting of live pathogenic cells with high host specificity (Smartt et al., 2012). Previous model showed that engineered M13 phage can prevent biofilm formation by drug resistant pathogens (Lu & Collins, 2009). Similarly, Lu and his team produced an engineered enzymatic T7 phage harbored dispersin B gene, which has been most effective than wild-type phage against *E. coli* biofilms (Lu & Collins, 2007). CRISPR-Cas system also recognized as a novel approach to treat drug-resistant pathogens (Y. Li & Peng, 2019; Shabbir et al., 2019). Recent studies reported that CRISPR-Cas system has been successfully examined to recognize antibiotic resistance genes of virulent strains. Moreover, this approach has the ability to sensitize and eradicate pathogens as well as treat diseased animal effectively (Pursey, Sunderhauf, Gaze, Westra, & van Houte, 2018).

Previous report revealed that phage endolysins have been successfully applied as antimicrobial agents in rat, mouse, and cow against *Pneumococcal meningitis* and Streptococcal infections (Grandgirard, Loeffler, Fischetti, & Leib, 2008; Schmelcher, Powell, Camp, Pohl, & Donovan, 2015). In another case like exogenous applications, endolysin may not work against Gram-negative bacteria, due to the lack the access of peptidoglycan on the bacterial outer membrane (Lim, Shin, Heu, & Ryu, 2014). Thus, a recent engineered endolysin combine with artilysins peptide stimulated the transportation of endolysin to the outer membrane (Briers et al., 2014).

SYNERGISTIC APPROACH OF NANO-BIOMATERIALS AND BACTERIOPHAGES

Bacteria can exhibit resistance against bacteriophages through restriction-modification systems (Tock & Dryden, 2005) and also CRISPR-Cas systems (Barrangou et al., 2007). Similarly, nano-biomaterials also have few limitations to prevent or control the behavior of pathogens. The significant physicochemical properties such as size, charge, solubility and surface chemistry have been induced the antimicrobial properties of NBMs (L. Wang et al., 2017). Similarly, the characteristic features of recombinant endolysin such as high binding specificity and a lower chance of developing resistance have provided promise for endolysins as novel antibacterial agents (Briers & Lavigne, 2015). In addition, the emergence of antibiotic tolerance against endolysins have been reported at low very levels as compared to other antimicrobials like antibiotics and exolysins (Nelson et al., 2012). Hence, synergistic activity of bacteriophage or phage endolysin and nano-biomaterials are burgeoning candidates for the control of drug resistant bacterial pathogens in recent years.

Optimistically, the functional mechanism of the synergistic effect will defeat the efficiency of antimicrobial resistance. The conjugation of nano-biomaterials with biomolecules has the potential with catalytic and recognition properties of biomaterials (Nagamune, 2017). Nano-biomaterial based approaches have been played a critical role to eliminate bacterial infections in case of creating antimicrobial nanotherapeutics. Nano-biomaterials can exhibit synergistic effect with bacteriophage and act as a promising tool against multi-drug resistant bacteria (Gupta et al., 2019). Bacterial resistance pattern has perfectly described in figure 3, it shows unlimited and moderate bacterial resistance due to the exploits of phage and phage lytic protein therapy. Moreover, synergistic effect of phage-endolysins and nano-biomaterials showed maximum lysis of target bacterial pathogen. Despite several applications of nanobiomaterials and phage endolysin in the medical and clinical sectors, there is an urgent need for the development of phage endolysin - based nanobiomaterials to control the spread of antibiotic resistance.

CONCLUSION WITH FUTURE PROSPECTS

Over time, treatment of infectious diseases could be a substantial challenge to global public health due to multidrug-resistant bacteria. Drug resistant pathogens can transfer to human beings through food animals and vice versa and cause infection. Therefore a perfect appliance should be demanded to recognize and eradicate the antibiotic resistant bacterial contaminants. Recently researchers got attention to encourage the synergistic approach using phage therapy and nano-biomaterials to overcome the limitations. Both features have been applied for the efficient action plan against specific drug resistant pathogens and minimize its intense in the animals, humans and other environment. The aim of this chapter is the development of a novel approach for the treatment of infectious pathogen harboring various drug resistances. Historical approach with antibiotics mainly involved in the creation of resistance in bacteria. In the future prospects, the phage endolysin-based nanobiomaterial approach would effectively reduce the existence of bacterial infection and successfully enhance healthy animal and human life and also heeded as a novel way of defense against various bacterial pathogens.

REFERENCES

Arakha, M., Pal, S., Samantarrai, D., Panigrahi, T. K., Mallick, B. C., Pramanik, K., Mallick, B., & Jha, S. (2015). Antimicrobial activity of iron oxide nanoparticle upon modulation of nanoparticle-bacteria interface. *Scientific Reports*, *5*(1), 14813. doi:10.1038rep14813 PMID:26437582

Atterbury, R. J. (2009). Bacteriophage biocontrol in animals and meat products. *Microbial Biotechnology*, *2*(6), 601–612. doi:10.1111/j.1751-7915.2009.00089.x PMID:21255295

Bai, J., Kim, Y. T., Ryu, S., & Lee, J. H. (2016). Biocontrol and Rapid Detection of Food-Borne Pathogens Using Bacteriophages and Endolysins. *Frontiers in Microbiology*, *7*, 474. doi:10.3389/fmicb.2016.00474 PMID:27092128

Baker, K. R., Jana, B., Hansen, A. M., Vissing, K. J., Nielsen, H. M., Franzyk, H., & Guardabassi, L. (2019). Repurposing azithromycin and rifampicin against Gram-negative pathogens by combination with peptide potentiators. *International Journal of Antimicrobial Agents*, *53*(6), 868–872. doi:10.1016/j.ijantimicag.2018.10.025 PMID:30447380

Balsells, E., Shi, T., Leese, C., Lyell, I., Burrows, J., Wiuff, C., Campbell, H., Kyaw, M. H., & Nair, H. (2019). Global burden of Clostridium difficile infections: A systematic review and meta-analysis. *Journal of Global Health*, *9*(1), 010407. doi:10.7189/jogh.09.010407 PMID:30603078

Barrangou, R., Fremaux, C., Deveau, H., Richards, M., Boyaval, P., Moineau, S., Romero, D. A., & Horvath, P. (2007). CRISPR provides acquired resistance against viruses in prokaryotes. *Science*, *315*(5819), 1709–1712. doi:10.1126cience.1138140 PMID:17379808

Bengtsson-Palme, J., Kristiansson, E., & Larsson, D. G. J. (2018). Environmental factors influencing the development and spread of antibiotic resistance. *FEMS Microbiology Reviews*, *42*(1). Advance online publication. doi:10.1093/femsre/fux053 PMID:29069382

Besinis, A., De Peralta, T., & Handy, R. D. (2014). The antibacterial effects of silver, titanium dioxide and silica dioxide nanoparticles compared to the dental disinfectant chlorhexidine on Streptococcus mutans using a suite of bioassays. *Nanotoxicology*, *8*(1), 1–16. doi:10.3109/17435390.2012.742935 PMID:23092443

Bloom, G., Merrett, G. B., Wilkinson, A., Lin, V., & Paulin, S. (2017). Antimicrobial resistance and universal health coverage. *BMJ Global Health*, *2*(4), e000518. doi:10.1136/bmjgh-2017-000518 PMID:29225955

Brayner, R., Ferrari-Iliou, R., Brivois, N., Djediat, S., Benedetti, M. F., & Fiévet, F. (2006). Toxicological impact studies based on Escherichia coli bacteria in ultrafine ZnO nanoparticles colloidal medium. *Nano Letters*, *6*(4), 866–870. doi:10.1021/nl052326h PMID:16608300

Briers, Y., & Lavigne, R. (2015). Breaking barriers: Expansion of the use of endolysins as novel antibacterials against Gram-negative bacteria. *Future Microbiology*, *10*(3), 377–390. doi:10.2217/fmb.15.8 PMID:25812461

Briers, Y., Walmagh, M., Van Puyenbroeck, V., Cornelissen, A., Cenens, W., Aertsen, A., Oliveira, H., Azeredo, J., Verween, G., Pirnay, J.-P., Miller, S., Volckaert, G., & Lavigne, R. (2014). Engineered endolysin-based "Artilysins" to combat multidrug-resistant gram-negative pathogens. *mBio*, *5*(4), e01379–e01314. doi:10.1128/mBio.01379-14 PMID:24987094

Cabello, F. C. (2006). Heavy use of prophylactic antibiotics in aquaculture: A growing problem for human and animal health and for the environment. *Environmental Microbiology*, *8*(7), 1137–1144. doi:10.1111/j.1462-2920.2006.01054.x PMID:16817922

Cantas, L., Shah, S. Q., Cavaco, L. M., Manaia, C. M., Walsh, F., Popowska, M., Garelick, H., Bürgmann, H., & Sorum, H. (2013). A brief multi-disciplinary review on antimicrobial resistance in medicine and its linkage to the global environmental microbiota. *Frontiers in Microbiology*, *4*, 96. doi:10.3389/fmicb.2013.00096 PMID:23675371

Carling, P. C., & Bartley, J. M. (2010). Evaluating hygienic cleaning in health care settings: What you do not know can harm your patients. *American Journal of Infection Control*, *38*(5, Suppl 1), S41–S50. doi:10.1016/j.ajic.2010.03.004 PMID:20569855

Casewell, M., Friis, C., Marco, E., McMullin, P., & Phillips, I. (2003). The European ban on growth-promoting antibiotics and emerging consequences for human and animal health. *The Journal of Antimicrobial Chemotherapy*, *52*(2), 159–161. doi:10.1093/jac/dkg313 PMID:12837737

Catalao, M. J., Gil, F., Moniz-Pereira, J., & Pimentel, M. (2011). Functional analysis of the holin-like proteins of mycobacteriophage Ms6. *Journal of Bacteriology*, *193*(11), 2793–2803. doi:10.1128/JB.01519-10 PMID:21441511

Chopra, I., & Roberts, M. (2001). Tetracycline antibiotics: mode of action, applications, molecular biology, and epidemiology of bacterial resistance. *Microbiol Mol Biol Rev*, *65*(2), 232-260. doi:10.1128/MMBR.65.2.232-260.2001

Coffey, B., Mills, S., Coffey, A., McAuliffe, O., & Ross, R. P. (2010). Phage and their lysins as biocontrol agents for food safety applications. *Annual Review of Food Science and Technology*, *1*(1), 449–468. doi:10.1146/annurev.food.102308.124046 PMID:22129344

Comeau, A. M., Tetart, F., Trojet, S. N., Prere, M. F., & Krisch, H. M. (2007). Phage-Antibiotic Synergy (PAS): Beta-lactam and quinolone antibiotics stimulate virulent phage growth. *PLoS One*, *2*(8), e799. doi:10.1371/journal.pone.0000799 PMID:17726529

de Kraker, M. E., Stewardson, A. J., & Harbarth, S. (2016). Will 10 Million People Die a Year due to Antimicrobial Resistance by 2050? *PLoS Medicine*, *13*(11), e1002184. doi:10.1371/journal.pmed.1002184 PMID:27898664

Diarra, M. S., Giguere, K., Malouin, F., Lefebvre, B., Bach, S., Delaquis, P., Aslam, M., Ziebell, K. A., & Roy, G. (2009). Genotype, serotype, and antibiotic resistance of sorbitol-negative Escherichia coli isolates from feedlot cattle. *Journal of Food Protection*, *72*(1), 28–36. doi:10.4315/0362-028X-72.1.28 PMID:19205460

Drulis-Kawa, Z., Majkowska-Skrobek, G., & Maciejewska, B. (2015). Bacteriophages and phage-derived proteins—Application approaches. *Current Medicinal Chemistry*, *22*(14), 1757–1773. doi:10.2174/0929867322666150209152851 PMID:25666799

Durante-Mangoni, E., Andini, R., & Zampino, R. (2019). Management of carbapenem-resistant Enterobacteriaceae infections. *Clinical Microbiology and Infection*, *25*(8), 943–950. doi:10.1016/j.cmi.2019.04.013 PMID:31004767

Fernández, L., Gutiérrez, D., García, P., & Rodríguez, A. (2019). The Perfect Bacteriophage for Therapeutic Applications-A Quick Guide. *Antibiotics (Basel, Switzerland)*, *8*(3), 126. Advance online publication. doi:10.3390/antibiotics8030126 PMID:31443585

Fernández, L., Gutiérrez, D., Rodríguez, A., & García, P. (2018). Application of Bacteriophages in the Agro-Food Sector: A Long Way Toward Approval. *Frontiers in Cellular and Infection Microbiology*, *8*, 296. doi:10.3389/fcimb.2018.00296 PMID:30186776

Fung, F., Wang, H. S., & Menon, S. (2018). Food safety in the 21st century. *Biomedical Journal*, *41*(2), 88–95. doi:10.1016/j.bj.2018.03.003 PMID:29866604

Gajbhiye, M., Kesharwani, J., Ingle, A., Gade, A., & Rai, M. (2009). Fungus-mediated synthesis of silver nanoparticles and their activity against pathogenic fungi in combination with fluconazole. *Nanomedicine (London)*, *5*(4), 382–386. doi:10.1016/j.nano.2009.06.005 PMID:19616127

Golkar, Z., Bagasra, O., & Pace, D. G. (2014). Bacteriophage therapy: A potential solution for the antibiotic resistance crisis. *Journal of Infection in Developing Countries*, *8*(2), 129–136. doi:10.3855/jidc.3573 PMID:24518621

Gradmann, C. (2016). Re-Inventing Infectious Disease: Antibiotic Resistance and Drug Development at the Bayer Company 1945-80. *Medical History*, *60*(2), 155–180. doi:10.1017/mdh.2016.2 PMID:26971595

Grandgirard, D., Loeffler, J. M., Fischetti, V. A., & Leib, S. L. (2008). Phage lytic enzyme Cpl-1 for antibacterial therapy in experimental pneumococcal meningitis. *The Journal of Infectious Diseases*, *197*(11), 1519–1522. doi:10.1086/587942 PMID:18471063

Gupta, A., Mumtaz, S., Li, C. H., Hussain, I., & Rotello, V. M. (2019). Combatting antibiotic-resistant bacteria using nanomaterials. *Chemical Society Reviews*, *48*(2), 415–427. doi:10.1039/C7CS00748E PMID:30462112

Hellberg, R. S., & Chu, E. (2016). Effects of climate change on the persistence and dispersal of foodborne bacterial pathogens in the outdoor environment: A review. *Critical Reviews in Microbiology*, *42*(4), 548–572. doi:10.3109/1040841X.2014.972335 PMID:25612827

Holban, A. M., Gestal, M. C., & Grumezescu, A. M. (2016). Control of biofilm-associated infections by signaling molecules and nanoparticles. *International Journal of Pharmaceutics*, *510*(2), 409–418. doi:10.1016/j.ijpharm.2016.02.044 PMID:26945736

Honarvar, Z., Hadian, Z., & Mashayekh, M. (2016). Nanocomposites in food packaging applications and their risk assessment for health. *Electronic Physician*, *8*(6), 2531–2538. doi:10.19082/2531 PMID:27504168

Hornish, R. E., & Kotarski, S. F. (2002). Cephalosporins in veterinary medicine - ceftiofur use in food animals. *Current Topics in Medicinal Chemistry*, *2*(7), 717–731. doi:10.2174/1568026023393679 PMID:12052187

Hu, Y., Liu, A., Vaudrey, J., Vaiciunaite, B., Moigboi, C., McTavish, S. M., Kearns, A., & Coates, A. (2015). Combinations of beta-lactam or aminoglycoside antibiotics with plectasin are synergistic against methicillin-sensitive and methicillin-resistant Staphylococcus aureus. *PLoS One*, *10*(2), e0117664. doi:10.1371/journal.pone.0117664 PMID:25692771

Huang, L., Li, D. Q., Lin, Y. J., Wei, M., Evans, D. G., & Duan, X. (2005). Controllable preparation of Nano-MgO and investigation of its bactericidal properties. *Journal of Inorganic Biochemistry*, *99*(5), 986–993. doi:10.1016/j.jinorgbio.2004.12.022 PMID:15833320

Javani, S., Lorca, R., Latorre, A., Flors, C., Cortajarena, A. L., & Somoza, Á. (2016). Antibacterial Activity of DNA-Stabilized Silver Nanoclusters Tuned by Oligonucleotide Sequence. *ACS Applied Materials & Interfaces*, *8*(16), 10147–10154. doi:10.1021/acsami.6b00670 PMID:27058628

Jiang, W., Mashayekhi, H., & Xing, B. (2009). Bacterial toxicity comparison between nano- and micro-scaled oxide particles. *Environmental Pollution*, *157*(5), 1619–1625. doi:10.1016/j.envpol.2008.12.025 PMID:19185963

Jiang, W., Rutherford, D., Vuong, T., & Liu, H. (2017). Nanomaterials for treating cardiovascular diseases: A review. *Bioactive Materials*, *2*(4), 185–198. doi:10.1016/j.bioactmat.2017.11.002 PMID:29744429

Jo, A., Ding, T., & Ahn, J. (2016). Synergistic antimicrobial activity of bacteriophages and antibiotics against Staphylococcus aureus. *Food Science and Biotechnology*, *25*(3), 935–940. doi:10.100710068-016-0153-0 PMID:30263357

Johanesen, P. A., Mackin, K. E., Hutton, M. L., Awad, M. M., Larcombe, S., Amy, J. M., & Lyras, D. (2015). Disruption of the Gut Microbiome: Clostridium difficile Infection and the Threat of Antibiotic Resistance. *Genes*, *6*(4), 1347–1360. doi:10.3390/genes6041347 PMID:26703737

Johnson, J. R., Johnston, B., Clabots, C., Kuskowski, M. A., & Castanheira, M. (2010). Escherichia coli sequence type ST131 as the major cause of serious multidrug-resistant E. coli infections in the United States. *Clinical Infectious Diseases*, *51*(3), 286–294. doi:10.1086/653932 PMID:20572763

Kaakoush, N. O., Castano-Rodriguez, N., Mitchell, H. M., & Man, S. M. (2015). Global Epidemiology of Campylobacter Infection. *Clinical Microbiology Reviews*, *28*(3), 687–720. doi:10.1128/CMR.00006-15 PMID:26062576

Kalan, L., & Wright, G. D. (2011). Antibiotic adjuvants: Multicomponent anti-infective strategies. *Expert Reviews in Molecular Medicine*, *13*, e5. doi:10.1017/S1462399410001766 PMID:21342612

Kamal, F., & Dennis, J. J. (2015). Burkholderia cepacia complex Phage-Antibiotic Synergy (PAS): Antibiotics stimulate lytic phage activity. *Applied and Environmental Microbiology*, *81*(3), 1132–1138. doi:10.1128/AEM.02850-14 PMID:25452284

Karmali, M. A., Gannon, V., & Sargeant, J. M. (2010). Verocytotoxin-producing Escherichia coli (VTEC). *Veterinary Microbiology*, *140*(3-4), 360–370. doi:10.1016/j.vetmic.2009.04.011 PMID:19410388

Katz, E., & Willner, I. (2004). Integrated nanoparticle-biomolecule hybrid systems: Synthesis, properties, and applications. *Angewandte Chemie International Edition in English*, *43*(45), 6042–6108. doi:10.1002/anie.200400651 PMID:15538757

Khandelia, H., & Kaznessis, Y. N. (2007). Cation-pi interactions stabilize the structure of the antimicrobial peptide indolicidin near membranes: Molecular dynamics simulations. *The Journal of Physical Chemistry B*, *111*(1), 242–250. doi:10.1021/jp064776j PMID:17201448

Kim, K. J., Sung, W. S., Suh, B. K., Moon, S. K., Choi, J. S., Kim, J. G., & Lee, D. G. (2009). Antifungal activity and mode of action of silver nano-particles on Candida albicans. *Biometals*, *22*(2), 235–242. doi:10.100710534-008-9159-2 PMID:18769871

Kim, M. H., Jeong, H., Sim, Y. M., Lee, S., Yong, D., Ryu, C. M., & Choi, J. Y. (2020). Using comparative genomics to understand molecular features of carbapenem-resistant Acinetobacter baumannii from South Korea causing invasive infections and their clinical implications. *PLoS One*, *15*(2), e0229416. doi:10.1371/journal.pone.0229416 PMID:32084241

Koper, O. B., Klabunde, J. S., Marchin, G. L., Klabunde, K. J., Stoimenov, P., & Bohra, L. (2002). Nanoscale powders and formulations with biocidal activity toward spores and vegetative cells of bacillus species, viruses, and toxins. *Current Microbiology*, *44*(1), 49–55. doi:10.100700284-001-0073-x PMID:11727041

Kutateladze, M., & Adamia, R. (2010). Bacteriophages as potential new therapeutics to replace or supplement antibiotics. *Trends in Biotechnology, 28*(12), 591–595. doi:10.1016/j.tibtech.2010.08.001 PMID:20810181

Labrie, S. J., Samson, J. E., & Moineau, S. (2010). Bacteriophage resistance mechanisms. *Nature Reviews. Microbiology, 8*(5), 317–327. doi:10.1038/nrmicro2315 PMID:20348932

Law, J. W., Ab Mutalib, N. S., Chan, K. G., & Lee, L. H. (2014). Rapid methods for the detection of foodborne bacterial pathogens: Principles, applications, advantages and limitations. *Frontiers in Microbiology, 5*, 770. doi:10.3389/fmicb.2014.00770 PMID:25628612

Li, Q., Mahendra, S., Lyon, D. Y., Brunet, L., Liga, M. V., Li, D., & Alvarez, P. J. (2008). Antimicrobial nanomaterials for water disinfection and microbial control: Potential applications and implications. *Water Research, 42*(18), 4591–4602. doi:10.1016/j.watres.2008.08.015 PMID:18804836

Li, Y., & Peng, N. (2019). Endogenous CRISPR-Cas System-Based Genome Editing and Antimicrobials: Review and Prospects. *Frontiers in Microbiology, 10*, 2471. doi:10.3389/fmicb.2019.02471 PMID:31708910

Lim, J. A., Shin, H., Heu, S., & Ryu, S. (2014). Exogenous lytic activity of SPN9CC endolysin against gram-negative bacteria. *Journal of Microbiology and Biotechnology, 24*(6), 803–811. doi:10.4014/jmb.1403.03035 PMID:24690638

Look, M., Bandyopadhyay, A., Blum, J. S., & Fahmy, T. M. (2010). Application of nanotechnologies for improved immune response against infectious diseases in the developing world. *Advanced Drug Delivery Reviews, 62*(4-5), 378–393. doi:10.1016/j.addr.2009.11.011 PMID:19922750

Lopez-Carrizales, M., Velasco, K. I., Castillo, C., Flores, A., Magana, M., Martinez-Castanon, G. A., & Martinez-Gutierrez, F. (2018). In Vitro Synergism of Silver Nanoparticles with Antibiotics as an Alternative Treatment in Multiresistant Uropathogens. *Antibiotics (Basel, Switzerland), 7*(2), 50. Advance online publication. doi:10.3390/antibiotics7020050 PMID:29921822

Love, M. J., Bhandari, D., Dobson, R. C. J., & Billington, C. (2018). Potential for Bacteriophage Endolysins to Supplement or Replace Antibiotics in Food Production and Clinical Care. *Antibiotics (Basel, Switzerland), 7*(1), 17. Advance online publication. doi:10.3390/antibiotics7010017 PMID:29495476

Lu, T. K., & Collins, J. J. (2007). Dispersing biofilms with engineered enzymatic bacteriophage. *Proceedings of the National Academy of Sciences of the United States of America, 104*(27), 11197–11202. doi:10.1073/pnas.0704624104 PMID:17592147

Lu, T. K., & Collins, J. J. (2009). Engineered bacteriophage targeting gene networks as adjuvants for antibiotic therapy. *Proceedings of the National Academy of Sciences of the United States of America, 106*(12), 4629–4634. doi:10.1073/pnas.0800442106 PMID:19255432

Ma, K., Dong, P., Liang, M., Yu, S., Chen, Y., & Wang, F. (2020). Facile Assembly of Multifunctional Antibacterial Nanoplatform Leveraging Synergistic Sensitization between Silver Nanostructure and Vancomycin. *ACS Applied Materials & Interfaces, 12*(6), 6955–6965. doi:10.1021/acsami.9b22043 PMID:31977179

Mahlapuu, M., Hakansson, J., Ringstad, L., & Bjorn, C. (2016). Antimicrobial Peptides: An Emerging Category of Therapeutic Agents. *Frontiers in Cellular and Infection Microbiology*, *6*, 194. doi:10.3389/fcimb.2016.00194 PMID:28083516

Man, S. M. (2011). The clinical importance of emerging Campylobacter species. *Nature Reviews. Gastroenterology & Hepatology*, *8*(12), 669–685. doi:10.1038/nrgastro.2011.191 PMID:22025030

Marshall, B. M., & Levy, S. B. (2011). Food animals and antimicrobials: Impacts on human health. *Clinical Microbiology Reviews*, *24*(4), 718–733. doi:10.1128/CMR.00002-11 PMID:21976606

McEwen, S. A. (2006). Antibiotic use in animal agriculture: What have we learned and where are we going? *Animal Biotechnology*, *17*(2), 239–250. doi:10.1080/10495390600957233 PMID:17127534

Miro, E., Verges, C., Garcia, I., Mirelis, B., Navarro, F., Coll, P., Prats, G., & Martinez-Martinez, L. (2004). Resistencia a quinolonas y betalactámicos en Salmonella enterica, y su relación con mutaciones en las topoisomerasas, alteraciones en la permeabilidad celular y expresión de un mecanismo de expulsión activa. *Enfermedades Infecciosas y Microbiologia Clinica*, *22*(4), 204–211. doi:10.1016/S0213-005X(04)73067-0 PMID:15056435

Mohamed Hamouda, I. (2012). Current perspectives of nanoparticles in medical and dental biomaterials. *Journal of Biomedical Research*, *26*(3), 143–151. doi:10.7555/JBR.26.20120027 PMID:23554743

Moye, Z. D., Woolston, J., & Sulakvelidze, A. (2018). Bacteriophage Applications for Food Production and Processing. *Viruses*, *10*(4), 205. Advance online publication. doi:10.3390/v10040205 PMID:29671810

Nagamune, T. (2017). Biomolecular engineering for nanobio/bionanotechnology. *Nano Convergence*, *4*(1), 9. doi:10.118640580-017-0103-4 PMID:28491487

Nelson, D. C., Schmelcher, M., Rodriguez-Rubio, L., Klumpp, J., Pritchard, D. G., Dong, S., & Donovan, D. M. (2012). Endolysins as antimicrobials. *Advances in Virus Research*, *83*, 299–365. doi:10.1016/B978-0-12-394438-2.00007-4 PMID:22748813

Newman, L., Rowley, J., Vander Hoorn, S., Wijesooriya, N. S., Unemo, M., Low, N., Stevens, G., Gottlieb, S., Kiarie, J., & Temmerman, M. (2015). Global Estimates of the Prevalence and Incidence of Four Curable Sexually Transmitted Infections in 2012 Based on Systematic Review and Global Reporting. *PLoS One*, *10*(12), e0143304. doi:10.1371/journal.pone.0143304 PMID:26646541

Nino-Martinez, N., Salas Orozco, M. F., Martinez-Castanon, G. A., Torres Mendez, F., & Ruiz, F. (2019). Molecular Mechanisms of Bacterial Resistance to Metal and Metal Oxide Nanoparticles. *International Journal of Molecular Sciences*, *20*(11), 2808. Advance online publication. doi:10.3390/ijms20112808 PMID:31181755

Niño-Martínez, N., Salas Orozco, M. F., Martínez-Castañón, G. A., Torres Méndez, F., & Ruiz, F. (2019). Molecular Mechanisms of Bacterial Resistance to Metal and Metal Oxide Nanoparticles. *International Journal of Molecular Sciences*, *20*(11), 2808. Advance online publication. doi:10.3390/ijms20112808 PMID:31181755

O'Sullivan, L., Buttimer, C., McAuliffe, O., Bolton, D., & Coffey, A. (2016). Bacteriophage-based tools: Recent advances and novel applications. *F1000 Research*, *5*, 2782. doi:10.12688/f1000research.9705.1 PMID:27990274

Peleg, A. Y., Seifert, H., & Paterson, D. L. (2008). Acinetobacter baumannii: Emergence of a successful pathogen. *Clinical Microbiology Reviews*, *21*(3), 538–582. doi:10.1128/CMR.00058-07 PMID:18625687

Pires, D. P., Cleto, S., Sillankorva, S., Azeredo, J., & Lu, T. K. (2016). Genetically Engineered Phages: A Review of Advances over the Last Decade. *Microbiology and Molecular Biology Reviews*, *80*(3), 523–543. doi:10.1128/MMBR.00069-15 PMID:27250768

Prestinaci, F., Pezzotti, P., & Pantosti, A. (2015). Antimicrobial resistance: A global multifaceted phenomenon. *Pathogens and Global Health*, *109*(7), 309–318. doi:10.1179/2047773215Y.0000000030 PMID:26343252

Pursey, E., Sunderhauf, D., Gaze, W. H., Westra, E. R., & van Houte, S. (2018). CRISPR-Cas antimicrobials: Challenges and future prospects. *PLoS Pathogens*, *14*(6), e1006990. doi:10.1371/journal.ppat.1006990 PMID:29902258

Qadir, M. I., & Chauhdary, Z. (2018). Antibacterial Activity of Novel Strains of Bacteriophages: An Experimental Approach. *Critical Reviews in Eukaryotic Gene Expression*, *28*(1), 1–12. doi:10.1615/CritRevEukaryotGeneExpr.2017019608 PMID:29773011

Rau, R. B., de Lima-Morales, D., Wink, P. L., Ribeiro, A. R., Martins, A. F., & Barth, A. L. (2018). Emergence of mcr-1 Producing Salmonella enterica serovar Typhimurium from Retail Meat: First Detection in Brazil. *Foodborne Pathogens and Disease*, *15*(1), 58–59. doi:10.1089/fpd.2017.2346 PMID:29048947

Reddy, K. M., Feris, K., Bell, J., Wingett, D. G., Hanley, C., & Punnoose, A. (2007). Selective toxicity of zinc oxide nanoparticles to prokaryotic and eukaryotic systems. *Applied Physics Letters*, *90*(213902), 2139021–2139023. doi:10.1063/1.2742324 PMID:18160973

Saleh, M. M., & Petri, W. A. Jr. (2019). Type 3 Immunity during Clostridioides difficile Infection: Too Much of a Good Thing? *Infection and Immunity*, *88*(1), e00306-19. Advance online publication. doi:10.1128/IAI.00306-19 PMID:31570564

Sandhu, B. K., & McBride, S. M. (2018). Clostridioides difficile. *Trends in Microbiology*, *26*(12), 1049–1050. doi:10.1016/j.tim.2018.09.004 PMID:30297117

Scallan, E., Hoekstra, R. M., Angulo, F. J., Tauxe, R. V., Widdowson, M. A., Roy, S. L., Jones, J. L., & Griffin, P. M. (2011). Foodborne illness acquired in the United States—Major pathogens. *Emerging Infectious Diseases*, *17*(1), 7–15. doi:10.3201/eid1701.P11101 PMID:21192848

Schmelcher, M., Donovan, D. M., & Loessner, M. J. (2012). Bacteriophage endolysins as novel antimicrobials. *Future Microbiology*, *7*(10), 1147–1171. doi:10.2217/fmb.12.97 PMID:23030422

Schmelcher, M., Powell, A. M., Camp, M. J., Pohl, C. S., & Donovan, D. M. (2015). Synergistic streptococcal phage lambdaSA2 and B30 endolysins kill streptococci in cow milk and in a mouse model of mastitis. *Applied Microbiology and Biotechnology*, *99*(20), 8475–8486. doi:10.100700253-015-6579-0 PMID:25895090

Shabbir, M. A. B., Shabbir, M. Z., Wu, Q., Mahmood, S., Sajid, A., Maan, M. K., Ahmed, S., Naveed, U., Hao, H., & Yuan, Z. (2019). CRISPR-cas system: Biological function in microbes and its use to treat antimicrobial resistant pathogens. *Annals of Clinical Microbiology and Antimicrobials, 18*(1), 21. doi:10.118612941-019-0317-x PMID:31277669

Shang, D., Liu, Y., Jiang, F., Ji, F., Wang, H., & Han, X. (2019). Synergistic Antibacterial Activity of Designed Trp-Containing Antibacterial Peptides in Combination With Antibiotics Against Multidrug-Resistant Staphylococcus epidermidis. *Frontiers in Microbiology, 10*, 2719. doi:10.3389/fmicb.2019.02719 PMID:31824473

Sharma, C., Dhiman, R., Rokana, N., & Panwar, H. (2017). Nanotechnology: An Untapped Resource for Food Packaging. *Frontiers in Microbiology, 8*, 1735. doi:10.3389/fmicb.2017.01735 PMID:28955314

Sharma, V. K., Johnson, N., Cizmas, L., McDonald, T. J., & Kim, H. (2016). A review of the influence of treatment strategies on antibiotic resistant bacteria and antibiotic resistance genes. *Chemosphere, 150*, 702–714. doi:10.1016/j.chemosphere.2015.12.084 PMID:26775188

Shimizu, A., Takada, H., Koike, T., Takeshita, A., Saha, M., Rinawati, Nakada, N., Murata, A., Suzuki, T., Suzuki, S., Chiem, N. H., Tuyen, B. C., Viet, P. H., Siringan, M. A., Kwan, C., Zakaria, M. P., & Reungsang, A. (2013). Ubiquitous occurrence of sulfonamides in tropical Asian waters. *The Science of the Total Environment, 452-453*, 108–115. doi:10.1016/j.scitotenv.2013.02.027 PMID:23500404

Silbergeld, E. K., Graham, J., & Price, L. B. (2008). Industrial food animal production, antimicrobial resistance, and human health. *Annual Review of Public Health, 29*(1), 151–169. doi:10.1146/annurev.publhealth.29.020907.090904 PMID:18348709

Singh, P., Garg, A., Pandit, S., Mokkapati, V., & Mijakovic, I. (2018). Antimicrobial Effects of Biogenic Nanoparticles. *Nanomaterials (Basel, Switzerland), 8*(12), 1009. Advance online publication. doi:10.3390/nano8121009 PMID:30563095

Singh, P., Kim, Y. J., Zhang, D., & Yang, D. C. (2016). Biological Synthesis of Nanoparticles from Plants and Microorganisms. *Trends in Biotechnology, 34*(7), 588–599. doi:10.1016/j.tibtech.2016.02.006 PMID:26944794

Smartt, A. E., Xu, T., Jegier, P., Carswell, J. J., Blount, S. A., Sayler, G. S., & Ripp, S. (2012). Pathogen detection using engineered bacteriophages. *Analytical and Bioanalytical Chemistry, 402*(10), 3127–3146. doi:10.100700216-011-5555-5 PMID:22101465

Soothill, G., Hu, Y., & Coates, A. (2013). Can we prevent antimicrobial resistance by using antimicrobials better? *Pathogens (Basel, Switzerland), 2*(2), 422–435. doi:10.3390/pathogens2020422 PMID:25437042

Stanković, A., Dimitrijević, S., & Uskoković, D. (2013). Influence of size scale and morphology on antibacterial properties of ZnO powders hydrothemally synthesized using different surface stabilizing agents. *Colloids and Surfaces. B, Biointerfaces, 102*, 21–28. doi:10.1016/j.colsurfb.2012.07.033 PMID:23010107

Strawn, L. K., Fortes, E. D., Bihn, E. A., Nightingale, K. K., Grohn, Y. T., Worobo, R. W., Wiedmann, M., & Bergholz, P. W. (2013). Landscape and meteorological factors affecting prevalence of three food-borne pathogens in fruit and vegetable farms. *Applied and Environmental Microbiology*, *79*(2), 588–600. doi:10.1128/AEM.02491-12 PMID:23144137

Teixeira, V., Feio, M. J., & Bastos, M. (2012). Role of lipids in the interaction of antimicrobial peptides with membranes. *Progress in Lipid Research*, *51*(2), 149–177. doi:10.1016/j.plipres.2011.12.005 PMID:22245454

Tock, M. R., & Dryden, D. T. (2005). The biology of restriction and anti-restriction. *Current Opinion in Microbiology*, *8*(4), 466–472. doi:10.1016/j.mib.2005.06.003 PMID:15979932

Tshokey, T., Tshering, T., Pradhan, A. R., Adhikari, D., Sharma, R., Gurung, K., Dorji, T., Wangmo, S., Dorji, U., & Wangdi, K. (2018). Antibiotic resistance in Neisseria gonorrhoea and treatment outcomes of gonococcal urethritis suspected patients in two large hospitals in Bhutan, 2015. *PLoS One*, *13*(8), e0201721. doi:10.1371/journal.pone.0201721 PMID:30067836

Tyers, M., & Wright, G. D. (2019). Drug combinations: A strategy to extend the life of antibiotics in the 21st century. *Nature Reviews. Microbiology*, *17*(3), 141–155. doi:10.103841579-018-0141-x PMID:30683887

van Duin, D., & Doi, Y. (2017). The global epidemiology of carbapenemase-producing Enterobacteriaceae. *Virulence*, *8*(4), 460–469. doi:10.1080/21505594.2016.1222343 PMID:27593176

Velusamy, V., Arshak, K., Korostynska, O., Oliwa, K., & Adley, C. (2010). An overview of food-borne pathogen detection: In the perspective of biosensors. *Biotechnology Advances*, *28*(2), 232–254. doi:10.1016/j.biotechadv.2009.12.004 PMID:20006978

Vieira, A., Ramesh, A., Seddon, A. M., & Karlyshev, A. V. (2017). CmeABC Multidrug Efflux Pump Contributes to Antibiotic Resistance and Promotes Campylobacter jejuni Survival and Multiplication in Acanthamoeba polyphaga. *Applied and Environmental Microbiology*, *83*(22), e01600-17. Advance online publication. doi:10.1128/AEM.01600-17 PMID:28916560

Wang, L., Hu, C., & Shao, L. (2017). The antimicrobial activity of nanoparticles: Present situation and prospects for the future. *International Journal of Nanomedicine*, *12*, 1227–1249. doi:10.2147/IJN.S121956 PMID:28243086

Wang, R., Schmidt, J. W., Harhay, D. M., Bosilevac, J. M., King, D. A., & Arthur, T. M. (2017). Biofilm Formation, Antimicrobial Resistance, and Sanitizer Tolerance of Salmonella enterica Strains Isolated from Beef Trim. *Foodborne Pathogens and Disease*, *14*(12), 687–695. doi:10.1089/fpd.2017.2319 PMID:29035101

Whaley, S. R., English, D. S., Hu, E. L., Barbara, P. F., & Belcher, A. M. (2000). Selection of peptides with semiconductor binding specificity for directed nanocrystal assembly. *Nature*, *405*(6787), 665–668. doi:10.1038/35015043 PMID:10864319

Wong, D., Nielsen, T. B., Bonomo, R. A., Pantapalangkoor, P., Luna, B., & Spellberg, B. (2017). Clinical and Pathophysiological Overview of Acinetobacter Infections: A Century of Challenges. *Clinical Microbiology Reviews*, *30*(1), 409–447. doi:10.1128/CMR.00058-16 PMID:27974412

Woolhouse, M. E., & Ward, M. J. (2013). Microbiology. Sources of antimicrobial resistance. *Science, 341*(6153), 1460–1461. doi:10.1126cience.1243444 PMID:24030495

Wu, X., Li, Z., Li, X., Tian, Y., Fan, Y., Yu, C., Zhou, B., Liu, Y., Xiang, R., & Yang, L. (2017). Synergistic effects of antimicrobial peptide DP7 combined with antibiotics against multidrug-resistant bacteria. *Drug Design, Development and Therapy, 11*, 939–946. doi:10.2147/DDDT.S107195 PMID:28356719

Yehl, K., Lemire, S., Yang, A. C., Ando, H., Mimee, M., Torres, M. T., de la Fuente-Nunez, C., & Lu, T. K. (2019). Engineering Phage Host-Range and Suppressing Bacterial Resistance through Phage Tail Fiber Mutagenesis. *Cell, 179*(2), 459–469 e459. doi:10.1016/j.cell.2019.09.015 PMID:31585083

Yonezawa, R., Kuwana, T., Kawamura, K., & Inamo, Y. (2015). Invasive Community-Acquired Methicillin-Resistant Staphylococcus aureus in a Japanese Girl with Disseminating Multiple Organ Infection: A Case Report and Review of Japanese Pediatric Cases. *Case Reports in Pediatrics, 291025*, 1–6. Advance online publication. doi:10.1155/2015/291025 PMID:26819794

Yosef, I., Manor, M., Kiro, R., & Qimron, U. (2015). Temperate and lytic bacteriophages programmed to sensitize and kill antibiotic-resistant bacteria. *Proceedings of the National Academy of Sciences of the United States of America, 112*(23), 7267–7272. doi:10.1073/pnas.1500107112 PMID:26060300

Zhao, X., Hilliard, L. R., Mechery, S. J., Wang, Y., Bagwe, R. P., Jin, S., & Tan, W. (2004). A rapid bioassay for single bacterial cell quantitation using bioconjugated nanoparticles. *Proceedings of the National Academy of Sciences of the United States of America, 101*(42), 15027–15032. doi:10.1073/pnas.0404806101 PMID:15477593

Zharkova, M. S., Orlov, D. S., Golubeva, O. Y., Chakchir, O. B., Eliseev, I. E., Grinchuk, T. M., & Shamova, O. V. (2019). Application of Antimicrobial Peptides of the Innate Immune System in Combination With Conventional Antibiotics-A Novel Way to Combat Antibiotic Resistance? *Frontiers in Cellular and Infection Microbiology, 9*, 128. doi:10.3389/fcimb.2019.00128 PMID:31114762

Chapter 6
Medicine at Nanoscale:
A New Paradigm for the Emerging Trend of Multidrug Resistance in Bacteria

Chockaiyan Usha
Lady Doak College, India

Pandi Sakthieaswari
Lady Doak College, India

Parameswaran Kiruthika Lakshmi
The Madura College, India

ABSTRACT

Antibiotic therapy for life-threatening bacterial infections has greatly enhanced the lifespan and health of human beings. But continued increase in the improper usage of antibiotics and emergence of multidrug resistant bacteria (MDR) remains a major reason for the morbidity and mortality worldwide. Delay in the detection of pathogens, poor penetration power of the antimicrobial agent, and susceptibility for mutational resistance also resulted in clinical failure of conventional antibiotics. Innovation of nanoparticles (NPs) serves as a promising strategy to overcome MDR bacteria. The site-specific action of NPs, high penetration power, minimal dosage, and multidimensional mechanism makes NPs highly efficient to tackle MDR bacteria than conventional drugs. This chapter presents a brief overview on various types of NPs, strategies to combat drug resistance, mode of action of NPs as antibacterial agents, applicability in various medical fields, clinical manifestation, challenges in clinical translation, and future prospects.

INTRODUCTION

Infectious diseases remain an important cause of mortality worldwide. In recent days, researchers have focused their attention towards the development of novel antibiotics for chemotherapeutic applications. Despite the intense research in discovery of antibiotics, bacterial resistance to antibiotics is also becoming increasingly important in clinical practice (Mann, 1999). Bacteria present along with infectious micro-

DOI: 10.4018/978-1-7998-5049-6.ch006

organisms have been reported to develop protection against bioactive molecules and develop resistance against their own antibiotics. Such antibiotic resistance mechanism by bacteria can provide protection against chemotherapeutics and thereby results in development of difficult situation to manage the infection.

Tenover and Hughes (1996) have reported that the reason for increased resistance of bacteria to antibiotics is due to inappropriate, indiscriminate and incomplete course of antibiotic therapy in humans. Russell *et al.,* (1998) and McBain and Gilbert (2001) have reported that there exists a possible link between bacterial antibiotic and biocide cross-resistance.

Annually 100,000 tons of antibiotics are produced commercially to treat bacterial infections. In spite of large production of antibiotics, there also exists a pressing need for the development of a novel antibiotic against multi-drug resistance (Alavi & Rai, 2019). In this context, developments in the field of nanotechnology have been found to offer new perspectives in combating antibacterial resistance. The nanoparticles (NPs) were found to be useful in almost all arenas like medicine, agriculture, cosmetics, industry due to the nano size of particles.

Sathishkumar *et al.,* (2018) reported that metal nanoparticles have antibacterial activities based on the physicochemical properties of the respective metal. Further, the elemental structure, size and shape of NPs play a key role in conferring the antibacterial activity. Nanoparticles are synthesized by top-down (physical) and bottom-up (chemical) approach (Alavi & Rai, 2019). In bottom-up method of fabrication of NPs capping agents (polymers, citrate, thiol) and reducing agents (sodium citrate, ascorbic acid) are used (Usha & Rachel, 2014). In green synthesis of NPs, plant extracts, bacteria, fungi, algae, yeasts are used as reducing and capping agents (Taran *et al.,* 2016). This chapter presents and discusses the various types of NPs, strategies to combat drug resistance, mode of action of NPs as antibacterial agents and current developments in medical arena.

EMERGENCE OF ANTIBIOTIC RESISTANCE

Knowledge on the antibiotic resistance mechanism by bacteria is essential in order to design better molecules to circumvent the resistance strategy of the bacteria. Further, Savjani *et al.,* (2009) also reported several mechanisms of antibiotic resistance by bacteria. The major antibiotic resistance mechanisms are by inactivating enzymes that modify the drug, antibiotic efflux pumps and alteration of drug-binding site to the bacterium.

BACTERIAL RESISTANCE MECHANISMS TO ANTIBIOTICS

Tenover (2006) reported that antibacterial agents act on bacteria by inhibiting their cell wall construction, disrupting the bacterial cell membrane, inhibiting the synthesis of protein and interfering the synthesis of nucleic acid. Based on the mechanism of action, the antibiotics are classified as bacteriostatic and bactericidal (Neu, 1992).

Antibiotic resistance occurs in three steps such as acquisition of resistance genes by bacteria, expression of resistance genes and selection for microbes expressing the resistance genes (Pelgrift and Friedman, 2013). Ganjian *et al.,* (2012) reported that antibiotic resistance genes in bacteria are also acquired by spontaneous mutation of existing genes.

The important bacterial resistance mechanism to antibacterials is increased efflux and reduced uptake of antibiotics. The mechanism of reduced uptake and increased efflux of antibiotics such as tetracyclines, sulfonamides, quinolones, aminoglycosides, chloramphenicol was also confirmed by Deck and Winston (2012). Pelgrift and Friedman (2013) reported that *E.coli* expresses multi-drug resistance by using the energy source derived by transmembrane proton gradient. Outer membrane present in Gram negative bacteria acts as an extra barrier against antibiotics (Poole, 2002). Jayaraman (2009) reported that the increased efflux is also a antibiotic resistance mechanism of *Enterococcus faecalis* against quinupristin and dalfopristin. Poole (2002) reported that TetA, TetB and TetK genes encoding tetracycline efflux pumps are being expressed in both gram positive and gram negative bacteria.

MODE OF ACTION OF NANOPARTICLES TO OVERCOME ANTIBACTERIAL RESISTANCE

A vast number of NPs have been synthesized from various materials for therapeutic applications. In comparison with other conventional medicines/antibiotics, the mode of action of each nanoparticle is highly distinct from one another (Figure 1). This complex antimicrobial mechanism and multi-target approach of NPs may serve as an efficient strategy to combat multi-drug resistance in bacteria (Zaidi *et al.*, 2017).

Figure 1.

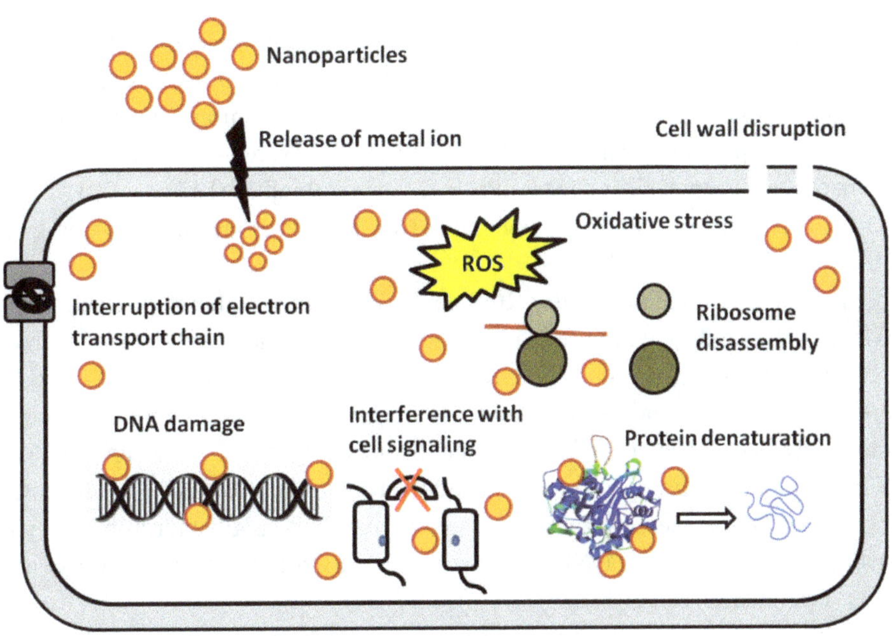

Generally, NPs exert its antibacterial activity by the following mechanisms 1) release of metal ions; 2) ROS-mediated stimulation of oxidative stress; 3) direct adhesion of NPs to bacterial cell wall; 4) inhibition of biofilm formation by interrupting bacterial communication system.

Release of Metal Ions

Metal and metal oxide NPs progressively release metal ions in aqueous environment which are absorbed through the bacterial cell membrane. The dissolved metal ions reacts with the functional groups [carboxyl (–COOH), amino (–NH), and mercapto (–SH) groups] found in nucleic acids, proteins and alters its cell structure and enzyme activities, which in turn disrupt normal physiological activity of the bacteria (Wang *et al.,* 2017). Efficiency of each metal ion is unique towards different microorganisms and its antibacterial potency depends on accomplishing the threshold NPs concentration. For instance, silver ions released by silver NPs interact with sulfhydryl groups in various enzymes leading to protein coagulation and ultimately resulted in altered physiology of the bacteria.

In Gram-positive bacteria, the silver ions prevent cell wall biosynthesis. It can also disrupt DNA replication and cell division in the bacterium (Dakal *et al.,* 2016). Similarly, copper ions can invade into the bacterium and can interact with carboxyl and amine groups in macromolecules resulting in alteration of enzymes involved in electron transport, nitrogen metabolism and can hinder denitrification process (Su *et al.,* 2015). Hemeg (2017) reported that the metal ions released by Au NPs can penetrate and disrupt the cell membrane of bacteria. However, the antibacterial activity mediated by metal ions depends on the pH of lipid vesicles and oxidation state of the metal being used (Tamayo *et al.,* 2014).

ROS-Mediated Stimulation of Oxidative Stress

ROS formation is the chief mechanism accountable for the antibacterial activity of NPs. Cellular oxidative metabolism of bacteria resulted in the formation of ROS, which is essential for its survival, proliferation, differentiation and signaling pathways. The generation and removal of ROS by the bacterial cells are stable under normal physiological conditions (Hemeg, 2017). But, excessive ROS production can destabilize the redox potential and creates oxidative stress which inhibits bacterial growth by limiting amino acid synthesis, lipid peroxidation and DNA replication (Wang *et al.,* 2017). Oxidative stress can also alter the membrane permeability that results in intense cell injury (Li *et al.,* 2012). Among the various types of ROS, singlet oxygen and hydroxyl radical attributes for extensive physiological damage and bacterial cell death (Zaidi *et al.,* 2017).

Each type of NPs can generate specific ROS depending on their chemical characteristics (Wang *et al.,* 2017). Metallic NPs exhibited increased antimicrobial efficiency as they are capable of producing ROS excessively and their surface-to-volume ratio was also higher when compared to other NPs. Metal oxide NPs like TiO_2 NPs and ZnO NPs induces ROS generation that interferes with cellular functions through photo catalysis (Ahmed *et al.,* 2017). After absorbing light, TiO_2 NPs can produce electron-hole pairs, which in turn react with water and air to produce ROS (Wu *et al.,* 2011). But, ZnO NPs generate minimal amounts of OH in the dark which is responsible for its antibacterial efficacy (Prasanna and Vijayaraghavan, 2015). Copper oxide NPs is known to produce all four types of ROS to induce oxidative stress in bacteria (Kumari *et al.,* 2017).

Zhang *et al.,* (2013) described the ability of uncoated Au NPs to generate ROS and metal ion with greater antibacterial efficacy, when subjected to UV irradiation at 365 nm. Similarly, Au NPs along with laser treatment also resulted in rapid loss of membrane permeability in bacterial cells. UV irradiation has been proved to be effective in enhancing the antibacterial potential of other metallic NPs. Li *et al.,* (2012) confirmed the ability of UV radiation to stimulate excessive production of free radicals by titanium oxide (TiO) and zinc oxide (ZnO) NPs, which resulted in enhanced oxidative stress and bacterial mortality. Ulloa- Ogaz *et al.,* (2017) demonstrated that copper oxide NPs stimulated intracellular signaling cascade within the bacteria that triggered oxidative stress and cell lysis.

Direct Adhesion of NPs to Bacterial Cell Wall

Direct interaction of NPs with bacterial cell wall accounts for non-oxidative mechanisms. This interaction can be achieved by vanderwaals, electrostatic, hydrophobic forces or receptor-ligand contact. Plasma membrane and cell wall of the bacterium act as defensive shield that protect the cell from adverse environment. The adsorption capacity of NPs differs among Gram-negative and Gram-positive bacteria as they have unique cell membrane components (Lesniak *et al.,* 2013).

In Gram-negative bacteria, presence of phospholipid bilayer and lipopolysaccharides (LPS) imparts highly negative-charge to the cells which attracts positively charged NPs. While, teichoic acid found in the cell wall of Gram-positive bacteria, attracts and binds the NPs evenly along its phosphate chain to avoid aggregation of NPs. Due to the variation in porosity; antibacterial activity of NPs is more prominent in Gram-positive bacteria when compared to Gram-negative bacteria (Wang *et al.,* 2017)

In addition, surface charge of the NPs plays a significant role in its cell adhesion capability. For instance, positively charged Ag NPs interacts with negatively charged bacterial cell membrane and causes alteration in cellular morphology of bacteria such as shrinkage of cytoplasm and cell wall disruption (Ghosh *et al.,* 2012). Whereas, the interaction of negatively charged ZnO NPs with both Gram-negative and Gram-positive cell membrane is accompanied via receptor-ligand contact (Jiang *et al.,* 2009). Functional groups such as amide, phosphate, carboxyl, hydroxyl groups found in the bacterial cell wall acts as binding site for the metal oxide NPs (Omoike *et al.,* 2004). After adhesion to the cell surface, the NPs inhibits activity of the enzymes involved in physiology and cell metabolism, hence leads to cellular dysfunction (Wang *et al.,* 2017). Aggregation of NPs may leads to formation of irregular shaped perforations and pits in cell surface and thereby affects metabolic reactions (Ansari *et al.,* 2014).

Inhibition of Biofilm Formation by Interrupting Bacterial Communication System

Infections caused by biofilm producing bacteria became a major medical challenge worldwide. Hetrick *et al.,* (2009) reported that the dosage of the antibiotic has to be increased upto 1000 times for effective control of biofilm producers. Bacterial cells embedded in the biofilm could signal each other by a ubiquitous regulatory system called quorum sensing and leads to phenotypic variations which aggravate its virulence towards the host cells (DeQueiroz and Day, 2007). In this context, recent research works have highlighted that the NPs can prevent and overcome biofilm formation.

Adherence of bacterial colonies together on any solid surface bounded by a slimy extracellular polymeric substance (EPS) matrix is termed as biofilm formation. Biofilm acts as a shield which blocks penetration of antibiotic into the bacteria and also helps the bacteria to evade host immune responses (Lebeaux *et al.,* 2014). NPs are known for their capacity to disrupt cell membrane and thereby preventing biofilm formation (Kulshrestha *et al.,* 2017). Moreover, NPs can also interact with EPS matrix formed by the bacteria and can block the bacterial communication-quorum sensing.

Inhibition of biofilm formation can be achieved by interrupting the quorum sensing system of bacteria (Singh *et al.,* 2017). Quorum sensing in bacteria involves production of signal molecules for cell-to-cell communication, by which the bacterial colonies coordinate their gene expression to combat adverse environmental conditions (Rutherford and Bassler, 2012).

Knetsch and Koole (2011) reported the biofilm inhibition capability of silver NPs on medical instruments. Hetrick *et al.,* (2009) revealed that the nitric oxide NPs were efficient in killing *Staphylococcus epidermidis, Staphylococcus aureus, Candida albicans* and *Pseudomonas aeruginosa* that were present on the surface of biofilm. Further Roe *et al.,* (2008) reported the utility of Ag NPs to control biofilm caused by *S. aureus* on catheters. Hajipour *et al.,* (2012) documented the anti- biofilm activity of zinc oxide NPs in glass surfaces.

An external magnetic field was applied to enhance the antibacterial activity of superparamagnetic iron oxide NPs (SPIONS), against biofilm producing bacteria (Hajipour *et al.,* 2012). According to a study of Anghel *et al.,* (2012), textile dressings coated with Fe_3O_4 NPs inhibited biofilm formation by *C. albicans*. Benzyl penicillin encapsulated in cationic liposomes was found to inhibit *S. aureus* biofilms efficiently (Huh and Kwon, 2011).

ANTIBACTERIAL MECHANISMS OF VARIOUS NANOPARTICLES

Chitosan Containing Nanoparticles

Chitosan is considered as a biodegradable, biocompatible polymer, observed as safe for human dietary use and for wound dressing purpose. Through various routes of administration it has been used as a carrier in polymeric NPs for drug delivery. In order to attain a tremendous range of potential applications, the chemical functional groups of chitosan can be modified to achieve specific goals by making it a polymer. NPs with chitosan and chitosan derivatives typically contain a positive charge on its surface and muco-adhesive properties with which it can adhere to the mucus membranes and releases the drug in a constant manner. Chitosan-based NPs possess various benefits (Figure 2) particularly in drug delivery for treating of cancer, ocular infections gastrointestinal, pulmonary diseases and brain related issues (Figure 3). Chitosan displays low toxicity both *in vitro* and in *vivo* conditions (Munawar *et al.*, 2017).

Metal and Non-Metallic Nanoparticles

Currently NPs are considered as an alternative to the antibiotics; hence it creates a huge interest in different research communities. They found to have the potential of combating multi-drug-resistant bacteria (Pelgrift and Friedman, 2013). However, NPs such as metal, metal oxide, and non-metallic NPs exhibit diverse nature in its chemical composition, surface and physicochemical properties, which enable to be have numerous action (Beyth *et al.*, 2015). Most of the NPs kill pathogenic bacteria in two main pathways (1) membrane integrity disruption and (2) ROS production. Commonly, NPs bind to the bacterial cell membrane electrostatically damages the membrane, which results in the changes of membrane depolarization, membrane potential and loss of membrane integrity (Beyth *et al.*, 2015). Finally, bacteria suffer from an imbalance in transport, interruption of energy transduction leading to impaired respiration and cell lysis resulting in cell death (Pelgrift and Friedman 2013). The increase in resistance of the microbes towards antibacterial agents has resulted in serious health issues in the recent years. Most of the infectious bacteria are resistant to a minimum of one of the antibiotics which are generally used in elimination of the infection.

Figure 2. Major properties and activity of Chitosan in biomedicine

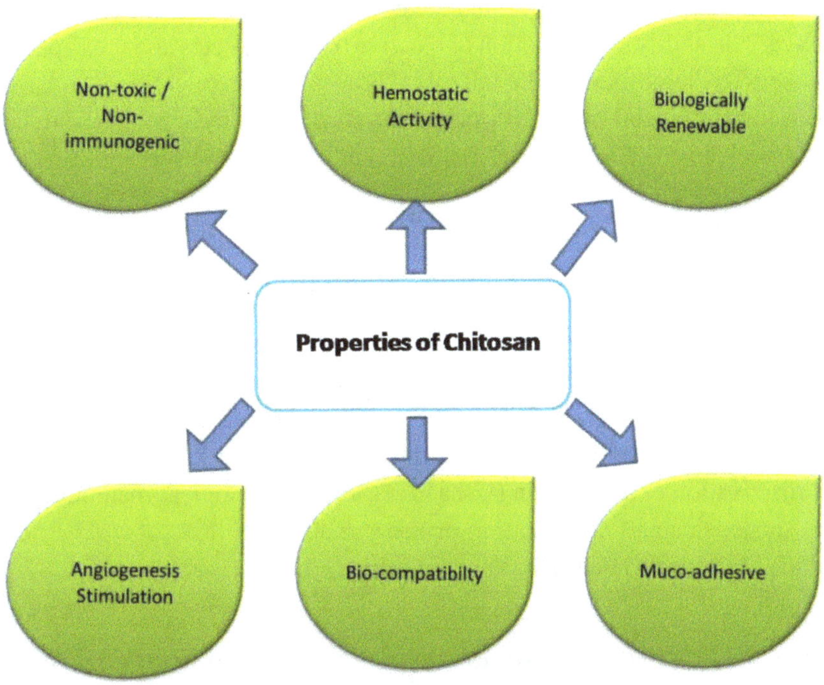

Figure 3. Therapeutic applications of Chitosan Nanoparticles

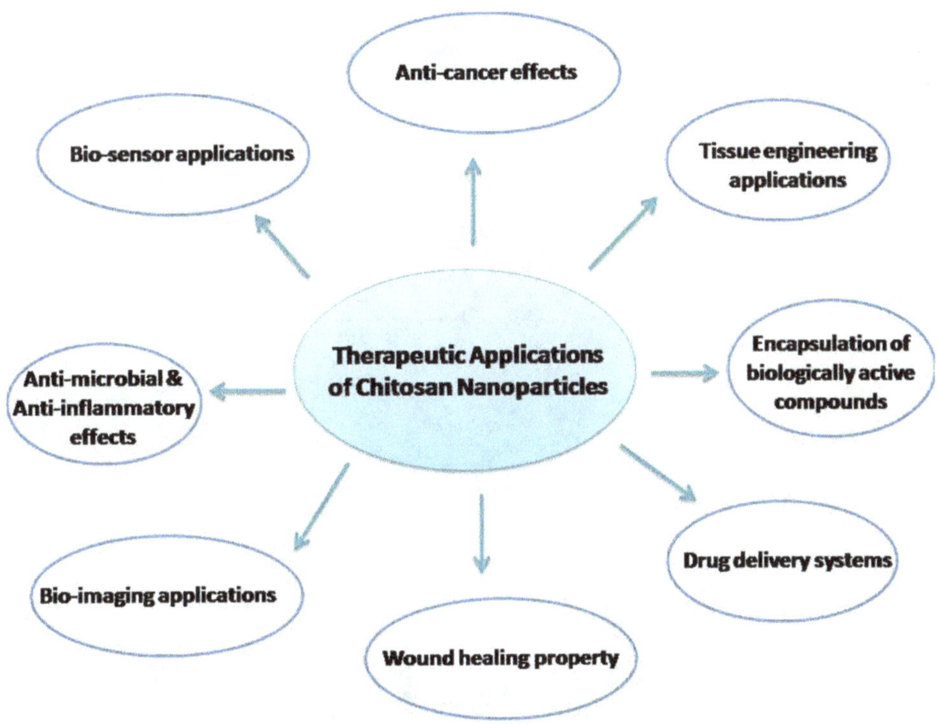

Metallic NPs are given much importance based on its small size, surface charge and large surface area. Some metallic NPs, increases their cellular interactions based on the different functional groups present on its surface. The ability of production of reactive oxygen species and the capability to bind with DNA or RNA by the NPs thereby, affect the microbial processes of replication which results in damage of bacteria. Various kinds of metal-based NPs have been recognized by several researchers in the recent years. The toxicity mechanism of metallic NPs differs within species to species (Singh *et al.*, 2019).

Silver Nanoparticles

Silver nanoparticles (Ag NPs) are considered to be the most important and fascinating nanoparticles among various other metallic NPs which are involved in the field of bio-nanomedical applications (Figure 4). The activity of Ag NPs depends mainly based on the morphology and structure of the Ag NPs which are controlled by shape and size of the particles (Zhang *et al.*, 2016). The oligo-dynamic effect of Ag NPs on microbes, results in the binding of ions to the reactive groups in bacterial cells causing precipitation and inactivation of a pathogen (Usha *et al.*, 2014). Biological methods are much effective in controlling the shape, size and optimization of the synthesis methods of Ag NPs compared to the chemical method of synthesis (Khodashenas and Ghorbani, 2015).

Figure 4. Application and Exposure of Silver Nanoparticles in various fields of applications

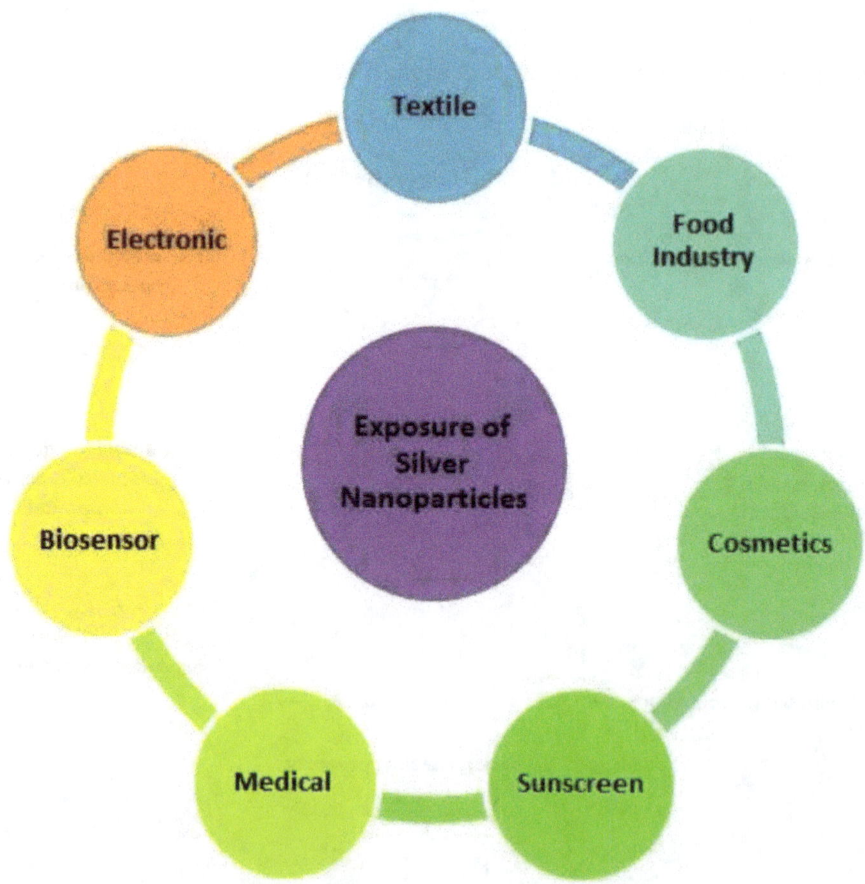

Zinc Oxide Nanoparticles (ZnO NPs)

Zinc oxide is an important ingredient of enzymes, sunscreens and ointments used for pain and itch relief. Microcrystals of ZnO NPs are effectual light absorbers in UV-A and UV-B region of spectra due to the presence of wide band gap. A biological function of ZnO NPs depends on its morphology, particle size, exposure time, concentration, pH and biocompatibility.

They are effective against *Bacillus subtilis*, *Bacillus megaterium*, *Staphylococcus aureus*, *Sarcina lutea*, *Escherichia coli*, *Pseudomonas aeruginosa*, *Klebsiella pneumonia*, *Pseudomonas vulgaris*, *Candida albicans*, and *Aspergillus niger.* ZnO NPs were activated by light, which which penetrate the bacterial cell wall through the process of diffusion. Based on the results of the images obtained from SEM and TEM, clearly depicts the disintegration of the cell membrane and accumulate in the cytoplasm, where they interact with biomolecules causing cell apoptosis which leads to bacterial cell death (Figure 5).

Figure 5. Action mechanism of Zinc Oxide Nanoparticles against a bacterial cell

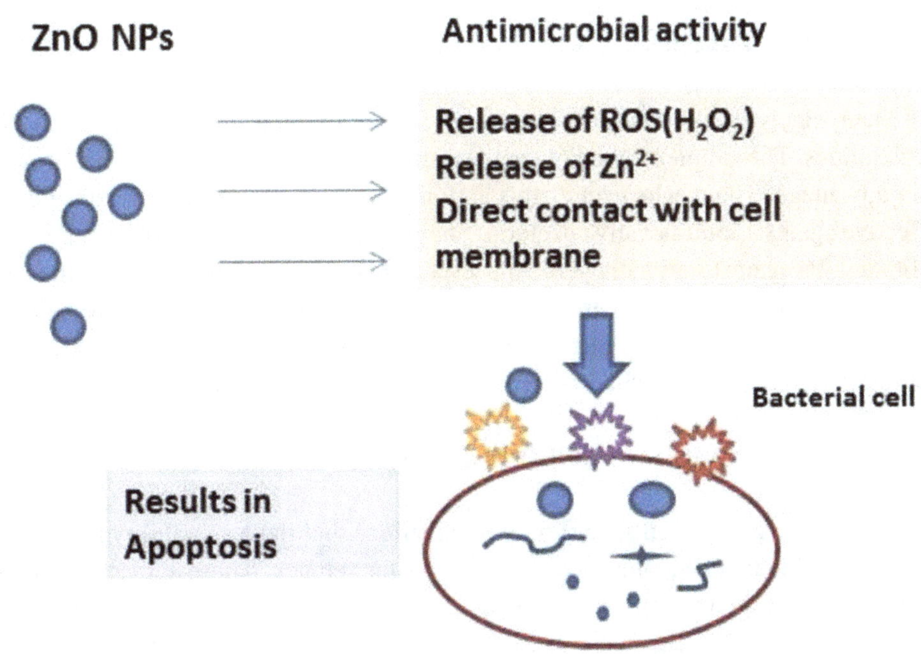

Recently, Pati *et al.*, (2014) observed that ZnO NPs disrupt bacterial cell membrane integrity, reduce cell surface hydrophobicity and disturbs the transcription mechanism of oxidative stress-resistance genes in bacteria. They enhance intracellular bacterial killing due the production of ROS. These NPs disrupt biofilm formation and inhibit hemolysis by hemolysin toxin produced that was produced by pathogens.

Copper Oxide Nanoparticles (CuO NPs)

CuO NPs possess numerous biological properties including effective antimicrobial action against a wide range of pathogens and drug resistant bacteria. These selective properties induced the scientists in conducting research with various approaches in field biomedical field and especially in modified textiles. It is believed that these nano systems could represent efficient alternatives in the development of smart systems which can be utilized for the detection of pathogens and for the treatment of infections (Grigore *et al.*, 2016). Zhang *et al.*, (2014) reported that the generation of ROS within bacterial cells is enhanced while using CuO water suspensions.

Magnesium Oxide Nanoparticles (MgO NPs)

MgO NPs is considered as a light metal based antimicrobial NPs, which can be metabolized and fully absorbed in the body. Nguyen *et al.*, (2018) stated the production of reactive oxygen species (ROS), Ca^{2+} ion concentrations, and quorum sensing contribute to the action mechanisms of magnesium oxide against planktonic bacteria, but transient alkaline pH of 7 to 10 or increased Mg^{2+} ion concentrations from 1 to 50 mM showed no inhibitory or killing effects on bacteria such as *S. epidermidis*. In general, while the concentrations of magnesium oxide increased, the adhesion densities of gram-negative bacteria, gram-positive bacteria, and yeasts get decreased.

Gold Nanoparticles (Au NPs)

Au NPs have been widely used in bio-nanotechnology, based on their unique properties and multiple surface functionalities. The simple method of synthesis by chemical reduction and exhibiting low toxicity is its main advantage. The mechanism of this activity was found to be size and dose-dependent. It was more effective against Gram-negative bacteria. Shamaila *et al.*, (2016) determined the antibacterial mechanism of Au NPs against four pathogenic bacteria and stated that Au NPs can be the next therapy against the enteric group of organisms. In biomedicine, Au NPs have been proved to be an important revolution for drug delivery and cancer therapy. They also perform as a safe and non-toxic antimicrobial agent due to their functional nature while compared with antibiotics.

Iron Oxide Magnetic Nanoparticles (FeO NPs)

FeO NPs are appropriate for a broad spectrum of applications, but the low stability and heterogeneous size distribution in an aqueous medium represent the major setback. These setbacks can however be reduced or diminished through the coating of FeO NPs with various polymers, such as polysaccharides. Sakthieaswari, (2016) explained that the polysaccharides are biocompatible, non- toxic and renewable. In addition, they possess chemical groups that permit further functionalization of the FeO NPs. Multi-functional entities can be created through decoration with specific molecules such as proteins, peptides, drugs, antibodies, bio-mimetic ligands, transfection agents, cells and other ligands. This development opens a whole range of applications for FeO NPs.

Aluminium Oxide Nanoparticles (AlO NPs)

Aluminium oxide NPs antimicrobial activity is due to the attachment of the NPs to the surface of the bacterial cell, resulting in cell death. There are few reports on the antimicrobial activity of AlO NPs. The NPs perforate in the Candida cells, and disrupt the morphological and physiological activity of the cells, resulting in cell death. Ansari *et al.*, (2015) prepared AlO NPs using leaf extracts of lemongrass.

Their antibacterial activity against extended-spectrum lactamases and metallo-lactamases of clinical isolates of *Pseudomonas aeruginosa* was found to be a significant activity of AlO NPs when compared with other nanoparticles which was reported by Singh *et al.*, (2019).

Carbon Nanotubes

NPs based on carbon possess an elevated antimicrobial activity. Prior studies showed that single-walled carbon nanotubes (SWCNTs), fullerenes, and graphene oxide (GO) NPs showed potential antimicrobial or bactericidal properties. Such novel forms of allotropic carbon have been discovered in the last 2 decades, and since then, they have been used in many fields of biosciences. Carbon nanotubes (CNTs) are a hollow cylindrical form of nanosize, which have been used in different fields of science and technology. SWCNTs with surface groups of eOH and eCOOH presented enhanced antimicrobial activity toward both gram-positive and gram-negative bacteria, whereas multi-walled carbon nanotubes (MWCNTs) with the same surface groups did not possess any notable antimicrobial effect. Their results showed that synthesis of cell-CNT aggregates affected the cell walls of bacteria (Figure 6) and then release of their DNA content (Kang *et al.*, 2007).

Figure 6. Antimicrobial mechanism of Carbon nanotubes in bacterial cell wall disruption and release of DNA content.

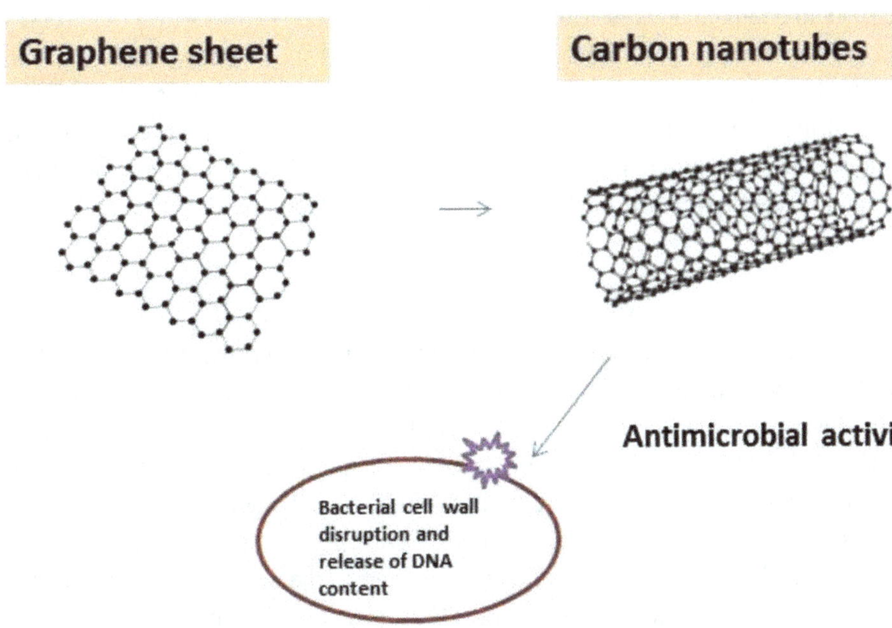

NANOPARTICLE-BASED DETECTION OF PATHOGENS

NPs can be used for rapid detection of pathogens with high accuracy and sensitivity. Superparamagnetic nanoparticles (SPM NPs) such as FeO_4-Au, Fe_3O_4-Ag and FePt-Ag, can be applied directly or as a part of electrochemical immunoassay to enhance the sensitivity. Kearns *et al.,* (2017) developed lectin-functionalised, antibody-coated silver NPs for detection of MRSA. A portable multiplex bar-chart SpinChip was fabricated by Wei *et al.,* (2018) for instrument-free detection of food-borne bacterial pathogens like *L. monocytogenes*, *Salmonella enterica* and *E. coli,* with a detection limit of 10 CFU/mL. Chung *et al.,* (2013) fabricated magnetic-DNA probe with nuclear magnetic resonance for detection of pathogenic bacterial RNA.

NANOVEHICLES FOR TARGETED DRUG DELIVERY

Besides bacterial detection, nanoparticles can also be utilized as delivery vehicles for the antimicrobial drugs to the specific infection site. NPs such as liposomal NPs, mesoporous silica NPs, polymer micelles, terpenoid-based NPs, carbon nanomaterials, polymer-based NPs, solid lipid NPs, magnetic NPs inorganic nanodrug carriers and dendrimer NPs can be used as a carrier for site-specific release of antibiotics and hence can be applied to combat microbial resistance (Ranghar, 2012). This target-oriented approach can be used to ensure the release of optimum concentration of drug at the infection site and to minimize its side effects (Liu *et al.,* 2016). NPs can be directed to the infection site by active or passive mechanism. In active system, NPs containing ligands such as antibodies are used for its specific binding to receptor at the infection site. While in passive system, NPs endure extravasations at the infection site and increased the blood vessel permeability. Moreover, NPs can be controlled through external stimuli like light, heat, magnetic field, pH and chemical agents (Lim *et al.,* 2018).

Aptamers are NPs containing DNA or RNA oligonucleotides that can bind specifically to the antigens at the infection site. Hence, aptamers can also be used as a carrier for delivering drugs to the exact site. Besides delivering drug, aptamers are known to have antibacterial and antiviral properties (Blecher *et al.,* 2011). Now-a-days, nanoparticles are being explored in the fields of disease diagnosis and site-specific delivery of therapeutic agents (precise medicines) in a well defined manner (Patra *et al.*, 2018).

Liposomes

Liposomes are phospholipid vesicles consisting of concentric lipid bilayers. Liposomes have been reported to offer improved therapy for biomedical applications. Liposomes have been proved to be advantageous as they overcome the tissue uptake and improves the biodistribution of medicinal compounds in *in vivo* condition. Liposomes are also biocompatible, ability to carry large size of drug molecules with controlled physicochemical and biophysical properties (Hua and Wu, 2013). The uniqueness of liposomes is that they entrap both lipophilic and hydrophilic compounds. Hydrophobic molecules get entrapped into the phosopholipid bilayer and hydrophilic molecules get inserted in the aqueous region which permits the delivery of a large range of macromolecules such as DNA, proteins and imaging agents.

Solid Lipid Nanoparticles

Solid lipid nanoparticles (SLNs) offer great promise for controlled and site specific drug delivery. SLNs have high drug loading capacity, large surface area and also increase the bioavailability of the drug in *in vivo* condition. Similar to liposomes, SLNs also acts as good candidate for the delivery of a large range of macromolecules such as DNA and proteins. Ezzati *et al.,* (2015) reported that SLNs offer an advantage of enhanced encapsulation efficiency with less active drug during the formulation process. Drug incorporation models of SLNs have been produced by using core shell model in which drug is dispersed in the lipid matrix. In the drug enriched shell model, the nanoemulsion results in a supersaturation of the drug.

Dendrimers

Dendrimers are three dimensional organized nanoscopic macromolecules with low polydispersity index. Dendrimers are derived from the Greek word 'dendron' meaning tree. Dendrimers have a unique tree like branching architecture. The architecture of dendrimers can be distinguished into inner core moiety followed by radially attached generations. The radially attached structures possess functional groups at their terminals. Dendrimers are highly preferred for biological applications due to their high water solubility, biocompatibility and polyvalent in nature. These features make dendrimers as a potent carrier for drug delivery and drug targeting applications (Madaan *et al.,* 2014).

NANOPARTICLES TO COMBAT INTRACELLULAR BACTERIA

Treating infections caused by facultative or obligate intracellular microbes are extremely difficult because most of the existing antibiotics exhibited poor penetration and retention power intracellularly. Moreover, the antimicrobial efficiency has been reduced greatly by the acidic pH in the lysosomes. This necessitates the need for the invention of novel therapeutics with greater intracellular activity and effective carrier molecules. It makes polymeric NPs as suitable weapon to overcome infections caused by intracellular bacteria. NPs, owing to its smaller size, have the capability to get engulfed by host cell phagocytes that contains intracellular bacteria. These NPs can also act as a carrier molecule which liberates required concentration of antibacterial drugs at the specific infection site with reduced toxicity and low side effects (Blecher *et al.,* 2011). This site-specific mechanism prevents the development of drug resistance by the microorganisms (Huang *et al.,* 2011).

The intracellular delivery and antimicrobial efficiency of the NPs depends on the surface charge and size of the NPs. Akbari *et al.,* (2013) demonstrated that ciprofloxacin-conjugated nano-niosomes of size 300–600 nm were readily phagocytosed and accumulated by macrophages than other nano-niosomes of size 600–1000 nm and 160–300 nm. NPs can be synthesized and modified to exhibit preferred surface charge and can be administered in combination along with other drug molecules.

Bacteria such as *M. tuberculosis, Chlamydophila pneumonia, L. monocytogenes* and *Legionella pneumophila* which can proliferate in the alveolar macrophages have been proven to be controlled by the use of NPs (Huh and Kwon, 2011). Alveolar macrophages are known to have more surface receptors for mannose, hence the addition of mannose to the NPs allowing them to target alveolar macrophages specifically. In case of type II pneumocytes, administration of liposomes combined with mannose leads to enhanced accumulation of antibacterial drugs in alveolar macrophages (Huh and Kwon, 2011). Similarly, accumulation of ZnO NPs in macrophages induced ROS, nitric oxide production by intracellular *Mycobacterium smegmatis* and resulted in death of the bacterium (Pati *et al.,* 2014).

COMBINATORIAL APPLICATION OF NANOPARTICLES WITH ANTIBIOTICS

Application of NPs with conventional antibiotics may lead to reduction in drug dosage and pose least toxicity to mammalian cells (Naqvi *et al.,* 2013). Panacek *et al.,* (2016) reported the synergistic activity of Ag NPs with other commercial antibiotics on selected Gram-positive and Gram-negative bacterial pathogens. Similar to this, antibacterial efficiency of Ag NPs was found to be higher when administered in combination with ampicillin (Franci *et al.,* 2015). Ag NPs when combined with ciprofloxacin (Naqvi *et al.,* 2013), clotrimazole (Kalhapure *et al.,* 2015) and vancomycin (Esmaeillou *et al.,* 2017) also exhibited synergistic inhibition activity against MRSA species. TiO_2 supplemented along with Ag NPs were found to exhibit efficient antifungal activity on *C. albicans* and *Aspergillus sp.* (Blecher *et al.,* 2011). Gao *et al.,* (2017) recorded the combinatorial effect of graphene with standard antibiotics on the control of *E.coli* and *Staphylococcus aureus*. Complete eradication of uropathogenic *Escherichia coli* using chitosan-conjugated antibiotics has been recorded previously by Erman *et al.,* (2017).

NANOTHERANOSTICS -A BUDDING TECHNOLOGY

NPs exhibited a wide range of applications in detection of pathogens (Yang *et al.,* 2018). Nanotheranostics is a newly emerging field that allows NPs based detection and real-time monitoring of therapy given to treat the infection. For instance, Au NPs -based Verigene R kit has been commercialized for diagnosis and identification of Gram-negative pathogens found in blood samples. Detection of resistance mechanism is also possible using this kit (Claeys *et al.,* 2018). Matrix-assisted magnetic iron oxide NPs have been used as a probe for collecting bacteria, which will be analyzed within 1 hour by using laser desorption or ionization mass spectroscopy (Reddy *et al.,* 2014). Similar work has been previously carried out by Galanzha *et al.,* (2012) where gold multi-layered magnetic NPs attached with specific antibody were used for detection of *Staphylococcus aureus* under *in vivo* condition and photothermal destruction of bacterial cells.

CHALLENGES IN CLINICAL TRANSLATION OF NANOANTIBIOTICS AND TACKLING STRATEGIES

NPs have promising antimicrobial efficiency even against MDR organisms and have several advantages over conventional antibiotics. NPs have also been extensively used in various allied medical fields such as medical diagnostics, solar cells, therapeutics, cosmetics, sensors and coatings. However, several challenges should be addressed before its application in clinical scale. Some of the bottlenecks include the interaction of NPs with cells, tissues and other organs for determining its cytotoxicity, dosage and proper route of administration (Sandhiya *et al.,* 2009).

In vitro assays using cell cultures has been employed for evaluating the biocompatibility of the NPs. But for clinical translation, *in vivo* models should be applied to understand the route of administration, mode of action, stability, release, toxicity, metabolism and clearance (Beyth *et al.,* 2015). But there are no specific guidelines to quantify biocompatibility of nanomaterials (Rai *et al.,* 2016). A more detailed long-term study is needed for determining the interaction of NPs on various tissues.

Cytotoxicity of Nanoparticles

Intravenous administration of NPs may lead to its accumulation in spleen, liver, lung, colon, lymphatic systems and bone marrow (Hagens *et al.,* 2007). Clinical administration of NPs has resulted in multi-organ nanotoxicity, like free radical-mediated oxidative stress which ultimately resulted in nephrotoxicity and hepatotoxicity (Lei *et al.,* 2008). Accumulation of Ag NPs has been observed in various organs like kidney, lungs, liver and spleen, specifically in brain owing to its capability to pass through the blood–brain barrier (Kim *et al.,* 2017). Exposure to NPs has also been resulted in increased heart rate, in few cases (Sharifi *et al.,* 2012). ZnO NPs accounted for toxicity via membrane injury and oxidative stress in mammalian cell lines (Huang *et al.,* 2008). Toxicity analysis of ZnO NPs and CuO NPs revealed that the size, duration and dosage of NPs determine its toxicity level (Dobrucka *et al.,* 2019). TiO_2 NPs exhibited toxic effects such as genetic toxicity, DNA damage and inflammation of lungs (Trouiller *et al.,* 2009). Naskar *at al.,* (2016) demonstrated the antibacterial efficiency of Ag–ZnO–graphene nanocomposite at lower dosage level and its cytotoxicity at higher concentrations. Aluminium oxide NPs revealed similar toxic effects like distortion in cell morphology and shrinkage at higher concentrations (Cha and Myung *et al.,* 2007).

The best implemented strategy for tackling the toxicity of NPs is the capping of NPs using biocompatible polymers like chitosan or PEG. Cinteza *et al.,* (2018) reported that capping of NPs using PEG enhanced its biocompatibility and reduced its toxic effects. Moreover, the synergistic effect of NPs with PEG resulted in effective killing of bacteria. Similarly, silica coating of ZnO NPs were found to reduce its toxicity (Chia and Leong, 2016). Doping is the second most important strategy applied for reducing toxicity of NPs. For example, doping of ZnO NPs with Fe decreased the dissolution of NPs and thereby reduced its toxic effect in zebra fish embryo and lungs of rodent (Xia *et al.,* 2011). In congruent with this, Sekar *et al.,* (2019) observed that the incorporation of Fe-doped ZnO NPs with polyvinyl alcohol nanofibers, exhibited effective antibacterial activity. Iqbal *et al.,* (2014) reported the antibacterial efficiency and biocompatibility of zinc–silver-doped hydroxyapatite NPs.

Pitfalls in Production and Characterization of Nanoparticles

Clinical translation of NPs necessitates proper guidelines for production, scaleup and characterization of NPs along with their biocompatibility (Zazo *et al.,* 2016). Currently, various physical, chemical and biological methods are being employed to synthesize NPs. Among these, bio-mediated NPs exhibited less toxicity in comparison with other techniques (Tyagi *et al.,* 2013). Hazarika *et al.,* (2016) described the effective antibacterial efficacy of Ag NPs synthesized from *Thalictrum foliolosum* root extract on RAW264.7 cells with no cytotoxic effects. Similar to this, Ag NPs synthesized from *Aloe vera* plant extract exhibited greater antibacterial potential against *S. epidermidis* and *P. aeruginosa* with zero cytotoxicity on human PBMCs (Tippayawat *et al.,* 2016).

Moreover, it is essential to address the economic affordability of large scale manufacturing of NPs (Bertrand and Leroux, 2012). The cost of production of NPs in industrial scale still remains high (Zazo *et al.,* 2016). Moreover, because of the site-specific action of NPs, no standard protocols are available to determine its dosage in terms of mass, surface area, number and biological samples (blood, urine, tissues). To overcome this issue, additional characterization techniques are needed which are independent of the properties of NPs (Lei *et al.,* 2008).

However, various NPs have been commercialized successfully to treat bacterial infections in humans. For instance, Ag NPs is widely used for treating burn wound and dental caries (Politano *et al.,* 2013). Magnetic NPs are being applied for antibiotic drug delivery (Tokajuk *et al.,* 2017). SiO_2 NPs are utilized for delivering drugs through oral route (Martínez-Carmona *et al.,* 2018). Nanomaterials conjugated with AMPs are used to treat skin infections to minimize the cyotototoxicity. For example, Lytixar (LTX-109) and pexiganan can be used as a topical application for treating skin infections (Cattoir and Felden, 2009). Brilacidin, murepavadin are also under phase 2 clinical trials to treat acute bacterial skin infections caused by resistant *P. aeruginosa* (WHO, 2019).

CONCLUSION

NPs have already proven its promising role in various medical fields and treatment of diseases like cancer. However, in depth knowledge on application of NPs in various other fields such as diagnosis, imaging, therapy and medicine is still in developmental stage. Eventhough nanoantibiotics exhibited several advantages over conventional antibiotics in treating multiple drug resistant pathogens, more interdisciplinary understanding from various disciplines such as microbiology, pathology, immunology, biomaterials, pharmacology, toxicology, and nanotechnology are needed. Metallic NP provides a promising approach when utilized as such or in combination with other antimicrobial agents to combat bacterial drug resistance. In addition, targeted or site specific action of NPs displays it's potential in controlling and mitigating various MDR strains. But, a systematic investigation becomes imperative before the clinical translation of NPs against MDR organisms. Pharmacology aspects of nanoantibiotics should be addressed as the dosage level to be administered is essential for clinical translation of NPs. More investigations on toxicity of NPs and its mechanism of action on other cells are needed before targeting multiple drug resistant bacteria.

The biosafety aspect of NPs has been addressed by NIEHS (National Institute of Environmental Health Sciences), an organization which supports research on nanoparticles to improve human health while examining the health hazards associated with it (Warheit, 2018). In order to avoid conflicting data regarding engineered nanoparticles, the standardization of *in vitro* methodologies for production, characterization of NPs becomes mandatory for establishing its toxicology profile. Construction of database specifically for classification and toxicological investigations of NPs can be of great use. Effective advancement in pre-clinical and clinical trials will help nanoantibiotics to enter the medicinal frame works in the near future.

REFERENCES

Ahmed, A. D., Mohammed, K. H., Lee, S. B., Kim, K., Saha, S. K., ... Gwang-Mo, Y. (2017). T.he rol.e of reactive oxygen species (ROS) in the biological activities of metallic nanoparticles. *International Journal of Molecular Sciences*, *18*(1), 120. doi:10.3390/ijms18010120

Akbari, V., Abedi, D., Pardakhty, A., & Sadeghi-Aliabadi, H. (2013). Ciprofloxacin nano niosomes for targeting intracellular infections: An *in vitro* evaluation. *Journal of Nanoparticle Research*, *15*(4), 1–14. doi:10.100711051-013-1556-y

Alavi, M., & Rai, M. (2019). Recent advances in antibacterial applications of metal nanoparticles (MNPs) and metal nanocomposites (MNCs) against multi-drug-resistant (MDR) bacteria. *Expert Review of Anti-Infective Therapy*, *17*(6), 419–428. doi:10.1080/14787210.2019.1614914 PMID:31046483

Anghel, I., Grumeezescu, A. M., Andronescu, E., ... Chifiriuc, M. C. (2012). Magnetite nanoparticles for functionalized textile dressing to prevent fungal biofilms development. *Nanoscale Research Letters*, *7*(1), 501. doi:10.1186/1556-276X-7-501 PMID:22950367

Ansari, M. A., Khan, H. M., Khan, A. A., Cameotra, S. S., Saquib, Q., & Musarrat, J. (2014). Interaction of Al(2)O(3) nanoparticles with *Escherichia coli* and their cell envelope biomolecules. *Journal of Applied Microbiology*, *116*(4), 772–783. doi:10.1111/jam.12423 PMID:24354999

Ansari, M. A., Khan, H. M., Alzohairy, M. A., Jalal, M., Ali, S. G., Pal, R., & Musarrat, J. (2015). Green synthesis of Al_2O_3 nanoparticles and their bactericidal potential against clinical isolates of multi-drug resistant *Pseudomonas aeruginosa. World Journal of Microbiology & Biotechnology*, *31*(1), 153–164. doi:10.100711274-014-1757-2 PMID:25304025

Bertrand, N., & Leroux, J. C. (2012). The journey of a drug-carrier in the body: An anatomo-physiological perspective. *Journal of Controlled Release*, *161*(2), 152–163. doi:10.1016/j.jconrel.2011.09.098 PMID:22001607

Beyth, N., Houri-Haddad, Y., Domb, A., Khan, W., & Hazan, R. (2015). Alternative antimicrobial approach: nano-antimicrobial materials. *Evid. Based Complement. Altern. Med.*, 1–16.

Blecher, K., Nasir, A., & Friedman, A. (2011). The growing role of nanotechnology in combating infectious disease. *Virulence*, *2*(5), 395–401. doi:10.4161/viru.2.5.17035 PMID:21921677

Cattoir, V., & Felden, B. (2009). Future antibacterial strategies: From basic concepts to clinical challenges. *The Journal of Infectious Diseases*, *220*(3), 350–360. doi:10.1093/infdis/jiz134 PMID:30893436

Cha, K. E., & Myung, H. (2007). Cytotoxic effects of nanoparticles assessed *in vitro* and *in vivo. Journal of Microbiology and Biotechnology*, *17*, 1573–1578. PMID:18062241

Chia, S. L., & Leong, D. T. (2016). Reducing ZnO nanoparticles toxicity through silica coating. *Heliyon*, *2*(10), e00177. doi:10.1016/j.heliyon.2016.e00177 PMID:27812550

Chung, H. J., Castro, C. M., Im, H., Lee, H., & Weissleder, R. (2013). A magneto-DNA nanoparticle system for rapid detection and phenotyping of bacteria. *Nature Nanotechnology*, *8*(5), 369–375. doi:10.1038/nnano.2013.70 PMID:23644570

Cinteza, L. O., Scomoroscenco, C., Voicu, S. N., Nistor, C. L., Nitu, S., Trica, B., Jecu, M.-L., & Petcu, C. (2018). Chitosa.n-stabilized Ag nanoparticles with superior biocompatibility and their synergistic antibacterial effect in mixtures with essential oils. *Nanomaterials (Basel, Switzerland)*, *8*(10), 826. doi:10.3390/nano8100826 PMID:30322127

Claeys, K. C., Heil, E. L., Pogue, J. M., Lephart, P. R., & Johnson, J. K. (2018). The Verigene dilemma: Gram-negative polymicrobial bloodstream infections and clinical decision making. *Diagnostic Microbiology and Infectious Disease*, *91*(2), 144–146. doi:10.1016/j.diagmicrobio.2018.01.012 PMID:29449043

Dakal, T. C., Kumar, A., Majumdar, R. S., & Yadav, V. (2016). Mechanistic basis of antimicrobial actions of silver nanoparticles. *Frontiers in Microbiology*, *7*, 1831. doi:10.3389/fmicb.2016.01831 PMID:27899918

Dale-Skinner, J. W., & Bonev, B. B. (2009). Molecular Mechanisms of Antibiotic Resistance: The need for novel antimicrobial therapies. Wiley-VCH verlag GmbH & Co.

Deck, D. H., & Winston, L. G. (2012). Sulfonamides, trimethhoprim and quinolones. In Basic and clinical pharmacology (12th ed.). McGraw-Hill.

DeQueiroz, G. A., & Day, D. F. (2007). Antimicrobial activity and effectiveness of a combination of sodium hypochlorite and hydrogen peroxide in killing and removing *Pseudomonas aeruginosa*biofilms from surfaces. *Journal of Applied Microbiology*, *103*(4), 794–802. doi:10.1111/j.1365-2672.2007.03299.x PMID:17897181

Dobrucka, R., Kaczmarek, M., Łagiedo, M., Kielan, A., & Dlugaszewska, J. (2019). Evaluation of biologically synthesized Au-CuO and CuO-ZnO nanoparticles against glioma cells and microorganisms. *Saudi Pharmaceutical Journal*, *27*(3), 373–383. doi:10.1016/j.jsps.2018.12.006 PMID:30976181

Erman, A., Hergouth, V. K., Blango, M. G., Kos, M. K., Mulvey, M. A., & Veranic, P. (2017). Repeated treatments with chitosan in combination with antibiotics completely eradicate uropathogenic Escherichia coli from infected mouse urinary bladders. *The Journal of Infectious Diseases*, *216*, 375–381. doi:10.1093/infdis/jix023 PMID:28119486

Esmaeillou, M., Zarrini, G., Ahangarzadeh, R. M., Shahbazi, M. J., & Bahadori, A. (2017). Vancomycin capped with silver nanoparticles as an antibacterial agent against multi-drug resistance bacteria. *Advanced Pharmaceutical Bulletin*, *7*(3), 479–483. doi:10.15171/apb.2017.058 PMID:29071232

Ezzati, J. N. D., Valizadeh, H., & Hamishehkar, H. (2015). Solid Lipid nanoparticles as Efficient Drug and Gene Delivery Systems: Recent Breakthroughs. *Advanced Pharmaceutical Bulletin*, *5*(2), 151–159. doi:10.15171/apb.2015.022 PMID:26236652

Franci, G., Falanga, A., Galdiero, S., Palomba, L., Rai, M., ... Morelli, G. (2015). Silver nanoparticles as potential antibacterial agents. *Molecules (Basel, Switzerland)*, *20*(5), 8856–8874. doi:10.3390/molecules20058856 PMID:25993417

Galanzha, E. I., Shashkov, E., Sarimollaoglu, M., Beenken, K. E., ... Shirtliff, M. E. (2012). *In vivo* magnetic enrichment, photoacoustic diagnosis, and photothermal purging of infected blood using multifunctional gold and magnetic nanoparticles. *PLoS One*, *7*(9), e45557. doi:10.1371/journal.pone.0045557 PMID:23049814

Ganjian, H., Nikokar, A., Tieshayar, A., Mostafaei, N., Amirmozafari, S., & Kiani, S. (2012). Effects of salt stress on the antimicrobial drug resistance and protein profile of *Staphylococcus aureus*. *Jundishapur Journal of Microbiology*, *5*(28), 328–331.

Gao, P., Xia, G., Bao, Z., Feng, C., Cheng, X., Kong, M., Liu, Y., & Chen, X. (2016). Chitosan based nanoparticles as protein carriers for efficient oral antigen delivery. *International Journal of Biological Macromolecules*, *91*, 716–723. doi:10.1016/j.ijbiomac.2016.06.015 PMID:27287772

Gao, Y., Wu, J., Ren, X., Tan, X., Hayat, T., Alsaedi, A., Cheng, C., & Chen, C. (2017). Impact of graphene oxide on the antibacterial activity of antibiotics against bacteria. *Environmental Science. Nano*, *4*(5), 1016–1024. doi:10.1039/C7EN00052A

Ghosh, S., Patil, S., Ahire, M., Kitture, R., Kale, S., ... Pardesi, K. (2012). Synthesis of silver nanoparticles using *Dioscorea bulbifera* tuber extract and evaluation of its synergistic potential in combination with antimicrobial agents. *International Journal of Nanomedicine*, *7*, 483–496. PMID:22334779

Grigore, M. E., Biscu, E., Holban, A., Gestal, M., & Grumezescu, A. (2016). Methods of Synthesis, Properties and Biomedical Applications of CuO Nanoparticles. *Pharmaceuticals (Basel, Switzerland)*, *9*(4), 75. doi:10.3390/ph9040075 PMID:27916867

Hagens, W. I., Oomen, A. G., de Jong, W. H., Cassee, F. R., & Sips, A. J. (2007). What do we (need to) know about the kinetic properties of nanoparticles in the body? *Regulatory Toxicology and Pharmacology*, *49*(3), 217–229. doi:10.1016/j.yrtph.2007.07.006 PMID:17868963

Hajipour, M.J., & Fromm, K.M., AkbarAshkarran, A., & Mahmoudi, M. (2012). Antibacterial properties of nanoparticles. *Trends*, *30*(10), 499–511. PMID:22884769

Hazarika, S. N., Gupta, K., Shamin, K. N. A. M., Bhardwaj, P., Boruah, R., Yadav, K. K., Naglot, A., Deb, P., Mandal, M., Doley, R., Veer, V., Baruah, I., & Namsa, N. D. (2016). One-pot facile green synthesis of biocidal silver nanoparticles. *Materials Research Express*, *3*(7), 075401. doi:10.1088/2053-1591/3/7/075401

Hemeg, H. A. (2017). Nanomaterials for alternative antibacterial therapy. *International Journal of Nanomedicine*, *12*, 8211–8225. doi:10.2147/IJN.S132163 PMID:29184409

Hetrick, E. M., Shin, J. H., Paul, H. S., & Schoenfisch, M. H. (2009). Antibiofilm efficacy of nitric oxide-releasing silica nanoparticles. *Biomaterials*, *30*(14), 2782–2789. doi:10.1016/j.biomaterials.2009.01.052 PMID:19233464

Hua, S., & Wu, S. Y. (2013). The use of lipid-based nanocarriers for targeted pain therapies. *Frontiers in Pharmacology*, *4*, 143. doi:10.3389/fphar.2013.00143 PMID:24319430

Huang, Z., Zheng, X., Yan, D., Yin, G., Liao, X., Kang, Y., Yao, Y., Huang, D., & Hao, B. (2008). Toxicological effect of ZnO nanoparticles based on bacteria. *Langmuir*, *24*(8), 4140–4144. doi:10.1021/la7035949 PMID:18341364

Huang, C. M., Chen, C. H., Pornpattananangkul, D., Zhang, L., Chan, M., Hsieh, M. F., & Zhang, L. (2011). Eradication of drug resistant *Staphylococcus aureus* by liposomal oleic acids. *Biomaterials*, *32*(1), 214–221. doi:10.1016/j.biomaterials.2010.08.076 PMID:20880576

Huang, N., Chen, X., Zhu, X., Xu, M., & Liu, J. (2017). Ruthenium complexes/polypeptide self-assembled nanoparticles for identification of bacterial infection and targeted antibacterial research. *Biomaterials*, *141*, 296–313. doi:10.1016/j.biomaterials.2017.07.005 PMID:28709020

Huh, A. J., & Kwon, Y. J. (2011). "Nanoantibiotics": A new paradigm for treating infectious diseases using nanomaterials in the antibiotics resistant era. *Journal of Controlled Release*, *156*(2), 128–145. doi:10.1016/j.jconrel.2011.07.002 PMID:21763369

Iqbal, N., Kadir, M. R. A., Mahmood, N. H., Salim, N., Froemming, G. R. A., Balaji, H. R., & Kamarul, T. (2014). Characterization, antibacterial and in vitro compatibility of zinc–silver doped hydroxyapatite nanoparticles prepared through microwave synthesis. *Ceramics International*, *40*(3), 4507–4513. doi:10.1016/j.ceramint.2013.08.125

Jayaraman, R. (2009). Antibiotic resistance: An overview of mechanisms and a paradigm shift. *Current Science*, *96*(11), 1475–1484.

Jiang, W., Mashayekhi, H., & Xing, B. (2009). Bacterial toxicity comparison between nano- and micro-scaled oxide particles. *Environmental Pollution*, *157*(5), 1619–1625. doi:10.1016/j.envpol.2008.12.025 PMID:19185963

Kalhapure, R. S., Sonawane, S. J., Sikwal, D. R., Jadhav, M., Rambharose, S., Mocktar, C., & Govender, T. (2015). Solid lipid nanoparticles of clotrimazole silver complex: An efficient nano antibacterial against *Staphylococcus aureus* and MRSA. *Colloids and Surfaces. B, Biointerfaces*, *136*, 651–658. doi:10.1016/j.colsurfb.2015.10.003 PMID:26492156

Kang, S., Pinault, M., Pfefferle, L. D., & Elimelech, M. (2007). Single-walled carbon nanotubes exhibit strong antimicrobial activity. *Langmuir*, *23*(17), 8670–8673. doi:10.1021/la701067r PMID:17658863

Kearns, H., Goodacre, R., Jamieson, L. E., Graham, D., & Faulds, K. (2017). SERS detection of multiple antimicrobial-resistant pathogens using nanosensors. *Analytical Chemistry*, *89*(23), 12666–12673. doi:10.1021/acs.analchem.7b02653 PMID:28985467

Kulshrestha, S., Qayyum, S., & Khan, A. U. (2017). Antibiofilm efficacy of green synthesized graphene oxide-silver nanocomposite using *Lagerstroemia speciosa* floral extract: A comparative study on inhibition of gram-positive and gram-negative biofilms. *Microbial Pathogenesis*, *103*, 167–177. doi:10.1016/j.micpath.2016.12.022 PMID:28027944

Khodashenas, B., & Ghorbani, H. R. (2015). Synthesis of silver nanoparticles with different shapes. *Arabian Journal of Chemistry*, 1–16.

Kim, M., Park, J. H., Jeong, H., Hong, J., Choi, W. S., Lee, B. H., & Park, C. Y. (2017). An evaluation of the in vivo safety of nonporous silica nanoparticles: Ocular topical administration versus oral administration. *Scientific Reports*, *7*(1), 8238. doi:10.103841598-017-08843-9 PMID:28811672

Knetsch, M. L., & Koole, L. H. (2011). New strategies in the development of antimicrobial coatings: The example of increasing usage of silver and silver nanoparticles. *Polymers*, *3*(1), 340–366. doi:10.3390/polym3010340

Kumari, P., Panda, P. K., Ealisha, J., Khushboo, K., & Mallick, A. M. (2017). Mechanistic insight to ROS and apoptosis regulated cytotoxicity inferred by green synthesized CuO nanoparticles from *Calotropis gigantea* to embryonic Zebraish. *Scientific Reports*, *7*(1), 16284. doi:10.103841598-017-16581-1 PMID:29176605

Lebeaux, D., Ghigo, J. M., & Beloin, C. (2014). Biofilm-related infections: Bridging the gap between clinical management and fundamental aspects of recalcitrance toward antibiotics. *Microbiology and Molecular Biology Reviews*, *78*(3), 510–543. doi:10.1128/MMBR.00013-14 PMID:25184564

Lei, R., Wu, C., Yang, B., Ma, H., Shi, C., Wang, Q., Wang, Q., Yuan, Y., & Liao, M. (2008). Integrated metabolomic analysis of the nano-sized copper particle-induced hepatotoxicity and nephrotoxicity in rats: A rapid in vivo screening method for nanotoxicity. *Toxicology and Applied Pharmacology*, *232*(2), 292–301. doi:10.1016/j.taap.2008.06.026 PMID:18706438

Lesniak, A., Salvati, A., Santos-Martinez, M. J., Radomski, M. W., Dawson, K. A., & Aberg, C. (2013). Nanoparticle adhesion to the cell membrane and its effect on nanoparticle uptake efficiency. *Journal of the American Chemical Society*, *135*(4), 1438–1444. doi:10.1021/ja309812z PMID:23301582

Li, Y., Zhang, W., Niu, J., & Chen, Y. (2012). Mechanism of photogenerated reactive oxygen species and correlation with the antibacterial properties of engineered metal oxide nanoparticles. *ACS Nano*, *6*(6), 1–22. doi:10.1021/nn300934k PMID:22587225

Lim, E. K., Chung, B. H., & Chung, S. J. (2018). Recent advances in pH-sensitive polymeric nanoparticles for smart drug delivery in cancer therapy. *Current Drug Targets*, *19*(4), 300–317. doi:10.2174/1389450117666160602202339 PMID:27262486

Liu, J. L., Zhang, W. J., Li, X. D., Yang, N., Pan, W. S., Kong, J., & … . (2016). Sustained-release genistein from nano-structured lipid carrier suppresses human lens epithelial cell growth. *International Journal of Ophthalmology*, *9*(5), 643–649. PMID:27275415

Madaan, K., Kumar, S., Poonia, N., Lather, V., & Pandita, D. (2014). Dendrimers in drug delivery and targeting: Drug-dendrimer interactions and toxicity issues. *Journal of Pharmacy & Bioallied Sciences*, *6*(3), 139–150. doi:10.4103/0975-7406.130965 PMID:25035633

Mann, J. (1999). Fighting bacteria: the mould that grew on sewage: evolution of the cephalosporins. In *The Elusive Magic Bullet: The Search for the Perfect Drug* (pp. 57–58). Oxford University Press.

Martínez-Carmona, M., Gun'ko, Y. K., & Vallet-Regí, M. (2018). Mesoporous silica materials as drug delivery: "The Nightmare" of bacterial infection. *Pharmaceutics*, *10*(4), 279. doi:10.3390/pharmaceutics10040279 PMID:30558308

McBain, A. J., & Gilbert, P. (2001). Biocide tolerance and the harbingers of doom. *International Biodeterioration & Biodegradation*, *47*(2), 55–61. doi:10.1016/S0964-8305(01)00037-3

Moken, M. C., Mc Murry, L. M., & Levy, S. B. (1997). Selection of multiple-antibiotic-resistant (mar) mutants of Escherichia coli by using the disinfectant pine oil: Roles of the mar and acrAB loci. *Antimicrobial Agents and Chemotherapy*, *41*(12), 2270–2772. doi:10.1128/AAC.41.12.2770 PMID:9420057

Munawar, A., Mohammed, J. T. M., Syeda, K. M. W., & Ellen, K. W. (2017). An Overview of Chitosan Nanoparticles and Its Application in Non-Parenteral Drug Delivery. *Pharmaceutics*, *9*(4), 53. doi:10.3390/pharmaceutics9040053 PMID:29156634

Naqvi, S. Z. H., Kiran, U., Ali, M. I., Jamal, A., Hameed, A., Ahmed, S., & Ali. (2013). Combined eficacy of biologically synthesized silver nanoparticles and different antibiotics against multi-drug-resistant bacteria. *International Journal of Nanomedicine*, *8*, 3187–3195. doi:10.2147/IJN.S49284 PMID:23986635

Naskar, A., Bera, S., Bhattacharya, R., Roy, S. S., & Jana, S. (2016). Synthesis, characterization and cytotoxicity of polyethylene glycol coupled zinc oxide–chemically converted graphene nanocomposite on human OAW42 ovarian cancer cells. *Polymers for Advanced Technologies*, *27*(4), 436–443. doi:10.1002/pat.3689

Neu, H. (1992). The Crisis in Antibiotic Resistance. *Science*, *257*(5073), 1064–1073. doi:10.1126cience.257.5073.1064 PMID:1509257

Nguyen, N. T., Grelling, N., Wetteland, C. L., Rosario, R., & Liu, H. (2018). Antimicrobial Activities and Mechanisms of Magnesium Oxide Nanoparticles (nMgO) against Pathogenic Bacteria, Yeasts, and Biofilms. *Scientific Reports*, *8*(1), 16260. doi:10.103841598-018-34567-5 PMID:30389984

Omoike, A., Chorover, J., Kwon, K. D., & Kubicki, J. D. (2004). Adhesion of bacterial exopolymers to α-FeOOH: Inner-sphere complexation of phosphodiester groups. *Langmuir*, *20*(25), 11108–11114. doi:10.1021/la048597+ PMID:15568864

Pal, S., Tak, Y. K., & Song, J. M. (2007). Does the antibacterial activity of silver nanoparticles depend on the shape of the nanoparticle? A study of the gram-negative bacterium *Escherichia coli. Applied and Environmental Microbiology*, *73*(6), 1712–1720. doi:10.1128/AEM.02218-06 PMID:17261510

Panacek, A., Smekalova, M., Kilianova, M., Prucek, R., & Bogdanova, K.,V ec erova, R., et al. (2016). Strong and nonspecific synergistic antibacterial efficiency of antibiotics combined with silver nanoparticles at very low concentrations showing no cytotoxic effect. *Molecules (Basel, Switzerland)*, *21*, 1–17. PMID:26729075

Pati, R., Mehta, R. K., Mohanty, S., Padhi, A., Sengupta, M., Vaseeharan, B., Goswami, C., & Sonawane, A. (2014). Topical application of zinc oxide nanoparticles reduces bacterial skin infection in mice and exhibits antibacterial activity by inducing oxidative stress response and cell membrane disintegration in macrophages. *Nanomedicine (London)*, *10*(6), 1195–1208. doi:10.1016/j.nano.2014.02.012 PMID:24607937

Patra, J. K., Das, G., Fraceto, L. F., Campos, E. V. R., Rodriguez-Torres, M. P., Acosta-Torres, L. S., Diaz-Torres, L. A., Grillo, R., Swamy, M. K., Sharma, S., Habtemariam, S., & Shin, H.-S. (2018). Nano based drug delivery systems: Recent developments and future prospects. *Journal of Nanobiotechnology*, *16*(1), 71. doi:10.118612951-018-0392-8 PMID:30231877

Pelgrift, R. Y., & Friedman, A. J. (2013). Nanotechnology as a therapeutic tool to combat microbial resistance. *Advanced Drug Delivery Reviews*, *65*(13-14), 1803–1815. doi:10.1016/j.addr.2013.07.011 PMID:23892192

Politano, A. D., Campbell, K. T., Rosenberger, L. H., & Sawyer, R. G. (2013). Use of silver in the prevention and treatment of infections. *Silver review. Surgical Infections*, *14*(1), 8–20. doi:10.1089ur.2011.097 PMID:23448590

Poole, K. (2002). Mechanisms of bacterial biocide and antibiotic resistance. *Journal of Applied Microbiology*, *92*(S1), 55S–64S. doi:10.1046/j.1365-2672.92.5s1.8.x PMID:12000613

Prasanna, V. L., & Vijayaraghavan, R. (2015). Insight into the mechanism of antibacterial activity of ZnO: Surface defects mediated reactive oxygen species even in the dark. *Langmuir*, *31*(33), 9155–9162. doi:10.1021/acs.langmuir.5b02266 PMID:26222950

Rai, M., Ingle, A. P., Gaikwad, S., Gupta, I., Gade, A., & Silvério da Silva, S. (2016). Nanotechnology based anti-infectives to fight microbial intrusions. *Journal of Applied Microbiology*, *120*(3), 527–542. doi:10.1111/jam.13010 PMID:26599354

Ranghar, S., Sirohi, P., Verma, P., & Agarwal, V. (2012). Nanoparticle-based drug delivery systems: Promising approaches against infections. *Brazilian Archives of Biology and Technology*, *57*(2), 209–222. doi:10.1590/S1516-89132013005000011

Reddy, P. M., Chang, K. C., Liu, Z. J., Chen, C. T., & Ho, Y. P. (2014). Functionalized magnetic iron oxide (Fe_3O_4) nanoparticles for capturing gram-positive and gram-negative bacteria. *Journal of Biomedical Nanotechnology*, *10*(8), 1429–1439. doi:10.1166/jbn.2014.1848 PMID:25016643

Roe, D., Karandikar, B., Bonn-Savage, N., Gibbins, B., & Roullet, J. B. (2008). Antimicrobial surface functionalization of plastic catheters by silver nanoparticles. *The Journal of Antimicrobial Chemotherapy*, *61*(4), 869–876. doi:10.1093/jac/dkn034 PMID:18305203

Russell, A. D., Tattawasart, U., Maillard, J. Y., & Furr, J. R. (1998). Possible link between bacterial resistance and use of antibiotics and biocides. *Antimicrobial Agents and Chemotherapy*, *42*(8), 2151. doi:10.1128/AAC.42.8.2151 PMID:9722471

Rutherford, S. T., & Bassler, B. L. (2012). Bacterial quorum sensing: Its role in virulence and possibilities for its control. *Cold Spring Harbor Perspectives in Medicine*, *2*(11), 12427. doi:10.1101/cshperspect.a012427 PMID:23125205

Sakthieaswari, P. (2016). *Biosynthesis and characterization of iron oxide nanoparticles produced from a few selected seaweeds collected from Mandapam Coast, Tamil Nadu* (Thesis). Madurai Kamaraj University.

Sandhiya, S., Dkhar, S. A., & Surendiran, A. (2009). Emerging trends of nanomedicine - an overview. *Fundamental & Clinical Pharmacology*, *23*(3), 263–269. doi:10.1111/j.1472-8206.2009.00692.x PMID:19527298

Sathishkumar, G., Logeshwaran, V., Sarathbabu, S., Jha, P. K., Jeyaraj, M., Rajkuberan, C., Senthilkumar, N., & Sivaramakrishnan, S. (2018). Green synthesis of magnetite Fe_3O_4 nanoparticles using *Couroupita guianensis* Aubl. fruit extract for their antibacterial and cytotoxicity activities. *Artificial Cells, Nanomedicine, and Biotechnology*, *46*(3), 589–598. doi:10.1080/21691401.2017.1332635 PMID:28554257

Savjani, J. K., Gajjar, A. K., & Savjani, K. T. (2009). Mechanisms of resistance: Useful tool to design antibacterial agents for drug-resistant bacteria. *Mini-Reviews in Medicinal Chemistry*, *9*(2), 194–205. doi:10.2174/138955709787316038 PMID:19200024

Sekar, A. D., Kumar, V., Muthukumar, H., Gopinath, P., & Matheswaran, M. (2019). Electrospinning of Fe-doped ZnO nanoparticles incorporated polyvinyl alcohol nanofibers for its antibacterial treatment and cytotoxic studies. *European Polymer Journal*, *118*, 27–35. doi:10.1016/j.eurpolymj.2019.05.038

Shamaila, S., Noshin, Z., Saira, R., Rehana, S., Jawad, N., & Shahzad, N. (2016). Gold Nanoparticles: An Efficient Antimicrobial Agent against Enteric Bacterial Human Pathogen. *Nanomaterials (Basel, Switzerland)*, *6*(4), 71. doi:10.3390/nano6040071 PMID:28335198

Sharifi, S., Behzadi, S., Laurent, S., Forrest, M. L., Stroeve, P., & Mahmoudi, M. (2012). Toxicity of nanomaterials. *Chemical Society Reviews*, *41*(6), 2323–2343. doi:10.1039/C1CS15188F PMID:22170510

Singh, B. N., Prateeksha, U. D. K., Singh, B. R., Defoirdt, T., Gupta, V. K., De Souza, A. O., & .... (2017). Bactericidal, quorum quenching and anti-biofilm nanofactories: A new niche for nanotechnologists. *Critical Reviews in Biotechnology*, *37*(4), 525–540. doi:10.1080/07388551.2016.1199010 PMID:27684212

Singh, K., Abha, M., Deepankar, S., & Kavita, S. (2019). *Antiviral and antimicrobial potentiality of Nano drugs*. Indian Institute of Technology, Banaras Hindu University. doi:10.1016/B978-0-12-814029-1.00013-2

Su, Y., Zheng, X., Chen, Y., Li, M., & Liu, K. (2015). Alteration of intracellular protein expressions as a key mechanism of the deterioration of bacterial denitrification caused by copper oxide nanoparticles. *Scientific Reports*, *5*(1), 15824. doi:10.1038rep15824 PMID:26508362

Tamayo, L. A., Zapata, P. A., Vejar, N. D., Azocar, M. I., Gulppi, M. A., Zhou, X., Thompson, G. E., Rabagliati, F. M., & Páez, M. A. (2014). Release of silver and copper nanoparticles from polyethylene nanocomposites and their penetration into *Listeria monocytogenes*. *Materials Science and Engineering*, *40*, 24–31. doi:10.1016/j.msec.2014.03.037 PMID:24857461

Taran, M., Rad, M., & Alavi, M. (2016). Characterization of Ag nanoparticles biosynthesized by *Bacillus* sp. HAI4 in different conditions and their antibacterial effects. *Journal of Applied Pharmaceutical Science, 6*(11), 94-99.

Tenover, F. C. (2006). Mechanisms of Antimicrobial resistance in bacteria. *The American Journal of Medicine*, *119*(6, 6A), 3–10. doi:10.1016/j.amjmed.2006.03.011 PMID:16735149

Tenover, F. C., & Hughes, J. M. (1996). Development and spread of multiply-resistant bacterial pathogens. *Journal of the American Medical Association*, *275*, 300–304. doi:10.1001/jama.1996.03530280052036 PMID:8544270

Tippayawat, P., Phromviyo, N., Boueroy, P., & Chompoosor, A. (2016). Green synthesis of silver nanoparticles in aloe vera plant extract prepared by a hydrothermal method and their synergistic antibacterial activity. *PeerJ*, *4*, 2589. doi:10.7717/peerj.2589 PMID:27781173

Tokajuk, G., Niemirowicz, K., Deptuła, P., Piktel, E., Cie'sluk, M., Wilczewska, A. Z., Dabrowski, J. R., & Bucki, R. (2017). Use of magnetic nanoparticles as a drug delivery system to improve chlorhexidine antimicrobial activity. *International Journal of Nanomedicine*, *12*, 7833–7846. doi:10.2147/IJN.S140661 PMID:29123396

Trouiller, B., Reliene, R., Westbrook, A., Solaimani, P., & Schiestl, R. H. (2009). Titanium dioxide nanoparticles induce DNA damage and genetic instability *in vivo* in mice. *Cancer Research*, *69*(22), 8784–8789. doi:10.1158/0008-5472.CAN-09-2496 PMID:19887611

Tyagi, P. K., Tyagi, S., Verma, C., & Rajpal, A. (2013). Estimation of toxic effects of chemically and biologically synthesized silver nanoparticles on human gut microflora containing *Bacillus subtilis. J. Toxicology and Environmental Health Sciences*, *5*(9), 172–177. doi:10.5897/JTEHS2013.0271

Ulloa-Ogaz, A. L., Pinon-Castillo, H. A., Munoz-Castellanos, L. N., Athie-Garcia, M. S., Ballinas-Casarrubias, M. D. L., Murillo-Ramirez, J. G., Flores-Ongay, L. Á., Duran, R., & Orrantia-Borunda, E. (2017). Oxidative damage to *Pseudomonas aeruginosa* ATCC 27833 and *Staphylococcusaureus* ATCC 24213 induced by CuO-NPs. *Environmental Science and Pollution Research International*, *24*(27), 22048–22060. doi:10.100711356-017-9718-6 PMID:28791555

Usha, C., & Rachel, G. A. (2014). Biogenic synthesis of silver nanoparticles by *Acacia nilotica* and their Antibacterial activity. *International Journal of Scientific Research*, *3*(6), 27–29. doi:10.15373/22778179/June2014/11

Wang, L., Hu, C., & Shao, L. (2017). The antimicrobial activity of nanoparticles: Present situation and prospects for the future. *International Journal of Nanomedicine*, *12*, 1227–1249. doi:10.2147/IJN.S121956 PMID:28243086

Warheit, D. B. (2018). Hazard and risk assessment strategies for nanoparticle exposures: How far have we come in the past 10 years? *F1000 Research*, *7*, 376. doi:10.12688/f1000research.12691.1 PMID:29636906

Wei, X., Zhou, W., Sanjay, S. T., Zhang, J., Jin, Q., Xu, F., Dominguez, D. C., & Li, X. J. (2018). Multiplexed instrument-free bar-chart spin chip integrated with nanoparticle-mediated magnetic apt sensors for visual quantitative detection of multiple pathogens. *Analytical Chemistry*, *90*(16), 9888–9896. doi:10.1021/acs.analchem.8b02055 PMID:30028601

World Health Organization. (n.d.). *Antibacterial Agents in Clinical Development: An Analysis of the Antibacterial Clinical Development Pipeline, Including Tuberculosis*. World Health Organization. https://apps.who.int/iris/handle/10665/258965

Xia, T., Zhao, Y., Sager, T., George, S., Pokhrel, S., Li, N., Schoenfeld, D., Meng, H., Lin, S., Wang, X., Wang, M., Ji, Z., Zink, J. I., Mädler, L., Castranova, V., Lin, S., & Nel, A. E. (2011). Decreased dissolution of ZnO by iron doping yields nanoparticles with reduced toxicity in the rodent lung and zebrafish embryos. *ACS Nano*, *5*(2), 1223–1235. doi:10.1021/nn1028482 PMID:21250651

Yang, X., Dang, Y., Lou, J., Shao, H., & Jiang, X. (2018). D-alanyl-D-alanine- modified gold nanoparticles form a broad-spectrum sensor for bacteria. *Theranostics*, *8*(5), 1449–1457. doi:10.7150/thno.22540 PMID:29507633

Young, L. S., & Hindler, J. (1986). Aminoglycoside resistance: A worldwide perspective. *The American Journal of Medicine*, *80*(6, Suppl 5C), 35–39. doi:10.1016/0002-9343(86)90474-2 PMID:3717187

Zaidi, S., Misba, L., & Khan, A. U. (2017). Nano-therapeutics : A revolution in infection control in post antibiotic era. *Nanomedicine; Nanotechnology, Biology, and Medicine*, *13*(7), 2281–2301. doi:10.1016/j.nano.2017.06.015 PMID:28673854

Zazo, H., Colino, C. I., & Lanao, J. M. (2016). Current applications of nanoparticles in infectious diseases. *Journal of Controlled Release*, *224*, 86–102. doi:10.1016/j.jconrel.2016.01.008 PMID:26772877

Zhang, W., Li, Y., Niu, J., & Chen, Y. (2013). Photogeneration of reactive oxygen species on uncoated silver, gold, nickel, and silicon nanoparticles and their antibacterial effects. *Langmuir*, *29*(15), 4647–4651. doi:10.1021/la400500t PMID:23544954

Zhang, Q., Zhang, K., Xu, D., Yang, G., Huang, H., Nie, F., Liu, C., & Yang, S. (2014). CuO nanostructures: Synthesis, characterization, growth mechanisms, fundamental properties, and applications. *Progress in Materials Science*, *4*(60), 208–337. doi:10.1016/j.pmatsci.2013.09.003

Zhang, X., Zhi-Guo, L., Wei, S., & Sangiliyandi, G. (2016). Silver Nanoparticles: Synthesis, Characterization, Properties, Applications, and Therapeutic Approaches. *International Journal of Molecular Sciences*, *7*(9), 1534. doi:10.3390/ijms17091534

Chapter 7
Encountering the Survival Strategies Using Various Nano Assemblages:
A Mechanistic Approach

Lakshmipathy Muthukrishnan
Saveetha Dental College and Hospitals, Saveetha Institute of Medical and Technical Sciences, India

ABSTRACT

The technological advancements have not only made humans more civilized but have also caused the micro-organisms to develop several survival strategies via antimicrobial resistance to keep pace. Such highly developed microbial systems have been classified as superbugs, exhibiting Trojan-horse mechanism. This uncertain behaviour in microbes has challenged humans to scour around novel moiety to shield themselves from the detrimental effects. One such natural phenomenon that has drawn the attention of researchers is the metal-microbe interaction where microbes were found to be controlled during their interaction with metals. Fine tuning could bestow them with enhanced physico-chemical properties capable of controlling life-threatening micro-organisms. Nano forms of metals (nanoparticles, quantum dots, polymeric nanostructures) exhibiting medicinal properties have been implied toward biomedical theranostics. This chapter highlights the mechanistic antimicrobial resistance and the containment strategy using various nano assemblage highlighting its fabrication and bio-molecular interaction.

INTRODUCTION

In the global scenario, the upsurge in the emergence of drug-resistant strains has been accelerated by both antibiotic overuse and reduced infection control. As a result, the most susceptible bacterial population is eliminated and replaced by the resistant strains to flourish (Baym et al., 2016; Holmes et al., 2016). This scenario is well explained by heteroresistance, where the bacterial sub-populations with reduced antibiotic susceptibility on a par with the main population. Tolerance is yet another factor that showcases the level of resistance exhibited by the antibiotic resistant sub-populations. They are capable

DOI: 10.4018/978-1-7998-5049-6.ch007

of surviving at concentrations close to their minimum inhibitory concentration (MIC). There are two principal mechanisms associated with the development of antibiotic resistance: Spontaneous mutation and Horizontal gene transfer (HGT) of resistance genes (Holmes et al., 2016). Mutations that occur spontaneously could modify the target or reduce the intake by efflux system. Notably, acquisition of resistant gene could lead to post-translational modification by providing alternative metabolic pathways pertaining to antibiotic inactivation (Blair et al., 2015).

Despite innumerable origins attributing antibiotic resistance, there exist some classical mechanisms for resistance development in bacterial populations (Alcalde-Rico et al., 2016). Intrinsic efflux mechanism is one among them which by mutation in energy-dependent transporters that most of the resistant phenotypes emerge (Blair et al., 2015; Bergmiller et al., 2015; Li et al., 2015). It is also the most rapid and effective drug resistance mechanism evidenced in the bacterial population challenged with the antibiotic or toxin capable of inducing stress. This oxidative stress triggers very complex adaptive responses in bacteria involving modifications at the gene level and their expression as investigated in *Escherichia coli* (Nichols et al., 2011). In the current scenario, almost 80% of all bacterial infections are empirically related to multidrug-resistance (Fraimow et al., 2011; Viale et al., 2015). In particular, the presence of bacterial envelope in Gram negative bacteria acts as a barrier in restricting the entry of antibiotics and related compounds.

The growing multidrug resistance in ESKAPE pathogens encompassing *Enterococcus faecium, Staphylococcus aureus, Klebsiella pneumoniae, Acinetobacter baumannii, Pseudomonas aeruginosa* and *Enterobacter spp.* have spurred the treatment regimen and containment strategies (Dantas et al., 2008; Nikaido, 2008; Fischbach et al., 2009; Laxminarayan et al., 2016). In order to combat the emergence of AMR pathogens, an intensive research to explore the molecular identity and mechanisms involved in the acquisition and dissemination of AMR genes in the microbial community is a mandate.

BACKGROUND

Despite improvement in the pharmacokinetics and pharmacodynamics driving novel drug design strategy, the difficulty in identifying the bacterial target sites for commercialized antibiotics is cumbersome. The emergence and dissemination of antimicrobial resistance could well be appreciated prior to the introduction of novel antimicrobials (Levy 2007; Chambers et al. 2009). Indeed resistance to antibiotic was documented even before the antibiotics were introduced into clinical practice (Abraham 1940). For instance, sulfonamides which were introduced in 1937 owing to its aggravated therapeutic usage led to the development of resistance in the late 1930s (Chopra et al. 2002). Subsequently, aminoglycoside-resistant *S. aureus* emerged within six years of its introduction (Gootz 1990; Katayama et al. 2000). To counterbalance this scenario, the very first semisynthetic antibiotic Methicillin was introduced in 1961 to target penicillinase producing strains of *S. aureus* but to the dismay, they too developed resistance to methicillin (Barber 1961).

In the antibiotic resistant sequel, fluoroquinolones which were introduced in the 1980s for the containing Gram negative strains of bacteria at a later stage developed resistance, but were capable enough to treat Gram-positive infections (Lowy 2003). Subsequently, in 2002 *S. aureus* strains resistant to vancomycin (VRSA) were reported after 44 years of introduction of vancomycin (Appelbaum 2006) as shown in timeline (Figure 1). Most recently, tigecycline-resistance has been identified during pre-clinical stages and even before its approval for human use (Moore et al. 2013).

Figure 1. Schematic representation of evolution of drug resistance by various clinical strains in due course of time

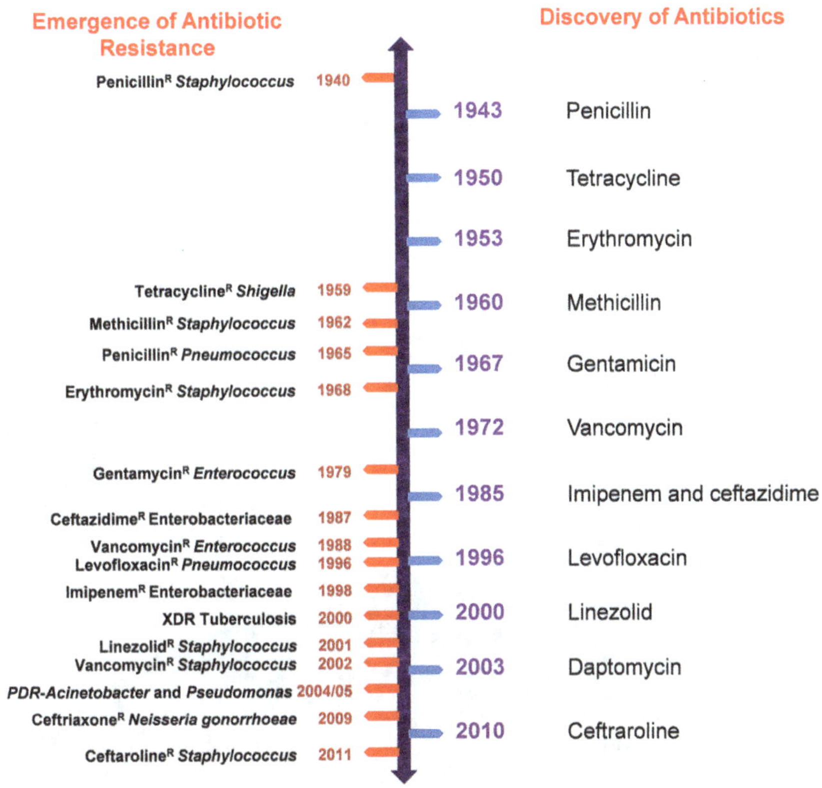

DEVELOPMENT OF ANTIMICROBIAL RESISTANCE (AMR)

Antimicrobial resistance (AMR), one of its kinds seems to co-exist in a complex relationship between the antimicrobials and the pathogens (both resistant and susceptible strains) based on its wheel of life. In a broader sense, those microbial pathogens viz., bacteria, viruses and protozoa with high rates of doubling period have short life-cycles, were more likely to become drug- resistant.

Drug resistance is acquired by the micro-organisms through natural selection. When a microbial pathogen grow and multiplies at a fast pace, the resulting ——progeny may contain errors, so called mutations which reduce its susceptibility to a given drug. When such a mutation gives the pathogen an advantage in a particular environment, it will grow better than its neighbors developing resistance and can increase its copy number for what it is selected. Another scenario is that Nature has pre-programmed all bacteria with a state of low resistance. For instance, a study reported from Thailand showed that the yesteryear antibiotics namely trimethoprim/sulfamethoxazole (TMP/SMZ), ampicillin and tetracycline are no longer used for the treatment of non-cholera diarrhea (NCD) (Hoge et al., 1998). In contrast, a study conducted in Bangladesh reported the effectiveness of the above mentioned antibiotics in treating NCD effectively (Rahman et al., 2017). Meanwhile without the use of antimicrobials, the susceptible strains will continue multiplying and challenge the resistant strains restricting their spread. Hence sufficient dosage of an antimicrobial administered during early infection augment microbicidal action toward susceptible strains and checks wide multiplication of newly-mutated pathogen for onward transmission (Figure 2).

Figure 2. Schematic representation of Natural selection for antibiotic resistance

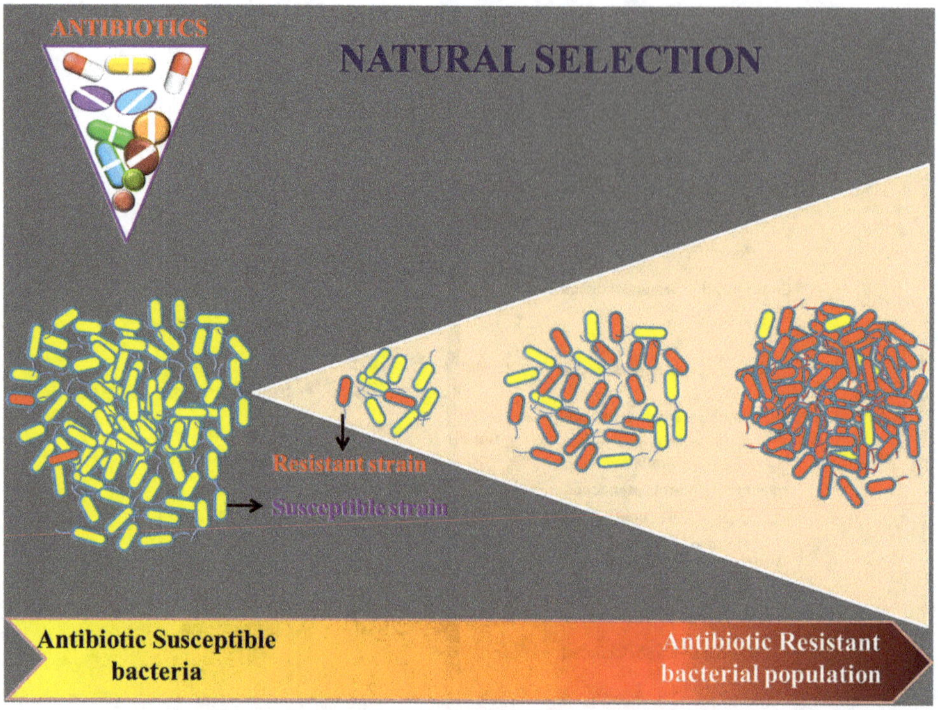

Poor Quality Medicines

According to World Health Organisation (WHO) poor quality medicines have been typified as SSFFC (Substandard, Spurious, Falsely-labeled, Falsified and Counterfeit) medical products. It denotes medicines that have not been formulated with correct ingredients or they might have degraded with time with reduced bioavailability. Although the definition sounds simple, the process encompasses problematic areas associated with antimicrobial resistance that overlooks different causes, implications with the regulatory non-compliance posing a threat to public health (Bate et al., 2014).

- High quality products formulated by legitimate manufacturers in compliance with all regulations
    - Quality products degradation due to inappropriate storage and transit.
    - Haphazard production or errors in packaging on a par with regulations
- Products with good quality formulations but not in compliance with all regulations
    - Unlicensed selling of medicines in markets
    - Innovative drug products under patent
    - Diversion of markets for selling quality medicines

Besides sub-standard medicines, it is from the manufacturers' standpoint, manufacturing errors and supply chain need great attention. As a result, the products switch over to right hand quartile. In general, three different types of manufacturers have been classified:

- Unregistered manufacturers (medicines produced without active ingredients)
- Licensed manufacturers (medicines produced with good manufacturing practises (GMP) but due to underinvestment leads to poor quality production).
- High quality manufacturers (medicines produced with highest quality standards)

The pathogenicity stands unique irrespective of manufacturing or regulatory or legal status of the drugs. It is only the ingredients and formulation and its synergistic interaction with patient characteristics determines the reliability. As discussed above these poor quality medicines aggravates development of resistance depending upon the circuition of the pathogen, mechanism of action and host metabolic rate at genetic level.

Sub-therapeutic Dose

On the other hand, sub-therapeutic doses of drugs create a selective pressure in killing the most vulnerable strain. This approach allows rapid multiplication of the most resistant strains on a par with the susceptible strain. As long as the administered antimicrobial stays in the blood, resistance factor gets amplified owing to the selective pressure that is built over and over by monotonous reproductive cycles. Upon acquiring, the resistance characteristics are transmitted via gene transfer mechanism. According to the population dynamics, the resistant strains tend to have developed higher fitness on a par with de novo resistance and may be selected. With diversified characteristics, the resistance pattern develops when an elevated pathogenic population encounters minimal inhibitory concentration of medicines capable of killing the most susceptible strains. This establishes the dissemination of resistant strains under lower selective pressure (Abdul-Aziz et al., 2015).

In order to reduce the prevalence of sub-standard or poor quality medicines, public health response is needed which include: rigorous improvement in the supply chain, ensuring compliance, detecting production errors and validating the quality of the product. These systems would support legitimate manufacturers incentivize to have a vigil on quality on a par with manufacturing errors.

Antimicrobial resistance - Mechanism

The increase in commercialization of antibiotics was proved to be fatal and led to the emergence of antibiotic resistance, one such strategy urging the need for survival of microbes. It is indeed the adaptation of microbes to the dynamic environments exposed to antibiotics by developing resistance through wide range of defense mechanisms (Figure 3). They constitute (1) increase in the permeability of cell, (2) active efflux, (3) catalytic activity on antibiotics, (4) nullifying effect, (5) alternative metabolic pathways, (6) excess deployment of the target enzyme and (7) alteration in target site.

Antimicrobial resistance in microorganisms was generally found to occur via mutation in the genomic DNA and HGT, the most important driving forces in phylogeny. This mutagenic influence brings about conformational changes in the secondary structures of proteins accounting for its virulence property. Subsequently, the transfer of acquired genes among bacterial species led to the wide spread emergence in antibiotic resistance. For instance, nucleotide base substitution(s) in 23S rRNA and ribosomal proteins (L4 and L22) led to macrolide resistance (Vester and Douthwaite 2001). Secondly, there exist a relationship between antibiotics consumption and emergence of resistance. In some European countries, antibiotics were used as antimicrobial growth promoters (AGPs) in animal husbandry mainly to enhance production. Increase in dosage of antibiotics and its continued exposure render the gut flora insensitive and increase the chance of survival. This led to the development of resistant bacteria in food products which when consumed by human results in food borne infections caused by resistant strains e.g. MRSA and Vancomycin-resistant Enterococci (VRE). The underlying mechanism of bacteria toward different classes of antibiotics has been clearly depicted in Table 1.

Figure 3. Schematic representation of antimicrobial resistance mechanism in bacteria

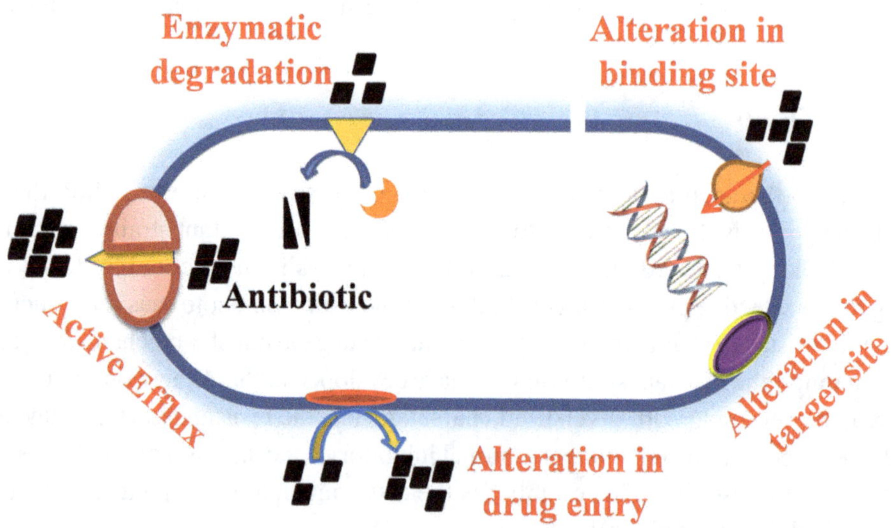

FACTORS DRIVING ACQUIRED ANTIMICROBIAL RESISTANCE

In acquired antimicrobial resistance (AAR), the microorganism develops a tendency to resist the effect of specific antimicrobial agent to which it had shown susceptibility at an earlier stage. This is the direct result of mutation in genes involved in normal metabolic processes and cellular structures, that may be acquired or synergistic effect.

Intrinsic Resistance

Intrinsic as the name implies is the resistance developed inherently in a bacterial species to a specific antimicrobial agent that can able to surpass either structurally or functionally. They may also be termed *insensitive* as it was evidenced in in microbes which have not been exposed or showed susceptibility toward particular class of antibiotic. There are various factors that determine the *insensitive* characteristics viz. drug-target affinity, inaccessibility, extrusion or efflux pumps and enzymatic inactivation (Table 2).

Table 1. Antimicrobial resistance Mechanism toward various classes of antibiotics (Zaman et al., 2017)

Classes of Antibiotics	Resistant genes	Mechanism of resistance
Beta-lactams – Penicillin derivatives (Penicillins, Cephalosporins, Monobactams etc.)	*TEM, SHV, CTX-M, OXA*	Drug hydrolysis, efflux pumps and alteration of target.
Aminoglycosides (Amikacin, Gentamicin, Streptomycin etc.)	*ArmA, RmtA-H, NpmA,*	Decreased cell permeability, Alterations at ribosomal binding site, Aminoglycoside modifying enzymes
Glycopeptides (Vancomycin)	*VanH, VanA, Van X*	Reprogramming cell wall biosynthesis
Tetracyclines (Tetracycline, Minocycline, Tigecycline, Oxytetracycline)	*tetA, tetB, tetD, tetE* and *tetG*	Monooxygenation, tetracycline efflux, ribosomal protection
Macrolides (Clarithromycin, Azithromycin)	*ermA, ermB, ermC, ermTR, mefA/E, msrA*	Ribosomal dimethylation, and Efflux.
Linosamides (Clindamycin)	*linB*	Enzymatic inactivation, Ribosomal protection by ABC-F proteins, Methylation.
Oxazolidinones (Linezolid)	*cfr, cfr(B), optrA, poxtA*	Efflux pumps and altered target
Quinolones (Ciprofloxacin, Norfloxacin, Ofloxacin)	*qnrA, qnrB, qnrS*	Mobile efflux pumps, altered target
Sulfonamides (Sulfadiazine, Sulfamethoxazole)	*sul1, sul2, sul3*	Efflux pumps, enzyme inactivation

Table 2. Intrinsic resistance and their respective mechanisms (Forbes et al., 1998; Giguere et al., 2006)

Intrinsic Resistance	Mechanism
Aminoglycosides	Deprived of oxidative metabolism
Metronidazole	Anaerobic reduction deficit
Aztreonam (β-lactam)	Paucity in PBPs
Vancomycin	Deficiency in uptake of drug
Ampicillin (β-lactam)	Enzymatic inactivation (Beta lactamases)
Imipenem (β-lactam)	Enzymatic inactivation (Beta lactamases)
Vancomycin	Unavailability of precursor molecules
Sulfonamides, trimethoprim, tetracycline etc.	Inefficiency in drug uptake
Aminoglycosides	Deprived of oxidative metabolism
All cephalosporins	Paucity in PBPs

Mobile Genes (Transposons)

The genes involved in AAR are confined within mobile DNA i.e. a fragment of DNA capable of translocating itself within same or between different genomes. The major players in HGT are the conjugative and mobilizing rudiments. Such mobilizing elements containing the entire genetic information are termed transposons. On the other hand, the use of conjugation machineries to transfer the genetic information to another host is termed conjugative plasmids or conjugative transposons.

Conjugative Transposons

The conjugative transposons behave in a similar fashion as conjugative plasmids possessing OriT (origin of transfer) and the genes responsible for various physiological activities pertaining to conjugation (Roberts et al., 2008). Alongside, they are far different from plasmids where the integration phase (either with plasmid or chromosome) is mandate to maintain the unique characteristic. This makes them more efficient in finding a larger host range over plasmids. Furthermore, they constitute a highly heterogeneous group with greater efficiency of integration of replicons conferring unique characteristics and broad host range (Frost and Koraimann 2010).

Insertion Sequences

One of the simplest among the mobile genes is the insertion sequence (IS). They are characterized by the presence of mobile genes with inverted repeats at either ends of the element. They may be short as 1Kb (Stefan Kohl and Ralph Bock 2009). IS besides comprehending mobile genetic element contains accessory genes which does not have any role in translocation of genetic elements are thus termed transposons. A typical transposon encompasses an accessory gene encoding antimicrobial resistance and an enzyme transposase.

Transposons are efficiently disseminated only when they are enclosed in a conjugative rudiment. For instance, integrative and conjugative elements were found efficient in transferring / translocating the transposons to a different genomic arena. Alongside, the specificity of integration depends on suitability of recombinases. In line, transposon Tn*916* although can use large number of integration sites, they are highly specific in their integration (Roberts and Mullany 2009; Wozniak and Waldor 2010).

Plasmids

They are the extra-chromosomal genetic elements with their replication origin found in almost all bacteria. The oriT and genes encoding physiological functions are transferred to another host via conjugation (Smillie et al., 2010). They are classified into conjugative and mobilizing plasmids. The former harbor genes responsible for conjugation and the latter with only an oriT while transferring genes among hosts. Most of the antimicrobial resistance (AMR) pattern is transferred via plasmids to new hosts. Plasmids capable of transferring AMR genes but incapable of replicating within the new host may lead to the loss of that plasmid. On the other hand when the same AMR is present on the transposons, they can be translocated within or between the genomes thereby maintaining the same even in the absence of plasmids. This implies that the plasmid also has a role in spreading the resistance characteristics besides coexisting plasmids (Couturier et al., 1988; Elisa Carloni et al., 2017).

Transduction

In general, it is the introduction of foreign DNA into a bacterial cell either by a bacteriophage or a vector. Although capable of translocation to new genome sites, they are not done by themselves but only upon integrating themselves with any conjugative element. Transposons and mobile introns are grouped under this category (Willi et al., 1997; Del Grosso et al., 2011).

Transformation

It is the process by which bacteria takes up naked DNA from the environment. In accord, there is a higher probability of acquiring antimicrobial resistant genes via transformation. This was also well studied in both Gram positive and negative bacteria. Bacteria capable of introducing a new strand of DNA into them are termed competent strains e.g. *Streptococci* group may show competence during a specific growth stage. Moreover, sequence specificity determines the efficiency of absorption as observed in *Neisseria sp.* (Smith et al., 1999) whereas in *Bacillus subtilis* there arise no such requirement. Once taken up, the foreign DNA is integrated with the host genome by homologous recombination acquiring the property of autonomous replication.

Integrons

Integrons works on the concept of gene cassettes. This process rapidly prepares the bacteria to adapt and evolve under adverse conditions. Genes responsible for antimicrobial resistance are stockpiled and embedded in a gene cassette lacking the functions of self-mobility (Fluit and Schmitz 2004; Depardieu et al., 2007). Typically the DNA fragments are exchanged via site-specific recombination system capable of integrating and expressing the typical gene characteristics (Mazel 2006; Labbate 2009). They differ from transposons in lacking the inverted terminal repeats and genes encoding proteins catalyzing their mobility. As they lack the self-mobilizing functionality, the gene exchange is made possible by plasmids or transposons (Carattoli 2001; Partridge et al., 2009).

BIOFILMS

Antimicrobial resistance, despite innumerous factors could also result from biofilm forming pathogenic micro-organisms owing to its inherent resistance to antimicrobials (Ribeiro SM et al., 2016). They have a unique architectural feature encompassing interstitial matrices assembled into macro and micro-colonies. These matrices facilitate diffusion of gases, nutrients and antimicrobials through them. The close proximity evidenced among cells establishes communication via quorum sensing, a remarkable extrachromosomal characteristic of biofilm.

Survival Strategy

One of the naturally acquired strategies in bacteria for survival is the formation of biofilm. They are syntrophic microbial consortia entrenched in a slimy matrix comprising of polymeric substances, proteins and extracellular DNA. This micro-architecture confers resistance characteristics to a wide range of anti-microbials, disinfectants, chemotherapeutic agents etc. interfering with host defence (Hoiby et al., 2011).

Glycocalyx or Capsule

It is the integral part of biofilms covering the cell membranes of bacteria (G +ve and G -ve) allowing them to adhere to inert surfaces having thickness ranging from 0.2 to 1.0μm (Reitsma 2007). To accomplish structural rigidity to the biofilm and help maturation, glycocalyx employs several weaker and stronger interaction forces (Pena et al., 2011) attributing for the development of resistance toward antimicrobials. Moreover, glycocalyx layer has the tendency to retain 25% w/w of the antimicrobial to which it is exposed. Secondly, the selective permeability of glycocalyx restricts movement of antimicrobials with the aid of embedded exoenzymes (Arciola et al., 2012; Sugano et al., 2016).

Heterogeneity in Metabolism

In biofilm assembly, the nutritional pyramid rolls from the top to bottom i.e. nutrient and oxygen rich areas show higher metabolic activity and those located at nutrient deficient areas shows reduced metabolic activity. It was presumed as a possible mechanism of action of biofilms. Poulsen et al., (1993) has reported on the heterogeneity associated with the cellular enzyme synthesis within the biofilm. Slow growing bacteria in the stationary phase shows reduced enzyme activity, metabolically active bacteria are killed by antimicrobial agent and the persistent bacteria remaining dormant show less susceptibility to various antimicrobials. For instance, *E. coli* synthesizes 1A-dependent ppGpp capable of suppressing autolysin thereby limiting anabolic process in cells. However, mutation induced in *rel*A gene does not necessarily affect the growth rate but makes such population more prone to antimicrobials by interfering with peptidoglycan synthesis (Lei Yang et al 2008).

Persistence Phenomenon in Biofilm

Bacterial cells (Persisters) constituting biofilm remain tolerant to various antimicrobial agents, disinfectants, solvents etc. augmenting severity and infectivity (Lewis K 2005) at global level. Persister cells comprises of late growing bacterial strains influenced by various growth stages of bacterial communities challenged with typical lethal doses of wide array of antimicrobial agents. They are the key contributors for the development of multidrug resistance and antibiotic tolerance. Bacteria that remain dormant duirng the stationary phase contributed much in the propagation of persister cells accounting for increased resistance inside biofilm (del Pozo and Patel R 2007). For instance, in patients with cystic fibrosis (CF), *P. aeruginosa* biofilms have been measured stating most cells at stationary phase. When they were exposed to single antibiotic, they were able to encounter actively dividing cells but were not efficient enough to kill a subset of population constituting biofilm. Conversely, the incredible tolerance exhibited by persisters augmented programmed cell death by partially damaging the cell (Keren et al 2004).

Acinetobacter baumannii (A. baumannii), a biofilm forming G -ve, opportunistic pathogen exhibited increased resistance toward antimicrobial agents. It was evidenced that the process of desiccating a low biofilm forming environmental MDR +ve strain showed 2.7 fold increase in risk of cell death compared with MDR-ve counterpart. Conversely, high biofilm forming MDR –ve environmental strain showed 60% decrease in acquiring the risk compared to low biofilm forming counterpart (Greene et al., 2016). In this line, *S. aureus* infections in community and health care settings pose yet another challenge (Kong et al., 2016) that demands development of anti-virulence therapeutics to regulate and contain the spread of infection and dissemination of antibiotic resistance.

OVERCOMING DRUG RESISTANCE

The discovery of antibiotics is one of the breakthroughs in the medical history that had revolutionized pharmaceutical industry in containing the most dreadful diseases caused by bacteria. However, large scale production and precarious application of antibiotics have contributed for the emergence of MDR strains which made the treatment more difficult. The only option to encounter them was through the use of broad spectrum antibiotics, which are toxic and more cost-intensive. Modification of existing antimicrobial compounds and / or formulating novel antimicrobial agents with bioactive potential is the need of the hour.

Metal Nanoparticles

In this line, Nanotechnology with its tremendous and widespread application plays a key role in manipulating materials at nanoscale to anticipate the already existing characteristics with new dimensions. There are indeed various methods of nanoparticle fabrication but green synthesis has gained popularity owing to its cheap, non-toxic, reliable, easy and eco-friendly approach devoid of chemicals. The nanofabricated metals have been employed for diagnosis and treatment of infectious diseases like cancer and tuberculosis. Slavin et al. (2017) opined on the tiny dimensions of NPs of various metals such as Ag, Au, ZnO, TiO_2, Pt etc and their role in bringing about an effective antimicrobial effect (Table 3). Their extremely small size, shape and surface charge enable them to accomplish contact with the bacterial cell membrane via electrostatic interaction, Vander Waals force, receptor-ligand interactions etc. Upon successful contact, these nanoparticles could efficiently cross the cell membrane and interfere with metabolic pathways. Their broad target mechanism is one of its kinds that facilitate the paralysis of bacterial cell by inducing oxidative stress (Xu et al., 2016) (Figure 4).

Figure 4. Schematic representation of Mechanism of action of various metal nanoparticles on bacterial cell

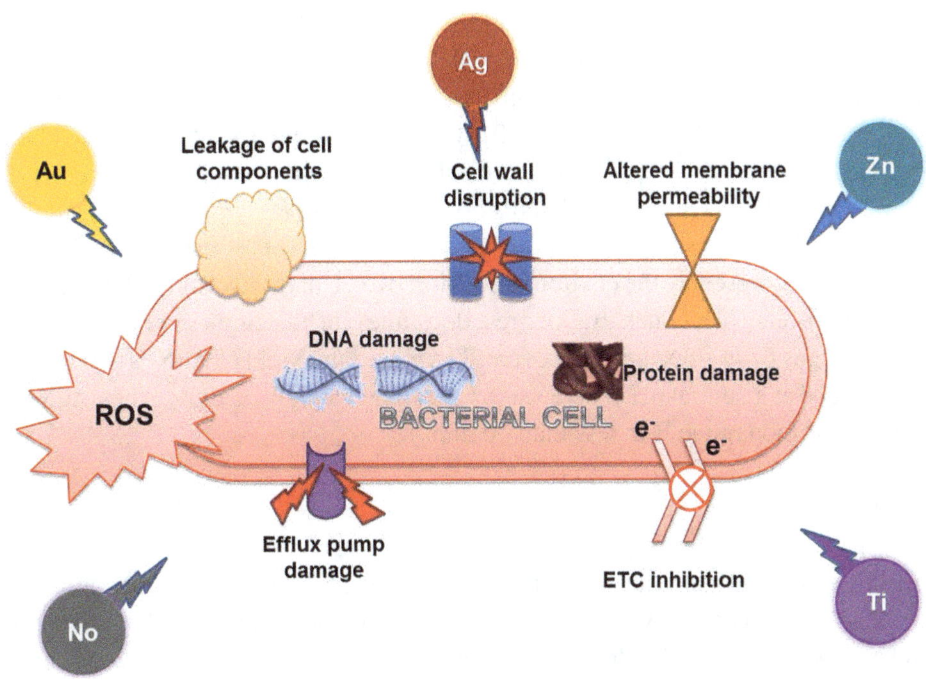

Although, MDR related cases have been surging, there have been reports on female infertility due to MDR infections. Gopinath et al. (2015) reported that MDR strains viz. *Enterobacter* sp., *P. aeruginosa*, *K. pneumoniae* and *E. coli* were responsible for infertility in females. Further, they demonstrated the synergistic antibacterial activity of AgNPs synthesized from *Fusarium oxysporum* along with antibiotics. It was inferred from the study that the resistant bacteria were found susceptible to AgNPs in a dose dependent fashion.

In recent years, bio-fabrication of various metal nanoparticles with significant biological activity comparable to other (physical and chemical) route has been reported. Moreover, extracellular synthesis of metal nanoparticles using bacteria with potential antimicrobial activity against Gram +ve and –ve strains including MDR strains has also been investigated (Panacek et al., 2006; Nanda et al., 2009). Furthermore, the correlation between size-dependent AgNPs and the inhibitory effect toward MRSA and Gram- negative bacteria have been well documented. Haq et al (2015) have demonstrated the antibacterial effect of AgNPs synthesized from higher fungi viz. *Agaricus bisporus*, *Helvella lacunosa*, *Ganoderma applanatum*, *Pleurotus florida* and *Fomes fomentarius* with potential activity exhibited by *A. bisporus*. Similarly, Behera and Nayak (2013) have prepared AgNPs using *Syzygium cumini* (jamun) extract and tested its activity against MRSA, ampicillin-resistant *E. coli* and erythromycin-resistant *Streptococcus pyogenes*. Alongside, surface-modified AgNPs fabricated using pluronic acid was also found to exhibit significant antimicrobial activity toward MRSA strains (Marta et al., 2014).

Further, Agarwal et al., (2013) have investigated on the potential inhibitory effect of bio-mediated AgNPs against *M. tuberculosis* and 26 various clinical isolates including MDR-TB and XDR-TB. Similarly, Ninganagouda et al., (2013) have reported on the susceptibility of MDR strains (*P. aeruginosa*, *E. coli* and *K. pneumoniae*) to AgNPs synthesized from microfungus, *Aspergillus flavus*. Saeb et al., (2014) have synthesized AgNPs using soil microbes viz. *Escherichia hermannii*, *Citrobacter sedlakii* and *Pseudomonas putida* and evaluated its antimicrobial potential against a panel of clinical isolates including MRSA. It was inferred that AgNPs significantly inhibited the growth of MRSA and whose activity was also enhanced by combining them with vancomycin antibiotic.

Beveridge and Murray (1980) for the first time reported the deposition of nano gold (AuNPs) on the cell wall of *B. subtilis* from gold chloride ($HAuCl_4$). Mishra et al., (2014) have optimized physico-chemical parameters for the biofabrication of AuNPs using the culture supernatant of *Hypocrea lixii* and *Trichoderma viride*. Similarly, Kitching et al., (2016) have extracellularly synthesized AuNPs using the membrane proteins of *Rhizopus oryzae* for biomedical and environmental applications.

Srinath et al., (2018) have synthesized AuNPs efficiently using *B. subtilis* isolated from Hatti Gold Mine, India. Further, the synthesized AuNPs was used as a biocatalyst for the degradation of methylene blue and other toxic dyes present in the environment. Moreover, Bhargava et al., (2016) have investigated on the process parameters (pH, concentration, reaction time) influencing the crystal size and yield of AuNPs using the fungus *Cladosporium oxysporum*. The maximum yield of AuNPs to biomass and water was found at 1:5:1 and at a neutral pH. There was an excellent catalytic activity when AuNPs were used for the degradation of rhodamine B in less than 10 min.

a. Mechanism

Among metals silver and gold have long been associated with human civilization as an ornament owing to its biocompatibility and non-toxic nature. Hitherto, in their nanoform they acquire remarkable properties and are considered safe for drug delivery applications. Mohamed et al., (2017) have evaluated the antimicrobial efficacy of AuNPs synthesized using co-precipitation method against *Corynebacterium pseudotuberculosis*. The MIC was determined to be 200 µg and 100 µg for AuNPs and AuNPs – laser combined therapy respectively. It was inferred that the antibacterial effect of AuNPs was due to the generation of ROS increasing the oxidative stress and thus leading to vacuole formation. Further, the activity was enhanced by the synergistic activity of AuNPs-laser therapy for possible application in the treatment of chronic caseous lymphadenitis.

Table 3. Modus operandi of Metallic nanoparticles

Nanoparticles	Antimicrobial mechanism
Silver (AgNPs)	Cell membrane disruption, DNA damage and electron transport chain inhibition
Gold (AuNPs)	Metal-microbe interaction facilitated by electrostatic attraction, accumulation and intrusion through cell membrane
Zinc oxide (ZnO)	Cell membrane disruption, mobilization & accumulation and ROS generation.
Titanium dioxide (TiO_2)	Cell membrane damage and ROS generation.
Nitric oxide-releasing NPs	Oxidative stress and Generation of ROS

Navale et al., (2015) have investigated on the antimicrobial and growth inhibitory effects induced by ZnONPs. The growth inhibitory study revealed that ZnO NPs of size 20 – 25 nm was found to possess significant bactericidal effects against fungi and bacteria when exposed to UV light. This event subsequently triggered generation of ROS inducing oxidative stress augmenting antibacterial effect. The possible mechanism of action of ZnO NPs in halting the cellular growth needs further research. It was presumed that ZnO NPs might disturb the cell integrity when in close contact with the bacteria, leading to the formation of ROS and releasing Zn^{2+} ions responsible for antimicrobial effect (Li et al., 2011; Lipovsky et al., 2011).

Furthermore, ZnO NPs exhibit remarkable photocatalytic activity and at the same time biocompatible. Nirmala et al., (2010) have synthesized ZnO NPs capable of degrading redox dye, methylene blue and also efficient enough inhibiting the growth of clinically significant bacteria viz. *B. subtilis, E. coli, S. typhi* and *S. aureus*. These nanoparticles upon UV irradiation generate ROS, H_2O_2, and superoxide (O^{2-}) capable of targeting broad spectrum of microbes (He et al., 2016; Siwinska et al., 2018) (Table 3).

Polymeric Nanoparticles

Nanoparticulate system with exorbitant biological properties is being extensively used in drug and gene delivery applications both *in vitro* and *in vivo*. For an efficient delivery, it is vital to understand the interaction between nanomaterials and biological system. There are various drug delivery systems bead-, liposome-, nanoparticle- and polymer-based systems for sustained and targeted delivery (Tiwari et al., 2012). Beside other systems, polymer-based carriers have been the most sought after delivery vehicle for ease of synthesis, manipulation and release of non-toxic products. Further, the tendency of cationic polymers in the formation of stable complexes made them more preferable than cationic lipids in surpassing the cell trafficking. On the other hand, nanoparticles were found to exhibit adjuvant properties facilitating efficient delivery of vaccines as they may be taken up by antigen-presenting cells (APCs).

Naturally occurring polymers (polysaccharides, alginates etc.) and fabricated polymers have shown promising candidates for vaccine delivery once converted to nanoforms. They form the base for the construction of nanoparticle-mediated drug delivery system. Peer et al. (2007) have reported on the use of polyalkyl cyanoacrylate nanoparticles for delivering anticancer drugs. It is noteworthy to mention that cationic polymers have been used for targeted drug delivery, imaging and detection of pathways leading to cell death. Furthermore, their gene delivery applications have also been explored *in vitro* and *in vivo* as well. The packaging of condensed and / or non-condensed DNA may greatly rely on the type and nature of polymer and the vector system. The most preferred ones are the biodegradable polymers owing to their reduced toxicity and accumulation in cells accounting for repeated administration. Upon degradation of polymer, the encapsulated drug / DNA / plasmid are efficiently released into the cytosol (Luten et al., 2008) (Table 4).

Table 4. Polymer based drug delivery system

Polymeric materials / Route of Administration	Disease / Clinical condition
PLGA (IM / ID / IN) and PLGA with CTAB	HIV, Measles virus, Hepatitis C Virus (HCV), Foot and Mouth Disease virus (FMDV) & Solid tumors
Amolimogene / ZYC101a (IM)	HPV cervical neoplasia
PLGA microspheres + PEI NPs (Oral / IM)	Model antigens
PLGA + PEI coating (ID / IM)	B-cell lymphoma
PLGA + PBAE (ID)	Tumor antigen
Chitosan NPs (IN / Oral)	Allergy / RSV / TB
PEI-mannose (Transdermal)	HIV

Poly (lactic-*co*-glycolic acid) (PLGA), one of the biodegradable polymers are at large used as an effective vehicle for delivering matrix antigen. In this line, fabrication of PLGA microspheres are proficient enough to be readily taken up by M cells and translocated into the lymphatic tissue in less than an hour. For instance, the core antigen of Hepatitis B virus loaded into PLGA NPs of size 300 nm triggered a strong immune response when compared to the core antigen without NPs as evidenced in mouse model. Most importantly, the size of the particle had a greater influence on eliciting an immune response. Administration of PLGA NPs of size range 200 – 600 nm induced IFN-γ at higher levels in response to Th1 cells. Alongside, PLGA microparticles in the size range of 2 – 8 μm stimulated the secretion of IL-4 in response to Th2 cells. These studies indicated that the nanoforms of both PLGA and liposomes were phagocytosed efficiently by the dendritic cells leading to their internalization and localization. However, acid hydrolytic degradation products of PLGA may restrict its use owing to loss of immunogenicity and posing detrimental effects to the entrapped protein / antigen when used along with organic solvents (Saroja et al., 2011). Most of the polymers have been used as carriers for delivering gene(s) with profound implications in the manufacture of DNA vaccines.

Quantum Dots (QDs)

QDs otherwise termed nanoscale semiconductor crystals exhibit unique optical and electronic properties in the form of bright fluorescence. QDs with their tunable optical properties have gained lot of interest on a par with the conventional labeling using dyes with limited emission probability at near infrared region. Moreover, their excellent photochemical stability, high quantum yield and size-dependent emission made them stand unique upon excitation with light of same wavelength. Secondly, their narrow range of emission facilitates detection of bands for multiple assays has drawn interest among researchers in the field of nano-theranostics for instantaneous probing, diagnosing, imaging and delivering therapy.

Kim et al. (2004) have proposed application of polydentate phosphine-coated QDs for cancer surgery in large animals through step-by-step mapping procedure in real-time, confirming complete resection. Similarly, Kobayashi et al. (2007) have developed Indium based InP@ZnS QDs for cellular imaging whereby complete visualization of lymphatic flow draining was investigated.

In recent years, heavy metal free (Cd, Hg, Pb) biocompatible QDs possessing high photoluminescence quantum yield were recently evaluated for their mapping capability of lymph node using *ex vivo* imaging model. However, indium-based QDs were the most preferred owing to strong retention, fast localization and photoluminescence stability confined to the regional lymph (Yaghini et al., 2016). Yu et al. (2016) have developed gadolinium-doped carbon dots for magnetic resonance imaging (MRI) imaging using one-pot hydrothermal method. The nanoprobes developed were in the size range of ~ 5.5 nm and spherical shaped with high spatial resolution for MRI with a broad emission spectra compared to that of only carbon dots.

Similarly, nitrogen and sulfur doped carbon dots of size 1.7 nm with a high fluorescence quantum yield (39.7%) were developed by Chen et al. (2017). Their low toxicity, good biocompatibility when incubated with HeLa cells was well documented. Such doped CDs were used to detect Cr(IV) by fluorescence quenching. Moreover, gadopentetic acid and glycine were employed to fabricate Gd-doped carbon dots of size ~18 nm for use in biomedical application (Du et al., 2017). As a theranostic agent they may act either as the chief nanocarrier or a part of a more complex structure as fluorescent labels (Chen et al. 2017). Cai et al. (2016) have proposed a pH sensitive ZnO QDs-Doxorubicin NPs targeting lung carcinoma. These nanocarriers sized ~3 nm were aminated and conjugated for specific binding with CD44 by cancer cells. Once triggered by pH, the nanocarrier dissociated and released DOX in a sustained fashion.

Hydrogels

Hydrogels formed by the network of crosslinked polymers that are hydrophilic capable of holding large amounts of water (Ahmed, 2015). Their tendency to extensively swell in an aqueous medium, make them the suitable candidate for biomedical applications (Daniele et al., 2014). Recently, hydrogels have been recognized to have a wide application in tissue engineering, drug delivery, biosensors and 3D bioprinting (Li et al., 2015). In comparison, these hydrogels exhibit biocompatibility, biodegradability, mechanical stability and high porosity. Alongside, their low strength, fragility and low interfacial tension with aqueous and other biological fluids represent them as a better alternative to natural tissues (Hamidi et al., 2008).

The highly porous stature of hydrogels forms a mesh that facilitates easy loading and protection of drugs from hostile environment during drug delivery. There exist a correlation between the porosity and the amount of drug loaded onto gel matrix. Yet another parameter that determines the drug delivery efficacy is the release rate that depends on the diffusion coefficient. These properties confer hydrogels a better candidate with a great potential for drug delivery applications (Qiu et al., 2001). The use of pluronic or poloxamer has seen a surge in bio-pharmaceutical field (Hoare et al., 2008) where an injectable gel-based poloxamer was fabricated for sustained release of lidocaine (Paavola et al., 1995). Such approach had the advantage of delivering psychotropic drugs in hospital environment to surpass the pitfalls encountered in the conventional method. Despite delivery, rapid diffusion and duration of drug release, need further tenability of gel matrix. Moreover, by adopting covalent crosslinking with other functional moieties viz. ethoxysilane, amine and / or carbohydrates, may help prevent the considerable loss of the drug when exposed to biological fluids (Kim et al., 2002).

LIMITATIONS

The recent years has seen a dramatic increase in the technological development and its implications in wide arena is a good sign of scientific refinement. As with the nanotechnological intervention, there is no doubt that we are progressing but at the same time we need to be prepared to run the risk while using manufactured nanomaterials. This demands a thorough understanding on the performance characteristics of the nanomaterials, their toxicity, environmental and human impacts. In this line, evaluating toxicity, dosing metrics and adverse effects must be closely monitored by pursuing *in vitro*, *in vivo* animal and human studies.

The remarkable physico-chemical properties acquired by the nanomaterials upon size reduction need extensive research on the mechanism of chemical reactivity and biological uncertainty. There is a pressing need for investigating the behavioral pattern of these nanoforms inside the biological system like activating immune response, inducing oxidative stress and damaging DNA / proteins. There is a great deal of efforts in bringing down the size and shape so as to explore the potential of nanomaterials to the fullest extent without affecting the surrounding environment. This requires a careful analysis of each and every particle for its interactions and tendency to form aggregates that may hinder from expressing its activity. There are other parameters that may contribute greatly for toxicity viz. chemical composition, surface charge and the role of functional groups on the nanoparticles.

As for the QDs are concerned, the toxicity may be due to surface functionalization or ROS generation which inturn damage the mitochondria, nucleus and plasma membranes. It is therefore essential to select a different type of nanoparticles based on the desired application. For instance, metal nanoparticles induce toxicity on its own and can have a broad target range whereas when they are functionalized with a polymer or capped with a protein moiety, the possibility of aggregation is reduced with no cytotoxic effects.

One of the major limiting factors dealing with nanotechnology and nanomaterials is the cost-effectiveness. The increased cost in health care settings would enable only the affordable to enjoy the privilege while making it difficult for the poor to access them (Hui Nicole Chia Poh, 2005). Furthermore, the ethical, societal and legal aspects of nanotechnology in medicine need critical analysis. Most importantly, the exposure of nanomaterials to humans need randomized clinical trials before being launched into the market pertaining to human exposure, environmental release and the typical half-life of the material concerned (Resnik et al., 2007).

CONCLUSION

Antimicrobial resistance (AMR) is considered as one of the major threat challenging the scientists across the globe for which humans are victimized. This phenomenon is predominantly evidenced not only in bacteria but also in fungi, viruses and protozoans. It is at this juncture that the world needs novel antimicrobials to counteract the current situation. In this context, the origin of antimicrobial resistance and the factors influencing the dissemination of AMR genes have been critically discussed. Further, the mechanism driving the antimicrobial resistance through HGT, plasmid, transposons etc. have also been described in brief. Besides antibiotics, novel nanotechnological approach in containing the spread of MDR strains has been reviewed. Metal nanoparticles that include Au, Ag, ZnO, TiO_2 etc. and their mechanism of action in the inhibition of MDR strains comprising bacteria have been elaborated with illustrations. Moreover, improving the antimicrobial efficacy of nanoparticulate system by conjugating with either polymer or with other antibiotics (synergistic effect) has also been addressed. Most importantly, the limitations of nanoparticulate system under various physiological conditions and lack of in vivo animal and human studies have been mentioned. In addition, the environmental impact, huge investment and high technical expertise made nanotechnology available to the affordable leaving behind the poor. Once these constraints are considerably reduced, the possibility of nanomedicine and its reach might have a significant impact on the healthcare system for building a disease-free society.

ACKNOWLEDGMENT

The author greatly thanks all the authors and their contributions that served as a base for drafting this book chapter. The timely support and the infrastructure facilities provided by Saveetha Institute of Medical and Technical Sciences are greatly acknowledged. This research received no specific grant from any funding agency in the public, commercial or not-for-profit sectors.

REFERENCES

Abdul-Aziz, M., Lipman, J., Mouton, J., Hope, W., & Roberts, J. (2015). Applying Pharmacokinetic / Pharmacodynamic Principles in Critically Ill Patients: Optimizing Efficacy and Reducing Resistance Development. *Seminars in Respiratory and Critical Care Medicine*, *36*(01), 136–153. doi:10.1055-0034-1398490 PMID:25643277

Abraham, E. P., & Chain, E. (1940). An enzyme from bacteria able to destroy penicillin. *Nature*, *146*(3713), 837. doi:10.1038/146837a0

Agarwal, P., Mehta, A., Kachhwaha, S., & Kothari, S. L. (2013). Green Synthesis of Silver Nanoparticles and Their Activity Against *Mycobacterium tuberculosis*. *Advanced Science, Engineering and Medicine*, *5*(7), 709–714.

Ahmed, E. M. (2015). Hydrogel: Preparation, characterization, and applications: A review. *Journal of Advanced Research*, *6*(2), 105–121. doi:10.1016/j.jare.2013.07.006 PMID:25750745

Alcalde-Rico, M., Hernando-Amado, S., Blanco, P., Martínez, J. L. (2016). Multidrug Efflux Pumps at the Crossroad between Antibiotic Resistance and Bacterial Virulence. *Frontiers in Microbiology, 21*(7), 1483.

Alexander, J. W. (2009). History of the medical use of silver. *Surgical Infections, 10*(3), 289–292. doi:10.1089ur.2008.9941 PMID:19566416

Appelbaum, P. (2006). The emergence of vancomycin-intermediate and vancomycin-resistant *Staphylococcus aureus. Clinical Microbiology and Infection, 12*, 16–23. doi:10.1111/j.1469-0691.2006.01344.x PMID:16445720

Arciola, C. R., Campoccia, D., Speziale, P., Montanaro, L., & Costerton, J. W. (2012). Biofilm formation in *Staphylococcus* implant infections. A review of molecular mechanisms and implications for biofilm-resistant materials. *Biomaterials, 33*(26), 5967–5982. doi:10.1016/j.biomaterials.2012.05.031 PMID:22695065

Arora, S., Jain, J., Rajwade, J. M., & Paknikar, K. M. (2009). Interactions of silver nanoparticles with primary mouse fibroblasts and liver cells. *Toxicology and Applied Pharmacology, 236*(3), 310–318. doi:10.1016/j.taap.2009.02.020 PMID:19269301

AshaRani, P. V., Low Kah Mun, G., Hande, M. P., & Valiyaveettil, S. (2009). Cytotoxicity and genotoxicity of siver nanoparticles in human cells. *ACS Nano, 3*(2), 279–290. doi:10.1021/nn800596w PMID:19236062

ATSDR (Agency for toxic substances and Disease Registry). (1990). *Toxicological profile for Silver.* Prepared by Clement international corporation, under Contract 205-88-0608. U.S. public Health Service. ATSDR/TP-90-24.

Barber, M. (1961). Methicillin-resistant staphylococci. *Journal of Clinical Pathology, 14*(4), 385–393. doi:10.1136/jcp.14.4.385 PMID:13686776

Bate, R., Jin, G. Z., Mathur, A., & Attaran, A. (2016). Poor Quality Drugs and Global Trade: A Pilot Study. *American Journal of Health Economics, 2*(3), 373–398. doi:10.1162/AJHE_a_00049

Baym, M., Stone, L. K., & Kishony, R. (2016). Multidrug evolutionary strategies to reverse antibiotic resistance. *Science, 351*(6268), 3292. doi:10.1126cience.aad3292 PMID:26722002

Bergmiller, T., Andersson, A. M. C., Tomasek, K., Balleza, E., Kiviet, D. J., Hauschild, R., & Guet, C. C. (2017). Biased partitioning of the multidrug efflux pump AcrAB-TolC underlies long-lived phenotypic heterogeneity. *Science, 356*(6335), 311–315. doi:10.1126cience.aaf4762 PMID:28428424

Bernardi, P., Scorrano, L., Colonna, R., Petronilli, V., & Di Lisa, F. (1999). Mitochondria and cell death. *European Journal of Biochemistry, 264*(3), 687–701. doi:10.1046/j.1432-1327.1999.00725.x PMID:10491114

Beveridge, T. J., & Murray, R. G. (1980). Sites of metal deposition in the cell wall of *Bacillus subtilis. Journal of Bacteriology, 141*(2), 876–887. PMID:6767692

Bhargava, A., Jain, N., Khan, M. A., Pareek, V., Dilip, R. V., & Panwar, J. (2016). Utilizing metal tolerance potential of soil fungus for efficient synthesis of gold nanoparticles with superior catalytic activity for degradation of rhodamine B. *Journal of Environmental Management*, *1*(183), 22–32. doi:10.1016/j.jenvman.2016.08.021 PMID:27567934

Blair, J. M. A., Bavro, V. N., Ricci, V., Modi, N., Cacciotto, P., Kleinekathöfer, U., & Piddock, L. J. V. (2015). AcrB drug-binding pocket substitution confers clinically relevant resistance and altered substrate specificity. *Proceedings of the National Academy of Sciences of the United States of America*, *112*(11), 3511–3516. doi:10.1073/pnas.1419939112 PMID:25737552

Blair, J. M. A., Webber, M. A., Baylay, A. J., Ogbolu, D. O., & Piddock, L. J. V. (2015). Molecular mechanisms of antibiotic resistance. *Nature Reviews. Microbiology*, *13*(1), 42–51. doi:10.1038/nrmicro3380 PMID:25435309

Brown, C. L., Bushell, G., Whitehouse, M. W., Agrawal, D., Tupe, S., Paknikar, K., & Tiekink, E. R. (2007). Nanogoldpharmaceutics. *Gold Bulletin*, *40*(3), 245–250. doi:10.1007/BF03215588

Cai, X., Luo, Y., Zhang, W., Du, D., & Lin, Y. (2016). pH-Sensitive ZnO quantum dots–doxorubicin nanoparticles for lung cancer targeted drug delivery. *ACS Applied Materials & Interfaces*, *8*(34), 22442–22450. doi:10.1021/acsami.6b04933 PMID:27463610

Carattoli, A. (2001). Importance of integrons in the diffusion of resistance. *Veterinary Research*, *32*(3-4), 243–259. doi:10.1051/vetres:2001122 PMID:11432416

Carlson, C., Hussain, S. M., Schrand, A. M. K., Braydich-Stolle, L., Hess, K. L., Jones, R. L., & Schlager, J. J. (2008). Unique cellular interaction of silver nanoparticles: Size-dependent generation of reactive oxygen species. *The Journal of Physical Chemistry B*, *112*(43), 13608–13619. doi:10.1021/jp712087m PMID:18831567

Castellano, J. J., Shafii, S. M., Ko, F., Donate, G., Wright, T. E., Mannari, R. J., & Robson, M. C. (2007). Comparative evaluation of silver-containing antimicrobial dressings and drugs. *International Wound Journal*, *4*(2), 114–122. doi:10.1111/j.1742-481X.2007.00316.x PMID:17651227

Chambers, H. F., & DeLeo, F. R. (2009). Waves of resistance: *Staphylococcus aureus* in the antibiotic era. *Nature Reviews. Microbiology*, *7*(9), 629–641. doi:10.1038/nrmicro2200 PMID:19680247

Chen, J., Liu, J., Li, J., Xu, L., & Qiao, Y. (2017). One-pot synthesis of nitrogen and sulphur co-doped carbon dots and its application for sensor and multi-color cellular imaging. *Journal of Colloid and Interface Science*, *485*, 167–174. doi:10.1016/j.jcis.2016.09.040 PMID:27662029

Chopra, I., Hesse, L., & O'Neill, A. J. (2002). Exploiting current understanding of antibiotic action for discovery of new drugs. *Journal of Applied Microbiology*, *92*, 4S–15S. doi:10.1046/j.1365-2672.92.5s1.13.x PMID:12000608

Costa, C. S., Ronconi, J. V. V., Daufenbach, J. F., Gonçalves, C. L., Rezin, G. T., Streck, E. L., & da Silva Paula, M. M. (2010). In vitro effects of silver nanoparticles on the mitochondrial respiratory chain. *Molecular and Cellular Biochemistry*, *342*(1-2), 51–56. PMID:20411305

Couturier, M., Bex, F., Bergquist, P. L., & Maas, W. K. (1988). Identification and classification of bacterial plasmids. *Clinical Microbiology Reviews*, *52*(3), 375–395. doi:10.1128/MR.52.3.375-395.1988 PMID:3054468

Daniele, M. A., Adams, A. A., Naciri, J., North, S. H., & Ligler, F. S. (2014). Interpenetrating networks based on gelatin methacrylamide and PEG formed using concurrent thiol click chemistries for hydrogel tissue engineering scaffolds. *Biomaterials*, *35*(6), 1845–1856. doi:10.1016/j.biomaterials.2013.11.009 PMID:24314597

Dantas, G., Sommer, M. O., Oluwasegun, R. D., & Church, G. M. (2008). Bacteria subsisting on antibiotics. *Science*, *320*(5872), 100–103. doi:10.1126cience.1155157 PMID:18388292

Del Grosso, M., Camilli, R., Barbabella, G., Blackman Northwood, J., Farrell, D. J., & Pantosti, A. (2011). Genetic resistance elements carrying mef subclasses other than mef(A) in *Streptococcus pyogenes*. *Antimicrobial Agents and Chemotherapy*, *55*(7), 3226–3230. doi:10.1128/AAC.01713-10 PMID:21502613

del Pozo, J. L., & Patel, R. (2007). The challenge of treating biofilm-associated bacterial infections. *Clinical Pharmacology and Therapeutics*, *82*(2), 204–209. doi:10.1038j.clpt.6100247 PMID:17538551

Depardieu, F., Podglajen, I., Leclercq, R., Collatz, E., & Courvalin, P. (2007). Modes and Modulations of Antibiotic Resistance Gene Expression. *Clinical Microbiology Reviews*, *20*(1), 79–114. doi:10.1128/CMR.00015-06 PMID:17223624

Dmitriev, B. A., Toukach, F. V., Holst, O., Rietschel, E. T., & Ehlers, S. (2004). Tertiary structure of *Staphylococcus aureus* cell wall murein. *Journal of Bacteriology*, *186*(21), 7141–7148. doi:10.1128/JB.186.21.7141-7148.2004 PMID:15489425

Dong, J., Signo, K. S. L., Vanderlinder, E. M., Yost, C. K., & Dahms, T. E. S. (2011). Atomic force microscopy of a ctpA mutant in *Rhizobium leguminosarum* reveals surface defects linking CtpA function to biofilm formation. *Microbiology*, *157*(11), 3049–3058. doi:10.1099/mic.0.051045-0 PMID:21852352

Du, F., Zhang, L., Zhang, L., Zhang, M., Gong, A., Tan, Y., & Zou, S. (2017). Engineered gadolinium-doped carbon dots for magnetic resonance imaging-guided radiotherapy of tumors. *Biomaterials*, *121*, 109–120. doi:10.1016/j.biomaterials.2016.07.008 PMID:28086179

Carloni, E., Andreoni, F., Omiccioli, E., Villa, L., Magnani, M., & Carattoli, A. (2017). Comparative analysis of the standard PCR-Based Replicon Typing (PBRT) with the commercial PBRT-KIT. *Plasmid*, *90*, 10–14. doi:10.1016/j.plasmid.2017.01.005 PMID:28137396

Pisani, E. (2015). Antimicrobial resistance: What does medicine quality have to do with it? *Antimicrobial Resistance Review,* 1-40. Retrieved August 14, 2020, from https://amr-review.org/sites/default/files/ElizabethPisaniMedicinesQualitypaper.pdf

Fayaz, A. M., Balaji, K., Girilal, M., Yadav, R., Kalaichelvan, P. T., & Venketesan, R. (2010). Biogenic synthesis of silver nanoparticles and their synergistic effect with antibiotics: A study against Gram-positive and Gram-negative bacteria. *Nanomedicine (London)*, *6*(1), 103–109. doi:10.1016/j.nano.2009.04.006 PMID:19447203

Fenchel, T., Blackburn, H., & King, G. M. (2012). *Bacterial Biogeochemistry: The ecophysiology of mineral cycling* (3rd ed.). Academic Press/Elsevier.

Feng, Q. L., Wu, J., Chen, G. Q., Cui, F. Z., Kim, T. N., & Kim, J. O. (2000). A mechanistic study of the antibacterial effect of silver ions on *Escherichia coli* and *Staphylococcus aureus. Journal of Biomedical Materials Research, 52*(4), 662–668. doi:10.1002/1097-4636(20001215)52:4<662::AID-JBM10>3.0.CO;2-3 PMID:11033548

Fischbach, M. A., & Walsh, C. T. (2009). Antibiotics for emerging pathogens. *Science, 325*(5944), 1089–1093. doi:10.1126cience.1176667 PMID:19713519

Fluit, A. C., & Schmitz, F. J. (2004). Resistance integrons and super-integrons. *Clinical Microbiology and Infection, 10*(4), 272–288. doi:10.1111/j.1198-743X.2004.00858.x PMID:15059115

Foldbjerg, R., Dang, D. A., & Autrup, H. (2011). Cytotoxicity and genotoxicity of silver nanoparticles in the human lung cancer cell line, A549. *Archives of Toxicology, 85*(7), 743–750. doi:10.100700204-010-0545-5 PMID:20428844

Forbes, B. A., Sahm, D. F., & Weissfeld, A. S. (1998). *Bailey And Scott's Diagnostic Microbiology* (10th ed.). Mosby Inc.

Fraimow, H. S., & Tsigrelis, C. (2011). Antimicrobial resistance in the intensive care unit: Mechanisms, epidemiology, and management of specific resistant pathogens. *Critical Care Clinics, 27*(1), 163–205. doi:10.1016/j.ccc.2010.11.002 PMID:21144992

Franco-Molina, M. A., Mendoza-Gamboa, E., Sierra-Rivera, C. A., Gómez-Flores, R. A., Zapata-Benavides, P., Castillo-Tello, P., & Rodríguez-Padilla, C. (2010). Antitumor activity of colloidal silver on MCF-7 human breast cancer cells. *Journal of Experimental & Clinical Cancer Research, 29*(1), 148. doi:10.1186/1756-9966-29-148 PMID:21080962

Frost, L. S., & Koraimann, G. (2010). Regulation of bacterial conjugation: Balancing opportunity with adversity. *Future Microbiology, 5*(7), 1057–1071. doi:10.2217/fmb.10.70 PMID:20632805

Ghosh, S., Patil, S., Ahire, M., Kitture, R., Kale, S., Pardesi, K., Cameotra, S. S., Bellare, J., Dhavale, D. D., Jabgunde, A., & Chopade, B. A. (2012). Synthesis of silver nanoparticles using *Dioscorea bulbifera* tuber extract and evaluation of its synergistic potential in combination with antimicrobial agents. *International Journal of Nanomedicine, 7*, 483–496. PMID:22334779

Giguère, S. (2006). Antimicrobial Drug Action and Interaction: An Introduction. Antimicrobial therapy in Veterinary Medicine (4th ed.). Blackwell Publishing.

Gootz, T. D. (1990). Discovery and development of new antimicrobial agents. *Clinical Microbiology Reviews, 3*(1), 13–31. doi:10.1128/CMR.3.1.13 PMID:2404566

Gopinath, P. M., Narchonai, G., Dhanasekaran, D., Ranjani, A., & Thajuddin, N. (2015). Mycosynthesis, characterization and antibacterial properties of AgNPs against multidrug resistant (MDR) bacterial pathogens of female infertility cases. *Asian Journal of Pharmaceutical Sciences, 10*(2), 138–145. doi:10.1016/j.ajps.2014.08.007

Greenblatt, M. S., Bennet, W. P., Hollstein, M., & Harris, C. C. (1994). Mutations in the p53 tumor suppressor gene: Clues to cancer etiology and molecular pathogenesis. *Cancer Research*, *54*(18), 4855. PMID:8069852

Greene, C., Vadlamudi, G., Newton, D., Foxman, B., & Xi, C. (2016). The influence of biofilm formation and multidrug resistance on environmental survival of clinical and environmental isolates of *Acinetbacter baumannii*. *American Journal of Infection Control*, *44*(5), e65–e71. doi:10.1016/j.ajic.2015.12.012 PMID:26851196

Hamidi, M., Azadi, A., & Rafiei, P. (2008). Hydrogel nanoparticles in drug delivery. *Advanced Drug Delivery Reviews*, *60*(15), 1638–1649. doi:10.1016/j.addr.2008.08.002 PMID:18840488

He, W., Jia, H., Cai, J., Han, X., Zheng, Z., Wamer, W. G., & Yin, J. J. (2016). Production of reactive oxygen species and electrons from photoexcited ZnO and ZnS nanoparticles: A comparative study for unraveling their distinct photocatalytic activities. *The Journal of Physical Chemistry C*, *120*(6), 3187–3195. doi:10.1021/acs.jpcc.5b11456

Hill, W. R., & Pillsbury, D. M. (1939). *Argyria–The Pharmacology of Silver Baltimore*. Williams and Wilkins.

Hoare, T. R., & Kohane, D. S. (2008). Hydrogels in drug delivery: Progress and challenges. *Polymer*, *49*(8), 1993–2007. doi:10.1016/j.polymer.2008.01.027

Hoge, C. W., Gambel, J. M., Srijan, A., Pitarangsi, C., & Echeverria, P. (1998). Trends in antibiotic resistance among diarrheal pathogens isolated in Thailand over 15 years. *Clinical Infectious Diseases*, *26*(2), 341–345. doi:10.1086/516303 PMID:9502453

Hoiby, N., Ciofu, O., Johansen, H. K., Song, Z. J., Moser, C., Jensen, P. O., Molin, S., Givskov, M., Tolker-Nielsen, T., & Bjarnsholt, T. (2011). The clinical impact of bacterial biofilms. *International Journal of Oral Science*, *3*(2), 55–65. doi:10.4248/IJOS11026 PMID:21485309

Holmes, A. H., Moore, L. S., Sundsfjord, A., Steinbakk, M., Regmi, S., Karkey, A., Guerin, P. J., & Piddock, L. J. (2015). Understanding the mechanisms and drivers of antimicrobial resistance. *Lancet*, *387*(10014), 176–187. doi:10.1016/S0140-6736(15)00473-0 PMID:26603922

Holt, K. B., & Bard, A. J. (2005). Interaction of silver (I) ions with the respiratory chain of *Escherichia coli*: An electrochemical and scanning electrochemical microscopy study of the antimicrobial mechanism of micromolar Ag^{+}. *Biochemie.*, *44*(39), 13214–13223. doi:10.1021/bi0508542 PMID:16185089

Hui Nicole Chia Poh. (2005). *Nanomedicine and Cancer. Science and Technology Studies, Section 84405*. Retrieved August 14, 2020, from http://www.tahan.com/charlie/nanosociety/course201/nanos/NH.pdf

Hussain, S. M., Hess, K. L., Gearhart, J. M., Geiss, K. T., & Schlager, J. J. (2005). In vitro toxicity of nanoparticles in BRL 3A rat liver cells. *Toxicology In Vitro*, *19*(7), 975–983. doi:10.1016/j.tiv.2005.06.034 PMID:16125895

Imran, M., Iqbal, J., & Mehmood, T. (2006). Synthesis, characterization and (*in vitro*) Screening of Amoxicillin and its complexes with Ag(I), Zn(II), Cu(II), Co(II), Ni(II). *Journal of Biosciences*, *5*, 946–949.

Katayama, Y., Ito, T., & Hiramatsu, K. (2000). A new class of genetic element, *Staphylococcus* cassette chromosome *mec*, encodes methicillin resistance in *Staphylococcus aureus. Antimicrobial Agents and Chemotherapy*, *44*(6), 1549–1555. doi:10.1128/AAC.44.6.1549-1555.2000 PMID:10817707

Lee, K. (2003). *Health impacts of globalization: towards global governance*. Palgrave Macmillan. doi:10.1057/9781403943828

Keren, I., Kaldalu, N., Spoering, A., Wang, Y., & Lewis, K. (2004). Persister cells and tolerance to antimicrobials. *FEMS Microbiology Letters*, *230*(1), 13–18. doi:10.1016/S0378-1097(03)00856-5 PMID:14734160

Kim, M. R., & Park, T. G. (2002). Temperature-responsive and degradable hyaluronic acid/Pluronic composite hydrogels for controlled release of human growth hormone. *Journal of Controlled Release*, *80*(1-3), 69–77. doi:10.1016/S0168-3659(01)00557-0 PMID:11943388

Kim, S., Lim, Y. T., Soltesz, E. G., De Grand, A. M., Lee, J., Nakayama, A., Parker, J. A., Mihalijevic, T., Laurence, R. G., Dor, D. M., Cohn, L. H., Bawendi, M. G., & Frangioni, J. V. (2004). Near-infrared fluorescent type II quantum dots for sentinel lymph node mapping. *Nature Biotechnology*, *22*(1), 93–97. doi:10.1038/nbt920 PMID:14661026

Kitching, M., Choudhary, P., Inguva, S., Guo, Y., Ramani, M., Das, S. K., & Marsili, E. (2016). Fungal surface protein mediated one-pot synthesis of stable and hemocompatible gold nanoparticles. *Enzyme and Microbial Technology*, *95*, 76–84. doi:10.1016/j.enzmictec.2016.08.007 PMID:27866629

Kobayashi, H., Hama, Y., Koyama, Y., Barrett, T., Regino, C. A. S., Urano, Y., & Choyke, P. L. (2007). Simultaneous multicolor imaging of five different lymphatic basins using quantum dots. *Nano Letters*, *7*(6), 1711–1716. doi:10.1021/nl0707003 PMID:17530812

Kong, C., Neoh, H. M., & Nathan, S. (2016). Targeting *Staphylococcus aureus* toxins: A potential form of anti-virulence therapy. *Toxins*, *8*(3), 72. doi:10.3390/toxins8030072 PMID:26999200

Kumar, R., Roopan, S. M., Prabhakar, A., Khanna, V. G., & Chakroborty, S. (2012b). Agricultural waste *Annona squamosa* peel extract: Biosynthesis of silver nanoparticles. *Spectrochimica Acta Part A*, *90*, 173–176. doi:10.1016/j.saa.2012.01.029 PMID:22336049

Kumari, M., Mukherjee, A., & Chandrasekaran, N. (2009). Genotoxicity of silver nanoparticles in *Allium cepa. The Science of the Total Environment*, *407*(19), 5243–5246. doi:10.1016/j.scitotenv.2009.06.024 PMID:19616276

Kuriyama, R., & Sakai, H. (1974). Role of tubulin-SH group in polymerization to microtubules. *Journal of Biochemistry*, *76*(3), 651–654. doi:10.1093/oxfordjournals.jbchem.a130609 PMID:4474165

Labbate, M., Case, R. J., & Stokes, H. W. (2009). The integrin /gene cassette system: An active player in bacterial adaptation. *Methods in Molecular Biology (Clifton, N.J.)*, *532*, 103–125. doi:10.1007/978-1-60327-853-9_6 PMID:19271181

Lara, H. H., Ayala-Nunez, N. V., Turrent, L. C. I., & Padilla, C. R. (2010). Bactericidal effect of silver nanoparticles against multi-drug resistant bacteria. *World Journal of Microbiology & Biotechnology*, *26*(4), 615–621. doi:10.100711274-009-0211-3

Laxminarayan, R., & Chaudhury, R. R. (2016). Antibiotic Resistance in India: Drivers and Opportunities for Action. *PLoS Medicine*, *13*(3), e1001974. doi:10.1371/journal.pmed.1001974 PMID:26934098

Levy, S. B. (2007). *Antibiotic Resistance: An Ecological Imbalance, in Ciba Foundation Symposium 207 - Antibiotic Resistance: Origins, Evolution, Selection and Spread*. John Wiley & Sons, Ltd.

Lewis, K. (2005). Persister cells and the riddle of biofilm survival. *Biochemistry (Moscow)*, *70*(2), 267–274. doi:10.100710541-005-0111-6 PMID:15807669

Li, C., Liu, H., Sun, Y., Wang, H., Guo, F., Rao, S., Deng, J., Zhang, Y., Miao, Y., Guo, C., Meng, J., Chen, X., Li, L., Li, D., Xu, H., Wang, H., Li, B., & Jiang, C. (2009). PAMAM nanoparticles promote acute lung injury by inducing autophagic cell death through the Akt- TSC2-mTOR signaling pathway. *Journal of Molecular Cell Biology*, *1*(1), 37–45. doi:10.1093/jmcb/mjp002 PMID:19516051

Li, L., Wang, Y., Pan, L., Shi, Y., Cheng, W., Shi, Y., & Yu, G. A. (2015). Nanostructured conductive hydrogels-based biosensor platform for human metabolite detection. *Nano Letters*, *15*(2), 1146–1151. doi:10.1021/nl504217p PMID:25569673

Li, M., Zhu, L., & Lin, D. (2011). Toxicity of ZnO nanoparticles to *Escherichia coli*: Mechanism and the influence of medium components. *Environmental Science & Technology*, *45*(5), 1977–1983. doi:10.1021/es102624t PMID:21280647

Li, X. Z., Plésiat, P., & Nikaido, H. (2015). The challenge of efflux-mediated antibiotic resistance in Gram-negative bacteria. *Clinical Microbiology Reviews*, *28*(2), 337–418. doi:10.1128/CMR.00117-14 PMID:25788514

Lipovsky, A., Nitzan, Y., Gedanken, A., & Lubart, R. (2011). Antifungal activity of ZnO nanoparticles - The role of ROS mediated cell injury. *Nanotechnology*, *22*(10), 105101. doi:10.1088/0957-4484/22/10/105101 PMID:21289395

Lowy, F. D. (2003). Antimicrobial resistance: The example of *Staphylococcus aureus. The Journal of Clinical Investigation*, *111*(9), 1265–1273. doi:10.1172/JCI18535 PMID:12727914

Luten, J., van Nostrum, C. F., De Smedt, S. C., & Hennink, W. E. (2008). Biodegradable polymers as non-viral carriers for plasmid DNA delivery. *Journal of Controlled Release*, *126*(2), 97–110. doi:10.1016/j.jconrel.2007.10.028 PMID:18201788

Manzoor-ul-Haq, Rathod, V., Singh, D., Singh, A. K., Ninganagouda, S., & Hiremath, J. (2015). Dried Mushroom *Agaricus bisporus* mediated synthesis of silver nanoparticles from Bandipora District (Jammu and Kashmir) and their efficacy against Methicillin Resistant *Staphylococcus aureus* (MRSA) strains. *Nanoscience and Nanotechnology*, *5*(1), 1–8. PMID:26328301

Marta, B., Jakab, E., Potara, M., Simon, T., Imre-Lucaci, F., Barbu-Tudoran, L., Popescu, O., & Astilean, S. (2014). Pluronic-coated silver nanoprisms: Synthesis, characterization and their antibacterial activity. *Colloids and Surfaces. A, Physicochemical and Engineering Aspects*, *441*, 77–83. doi:10.1016/j.colsurfa.2013.08.076

Mazel, D. (2006). Integrons: Agents of bacterial evolution. *Nature Reviews. Microbiology*, *4*(8), 608–620. doi:10.1038/nrmicro1462 PMID:16845431

McDonnell, G. E. (2007). Chemical Disinfection. In *Antisepsis, disinfection and sterilization* (pp. 111–115). STERIS Limited.

McNeill, W. H. (1976). Plagues and peoples. Bantam, Doubleday Dell Publishing Group, Inc.

Merino, S., Camprubi, S., Alberti, S., Benedi, V. J., & Tomás, J. M. (1992). Mechanisms of *Klebsiella pneumoniae* resistance to complement-mediated killing. *Infection and Immunity*, *60*(6), 2529–2535. doi:10.1128/IAI.60.6.2529-2535.1992 PMID:1587619

Mishra, A., Kumari, M., Pandey, S., Chaudhry, V., Gupta, K. C., & Nautiyal, C. S. (2014). Biocatalytic and antimicrobial activities of gold nanoparticles synthesized by *Trichoderma* sp. *Bioresource Technology*, *166*, 235–242. doi:10.1016/j.biortech.2014.04.085 PMID:24914997

Mohamed, M. M., Fouad, S. A., Elshoky, H. A., Mohammed, G. M., & Salaheldin, T. A. (2017). Antibacterial effect of gold nanoparticles against *Corynebacterium pseudotuberculosis*. *International Journal of Veterinary Science and Medicine*, *5*(1), 23–29. doi:10.1016/j.ijvsm.2017.02.003 PMID:30255044

Moore, A. M., Patel, S., Forsberg, K. J., Wang, B., Bentley, G., Razia, Y., Qin, X., Tarr, P. I., & Dantas, G. (2013). Pediatric fecal microbiota harbor diverse and novel antibiotic resistance genes. *PLoS One*, *8*(11), e78822. doi:10.1371/journal.pone.0078822 PMID:24236055

Nanda, A., & Saravanan, M. (2009). Biosynthesis of silver nanoparticles from *Staphylococcus aureus* and its antimicrobial activity against MRSA and MRSE. *Nanomedicine (London)*, *5*(4), 452–456. doi:10.1016/j.nano.2009.01.012 PMID:19523420

Navale, G. R., Thripuranthaka, M., Late, D. J., & Shinde, S. S. (2015). Antimicrobial activity of ZnO nanoparticles against pathogenic bacteria and fungi. *JSM Nanotechnology and Nanomedicine*, *3*, 1033.

Nichols, R. J., Sen, S., Choo, Y. J., Beltrao, P., Zietek, M., Chaba, R., Lee, S., Kazmierczak, K. M., Lee, K. J., Wong, A., Shales, M., Lovett, S., Winkler, M. E., Krogan, N. J., Typas, A., & Gross, C. A. (2011). Phenotypic landscape of a bacterial cell. *Cell*, *144*(1), 143–156. doi:10.1016/j.cell.2010.11.052 PMID:21185072

Nikaido, H. (2009). Multidrug resistance in bacteria. *Annual Review of Biochemistry*, *78*(1), 119–146. doi:10.1146/annurev.biochem.78.082907.145923 PMID:19231985

Ninganagouda, S., Rathod, V., Jyoti, H., Singh, D., & Prema, K., & Manzoor-Ul-Haq. (2013). Extracellular biosynthesis of silver nanoparticles using *Aspergillus flavus* and their antimicrobial activity against gram negative MDR strains. *International Journal of Pharma and Bio Sciences*, *4*(2), 222–229.

Nirmala, M., Nair, M. G., Rekha, K., Anukaliani, A., Samdarshi, S., & Nair, R. G. (2010). Photocatalytic activity of ZnO nanopowders synthesized by DC thermal plasma. *African Journal of Basic & Applied Sciences*, *2*, 161–166.

Oda, K., Arakawa, H., Tanaka, T., Matsuda, K., Tanikawa, C., Mori, T., Nishimori, H., Tamai, K., Tokino, T., Nakamura, Y., & Taya, Y. (2000). p53AIP1, a potential mediator of p53-dependent apoptosis, and its regulation by Ser-46- phosphorylated p53. *Cell*, *15*(102(6)), 849–862. doi:10.1016/S0092-8674(00)00073-8 PMID:11030628

Paavola, A., Yliruusi, J., Kajimoto, Y., Kalso, E., Wahlström, T., & Rosenberg, P. (1995). Controlled release of lidocaine from injectable gels and efficacy in rat sciatic nerve block. *Pharmaceutical Research*, *12*(12), 1997–2002. doi:10.1023/A:1016264527738 PMID:8786979

Pal, S., Tak, Y. K., & Song, J. M. (2007). Does the antibacterial activity of silver nanoparticles depend on the shape of the nanoparticle? A study of the Gram-negative bacterium *Escherichia coli*. *Applied and Environmental Microbiology*, *73*(6), 1712–1720. doi:10.1128/AEM.02218-06 PMID:17261510

Panacek, A., Kvítek, L., Prucek, R., Kolar, M., Vecerova, R., Pizúrova, N., Sharma, V. K., Nevecna, T., & Zboril, R. (2006). Silver colloid nanoparticles: Synthesis, characterization and their antibacterial activity. *The Journal of Physical Chemistry B*, *110*(33), 16248–16253. doi:10.1021/jp063826h PMID:16913750

Papageorgiou, I., Brown, C., Schins, R., Singh, S., Newson, R., Davis, S., Fisher, J., Ingham, E., & Case, C. P. (2007). The effect of nano- and micron-sized particles of cobalt-chromium alloy on human fibroblasts in vitro. *Biomaterials*, *28*(19), 2946–2958. doi:10.1016/j.biomaterials.2007.02.034 PMID:17379299

Poulsen, L. K., Ballard, G., & Stahl, D. A. (1993). Use of rRNA fluorescence in situ hybridization for measuring the activity of single cells in young and established biofilms. *Applied and Environmental Microbiology*, *59*(5), 1354–1360. doi:10.1128/AEM.59.5.1354-1360.1993 PMID:7685999

Payne, D. J., Gwynn, M. N., Holmes, D. J., & Pompliano, D. L. (2007). Drugs for bad bugs: Confronting the challenges of antibacterial discovery. *Nature Reviews. Drug Discovery*, *6*(1), 29–40. doi:10.1038/nrd2201 PMID:17159923

Peer, D., Karp, J. M., Hong, S., Farokhzad, O. C., Margalit, R., & Langer, R. (2007). Nanocarriers as an emerging platform for cancer therapy. *Nature Nanotechnology*, *2*(12), 751–760. doi:10.1038/nnano.2007.387 PMID:18654426

Peña, J., Bargar, J. R., & Sposito, G. (2011). Role of bacterial biomass in the sorption of Ni by biomass-birnessite assemblages. *Environmental Science & Technology*, *45*(17), 7338–7344. doi:10.1021/es201446r PMID:21780745

Qiu, Y., & Park, K. (2001). Environment-sensitive hydrogels for drug delivery. *Advanced Drug Delivery Reviews*, *53*(3), 321–339. doi:10.1016/S0169-409X(01)00203-4 PMID:11744175

Rahman, A. E., Iqbal, A., Hoque, D. E., Moinuddin, M., Zaman, S. B., Rahman, Q., Begum, T., Chowdhury, A. I., Haider, R., Arifeen, S. E., Kissoon, N., & Larson, C. P. (2017). Managing neonatal and early childhood syndromic sepsis in sub-district hospitals in resource poor settings: Improvement in quality of care through introduction of a package of interventions in rural Bangladesh. *PLoS One*, *12*(1), e0170267. doi:10.1371/journal.pone.0170267 PMID:28114415

Sietze, R. (2007). The endothelial glycocalyx: Composition, functions, and visualization. *European Journal of Phycology*, *454*(3), 345–359. PMID:17256154

Resnik, D., & Tinkle, S. (2007). Ethics in Nanomedicine. *Nanomedicine (London)*, *2*(3), 345–350. doi:10.2217/17435889.2.3.345 PMID:17716179

Ribeiro, S. M., Felício, M. R., Boas, E. V., Goncalves, S., Costa, F. F., Samy, R. P., Santos, N. C., & Franco, O. L. (2016). New frontiers for anti-biofilm drug development. *Pharmacology & Therapeutics*, *160*, 133–144. doi:10.1016/j.pharmthera.2016.02.006 PMID:26896562

Richard, J. W., Spencer, B. A., McCoy, L. F., Carina, E., Washington, J., Edgar, P., Rosenblatt, J., & Goodheart, R. (2002). Acticoat versus silverlon: The truth. *Journal of Burns and Surgical Wound Care*, *1*, 11–20.

Roberts, A. P., & Mullany, P. (2009). A modular master on the move: The Tn916 family of mobile genetic elements. *Trends in Microbiology*, *6*(6), 251–258. doi:10.1016/j.tim.2009.03.002 PMID:19464182

Roberts, A. P., Chandler, M., Courvalin, P., Guédon, G., Mullany, P., Pembroke, T., Rood, J. I., Jeffery Smith, C., Summers, A. O., Tsuda, M., & Berg, D. E. (2008). Revised nomenclature for transposable genetic elements. *Plasmid*, *60*(3), 167–173. doi:10.1016/j.plasmid.2008.08.001 PMID:18778731

Saeb, A. T. M., Alshammari, A. S., Al-Brahim, H., & Al-Rubeaan, K. A. (2014). Production of silver nanoparticles with strong and stable antimicrobial activity against highly pathogenic and multidrug resistant bacteria. *TheScientificWorldJournal*, *2014*, 1–9. doi:10.1155/2014/704708 PMID:25093206

Partridge, S., Tsafnat, G., Coiera, E., & Iredell, J. (2009). Gene cassettes and cassette arrays in mobile resistance integrons. *FEMS Microbiology Reviews*, *33*(4), 757–784. doi:10.1111/j.1574-6976.2009.00175.x PMID:19416365

Saroja, Ch., Lakshmi, P. K., & Bhaskaran, S. (2011). Recent trends in vaccine delivery systems: A review. *International Journal of Pharmaceutical Investigation*, *1*(2), 64–74. doi:10.4103/2230-973X.82384 PMID:23071924

Schneider, G. (1984). Silver nitrate prophylaxis. *Canadian Medical Association Journal*, *131*, 193–196. PMID:6378348

Schuler, M., Bossy-Wetzel, E., Goldstein, J. C., Fitzgerald, P., & Green, D. R. (2000). p53 induces apoptosis by caspase activation through mitochondrial cytochrome c release. *The Journal of Biological Chemistry*, *275*(10), 7337–7342. doi:10.1074/jbc.275.10.7337 PMID:10702305

Shahverdi, A. R., Fakhimi, A., Shahverdi, H. R., & Minaian, S. (2007). Synthesis and effect of silver nanoparticles on the antibacterial activity of different antibiotics against *Staphylococcus aureus* and *Escherichia coli*. *Nanomedicine (London)*, *3*(2), 168–171. doi:10.1016/j.nano.2007.02.001 PMID:17468052

Silhavy, T. J., Kahne, D., & Walker, S. (2010). The bacterial cell envelope. *Cold Spring Harbor Perspectives in Biology*, *5*, 1–16. PMID:20452953

Sims, M. J. (1884). *The Story of My Life Marion-Sims* (Vol. H). D. Appleton & Co.

Singh, S. K., Chidrawar, V. R., Ushir, Y. V., Vadalia, K. R., Sheth, N. R., & Singh, S. (2010). Pharmaceutical characterization of amoxicillin trihydrate as mucoadhesive microspheres in management of *H. pylori*. *International Journal of Pharm Tech Research*, *2*(1), 348–358.

Siwinska-Stefanska, K., Kubiaka, A., Piasecki, A., Goscianska, J., Nowaczyk, G., Jurga, S., & Jesionowski, T. (2018). TiO_2-ZnO binary oxide systems: Comprehensive characterization and tests of photocatalytic activity. *Materials (Basel)*, *11*(5), 841. doi:10.3390/ma11050841 PMID:29783700

Slavin, Y. N., Asnis, J., Hafeli, U. O., & Bach, H. (2017). Metal nanoparticles: Understanding the mechanisms behind antibacterial activity. *Journal of Nanobiotechnology*, *15*(1), 65. doi:10.118612951-017-0308-z PMID:28974225

Smetana, A., Klabunde, K., Marchin, G., & Sorenson, C. (2008). Biocidal activity of nanocrystalline silver powders and particles. *Langmuir*, *24*(14), 7457–7464. doi:10.1021/la800091y PMID:18543995

Smillie, C., Garcillán-Barcia, M. P., Francia, M. V., Rocha, E. P. C., & de la Cruz, F. (2010). Mobility of plasmids. *Microbiology and Molecular Biology Reviews*, *74*(3), 434–452. doi:10.1128/MMBR.00020-10 PMID:20805406

Smith, H. O., Gwinn, M. L., & Salzberg, S. L. (1999). DNA uptake signal sequences in naturally transformable bacteria. *Research in Microbiology*, *150*(9-10), 603–616. doi:10.1016/S0923-2508(99)00130-8 PMID:10673000

Sondi, I., & Salopek-Sondi, B. (2004). Silver nanoparticles as antimicrobial agent: A case study on *E. coli* as a model for Gram-negative bacteria. *Journal of Colloid and Interface Science*, *275*(1), 177–182. doi:10.1016/j.jcis.2004.02.012 PMID:15158396

Srinath, B. S., Namratha, K., & Byrappa, K. (2018). Eco-Friendly Synthesis of Gold Nanoparticles by *Bacillus subtilis* and Their Environmental Applications. *Advanced Science Letters*, *24*(8), 5942–5946. doi:10.1166/asl.2018.12224

Starodub, M. E., & Trevors, J. T. (1989). Silver resistance in *Escherichia coli* R1. *Journal of Medical Microbiology*, *29*(2), 101–110. doi:10.1099/00222615-29-2-101 PMID:2659795

Kohl, S., & Bock, R. (2009). Transposition of a bacterial insertion sequence in chloroplasts. *The Plant Journal*, *58*(3), 423–436. doi:10.1111/j.1365-313X.2009.03787.x PMID:19144000

Stout, R. D., Ferguson, K. P., Li, Y. N., & Lambe, D. W. Jr. (1992). Staphylococcal exopolysaccharides inhibit lymphocyte proliferative responses by activation of monocyte prostaglandin production. *Infection and Immunity*, *60*(3), 922–927. doi:10.1128/IAI.60.3.922-927.1992 PMID:1541565

Sugano, M., Morisaki, H., Negishi, Y., Endo-Takahashi, Y., Kuwata, H., Miyazaki, T., & Yamamoto, M. (2016). Potential effect of cationic liposomes on interactions with oral bacterial cells and biofilms. *Journal of Liposome Research*, *26*(2), 156–162. PMID:26152278

Tiwari, G., Tiwari, R., Sriwastawa, B., Bhati, L., Pandey, S., Pandey, P., & Bannerjee, S. K. (2012). Drug delivery systems: An updated review. *International Journal of Pharmaceutical Investigation*, *2*(1), 2–11. doi:10.4103/2230-973X.96920 PMID:23071954

Vester, B., & Douthwaite, S. (2001). Macrolide resistance conferred by base substitutions in 23S rRNA. *Antimicrobial Agents and Chemotherapy*, *45*(1), 1–12. doi:10.1128/AAC.45.1.1-12.2001 PMID:11120937

Viale, P., Giannella, M., Tedeschi, S., & Lewis, R. (2015). Treatment of MDR-Gram negative infections in the 21st century: A never ending threat for clinicians. *Current Opinion in Pharmacology*, *24*, 30–37. doi:10.1016/j.coph.2015.07.001 PMID:26210268

Wei, D., Sun, W., Qian, W., Ye, Y., & Ma, X. (2009). The synthesis of chitosan-based silver nanoparticles and their antibacterial activity. *Carbohydrate Research*, *344*(17), 2375–2382. doi:10.1016/j.carres.2009.09.001 PMID:19800053

Willi, K., Sandmeier, H., Kulik, E. M., & Meyer, J. (1997). Transduction of antibiotic resistance markers among *Actinobacillus actinomycetemcomitans* strains by temperate bacteriophages Aaϕ23. *Cellular and Molecular Life Sciences*, *53*(11-12), 904–910. doi:10.1007000180050109 PMID:9447241

Wozniak, R. A. F., & Waldor, M. K. (2010). Integrative and conjugative elements: Mosaic mobile genetic elements enabling dynamic lateral gene flow. *Nature Reviews. Microbiology*, *8*(8), 552–563. doi:10.1038/nrmicro2382 PMID:20601965

Wyllie, A. H. (1980). Glucocorticoid-induced thymocyte apoptosis is associated with endogenous endonuclease activation. *Nature*, *284*(5756), 555–556. doi:10.1038/284555a0 PMID:6245367

Xu, Y., Wei, M. T., Ou-Yang, H. D., Walker, S. G., Wang, H. Z., Gordon, C. R., Guterman, S., Zawacki, E., Applebaum, E., Brink, P. R., Rafailovich, M., & Mironava, T. (2016). Exposure to TiO_2 nanoparticles increases *Staphylococcus aureus* infection of HeLa cells. *Journal of Nanobiotechnology*, *14*(1), 34. doi:10.118612951-016-0184-y PMID:27102228

Yaghini, E., Turner, H. D., LeMarois, A. M., Suhling, K., Naasani, I., & MacRobert, A. J. (2016). *In vivo* bio-distribution studies and ex vivo lymph node imaging using heavy metal-free quantum dots. *Biomaterials*, *104*, 182–191. doi:10.1016/j.biomaterials.2016.07.014 PMID:27454064

Yang, L., Haagensen, J. A., Jelsbak, L., Johansen, H. K., Sternberg, C., Høiby, N., & Molin, S. (2008). In situ growth rates and biofilm development of *Pseudomonas aeruginosa* populations in chronic lung infections. *Journal of Bacteriology*, *190*(8), 2767–2776. doi:10.1128/JB.01581-07 PMID:18156255

Yu, C., Xuan, T., Chen, Y., Zhao, Z., Liu, X., Lian, G., & Li, H. (2016). Gadolinium-doped carbon dots with high quantum yield as an effective fluorescence and magnetic resonance bimodal imaging probe. *Journal of Alloys and Compounds*, *688*, 611–619. doi:10.1016/j.jallcom.2016.07.226

Zaman, S., Hussain, M., Nye, R., Mehta, V., Mamun, K. T., & Hossain, N. (2017). A Review on Antibiotic Resistance: Alarm Bells are Ringing. *Cureus Journal of Medical Science*, *9*(6), e1403. doi:10.7759/cureus.1403 PMID:28852600

Chapter 8
Nanotechnology Based Emerging Approaches to Combat Malaria and Dengue Fever

Toqeer Ahmed
Centre for Climate Research and Development (CCRD), COMSATS University Islamabad, Pakistan

Irfan Liaqat
Centre for Climate Research and Development (CCRD), COMSATS University Islamabad, Pakistan

Muhammad Zeeshan Hyder
https://orcid.org/0000-0002-9172-499X
Department of Biosciences, COMSATS University Islamabad, Pakistan

ABSTRACT

Nanotechnology has vast applications in virtually all the fields including health and environment. Nanomaterials have different origins like plants, animals, metals, and microorganisms. They have larvicidal activities against mosquito larvae that cause malaria and dengue fever which are most malicious parasitic diseases of human beings. Multidrug resistance of malaria has been reported in different parts of Southeast Asia especially to the first line of antimalarial drugs. Nanomaterials having different origin, sizes, concentrations, and shapes, have varied efficacy against the vectors which are comparable with the commercially available insecticides and even have better results than the same. Nanomaterials can be used as an alternative to commercially available insecticides for the eradication of vectors causing both the diseases. In this chapter, types, sources, composition, larvicidal potential against vectors of nanomaterials have been discussed along with possible toxicity and future recommendations on the eco-friendly approach for the control of mosquito-borne diseases.

DOI: 10.4018/978-1-7998-5049-6.ch008

INTRODUCTION

Nanotechnology is not definite and restricted to a single discipline, instead it is interdisciplinary field including different disciplines like natural sciences, engineering science and lately toxicology field. Through application of molecular biology with engineering have produced new multifunctional structures for higher biological assessment with significant qualities such as higher specificity, sensitivity and recognition rate. The significant study of nanotechnology is the incorporation of several types of nanoparticles (NPs) in diverse compositions, sizes and shapes with specific differences (Ullah Khan et al., 2018). Lately, the production of metallic NPs, i.e., silver, gold, iron, titanium, palladium, copper, zinc, platinum, and cobalt from natural sources has obtained significance (Goodsell, 2004; Rahman et al., 2019). The biosynthesis of metallic NPs is commonly based on the reducing and stabilizing potential of plant extracts and other metabolites. The key factors that influenced the NPs size, shape, and their stability; are the concentration of metal ions and the plant extract (Rajan et al., 2015).

Recently, the awareness of researchers in nanotechnology has increased exaggeratedly to synthesized nanoparticles for their biomedical and environmental applications like drug delivery, gene delivery, diagnostics, tissue engineering, imaging, artificial implants and especially for the mosquito-borne diseases and other pest management (Amerasan et al., 2016). Several reliable plant-mediated and chemical composites have been anticipated for effective and prompt synthesis of metallic nanoparticles that exhibit exceptional anti-plasmodial activity, as well as larvicidal potential, in laboratory and field conditions.

Mosquito is an insect belongs to Family Culicidae, order Diptera. More than 3500 species of mosquito have been identified throughout the world (Harbach, 2013). In these species the major genus is *Aedes* which has more than 900 species. Some of them are vectors for several pathogens including arboviruses, protozoans, filariae and few bacteria that cause serious infectious diseases (Becker et al., 2010; Dieme et al., 2015). Pathologically most significant genera are *Aedes, Anopheles, Culex* (Ghosh et al., 2012). *Aedes* mosquitoes are the best-known vectors of dengue and yellow fever. Several *Aedes* species also act as vectors of other viral and filariasis diseases. *Ae. aegypti* and *Ae. albopictus* are causal agents of diseases like dengue and hemorrhagic fever (Gubler, 2002).

The four types of viruses (DENV-1, DENV-2, DENV-3, DENV-4) belonging to the family *Flaviviridae* causes viral infection, dengue fever. It is either transmitted by biting the *Ae. albopictus* and *Ae. aegypti* (female mosquitoes) from dawn to desk (ECDC, 2019a, 2019b). Malaria is often caused by parasites named, *Plasmodium*. The parasite is transferred to other people and animals through the biting of infected *Anopheles* mosquitoes, the *Plasmodium* mainly bite during dawn and dusk (Jensen & Mehlhorn, 2009). According to malaria control program, 177 million people in Pakistan are at danger of malaria, and 3.5 million are confirmed cases annually. It was aimed to reduce the disease burden by 75% in 2020 (DMC, 2020) but still there is need to work more to control the spread of this disease along with dengue fever as 998 cases of dengue fever appeared in 48 week of 2019 (WHO, 2019). In 2019, 52485 cases of dengue fever were reported including 91 related deaths in almost all provinces of Pakistan(WHO, 2019).

Various conventional methods are used to control population of vector like mosquitoes including chemicals (insecticides), physical, biological, environmental management etc. (Ahmed, 2018) along with the medicine to prevent the malaria and dengue fever. The prevention and control of vectors through these conventional methods has caused the development of resistance in mosquitoes against antiparasitic drugs and insecticides. The other factor in limiting insecticidal approach for mosquito control is; their unfavorable impacts on the environment including plants and human health. Among the emerging approaches, nanomaterials played a tremendous role to control the mosquito vectors. Empirical studies have reported the impact of Metallic and other Nanoparticles (MNPs) against vector causing dengue and malaria. Plant synthesized nanoparticles has received great attention due to their safe, cheap and environment friendly nature. Limited work has been done on fate, transport and toxicity of metallic and other nanoparticles. Mostly review studies have emphasized on supplementary work to address the issues related to toxicity of NPs. In this chapter, we focused on the different types of nanomaterials, their sources and composition, nanomaterials been tested for larvicidal potential against vectors of dengue and malaria, the possible toxicity of those nanomaterials and future recommendations on the eco-friendly approach for the control of mosquito-borne diseases.

TYPES, SOURCES AND COMPOSITION OF NANOMATERIALS

The types of nanomaterials applied for environmental and health applications and specifically applied for insecticides are:

Carbon Based Nanomaterials

Commonly, this type of nanomaterials NMs comprise carbon, and are observed in spheres, hollow tubes, or ellipsoids morphologies. Normally carbon-nanotubes (CNTs), carbon nanofibers (CNFs), Fullerenes (C60), graphene (Gr), Nanocones, carbon black, and carbon onions are contained within this group. The synthesis methods used for these NMs are laser ablation, arc discharge, and chemical vapor deposition (CVD) except carbon black (Kumar et al., 2016).

Inorganic Based Nanomaterials

Inorganic-based NMs comprised of metal and metal oxide nanoparticles. These nanomaterials can be shaped into metals such as Fe, Au, Co, Cu, Ag, Ti or Zn NPs, their metal oxides i.e., CuO, TiO_2 and ZnO NPs, or binary compounds when including carbides, nitrides, etc. and in the form of semi-conductors like silicon and ceramics (Jeevanandam et al., 2018).

Organic Based Nanomaterials

In this group, nanomaterials mostly comprised of organic matter, exclusive of carbon and inorganic based NMs. The weak (noncovalent) interactions of molecules for the design and self-assembly facilitates to alter the organic nanomaterials into preferred nanostructures including liposomes, micelles, dendrimers, and polymer NPs (polystyrene) (Jeevanandam et al., 2018; Kumar et al., 2016).

Composite Based Nanomaterials

Composite based nanomaterials are multi-phase nanoparticles in which one phase is in nanoscale either combine one NPs with another NPs or any NPs combined with large bulky-materials such as alloys, hybrid nanofibers or more complex structures (metal-organic structure). In other words, composites are any combinations of carbon, metal or organic based nanomaterials with any metal, ceramic, and polymer bulk material (Jeevanandam et al., 2018).

Quantum Dots

Quantum dots are small particles of any semi-conducting material, display unique electronic properties and produce distinct colors based on particle's size (e.g. LEDs, solar cells)

Different sources are used for the synthesis of nanoparticles like plants (Bianca Morejón et al., 2018; Morejon, 2019) including herbs herbal essential oils (Adhikari et al., 2013; Sutthanont et al., 2019) metallic NPs (Kumar et al., 2016). Different methods like gas phase processes, liquid phase synthesis, mechanical synthesis and biological synthesis are used to synthesize MNPs (Adhikari et al., 2013) but ecofriendly, effective and more economic methods are widely adopted for green synthesis (Aitken et al., 2004). Schematic diagram of nanomaterials and their applications against larvae of mosquitoes (both malaria and dengue causing mosquitoes) are presented in Figure. 1.

Figure 1. Schematic diagram of Nanotechnology impacts on malaria and dengue vectors.

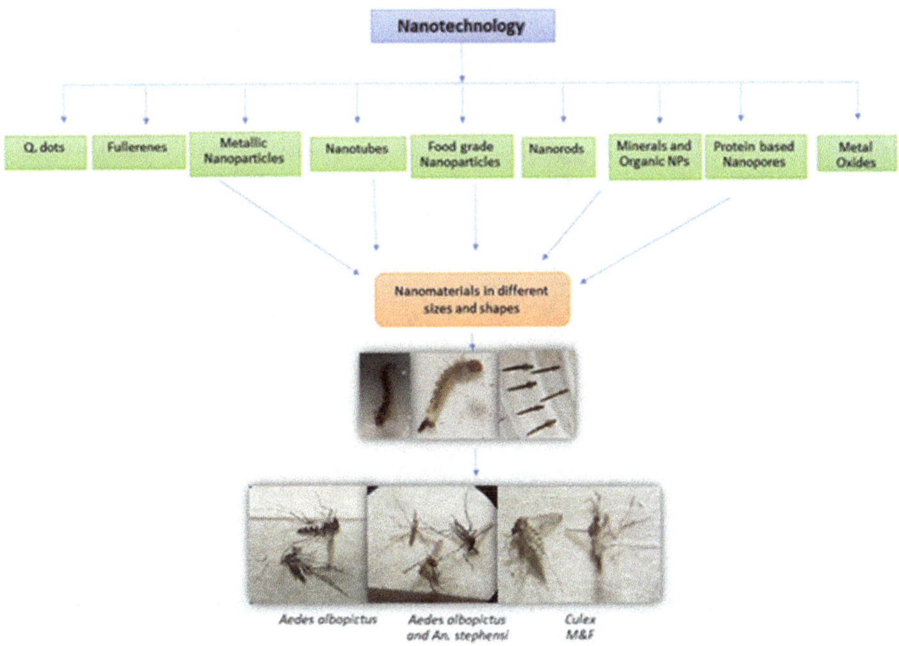

Schematic diagram of Nanoparticles used against mosquito vector and how they act against *Plasmodium* is given in Figure 2.

NANOMATERIALS FOR THE CONTROL OF DENGUE AND MALARIA VECTORS

Nanobiotechnology has vast applications in the field of tropical diseases including use of nano-insecticides, nano-biodevices, nanowires-based sensors, nanoparticles etc. Nano-biosensors can help in the detection of dengue virus which are inexpensive, easy to correlate and use (Durán et al., 2016). In recent times, few plant extracts have been discovered for their effectiveness against *Ae. aegypti.* larvae and pupae. Plants as a reservoir of rich bio-active organic compounds, produce different secondary metabolites that serve as defense mechanism against pests (arthropod). These secondary metabolites act as repellents, insecticides, attractants, growth inhibitors, antifeedants, mimics of juvenile hormone, oviposition deterrents, and moulting hormones (Benelli, 2015b; Pavela, 2015a).

Figure 2.. Nanoparticles scheme used against mosquito vector (Plasmodium and mosquitoes' vector) (Barabadi et al., 2019).

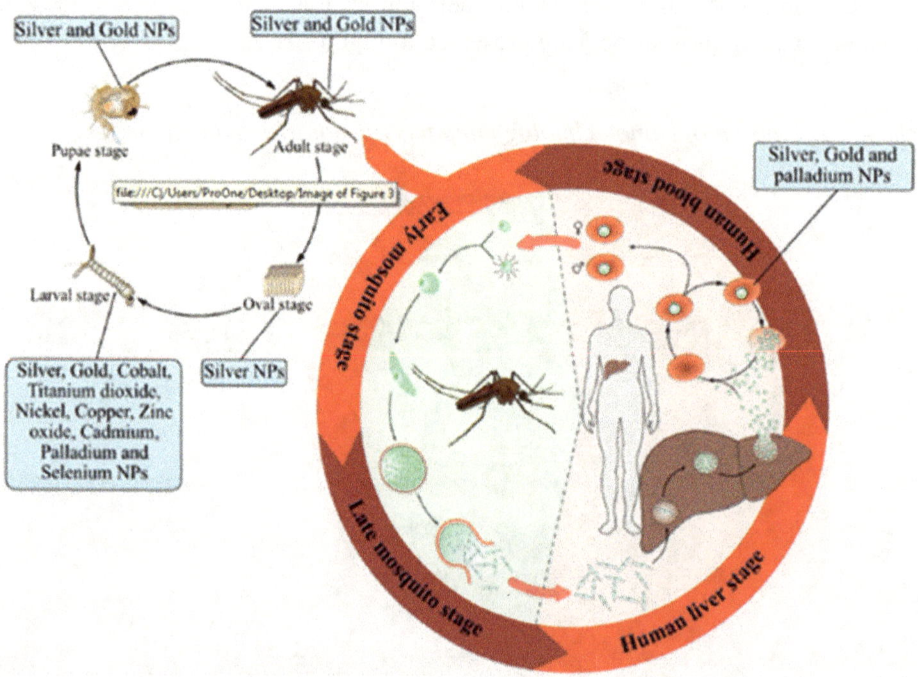

Biosynthesized silver nanoparticles (AgNPs) from the aqueous leaf extracts of *Ambrosia arborescens* were evaluated against third instar larvae of *Ae. aegypti.* The larvicidal activity shown by these nanoparticles were more higher (LC50) than the plant extract with LC50 and LC90 of (0.28 ppm, 0.43 ppm; 1844.61 ppm, 6043.95 ppm) after 24 h exposure (Morejon, 2019). The water extract of rhizome (*H. coronarium)* was recognized as more fatal against 1st to 4th instar *Ae. Aegypti* larvae. The calculated dose-dependent effect was found in this study with LC50 values of 24.264 ppm (1st instar), 39.769 ppm (2nd instar), 52.715ppm (3rd instar), and 72.618ppm (4th instar) (Kalimuthu et al., 2017). In previous studies, Kalimuthu et. al., 2013 noted the high larvicidal activity of methanol extract of the seaweed *Gracilaria firma* against larvae of *Ae. Aegypti* (Kalimuthu et al., 2013). More, Murugan et. al., 2015c described the exposure of leaf extract *Artemisia vulgaris* as an effective larvicidal and pupicidal chemical against *Ae. aegypti*, with LC50 of 240.8 ppm (1st instar), 269.3 (2nd instar), 288.2 (3rd instar), 322.7 (4th instar), and 342.2 (pupa) (Murugan et al., 2015c).

Table 1. Nanomaterials for the control of malaria and dengue vectors

Sr. no	Sources	NPs Type	Shape & Size	Characterization technique used	Optimal/ Incubation temperature	Stage	Conc. used	Organism applied	Exposure time	Results (LC50)	Reference
1	Rhizome extract of *Hedychium coronarium*	Silver nanoparticles (AgNPs)	Spherical and Oval shaped NPs 9.54-49.0 nm	UV-vis spectroscopy, EDX, TEM, FTIR	Room temperature	1st, 2nd, 3rd, 4th Instar larvae and Pupa	20, 40, 80, 160, and 320 ppm	*Ae. aegypti*	24 h	24.264 ppm, 39.769 ppm, 52.715 ppm, 72.618 ppm, 348.68 ppm	(Kalimuthu et al., 2017)
2	Aqueous leaf extract of *Artemisia vulgaris*	Silver nanoparticles (AgNPs)	Crystalline Irregular shaped NPs 30-70nm	UV-vis spectroscopy, EDX, SEM, FTIR, XRD	Room temperature	1st, 2nd, 3rd, 4th Instar larvae and Pupa	2,4,8,16, and 32 ppm	*Ae. aegypti*	24 h	4.429 ppm, 7.209 ppm, 8.273 ppm, 10.776ppm, 13.089 ppm	(Murugan et al., 2015c)
3	Aqueous leaf extract of *Pteridium aquilinum*	Silver nanoparticles (AgNPs)	Spherical shaped NPs 35-65nm	UV-vis spectroscopy, EDX, SEM, FTIR, XRD	Room temperature	1st, 2nd 3rd, 4th Instar larvae and Pupa	3.125, 6.25, 12.5, 25, and 50 ppm	*An. stephensi*	24 h	7.48 ppm, 10.68 ppm, 13.77 ppm, 18.45 ppm, 31.51 ppm	(Panneerselvam et al., 2016)
4	Aqueous extract of *Centroceras clavulatum*	Silver nanoparticles (AgNPs)	spherical and cubic shaped NPs 35-65nm	UV-vis spectroscopy, EDX, SEM, FTIR, XRD	Room temperature	1st, 2nd 3rd, 4th Instar larvae and Pupa	10, 20, 30, 40, and 50 ppm	*Ae. aegypti*	-	21.460 ppm, 23.579 ppm, 25.912 ppm, 29.155 ppm, 33.877 ppm	(Murugan et al., 2016b)
5	Aqueous extract of the seaweed *Sargassum muticum*	Silver nanoparticles (AgNPs)	Spherical shaped NPs 43-79nm	UV-vis spectroscopy, EDX, SEM, FTIR, XRD	Room temperature	1st, 2nd, 3rd, 4th Instar larvae and Pupa	10, 20, 30, 40, and 50 ppm	*Ae. aegypti*	24 h	20.237 ppm, 21.802 ppm, 23.732 ppm, 28.099 ppm, 30.579 ppm	(Madhiyazhagan et al., 2015)
								An. stephensi	24 h	16.156 ppm, 17.809 ppm, 22.068 ppm, 26.086 ppm, 28.881 ppm	
6	Leaf extract of *Avicennia marina*	Silver nanoparticles (AgNPs)	-	UV-vis spectroscopy, FTIR	Room temperature	1st, 2nd 3rd, 4th Instar larvae and Pupa	1.25, 2.50, 5.0, 10.0, and 20.0 mg/L	*Ae. aegypti* and *An. stephensi*	-	4.374 mg/L and 7.406 mg/L	(Balakrishnan et al., 2016)
7	Aqueous extract of *Syzgium cumini*	Zinc Oxide nanoparticles (ZnONPs)	Spherical shaped NPs 50–60 nm	UV-vis spectroscopy, SEM, TEM, FTIR, XRD	450 °C	Larvae	15, 30, 60 and 120 ppm	*Ae. aegypti*	24 h	51.94 ppm	(Roopan et al., 2018)
8	Exopolysaccharide strain of *Bacillus licheniformis*	Zinc Oxide nanoparticles (ZnONPs)	Hexagonal shaped NPs 100 nm	UV-vis spectroscopy, SEM, TEM, EDX	37 °C	4th instar larvae	20,40,60,80 and 100 μg/ml	*Ae. aegypti* and *An. stephensi*	24 h	43.57 μg/ml and 50.20 μg/ml	(Abinaya et al., 2018)
9	Leaf extract of *Tridax procumbens*	Copper oxide nanoparticles (CuONPs)	Monoclinic structure NPs 16 nm	UV-vis spectroscopy, Fluorescence spectroscopy, FE-SEM, EDAX, XRD	80 °C	larvae	1, 2.5, 5,7.5, and 10 mg/L	*Ae. aegypti*	24 h	4.209 mg/L	(Muthamil Selvan et al., 2018)

continues on followng page

Table 1. Continued

Sr. no	Sources	NPs Type	Shape & Size	Characterization technique used	Optimal/ Incubation temperature	Stage	Conc. used	Organism applied	Exposure time	Results (LC50)	Reference
10	Root extract of *Morinda citrifolia*	Titanium dioxide nanoparticles (TiO_2NPs)	Spherical, oval and triangle shaped NPs 20.46-39.20 nm	FTIR, XRD, FE-SEM, EDAX, TEM	50 °C	larvae	5,12,25,50 and 100 mg/L	*Ae. aegypti and An. stephensi*	24 h & 48 h	23.711 mg/L, 16.292 mg/L and 13.620 mg/L, 5.032 mg/L	(Suman et al., 2015)
11	Fibroin polymer from *Bombyx mori*	Fibroin-silver nano-composites (FSNC), Fibroin-gold nano-composites (FGNC),	Spherical and elongated FSNC, Evenly distributed smaller FGNC, 20-30nm	UV-vis spectroscopy, AFM,	Room temperature	1st, 2nd, 3rd, 4th Instar larvae and Pupa	50,100,150, and 200 µl/lit	*Ae. aegypti*	-	100% mortality achieved by using 200 µl/lit FSNC, and 86–98% mortality achieved by using 200 µl/lit FGNC	(Mane et al., 2017)
12	*Penaeus monodon* Fabricius shells	Ag NPs embedded α-chitin nanocomposite (CNP/AgNP)	Spherical shaped Chitin and silver NPs 29 nm AgNPs 115 nm CNPs	UV-vis spectroscopy, FTIR, XRD, SEM, EDX, High resolution TEM, DLS	70 °C	1st, 2nd, 3rd, 4th Instar larvae	10 to 100 ppm	*Ae. aegypti*	24 h & 48 h	9.72 ppm, & 5.96 ppm, 10.41 ppm, & 7.16 ppm, 13.93 ppm, & 9.87 ppm, 23.85 ppm, & 13.58 ppm	(Solairaj & Rameshthangam, 2017)
13	Leaf Aqueous Extract of *Fumaria indica*	Silver (Ag) Nano-composites	Spherical shaped nano-composite 13-38 nm	UV-vis spectroscopy, FTIR, XRD, SEM, TEM	-	3rd instar larvae	5–25 µg/mL	*Ae. albopictus*, and *An. subpictus*	-	11.42 µg/mL, 10.34 µg/mL	(Benelli et al., 2017)
14	Lichen *Parmelia sulcata* (PSE) extract	Gold Nanoparticles (AuNPs)	Spherical shaped NPs 54.18 nm	UV-vis spectroscopy, FTIR, XRD, High resolution (HRTEM), DLS, EDS	60 °C	1st, 2nd, 3rd, 4th Instar larvae and Pupa	75, 150, 225, 300, & 375 ppm	*Ae. aegypti*	-	34.49 ppm, 38.72 ppm, 44.72 ppm, 51.41 ppm, 59.00 ppm	(Gandhi et al., 2019)
								An. stephensi	-	29.82 ppm, 33.88 ppm, 37.55 ppm, 44.26 ppm, 50.44 ppm	
15	Sea anemone *Stichodactyla haddoni* (ShAp)	ShAp-Zinc Oxide nanoparticles (ShAp ZnO NPs)	Hexagonal shaped NPs 10 nm	UV-vis spectroscopy, FTIR, XRD, High resolution (HRTEM), SAED analysis	60 °C	3rd instar larvae	20, 40, 60, 80, and 100 µg/mL	*Ae. aegypti*	24 h	31.49 µg/mL	(Abinaya et al., 2019)

Panneerselvam et. al., 2016 explained the *Pteridium aquilinum*-synthesized silver nanoparticles as toxic larvicidal agent to *Anopheles stephensi* larvae, with LC50 starting from 7.48 (1st instar) to 31.51 ppm (pupae) (Panneerselvam et al., 2016). The green synthesized AgNPs using an aqueous extract of *Centrocerus clavulatum* were found highly toxic to both pupae and larvae of *Ae. Aegypti* (Murugan et al., 2016b). The LC50 of synthesized AgNPs were 21.460 ppm against 1st instar larvae and 33.877 ppm against pupae of *Ae. aegypti*. Madhiyazhagan, P., et.al., 2015 experimented the larvicidal effectiveness of *Sargassum muticum* synthesized AgNPs against *Ae. aegypti,* and *An. stephensi.* The calculated LC50 against *ae. aegypti* were 20.237 ppm (1st instar), 21.802 ppm (2nd instar), 23.732 ppm (3rd instar), 28.099 ppm (4th instar), 30.579 ppm (pupa) and LC50 against *An. stephensi* were 16.156 ppm (1st instar), 17.809 ppm (2nd instar), 22.068 ppm (3rd instar), 26.086 ppm (4th instar), and 28.881 ppm (pupa) (Madhiyazhagan et al., 2015). Similarly, Balakrishnan et. al., 2016 examined that the larvicidal potential of *Avicennia marina* synthesized NPs compared to *Ae. aegypti* and *An. stephensi* through LC50 of 7.406 mg/L and 4.374 mg/L, respectively (Balakrishnan et al., 2016). Field assays probing the mosquitocidal potential of green-synthesized AgNPs are currently constrained (Dinesh et al., 2015; Subramaniam et al., 2016). Suresh et al. 2015, stated that the field application of the extract of *P. niruri* (10 * LC50) resulted in larval reduction of *Ae. aegypti* to 39.9, 69.2, and 100%, after exposure of 24 h, 48 h, and 72 h, respectively (Suresh et al., 2015).

Roopan, S. et.al., 2018 depicted the larvicidal efficacy of the synthesized ZnO NPs for different concentrations ranging from 15 ppm to 120 ppm against the *Ae. Aegypti* (Roopan et al., 2018). Mortality rate resulted from ZnO NPs exposure was 21.4±2.3 to 88.6±1.2. Abinaya, M., et.al., 2018 investigated the EPS coated zinc oxide nanoparticle (EPS ZnO NPs) for their larvicidal activity. The calculated LC50 and LC90 against *Ae. aegypti* were 50.20 μg/ml and 93.13 μg/ml; and LC50 and LC90 against *An. stephensi* were 43.57 μg/ml and 82.49 μg/ml, respectively (Abinaya et al., 2018). In 2018, larvicidal potential of leaf extract of *T. procumbens* and synthesized CuO nanoparticles against *Ae. aegypti* were explored by (Muthamil Selvan et al., 2018). The highest larvicidal potential found in leaf extract of *T. procumbens* and CuO NPs against the *Ae. aegypti* larvae with LC50 of 60.965 mg/L and 4.209 mg/L, respectively. Suman, TY., et.al., 2015 demonstrated the larvicidal potential of biosynthesized titanium dioxide ($Ti\ O_2$) NPs with LC50 of 23.711 mg/L, 16.292 mg/L and 13.620 mg/L, 5.032 mg/L against *ae. aegypti* and *an. stephensi* after 24 h and 48 h, respectively (Suman et al., 2015). Further detail of studies of nanomaterials against vectors and their control is given in Table-1.

TOXICITY AND ECOTOXICITY OF NANOMATERIALS

Although, nanomaterials have fruitful applications but toxic effects cannot be ignored like damage DNA, membrane, mitochondrial damage, oxidative stress, neurotoxicity, penetration through exoskeleton and intercellular damage (Tunçsoy, 2018). Although the intact skin is thought to be an effective restraint; but silver absorbed through the mucosal exteriors and compromised skin. Widespread argyria was seen with nanocrystalline ionic silver in humans with increasing doses in between 70 to 1500 mg silver/kg body weight. Silver is measured to have a low risk of skin irritation. Few cases of allergic contact dermatitis and eye irritation have been informed. Silver can cause genotoxicity, but supplementary data is necessary to evaluate its oncogenic potential (Hadrup et al., 2018). A 100 cm^2 gauze of cotton fabric comprising a 2% AgNPs dispersal was dermally employed in rabbits. The silver formulation was categorized as a scarcely traceable irritant (Zelga et al., 2016). The toxic effect of SiO_2 and Al_2O_3 NPs reported due to binding with cuticle and later physio-sorption of waxes and lipids, leading to insect dehydration (Benelli, 2018). They can enter through lungs and accumulate in the liver and damage brain (Elsaesser & Howard, 2012) but some studies reported that food grade nanoparticles like ZnO and Curcumin NPs are safe and may be present in nature. Case to case study is recommended for the food grade nanoparticles and more research has been suggested to understand the behavior of organic NPs (McClements & Xiao, 2017).

Mitochondria specific effects of AgNPs exposures have been recognized in numerous cell types, together with membrane potential loss, decreased MTT reduction, decreased ATP content, release of cytochrome c, (Gerencser et al., 2012; Jaiswal et al., 2015; Sanpui et al., 2011) enzymes inhibition embraced in oxidative phosphorylation, and alteration in calcium sequestration. Still, the biological importance of mitochondrial toxicity because of AgNPs exposure is formerly partly understood (Maurer & Meyer, 2016). Applications of nanomaterial on environment and applications against insecticides may harm other useful species like Daphnia. A study on photoinduced toxicity of TiO_2 in natural sunshine reported 139 ppb under full strength ambient natural sunshine at 8h LC50 of anatase TiO_2 NP and 778 ppb was observed at 50%, while more than 500 ppm under 10% natural sunshine (Mansfield et al., 2015). Similarly, NPs may have some low impacts on longer exposure to other beneficial soil organisms like BNF bacteria and nitrifying bacteria (McKee & Filser, 2016).

Some nanoparticles line mineral NPs and organic polymer-based NPs like starch derivatives and dendrimers etc. are highly degradable and may dissolve over time in the environment (Rana & Kalaichelvan, 2013). A study reported that non targeted effects had been observed with no genotoxicity or acute toxicity on other aquatic organisms at the recommended lethal dose (Benelli, 2016), however, further studies on behavior and fate of Ag ions and other MNPs were suggested before its applications. To understand the impact of MNPs on biological systems, there is need to identify the biomarkers and further tests have been suggested (Sahu & Hayes, 2017). Schematic diagram of human body with different exposure pathways to nanoparticles, affected organs and their associated diseases from epidemiological in-vitro and in-vivo studies is given in Fig. 3.

NANOMATERIALS VS ANTIPARASITIC DRUGS

Like the antibiotic resistance, antiparasitic drug against malaria and dengue fever has been reported. On the basis of *Plasmodium* species, the geographic region of the infecting species, and the severity of the patient's infection, drug is suggested (Hill SR & GK. , 2019). Chloroquine drug is used to treat uncomplicated malaria caused by all species of *Plasmodium* except *P. falciparum*, which are becoming resistant increasingly. According to WHO, Amodiaquine in combination with artesunate used to treat uncomplicated chloroquine-resistant *P. falciparum infections* (WHO, 2015). An artemether-lumefantrine combination or an atovaquone-proguanil combination may be used for the treatment for chloroquine-resistant *P. falciparum* (Campbell S & K, 2019). A study conducted in 2004 suggested the use of anti-malarial drug combination like Artemisinin derivatives are effective against malaria (White, 2004). A case of early treatment failure was reported in coexisting malaria and dengue infections with resistance to artemisinin therapy and detected pfcrt, pfmdr-1, pfdhfr, and pfdhps genes as a probable cause of treatment failure (Saksena et al., 2017). According to a recent news, it has been reported that first two-line antimalarial drug resistance has spread from Cambodia to Southeast Asia (Hamilton et al., 2019).

Figure 3. Schematic diagram of human body with different exposure pathways to nanoparticles, affected organs and their associated diseases from epidemiological in-vitro and in-vivo studies (Cristina Buzea et al., 2007).

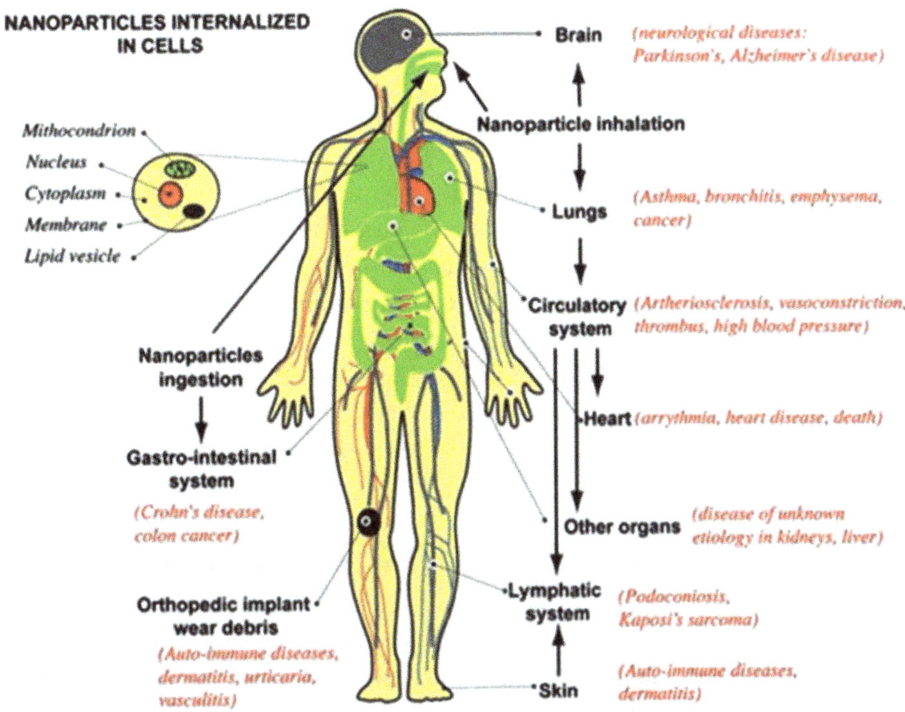

For dengue fever, Chloroquine has some antiviral effects along with antimalarial and helps in improving the dengue related symptoms (Borges et al., 2013). The antiparasitic drug niclosamide deliberates antiviral activity, together with effects on release of viral genome, expression of viral protein, dsRNA replication, and release of virus *in vitro* in numerous DENV infected cell lines (Kao et al., 2018; Li et al., 2017). A study conducted by T-L Xu et.al., 2018 reported that Ivermectin can inhibit directly or indirectly the DENV-2 multiplication in *Aedes albopictus* (Xu et al., 2018). The antiviral mechanism of ivermectin against DENV-2 is still unknown but it is assumed from the in-vitro studies that it inhibits the activity of NS3 helicase, NS5 polymerase proteins, altering mosquito physiology and interrelating with glutamate-gated chloride channels or γ-aminobutyric acid-gated chloride channel (Croci et al., 2016; Mastrangelo et al., 2012). NPs effectively assists in targeted delivery of medicines as it is one of the applications of NPs in targeted drug delivery. A study reported the efficient delivery of small interfering RNAs through AuNPs and helped in the control of dengue viral infection (Amber M. Paul, 2014). Similarly, experiments on mice demonstrated that tetravalent DENV subunit vaccines exhibited on nanoparticles had the potential to overcome unbalanced immunity (Metz et al., 2018). Nanoparticles may be use in targeted delivery of developed vaccines for dengue and malaria and against resistance drugs to make it more effective. As effective drugs are available for malaria (Ahmed et al., 2019) but to decease drug resistance, NPs may be helpful. Another useful application of nanotechnology is the detection and diagnosis of diseases. Pragmatic studies have reported that different types of nanocarriers are used for drug delivery like micelles, quantum dots, dendrimers, zeolites, hydrogels, fullerenes, liposomes, polymeric, magnetic and metallic nanoparticles, like silica, gold etc. (Salouti, 2014).

CONCLUSION

Nanomaterials have been found effective in combating the diseases including malaria and dengue fever. It is important to understand that these should be nontoxic to both environment and human health. Metallic NPs have been found effective, but they can be toxic to human health. Food grade NPs like Curcumin and others have been found comparable to metallic NPs and they are non-toxic to human body. Such kind of NPs can be more useful, cost effective, applicable to control larvae of mosquitoes and helpful to reduce the diseases burden. Comprehensive studies on the toxicity and ecotoxicity should be conducted in future to address the effectiveness of NPs against the diseases. Similarly, NPs can be used as carrier material for vaccines for the effective delivery of medicines.

RECOMMENDATIONS

The following are the important recommendations to reduce the malaria and dengue fever by using nanoparticles:

- There is need to address the knowledge gaps of toxicity assessment especially the fate and impact of NMs against other beneficial species in the complex ecosystem.
- There is need to study behavior and characteristics of NPs under changing conditions before its application to the environment.
- Use of economic, ecofriendly and health beneficial food grade NPs are recommended to use against the vectors causing malaria and dengue fever. However, their toxicity must be tested before use.
- To overcome the development of resistance to available insecticides, use alternate food grade NMs is suggested.

ACKNOWLEDGMENT

Authors gratefully acknowledge the fund provided by the Higher Education Commission (HEC), Islamabad under the project No. 5424/Federal/NRPU/R&D/HEC/2016 and resources of Centre for Climate Research and Development (CCRD), COMSATS University Islamabad, used for this work.

REFERENCES

Abinaya, M., Rekha, R., Sivakumar, S., Govindarajan, M., Alharbi, N. S., Kadaikunnan, S., Khaled, J. M., Alobaidi, A. S., Al-Anbr, M. N., & Vaseeharan, B. (2019). Novel and Facile Synthesis of Sea Anemone Adhesive Protein-Coated ZnO Nanoparticles: Antioxidant, Antibiofilm, and Mosquito Larvicidal Activity Against Aedes aegypti. *Journal of Cluster Science*, *30*(6), 1393–1402. doi:10.100710876-019-01581-0

Abinaya, M., Vaseeharan, B., Divya, M., Sharmili, A., Govindarajan, M., Alharbi, N. S., Kadaikunnan, S., Khaled, J. M., & Benelli, G. (2018). Bacterial exopolysaccharide (EPS)-coated ZnO nanoparticles showed high antibiofilm activity and larvicidal toxicity against malaria and Zika virus vectors. *Journal of Trace Elements in Medicine and Biology*, *45*, 93–103. doi:10.1016/j.jtemb.2017.10.002 PMID:29173489

Adhikari, U., Ghosh, A., & Chandra, G. (2013). Nano particles of herbal origin: A recent eco-friend trend in mosquito control. *Asian Pacific Journal of Tropical Disease*, *3*(2), 167–168. doi:10.1016/S2222-1808(13)60065-1

Ahmed, T., Hyder, M. Z., Liaqat, I., & Scholz, M. (2019). Climatic Conditions: Conventional and Nanotechnology-Based Methods for the Control of Mosquito Vectors Causing Human Health Issues. *International Journal of Environmental Research and Public Health*, *16*(17), 3165. doi:10.3390/ijerph16173165 PMID:31480254

Ahmed, T., & Mukhtar, M. (2018). *Climate Change and mosquitoes: need and ways to diminish mosquito related diseases in Pakistan*. Academic Press. Technology Times, 24 May 2018. https://www.technologytimes.pk/2018/05/24/climate-change-mosquitoes/

Aitken, R., Ks, C., & Tran, L. (2004). Nanoparticle: An Occupational Hygiene Review. *Institute of Occupational Medicine for the Health and Safety Executive*, 1-102. doi:10.1055os-SD-013-01214

Amber, M. (2014). Delivery of antiviral small interfering RNA with gold nanoparticles inhibits dengue virus infection in vitro. *The Journal of General Virology*, *95*(8), 1712–1722. doi:10.1099/vir.0.066084-0 PMID:24828333

Amerasan, D., Nataraj, T., Murugan, K., Panneerselvam, C., Madhiyazhagan, P., Nicoletti, M., & Benelli, G. (2016). Myco-synthesis of silver nanoparticles using Metarhizium anisopliae against the rural malaria vector Anopheles culicifacies Giles (Diptera: Culicidae). *Journal of Pest Science*, *89*(1), 249–256. doi:10.100710340-015-0675-x PMID:28217039

Balakrishnan, S., Srinivasan, M., & Mohanraj, J. (2016). Biosynthesis of silver nanoparticles from mangrove plant (Avicennia marina) extract and their potential mosquito larvicidal property. *Journal of Parasitic Diseases : Official Organ of the Indian Society for Parasitology*, *40*(3), 991–996. doi:10.100712639-014-0621-5 PMID:27605825

Barabadi, H., Alizadeh, Z., Rahimi, M. T., Barac, A., Maraolo, A. E., Robertson, L. J., Masjedi, A., Shahrivar, F., & Ahmadpour, E. (2019). Nanobiotechnology as an emerging approach to combat malaria: A systematic review. *Nanomedicine; Nanotechnology, Biology, and Medicine*, *18*, 221–233. doi:10.1016/j.nano.2019.02.017 PMID:30904586

Becker, N., Petric, D., Zgomba, M., Boase, C., Madon, M., Dahl, C., & Kaiser, A. (2010). *Mosquitoes and their control*. Springer Science & Business Media. doi:10.1007/978-3-540-92874-4

Benelli, G. (2015b). Plant-borne ovicides in the fight against mosquito vectors of medical and veterinary importance: A systematic review. *Parasitology Research, 114*(9), 3201–3212. doi:10.100700436-015-4656-z PMID:26239801

Benelli, G. (2016). Green synthesized nanoparticles in the fight against mosquito-borne diseases and cancer—A brief review. *Enzyme and Microbial Technology, 95*, 58–68. doi:10.1016/j.enzmictec.2016.08.022 PMID:27866627

Benelli, G. (2018). Mode of action of nanoparticles against insects. *Environmental Science and Pollution Research International, 25*(13), 12329–12341. doi:10.100711356-018-1850-4 PMID:29611126

Benelli, G., Govindarajan, M., Kadaikunnan, S., & Alharbi, N. S. (2017). What Kind of Reducing Botanical? High Mosquitocidal Efficacy of a Silver Nanocomposite Synthesized Using a Leaf Aqueous Extract of Fumaria indica. *Journal of Cluster Science, 28*(1), 637–643. doi:10.100710876-017-1159-3

Morejón, Pilaquinga, Domenech, Ganchala, & & Debut. (2018). *Larvicidal Activity of Silver Nanoparticles Synthesized Using Extracts of Ambrosia arborescens (Asteraceae) to Control Aedes aegypti L. (Diptera: Culicidae).* Article 6917938.

Borges, M. C., Castro, L. A., & Fonseca, B. A. L. (2013). Chloroquine use improves dengue-related symptoms. *Memorias do Instituto Oswaldo Cruz, 108*(5), 596–599. doi:10.1590/S0074-02762013000500010 PMID:23903975

Campbell, S., & K, S.-F. (2019). *Antiparasitic Drugs.* StatPearls Publishing.

Cristina Buzea, I. I. P., & Robbie, K. (2007). Nanomaterials and nanoparticles: Sources and toxicity. *Biointerphases, 2*(4), 17–71. doi:10.1116/1.2815690 PMID:20419892

Croci, R., Bottaro, E., Chan, K., Watanabe, S., Pezzullo, M., Mastrangelo, E., & Nastruzzi, C. (2016). Liposomal Systems as Nanocarriers for the Antiviral Agent Ivermectin. *International Journal of Biomaterials, 2016*, 1–15. Advance online publication. doi:10.1155/2016/8043983 PMID:27242902

Dieme, C., Bechah, Y., Socolovschi, C., Audoly, G., Berenger, J. M., Faye, O., Raoult, D., & Parola, P. (2015). Transmission potential of Rickettsia felis infection by Anopheles gambiae mosquitoes. *Proceedings of the National Academy of Sciences of the United States of America, 112*(26), 8088–8093. doi:10.1073/pnas.1413835112 PMID:26056256

Dinesh, D., Murugan, K., Madhiyazhagan, P., Panneerselvam, C., Kumar, P. M., Nicoletti, M., Jiang, W., Benelli, G., Chandramohan, B., & Suresh, U. (2015). Mosquitocidal and antibacterial activity of green-synthesized silver nanoparticles from Aloe vera extracts: Towards an effective tool against the malaria vector Anopheles stephensi? *Parasitology Research, 114*(4), 1519–1529. doi:10.100700436-015-4336-z PMID:25653031

DMC. (2020). *Malaria control Program.* DMC.

Durán, N., Islan, G. A., Durán, M., & Castro, G. R. (2016). Nanobiotechnology Solutions against Aedes aegypti %J. *Journal of the Brazilian Chemical Society, 27*, 1139–1149. http://www.scielo.br/scielo.php?script=sci_arttext&pid=S0103-50532016000701139&nrm=iso

ECDC. (2019a). *Communicable Disease Threats Report*. European Centre for Disease Prevention and Control Issue. www.ecdc.europa.eu/en/publications-data/communicable-disease-threats-report-15-21-december-2019-week-51

ECDC. (2019b). *Communicable Disease Threats Report, 15-21 December 2019, week 51*. ECDC.

Elsaesser, A., & Howard, C. V. (2012). Toxicology of nanoparticles. *Advanced Drug Delivery Reviews*, *64*(2), 129–137. doi:10.1016/j.addr.2011.09.001 PMID:21925220

Gandhi, A. D., Murugan, K., Umamahesh, K., Babujanarthanam, R., Kavitha, P., & Selvi, A. (2019). Lichen Parmelia sulcata mediated synthesis of gold nanoparticles: An eco-friendly tool against Anopheles stephensi and Aedes aegypti. *Environmental Science and Pollution Research International*, *26*(23), 23886–23898. doi:10.100711356-019-05726-6 PMID:31218582

Gerencser, A. A., Chinopoulos, C., Birket, M. J., Jastroch, M., Vitelli, C., Nicholls, D. G., & Brand, M. D. (2012). Quantitative measurement of mitochondrial membrane potential in cultured cells: Calcium-induced de- and hyperpolarization of neuronal mitochondria. *The Journal of Physiology*, *590*(12), 2845–2871. doi:10.1113/jphysiol.2012.228387 PMID:22495585

Ghosh, A., Chowdhury, N., & Chandra, G. (2012). Plant extracts as potential mosquito larvicides. *The Indian Journal of Medical Research*, *135*(5), 581–598. PMID:22771587

Goodsell, D. S. (2004). Bionanotechnology: Lessons from Nature (illustrated ed.). John Wiley & Sons.

Gubler, D. J. (2002). The Global Emergence/Resurgence of Arboviral Diseases As Public Health Problems. *Archives of Medical Research*, *33*(4), 330–342. doi:10.1016/S0188-4409(02)00378-8 PMID:12234522

Hadrup, N., Sharma, A. K., & Loeschner, K. (2018). Toxicity of silver ions, metallic silver, and silver nanoparticle materials after in vivo dermal and mucosal surface exposure: A review. *Regulatory Toxicology and Pharmacology*, *98*, 257–267. doi:10.1016/j.yrtph.2018.08.007 PMID:30125612

Hamilton, W. L., Amato, R., van der Pluijm, R. W., Jacob, C. G., Quang, H. H., Thuy-Nhien, N. T., Hien, T. T., Hongvanthong, B., Chindavongsa, K., Mayxay, M., Huy, R., Leang, R., Huch, C., Dysoley, L., Amaratunga, C., Suon, S., Fairhurst, R. M., Tripura, R., Peto, T. J., ... Miotto, O. (2019). Evolution and expansion of multidrug-resistant malaria in southeast Asia: A genomic epidemiology study. *The Lancet. Infectious Diseases*, *19*(9), 943–951. doi:10.1016/S1473-3099(19)30392-5 PMID:31345709

Harbach, R. E. (2013). *Mosquito taxonomic inventory*. Academic Press.

Hill, S.R., & GK, S. (2019). *Antiparasitic Drugs*. StatPearls Publishing.

Jaiswal, S., Bhattacharya, K., McHale, P., & Duffy, B. (2015). Dual effects of beta-cyclodextrin-stabilised silver nanoparticles: Enhanced biofilm inhibition and reduced cytotoxicity. *Journal of Materials Science. Materials in Medicine*, *26*(1), 5367. doi:10.100710856-014-5367-1 PMID:25596861

Jeevanandam, J., Barhoum, A., Chan, Y. S., Dufresne, A., & Danquah, M. K. B. J. (2018). Review on nanoparticles and nanostructured materials: History, sources, toxicity and regulations. *Beilstein Journal of Nanotechnology*, *9*, 1050–1074. doi:10.3762/bjnano.9.98 PMID:29719757

Jensen, M., & Mehlhorn, H. (2009). Seventy-five years of Resochin in the fight against malaria. *Parasitology Research, 105*(3), 609–627. doi:10.100700436-009-1524-8 PMID:19593586

Kalimuthu, K., Lin, S., Tseng, L., Murugan, K., & Hwang, J.-S. (2013). Bio-efficacy potential of seaweed Gracilaria firma with copepod, Megacyclops formosanus for the control larvae of dengue vector Aedes aegypti. *Hydrobiologia, 741*(1), 113–123. doi:10.100710750-013-1745-9

Kalimuthu, K., Panneerselvam, C., Chou, C., Tseng, L.-C., Murugan, K., Tsai, K.-H., Alarfaj, A. A., Higuchi, A., Canale, A., Hwang, J.-S., & Benelli, G. (2017). Control of dengue and Zika virus vector Aedes aegypti using the predatory copepod Megacyclops formosanus: Synergy with Hedychium coronarium-synthesized silver nanoparticles and related histological changes in targeted mosquitoes. *Process Safety and Environmental Protection, 109*, 82–96. doi:10.1016/j.psep.2017.03.027

Kao, J. C., HuangFu, W.-C., Tsai, T.-T., Ho, M.-R., Jhan, M.-K., Shen, T.-J., Tseng, P.-C., Wang, Y.-T., & Lin, C.-F. (2018). The antiparasitic drug niclosamide inhibits dengue virus infection by interfering with endosomal acidification independent of mTOR. *PLoS Neglected Tropical Diseases, 12*(8), e0006715. doi:10.1371/journal.pntd.0006715 PMID:30125275

Kumar, P. M., Murugan, K., Madhiyazhagan, P., Kovendan, K., Amerasan, D., Chandramohan, B., Dinesh, D., Suresh, U., Nicoletti, M., Alsalhi, M. S., Devanesan, S., Wei, H., Kalimuthu, K., Hwang, J. S., Lo Iacono, A., & Benelli, G. (2016). Biosynthesis, characterization, and acute toxicity of Berberis tinctoria-fabricated silver nanoparticles against the Asian tiger mosquito, Aedes albopictus, and the mosquito predators Toxorhynchites splendens and Mesocyclops thermocyclopoides. *Parasitology Research, 115*(2), 751–759. doi:10.100700436-015-4799-y PMID:26499804

Li, Z., Brecher, M., Deng, Y. Q., Zhang, J., Sakamuru, S., Liu, B., Huang, R., Koetzner, C. A., Allen, C. A., Jones, S. A., Chen, H., Zhang, N. N., Tian, M., Gao, F., Lin, Q., Banavali, N., Zhou, J., Boles, N., Xia, M., ... Li, H. (2017). Existing drugs as broad-spectrum and potent inhibitors for Zika virus by targeting NS2B-NS3 interaction. *Cell Research, 27*(8), 1046–1064. doi:10.1038/cr.2017.88 PMID:28685770

Madhiyazhagan, P., Murugan, K., Kumar, A. N., Nataraj, T., Dinesh, D., Panneerselvam, C., Subramaniam, J., Mahesh Kumar, P., Suresh, U., Roni, M., Nicoletti, M., Alarfaj, A. A., Higuchi, A., Munusamy, M. A., & Benelli, G. (2015). S argassum muticum-synthesized silver nanoparticles: An effective control tool against mosquito vectors and bacterial pathogens. *Parasitology Research, 114*(11), 4305–4317. doi:10.100700436-015-4671-0 PMID:26281786

Mane, P. C., Chaudhari, R. D., Shinde, M. D., Kadam, D. D., Song, C. K., Amalnerkar, D. P., & Lee, H. (2017). Designing Ecofriendly Bionanocomposite Assembly with Improved Antimicrobial and Potent on-site Zika Virus Vector Larvicidal Activities with its Mode of Action. *Scientific Reports, 7*(1), 15531. doi:10.103841598-017-15537-9 PMID:29138496

Mansfield, C. M., Alloy, M. M., Hamilton, J., Verbeck, G. F., Newton, K., Klaine, S. J., & Roberts, A. P. (2015). Photo-induced toxicity of titanium dioxide nanoparticles to Daphnia magna under natural sunlight. *Chemosphere, 120*, 206–210. doi:10.1016/j.chemosphere.2014.06.075 PMID:25062026

Mastrangelo, E., Pezzullo, M., De Burghgraeve, T., Kaptein, S., Pastorino, B., Dallmeier, K., de Lamballerie, X., Neyts, J., Hanson, A. M., Frick, D. N., Bolognesi, M., & Milani, M. (2012). Ivermectin is a potent inhibitor of flavivirus replication specifically targeting NS3 helicase activity: New prospects for an old drug. *The Journal of Antimicrobial Chemotherapy*, *67*(8), 1884–1894. doi:10.1093/jac/dks147 PMID:22535622

Maurer, L., & Meyer, J. (2016). A Systematic Review of Evidence for Silver Nanoparticle-Induced Mitochondrial Toxicity. *Environmental Science. Nano*, *3*(2), 311–322. Advance online publication. doi:10.1039/C5EN00187K

McClements, D., & Xiao, H. (2017). Is nano safe in foods? Establishing the factors impacting the gastrointestinal fate and toxicity of organic and inorganic food-grade nanoparticles. *NPJ Science of Food, 1*. doi:10.103841538-017-0005-1

McKee, M. S., & Filser, J. (2016). Impacts of metal-based engineered nanomaterials on soil communities. *Environmental Science: Nano, 3*(3), 506-533. doi:10.1039/C6EN00007J

Metz, S. W., Thomas, A., Brackbill, A., Xianwen, Y., Stone, M., Horvath, K., Miley, M. J., Luft, C., DeSimone, J. M., Tian, S., & de Silva, A. M. (2018). Nanoparticle delivery of a tetravalent E protein subunit vaccine induces balanced, type-specific neutralizing antibodies to each dengue virus serotype. *PLoS Neglected Tropical Diseases*, *12*(9), e0006793. doi:10.1371/journal.pntd.0006793 PMID:30248097

Morejon, B., Pilaquinga, F., Domenech, F., Ganchala, D., Debut, A., & Neira, M. (2019). Larvicidal Activity of Silver Nanoparticles SynthesizedUsing Extracts ofAmbrosia arborescens(Asteraceae) to-ControlAedes aegyptiL. (Diptera: Culicidae). *Journal of Nanotechnology*.

Murugan, K., Aruna, P., Panneerselvam, C., Madhiyazhagan, P., Paulpandi, M., Subramaniam, J., Rajaganesh, R., Wei, H., Alsalhi, M. S., Devanesan, S., Nicoletti, M., Syuhei, B., Canale, A., & Benelli, G. (2016b). Fighting arboviral diseases: Low toxicity on mammalian cells, dengue growth inhibition (in vitro), and mosquitocidal activity of Centroceras clavulatum-synthesized silver nanoparticles. *Parasitology Research*, *115*(2), 651–662. doi:10.100700436-015-4783-6 PMID:26462804

Murugan, K., Priyanka, V., Dinesh, D., Madhiyazhagan, P., Panneerselvam, C., Subramaniam, J., Suresh, U., Chandramohan, B., Roni, M., Nicoletti, M., Alarfaj, A. A., Higuchi, A., Munusamy, M. A., Khater, H. F., Messing, R. H., & Benelli, G. (2015c). Predation by Asian bullfrog tadpoles, Hoplobatrachus tigerinus, against the dengue vector, Aedes aegypti, in an aquatic environment treated with mosquitocidal nanoparticles. *Parasitology Research*, *114*(10), 3601–3610. doi:10.100700436-015-4582-0 PMID:26091763

Muthamil Selvan, S., Vijai Anand, K., Govindaraju, K., Tamilselvan, S., Kumar, V. G., Subramanian, K. S., Kannan, M., & Raja, K. (2018). Green synthesis of copper oxide nanoparticles and mosquito larvicidal activity against dengue, zika and chikungunya causing vector Aedes aegypti. *IET Nanobiotechnology / IET*, *12*(8), 1042–1046. doi:10.1049/iet-nbt.2018.5083 PMID:30964011

Panneerselvam, C., Murugan, K., Roni, M., Aziz, A. T., Suresh, U., Rajaganesh, R., Madhiyazhagan, P., Subramaniam, J., Dinesh, D., Nicoletti, M., Higuchi, A., Alarfaj, A. A., Munusamy, M. A., Kumar, S., Desneux, N., & Benelli, G. (2016). Fern-synthesized nanoparticles in the fight against malaria: LC/MS analysis of Pteridium aquilinum leaf extract and biosynthesis of silver nanoparticles with high mosquitocidal and antiplasmodial activity. *Parasitology Research, 115*(3), 997–1013. doi:10.100700436-015-4828-x PMID:26612497

Pavela, R. (2015a). Essential oils for the development of eco-friendly mosquito larvicides: A review. *Industrial Crops and Products, 76*, 174–187. doi:10.1016/j.indcrop.2015.06.050

Rahman, K., Khan, S. U., Fahad, S., Chang, M. X., Abbas, A., Khan, W. U., Rahman, L., Haq, Z. U., Nabi, G., & Khan, D. (2019). Nano-biotechnology: A new approach to treat and prevent malaria. *International Journal of Nanomedicine, 14*, 1401–1410. doi:10.2147/IJN.S190692 PMID:30863068

Rajan, R., Chandran, K., Harper, S. L., Yun, S.-I., & Kalaichelvan, P. T. (2015). Plant extract synthesized silver nanoparticles: An ongoing source of novel biocompatible materials. *Industrial Crops and Products, 70*, 356–373. doi:10.1016/j.indcrop.2015.03.015

Rana, S., & Kalaichelvan, P. (2013). Ecotoxicity of Nanoparticles. *ISRN Toxicology, 574648*, 1–11. Advance online publication. doi:10.1155/2013/574648 PMID:23724300

Roopan, S., Mathew, R., Mahesh, S., Titus, D., Aggarwal, K., Bhatia, N., Kasinathan, D., Elumalai, K., & Samuel, J. (2018). Environmental friendly synthesis of zinc oxide nanoparticles and estimation of its larvicidal activity against Aedes aegypti. *International Journal of Environmental Science and Technology, 16*(12), 8053–8060. Advance online publication. doi:10.100713762-018-2175-z

Sahu, S. C., & Hayes, A. W. (2017). *Toxicity of nanomaterials found in human environment: A literature review*. doi:10.1177/2397847317726352

Saksena, R., Matlani, M., Singh, V., Kumar, A., Anveshi, A., Kumar, D., & Gaind, R. (2017). Early treatment failure in concurrent dengue and mixed malaria species infection with suspected resistance to artemisinin combination therapy from a tertiary care center in Delhi: A case report. *International Medical Case Reports Journal, 10*, 289–294. doi:10.2147/IMCRJ.S139729 PMID:28860870

Salouti, M. (2014). *Nanoparticle based Drug Delivery Systems for Treatment of Infectious Diseases in book Application of Nanotechnology in Drug Delivery*. Academic Press.

Sanpui, P., Chattopadhyay, A., & Ghosh, S. S. (2011). Induction of apoptosis in cancer cells at low silver nanoparticle concentrations using chitosan nanocarrier. *ACS Applied Materials & Interfaces, 3*(2), 218–228. doi:10.1021/am100840c PMID:21280584

Solairaj, D., & Rameshthangam, P. (2017). Silver Nanoparticle Embedded α-Chitin Nanocomposite for Enhanced Antimicrobial and Mosquito Larvicidal Activity. *Journal of Polymers and the Environment, 25*(2), 435–452. doi:10.100710924-016-0822-3

Subramaniam, J., Murugan, K., Jebanesan, A., Pontheckan, P., Dinesh, D., Nicoletti, M., Wei, H., Higuchi, A., Kumar, S., Canale, A., & Benelli, G. (2016). Do Chenopodium ambrosioides-Synthesized Silver Nanoparticles Impact Oryzias melastigma Predation Against Aedes albopictus Larvae? *Journal of Cluster Science, 28*(1), 413–436. doi:10.100710876-016-1113-9

Suman, T. Y., Ravindranath, R. R. S., Elumalai, D., Kaleena, P. K., Ramkumar, R., Perumal, P., Aranganathan, L., & Chitrarasu, P. S. (2015). Larvicidal activity of titanium dioxide nanoparticles synthesized using Morinda citrifolia root extract against Anopheles stephensi, Aedes aegypti and Culex quinquefasciatus and its other effect on non-target fish. *Asian Pacific Journal of Tropical Disease*, *5*(3), 224–230. doi:10.1016/S2222-1808(14)60658-7

Suresh, U., Murugan, K., Benelli, G., Nicoletti, M., Barnard, D. R., Panneerselvam, C., Kumar, P. M., Subramaniam, J., Dinesh, D., & Chandramohan, B. (2015). Tackling the growing threat of dengue: Phyllanthus niruri-mediated synthesis of silver nanoparticles and their mosquitocidal properties against the dengue vector Aedes aegypti (Diptera: Culicidae). *Parasitology Research*, *114*(4), 1551–1562. doi:10.100700436-015-4339-9 PMID:25669140

Sutthanont, N., Attrapadung, S., & Nuchprayoon, S. (2019). Larvicidal Activity of Synthesized Silver Nanoparticles from Curcuma zedoaria Essential Oil against Culex quinquefasciatus. *Insects*, *10*(1), 27. doi:10.3390/insects10010027 PMID:30641859

Tunçsoy, B. S. (2018). Toxicity of nanoparticles on insects: A Review. *Adana Science and Technology University Journal of Science, 49.*

Ullah Khan, S., Saleh, T. A., Wahab, A., Khan, M. H. U., Khan, D., Ullah Khan, W., Rahim, A., Kamal, S., Ullah Khan, F., & Fahad, S. (2018). Nanosilver: New ageless and versatile biomedical therapeutic scaffold. *International Journal of Nanomedicine*, *13*, 733–762. doi:10.2147/IJN.S153167 PMID:29440898

White, N. J. (2004). Antimalarial drug resistance. *The Journal of Clinical Investigation*, *113*(8), 1084–1092. doi:10.1172/JCI21682 PMID:15085184

WHO. (2015). *Guidelines for the Treatment of Malaria*. WHO.

WHO. (2019). *Outbreak update – Dengue in Pakistan*. WHO.

Xu, T. L., Han, Y., Liu, W., Pang, X. Y., Zheng, B., Zhang, Y., & Zhou, X. N. (2018). Antivirus effectiveness of ivermectin on dengue virus type 2 in Aedes albopictus. *PLoS Neglected Tropical Diseases*, *12*(11), e0006934. doi:10.1371/journal.pntd.0006934 PMID:30452439

Zelga, P. J., Gornicz, M. M., Gluszkiewicz, J. M., & Piasecka-Zelga, J. (2016). Outcomes of acute dermal irritation and sensitisation tests on active dressings for chronic wounds: A comparative study. *Journal of Wound Care*, *25*(12), 722–729. doi:10.12968/jowc.2016.25.12.722 PMID:27974010

Chapter 9
Current Strategies in Peptide Conjugated Nanoparticles:
A Novel Approach in the Field of Biomedicine

Palanisamy Priya Dharsini
Anna University, India

Palanisamy Selvamani
Anna University, India

Subbiah Latha
Anna University, India

ABSTRACT

In the arena of biomedicine, peptide-nanoparticle conjugates (PNCs) have been newly arising as a multifunctional tool. The synergistic effect of the two groups was directly heightened the various biological properties. There are thousands of PNCs that have been developed for both the narrow and broad bioactivities. In this review, the authors highlighted the novel technologies like single and multi-walled carbon nanotubes conjugates, trivalent, and multivalent conjugates and their importance. This study provides a clear picture about the emerging technologies in the field of PNCs and also offers a guideline for the budding researchers, particularly who are new to this field.

INTRODUCTION

There is an increasing resistance phase in the field of antibiotics, which poses a reverse reaction in global health sector and also hindered the development of modern medicine (Aloush et al., 2006; Manchanda et al., 2010; Guilhelmelli et al., 2013). The growth and emergence of new and novel antibiotic drugs is slow when compared to the increased activity in the resistance. The bacterial candidate those resist the drugs are biofilm producers i.e.) there is a protection by the development of extracellular polymer matrixes.

DOI: 10.4018/978-1-7998-5049-6.ch009

Biofilm formation is considered as a serious issue in various sectors like health, food industry and marine and so on (Hoiby et al., 2010; Otter et al., 2015). In order to battle from multidrug-resistant (MDR) pathogens, there is a pressing need for novel alternatives with little chance for resistance development.

To overcome this issue antimicrobial peptides (AMPs) are scrutinized as gorgeous option to combat MDR pathogens (Hancock and Chapple, 1999). Nowadays AMPs are evolving as a vital element for wiped out the pathogenic microorganisms due to their broad band activity with low rate of resistance formation (Wang et al., 2016; Zhang and Gallo, 2016). Even though AMPs also have some constraint attributes during their development as therapeutics due to their low permeability rate in the biological barriers and poor stability in the enzymatic action (Brandelli, 2012; Kang et al., 2014). These drawbacks can be overcome by some alterations like peptide cyclisation, use of non-protein amino acids, peptidomimetics, and lipidation (Fosgerau and Hoffmann, 2015; Kaspar and Reichert, 2013; White and Yudin, 2011).

Nanoparticles provide a hidden platform to conflict multidrug-resistant pathogens whereas it holds them the antimicrobial activities and many workers proved their mechanism of action (Raghunath and Perumal 2017; Azam et al., 2012, Wang et al., 2017). It can rupture the cell membrane of bacteria by cell penetration and retort with intracellular targets which leads to toxicity (Zhao et al., 2010). The AMPs immobilization to metallic nanoparticles might be an alternate to battle against the multi-drug resistant pathogens which leads to increase the antimicrobial activity (Allahverdiyev et al., 2011).This criteria comforts few AMPs to overwhelmed the downside such as proneness to proteases and deprived penetrability in the biological barriers, since nanoparticles showed tremendous effects in targeted drug delivery systems (Brandelli, 2012). This review lightened the nano-peptide conjugates (NPCs) and their uses in theranostic and imaging purposes.

POTENTIAL CANDIDATES FOR CONJUGATION (NANO-PEPTIDE PARTICLES)

Nanoparticles are magnificent candidate for drug carriers to the particular target sites. The combination of nano-peptides generates, there is a decline in their toxicity due to its synergistically heightened antimicrobial activities and low requirement of high dosages (Allahverdiyev et al., 2011). In antibiotic therapy, nanoparticles as peptide conjugates and carries to the infection site are evolving as a promising approach (Gao et al., 2014).NPCs can endure a robust light plasmon resonance, which used for drug release by photo activation (Pissuwan et al.,2011). In biological structures, mostly interface are centred on non-covalent interactions and hydrophobic interactions. The weak discrete bands interrupt the interaction meanwhile the supportive action allows vigorously bonding in binding kinetics grounded on the formula that the number of distinct binding pair is directly proportional to the collective binding strength.

TYPES AND SHAPES OF NANO-PEPTIDE CONJUGATES

The exclusive features of the conjugates depends upon the tiny molecules and greater micrometer shape particles which have been commonly used in the field of biomedical engineering (Bianco et al., 2015, Bozzuto and Molinari,2015, Banik et al., 2016, Bhatia,2016, Giner-Casares et al., 2016, McNamara and Tofail,2017).The distinctive assets of nanoparticles gathering them with bioactive groups which aid to massive biomedical applications. The probable effects of well - studied nanoparticles are tabulated in Table 1.

Table 1. Nanomaterial types and its credible effects

S.No	Types of Nanomaterial	Probable Effects	References
1	Metallic nanoparticles	Antimicrobial, biodistribution, photoablation and hyperthermic therapies	Versiani et al., 2016, Dreaden et al., 2012, Gao et al., 2012, Howes et al., 2014
2	Carbon nanotubes	invasive stimuli for engendering hyperthermia	Akbarzadeh et al., 2012, Shen et al., 2017
3	Liposome	Drug delivery agents and functionalised with lipids and ligands	Bozzuto et al., 2015
4	Quantum dots	In imaging system (*in vitro* and *in vivo)* targeted diagnosis or therapy	Howes et al., 2014, Zhao, et al., 2017, Argyo et al., 2014, Bharti et al.,2015, Croissant et al., 2017, Mehmood et al., 2017
5	Polymeric nanoparticles	Precise anticancer drug delivery	Ahmad et al., 2014, Zhang and Zhang, 2017, Moreno-Vega et al., 2012

The FDA approved nanoparticles are used for clinical purposes (Anselmo and Mitragotri, 2016). Amongst all, Gold nanoparticles were found to be the potential for its biocompatibility and stability. These can be synthesized by a precise mechanism in order to achieve the perfect shapes and sizes for particular application, which is tabulated in table .2

Table 2. Shapes and Size of the distant NPCs and its application

S.No	Shape	Size	Application
1	Clusters	< 2nm	Enhanced laser-induced bacterial killing, cancer therapy
2	Nanospheres	2-100 nm	Drug delivery, enhanced Raman imaging for small living cell
3	Nano rods	>2 nm	Drug delivery, cell imaging, *in vivo* imaging and cancer therapy
4	Nanoshells	10-400 nm	*In vitro* and *in vivo* imaging, cancer therapy,
5	Nanocage	20-200 nm	Photothermal cancer therapy

METHODS FOR CONJUGATING NANOPARTICLES TO PEPTIDE

The procedure used to conjugate nanoparticles and peptides can be achieved either through physisorption or chemisorption (Katz and Willner, 2004). Physisorption method is facilitated by intermolecular forces via van der waals force. The latter one is mediated through the covalent binding of the biomolecules with various conjugated chemistries in the thiolate bonds. For example there is a direct conjugation takes place to the gold and silver nanoparticles, where the thiol group containing peptides (Jazayeri et al., 2016; Ravindran et al., 2013). Another example for covalent chemistry is vancomycin antibiotic glycopeptide with gold nanoparticle as bis vancomycin cystamide (Gu et al., 2003). Many parameters may influence the conjugation like concentration of peptide, reaction time, conjugation buffer used to produce low Polydispersity along with peptide conjugates in high density. The single step process was taken place while in formation of conjugates, due to presence of cysteine which has numerous noteworthy advantages like peptide density, uniform particle size and aggregation behaviour.

MECHANISM OF NANO-PEPTIDE CONJUGATES

The mechanism for the significant antimicrobial activity is facilitating the drug concentration at the specific site of infection. Additionally, it also interacts with carbohydrates and proteins which are present in the outer membrane of the gram negative bacterial cell wall and left over in the membrane (Verma et al., 2013). The porin ion channels in the plasma membrane allowed the deposited proteins or the drugs are diffused by the phospholipid layer (Katz and Willner, 2004). The particular action is beneficial for biofilms producing bacteria for the interaction between the therapeutic as well as the conjugated peptide when compared to non-conjugated peptides.

STABILITY ENHANCEMENT

It is a well-known fact that there is an huge amount of NPCs have been showed positive results in the field of biomedicine, but still the NPCs are in infancy stage which needs to constructs the NPCs under controlled environment. There are many obstacles in translating NPCs stability under venerated situation in the multifacinated biological environments (Park, 2013). Coating of drugs such as peptides or proteins played a vital role in facilitating NPCs centered technologies (Bjornmalam et al.,2016).NPCs clumping was prevented by coating which results the drugs was cleared from the blood circulation. So, constructions of NPCs are multidimensional criteria, which were designed carefully with utmost care that safeguards the final application of the technology.

The attachment of opsonin proteins to the NPCs showed a vital role in eliciting clearance, provoking macrophage identification and finally exclusion takes place by phagocytosis (Peng et al., 2013, Gao and He, 2014). Apart from that, physical stability is also considered as an important factor while construct NPCs in the field of biomedical science (Moore et al., 2015). The discharge of tiny particles and accretion of larger aggregates leads to the reduction in the toxicity of the molecules (Herce et al., 2014, Bechara and sagan,2013, Kettier et al., 2014, Guidotti et al., 2017, Zhu et al., 2013). Several studies proved the ability of NPCs under in vitro and in vivo conditions, produce a disseminated and stable in various controlled conditions (Cui et al., 2016, Feliu et al., 2017, Pamies et al., 2014, Neupane et al., 2017). Additionally, the studies showed agglomerated nature which alters the assets mainly in the lethal effects. The accumulated molecules may significantly decline the surface area of the particles and also low uptake of cellular components (Gupta et al., 2016, Wu et al., 2014).

BIOMARKERS AND BIO IMAGING

Bio imaging at molecular level depicts the clear pictorial representation on evidence about biological process at high resolution (Cassidy and Radda, 2005, Prabhakar et al., 2013, Nichols and Bae, 2014, Danhier,2016, Anchordoquy et al., 2017). In the field of nanobiotechnology, there is an advanced growth in the molecular imaging technique by using the imaging probes (Kim et al., 2018, Yao et al., 2016, Gupta et al., 2014, Kunjachan et al., 2014, Curnis et al., 2016). Novel probe was preferred for bio imaging due to the fact that prolonged stability, accurate target site and rapid clearance property of the peptides (Spicer et al., 2018, Geng et al., 2016, Ding et al., 2017, Mu et al., 2015, Yeh et al., 2016). The integrated and prefabricated nature of the peptides helps to study the imaging modalities, which resulted in fabulous outcomes in the *in vivo* models and preclinical studies. The facility to enrich the target to background signal offered by peptide nanoparticle conjugates is a key lead in this study (Chapman et al., 2013, Mann et al., 2016, Agemy et al., 2016, Chen et al., 2011, Yang et al., 2013).

NPCs biological dispersal may epitomize the accumulation of the particles in some types of tumours through EPR effect (Bae and Park,2011, Lee and Yu,2017, Schwendener, 2014, Liu et al., 2015).NPCs targeting the liver and spleen which obviously dropped clearance from blood and other parts of the organs needs active mode of guiding drug delivery (Weis,2008).The recognition motif nearby the corona of NPCs is the mostly employed technique which muddle the cell surface receptors (Panagi et al.,2001).

The emission spectra at 700 and 1100 nm in infra-red (NIR) fluorescence imaging consumes imaging agents which breaches the tissue and heightens the tissue imaging (Owens et al., 2016).Now a days NIR imaging was used with the aid of NPCs for the detection of abnormal cancerous tissue at high specificity (Cao et al., 2011, Estelrich et al., 2015). The fluorescent nanoparticles conjugate with cyclic peptides were co assembled with Zn^2+ ions to create strong NIR fluorescent signals (Fan et al., 2018, Thakor et al.,2016, Yun and kwok,2017). Even though many advantages, still the toxicity and biological stability could be addressed for the effective scientific translation and NPCs based bio imaging paves a new attention in the biomedical field for current diagnostic and therapeutic platforms.

CONCLUSION

Nature takes millions of years to gradually develop the proteins with bioactivity but still there is a challenging task to partitioned, constructed the peptide from the resources. The bioactive domain with flexible orientation was responsible for the powerful NPCs or Nano protein conjugates (Parolo et al., 2013, Heinis and Winter, 2015).The NPCs provides the stable antimicrobial activity when compare to free antimicrobial peptides against the drug resistant bacterial pathogens. It is a novel group of biomaterial to those dragging MDR pathogens, as biomarkers from human body fluids and as submicron bio imaging agents. Generally along with the prompt developments in the field of nanotechnology, the NPCs will probable become an innovative platform for typical therapeutic and investigative structures.

REFERENCES

Aloush, V., Navon-Venezia, S., Seigman-Igra, Y., Cabili, S., & Carmeli, Y. (2006). Multidrug-resistant Pseudomonas aeruginosa: Risk factors and clinical impact. *Antimicrobial Agents and Chemotherapy*, *50*(1), 43–48. doi:10.1128/AAC.50.1.43-48.2006 PMID:16377665

Manchanda, V., Sanchaita, S., & Singh, N. P. (2010). Multidrug resistant acinetobacter. *Journal of Global Infectious Diseases*, *2*(3), 291. doi:10.4103/0974-777X.68538 PMID:20927292

Guilhelmelli, F., Vilela, N., Albuquerque, P., Derengowski, L., Silva-Pereira, I., & Kyaw, C. (2013). Antibiotic development challenges: The various mechanisms of action of antimicrobial peptides and of bacterial resistance. *Frontiers in Microbiology*, *4*, 353. doi:10.3389/fmicb.2013.00353 PMID:24367355

Høiby, N., Bjarnsholt, T., Givskov, M., Molin, S., & Ciofu, O. (2010). Antibiotic resistance of bacterial biofilms. *International Journal of Antimicrobial Agents*, *35*(4), 322–332. doi:10.1016/j.ijantimicag.2009.12.011 PMID:20149602

Otter, J. A., Vickery, K., Walker, J. D., Pulcini, E. D., Stoodley, P., Goldenberg, S. D., ... Edgeworth, J. D. (2015). Surface-attached cells, biofilms and biocide susceptibility: Implications for hospital cleaning and disinfection. *The Journal of Hospital Infection*, *89*(1), 16–27. doi:10.1016/j.jhin.2014.09.008 PMID:25447198

Hancock, R. E., & Chapple, D. S. (1999). Peptide antibiotics. *Antimicrobial Agents and Chemotherapy*, *43*(6), 1317–1323. doi:10.1128/AAC.43.6.1317 PMID:10348745

Wang, S., Zeng, X., Yang, Q., & Qiao, S. (2016). Antimicrobial peptides as potential alternatives to antibiotics in food animal industry. *International Journal of Molecular Sciences*, *17*(5), 603. doi:10.3390/ijms17050603 PMID:27153059

Zhang, L., & Gallo, R. L. (2016). Antimicrobial peptides. *Current Biology*, *26*(1), R14–R19. doi:10.1016/j.cub.2015.11.017 PMID:26766224

Brandelli, A. (2012). Nanostructures as promising tools for delivery of antimicrobial peptides. *Mini-Reviews in Medicinal Chemistry*, *12*(8), 731–741. doi:10.2174/138955712801264774 PMID:22512554

Kang, S. J., Park, S. J., Mishig-Ochir, T., & Lee, B. J. (2014). Antimicrobial peptides: Therapeutic potentials. *Expert Review of Anti-Infective Therapy*, *12*(12), 1477–1486. doi:10.1586/14787210.2014.976613 PMID:25371141

Fosgerau, K., & Hoffmann, T. (2015). Peptide therapeutics: Current status and future directions. *Drug Discovery Today*, *20*(1), 122–128. doi:10.1016/j.drudis.2014.10.003 PMID:25450771

Kaspar, A. A., & Reichert, J. M. (2013). Future directions for peptide therapeutics development. *Drug Discovery Today*, *18*(17-18), 807–817. doi:10.1016/j.drudis.2013.05.011 PMID:23726889

White, C. J., & Yudin, A. K. (2011). Contemporary strategies for peptide macrocyclization. *Nature Chemistry*, *3*(7), 509. doi:10.1038/nchem.1062 PMID:21697871

Raghunath, A., & Perumal, E. (2017). Metal oxide nanoparticles as antimicrobial agents: A promise for the future. *International Journal of Antimicrobial Agents*, *49*(2), 137–152. doi:10.1016/j.ijantimicag.2016.11.011 PMID:28089172

Azam, A., Ahmed, A. S., Oves, M., Khan, M. S., Habib, S. S., & Memic, A. (2012). Antimicrobial activity of metal oxide nanoparticles against Gram-positive and Gram-negative bacteria: A comparative study. *International Journal of Nanomedicine*, *7*, 6003. doi:10.2147/IJN.S35347 PMID:23233805

Wang, L., Hu, C., & Shao, L. (2017). The antimicrobial activity of nanoparticles: Present situation and prospects for the future. *International Journal of Nanomedicine*, *12*, 1227–1249. doi:10.2147/IJN.S121956 PMID:28243086

Zhao, Y., Tian, Y., Cui, Y., Liu, W., Ma, W., & Jiang, X. (2010). Small molecule-capped gold nanoparticles as potent antibacterial agents that target gram-negative bacteria. *Journal of the American Chemical Society*, *132*(35), 12349–12356. doi:10.1021/ja1028843 PMID:20707350

Allahverdiyev, A. M., Kon, K. V., Abamor, E. S., Bagirova, M., & Rafailovich, M. (2011). Coping with antibiotic resistance: Combining nanoparticles with antibiotics and other antimicrobial agents. *Expert Review of Anti-Infective Therapy*, *9*(11), 1035–1052. doi:10.1586/eri.11.121 PMID:22029522

Gao, W., Thamphiwatana, S., Angsantikul, P., & Zhang, L. (2014). Nanoparticle approaches against bacterial infections. *Wiley Interdisciplinary Reviews. Nanomedicine and Nanobiotechnology*, *6*(6), 532–547. doi:10.1002/wnan.1282 PMID:25044325

Pissuwan, D., Niidome, T., & Cortie, M. B. (2011). The forthcoming applications of gold nanoparticles in drug and gene delivery systems. *Journal of Controlled Release*, *149*(1), 65–71. doi:10.1016/j.jconrel.2009.12.006 PMID:20004222

Anselmo, A. C., & Mitragotri, S. (2016). Nanoparticles in the clinic. *Bioengineering & Translational Medicine*, *1*(1), 10–29. doi:10.1002/btm2.10003 PMID:29313004

Katz, E., & Willner, I. (2004). Integrated nanoparticle–biomolecule hybrid systems: Synthesis, properties, and applications. *Angewandte Chemie International Edition*, *43*(45), 6042–6108. doi:10.1002/anie.200400651 PMID:15538757

Jazayeri, M. H., Amani, H., Pourfatollah, A. A., Pazoki-Toroudi, H., & Sedighimoghaddam, B. (2016). Various methods of gold nanoparticles (GNPs) conjugation to antibodies. *Sensing and Bio-Sensing Research*, *9*, 17–22. doi:10.1016/j.sbsr.2016.04.002

Ravindran, A., Chandran, P., & Khan, S. S. (2013). Biofunctionalized silver nanoparticles: Advances and prospects. *Colloids and Surfaces. B, Biointerfaces*, *105*, 342–352. doi:10.1016/j.colsurfb.2012.07.036 PMID:23411404

Gu, H., Ho, P. L., Tong, E., Wang, L., & Xu, B. (2003). Presenting vancomycin on nanoparticles to enhance antimicrobial activities. *Nano Letters*, *3*(9), 1261–1263. doi:10.1021/nl034396z

Verma, A., Uzun, O., Hu, Y., Hu, Y., Han, H. S., Watson, N., Chen, S., Irvine, D. J., & Stellacci, F. (2008). Surface-structure-regulated cell-membrane penetration by monolayer-protected nanoparticles. *Nature Materials*, *7*(7), 588–595. doi:10.1038/nmat2202 PMID:18500347

Cassidy, P. J., & Radda, G. K. (2005). Molecular imaging perspectives. *Journal of the Royal Society, Interface*, *2*(3), 133–144. doi:10.1098/rsif.2005.0040 PMID:16849174

Kim, D., Kim, J., Park, Y. I., Lee, N., & Hyeon, T. (2018). Recent development of inorganic nanoparticles for biomedical imaging. *ACS Central Science*, *4*(3), 324–336. doi:10.1021/acscentsci.7b00574 PMID:29632878

Spicer, C. D., Jumeaux, C., Gupta, B., & Stevens, M. M. (2018). Peptide and protein nanoparticle conjugates: Versatile platforms for biomedical applications. *Chemical Society Reviews*, *47*(10), 3574–3620. doi:10.1039/C7CS00877E PMID:29479622

Chapman, S., Dobrovolskaia, M., Farahani, K., Goodwin, A., Joshi, A., Lee, H., ... Singh, R. (2013). Nanoparticles for cancer imaging: The good, the bad, and the promise. *Nano Today*, *8*(5), 454–460. doi:10.1016/j.nantod.2013.06.001 PMID:25419228

Owens, E. A., Henary, M., El Fakhri, G., & Choi, H. S. (2016). Tissue-specific near-infrared fluorescence imaging. *Accounts of Chemical Research*, *49*(9), 1731–1740. doi:10.1021/acs.accounts.6b00239 PMID:27564418

Cao, C. Y., Chen, Y., Wu, F. Z., Deng, Y., & Liang, G. L. (2011). Caspase-3 controlled assembly of nanoparticles for fluorescence turn on. *Chemical Communications*, *47*(37), 10320–10322. doi:10.1039/c1cc14112k PMID:21853204

Fan, Z., Chang, Y., Cui, C., Sun, L., Wang, D. H., Pan, Z., & Zhang, M. (2018). Near infrared fluorescent peptide nanoparticles for enhancing esophageal cancer therapeutic efficacy. *Nature Communications*, *9*(1), 1–11. doi:10.103841467-018-04763-y PMID:29973582

Blanco, E., Shen, H., & Ferrari, M. (2015). Principles of nanoparticle design for overcoming biological barriers to drug delivery. *Nature Biotechnology*, *33*(9), 941–951. doi:10.1038/nbt.3330 PMID:26348965

Bozzuto, G., & Molinari, A. (2015). Liposomes as nanomedical devices. *International Journal of Nanomedicine*, *10*, 975. doi:10.2147/IJN.S68861 PMID:25678787

Banik, B. L., Fattahi, P., & Brown, J. L. (2016). Polymeric nanoparticles: The future of nanomedicine. *Wiley Interdisciplinary Reviews. Nanomedicine and Nanobiotechnology*, *8*(2), 271–299. doi:10.1002/wnan.1364 PMID:26314803

Bhatia, S. (2016). Nanoparticles types, classification, characterization, fabrication methods and drug delivery applications. In *Natural polymer drug delivery systems* (pp. 33–93). Springer. doi:10.1007/978-3-319-41129-3_2

Giner-Casares, J. J., Henriksen-Lacey, M., Coronado-Puchau, M., & Liz-Marzán, L. M. (2016). Inorganic nanoparticles for biomedicine: Where materials scientists meet medical research. *Materials Today*, *19*(1), 19–28. doi:10.1016/j.mattod.2015.07.004

McNamara, K., & Tofail, S. A. (2017). Nanoparticles in biomedical applications. *Advances in Physics: X*, *2*(1), 54–88. doi:10.1080/23746149.2016.1254570

Versiani, A. F., Andrade, L. M., Estefania, M. N., Sergio, S., Geraldo, J. M., Claudilne, R. C., ... Fonseca, F. G. (2016). Gold nanoparticles and their applications in biomedicine. *Future Virology*, *11*(4), 293–309. doi:10.2217/fvl-2015-0010

Dreaden, E. C., Alkilany, A. M., Huang, X., Murphy, C. J., & El-Sayed, M. A. (2012). The golden age: Gold nanoparticles for biomedicine. *Chemical Society Reviews*, *41*(7), 2740–2779. doi:10.1039/C1CS15237H PMID:22109657

Gao, J., Huang, X., Liu, H., Zan, F., & Ren, J. (2012). Colloidal stability of gold nanoparticles modified with thiol compounds: Bioconjugation and application in cancer cell imaging. *Langmuir*, *28*(9), 4464–4471. doi:10.1021/la204289k PMID:22276658

Howes, P. D., Rana, S., & Stevens, M. M. (2014). Plasmonic nanomaterials for biodiagnostics. *Chemical Society Reviews*, *43*(11), 3835–3853. doi:10.1039/C3CS60346F PMID:24323079

Akbarzadeh, A., Samiei, M., & Davaran, S. (2012). Magnetic nanoparticles: Preparation, physical properties, and applications in biomedicine. *Nanoscale Research Letters*, *7*(1), 144. doi:10.1186/1556-276X-7-144 PMID:22348683

Shen, Z., Wu, A., & Chen, X. (2017). Iron oxide nanoparticle based contrast agents for magnetic resonance imaging. *Molecular Pharmaceutics*, *14*(5), 1352–1364. doi:10.1021/acs.molpharmaceut.6b00839 PMID:27776215

Lee, D. E., Koo, H., Sun, I. C., Ryu, J. H., Kim, K., & Kwon, I. C. (2012). Multifunctional nanoparticles for multimodal imaging and theragnosis. *Chemical Society Reviews*, *41*(7), 2656–2672. doi:10.1039/C2CS15261D PMID:22189429

Ulbrich, K., Hola, K., Subr, V., Bakandritsos, A., Tucek, J., & Zboril, R. (2016). Targeted drug delivery with polymers and magnetic nanoparticles: Covalent and noncovalent approaches, release control, and clinical studies. *Chemical Reviews*, *116*(9), 5338–5431. doi:10.1021/acs.chemrev.5b00589 PMID:27109701

Ahmad, Z., Shah, A., Siddiq, M., & Kraatz, H. B. (2014). Polymeric micelles as drug delivery vehicles. *RSC Advances*, *4*(33), 17028–17038. doi:10.1039/C3RA47370H

Zhang, X. Y., & Zhang, P. Y. (2017). Polymersomes in nanomedicine-A review. *Current Nanoscience*, *13*(2), 124–129. doi:10.2174/1573413712666161018144519

Moreno-Vega, A. I., Gomez-Quintero, T., Nunez-Anita, R. E., Acosta-Torres, L. S., & Castaño, V. (2012). Polymeric and ceramic nanoparticles in biomedical applications. *Journal of Nanotechnology*, *2012*, 2012. doi:10.1155/2012/936041

Bozzuto, G., & Molinari, A. (2015). Liposomes as nanomedical devices. *International Journal of Nanomedicine*, *10*, 975. doi:10.2147/IJN.S68861 PMID:25678787

Howes, P. D., Chandrawati, R., & Stevens, M. M. (2014). Colloidal nanoparticles as advanced biological sensors. *Science*, *346*(6205), 1247390. doi:10.1126cience.1247390 PMID:25278614

Zhao, P., Xu, Q., Tao, J., Jin, Z., Pan, Y., Yu, C., & Yu, Z. (2018). Near infrared quantum dots in biomedical applications: Current status and future perspective. *Wiley Interdisciplinary Reviews. Nanomedicine and Nanobiotechnology*, *10*(3), e1483. doi:10.1002/wnan.1483 PMID:28719080

Argyo, C., Weiss, V., Bräuchle, C., & Bein, T. (2014). Multifunctional mesoporous silica nanoparticles as a universal platform for drug delivery. *Chemistry of Materials*, *26*(1), 435–451. doi:10.1021/cm402592t

Bharti, C., Nagaich, U., Pal, A. K., & Gulati, N. (2015). Mesoporous silica nanoparticles in target drug delivery system: A review. *International Journal of Pharmaceutical Investigation*, *5*(3), 124. doi:10.4103/2230-973X.160844 PMID:26258053

Croissant, J. G., Fatieiev, Y., & Khashab, N. M. (2017). Degradability and clearance of silicon, organosilica, silsesquioxane, silica mixed oxide, and mesoporous silica nanoparticles. *Advanced Materials*, *29*(9), 1604634. doi:10.1002/adma.201604634 PMID:28084658

Mehmood, A., Ghafar, H., Yaqoob, S., Gohar, U. F., & Ahmad, B. (2017). Mesoporous silica nanoparticles: A review. *Journal of Developing Drugs*, *6*(02). Advance online publication. doi:10.4172/2329-6631.1000174

Park, K. (2013). Facing the truth about nanotechnology in drug delivery. *ACS Nano*, *7*(9), 7442–7447. doi:10.1021/nn404501g PMID:24490875

Björnmalm, M., Faria, M., & Caruso, F. (2016). Increasing the impact of materials in and beyond bio-nano science. *Journal of the American Chemical Society*, *138*(41), 13449–13456. doi:10.1021/jacs.6b08673 PMID:27672703

Bae, Y. H., & Park, K. (2011). Targeted drug delivery to tumors: Myths, reality and possibility. *Journal of Controlled Release*, *153*(3), 198–205. doi:10.1016/j.jconrel.2011.06.001 PMID:21663778

Grimsey, N. J., Lin, Y., Narala, R., Rada, C. C., Mejia-Pena, H., & Trejo, J. (2019). G protein–coupled receptors activate p38 MAPK via a non-canonical TAB1–TAB2–and TAB1–TAB3–dependent pathway in endothelial cells. *The Journal of Biological Chemistry*, *294*(15), 5867–5878. doi:10.1074/jbc.RA119.007495 PMID:30760523

Panagi, Z., Beletsi, A., Evangelatos, G., Livaniou, E., Ithakissios, D. S., & Avgoustakis, K. (2001). Effect of dose on the biodistribution and pharmacokinetics of PLGA and PLGA–mPEG nanoparticles. *International Journal of Pharmaceutics*, *221*(1-2), 143–152. doi:10.1016/S0378-5173(01)00676-7 PMID:11397575

Gao, H., & He, Q. (2014). The interaction of nanoparticles with plasma proteins and the consequent influence on nanoparticles behavior. *Expert Opinion on Drug Delivery*, *11*(3), 409–420. doi:10.1517/17425247.2014.877442 PMID:24397260

Peng, Q., Zhang, S., Yang, Q., Zhang, T., Wei, X. Q., Jiang, L., Zhang, C.-L., Chen, Q.-M., Zhang, Z.-R., & Lin, Y. F. (2013). Preformed albumin corona, a protective coating for nanoparticles based drug delivery system. *Biomaterials*, *34*(33), 8521–8530. doi:10.1016/j.biomaterials.2013.07.102 PMID:23932500

Moore, T. L., Rodriguez-Lorenzo, L., Hirsch, V., Balog, S., Urban, D., Jud, C., Rothen-Rutishauser, B., Lattuada, M., & Petri-Fink, A. (2015). Nanoparticle colloidal stability in cell culture media and impact on cellular interactions. *Chemical Society Reviews*, *44*(17), 6287–6305. doi:10.1039/C4CS00487F PMID:26056687

Cui, J., Faria, M., Björnmalm, M., Ju, Y., Suma, T., Gunawan, S. T., Richardson, J. J., Heidari, H., Bals, S., Crampin, E. J., & Caruso, F. (2016). A framework to account for sedimentation and diffusion in particle–cell interactions. *Langmuir, 32*(47), 12394–12402. doi:10.1021/acs.langmuir.6b01634 PMID:27384770

Feliu, N., Sun, X., Alvarez Puebla, R. A., & Parak, W. J. (2017). Quantitative particle–cell interaction: Some basic physicochemical pitfalls. *Langmuir, 33*(27), 6639–6646. doi:10.1021/acs.langmuir.6b04629 PMID:28379704

Pamies, R., Cifre, J. G. H., Espín, V. F., Collado-González, M., Baños, F. G. D., & de la Torre, J. G. (2014). Aggregation behaviour of gold nanoparticles in saline aqueous media. *Journal of Nanoparticle Research, 16*(4), 2376. doi:10.100711051-014-2376-4

Neupane, S., Pan, Y., Takalkar, S., Bentz, K., Farmakes, J., Xu, Y., Chen, B., Liu, G., Qian, S. Y., & Yang, Z. (2017). Probing the aggregation mechanism of gold nanoparticles triggered by a globular protein. *The Journal of Physical Chemistry C, 121*(2), 1377–1386. doi:10.1021/acs.jpcc.6b11963

Gupta, A., Moyano, D. F., Parnsubsakul, A., Papadopoulos, A., Wang, L. S., Landis, R. F., Das, R., & Rotello, V. M. (2016). Ultrastable and biofunctionalizable gold nanoparticles. *ACS Applied Materials & Interfaces, 8*(22), 14096–14101. doi:10.1021/acsami.6b02548 PMID:27191946

Wu, R. H., Nguyen, T. P., Marquart, G. W., Miesen, T. J., Mau, T., & Mackiewicz, M. R. (2014). A facile route to tailoring peptide-stabilized gold nanoparticles using glutathione as a synthon. *Molecules (Basel, Switzerland), 19*(5), 6754–6775. doi:10.3390/molecules19056754 PMID:24858266

Parolo, C., de la Escosura-Muñiz, A., Polo, E., Grazú, V., De La Fuente, J. M., & Merkoçi, A. (2013). Design, preparation, and evaluation of a fixed-orientation antibody/gold-nanoparticle conjugate as an immunosensing label. *ACS Applied Materials & Interfaces, 5*(21), 10753–10759. doi:10.1021/am4029153 PMID:24095174

Estelrich, J., Sánchez-Martín, M. J., & Busquets, M. A. (2015). Nanoparticles in magnetic resonance imaging: From simple to dual contrast agents. *International Journal of Nanomedicine, 10*, 1727. PMID:25834422

Thakor, A. S., Jokerst, J. V., Ghanouni, P., Campbell, J. L., Mittra, E., & Gambhir, S. S. (2016). Clinically approved nanoparticle imaging agents. *Journal of Nuclear Medicine, 57*(12), 1833–1837. doi:10.2967/jnumed.116.181362 PMID:27738007

Yun, S. H., & Kwok, S. J. (2017). Light in diagnosis, therapy and surgery. *Nature Biomedical Engineering, 1*(1), 1–16. doi:10.103841551-016-0008 PMID:28649464

Herce, H. D., Garcia, A. E., & Cardoso, M. C. (2014). Fundamental molecular mechanism for the cellular uptake of guanidinium-rich molecules. *Journal of the American Chemical Society, 136*(50), 17459–17467. doi:10.1021/ja507790z PMID:25405895

Bechara, C., & Sagan, S. (2013). Cell-penetrating peptides: 20 years later, where do we stand? *FEBS Letters, 587*(12), 1693–1702. doi:10.1016/j.febslet.2013.04.031 PMID:23669356

Kettler, K., Veltman, K., van de Meent, D., van Wezel, A., & Hendriks, A. J. (2014). Cellular uptake of nanoparticles as determined by particle properties, experimental conditions, and cell type. *Environmental Toxicology and Chemistry*, *33*(3), 481–492. doi:10.1002/etc.2470 PMID:24273100

Guidotti, G., Brambilla, L., & Rossi, D. (2017). Cell-penetrating peptides: From basic research to clinics. *Trends in Pharmacological Sciences*, *38*(4), 406–424. doi:10.1016/j.tips.2017.01.003 PMID:28209404

Zhu, M., Nie, G., Meng, H., Xia, T., Nel, A., & Zhao, Y. (2013). Physicochemical properties determine nanomaterial cellular uptake, transport, and fate. *Accounts of Chemical Research*, *46*(3), 622–631. doi:10.1021/ar300031y PMID:22891796

Prabhakar, U., Maeda, H., Jain, R. K., Sevick-Muraca, E. M., Zamboni, W., Farokhzad, O. C., ... Blakey, D. C. (2013). *Challenges and key considerations of the enhanced permeability and retention effect for nanomedicine drug delivery in oncology*. Academic Press.

Nichols, J. W., & Bae, Y. H. (2014). EPR: Evidence and fallacy. *Journal of Controlled Release*, *190*, 451–464. doi:10.1016/j.jconrel.2014.03.057 PMID:24794900

Danhier, F. (2016). To exploit the tumor microenvironment: Since the EPR effect fails in the clinic, what is the future of nanomedicine? *Journal of Controlled Release*, *244*, 108–121. doi:10.1016/j.jconrel.2016.11.015 PMID:27871992

Anchordoquy, T. J., Barenholz, Y., Boraschi, D., Chorny, M., Decuzzi, P., Dobrovolskaia, M. A., ... Godin, B. (2017). *Mechanisms and barriers in cancer nanomedicine: addressing challenges, looking for solutions*. Academic Press.

Yao, V. J., D'Angelo, S., Butler, K. S., Theron, C., Smith, T. L., Marchiò, S., ... Bradbury, A. R. (2016). Ligand-targeted theranostic nanomedicines against cancer. *Journal of Controlled Release*, *240*, 267–286. doi:10.1016/j.jconrel.2016.01.002 PMID:26772878

Gupta, M., Chashoo, G., Sharma, P. R., Saxena, A. K., Gupta, P. N., Agrawal, G. P., & Vyas, S. P. (2014). Dual targeted polymeric nanoparticles based on tumor endothelium and tumor cells for enhanced antitumor drug delivery. *Molecular Pharmaceutics*, *11*(3), 697–715. doi:10.1021/mp400404p PMID:24512060

Kunjachan, S., Pola, R., Gremse, F., Theek, B., Ehling, J., Moeckel, D., ... Storm, G. (2014). Passive versus active tumor targeting using RGD-and NGR-modified polymeric nanomedicines. *Nano Letters*, *14*(2), 972–981. doi:10.1021/nl404391r PMID:24422585

Curnis, F., Fiocchi, M., Sacchi, A., Gori, A., Gasparri, A., & Corti, A. (2016). NGR-tagged nano-gold: A new CD13-selective carrier for cytokine delivery to tumors. *Nano Research*, *9*(5), 1393–1408. doi:10.100712274-016-1035-8 PMID:27226823

Geng, L., Wang, Z., Jia, X., Han, Q., Xiang, Z., Li, D., ... Hu, Z. (2016). HER2 targeting peptides screening and applications in tumor imaging and drug delivery. *Theranostics*, *6*(8), 1261–1273. doi:10.7150/thno.14302 PMID:27279916

Ding, H., Gangalum, P. R., Galstyan, A., Fox, I., Patil, R., Hubbard, P., Murali, R., Ljubimova, J. Y., & Holler, E. (2017). HER2-positive breast cancer targeting and treatment by a peptide-conjugated mini nanodrug. *Nanomedicine; Nanotechnology, Biology, and Medicine*, *13*(2), 631–639. doi:10.1016/j.nano.2016.07.013 PMID:27520726

Mu, Q., Kievit, F. M., Kant, R. J., Lin, G., Jeon, M., & Zhang, M. (2015). Anti-HER2/neu peptide-conjugated iron oxide nanoparticles for targeted delivery of paclitaxel to breast cancer cells. *Nanoscale*, *7*(43), 18010–18014. doi:10.1039/C5NR04867B PMID:26469772

Yeh, C. Y., Hsiao, J. K., Wang, Y. P., Lan, C. H., & Wu, H. C. (2016). Peptide-conjugated nanoparticles for targeted imaging and therapy of prostate cancer. *Biomaterials*, *99*, 1–15. doi:10.1016/j.biomaterials.2016.05.015 PMID:27209258

Mann, A. P., Scodeller, P., Hussain, S., Joo, J., Kwon, E., Braun, G. B., ... Krajewski, S. (2016). A peptide for targeted, systemic delivery of imaging and therapeutic compounds into acute brain injuries. *Nature Communications*, *7*(1), 1–11. doi:10.1038/ncomms11980 PMID:27351915

Agemy, L., Friedmann-Morvinski, D., Kotamraju, V. R., Roth, L., Sugahara, K. N., Girard, O. M., Mattrey, R. F., Verma, I. M., & Ruoslahti, E. (2014). Targeted nanoparticle enhanced proapoptotic peptide as potential therapy for glioblastoma. *Proceedings of the National Academy of Sciences of the United States of America*, *108*(42), 17450–17455. doi:10.1073/pnas.1114518108 PMID:21969599

Chen, X., Wong, R., Khalidov, I., Wang, A. Y., Leelawattanachai, J., Wang, Y., & Jin, M. M. (2011). Inflamed leukocyte-mimetic nanoparticles for molecular imaging of inflammation. *Biomaterials*, *32*(30), 7651–7661. doi:10.1016/j.biomaterials.2011.06.030 PMID:21783245

Yang, S., Damiano, M. G., Zhang, H., Tripathy, S., Luthi, A. J., Rink, J. S., ... Thaxton, C. S. (2013). Biomimetic, synthetic HDL nanostructures for lymphoma. *Proceedings of the National Academy of Sciences of the United States of America*, *110*(7), 2511–2516. doi:10.1073/pnas.1213657110 PMID:23345442

Lee, K., & Yu, Y. (2017). Janus nanoparticles for T cell activation: Clustering ligands to enhance stimulation. *Journal of Materials Chemistry. B, Materials for Biology and Medicine*, *5*(23), 4410–4415. doi:10.1039/C7TB00150A PMID:28966791

Schwendener, R. A. (2014). Liposomes as vaccine delivery systems: A review of the recent advances. *Therapeutic Advances in Vaccines*, *2*(6), 159–182. doi:10.1177/2051013614541440 PMID:25364509

Liu, H., Tu, Z., Feng, F., Shi, H., Chen, K., & Xu, X. (2015). Virosome, a hybrid vehicle for efficient and safe drug delivery and its emerging application in cancer treatment. *Acta Pharmaceutica (Zagreb, Croatia)*, *65*(2), 105–116. doi:10.1515/acph-2015-0019 PMID:26011928

Heinis, C., & Winter, G. (2015). Encoded libraries of chemically modified peptides. *Current Opinion in Chemical Biology*, *26*, 89–98. doi:10.1016/j.cbpa.2015.02.008 PMID:25768886

Chapter 10
Modern Nanomaterials Extraction and Characterization Techniques Using Plant Samples and Their Biomedical Potential

Biswajit Patra
https://orcid.org/0000-0002-8248-0487
Sambalpur University, India

Paulraj R.
Jawaharlal Nehru University, India

Muthupandian Saravanan
https://orcid.org/0000-0002-1480-3555
Mekelle University, Ethiopia & Saveetha Dental College, Saveetha Institute of Medical and Technical Sciences (SIMATS), India

Ramovatar Meena
Jawaharlal Nehru University, India

Surya Narayan Pradhan
https://orcid.org/0000-0003-1672-855X
Sambalpur University, India

ABSTRACT

Nanotechnology has explored tremendous development during the last decade because of its specific properties. Nanotechnology offers generous prospects in the improvement of agriculture, water treatment, and food industry. In this study, modern extraction techniques of agricultural and plant samples with advanced characterization methods were discussed. Additionally, various factors affecting during synthesis of NPs are also briefly described. The assessments of NPs in these samples are very complex and various techniques are to be used to get essential data. The outcomes estimated by these various techniques and methodologies are not constantly identical because of different samples different standardization methodology. A new challenge emerges when testing samples with low concentration. For this situation, expository techniques with high affectability are wanted to gauge low convergences of NPs. A perfect analytical technique should be able to detect plant-NP association, for example, structure, morphology, natural speciation, size, mass concentration, etc.

DOI: 10.4018/978-1-7998-5049-6.ch010

INTRODUCTION

The expanding utilization of nanoparticles (NPs) on the planet has raised huge worries about their potential effects on biological systems, sanitation and human wellbeing, prompting for thorough examination regarding the association between NPs and crop plants (Chandran et al. 2006). In this manner, idea of plant-NPs interaction component is necessary for precise hazard appraisal to safe guard the adequate utilization of NPs (Handy et al. 2008). Nanotechnology has been generally applied on innovative products, biosensing techniques, optics, germicidal agents, memory chips, remote electrical devices and electrometers (Maruyama et al. 2016). Furthermore, in agrifarming perspectives, NPs are frequently consolidated to nano-figured pesticides, composts and nanosensors for plant safety (Serrano et al. 2014). Ascent of sap drives the transport of solutes across plants. Research confirms that built NPs accumulates in tissues of plant (Bandyopadhyay et al. 2013). There are different customized NPs with various morphology, properties and size. Designed NPs additionally show unique concoction and physical characteristics with various ecological practices and quality in correlation to mass partners, of 1 to 100 nm and high surface volume (Chaudhry et al. 2008; Joginder et al. 2017). Biosynthesis techniques follow conventional methods because of the accessibility of progressively natural substances and eco-accommodating methodology. The rich biodiversity and simplicity of plant elements have been profoundly investigated for the nanomaterials combination (Maruyama et al. 2016). Characterization of NPs during plant – NPs association with advanced instrumentation techniques have been developed to give important information about NPs association by the help of advanced microscopic imaging, chromatography, spectroscopy techniques etc. In this review, we portray the various synthesis techniques, approaches, factor affecting for the synthesis and modern characterization techniques for the synthesized NPs. Additionally, Green synthesis of leaf based NPs used in biomedical application and their nano-antimicrobial properties also briefly discussed.

Synthesis of NPs Different methods (biological, chemical and physical) are utilized for NPs synthesis. Physical strategies consist of including ball milling, thermal evaporation, layer by layer growth, plasma arcing, ultra thin films, spray pyrolysis, lithographic techniques, molecular beam epitasis, pulsed laser desorption, sputter deposition, and diffusion flame synthesis of NPs (Joerger et al. 2000). Likewise, the chemical chacaterisation protocol involves vapour deposition, electro deposition, sol gel method (Panigrahi et al. 2004), Langmuir Blodgett approaches, soft chemical techniques, catalytic methods, hydrolysed method (Pileni 1997), wet chemical process and co-precipitation techniques (Gan et al. 2012). Chemically, physical techniques are concentrated with high radiation stabilizing agents which are destructive to ecological and to human wellbeing. Thus, organic combination and green synthesis of NPs are easy steps towards bio reduction technique and lesser energy utilization during synthesis (Sathishkumar et al. 2009; Iravani 2011). So, adoption of bioresources like extracts of plant parts, fungi, bacteria, microalga are used for the NPs synthesis. In view of these special physiochemical interactions, nanoparticles are specifically noteworthy for various applications compound sensors, electronic parts, medicoanalytical imaging, pharmaceutics and therapeutics. For instance, metallic NPs like silver, gold, palladium and platinum are generally utilized from items going from restorative to pharmaceuticals and therapeutic. Gold NPs widely utilized in biomedical engineering (Sperling et al. 2008; Puvanakrishnan et al. 2012), chromatography sciences (Sykora et al. 2010), and pharmaceuticals (Cai et al. 2008; Bhumkar et al. 2007). Silver NPs have been found to contain antibacterial and mitigating properties that can promote quick treatment (Huang et al. 2007; Li et al. 2011). Platinum nanoparticles have been generally utilized in biomedical therapeutics being pure and alloyed with different NPs (Hrapovic et al. 2004) and NPs with palladium in catalyzed electro-chemical reactions (Gopidas et al. 2003; Mandali and Chand 2011), concoction biosensors (Coccia et al. 2012), electro-optical, and hostile to microbialmanagement (Brice-Profeta et al. 2005). Copper, iron, zinc oxide, and selenium (Pankhurst et al. 2003; Njagi et al. 2011; Lee et al. 2011; Brayner et al. 2006) have additionally been adopted in therapeutic medications and against bactericidal formulations (Prasad et al. 2012).

Figure 1. Graphical representation of (a) various methods for synthesis of NPs and (b) Bio-synthesis of NPs in plant samples.

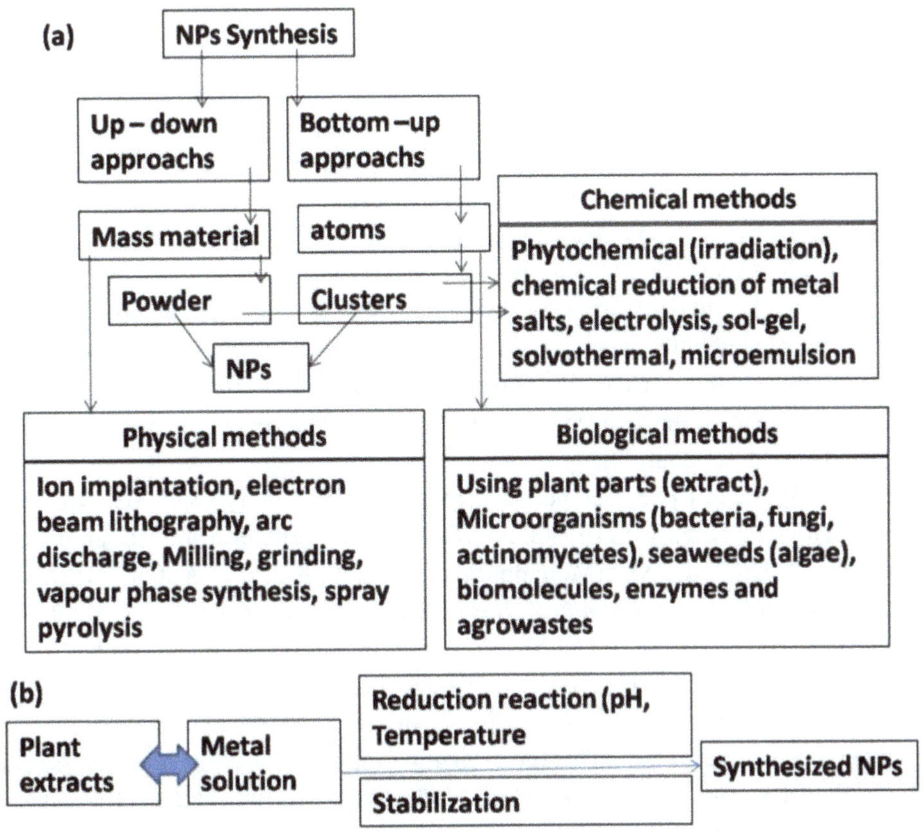

Factor affecting during synthesis of NPs Different factors that influence the NPs synthesis and characterization include temperature, pH of solutions, concentration, incubation period etc. (Kharissova et al. 2013; Narayanan and Sakthivel 2011; Ajayan 2004; Somorjai and Park 2008; Pennycook et al. 2012; Baker et al. 2013). The pH of samples affects the nano-synthesis. Specialists have found that pH of the solutions impacts the size and surface of the incorporated NPs (Kharissova et al. 2013). Accordingly, size of NPs could be constrained by alteration of pH of the media used. Temperature also influences the union of NPs utilizing every one of the three strategies. During the physical method, it requires the temperature greater than 350° C, while chemical method requires the temperature under 350°C but green biosynthesis of NPs required temperatures under 100°C and sometimes encompassing heat. The standardized temperature of response medium decides the characteristics of the formation of NPs (Pankhurst et al. 2003). However, the pressure also influences the NPs formation and synthesis. On the other hand, medium also influences the size and shape of synthesized NPs (Baker et al. 2013). Metal particles reduction speed with bio-agent has been seen faster at surrounding pressure conditions (Somorjai and park 2008). The variation and good qualities of synthesized NPs are attributed to incubation period, because variation occurs due to long storage and particles may shrink (Akbari et al. 2011; Yacaman et al. 2001). Size of particles also determine the properties of NPs. Composition of metabolites differ depending on plant species and their extraction procedures because many plant parts and their extracts show huge amount of secondary metabolites (Lynch et al. 2007).

CHARACTERIZATION OF SYNTHESIZED NPS

The study by Daniel and Astruc (2004) highlighted the physicochemical parameters like surface porosity and surface area, solubility, size, particle size distribution, zeta potential, hydrated surface investigation, adsorption potential, wettability, size, fractal measurements, surface shape, crystallinity, orientation, scattering of NPs and nanotubes in nanocomposite materials. The findings of Joerger et al. (2000) emphasize adoption of electron microscopy, atomic force microscopy (AFM), scanning electron microscopy (SEM). Metabolomics holds the key to confirm the presence of multi elements in a unknown compound by use of Gas Chromatography and mass spectroscopy (GC-MS), matrix-assisted laser desorption/ionization time-of-flight mass spectrometry (MALDI-TOF), dual polarization interferometry, nuclear magnetic resonance (NMR), nanoparticle tracking analysis (NTA) for evaluation of Brownian motion, and particle size analysis (Ingale and Chaudhari 2013).

Detection of NPs formation The inference of formation of different types of NPs is facilitated by UV-Vis spectroscopy with scan ranging 200 to 800 nm. This gives information about aggregation formation, stability, structure and size of the NPs (Mohanpuria et al. 2008; Kharissova et al. 2013).

Determination of particle size and morphology Particle size with surface morphology of NPs are imaging by microscopic technology like SEM, AFM and TEM (Ingale and Chaudhari 2013). Many experiments suggested that small micro-particles (MPs) are lesser effective for drug rescue in comparison to nano particles (NPs) due to their large size (Mohanpuria et al. 2008; Kreuter 1991).

TEM It is the most prevalent method for the determination of surface morphology, size and shape of NPs (Vimani et al. 2019; Patra et al. 2019). Media and sample preparation for TEM imaging are time consuming phenomenon because of preparing very ultrathin sample for transmittance of electron. So thin films samples are adsorbed with carbon coated copper grid (Tiwari et al. 2008; Suzuki 2002). After that samples were allowed to dry following mercury lamp and exposed under beam of monochromatic electron and projected onto display screen for viewing image. Surface atoms arrangement and nearby microstructures like screw axes, defects and lattice vacancies, glide plane, lattice fringe of crystalline NPs can be detected by using HRTEM (high resolution transmission electron microscopy) (Patra et al. 2019).

SEM This microscopic technique aims to visualize the surface morphology of NPs by direct observation. The protocol depends on electron microscopy and provides favorable circumstances for morphological and size investigation (Narayanan and sakthivel 2011). The instrument has an electron gun, condensed lenses and a vacuum chamber. SEM gives three sorts of chief images: secondary electron pictures, backscattered electron pictures and external X-ray maps (Patra et al. 2019). In SEM images, NPs must be media dried into powder, mounted on a sample holder, and covered with conductive metals like gold-palladium composite, gold, osmium, iridium, tungsten, platinum, chromiumusing a spit coater

(Suzuki 2002). High voltage electron is bombarded on the sample to view different emergent signals on the surface of the specimens (Jores et al. 2004). These techniques also time consuming, expensive and require integral data about the size distribution.

AFM This technique aims contemplate the three dimensional images of NPs and biomolecules. Dissimilar to transmission electron microscopy and scanning electron microscopy, atomic force microscopy gives three-dimensional pictures with the goal that height and molecular volume can be investigated (Jores et al. 2004). This method suits for high resolution NPs size evaluation and depends on physical scanning of the sample at the submicron plane utilizing a probe tip (Muhlen et al. 1996). Utilizing AFM, quantitative data in relation to single NPs and group of particles size (tallness, length and width), surface texture and morphology could be accessed by software based imaging solutions (Shi et al. 2003). It can be conducted either in gas or liquid medium. For this technique, a little volume of the nano-particles is spread on a glass spread, slip mounted on the stub and at room temperature, the sample are dried by nitrogen gas. Six to ten pictures then taken for a solitary sample to empower superior interpretation. These techniques also produce a topographical map of the samples dependent on the strength between the tip and the sample surface (Muhlen et al. 1996).

DLS These techniques are adopted for particle size distribution. Broadly, it does quantify the size of the Brownian NPs in colloidal suspensions. It is also called photon correlation spectroscopy (Mudunkotuwa et al. 2012). Monochromatic light emission corresponds to a solution of circular particles in Brownian movement and a Doppler shifting happens when the light hits these versatile particles. Thus, changing the wavelength of light emission is done by a value identified with particle or molecule size. Likewise, DLS empowers nanoparticle movement and size distribution in the sample. This is also processed by the estimating diffusion coefficient of the particle (Muhlen et al. 1996).

NTA The Nanoparticle Tracking Analysis is a developed frame work,designed to classify various kinds of NPs based on their size ranging from 30 - 1000 nm and the lower detection limit contingent upon its refractive index. By this technique, the liquid NPs suspension can be directly visualized (Mudunkotuwa et al. 2012).

XPS This x-ray photoelectron spectroscopy is used to consider the component of magnetic NPs surface and different element included and affirm the structure and various components present in the magnetic NPs (Faraji et al. 2010).

FT-IR This spectroscopic technique is performed to identify the functional groups available on NPs. Utilizing Fourier transform infrared spectroscopy investigation, the IR emission spectrum, photoconductivity, absorption of solid, gas and liquid can be investigated. The range corresponds to a unique finger impression of the NPs comprising of retention tops that compare to the frequencies of sensations within the obligations of molecules in the NPs. The transmission spectra of NPs are acquired through the arrangement of slim, straightforward KBr pellets (Potassium Bromide). The KBr blends were put in vacuum line to make pellets and pellets are again put in the vacuum line before use. The transmission spectra are acquired subsequent to cleansing in dry air and background comparative with blank KBr sample for reference (Virmani et al. 2019).

Zeta Potential The surface charge and stability of NPs are assessed indirectly by zeta potential utilizing a Zeta size nanomachine. Zeta potential investigation compares the distinction among the surface of shear and the external Helmholtz plane. Estimation of the zeta potential indicated the deposition capacity of the colloidal scattering. The characters of the particles embodied within the NPs or coated on the particle surface are additionally examined dependent on zeta potential (Otsuka et al. 2003).

TGA This is utilized to validate the composition of polymers and coating of surfactants to analyzed the binding competence on surface of magnetic NPs (Kreuter 1991).

XRD Crystallinity analysis is utilized to identify the different crystalline forms and elemental composition of natural and affected material of NPs. Size of the particles determined on the width of the peaks utilizing Scherrer equation (Virmani et al. 2019; Patra et al. 2019).

Hydrophobicity Analysis Surface hydro-phobicity assessment of NPs are evaluated by using different techniques like contact angle measurements, hydrophobic interaction chromatography, probe adsorption and biphasic partition. X-ray photon correlation spectroscopy mainly useful to recognize particular chemical groups on the NPs surface (Valsesia et al. 2018).

Magnetometry properties analysis Various strategies are available to search the characteristics of magnetic NPs including VSM (vibrating sample magnetometry) and SQUID (superconducting quantum interference device magnetometry) (Woodard et al. 2018). These methods are component explicit and it can evaluate only general polarization. SQUID magnetometry used for the evaluation of properties of magnetic NPs. In this method, NPs with magnetic field applied and warmed (Biehl et al. 2018). Polarization is functioned as temperature. VSM directed to assess the charge of magnetic NPs as an element of an applied outer magnetic field (H), range within −3 and 3 Tesla. In view of the VSM bend observed at low and room temperature, the magnetic behavior of NPs could be watched. VSM is a decent system for evaluating the impacts of a shell on saturation magnetization (Woodard et al. 2018).

EDX Energy dispersive X-ray spectra used to characterize the elemental composition of the NPs. EDS detector can be connected to filtering electron magnifying instrument could distinguish the quantity of X-beams which discharged to adjust the vitality contrast within two electrons. The vitality of the produced X-beam is a trademark highlight of the component, and the component is in this manner broke down subjectively and quantitatively (Virmani et al. 2019; Patra et al. 2019).

Chromatography and NMR techniques Many chromatographic techniques are used to separate NPs from the solution. Size exclusion chromatography (SEC) is used to separating NPs like quantum dots, single-walled carbon nanotubes and polystyrene NPs dependent on shape and size. Hydrodynamic chromatography (HDC) can separate particles according to their hydrodynamic radius ranging from 5 to 1200nm depending on column length (Faraji et al. 2010; Biehl et al. 2018). Gas chromatography and mass spectroscopy (GC-MS) are also adopted to discriminate the unknown and known metabolites in NPs. This technique also successfully implemented in agribusiness management system for pesticide detection (Biehl et al. 2018). NMR is also used to structural analysis of the metabolites found by GCMS and LCMS (Zhao et al. 2017). Changes in mineral supplement digestion actuated by NPs were controlled by inductively coupled plasma-mass spectrometry (ICP-MS) (Mo et al. 2017; Zhao et al. 2017). Blend of LCMS and GCMS based metabolomics investigation of hepatotoxic impact of nanoparticles in rodents were considered by studies MO et al., (2017). Zhao et al., (2017) concentrated the NMR and GCMS based metabolomics uncover NPs adjusted to organic product for healthful stock.

MFFF Sometimes mono dispersed particulate materials were difficult to detect. So, Magnetic field flow fractionation technique utilized to identify the poly-dispersed magnetic NPs. In this case, species are separated by magnetic susceptibility with size in nanometer to micrometer range (patra and Baek 2014).

LIBD Laser induced breakdown detection technique used to measure the breakdown probability. Other laser based detection methods are Raman spectrometry, laser induced fluorescence (LIF) and mass spectrometry (MS) (Bundschuh et al. 2001).

SAXS By this technique, structural based fluid and solid materials in nanometer range are investigated (Patra and Baek 2014).

ICPMS It is utilized to the detection of ultra-stress metals in wider range of NPs. This resolved the toxicity of NPs. various types of NPs like carbon nanotubes, nickel, gold are quantify by this Inductively Coupled Plasma Mass Spectrometry (Fabricius et al. 2014; Aznar et al. 2017; Lee et al. 2014).

Table 1. common characterization techniques used for NPs detection.

Types	Techniques
Formation of NPs	UV-Visible spectroscopy (UV-Vis)
Particle size and morphology	Atomic force microscopy (AFM), Scanning electron microscopy (SEM), High-resolution transmission electron microscopy (HRTEM), Dynamic light Scattering (DLS)
Surface charge	Thermal gravimetric analysis (TGA), X-ray photoelectron spectroscopy (XRP), Fourier transform infrared spectroscopy (FTIR), Zeta potential (ZEP)
Crystallinity	X-ray diffraction (XRD)
Magnetic Properties	Vibrating sample magnetometry (VSM), Superconducting quantum interference device magnetometry (SQID)
Other Techniques	Gas chromatography (GC), Mass spectroscopy (MS), X-ray fluoroscence spectroscopy (XRF), Nuclear Magnetic Resonance (NMR), Hyperspectral imaging, inductively coupled plasma mass spectrometry (ICPMS)

LEAVE BASED NPS USED IN BIOMEDICAL APPLICATION

MTT assay Evaluation of cell viability assumes a great job in different aspects of biomedical research. It gives critical starting information before performing clinical and in vivo investigations. Different kinds of test are accessible for the recognition of cell viability which extends from the routine trypan blue dye exclusion assay to exceptionally complex investigation of single cells. Di Methyl Thiazoldiphenyl-Tetrazoliumbromide (MTT) based test is one of the most abused methodologies in cancer research for measuring drug cytotoxicity, cell proliferation and viability. MTT assay involved the conversion of MTT tetrazolium salt to purple formazon by feasible multiplying cells. MTT test is known for its fast and sensitive examine (Virmani et al. 2019).

ROS This ROS estimation assumes a significant role to control development and growth, abiotic and biotic natural improvements in plants. The ROS creation in plants is mostly localized in the mitochondria, chloroplast and peroxisomes. The reactive oxygen species can adjust the cell signaling proteins and progressively intercede neurotic actions, like diabetes, atherosclerosis, unchecked development, aggravation, neuro degeneration, and maturing (Das and Roychoudhury 2014; Benoit and Michel 2007).

MMP The mitochondrial membrane potential regulate ATP synthesis, ROS, and respiratory rate. It also regulates the proton leaks and electron transport (Zorova et al. 2018).

Western Blotting Immuno-assays were accepted standard to determine the subcellular protein confinement in single cell. To upgrade the selective when giving single-cell resolution, a subcellular western blotting procedure accomplished to separate proteins in 2 pL core of single cells and 14 pL cytoplasm. The chemical cytometry assay quantifies the nucleo- cytoplasmic protein circulation in unfixed single cells; provide the protein signaling in assorted cells (Yamauchi and Herr 2017). Nanotechnology makes peptide discovery conceivable by making a metal-covered film of immobilin-poly-vinylidene difluoride layers.

FACS Flow cytometry is a routine procedure used to identify and measure the chemical and physical qualities of a heterogeneous cell mixture or particles. Fluorescence-actuated cell sorting (FACS) is also examine the GPI (Glycosyl-phosphatidylinositol) anchored protein expression. The mainly investigated GPI-anchored proteins include CD16b, CD59, CD55and fluoresce in labeled pro-aerolys in (FLAER) that binds to the GPI anchor (Peng et al. 2019).

Cellular Uptake The blend of micro bubbles and ultrasound can encourage cellular uptake of model drugs by intercellular localization. By utilizing fluorescent particles, this procedure can be examined by confocal fluorescence microscopy. This examination is intended to explore the connection between cellular uptake and fluorescence intensity of experimental model medications (Felix et al. 2017). Confocal Raman Microscopy, a non-obtrusive, non-damaging technique is adopted to detect the confinement and actual localization of NPs in the cell line set ups. The cells were presented in carbon nanotubes (CNTs) which were designed with poly-electrolytes and lipid bilayers, cerium dioxide nano particles and aluminum oxide. Raman spectra were applied to acquire the spatial appropriations of NPs together with proteins or lipids in cells. The association of the NPs with various lipid bodies, DNA, and protein were gathered. Lipid covered CNTs in relation to the specially designed lipid loaded areas are used, though poly-electrolyte covered CNTs are avoided from lipid. Al_2O_3 NPs were originated in the protoplasm. CeO_2 NPs were promptly taken and could be seen everywhere throughout the cells. Raman z-scans exhibit the intracellular circulation of the separate NPs (Plant et al. 1985).

Invivo study Most cancer nano medicines are intended to eliminate disease, the nanomaterial can prompt the development of micrometer-sized holes in the vein and blood vessel of endothelial walls. Nanomaterials-incited endothelial leakiness may support intravasation of enduring disease cells into the surrounding cells. Chick chorioallantoic layer test as an in vivo model to think about the impact of nanoparticle based anticancer medications in ovarian disease (Vu et al. 2018). Numerous surveys have condensed the techniques in both in vitro and in vivo investigations in certain nanostructures in various model frameworks (Potter et al. 2018).

Nano-antimicrobial activity Nanomaterial activities in antimicrobial fields have the prime enthusiasm by numerous specialists and scientists. Accessible reports showed that a portion of metal oxide NPs including ZnO, TiO_2,Al_2O_3, Co_3O_4,CuO, MgO,In_2O_3, ZrO_2,SiO_2, Ni_2O_3, Cr_2O_3,CoO and Mn_2O_3 with toxicity in the direction of a few microorganisms and they effectively kill various microscopic organisms (Hoseinzadeh et al. 2017). Some literature suggest the impact and the capacity of nanomaterials against some microbes, which are recorded as shape, varieties, grouping of nanomaterial and surface charge of the metal nanomaterial scattering and make contact with the nanomaterial to microbial cell, freedom of antimicrobial particles, medium segments and pH, physiochemical properties, development rate, cell mass of microorganisms, and impact of UV brightening (Wang et al. 2017; Dar et al. 2013). NPs were progressively utilized to targeted microorganisms as the alternative in contrast to anti-infection agents. Nanotechnology might be especially beneficial in treating bacterial diseases. Models remember the usage of NPs for antibacterial coatings for implantable gadgets and restorative materials to counteract contamination and advance injury recuperating, in anti-microbial conveyance frameworks to treatment malady, in microbial recognition frameworks to produce bacterial and fungal diagnostics, and in antibacterial antibodies to control microbial diseases (Kim et al. 2019). Some papers suggested that NPs have been explicitly utilized in agribusiness and drug as antibacterial, antifungal and cancer prevention agents. It has been exhibited that the development and augmentation of numerous microscopic organisms like *Citrobacter koseri, Staphylococcus aureus, Escherichia coli, Klebsiella pneumonia, Bacillus cereus, Vibrio parahaemolyticus, Salmonella typhii, Pseudomonas aeruginosa,* and fungus *Candida* *albicans* by restricting biomolecules present in the microbial cells (Wang et al. 2017; Kim et al. 2019; Mourdikoudis et al. 2018). Since NPs are smaller than the micro organisms and diffuse into cell and crack the cell wall which can be again characterized by SEM and TEM pictures of the suspension containing nanoparticles and pathogens.

CONCLUSION

The utilization of NPs in the therapeutics, pharmaceuticals, and horticultural ventures has earned a lot of enthusiasm, with an attention on improvement of increasingly advantageous methods utilizing green nano-biotechnology tools to generate eco-accommodating, non-toxic and naturally sustainable NPs (Serrano et al. 2014; Bandyopadhyay et al. 2013). There are numerous variables that impact the value and amount of blended NPs for their possible usage in different operations. Reasonable portrayal systems can exhibit the incorporated nanoparticles proficiently and augment their effective use in ecological, electronic, and bio-medical applications. However, biomedical innovations are also required. This review, various biomedical techniques used in disease prophylaxis by the administration of NPs was depicted in detail. Present and future examinations of green nano-biotechnology will be usher a progressively complete information base in relation to various factors that impact the combination of NPs and complex innovation which can be utilized to exhibit the incorporation of NPs for its increasingly productive future applications in pharma industry and biomedical applications (Mourdikoudis et al. 2018). So, more studies were prompted to design a novel system for synergistic effects for treatment of various diseases. Further research is also necessitated in animal models to confirm the mechanisms to gain clear cut information of biocompatibility and toxicity (Sathishkumar et al. 2009; Puvanakrishnan et al. 2012). Eventually, study on bio-safety of NPs and their interactions with cells and tissues are to be focused upon. The best concern is to develop tailor made nano based precision therapy for complex disease set ups.

REFERENCES

Ajayan, P. M. (2004). Nanotechnology: How does a nanofibre grow? *Nature*, *427*(6973), 402–403. doi:10.1038/427402a PMID:14749810

Akbari, B., Tavandashti, M. P., & Zandrahimi, M. (2011). Particle size characterization of nanoparticles—A practicalapproach. *Iranian Journal of Materials Science and Engineering*, *8*, 48–56.

Aznar, R., Barahona, F., Geiss, O., Ponti, J., Luis, T. J., & Moreno, J. B. (2017). Quantification and size characterization of silver nanoparticles in environmental aquous samples and consumer products by single particle ICPMS. *Talanta*, *175*, 200–208. doi:10.1016/j.talanta.2017.07.048 PMID:28841979

Baker, S., Rakshith, D., & Kavitha, K. S. (2013). Plants: Emerging as nanofactories towards facile route in synthesis of nanoparticles. *BioImpacts*, *3*, 111–117. PMID:24163802

Bandyopadhyay, S., Jose, R. P., & Jorge, L. G. (2013). Advanced analytical techniques for the measurement of nanomaterials in food and agricultural samples: A review. *Environmental Engineering Science*, *30*(3), 118–125. doi:10.1089/ees.2012.0325 PMID:23483065

Benoit, D. A., & Michel, B. T. (2007). ROS as signaling molecules: Mechanisms that generate specificity in ROS homeostasis. *Nature Reviews. Molecular Cell Biology*, *8*(10), 813–824. doi:10.1038/nrm2256 PMID:17848967

Bhumkar, D. R., Joshi, H. M., Sastry, M., & Pokharkar, V. B. (2007). Chitosan reduced gold nanoparticles as novel carriers for transmucosal delivery of insulin. *Pharmaceutical Research*, *24*(8), 1415–1426. doi:10.100711095-007-9257-9 PMID:17380266

Biehl, P., Luhe, M. V., Dutz, S., & Schacher, F. H. (2018). Synthesis, characterization and applications of magnetic nanoparticles featuring polyzwitterionic coatings. *Polymers*, *10*(1), 1–28. doi:10.3390/polym10010091 PMID:30966126

Brayner, R., Ferrari-Iliou, R., Brivois, N., Djediat, S., Benedetti, M., & Fiévet, F. (2006). Toxicological impact studies based on Escherichia coli bacteria in ultrafine ZnO nanoparticles colloidal medium. *Nano Letters*, *6*(4), 866–870. doi:10.1021/nl052326h PMID:16608300

Brice-Profeta, S., Arrio, M. A., Tronc, E., Menguy, N., Letard, I., Cartier dit Moulin, C., Noguès, M., Chanéac, C., Jolivet, J.-P., & Sainctavit, P. (2005). Magnetic order in γ-Fe2O3 nanoparticles: A XMCD study. J*ournal of Magnetism and Magnetic Materials, 288,* 354–365. doi:10.1016/j.jmmm.2004.09.120

Bundschuh, T., Knopp, R., & Kim, J. I. (2001). Laser-induced breakdown detection (LIBD) of aquatic colloids with different laser systems. *Colloids and Surfaces. A, Physicochemical and Engineering Aspects*, *177*(1), 47–55. doi:10.1016/S0927-7757(99)00497-5

Cai, W., Gao, T., Hong, H., & Sun, J. (2008). Applications of gold nanoparticles in cancer nanotechnology. *Nanotechnology, Science and Applications*, *1*, 17–32. doi:10.2147/NSA.S3788 PMID:24198458

Chandran, S. P., Chaudhary, M., Pasricha, R., Ahmad, A., & Sastry, M. (2006). Synthesis of gold nanotriangles and silver nanoparticles using Aloe vera plant extract. *Biotechnology Progress*, *22*(2), 577–583. doi:10.1021/bp0501423 PMID:16599579

Chaudhry, Q., Scotter, M., Blackburn, J., Ross, B., Boxall, A., Castle, L., Aitken, R., & Watkins, R. (2008). Applications and implications of nanotechnologies for the food sector. *Food Additives and Contaminants*, *25*(3), 241–244. doi:10.1080/02652030701744538 PMID:18311618

Coccia, F., Tonucci, L., Bosco, D., Bressan, M., & Alessandro, N. (2012). One pot synthesis of lignin-stabilized platinum and palladium nanoparticles and their catalytic behaviors in oxidation and reduction reactions. *Green Chemistry*, *14*(4), 1073–1078. doi:10.1039/c2gc16524d

Daniel, M. C., & Astruc, D. (2004). Gold nanoparticles: Assembly, supramolecular chemistry, quantum-size-related properties, and applications toward biology, catalysis, and nanotechnology. *Chemical Reviews*, *104*(1), 293–346. doi:10.1021/cr030698+ PMID:14719978

Dar, M. A., Ingle, A., & Rai, M. (2013). Enhanced antimicrobial activity of silver nanoparticles synthesized by cryphonectria sp. Evaluated singly and in combination with antibiotics. *Nanomedicine (London)*, *9*(1), 105–110. doi:10.1016/j.nano.2012.04.007 PMID:22633901

Das, K., & Roychoudhury, A. (2014). Reactive oxygen species (ROS) and response of antioxidants as ROS-scavengers during environmental stress plants. *Frontiers in Environmental Science*, *2*, 1–13. doi:10.3389/fenvs.2014.00053

Fabricius, A. L., Duester, L., Meermann, B., & Ternes, T. (2014). ICP-MS based characterization of inorganic nanoparticles-sample preparation and off line fractionation strategies. *Analytical and Bioanalytical Chemistry*, *406*(2), 467–479. doi:10.100700216-013-7480-2 PMID:24292431

Faraji, M., Yamini, Y., & Rezaee, M. (2010). Magnetic nanoparticles: Synthesis, stabilization, functionalization, characterization, and applications. *Journal of the Indian Chemical Society*, *7*(1), 1–37. doi:10.1007/BF03245856

Felix, L. C., Ortega, V. A., & Goss, G. G. (2017). Cellular uptake and intracellular localization of poly (acrylic acid) nanoparticles in a rainbow trout (Oncorhynchus mykiss) gill epithelial cell line RTgill-W1. *Aquatic Toxicology (Amsterdam, Netherlands)*, *192*, 58–68. doi:10.1016/j.aquatox.2017.09.008 PMID:28917946

Gan, P. P., Ng, S. H., Huang, Y., & Li, S. F. Y. (2012). Green synthesis of gold nanoparticles using palm oil mill effluent (POME): A low-cost and eco-friendly viable approach. *Bioresource Technology*, *113*, 132–135. doi:10.1016/j.biortech.2012.01.015 PMID:22297042

Gopidas, K. R., Whitesell, J. K., & Fox, M. A. (2003). Synthesis, characterization and catalytic activity of a palladium nanoparticle cored dendrimier. *Nano Letters*, *3*(12), 1757–1760. doi:10.1021/nl0348490

Handy, R. D., Owen, R., & Valsami-Jones, E. (2008). The ecotoxicology of nanoparticles and nanomaterials: Current status, knowledge gaps, challenges, and future needs. *Ecotoxicology (London, England)*, *17*(5), 315–325. doi:10.100710646-008-0206-0 PMID:18408994

Hoseinzadeh, E., Makhdoumi, P., Taha, P., Hossini, H., Stelling, J., Kamal, M. A., & Ashraf, G. M. (2017). A review on nano-antimicrobials: Metal nanoparticles, methods and mechanisms. *Current Drug Metabolism*, *18*, 120–128. doi:10.2174/1389200217666161201111146 PMID:27908256

Hrapovic, S., Liu, Y., Male, K. B., & Luong, J. H. T. (2004). Electrochemical biosensingplateform using platinum nanoparticles and carbon nanotubes. *Analytical Chemistry*, *76*(4), 1083–1088. doi:10.1021/ac035143t PMID:14961742

Huang, Y., Li, X., Liao, Z., Zhang, G., Liu, Q., Tang, J., Peng, Y., Liu, X., & Luo, Q. (2007). A randomized comparative trial between Acticoat and SD-Ag in the treatment of residual burn wounds including safety analysis. *Burns*, *33*(2), 161–166. doi:10.1016/j.burns.2006.06.020 PMID:17175106

Ingale, A. G., & Chaudhari, A. N. (2013). Biogenic synthesis of nanoparticles and potential applications: An eco-friendly approach. *Journal of Nanomedicine & Nanotechnology*, *4*, 2–4. doi:10.4172/2157-7439.1000165

Iravani, S. (2011). Critical review – green synthesis of metal nanoparticles using plants. *Green Chemistry*, *13*(10), 2638–2650. doi:10.1039/c1gc15386b

Joerger, R., Klaus, T., & Granqvist, C. G. (2000). Biologically produced silver–carbon composite materials for optically functional thin-film coatings. *Advanced Materials*, *12*(6), 407–409. doi:10.1002/(SICI)1521-4095(200003)12:6<407::AID-ADMA407>3.0.CO;2-O

Joerger, R., Klaus, T., & Granqvist, C. G. (2000). Biologically produced silver carbon composite materials for optically functional thinfilm coatings. *Advanced Materials*, *12*(6), 407–409. doi:10.1002/(SICI)1521-4095(200003)12:6<407::AID-ADMA407>3.0.CO;2-O

Joginder, S. D., Ravinder, K., Naresh, K., Pawan, K., Kiran, N., & Surekha, D. (2017). Nanotechnology: The new perspective in precision agriculture. *Biotechnology Reports (Amsterdam, Netherlands)*, *15*, 11–13. doi:10.1016/j.btre.2017.03.002 PMID:28603692

Jores, K., Mehnert, W., Drechsler, M., Bunjes, H., Johann, C., & Mader, K. (2004). Investigations on the structure of solid lipid ¨ nanoparticles (SLN) and oil-loaded solid lipid nanoparticles by photon correlation spectroscopy, field-flow fractionation and transmission electron microscopy. *Journal of Controlled Release*, *95*(2), 217–227. doi:10.1016/j.jconrel.2003.11.012 PMID:14980770

Kharissova, O. V., Dias, H. V. R., Kharisov, B. I., Perez, B. O., & Perez, V. M. J. (2013). The greener synthesis of nanoparticles. *Trends in Biotechnology*, *31*(4), 240–248. doi:10.1016/j.tibtech.2013.01.003 PMID:23434153

Kim, E. J., Choi, M., Park, H. Y., Hwang, J. Y., Kim, H. E., Hong, S. W., Lee, J., Yong, K., & Kim, W. (2019). Thorn like TiO2 nanoarrays with broad spectrum antimicrobial activity through physical puncture and photocatalytic action. *Scientific Reports*, *9*(1), 13697. doi:10.103841598-019-50116-0 PMID:31548584

Kreuter, J. (1991). Peroral administration of nanoparticles. *Advanced Drug Delivery Reviews*, *7*(1), 71–86. doi:10.1016/0169-409X(91)90048-H

Lee, H. J., Lee, G., Jang, N. R., Yun, J. H., Song, J. Y., & Kim, B. S. (2011). Biological synthesis of copper nanoparticles using plants extract. *Nanotechnology*, *1*, 371–374.

Lee, S., Bi, X., Reed, R. B., Ranville, J. F., Herckes, P., & Westerhoff, P. (2014). Nanoparticle size detection limits by single particle ICPMS for 40 elements. *Environmental Science & Technology*, *48*(17), 10291–10300. doi:10.1021/es502422v PMID:25122540

Li, W. R., Xie, X. B., Shi, Q. S., Duan, S. S., Ouyang, Y. S., & Chen, Y. B. (2011). Antibacterial effect of silver nanoparticles on *Staphylococcus aureus*. *Biometals*, *24*(1), 135–141. doi:10.100710534-010-9381-6 PMID:20938718

Lynch, I., Cedervall, T., Lundqvist, M., Cabaleiro-Lago, C., Linse, S., & Dawson, K. A. (2007). The nanoparticle-protein complex as a biological entity; a complex fluids and surface science challenge for the 21st century. *Advances in Colloid and Interface Science*, *135*, 167–174. doi:10.1016/j.cis.2007.04.021 PMID:17574200

Mandali, P. K., & Chand, D. K. (2011). Palladium nanoparticles catalysed Suzuki cross-coupling reactions in ambient conditions. *Catalysis Communications*, *31*, 16–20. doi:10.1016/j.catcom.2012.10.020

Maruyama, C. R., Guilger, M., Pascoli, M., Bileshy-José, N., Abhilash, P., Fraceto, L. F., & de Lima, R. (2016). Nanoparticles based on chitosan as carriers for the combined herbicides imazapic and imazapyr. *Scientific Reports*, *6*(1), 19768. doi:10.1038rep19768 PMID:26813942

Mo, H. Z., Jia, Q. C., Hui, M. G., Li, R. T., Yi, Q. G., Yuan, T., Zun, J. Z., & Yin, H. (2017). Combination of LCMS and GCMS based metabolomics to study the hepatotoxic effect of regular nanoparticles in rats. *Chinese Journal of Natural Medicines*, *15*(9), 684–694. doi:10.1016/S1875-5364(17)30098-5 PMID:28991530

Mohanpuria, P., Rana, N. K., & Yadav, S. K. (2008). Biosynthesis of nanoparticles: Technological concepts and future applications. *Journal of Nanoparticle Research*, *10*(3), 507–517. doi:10.100711051-007-9275-x

Mourdikoudis, S., Pallares, R. M., & Thanh, N. T. K. (2018). Characterizationtechniques for nanoparticles: Comparison and complementarity upon studying nanoparticle properties. *Nanoscale*, *10*(27), 12871–12934. doi:10.1039/C8NR02278J PMID:29926865

Mudunkotuwa, I. A., Pettibone, J. M., & Grassian, V. H. (2012). Environmental implications of nanoparticle aging in the processing and fate of copper-based nanomaterials. *Environmental Science & Technology*, *46*(13), 7001–7010. doi:10.1021/es203851d PMID:22280489

Muhlen, A. Z., Uhlen, E. Z. M., Niehus, H., & Mehnert, W. (1996). Atomic force microscopy studies of solid lipid nanoparticles. *Pharmaceutical Research*, *13*(9), 1411–1416. doi:10.1023/A:1016042504830 PMID:8893284

Narayanan, K. B., & Sakthivel, N. (2011). Green synthesis of biogenic metal nanoparticles by terrestrial and aquatic phototrophic and heterotrophic eukaryotes and biocompatible agents. *Advances in Colloid and Interface Science*, *169*(2), 59–79. doi:10.1016/j.cis.2011.08.004 PMID:21981929

Njagi, E. C., Huang, H., Stafford, L., Genuino, H., Galindo, H. M., Collins, J. B., Hoag, G. E., & Suib, S. L. (2011). Biosynthesis of iron and silver nanoparticles at room temperature using aqueous sorghum bran extracts. *Langmuir*, *27*(1), 264–271. doi:10.1021/la103190n PMID:21133391

Otsuka, H., Nagasaki, Y., & Kataoka, K. (2003). PEGylated nanoparticles for biological and pharmaceutical applications. *Advanced Drug Delivery Reviews*, *55*(3), 403–419. doi:10.1016/S0169-409X(02)00226-0 PMID:12628324

Panigrahi, S., Kundu, S., Ghosh, S., Nath, S., & Pal, T. (2004). General method of synthesis for metal nanoparticles. *Journal of Nanoparticle Research*, *6*(4), 411–414. doi:10.100711051-004-6575-2

Pankhurst, Q. A., Connolly, J., Jones, S. K., & Dobson, J. (2003). Applications of magnetic nanoparticles in biomedicine. *Journal of Physics. D, Applied Physics*, *36*(13), 167–181. doi:10.1088/0022-3727/36/13/201

Patra, B., Gautam, R., Priyadarsini, E., Rajamani, P., Pradhan, S. N., Saravanan, M., & Meena, R. (2019). Piper betle : Augmented synthesis of gold nanoparticles and its invitro cytotoxicity assessment on hela and hek 293 cells. *Journal of Cluster Science*, 1–13.

Patra, J.K., & Baek, K.H. (2014). Green nanobiotechnology: factor affecting synthesis and characterization techniques. *Journal of Nanomaterials,* 1-12. doi:10.1155/2014/417305

Peng, F., Setyawati, M. I., Tee, J. K., Ding, X., Wang, J., Nga, M. E., Ho, H. K., & Leong, D. T. (2019). Nanoparticles promote in vivo breast cancer cell intravasation and extravasation by inducing endothelial leakiness. *Nature Nanotechnology, 14*(3), 279–286. doi:10.103841565-018-0356-z PMID:30692675

Pennycook, T. J., McBride, J. R., Rosenthal, S. J., Pennycook, S. J., & Pantelides, S. T. (2012). Dynamic fluctuations in ultrasmall nanocrystals induce white light emission. *Nano Letters, 12*(6), 3038–3042. doi:10.1021/nl3008727 PMID:22568665

Pileni, M. P. (1997). Nanosized particles made in colloidal assemblies. *Langmuir, 13*(13), 3266–3276. doi:10.1021/la960319q

Plant, A. L., Benson, D. M., & Smith, L. C. (1985). Cellular uptake and intercellular localization of benzo (a) pyrene by digital fluorescence imaging microscopy. *The Journal of Cell Biology, 100*(4), 1295–1308. doi:10.1083/jcb.100.4.1295 PMID:3980583

Potter, T. M., Neun, B. W., & Dobrovolskaia, M. A. (2018). Invitro and invivo methods for analysis of nanoparticle potential to induce delayed type hypersensitivity reactions. *Methods in Molecular Biology (Clifton, N.J.), 1682*, 197–210. doi:10.1007/978-1-4939-7352-1_17 PMID:29039104

Prasad, K. S., Patel, H., Patel, T., Patel, K., & Selvaraj, K. (2012). Biosynthesis of Se nanoparticles and its effect on UV-induced DNA damage. *Colloids and Surfaces. B, Biointerfaces, 103*, 261–266. doi:10.1016/j.colsurfb.2012.10.029 PMID:23201746

Puvanakrishnan, P., Park, J., Chatterjee, D., Krishnan, S., & Tunnel, J. W. (2012). In vivo tumor targeting of gold nanoparticles: Effect of particle type and dosing strategy. *International Journal of Nanomedicine, 7*, 1251–1258. doi:10.2147/IJN.S29147 PMID:22419872

Sathishkumar, M., Sneha, K., Won, S. W., Cho, C. W., Kim, S., & Yun, Y. S. (2009). Cinnamon zeylanicum bark extract and powder mediated green synthesis of nano-crystalline silver particles and its bactericidal activity. *Colloids and Surfaces, 73*(2), 332–338. doi:10.1016/j.colsurfb.2009.06.005 PMID:19576733

Serrano, A. L., Olivas, R. M., Landaluze, J. S., & Camara, C. (2014). Nanoparticles: A global vision. Characterization, separation and quantification methods. Potential environmental and health impact. *Analytical Methods, 6*(1), 38–56. doi:10.1039/C3AY40517F

Shi, H. G., Farber, L., Michaels, J. N., Dickey, A., Thompson, K. C., Shelukar, S. D., Hurter, P. N., Reynolds, S. D., & Kaufman, M. J. (2003). Characterization of crystalline drug nanoparticles using atomic force microscopy and complementary techniques. *Pharmaceutical Research, 20*(3), 479–484. doi:10.1023/A:1022676709565 PMID:12669972

Somorjai, G. A., & Park, J. Y. (2008). Colloid science of metal nanoparticle catalysts in 2D and 3D structures. Challenges of nucleation, growth, composition, particle shape, size control and their influence on activity and selectivity. *Topics in Catalysis, 49*(3-4), 126–135. doi:10.100711244-008-9077-0

Sperling, R. A., Gil, P. R., Zhang, F., Zanella, M., & Parak, W. J. (2008). Biological applications of gold nanoparticles. *Chemical Society Reviews, 37*(9), 1896–1908. doi:10.1039/b712170a PMID:18762838

Suzuki, E. (2002). High-resolution scanning electron microscopy of immunogold-labelled cells by the use of thin plasma coating of osmium. *Journal of Microscopy*, *208*(3), 153–157. doi:10.1046/j.1365-2818.2002.01082.x PMID:12460446

Sykora, D., Kasicka, V., Miksik, I., Rezanka, P., Zaruba, K., Matejka, P., & Kral, V. (2010). Applications of gold nanoparticles in separation sciences. *Journal of Separation Science*, *30*(3), 372–387. doi:10.1002/jssc.200900677 PMID:20099261

Tiwari, D. K., Behari, J., & Sen, P. (2008). Time and dose-dependent antimicrobial potential of Ag nanoparticles synthesized by top down approach. *Current Science*, *95*, 647–655.

Valsesia, A., Desmet, C., Jimenez, I. O., Oddo, A., Capomaccio, R., Rossi, F., & Colpo, P. (2018). Direct quantification of nanoparticle surface hydrophobicity. *Communications Chemistry*, *1*(1), 53. doi:10.103842004-018-0054-7

Virmani, I., Sasi, C., Priyadarshini, E., Kumar, R., Sharma, S.K., Singh, G.P., Pachwarya, R.B., Paulraj, R., Barabadi, H., Saravanan, M., &Meena, R. (2019). Comparative anticancer potential of biologically and chemically synthesized gold nanoparticles. *Journal of Cluster Science*, 1-10.

Vu, B. T., Shahin, S. A., Croissant, J., Fatieiev, Y., Matsumoto, K., Doan, T. L., Yik, T., Simargi, S., Conteras, A., Ratliff, L., Jimenez, C. M., Raehm, L., Khashab, N., Durand, J. O., Glackin, C., & Tamanoi, F. (2018). Chick chorioallantoic membrane assay as an in vivo model to study the effect of nanoparticle based anticancer drugs in ovarian cancer. *Scientific Reports*, *8*(1), 8524. doi:10.103841598-018-25573-8 PMID:29867159

Wang, L., Hu, C., & Shao, L. (2017). The antimicrobial activity of nanoparticles: Present situation and prospects for the future. *International Journal of Nanomedicine*, *12*, 1227–1249. doi:10.2147/IJN.S121956 PMID:28243086

Woodard, L. E., Dennis, C. L., Borchers, J. A., Attaluri, A., Velarde, E., Dawidczyk, C., Searson, P. C., Pomper, M. G., & Ivkov, R. (2018). Nanoparticle architecture preserves magnetic properties during coating to enable robust multi-model functionality. *Scientific Reports*, *8*(12706), 1–13. doi:10.103841598-018-29711-0 PMID:30139940

Yacaman, M. J., Ascencio, J. A., Liu, H. B., & Torresdey, G. J. (2001). Structure shape and stability of nanometric sized particles. *Journal of Vacuum Science and Technology*, *19*(4), 1091–1103. doi:10.1116/1.1387089

Yamauchi, K. A., & Herr, A. E. (2017). Subcellular western blotting of single cells. *Microsystems & Nanoengineering*, *3*(1), 16079. doi:10.1038/micronano.2016.79 PMID:29333327

Zhao, L., Hu, J., Wang, H., Adeleye, A., Ortiz, C., & Keller, A. A. (2017). 1H NMR and GCMS based metabolomics reveal nano-cu altered cucumber fruit nutritional supply. *Plant Physiology and Biochemistry*, *110*, 138–146. doi:10.1016/j.plaphy.2016.02.010 PMID:26922143

Zorova, L. D., Popkov, V. A., Plotnikov, E. J., Silachev, D. N., Pevzner, I. B., Jankauskas, S. S., Zorov, S. D., Babenko, V. A., & Zorov, D. B. (2018). Mitochondrial membrane potential. *AnnalBiochem*, *552*, 50–59. PMID:28711444

Chapter 11
Role of Micro and Nano Motors in Therapeutics and Diagnostics

Karthik Vivekanandhan
K. S. Rangasamy College of Technology, India

Poornima Shanmugam
K. S. Rangasamy College of Technology, India

Balasubramanian Mythili Gnanamangai
K. S. Rangasamy College of Technology, India

Subbaiya Ramasamy
The Copperbelt University, Zambia

ABSTRACT

Nanomotors are the new generation drug delivery devices that find their major application in the pharmaceutical sector. This chapter highlights the basic aspects, mechanisms, and applications of nanomotors in antibacterial therapy, cancer therapy, nano-surgery, and imaging. Because of their higher penetration ability, rapid transportation, and controlled mobility, micro and nanomotors are referred as new generation targeted drug delivery devices. The detecting and sensing ability of nanomotors find potential applications in diagnostics and therapeutics. Bio-hybrid nanomotors are most attracting candidates for effective drug delivery. Bio-functionalized nanomotor can be used for detection and identification of tumor cells. Thus, nanomotors and micromotors can pave way for development of diagnostic and therapeutic tools for the future. Further development in these areas facilitate the discovery of lab-on-chip devices that can be used for super-fast screening and clinical diagnostic applications.

DOI: 10.4018/978-1-7998-5049-6.ch011

INTRODUCTION

Micro and nanomotors are the devices capable of converting energy into movement to propel themselves in the biological environment. The locomotion of nanomotors are due to fuel or energy (Baeza & Vallet-Regí, 2018). The integral components of nanomotors are engine and work unit which are responsible for providing kinetic energy and controlled response respectively. Based on the mechanism used for propulsion, micro and nanomotors can be classified as fuel powered motors and fuel free motors. If the energy required for propulsion from using chemicals it is called fuel powered motors. Fuel free motors are driven my magnetic, electric, light, acoustic and living organism driven. Ultrasound propelled nanomotors are the attractive alternates with high precision delivery (Kiristi et al., 2015). The nanomotors can also be specifically used for detection and sensing. A self-propelled gold/ nickel/ polyaniline/ platinum nanomotor, coupled with concanavalin A lecitin bioreceptor is proved to be useful for isolation of *E. coli* (Campuzano et al., 2012). A nanomotor functionalized with anti-carcinoembryogenic antigen monoclonal antibody is used for detection and identification of tumor cells, as CEA antigen is highly found antigen in gastric, colorectal and pancreatic tumour (Balasubramanian et al., 2011). Moreover, functionalized nanomachines with ssDNA can been used to clearly isolate the required molecule from biological fluid. Further development in these topics can help in the development of Lab on chip devices that can be used for super-fast screening and clinical diagnostic applications (Daniel et al., 2011). In the above cases the sensing capability of nanomotors and micromotors have been exploited to develop diagnostics and sensing tools.

Multiple nanomachines have been investigated for their properties to transport drugs, acting as carrier. Microspheres, nanospheres and lipospheres have already been subjected to studies regarding drug delivery. These nanocarriers requires fuels to propel to site of delivery. Diallyldimethyl ammonium chloride stabilized PtN's was coated in nanorockets at the inner layer and the outer layer was loaded with doxorubicin. Potential propulsion force of the nanomotor was able to effectively travel up to 30 cm (Zhiguang et al., 2013). In the presence of hydrogen peroxide as fuel, rolled up nanojets with a sharp tip has shown corkscrew like movement (Alexander, 2012), but the toxicity to the biological system eliminates its use. Thus, magnetically controlled nanojets are now under research to develop a fuel free motor that is also non-toxic (Xi et al., 2013). The nanorockets were also found to be as efficient as to penetrate Hela cells and release doxorubicin inside. Use of nanomotors in imaging will help us to obtain images of high details using the existing imaging systems. Catalase coated silica nanospheres was used for the inflammation relating to hydrogen peroxide can be converted to oxygen microbubbles under ultrasound, thus pinpointing the location of microbial abscess.(Olson et al., 2013). Employing metallic nanomotors has proved to be efficient in imaging as high detailed images which can used to generate more data can be generated with existing imaging systems (Martel et al., 2009). The motion of micro and nano motors are greatly influenzed by their geometric structures. Nano tubular structure, microspheres, Nanorods and asymmetric branches are the commonly employed structures (Abid et al., 2011; Cameron et al., 2018; Evans et al., 2013; Li, Wu, Qin et al., 2016; Moo et al., 2016; Tian et al., 2018; Zeng et al., 2015). Microspheres vary from 0.1 to 100 micrometers. The velocity of microsphere is higher than the other geometric structures. The mechanism of nanorod is self-electrophoretic propulsion. Tubular structures follow bubble induced propulsion mechanism (Braun et al., 2018) and symmetric structures follow light driven photo-electro chemical reaction mechanism (Zhan et al., 2018).

Nanomotors can be used to differentiate and localize cell using their capacity to sense hydrogen peroxide, temperature, and water. Microspheres, nanospheres and lipospheres have already been subjected to studies regarding drug delivery. These nano carriers require fuels to propel to site of delivery. Further, nanorockets are also found to have efficient penetration ability and release drug into the targeted site. Moreover, multiple nanomachines have been investigated for their properties to transport drugs and acting as a carrier.

MECHANISM OF NANOMOTORS

The mechanism of propulsion in macro motors does not apply to the propulsion of nanomotors, because the mechanism of energy conversion totally varies between macro and nanomotors. In macroscale motors the propulsion is maintained with the help of laws of inertia, but as the size decreases to nano, laws of inertia fail to be applicable, as the viscous force is multiple times higher than it. The only equation that can relate both size and velocity of the nanomotor in an environment, where propulsion is opposed by viscous force is Reynold's number. It is a dimensionless value which is given by,

$$Re = \frac{\rho v L}{\mu}$$

where, ρ= density
v= velocity of the particle
L= Length of the particle
μ= dynamic viscosity of the medium

This equation relates both the length and velocity of the particle. But development of nanomotors whose length are in nanoscale are difficult, not only to fabricate but also in designing it for propulsion. Thus, nanomotors have very low Reynolds around the value 10^{-4}. In such cases the law of inertia is not applicable for the motion of the particle and only instantaneous force can cause propulsion. Also designing of motor that has a unidirectional propulsion is very difficult. The particle can also exhibit Brownian motion as at nano size collision with molecules in aqueous phase can result in Brownian motion. But in biological system there are naturally established mechanism for directional propulsions. Brownian movement can be characterized by the use of diffusion coefficient *D,* it is given by the equation,

$$D= \frac{Tk_B}{6\pi r\eta}$$

Where, T = absolute temperature
K_B = Boltzmann constant
r = radius of the particle
η = viscosity

Nanomotors can be classified based on fuel type and mechanism of propulsion.

MECHANISM OF PROPULSION

Bubble Propulsion

Propulsion of nanomotors by the bubbles that arises from the catalytic decomposition of fuel into gas and other by-product is called as bubble propulsion mechanism. These nanomotors can propel until the fuel is consumed completely. A best example for bubble propulsion is decomposition of Hydrogen peroxide to water and oxygen by Pt (Figure 1). The gas released helps in propulsion of the nanomotor, but the directionality is not controlled, and is suspected to be along the fuel gradient (Mei et al., 2011).

Figure 1. Schematic illustration of bubble propulsion mechanism observed in Nanomotor that decomposes hydrogen peroxide using Pt.

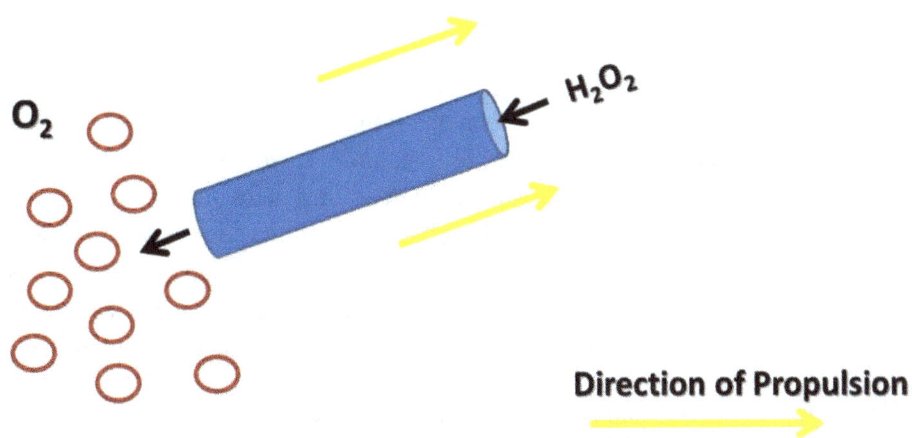

Diffusophoresis Propulsion

Asymmetric nanomotors with a fuel decomposing end on side, continuously decomposes the fuel, while on the other end, the decomposition of by-products get accumulated and transformed on the other side, subsequently creating a concentration gradient at the surface of the nanomotor. When the decomposed products concentration gets higher it starts diffusing into the environment causing propulsion (Gibbs & Zhao, 2011).

Self-Electrophoresis Propulsion

Asymmetric nanomotors with bimetallic end, where one end is anode and the other end is cathode are propelled autonomously by the proton gradient generated along the axis of the nanomotor. A Au/Pt nanomotor using hydrogen peroxide fuel, can be a good example for self-electrophoretically powered nanomotor. The hydrogen peroxide decomposes by oxidation at the Pt side, resulting proton and electron. These proton and electron flow to the Au side to form water molecule from hydrogen peroxide by reduction. Thus flow of electron towards the Au side and electromigration of proton to the Au side will result with gradient of proton along the entire length of nanomotor. The electron flow results in net negative charge of the nanomotor, resulting in propulsion by the attraction to the proton rich side that was passed by the anode (Gibbs & Zhao, 2011).

Propulsion by Interfacial Tension

When the catalytic end of a nanomotor converts the components from the surrounding, and produces a by-product, due to the interfacial tension created the motor move towards the direction of the catalyst (John & Yiping, 2011). Sen *et al.,* studied the propulsion by interfacial tension in a Platinum-Gold nanorod (Pt-Au). The gold side of the nanorod decomposes the hydrogen peroxide in the surrounding causing the environment around it to be filled with oxygen bubble. Due to the hydrophobicity created at the Au end, the nanorod was puhed along the long axis to move to the platinum end, causing propulsion. Thus due to the interfacial tension that was created, the nanorod propelled in the medium (Paxton et al., 2004).

Chemically Powered Nanomotors

Chemically powered nanomotors carry a catalyst on board that can interact with the chemical in the surrounding causing the nanomotor to propel due to the instantaneous force generated during the catalytic reaction. The generated instantaneous force must greater, so that the naomotor can propel along with the carrier weight. But for use in biomedical applications the fuel used must be non-toxic to humans.

The best example of a chemically powered nanomotors can be Pt–Au nanorods synthesized (Fournier-Bidoz et al., 2005). The nanorods are propelled along with their long axis using the interfacial tension stimulated motion by the decomposition of hydrogen peroxide. A propulsion speed of ~`3-8 μm s^{-1} was achieved.

Different bimetallic nanowires that can propel using hydrogen peroxide as fuel (Li, Wu, Qin et al., 2016). The Pt–Au nanowire created by them showed a propulsion speed of ~20 μm s^{-1} and the Pt–Ru nanowire synthesized by them showed the highest propulsion speed of ~30 μm s^{-1}(Wang et al., 2006) .

In the above cases the propulsion direction of the nanomotors is not controlled but to negate this disadvantage, self-assembled nanomotors were designed. The nanoparticles were engulfed in a bowl shape copolymer with a narrow outlet so the gas expulsion via the narrow opening can guide the propulsion of the nanomotor in a single direction. This design is called as a stomatocyte nanomotors (Wu et al., 2012).

The advantages of these self-assembled nanomotors is that, they can propel under lower fuel concentrations as low as 0.0005% (v/v), and this can be a huge advantage in biomedical applications. But chemically powered are less powered in case for biomedical applications as the use of fuel can prove to be toxic to the biological system. But at the same time bio-hybrid nanomotors are studies more to find a mechanism that uses the components in a biological system for its autonomous propulsion.

Chemically Powered Biohybrid Nanomotors

A carbon microfiber bound with glucose oxidase (GOx) on one side and bilirubin oxidase (BOD) on the other side, where GOx oxidises glucose to glucolactone on one end and subsequent reduction of peroxide to water on the BOD end resulting in flux of protons from anode end (GOx) to cathode end (BOD) resulting in propulsion speed upto 1000 μm s^{-1} (Mano & Heller, 2005). Similarly, enzymes such as catalase can be bound in nanotubes, which can decompose hydrogen peroxide resulting in propulsion (Abdelmohsen et al., 2014).

Physically Powered Nanomotors

Nanomotors that are propelled in an environment using an external force such as light, magnetic force and sound waves are physically powered nanomotors. These nanomotors do not depend on a fuel for propulsion, but rather depend on the external force applied. Magnetic and acoustic driven nanomotors have high biomedical applications than electric and photo driven nanomotors. Based on the type of external force used they are classified as follows.

2.3.1. Magnetic field driven nanomotors

The magnetic nanomotors can be driven using a homogenous, non-homogenous or gradient and oscillating magnetic field. The nanomotor must have ferromagnetic, diamagnetic or paramagnetic property for propulsion. The nanomotor be spherical, wire or helical shaped for better propulsion. The artificial bacterial flagella consist of a helical tail and ferromagnetic head. The helical tail is fabricated using indium–gallium–arsenic/gallium–arsenic (InGaAs/GaAs) or indium–gallium–arsenic/gallium–arsenic/chromium (InGaAs/GaAs/Cr). The head portion of ferromagnet is composed of chromium/nickel/gold (Cr/Ni/Au), fabricated by self-scrolling method. Under exposure to magnetic field, the induced rotational magnetic field results in translational motion on the direction of the chirality of the helix tail (Zhang et al., 2009).

In the design of a nanowire with a flexible Ag bridge in the middle of Au head and Ni tail. The hydrogen peroxide cause dissolution of Ag bridge resulting in a flexible nanowire. Complete dissolution is halted by silver oxide and hydroxide formation on the flexible bridge (Gao et al., 2010). Without the partial dissolution rigid Au/Ag/Ni doesn't show any free movement in the solution as the flexibility of the nanowire is also a reason for the propulsion.

Similarly, microjets and nanojets that are magnetically powered are of much importance for biomedical applications. Introduction of Ni between Au/Pt nanotube i.e., by fabricating a Au/Ni/Pt nanotube (Figure 2) and magnetizing it results in a permanent change in the magnetic domain of Ni. Such nanojets orients themselves along the magnetic field and propels itself towards or away the magnetic field (Zhao & Pumera, 2012).

Light Driven Nanomotors

Light is one of the common external stimuli and can be used to propel nanomotors, by photocatalysis, photolysis, photoisomerization, photo thermophoresis and photochromic mechanism. But photocatalysis and photo induced thermphophoresis is the highly studied propulsion mechanism as the propulsion speed achieved is high.

Titanium dioxide nanomotors driven by photocatalysis. When exposed to UV, due to self-diffusophoresis, the TiO_2 micromotors can achieve a propulsion speed of ~ 10 µm S^{-1} (Gibbs & Zhao, 2011). The

photosensitive property of AgCl was also used to fabricate light driven micromotors(Hong et al., 2010), which had a propulsion speed of 100 µm s^{-1}. The propulsion of the micromotor is by self-diffusophoresis, as under UV the AgCl is dissolute to proton and chloride ion, of which the movement of proton is faster compared to chloride ion, which results in an electrolyte gradient resulting in self-diffusophoresis propulsion (Ibele et al., 2009). Further research is being done on fabrication and modelling of light driven nanomotors. Geometric structures of light driven micro and nanomotors is given in Table 1.

Figure 2. Schematic illustration of magnetically powered propulsion mechanism

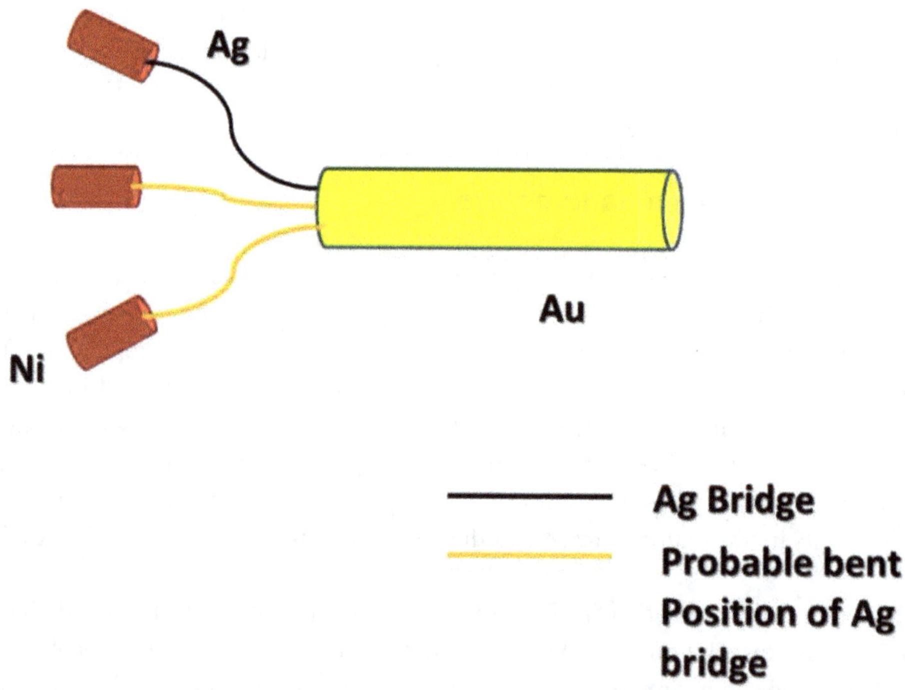

Table 1. Comparison of light driven micro/nano motors with respect to geometric structures and mechanism

Geometric structures	Mechanism	Light	Reference
Nanotubes	Photothermal-effect propulsion	NIR	(Wu, Si, Gao et al., 2016)
		NIR	(Wu, Lin, Zou et al., 2015)
Microspheres	Bubble Propulsion	UV-Vis	(Lee et al., 2018)
		UV	(Li, Mou, Chen et al., 2016)
	Osmotic propulsion	UV	(Chen et al., 2017)
	Photo-thermal effect propulsion	NIR	(Wu, Si, Shao et al., 2016)
	Bubble Propulsion	UV	(Mou et al., 2015)
Nanorods	Self-electrophoretic propulsion	Vis	(Palacci et al., 2013)
		UV-Vis	(Zhou, Li, Xu et al., 2017)
	Photo-thermal effect propulsion	NIR	(Xuan et al., 2015)
	Self-electrophoretic propulsion	UV	(Zhou, Ren, Li et al., 2017)
Asymmetric branches	Self-electrophoretic propulsion	UV-Vis	(Dai et al., 2016)

Electrically Driven Nanomotors

Here nanomotors are made to propel using an electric field. Millimetre scale semi-conductor diodes can move in an alternate electric field, due to electro-osmotic flow that is created around the diodes. Following this, many micromotors and nanomotors have been designed by designed researchers, and is yet to be perfected. Wang *et al* designed a poly(pyrrole)- cadmium nanowire and cadmium selenide – gold – cadmium selenide (CdSe –Au –CdSe) nanowire and observed their propulsion. Under uniform AC field the nanowires propulsion speed was found to be ~ 17.2 μm s^{-1}. They also noticed that the ionic strength of the medium interfered with the propulsion, indicating that electrochemistry of the medium also plays a major role in electrically driven nanomotors (Calvo-Marzal et al., 2010).

Acoustic Powered Nanomotors

Acoustic powered nanomotors uses sound waves especially ultrasound to propel the nanomotors. In these case the nanomotor should be of asymmetric shape for propulsion. A microrod was designed by template electrode deposition. The microrod consisted of concave and convex end. Thus when exposed to continuous or pulsed ultrasound due to the asymmetry in rod, a pressure gradient was induced on the surface of the rod, resulting in a propulsion along the direction of the sound wave (Wang et al., 2012).

A multifunctional nanomotor with three segments, Au-Ni-Au was designed. The nanomotor was fabricated using sphere lithography so that a concave end can be constructed (Figure 3). With magnetized Ni and ultra sound as external stimuli this nanomotor was driven with control on direction and speed (Garcia-Gradilla et al., 2014).

Figure 3. Illustration of Au/Ni/Au nanomotor driven using ultrasound and steadily guided by magnetic field.

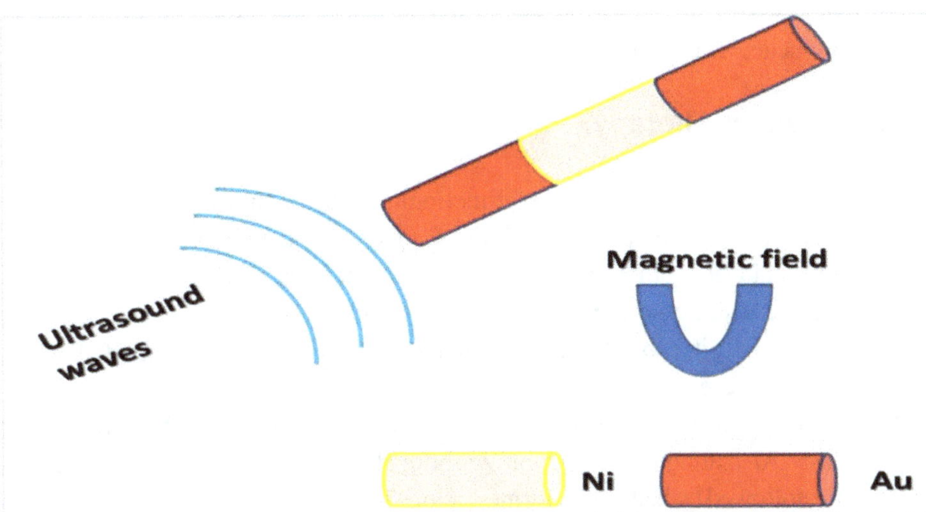

Cellular Driven Nanomotors

Unicellular units such a bacteria and sperm can be used to transport the nanoparticle to the target location. They are not energy-free, but an external energy is not required, and the components inside the body is just enough for their locomotion. Schmidt *et al.,* (Magdanz et al., 2013) used spermatozoa (sperm cell), for dragging a nanotube containing a specific genetic code isolated from a sperm cell, so that spermatozoa can drive it to ovum for spermatization. Another research uses magnetotactic bacteria for the delivery of drug to target cells. Magnetotactic bacteria are those bacteria that move along the magnetic field, with the help of magnetic crystal in them to guide along the field. Pouponneau *et al.,* (Martel et al., 2009) used a magnetotactic bacteria loaded with biological component in a microvasculature operation. Thus for specific purposes, cells/ microbes can be used as vehicle carrying a cargo especially in biological system.

APPLICATIONS OF NANOMOTORS

Micro and nanomotors, though it has a lot of space for research and development, they have high bio-medical applications. These nanomotors can be used for targeted drug delivery for effective treatments. They can also be used for enhanced bioimaging, specific cell isolation and other applications.

Nanomotors in Combatting Antibacterial Resistance

Antibacterial research is most important now-a- days due to multi drug resistant bacteria and inefficient conventional antibiotic therapies. To eradicate bacterial infections and to overcome the limitations of traditional antibiotic drugs, there is a need for development of novel and efficient technology. Nanomotors enhance destruction of bacteria with greater acceleration and efficiency. Lysozyme based nanomotors are developed for enhanced antibacterial effect which has been demonstrated against *Micrococcus lyso-deikticus* and *E. coli* (Kiristi et al., 2015). UV exposed TiO_2 and AgCl based nanomotors are effective against bacteria (Simmchen et al., 2017).

Nanomotors in Cancer Therapy

Cancer is the disease that arises when a normal functioning cell, malfunctions becoming an abnormal cell that grows uncontrollably in any part of the body. Cancer can be malignant or metastatic (when the cancer cell can spread and affect other parts it is called as metastasis). Generally, in case of malignant tumour, surgery is opted, but in case of metastatic cancer chemotherapy is advised. But the prime dis-advantage of chemotherapy is it also affects the normal cells. Because of this, targeted drug delivery to cancer cells is much under research.

Cancer cells produce abnormal levels of hydrogen peroxide causing oxidative stress, which helps in invasion and metastasis. Also tumour cells exhibit Warburg effect, where cancer cells obtain its energy via higher glycolysis rate, followed by lactic acid fermentation. Thus cancer cells exhibit a range of acidic pH.

Thus using this nature of cancer cells, chemically powered nanomotors where hydrogen peroxide can be used up as fuel can be used for drug delivery and cancer cell targeting. A tubular nano-rocket fabricated by template assisted layer by layer assembly, was incorporated with platinum nanoparticles, to catalytically decompose hydrogen peroxide causing propulsion. The tubular nanomotor was fabricated using positively charged chitosan and negatively charged sodium alginate by LBL assembly method using an etched template. The nano-rocket was loaded with doxorubicin was found to effectively penetrate HeLa cell lines under *In vitro* conditions (Wu et al., 2013).

Since the cancer cell microenvironment is acidic a Janus micromotor with calcium carbonate core was designed to exploit the lactic acid in microenvironment as fuel. The calcium carbonate core was shielded on one side of the hemisphere by thin cobalt coating, leaving the other half being able to interact with the lactic acid in environment. Calcium carbonate interacts with lactic acid to for calcium lactate and carbon dioxide, where the released carbon dioxide propels the micromotor resulting in a slow and continuous Brownian motion. Thus this micromotor effectively exploits the components around the cancer cell microenvironment as fuel for its propulsion (Guix et al., 2016).

The nanomotor used should be biodegradable and must not accumulate and cause toxicity to the cells. A biodegradable poly L lysine/ bovine serum albumin multilayer rocket was fabricated by LBL technique using a template. Then the micro rocket was loaded with a thermal sensitive gelatin that encapsulates catalase, Au nanoparticles and doxorubicin. Here the Au nanoparticles are used for photo thermal degradation or phase change of the gelatin hydrogel by external application of Near infrared irradiation. The released catalase will cause propulsion upon interaction with hydrogen peroxide in the cancer cell microenvironment and the doxorubicin drug treats the cancer cells (Wu et al., 2014).

Similarly, micro/Nano rockets can be synthesized for drug delivery using external physical force. A multilayer micro rocket was fabricated using alternate assembly of poly (styrene sulfonic acid) and poly (allyl amine hydrochloride) using a polycarbonate membrane as template. Then citrate stabilized gold nanoparticles were aligned in the microrocket by electrostatic interaction, where the gold nanoparticle diameter is of 20nm and doxorubicin was also loaded into the microrocket. Upon interaction with near infrared irradiation, the nanomotors propels by self-thermophoresis (Wu, Si, Gao et al., 2015). Similarly, gold nanowire loaded with anti-cancer drug can also be used for drug delivery. The gold nanowire is asymmetrically fabricated with a concave end so that the propulsion can be achieved using ultrasound and self-thermophoresis by near infrared irradiation (Garcia-Gradilla et al., 2014).

Nanomotors in Nano-Surgery

From invasive scalpel suture surgery to laparoscopy, the surgery techniques have evolved from invasive to minimally invasive techniques to reduce scarring and wound size. Now the nanoscale surgery is being researched upon. Nano surgery refers to manipulation of surgery at nanoscale which has an advantage as being minimally invasive surgery ever and the wound healing rate will also be faster. High powered lasers to micro drillers and nanorockets are employed for Nano surgery.

A Ti/Cr/Fe micro driller, fabricated using photolithography over a trapezia shaped silica as base with sharp ends is used as a micro driller in nano-surgery. The iron layer was added, so device can be controlled magnetically and the Ti and Cr layer facilitated the roll up process for drilling. The micro driller can be propelled magnetically and the axis of drilling can be altered by changing the magnetic field frequency (Xi et al., 2013). A InGaAs/GaAs/(Cr)Pt microjets with sharp ends that exhibited corkscrew movement using hydrogen peroxide as a fuel which is found in the microenvironment of HeLa cell *in vitro* (Solovev et al., 2012).

Nanomotors in Imaging

Nanomotors based on their unique sensitivity, and minimal invasive property of physiological or pathological environment they are gaining scope as a promising bio-imaging agent. These naomotors are sensitive to water, hydrogen peroxide, temperature and other such factors.

Catalase-coated silica nanosphere converters (NSCs) for imaging (Olson et al., 2013). The microbubble produced by decomposition of hydrogen peroxide by catalase in the NSC's results in reduction of inflammation. Thus the NSC's were able to produce high signal in region of bacterial abscess resulting in better pathological evaluation during ultra sound imaging (Olson et al., 2013). Similarly, magnetic nanomotors are under development as bioimaging agent in MRI due to their alignment and arrangement under magnetic field.

Nanomotors in Isolating Circulating Tumour Cells

Circulating tumour cells are cancer cells that have separated from the primary source and have entered the blood circulation. These CTC's are extravasate of the primary source of cancer and have the ability to localize in an area and develop into tumour in the new site. A rolled up micro machine with a platinum sheet inside a ferrous sheet in the middle and a gold sheet outside. The Pt layer decomposes the hydrogen peroxide enabling the micromachine to propel in the media, while the Fe layer helps in controlled motion of the micromachine under a magnetic field, while on the Au sheet is incorporated with antibody that can detect carcinogenic antigen found in cancer cells. Thus CTC targeted isolation is achievable using micro and nanomotors, but the efficiency of isolation depends on the design of the micromachine, targeted cell and the fluidic environment in which the cell is going to be isolated from (Balasubramanian et al., 2011).

FUTURE PROSPECTS

Lot of difficulties are to be crossed to fabricate a nanomotor for biomedical purposes, from biocompatibility to controlling the motion and action of the nanomotor inside the body. Most studies have been done i*n vitro*, and the effect of these nanomotors in an actual body is still unknown. Still a lot of research is going on development of nano-drug carriers, and when this concept becomes practically feasible, the phase of medical field will be open to lots of opportunities and technologies to treat complicated and complex disease. Moreover, there is a lot to study and explore about propulsion mechanisms, manipulation of the nanomotors, and the ability to fabricate a nanomotor with sensing ability to address specific requirements for, treatment of diseases, development of rapid detection kits for cancer/ diseases, specific cell targeting for high resolution imaging and many more applications that are yet to be found. Another critical criterion for a nanomotor, that has lot of space for development, is to be biodegradable, and biocompatible so that it is feasible for application in biological systems. The immune response the human body will give rise to face the nanomotor in its system is another question, that is yet to be answered.

Yet among all these unanswered questions, nanotechnology has been growing at a rapid pace in the last two decades, and we are coming to realise the possibilities and the extent of feasibility of the nanomotor for medical, environmental and other unknown applications yet to be discovered. The possibility use of nanomotor in surgery and biopsies could prove to be the most non-invasive or minimally invasive techniques developed as of yet. The possibility of feasible targeted drug delivery and targeted gene delivery could prove to be the ultimate cure for cancer, hereditary diseases and high resolution human body imaging technologies. The possibility of nanomotors in sensing can lead to miniaturisation of instruments and testing kits. All these possibilities can only be realised with time and continual effort of researchers from multiple discipline come together to answer the questions we face, one question at a time so that we can unlock the limitless possibilities of nanomotors.

CONCLUSION

Nanomotors emerge as a novel new generation drug delivery device with versatile applications. In the medical field, non-invasive and minimally invasive surgery techniques are very arduous and is highly preferred. The recent focus is on nanojets and microdrillers which can be used to perform minimally invasive and precise surgery. These techniques are still under development as the materials currently used for the propulsion of the motors are toxic to biological system. Magnetically controlled nanojets are now under research to develop a fuel free motor that is also non-toxic. Nanomotors find their applications in environmental remediation also. Studying immune response of these devices are also the promising area of research. The robust design, suitable material selection and bio-compatible nature outweigh the drawback of using nanomotors in physiological environment, hence these areas will drive the future of the nanomotor research.

ACKNOWLEDGMENT

The authors are thankful to the Management and Principal of K.S.Rangasamy College of Technology, Tiruchengode, Namakkal District, Tamil Nadu, India and also gratefully acknowledge DST FIST (Fund for Infrastructure for Science and Technology) FIST No: 368 (SR/FST/College – 235/2014) and DBT-STAR College Scheme (BT/HRD/11/09/2018) for their support for the instrumentation.

REFERENCES

Abdelmohsen, L. K. E. A., Peng, F., Tu, Y., & Wilson, D. A. (2014). Micro- and nano-motors for biomedical applications. *Journal of Materials Chemistry. B, Materials for Biology and Medicine*, *2*(17), 2395–2408. doi:10.1039/C3TB21451F PMID:32261411

Abid, J. P., Frigoli, M., Pansu, R., Szeftel, J., Zyss, J., Larpent, C., & Brasselet, S. (2011). Light-driven directed motion of azobenzene-coated polymer nanoparticles in an aqueous medium. *Langmuir*, *27*(13), 7967–7971. doi:10.1021/la200682p PMID:21651298

Alexander, A. (2012). Self-propelled nanotubes. *ACS Nano*, *6*(2), 1751–1756. doi:10.1021/nn204762w PMID:22233271

Baeza, A., & Vallet-Regí, M. (2018). Nanomotors for Nucleic Acid, Proteins, Pollutants and Cells Detection. *International Journal of Molecular Sciences*, *19*(6), 1579. doi:10.3390/ijms19061579 PMID:29799489

Balasubramanian, S., Kagan, D., Hu, C.M.J., Campuzano, S., Castañon, M.J.L., Dae, N.L., Zimmerman,Y.K.M., Zhang, L., & Wang, J. (2011). *Micromachine-Enabled Capture and Isolation of Cancer Cells in Complex Media*. doi:.1002/anie.20110011

Braun, L. B., Hessberger, T., Pütz, E., Müller, C., Giesselmann, F., Serra, C. A., & Zentel, R. (2018). Actuating thermo- and photo-responsive tubes from liquid crystalline elastomers. *Journal of Materials Chemistry. C, Materials for Optical and Electronic Devices*, *6*(34), 9093–9101. doi:10.1039/C8TC02873G

Calvo-Marzal, P., Sattayasamitsathit, S., Balasubramanian, S., Windmiller, J. R., Dao, C., & Wang, J. (2010). Propulsion of nanowire diodes. *Chemical Communications*, *46*(10), 1623. doi:10.1039/b925568k PMID:20177595

Cameron, J. M., Wales, D. J., & Newton, G. N. (2018). Shining a light on the photo-sensitisation of organic-inorganic hybrid polyoxometalates. *Dalton Transactions (Cambridge, England)*, *47*(15), 5120–5136. doi:10.1039/C8DT00400E PMID:29517788

Campuzano, U., Orozco, J., Kagan, D., Guix, M., Gao, W., Sattayasamitsathit, S., Jonathan, C. C., Merkoçi, A., & Wang, J. (2012). Bacterial isolation by lectin-modified microengines. *Nano Letters*, *12*(1), 396–401. doi:10.1021/nl203717q PMID:22136558

Chen, C. R., Mou, F. Z., Xu, L. L., Wang, S. F., Guan, J. G., Feng, Z. P., Wang, Q. W., Kong, L., Li, W., Wang, J., & Zhang, Q. J. (2017). Light-steered isotropic semiconductor micromotors. *Advanced Materials*, *29*(3), 1603374. doi:10.1002/adma.201603374 PMID:27748536

Dai, B. H., Wang, J. Z., Xiong, Z., Zhan, X. J., Dai, W., Li, C. C., Feng, S. P., & Tang, J. Y. (2016). Programmable artificial phototactic microswimmer. *Nature Nanotechnology*, *11*(12), 1087–1092. doi:10.1038/nnano.2016.187 PMID:27749832

Daniel, K., Susana, C., & Shankar, B. (2011). Functionalized Micromachines for Selective and Rapid Isolation of Nucleic Acid Targets from Complex Samples. *Nano Letters*, *11*(5), 2083–2087. doi:10.1021/nl2005687 PMID:21491941

Evans, J. S., Sun, Y., Senyuk, B., Keller, P., Pergamenshchik, V. M., Lee, T., & Smalyukh, I. I. (2013). Active shape-morphing elastomeric colloids in short-pitch cholesteric liquid crystals. *Physical Review Letters*, *110*(18), 187802. doi:10.1103/PhysRevLett.110.187802 PMID:23683245

Fournier-Bidoz, S., Arsenault, A. C., Manners, I., & Ozin, G. A. (2005). Synthetic self-propelled nanorotors. *Chemical Communications*, *441*(4), 441. doi:10.1039/b414896g PMID:15654363

Gao, W., Sattayasamitsathit, S., Manesh, K. M., Weihs, D., & Wang, J. (2010). Magnetically Powered Flexible Metal Nanowire Motors. *Journal of the American Chemical Society*, *132*(41), 14403–14405. doi:10.1021/ja1072349 PMID:20879711

Garcia-Gradilla, V., Sattayasamitsathit, S., Soto, F., Kuralay, F., Yardımcı, C., Wiitala, D., & Wang, J. (2014). Ultrasound-Propelled Nanoporous Gold Wire for Efficient Drug Loading and Release. *Small*, n/a. Advance online publication. doi:10.1002mll.201401013 PMID:24995778

Gibbs, J., & Zhao, Y. (2011). Catalytic nanomotors: Fabrication, mechanism, and applications. *Frontiers of Materials Science*, *5*(1), 25–39. doi:10.100711706-011-0120-x

Guix, M., Meyer, A. K., Koch, B., & Schmidt, O. G. (2016). Carbonate-based Janus micromotors moving in ultra-light acidic environment generated by HeLa cells in situ. *Scientific Reports*, *6*(1), 21701. Advance online publication. doi:10.1038rep21701 PMID:26905939

Hong, Y., Diaz, M., Córdova-Figueroa, U. M., & Sen, A. (2010). Light-Driven Titanium-Dioxide-Based Reversible Microfireworks and Micromotor/Micropump Systems. *Advanced Functional Materials*, *20*(10), 1568–1576. doi:10.1002/adfm.201000063

Ibele, M., Mallouk, T., & Sen, A. (2009). Schooling Behavior of Light-Powered Autonomous Micromotors in Water. *Angewandte Chemie International Edition*, *48*(18), 3308–3312. doi:10.1002/anie.200804704 PMID:19338004

John, G., & Yiping, Z. (2011). Catalytic nanomotors: Fabrication, mechanism, and applications. *Frontiers of Materials Science*, *5*(1), 25–39. doi:10.100711706-011-0120-x

Kiristi, M., Singh, V. V., Esteban-Fernández de Ávila, B., Uygun, M., Soto, F., Aktaş Uygun, D., & Wang, J. (2015). Lysozyme-Based Antibacterial Nanomotors. *ACS Nano*, *9*(9), 9252–9259. doi:10.1021/acsnano.5b04142 PMID:26308491

Lee, C.-S., Gong, J., Oh, D.-S., Jeon, J.-R., & Chang, Y.-S. (2018). Zerovalent-iron/platinum Janus micromotors with spatially separated functionalities for efficient water decontamination. *ACS Applied Nano Materials*, *1*(2), 768–776. doi:10.1021/acsanm.7b00223

Li, W., Wu, X. R., Qin, H., Zhao, Z. Q., & Liu, H. W. (2016). Light-driven and lightguided microswimmers. *Advanced Functional Materials*, *26*(18), 3164–3171. doi:10.1002/adfm.201505378

Li, Y., Mou, F. Z., Chen, C. R., You, M., Yin, Y. X., Xu, L. L., & Guan, J. G. (2016). Light controlled bubble propulsion of amorphous TiO2/Au Janus micromotors. *RSC Advances*, *6*(13), 10697–10703. doi:10.1039/C5RA26798F

Magdanz, V., Sanchez, S., & Schmidt, O. G. (2013). Development of a sperm-flagella driven micro-bio-robot. *Advanced materials (Deerfield Beach, Fla.)*, *25*(45), 6581–6588. doi:10.1002/adma.201302544 PMID:23996782

Mano, N., & Heller, A. (2005). Bioelectrochemical Propulsion. *Journal of the American Chemical Society*, *127*(33), 11574–11575. doi:10.1021/ja053937e PMID:16104713

Martel, S., Mohammadi, M., Felfoul, O., Lu, Z., & Pouponneau, P. (2009). Flagellated Magnetotactic Bacteria as Controlled MRI-trackable Propulsion and Steering Systems for Medical Nanorobots Operating in the Human Microvasculature. *The International Journal of Robotics Research*, *28*(4), 571–582. doi:10.1177/0278364908100924 PMID:19890435

Mei, Y., Solovev, A. A., Sanchez, S., & Schmidt, O. G. (2011). Rolled-up nanotech on polymers: From basic perception to self-propelled catalytic microengines. *Chemical Society Reviews*, *40*(5), 2109. doi:10.1039/c0cs00078g PMID:21340080

Moo, J. G. S., Presolski, S., & Pumera, M. (2016). Photochromic spatiotemporal control of bubble-propelled micromotors by a spiropyran molecular switch. *ACS Nano*, *10*(3), 3543–3552. doi:10.1021/acsnano.5b07847 PMID:26919161

Mou, F. Z., Li, Y., Chen, C. R., Li, W., Yin, Y. X., Ma, H. R., & Guan, J. G. (2015). Single component TiO2 tubular micro engines with motion controlled by light-induced bubbles. *Small, 11*(21), 2564–2570. doi:10.1002mll.201403372 PMID:25627213

Olson, E. S., Orozco, J., Wu, Z., Malone, C. D., Yi, B., Gao, W., Eghtedari, M., Wang, J., & Mattrey, R. F. (2013). Toward *in vivo* detection of hydrogen peroxide with ultrasound molecular imaging. *Biomaterials, 34*(35), 8918–8924. doi:10.1016/j.biomaterials.2013.06.055 PMID:23958028

Palacci, J., Sacanna, S., Vatchinsky, A., Chaikin, P. M., & Pine, D. J. (2013). Photoactivated colloidal dockers for cargo transportation. *Journal of the American Chemical Society, 135*(43), 15978–159781. doi:10.1021/ja406090s PMID:24131488

Paxton, W. F., Kistler, K. C., Olmeda, C. C., Sen, A., St Angelo, S. K., Cao, Y., Mallouk, T. E., Lammert, P. E., & Crespi, V. H. (2004). Catalytic nanomotors: Autonomous movement of striped nanorods. *Journal of the American Chemical Society, 126*(41), 13424–13431. doi:10.1021/ja047697z PMID:15479099

Simmchen, J., Baeza, A., Miguel-Lopez, A., Stanton, M. M., Vallet-Regi, M., Ruiz-Molina, D., & Sánchez, S. (2017). Dynamics of Novel Photoactive AgCl Microstars and Their Environmental Applications. *ChemNanoMat: Chemistry of Nanomaterials for Energy, Biology and More, 3*(1), 65–71. doi:10.1002/cnma.201600300

Solovev, A. A., Xi, W., Gracias, D. H., Harazim, S. M., Deneke, C., Sanchez, S., & Schmidt, O. G. (2012). Self-Propelled Nanotools. *ACS Nano, 6*(2), 1751–1756. doi:10.1021/nn204762w PMID:22233271

Tian, Z. M., He, J., Chen, X., Wen, T., Zhai, C., Zhang, Z. X., Cho, J. D., Chou, X. J., & Xue, C. Y. (2018). Core-shell coaxially structured triboelectric nanogenerator for energy harvesting and motion sensing. *Rsc Adv., 8*(6), 2950–2957. doi:10.1039/C7RA12739A

Wang, W., Castro, L. A., Hoyos, M., & Mallouk, T. E. (2012). Autonomous Motion of Metallic Microrods Propelled by Ultrasound. *ACS Nano, 6*(7), 6122–6132. doi:10.1021/nn301312z PMID:22631222

Wang, Y., Hernandez, R. M., Bartlett, D. J., Bingham, J. M., Kline, T. R., Sen, A., & Mallouk, T. E. (2006). Bipolar Electrochemical Mechanism for the Propulsion of Catalytic Nanomotors in Hydrogen Peroxide Solutions. *Langmuir, 22*(25), 10451–10456. doi:10.1021/la0615950 PMID:17129015

Wu, Y., Wu, Z., Lin, X., He, Q., & Li, J. (2012). Autonomous Movement of Controllable Assembled Janus Capsule Motors. *ACS Nano, 6*(12), 10910–10916. doi:10.1021/nn304335x PMID:23153409

Wu, Y. J., Si, T. Y., Shao, J. X., Wu, Z. G., & He, Q. (2016). Near-infrared light-driven Janus capsule motors: Fabrication, propulsion, and simulation. *Nano Research, 9*(12), 3747–3756. doi:10.100712274-016-1245-0

Wu, Z., Lin, X., Zou, X., Sun, J., & He, Q. (2014). Biodegradable Protein-Based Rockets for Drug Transportation and Light-Triggered Release. *ACS Applied Materials & Interfaces, 7*(1), 250–255. doi:10.1021/am507680u PMID:25496011

Wu, Z., Si, T., Gao, W., Lin, X., Wang, J., & He, Q. (2015). Superfast Near-Infrared Light-Driven Polymer Multilayer Rockets. *Small, 12*(5), 577–582. doi:10.1002mll.201502605 PMID:26690728

Wu, Z., Wu, Y., He, W., Lin, X., Sun, J., & He, Q. (2013). Self-Propelled Polymer-Based Multilayer Nanorockets for Transportation and Drug Release. *Angewandte Chemie International Edition*, *52*(27), 7000–7003. doi:10.1002/anie.201301643 PMID:23703837

Wu, Z. G., Lin, X. K., Zou, X., Sun, J. M., & He, Q. (2015). Biodegradable protein-based rockets for drug transportation and light-triggered release. *ACS Applied Materials & Interfaces*, *7*(1), 250–255. doi:10.1021/am507680u PMID:25496011

Wu, Z. G., Si, T. Y., Gao, W., Lin, X. K., Wang, J., & He, Q. (2016). Superfast near infrared light-driven polymer multilayer rockets. *Small*, *12*(5), 577–582. doi:10.1002mll.201502605 PMID:26690728

Xi, A., Solovev, A., Ananth, N., Gracias, D. H., Sanchez, S., & Schmidt, O. G. (2013). Rolled-up magnetic microdrillers: Towards remotely controlled minimally invasive surgery. *Nanoscale*, *5*(4), 1294–1297. doi:10.1039/C2NR32798H PMID:23154823

Xuan, M. J., Shao, J. X., Lin, X. K., Dai, L. R., & He, Q. (2015). Light-activated Janus self-assembled capsule micromotors. *Colloids and Surfaces. A, Physicochemical and Engineering Aspects*, *482*, 92–99. doi:10.1016/j.colsurfa.2015.04.032

Zeng, H., Wasylczyk, P., Parmeggiani, C., Martella, D., Burresi, M., & Wiersma, D. S. (2015). Light-fueled microscopic walkers. *Advanced Materials*, *27*(26), 3883–3887. doi:10.1002/adma.201501446 PMID:26033690

Zhan, Z., Wei, F., Zheng, J., Yang, W., Luo, J., & Yao, L. (2018). Recent advances of light-driven micro/nanomotors: Toward powerful thrust and precise. control. *Nanotechnology Reviews*, *7*(6), 555–581. doi:10.1515/ntrev-2018-0106

Zhang, L., Abbott, J. J., Dong, L., Kratochvil, B. E., Bell, D., & Nelson, B. J. (2009). Artificial bacterial flagella: Fabrication and magnetic control. *Applied Physics Letters*, *94*(6), 064107. doi:10.1063/1.3079655

Zhao, G., & Pumera, M. (2012). Magnetotactic Artificial Self-Propelled Nanojets. *Langmuir*, *29*(24), 7411–7415. doi:10.1021/la303762a PMID:23078532

Zhiguang, W., Yingjie, W., Wenping, H., Xiankun, L., Jianmin, S., & Qiang, H. (2013)... *Self-Propelled Polymer-Based Multilayer Nanorockets for Transportation and Drug Release, Wiley VCH*, *52*(27), 7000–7003. doi:10.1002/anie.201301643

Zhou, D. K., Li, Y. C. G., Xu, P. T., Ren, L. Q., Zhang, G. Y., Mallouk, T. E., & Li, L. Q. (2017). Visible-light driven Si-Au micromotors in water and organic solvents. *Nanoscale*, *9*(32), 11434–11438. doi:10.1039/C7NR04161F PMID:28786464

Zhou, D. K., Ren, L. Q., Li, Y. G. C., Xu, P. T., Gao, Y., Zhang, G. Y., Wang, W., Mallouk, T. E., & Li, L. Q. (2017). Visible light-driven, magnetically Steerable gold/iron oxide nanomotors. *Chemical Communications*, *53*(83), 11465–11468. doi:10.1039/C7CC06327J PMID:28983536

Chapter 12
Overview of Nano-Strategies for Combating Cancer

Yasodha Krishna Janapati
United States International University Africa, Kenya

Sunil Junapudi
Geethanjali College of Pharmacy, India

Sudharshan Reddy Dachani
College of Pharmacy, Al-Dawadmi, Saudi Arabia

ABSTRACT

Cancer is one of the prime rationales for mortality in humanity and remains a difficult disease to treat. Contemporary problems allied with conventional cancer chemotherapies embrace the insolubility of drugs in an aqueous medium, delivery of sub-therapeutic doses to target cells, lack of bioavailability, and most importantly, non-specific toxicity to normal tissues. Recent advances in nanotechnology investigation tackle potential solutions to these riddles. However, there are challenges regarding targeting specific sites, tracking the delivery system and control over the release of the drug to the target site. The nanodevices are 100 to 1000 times smaller than cells in humans; their size is comparable to the enzymes, the receptors. This enables them to have a large surface area and ability to interact with biomolecules on both the surface and inside cells. Nanomedicines between 8-100 nm have an enhanced permeability and retention (EPR) effect, which make these medicines to target passively the solid tumours.

INTRODUCTION

Nanotechnology has been a well-known research field since the last century. In 1959, the physicist Richard Feynmann (Richard, 1959) suggested a new area of research for science & technology on an atomic and molecular scale in a paper entitled "There's Plenty of Room at the Bottom." Since "nanotechnology" was named. Nanotechnology developed nanoscale-level materials of different forms. Nanoparticles (NPs) are broad nanomaterials with dimension of 1-100 nm (Laurent et al., 2008). Those materials can be 0D,

DOI: 10.4018/978-1-7998-5049-6.ch012

1D, 2D, or 3D depending on the overall shape (Tiwari et al, 2012). The importance of these materials became conscious when researchers exposed that dimensions can influence a material physicochemical properties, e.g. the optical properties. NPs of 20-nm Au (gold), Ag (silver) Pt (platinum), and Pd (palladium) have characteristic colour of red wine, yellowish gray, black, and dark black, respectively. Figure- 1 indicates that the diagram is an example, in which gold NPs synthesis with different sizes. Such NPs exhibited significant colors and characteristics with the variability in size and shape that can be used in bioimaging applications (Dreaden et al., 2012). Figure-1 shows that the colour of the solution varies because of difference in aspect ratio, nanoshell thickness and gold concentration percentage. Any modification in an element's physical properties affects the absorption properties of the NPs, and also has different colours.

Figure 1. Colour variation of Gold NPs with respect to size and shape (Dreaden et al, 2012)

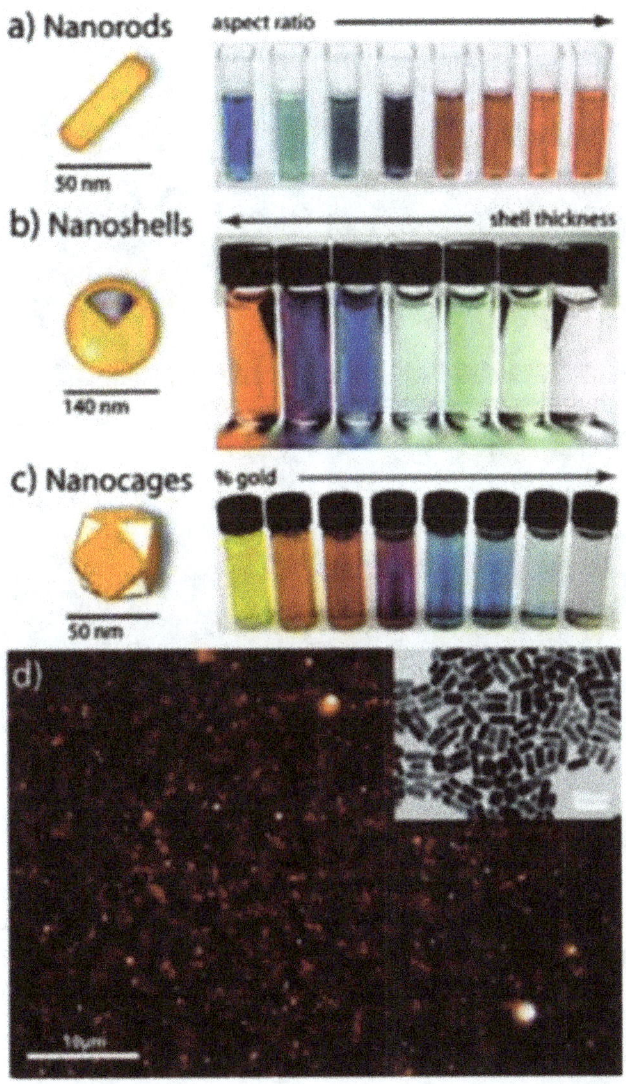

NPs are composed of three layers i.e. outer layer, shell layer or inter intermediate layer and the inner core. The outer layer is composed with diverse molecules like metal ions, polymers and surfactants, he shell layer or intermediate layer, which is distinct from the core material, and the core; which is fundamentally central part of a NP and refers specifically to the NP as its own (Shin et al., 2016).

The above materials got tremendous interest among researchers in different disciplines because of such unique properties. Figure-2 Shows images of mesoporous and non-porous methacrylate functionalized silica (MA-SiO_2) by scanning electron microscopy (SEM) and transmission electron microscope (TEM). Other characteristics of mesoporosity are imparted in NPs. The NPs are used for the delivery of drugs (Yamauchi et al, 2008), chemical and biological sensing (Barrak et al., 2019), gas sensing (Mansha et al., 2016), CO_2capture (Rawal et al., 2013), and other related applications (Khan et al., 2017).

Figure 2. FE-SEM micrographs of (a) nonporous MA-SiO_2 NPs, (b) mesoporous MA-SiO_2 NPs. TEM images of (c) nonporous MASiO_2NPs and (d) mesoporous MA-SiO_2 NPs (Lee et al., 2011)

Classification of NPs

NPs are commonly classified into different groups according to their size, and chemical properties and morphology. Below are the possibly the best-known classes of NPs, focusing on physical and chemical properties.

1. Carbon-based NPs
2. Metal NPs

3. Ceramics NPs
4. Polymeric NPs
5. Lipid-based NPs

Carbon-Based NPs

Two key carbon-based NP groups are fullerenes and carbon nanotubes (CNTs). Fullerenes contain nanomaterial, such as allotropic carbon derivatives, made from hollow globular container. Notable commercial interest has been generated by electrical conductivity, high strength, structure, electron affinity, and flexibility (Ganesh et al., 2017). Such materials have arranged in pentagonal and hexagonal units of carbon by SP^2 hybridisation.

CNTs are elongated, tubular structure with a diameter of 1–2 nm (Ramacharyulu et al., 2015). These can be predicted based on their diameter telicity as metallic or semi-conductive (Shaalan et al., 2016). The CNTs are structurally similar to the sheet of graphite rolled onto itself (Figure-3). The rolled sheets could be single, double or multiple walls and are therefore known as single-walled (SWNTs), double-walled (DWNTs), or multi-walled carbon nanotubes (MWNTs). CNTs are mostly synthesized by the accumulation of carbon precursors material, especially atomic carbons, vaporized by laser light or electric arc from graphite to metal particles Lately, they were synthesized using the technique of chemical vapor deposition (CVD) (Astefanei et al., 2015). These components are not only used in natural state due to their unique physical, chemical and mechanical features, but also nanostructured materials for several commercial products like fillers (Ibrahim KS., 2013), efficient gaseous adsorbents for environmental remediation, and as a support medium for various inorganic and organic catalysts (Aqel et al., 2012). For example, CNTs may stimulate inflammatory and apoptosis reactions in human T-cells (Gad 2017). The water-soluble CNTs are made of polyethylene glycol chains and have no adverse effects when tested in a wide variety of immune systems.

Figure 3. Rolling of graphite layer into single-walled and multi-walled CNTs. (Mashaghi et al., 2013)

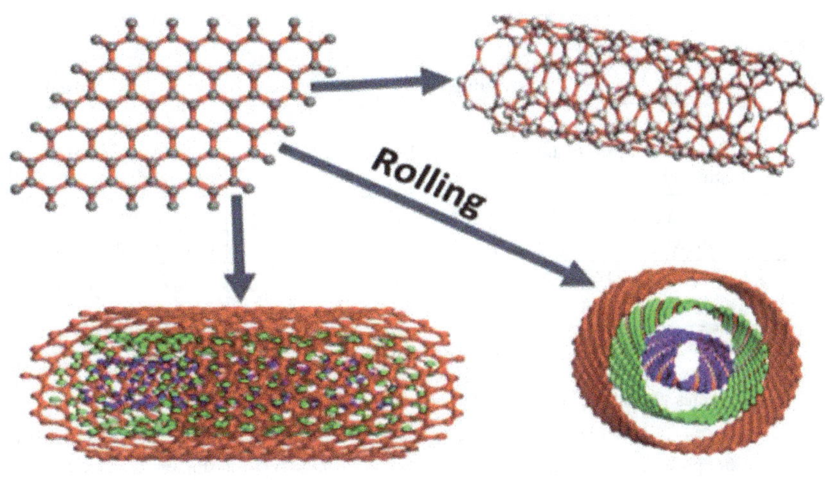

Metal NPs

Metal NPs are made purely of the precursors to the metal. The NPs exhibit special optoelectrical properties due to the well-known localized surface plasmon resonance (LSPR) characteristics. NPs of the alkali and noble metals i.e. Cu, Ag, and Au have a broad absorption band in the visible zone of the electromagnetic solar spectrum. Metal NPs find applications in many research areas, due to their advanced optical properties like Gold NPs coating is commonly used for SEM sampling, improving the electronic stream, helping to obtain high-quality (Elliott et al., 2013).

Ceramics NPs

Ceramics NPs are non-metallic inorganic solids which are synthesized by heat and by successive cooling. They are amorphous, polycrystalline, thick, porous, or hollow (Saeed et al., 2016). Because of their use in applications such as catalysis, photocatalysis, photodegradation of dyes, and imaging applications (Ngoy et al., 2014), these NPs receive great attention from researchers in the field of energy supply and storage, communication, transportation systems, construction, and medical technology. Their electrical properties may allow energy to be transferred with efficiencies approaching 100%.

Polymeric NPs

These are normally organic-based NPs and in the literature a special term polymer nanoparticle (PNP) collective used for it. They are mostly shaped was nanospheres or nanocapsules. The former are matrix particles whose overall mass is generally solid and the drug is dispersed. In the latter case the solid mass is encapsulated within the particle completely (Thomas et al., 2015). Applications of polymeric nanoparticles are controlled release; protect drug molecules, ability to combine therapy and imaging, targeting therapy, etc. They are highly biodegradable and biocompatible as in Table 1.

Table 1. The potential advantages and disadvantages of the polymers used in preparation of NPs (Suphiya et al., 2008).

S. no	Polymers	Advantages	Disadvantages
1	PLGA/PLA	The most commonly used polymers for the past 20 years are PLA and PLGA and these are bio-degradable, bio-compatible and non-toxic	The polymer is lipophilic and colloidal suspension stability is a problem. The coating process is laborious & expensive
2	PCL	Bio-compatible, bio-degradable and appropriate for long-term delivery	Very brittle, slow degradation, lack of side chain groups because of inert nature.
3	Chitosan	Bio-compatible, bio-degradable, lipophobic and has excellent mucoadhesive properties	Beads and agglomerates may form during synthesis
4	HAS	Bio-degradable, non-toxic, and non-antigenic. Allows electrostatic adsorption of positively or negatively charged species without the requirement of other compounds	Difficult to purify and to produce on a large scale
5	SLNs	Improved drug stability, high drug payload, and incorporate lipophilic and hydrophilic drugs.	Gel formation on solidification and upon storage

PLGA (Poly lactide-co-glycolide); PLA (polylactide acid); PCL (Poly-Ɛ-caprolactone); HAS (Human serum albumin); SLNs (Solid lipid nanoparticles)

Lipid-Based NPs

NPs include lipid moieties and are used in numerous biological devices. A lipid NP is typically spherical, with a diameter between 10 and 1000 nm.Lipid NPs have a solid lipid core, and a matrix contains soluble lipophilic molecules. Surfactants or emulsifiers have stabilised the external core of these NPs (Khan et al., 2017). Lipid nanotechnology focus on the design and synthesised of lipid NPs for various utilization like drug carriers and delivery (Hisatomi et al., 2014) (Sun et al., 1999) and the release of RNA for cancer therapy (Mansha et al., 2017).

NANOTECHNOLOGY IN CANCER TREATMENT

Nanotechnological principles improve the conditions of patients with cancer; because of better diagnosis of cancer, more efficient delivery of drugs to tumor cells, and molecular targeted cancer treatment. Nanotechnology can be applied to better diagnosis of cancer. One of the main usage fields of optical nanoparticles is to allow better cancer detection. To begin with, there are drawbacks to traditional approaches that are used in diagnosis.

Classified approaches like X-rays, tomography, or mammography often include the use of mutagenic agents on cancer-causing cells expresses harmful substances, and X-rays in the diagnosis of cancer is also linked to the causes of cancer (Prasad Rao et al., 2011). Diagnostic optical nanoparticles technique can be used to curtail those grievances. This approach works to interact with tumor cells with special dyes, and can be detected with optical nanoparticles. Such observations show that cancer detection with optical nanoparticles is an evolving topic, but it does have significant diagnostic advantages.

Recently, the Food and Drug Administration ratify several Investigational New Drug (IND) applications for nano-formulation or nano devices such as quantum dots, nanopores, nano cantilveres, naotubes, nanoshells, dendrimers and nanoparticles empowered clinical trials for breast, gynecological, lung, central nervous system, solid tumor, lymphomas, mesenchymal tissue, genitor-urinary cancer treatment and solid tumors. The examples of the drugs that have been approved by nanotechnology are paclitaxel, daunorubicine, daunorubicine, cytarabine, vincritine, mifamurtide, L-asparaginase and irinotecan. (www.fda.gov)

EMERGING STRATEGIES IN CANCER NANOTECHNOLOGY

Early to Late Stage Diagnosis: Nanotechnology Based Intercession

The greatest hope to overcome the adversity of cancer is indeed to identify cancer before the primary site of origin rises in cell proliferation and physical distribution of cancer cells. It is especially challenging to find cancer early because there will be few cancer cells, super low levels of biomarkers at the tumour site, and in initial phase body fluids. Furthermore, as the majority of cancers are diagnosed lately because the characteristics of the cancerous cells are not well established. But in the advanced stage, abundant cancer cells increased instantly, more genome, epigenome, proteome, and transcriptome changes are likely to occur in a similar fashion when characterized *ex vivo*, as well as more protein targets for *in vivo* molecular imaging probes. Ideally, all these issues can be solved as part of the nanomedicine sector by nanotechnology-based medical diagnostics. Nanomedicine, for its part, promises ground-breaking advances for early diagnosis, planning, and treatment.

Both *in vitro* nanoenabled diagnostic tools and nanoimaging, preoperative molecular diagnostic screening can spot and locate the tumor. These procedure exempt patients who have been metastasised beyond resection eligibility and recognize molecular targets that used direct elective surgeries, and screen appropriate cases with the most relevant biology and guide the surgeon. Nanotechnology provides advantages, for instance, extensive single cancer molecule can be detected in place of bulk measurements.

Nanotechnology offers

1. Sensitivity to an analysis;
2. Massive capacity for multiplexing of biomarkers / analytes;
3. Low operability of the clinical samples;
4. Comprehensive patient monitoring and the recognition by implantable sensors of any deviations from it;
5. Solutions for visualizing and responding to oncological pathogenesis
6. Concurrently trying to detect and treating cancer (theranostics);
7. Cost benefits for patient populations and the overall health-care system.

Recent Trends in Early to Late-Stage Nanotechnology-Based Diagnostic Testing

A wide array of nano-biosensors of preclinical research grade were already established but the main objective of the clinical treatment, inexpensive, high-throughput, powerful diagnostic gadget has yet to be properly addressed. Providing such a capacity in the clinical would undoubtedly enable improved cancer detection with potential aid for patients and the overall health protection. Often, on initiation of the diagnosis, the majority of long-term cancer survivors have tumours that appear to be confined to the primary site and can thus promote significantly from curative surgery, assisting primitive cancer recognition and intervention to augment patient's endurance. Till today, there are no serum bio-markers with the accuracy and precision to trace early pre-cancer cells. Our technologies are in many respects ahead of our compassionate of the fundamental biology of cancer. Moreover, heterogeneity of tumour and the inherently complex stromal micro-environment also pose a challenge to acknowledge impending biomarkers.

A recent mathematic study (Hori et al., 2011), it was discovered that cancer can develop undetected for over 10 years to reach a globular diameter 25 mm prior to actually identification by existing medical blood tests. These forecasts are well-aligned with clinical results. Currently, there are no suitable biomarkers for evaluating healthy people for the incidence of pre-cancerous events. Actually, cancer screening is carried out using traditional cancer discovery technologies like x-ray imaging based mammography for breast cancer, visible light colonoscopy for colorectal cancer, etc., none of which are currently allowed by nanotechnologies. Many pre-clinical medical diagnostic imaging devices are currently undergoing evaluation for their worthiness as supporting methods to present-day diagnostic approaches to cancer.

A few of these technologies include magnetic nanoparticles/ gadolinium chelate functional nanoparticles facilitate for higher solution MRI (Bennett et al., 2014), nanoparticles and inherent contrast based surface-enhanced Raman spectroscopy-based endoscopy, photoacoustic imaging, cancer-triggered self-assembled smart optical and MRI nano-imaging agents, micro-nuclear magnetic resonance imaging, dual (e.g., PET-Near Infrared) the related clinical trials with those methods. Currently, two very broad sets of devices and instruments are evolving in the field of cancer nanotechnology focused diagnostics, and there is a rapid and continuous investigation in both.

Such classes are

1. Bench-top or larger scale medical diagnostic instruments;
2. Miniaturized nano-based or nano-enabled diagnostic assays/devices designed for point of care or home use directly or suitable for inhalable, ingestible, implantable, and wearable uses.

The medical requirements of the 1st cohort of tools are that they will be exceedingly strong, flexible, error free and ideal for validating of clinical cancer management. Examples of this category are imaging agents based on nanoparticles like paramagnetic iron oxide, gold, silica-based nanoparticles, carbon nanotubes, surface enhanced Raman nanoparticles, etc., and their allied instrumentation detection or analysis and nano-imaging devices (e.g., Raman spectroscopy, photoacoustic imaging, MRI abet nanoparticles). These techniques usually better suited to later stage cancer detection.

The 2nd cohort of nanodiagnostic tools for cancer embraces nanocantilever, nano-pore, nano-wire, quantum dot, nanofluidic devices. These devices are relatively inexpensive, yield fast and simple results, often during the same visit to the office, and produce effective insights for the next round of risk evaluation of cancer. These screening tools are more appropriate to early phase cancer diagnosis, testing relapse, monitoring of therapeutic efficacy and overall monitoring. Persistent research into cancer nano-technology is needed to improve and innovate these two cohorts of medical diagnostic instruments.

EARLY TO LATE STAGE DIAGNOSIS: DETECTING AND ANALYSING CIRCULATING TUMOR CELLS

Circulating Tumor Cells (CTC)

Circulating tumor cells (CTCs) are released into the peripheral blood from the primary tumor microenvironment, and have the ability to invade other body parts, triggering metastasis (Alix et al., 2014). The tissue based assessment of biopsy specimens remains the mainstay in clinical care and research for diagnosis and prognosis. The research focuses on samples of tissue collected by biopsy or radiographically guided needle fractions. These procedures have several issues like metastases, ultimately expensive, several biopsies are required but usually excluded from the protocol because of paranoia, yet important in acquiring complex insight of the progress of cancer. Here research has focused on the common primary tissues of cancers where metastatic tissue biopsies are problematic.Circulating tumor cells (CTCs) most often suggested as a tool for resolving these issues, because it imparts a mode of sampling tumors throughout the (systemically perfused in the blood), together the with primary tumor and metastases (Lohr et al., 2014). Traditional diagnostic imaging and serum bio-marker identification in cancer, the detection, and characterisation of CTCs in cancer patients during the treatment provides new opportunities for individual cancer care by (Cohen et al., 2009)

1. Tracking the progression of cancer;

2. Recognizing the pathogenic pathways that cause fatal cancer and the complexities emerging biology, and
3. Assisting the effectiveness of the treatment and re-strategizing on the development of resistance (Wallwiener et al., 2014)

Substantial development has been achieved within the field of CTC diagnosis, isolation, and characterisation over the last decade, largely driven by teamwork and integrative research in chemistry, materials science, bioengineering, and oncology.

New technologies in nanomaterials offer efficient microfluidic systems as well as distinctive nanoparticles that will allow for a diversity of in-depth CTC characterization studies with significantly lower expenses, and eventually bring the oncology field closer to achieving of patient services.

Conventional CTC Assays

The predictive validity of CTCs in the clinical is limited by their low concentration (1–10 CTCs / billion of blood cells) (Jackson et al., 2017). The most commonly used assays for spotting CTC include:

1. Immunomagnetic separation: The method uses magnetic nanoparticles and captured agent (conjugated with protein antibody) to selective CTCs using a cell surface marker like anti-EpCAM (Epithelial cell adhesion molecule) or use anti-CD45 of white blood cells. The CellSearchTM screening test is the only CTC interventional technique approved by the FDA for breast, prostate, and colorectal metastatic cancers. CellSearch™ Assay harvests CTCs with anti-EpCAM coated magnetic beads, and the consequent immunocytochemistry (ICC) procedure aims to recognize CTCs from non-specifically collected white blood cells. Numerous new technologies have been introduced such as MagSweeper, IsoFlux, Cynvenio, magnetic sifters, VerIFAST and AdnaGe (Cristofanilli et al., 2004).
2. Flow Cytometry: It is one of the mainly sophisticated techniques for analyzing and sorting cell subpopulations, in conjunction with the use of fluorescent probes. This method, however, cannot provide the geometric features of CTCs in order to access pathogenesis. To address this weakness, a superior technique called the ensemble-decision aliquot ranking (eDAR), was developed which automatically sort and collects CTCs (Schiro et al, 2012).
3. Imaging by microscopy: Microscopic visualization of blood samples treated with ICC enables exceedingly sensitive identification of CTCs, followed by their morphometric features and protein expression.
    1. The following 'two' approaches identify CTCs as label free techniques.
4. CTC filters: In this method use micro filters to trap CTCs according to dimension.
5. Dielectrophoresis: CTCs can be sorted from WBCs according to dielectrical properties of the CTC's. ApoCell's technologies, Silicon Biosystems are two examples under this category.

Micro Fluidics-Enabled CTC Assays

Toner et al. developed microfluidic affinity capture devices which triggered notable research programs aimed at the progression of nanotechnology enabled CTC assays. The first-generation device is CTC-Chip portrayed chemically engraved microposts on the silicon substrate, the anti-EpCAM antibodies are attached to it by covalent bonding. These engraved microposts have been designed to enhance contact between device surfaces and cell flow (Nagrath et al., 2007).

Immunocytochemistry is analyzed to recognise CTCs after the detection of CTC. The CTC-Chips showed much more accomplishments in the validation than traditional CTC assays. Analogous equipment setups were developed to create new microfluidic chips such as geometrically improved differential immunocapture, GEDI approach (Gleghorn et al., 2010), and Biocept's CTC assay (Pecot et al., 2011).

The Ephesia system is a further promising example. The sample shall flow via a diamond-like chip into zone of the capture columns. Supermagnetic beads coated with EpCAM antibodies self-assemble into a periodic array under a high magnetic field that creates a dense sieve which, in turn, catches the EpCAM positive cells that passes across. It is used in the prognosis and diagnosis of breast, NSCLC, Prostate Colorectal cancers (Svobodova et al., 2014). Microscale herringbone patterns were designed into PDMS portion to create microvortices, resulting in increased interaction between both CTCs and chip surfaces coated with antibodies. n addition to frequently utilised ICC methodology, the HB-Chip 's transparent nature allowed the sized CTCs to be studied by histopathological stains such as H & E stain (haematoxylin and eosin stain).

Microfluidic setup enhances CTC-capture efficiency, due to vertical depths of 3-dimensional device configurations most microfluidic CTC assessments endure from depth of field problems when conducting microscopy imaging. Time is consumed for multiple cross-sectional imaging scans that produce bulky image files are required to avoid superimposed micrographs. By integrating, a couple of microelectrodes at plastic microfluidic chip end (Adams et al., 2008), the enzymatic discharge of the captured CTCs can be recorded electrically without the microscopy imaging.

3rd generation iChip describes (Yu et al., 2012) a pioneering tag-free method, where combining negative immunomagnetic depletion processes with an integrated microchip inertial focus setting. Most interestingly, this method permits the retrieval of unstaged CTCs with intended biochemical integrity and viability, enabling downstream expressional profiling, as well as ex-vivo culture and drug sensitivity validation (Yu et al., 2014). Other microfluidic CTC assays include micro-nuclear magnetic resonance (μNMR) platform, cell rolling, and Vortex technology have also been developed, tested, and validated.

Nanomaterials-enabled CTC Assays (Nano Velcro CTC Assay)

Nano Velcro is recent nanotechnology advances enabled powerful solutions (Yoon et al., 2014) involved in a wide range and in-depth CTC analysis, and drastically lowering prices. The development of such evolving innovations will inevitably bring oncology closer to achieving of personalized patient care. The cellular components, such as extracellular matrix, and the cellular membrane, exist in the tissue micro-environment which furnish structural and biochemical support and regulate the behaviour, fate of cells.

As interactions were observed throughout the tissue micro - environment, UCLA research team headed by Dr. Tseng initiated the concept of cell-affinity substrates "NanoVelcro". In this unique approach, capture agent-coated nanostructured substrates are used to immobilize CTC from the microenvironment with superior precision (Wang et al., 2009).

The mechanism of NanoVelcro mimics VelcroTM – when the two fabric strips of a Velcro fastener are pressurized around each other, there is interactions between the hairy surfaces on 'two' strips result in high affinity between cells and nanosubstrates containing SiNS (silicon nanowire substratum). NanoVelcro cell-affinity assay is supported by various nanomaterials such as polymer dots, nanopillars, TiO_2 nanowires, nanoparticles, layer-by-layer assembled nanostructures, silicon nanowire, gold clusters,

Fe_3O_4 nanoparticles, DNA networks, and graphene oxide nanosheets to achieve the highest affinity CTC capture.

The 1st generation NanoVelcro Chip, consist SiNS and an overlaid microfluidic chaotic mixer, was developed for CTC enumeration. The success of these NanoVelcro Chips above 85 percent of CTC capture efficiency was observed using artificial CTC samples. Studies of side-by-side analytical reassurance utilize clinical blood samples show that the susceptibility of the 1st generation NanoVelcro Chip best than CellSearch™ Assay approved by the FDA (Lu et al., 2013).

The NanoVelcro approach permits the immobilization of CTCs on a flat and small surface, thus facilitating following high-resolution immunofluorescence microscopy imaging of CTCs without multiple cross-sectional image analysis which is required for most current microfluidic CTC assays. The first generation NanoVelcro is used for detection of CTC of solid tumours, prostate cancer and pancreatic cancer.

The 2nd generation NanoVelcro Chips identified as the NanoVelcro-LCM approach, have been constructed by substituting SiNS with a transparent nanofibre-covered substratum made from PLGA, i.e. poly (lactic-co-glycolic acid). The PLGA NanoVelcro transparent substrate holds the intended CTC capture and enables complete connection with a laser microdissection (LCM) method to segregate single-cell immobilized CTCs with high resolution. To evaluate the activities of CTC as a tumor liquid biopsy, single-CTC genotyping is evaluated by Sanger sequencing and NGS (next-generation sequencing) may be conducted on individually isolated CTCs. The second generation NanoVelcro is used for the detection of CTC of solid tumours, metastatic melanoma, and pancreatic cancer

The 3rd generation Thermo responsive NanoVelcro Chip from UCLA has shown the feasibility of trapping, and releasing CTCs at 4 and 37°C, respectively. The temperature dependent conformation alters in the polymer brushes will change the availability of the capture agent on SiNS by grafting thermo responsive polymer brushes onto SiNS, permit quick CTC purification with required feasibility and biochemical integrity. This generation is used for identification of CLC of NSCLC (Hou et al., 2013).

TUMOUR MICROENVIRONMENT TARGETED NANOTHERAPY

Handling cancer worldwide remains a formidable job for physicians, oncologists, and scientists (Ferlay et al., 2010). Lung cancer is notably linked with the highest proportion of mortality accompanied by hepatic and stomach cancer among the various types of cancer. In recent times, patients suffering from colorectal and breast cancers have increased. By 2025, the number of cancer patients may be more than 20 million annually is estimated to rise globally (Fernandes et al., 2018). These alarming statistics have forced researchers around the globe to speed up research into new and powerful drug molecules to alleviate the acquired confrontation and eliminate the cancerous cells from the biological environment. The nature of the disease, however, needs systematic efforts to devise chemotherapy to curb tumor growth (Raave et al., 2018).

Nonetheless, these activities have been interpreted into cancer molecules capable, albeit in preclinical settings, of fighting cancer progression. Their clinical implementation is still burdened particularly with unwanted toxic effects (Dai et al., 2016). Another devastating problem with the chemotherapeutic field is the advancement of chemo-resistance, which is again disturbing fact for both individual and combination cancer therapy after the initial response period (Cree et al., 2017). Clinical manifestations persuasive implicate the presence of malignant and metastatic components in the tumor microenvironment are a fundamental truth of tumor resistance to chemotherapy (Cheng et al., 2016).

Figure 4. Semantic depiction of the 'three' generations of nanovelcroCTC Assays invented by the UCLA group of scientists to reach various clinical aspects.

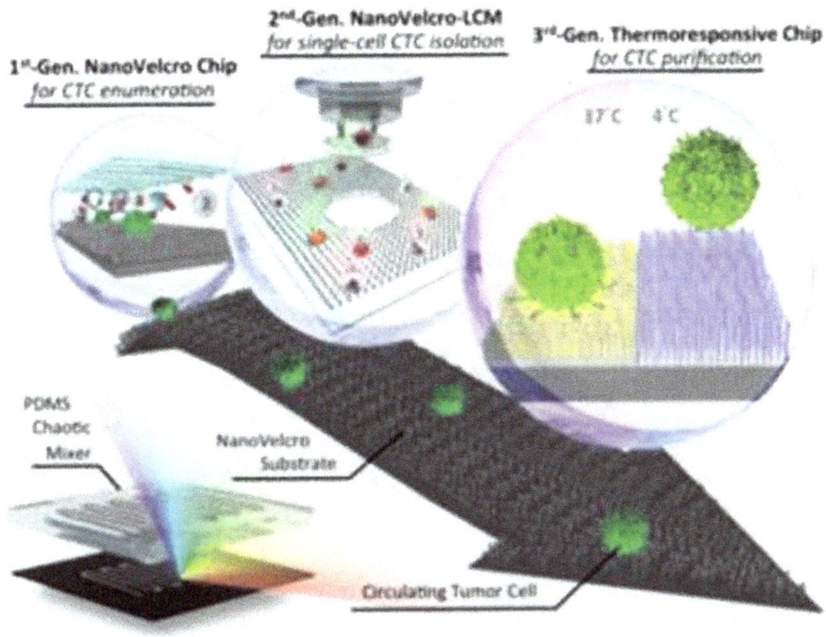

Because of the intricacy of the tumor microenvironment showing figure-5, the conventional drug delivery system does not distribute the chemo-drug in a cancer cell to destroy it due to concentration and also connected with serious side effects. This urged the use of the nanotechnology technique to accomplish tumor specificity to improve the therapeutic index and the pharmacokinetic profile of the chemotherapeutic agent (Danhier F., 2016). Passive diffusion was found through EPR-effect (enhanced permeability and retention) to allow nano chemotherapeutics tumor localization. Due to inadequate knowledge of tumor microenvironment and accumulated preliminary success exploiting the EPR effect; previous investigation mainly emphasized on constructing sustainable long-circulating nanocarriers to enable making tremendous local effect of drugs with negligible drug loss in the circulatory system. Today, these efforts resulted in the commercialization of FDA, approved nanochemotherapeutics; the liposomal formulation of doxorubicin, Daunorubicin, and albumin-bound Paclitaxel (Jokerst et al., 2011). Clinically, however, these formulations were found to be moderately successful because of insufficient drug delivery to the site of the tumor (Primeau et al, 2005). The likely explanations for such a disparity could be attributed to the containment of nano-chemotherapeutics in extremely infused regions, often depriving the tumor regions of reduced perfused or avascular. Such an inhomogeneity in tumor distribution has been shown to relate to suboptimal clinical effects, acquired resistance, cancer cell relapse, and therefore requires a high dosage of drugs. Consequently, this results in unwanted adverse/toxic effects (Waite et al, 2012).

The localization of nanochemotherapeutics within the tumor microenvironment may obstruct by larger quantities of interstitial fluid pressure, modified extracellular matrix (ECM) structure, enhanced cell division and impeded lymphatic drainage (Ozcelikkale et al., 2013). Thus, there is no complete knowledge about these complicated variables controlling such transport mechanisms, and tumor localization poses enormous bottlenecks to design an effective nano-strategy for tumor eradication.

Figure 5. **Demonstrating tumor microenvironment** *(Fernandes et al., 2018)*

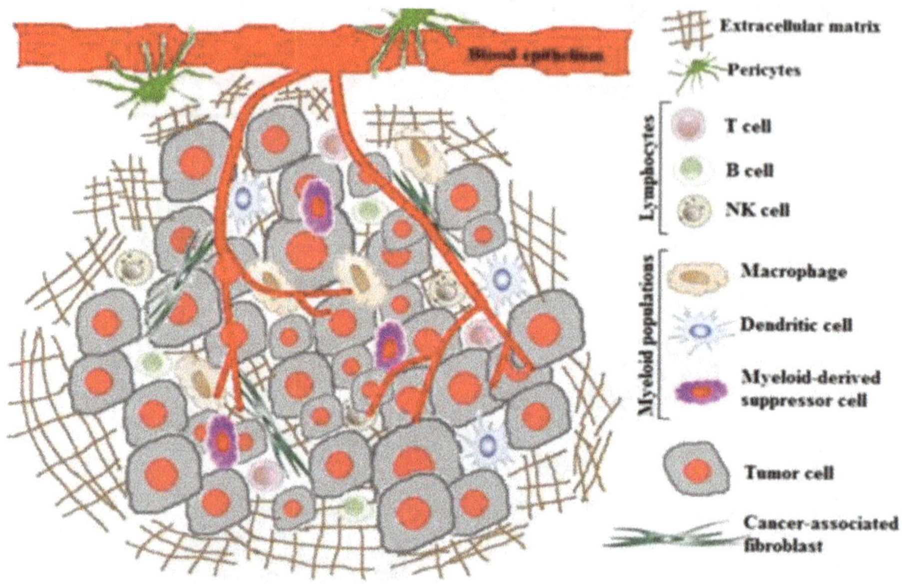

TOXICITY EFFECTS OF NANOMATERIALS ON HUMANS AND ENVIRONMENT

Nanomedicine is seen by researchers as another logical choice in science, integrating engineering with physics, chemistry, biology, and medicine (Thomas et al., 2003). When material sizes become quite tiny, its physicochemical characteristics can vary greatly from that of the same material in bulk form. The latest nanomaterials construct microscopic or even molecular devices capable of supporting medicine, environmental conservation, energy, and space excavation (Stewart et al., 2008). With increasing knowledge of nanomaterials as well as the capacity to architect innovative products, it wouldn't be long before the whole history could be squeezed within our pockets or widened by designed specifically molecules that imitate living systems. The idiom "Nanotechnology" has been exaggerated in the last few years and has become almost synonymous with items that are revolutionary and very encouraging (Teo et al., 2007). Nanotechnology allows us to manipulate atomic and molecular scales, and to exploit new properties and phenomena to develop minute materials, devices, and systems. Considerably smaller size, lesser mass, extra moderate power specifications, greater accuracy, and improved precision are few advantages for humans use in the construction of nanomedicine, nanoimagine, sensors, etc. For the past three decades the smaller and faster transistors were a guiding factor for the central processing units. When transistor dimensions are decreased to the nanometer, designers have been getting close to the point in which nano lithography could accomplish the necessary resolution to construct such nanometer-sized devices.

Nanomaterials have applications in the areas of individual endeavors, as well as manufacturing, agriculture, health care, public health, and public interest (Mirkin et al., 1996). Nanomaterials are integral part of society in the form of beauty products, packaged food material, targeted drug delivery, therapeutics, biosensors, bioimaging, etc. The size, and shape is comparable to that of biomolecules, and also due of its antibacterial, odor fighting properties, these materials are comprehensively used in variety of commercial materials like wound dressing, detergents, deodorants, and antibacterial adhesives.

As per the National Nanotechnology Initiative (America) (Zhang et al, 2005) every year thousands of tons of silica-dioxide (SiO_2), alumina-dioxide, cerium are used in slurries for the precise scrubbing of silicon wafers in the form of the ultra-fine abrasive particulate mixture (Lü et al., 2009, Martins et al., 2007),.

Therefore, the population exposed to nanomaterials keeps growing because of the surge in their usage. Amidst clear benefits due to small size, there are numerous questions as raised to how the nanoparticles used during daily life can impact the environment. Another crucial issue to be addressed in the coming years, before the huge manufacture of nanomaterials, is their human toxicity and environmental consequences. There are significant discussions about how nanoparticles' unique characteristics can result in detrimental health impacts, with the ability to cause harmful effects. One needs to recognize what the cellular responses will be when nanomaterials undertake biotransformation in the cell membrane? Biodegraded nanoparticles, for example, Could build up in cells and cause intracellular transformation like disordering in organelle composition, gene alternation. A few critical questions are:

1. Are nanomaterials more toxic than their non-nano counterparts?
2. Will the nanoparticles transform into more toxic forms in the environment?

It is essential for nanotoxicology investigation to investigate and acknowledge how nanoparticles impact the ecosystem so that their harmful features could be spared before nanomaterials are permitted for use in everyday lives. The potential health effects of emerging nanotechnologies on humans is discussed in Table 2,

Table 2. **Potential toxic effects of nanomaterials on humans** *(www.sigmaaldrich.com)*

S.No	Toxic effects	Examples of Nanomaterial
1	Cell membrane damage by leaking or thinning	Cationic NPs.
2	Blocking or change in protein function	Metal oxide NPs, Polystyrene, Dendrimer, Carbon material.
3	DNA strand cleavage or mutation	Sliver NPs.
4	Mitochondrial impairment by blocking electron transport chain or ATP production	Ultrafine particles, Cationic NPs.
5	Lysosomal damage	Ultrafine particles, Cationic NPs, Carbon nanotubes
6	Induction of inflammatory signal sequence and apoptosis response in T cells	Metal oxide NPs, Carbon nanotubes
7	Fibrinogensis and tissue remodeling cell injury	Carbon nanotubes
8	Blood platelet, vascular endothelial and clotting abnormalities	Colloidal silica nanoparticles
9	Free radical activation and production	Ultrafine particles, Metal oxide NPs, Cationic NPs.
10	Immune response activation	Gold NPs, Poly NPs Liposomes, Dendrimers,

7. Anticipated Economic Impact on Nanotechnology

Nano-medicine is vital not only in terms of social and health care sector but also in terms of economic capability. The economic growth is forecasted to expand up to $ 1 trillion and reach $3,231.4 trillion by the National Science Foundation's studies by 2025 (Parveen et al., 2006). In 2013, incomes from product categories that integrate nanotechnology into the global industry studies approached $1014 billion, with 31.8 percent in Europe, 31.3 percent in USA, 30.5 percent in Asia, and 6.4 percent for the rest of the globe. The key segments recognized mostly in census are 2 percent towards of energy and environment,10 percent include for health and life sciences sector, 29 percentage for electronics and information technology, and finally 59 percentage in the sector of materials and manufacturing (Lux Research 2016).

Nanotechnology is luring fast-growing investment funds from government and other relevant corporate communities. The estimated worldwide investment in this sector is currently around 5 billion Euros, of which 2 billion Euros originates from private industry (Parveen et al., 2006). In 2005, the National Institutes of Health spent $89 million on nanotechnology, covering $30 million for the Nanomedicine in Cancer at the National Cancer Institute. The quantity of published patents claims in nanotechnology augmented 4 times from 531 patent claims in 1995 to 24,206 patents claims in the last two decades. As most people do believe that nanotechnologies would have an influence along a broad variety of sectors, a study of nanotechnology experts recognized hype, i.e. "misguided promises that nanotechnology can fix everything" as the most likely issue to lead to a negative backlash" (Parveen et al., 2006). Switzerland, South Korea, Germany, France, the USA and Japan play a significant role in the nanotechnology science. The BRICS (Brazil, the Russian Federation, India, China and South Africa) strive to be excellence in the field of nanotechnology. Nanotechnology is now seen as a research priority by an increasing number of countries like Argentina, Azerbaijan, Chile, Croatia, Jordan, Kazakhstan, Mexico, Morocco, Nepal, the Philippines, Saudi Arabia, Serbia, Slovenia, Sri Lanka and Tunisia (http://www.unesco.org/).

CONCLUSION

This chapter is detailed with classification of NPs, Nanotechnology in cancer, its emerging strategies, and targeted nanotherapy, its effects on humans and environment, and finally anticipated economic impact. Nanotechnologies are promising tools for augmenting the sensitivity and accuracy of both cancer diagnosis and treatment. Different nanomaterials are used as contrast media for MRI and CT imaging, providing better chances of premature cancer cells being identified. The fundamental part of oncologists who practices cancer treatment monitors the biological environment by tracking the dynamic revisions in the patient body, which allows the clinicians to follow a distinct and biologically relevant rhythm. CTCs associated with nanoparticles should be replaced by the existing biopsies or testing limited biomarkers. But, there is lot of concerning effects of nanotechnology on human health as well as environmental effects. Thus an efficient understanding of the importance of the nanotechnology for our life will be greater. So, nanotechnology must be improved and this must be continued for the future endeavour.

REFERENCES

Adams, A. A., Okagbare, P. I., Feng, J., Hupert, M. L., Patterson, D., Gottert, J., McCarley, R. L., Nikitopoulos, D., Murphy, M. C., & Soper, S. A. (2008). Highly efficient circulating tumor cell isolation from whole blood and label-free enumeration using polymer-based microfluidics with an integrated conductivity sensor. *Journal of the American Chemical Society*, *130*(27), 8633–8641. doi:10.1021/ja8015022 PMID:18557614

Alix Panabieres, C., & Pantel, K. (2014). Challenges in circulating tumour cell research. *Nature Reviews. Cancer*, *14*(9), 623–631. doi:10.1038/nrc3820 PMID:25154812

Aqel, M., El Nour, R.A., Ammar, A., & Warthan, A. (2012). Carbon nanotubes, science and technology part (I) structure, synthesis and characterisation. *Arabian Journal of Chemistry, 5*(1), 1-23.

Astefanei, A., Núñez, O., & Galceran, M. T. (2015). Characterisation and determination of fullerenes: A critical review. *Analytica Chimica Acta*, *882*, 1–21. doi:10.1016/j.aca.2015.03.025 PMID:26043086

Barrak, H., Saied, T., Chevallier, P., Laroche, G., M'nif, A., & Hamzaoui, A. H. (2019). Synthesis, characterization, and functionalization of ZnO nanoparticles by N-(trimethoxysilylpropyl) ethylenediamine triacetic acid (TMSEDTA). *Arabian Journal of Chemistry*, *12*(8), 4340–4347. doi:10.1016/j.arabjc.2016.04.019

Bennett, K.M., Jo, J., Cabral, H., Bakalova, R., & Aoki, I. (2014). MR imaging techniques for nano-pathophysiology and theranostics. *Adv. Drug Deliv*, 75-94.

Cheng, Y. T., Yang, C. C., & Shyur, L. F. (2016). Phytomedicine-Modulating oxidative stress and the tumor microenvironment for cancer therapy. *Pharmacological Research*, *114*, 128–143. doi:10.1016/j.phrs.2016.10.022 PMID:27794498

Cohen, S. J., Punt, C. J. A., Iannotti, N., Saidman, B. H., Sabbath, K. D., Gabrail, N. Y., Picus, J., Morse, M., Mitchell, E., Miller, M. C., Doyle, G. V., Tissing, H., Terstappen, L. W. M. M., & Meropol, N. J. (2009). Relationship of Circulating Tumor Cells to Tumor Response, Progression-Free Survival, and Overall Survival in Patients With Metastatic Colorectal Cancer. *Journal of Clinical Oncology*, *26*(19), 3213–3221. doi:10.1200/JCO.2007.15.8923 PMID:18591556

Cree, I. A., & Charlton, P. (2017). Molecular chess? Hallmarks of anti-cancer drug resistance. *BMC Cancer*, *17*(1), 10. doi:10.118612885-016-2999-1 PMID:28056859

Cristofanilli, M., Budd, G. T., Ellis, M. J., Stopeck, A., Matera, J., Miller, M. C., Reuben, J. M., Doyle, G. V., Allard, W. J., Terstappen, L. W. M. M., & Hayes, D. F. (2004). Circulating Tumor Cells, Disease Progression, and Survival in Metastatic Breast Cancer. *The New England Journal of Medicine*, *351*(8), 781–791. doi:10.1056/NEJMoa040766 PMID:15317891

Dai, L., Liu, J., Luo, Z., Li, M., & Cai, K. (2016). Tumor therapy: Targeted drug delivery systems. *Journal of Materials Chemistry. B, Materials for Biology and Medicine*, *4*(42), 6758–6772. doi:10.1039/C6TB01743F PMID:32263571

Danhier, F. (2016). To exploit the tumor microenvironment: Since the EPR effect fails in the clinic, what is the future of nanomedicine. *Journal of Controlled Release, 244*, 108–121. doi:10.1016/j.jconrel.2016.11.015 PMID:27871992

Elliott, J. A., Shibuta, Y., Amara, H., Bichara, C., & Neyts, E. C. (2013). Atomistic modelling of CVD synthesis of carbon nanotubes and graphene. *Nanoscale, 5*(15), 6662–6676. doi:10.1039/c3nr01925j PMID:23774798

Ferlay, J., Shin, H.R., Bray, F., Forman, D., Mathers, C., & Parkin, D.M. (2010). GLOBOCAN 2008; cancer incidence and mortality worldwide. *IArC CancerBase*, 10.

Fernandes, C., Suares, D., & Yergeri, M. C. (2018). Tumor microenvironment targeted nanotherapy. *Frontiers in Pharmacology, 9*, 1230. doi:10.3389/fphar.2018.01230 PMID:30429787

Gad, S.C. (2017). Oral drug formulation development in pharmaceutical lead selection stage. *Oral Formulation Roadmap from Early Drug Discovery to Development*, 39-88.

Ganesh, M., Hemalatha, P., Peng, M. M., & Jang, H. T. (2017). One pot synthesized Li, Zr doped porous silica nanoparticle for low temperature CO2 adsorption. *Arabian Journal of Chemistry, 10*, S1501–S1505. doi:10.1016/j.arabjc.2013.04.031

Gleghorn, J. P., Pratt, E. D., Denning, D., Liu, H., Bander, N. H., Tagawa, S. T., Nanus, D. M., Giannakakou, P. A., & Kirby, B. J. (2010). Capture of circulating tumor cells from whole blood of prostate cancer patients using geometrically enhanced differential immunocapture (GEDI) and a prostate-specific antibody. *Lab on a Chip, 10*(1), 27–29. doi:10.1039/B917959C PMID:20024046

Hisatomi, T., Kubota, J., & Domen, K. (2014). Recent advances in semiconductors for photocatalytic and photoelectrochemical water splitting. *Chemical Society Reviews, 43*(22), 7520–7535. doi:10.1039/C3CS60378D PMID:24413305

Hori, S.S., & Gambhir, S.S. (2011). Mathematical model identifies blood biomarker-based early cancer detection strategies and limitations. *Sci. Transl. Med.*, 109ra116–109ra116.

Hou, S., Zhao, H., Zhao, L., Shen, Q., Wei, K. S., Suh, D. Y., Nakao, A., Garcia, M. A., Song, M., Lee, T., Xiong, B., Luo, S.-C., Tseng, H.-R., & Yu, H. (2013). Capture and stimulated release of circulating tumor cells on polymer-grafted silicon nanostructures. *Advanced Materials, 25*(11), 1547–1551. doi:10.1002/adma.201203185 PMID:23255101

Ibrahim, K.S. (2013). Carbon nanotubes-properties and applications: a review. *Carbon Letters, 14*(3), 131-144.

Jackson, J. M., Witek, M. A., Kamande, J. W., & Soper, S. A. (2017). Materials and microfluidics: Enabling the efficient isolation and analysis of circulating tumour cells. *Chemical Society Reviews, 46*(14), 4245–4280. doi:10.1039/C7CS00016B PMID:28632258

Jan, Y. J., Chen, J. F., Zhu, Y., Lu, Y. T., Chen, S. H., Chung, H., Smalley, M., Huang, Y.-W., Dong, J., Chen, L.-C., Yu, H.-H., Tomlinson, J. S., Hou, S., Agopian, V. G., Posadas, E. M., & Tseng, H.-R. (2018). NanoVelcro rare-cell assays for detection and characterization of circulating tumor cells. *Advanced Drug Delivery Reviews, 125*, 78–93. doi:10.1016/j.addr.2018.03.006 PMID:29551650

Jokerst, J. V., Lobovkina, T., Zare, R. N., & Gambhir, S. S. (2011). Nanoparticle PEGylation for imaging and therapy. *Nanomedicine (London)*, *6*(4), 15–28. doi:10.2217/nnm.11.19 PMID:21718180

Khan, I., Abdalla, A., & Qurashi, A. (2017). Synthesis of hierarchical WO3 and Bi2O3/WO3 nanocomposite for solar-driven water splitting applications. *International Journal of Hydrogen Energy*, *42*(5), 3431–3439. doi:10.1016/j.ijhydene.2016.11.105

Khan, I., Yamani, Z. H., & Qurashi, A. (2017). Sonochemical-driven ultrafast facile synthesis of SnO2 nanoparticles: Growth mechanism structural electrical and hydrogen gas sensing properties. *Ultrasonics Sonochemistry*, *34*, 484–490. doi:10.1016/j.ultsonch.2016.06.025 PMID:27773272

Laurent, S., Forge, D., Port, M., Roch, A., Robic, C., Vander Elst, L., & Muller, R. N. (2008). Magnetic iron oxide nanoparticles synthesis, stabilization, vectorization, physicochemical characterizations, and biological applications. *Chemical Reviews*, *108*(6), 2064–2110. doi:10.1021/cr068445e PMID:18543879

Lee, J. E., Lee, N., Kim, T., Kim, J., & Hyeon, T. (2011). Multifunctional mesoporous silica nanocomposite nanoparticles for thernaoustic applications. *Accounts of Chemical Research*, *44*(10), 893–902. doi:10.1021/ar2000259 PMID:21848274

Lohr, J. G., Adalsteinsson, V. A., Cibulskis, K., Choudhury, A. D., Rosenberg, M., Cruz-Gordillo, P., Francis, J. M., Zhang, C.-Z., Shalek, A. K., Satija, R., Trombetta, J. J., Lu, D., Tallapragada, N., Tahirova, N., Kim, S., Blumenstiel, B., Sougnez, C., Lowe, A., Wong, B., ... Love, J. C. (2014). Whole-exome sequencing of circulating tumor cells provides a window into metastatic prostate cancer. *Nature Biotechnology*, *32*(5), 479–484. doi:10.1038/nbt.2892 PMID:24752078

Lü, J. M., Wang, X., Marin-Muller, C., Wang, H., Lin, P. H., Yao, Q., & Chen, C. (2009). Current advances in research and clinical applications of PLGA-based nanotechnology. *Expert Review of Molecular Diagnostics*, *9*(4), 325–341. doi:10.1586/erm.09.15 PMID:19435455

Lu, Y. T., Zhao, L., Shen, Q., Garcia, M. A., Wu, D., Hou, S., Song, M., Xu, X., OuYang, W.-H., OuYang, W. W.-L., Lichterman, J., Luo, Z., Xuan, X., Huang, J., Chung, L. W. K., Rettig, M., Tseng, H.-R., Shao, C., & Posadas, E. M. (2013). NanoVelcro Chip for CTC enumeration in prostate cancer patients. *Methods (San Diego, Calif.)*, *64*(2), 144–152. doi:10.1016/j.ymeth.2013.06.019 PMID:23816790

Lux Research. (2016). Revenue from nanotechnology (Figure 11). In Nanotechnology Update: U.S. Leads in Government Spending Amidst Increased Spending Across Asia. Lux Research, Inc.

Mabena, L. F., Ray, S. S., Mhlanga, S. D., & Coville, N. J. (2011). Nitrogen-doped carbon nanotubes as a metal catalyst support. *Applied Nanoscience*, *1*(2), 67–77. doi:10.100713204-011-0013-4

Mansha, M., Qurashi, A., Ullah, N., Bakare, F. O., Khan, I., & Yamani, Z. H. (2016). Synthesis of In2O3/graphene heterostructure and their hydrogen gas sensing properties. *Ceramics International*, *42*(9), 11490–11495. doi:10.1016/j.ceramint.2016.04.035

Mansha, M., Qurashi, A., Ullah, N., Bakare, F. O., Khan, I., & Yamani, Z. H. (2017). Synthesis, characterization and visible-light-driven photoelectrochemical hydrogen evolution reaction of carbazole-containing conjugated polymers. *International Journal of Hydrogen Energy*, *42*(16), 10952–10961. doi:10.1016/j.ijhydene.2017.02.053

Martins, R., Baptista, P., Raniero, L., Doria, G., Silva, L., Franco, R., & Fortunato, E. (2007). Amorphous/nanocrystalline silicon biosensor for the specific identification of unamplified nucleic acid sequences using gold nanoparticle probes. *Applied Physics Letters*, *90*(2), 023903. doi:10.1063/1.2431449

Mashaghi, S., Jadidi, T., Koenderink, G., & Mashaghi, A. (2013). Lipid Nanotechnology. *International Journal of Molecular Sciences*, *14*(2), 4242–4282. doi:10.3390/ijms14024242 PMID:23429269

Mirkin, C. A., Letsinger, R. L., Mucic, R. C., & Storhoff, J. J. (1996). A DNA-based method for rationally assembling nanoparticles into macroscopic materials. *Nature*, *382*(6592), 607–609. doi:10.1038/382607a0 PMID:8757129

Nagrath, S., Sequist, L. V., Maheswaran, S., Bell, D. W., Irimia, D., Ulkus, L., Smith, M. R., Kwak, E. L., Digumarthy, S., Muzikansky, A., Ryan, P., Balis, U. J., Tompkins, R. G., Haber, D. A., & Toner, M. (2007). Isolation of rare circulating tumour cells in cancer patients by microchip technology. *Nature*, *450*(7173), 1235–1239. doi:10.1038/nature06385 PMID:18097410

Ngoy, J. M., Wagner, N., Riboldi, L., & Bolland, O. (2014). A CO2 capture technology using multi-walled carbon nanotubes with polyaspartamide surfactant. *Energy Procedia*, *63*, 2230–2248. doi:10.1016/j.egypro.2014.11.242

Ozcelikkale, A., Ghosh, S., & Han, B. (2013). Multifaceted transport characteristics of nanomedicine: Needs for characterization in dynamic environment. *Molecular Pharmaceutics*, *10*(6), 2111–2126. doi:10.1021/mp3005947 PMID:23517188

Parveen, S., & Sahoo, S. K. (2006). Nanomedicine. *Clinical Pharmacokinetics*, *45*(10), 965–988. doi:10.2165/00003088-200645100-00002 PMID:16984211

Parveen, S., & Sahoo, S. K. (2008). Polymeric nanoparticles forcancer therapy. *Journal of Drug Targeting*, *16*(2), 108–123. doi:10.1080/10611860701794353 PMID:18274932

Pecot, C. V., Bischoff, F. Z., Mayer, J. A., Wong, K. L., Pham, T., Bottsford-Miller, J., Stone, R. L., Lin, Y. G., Jaladurgam, P., Roh, J. W., Goodman, B. W., Merritt, W. M., Pircher, T. J., Mikolajczyk, S. D., Nick, A. M., Celestino, J., Eng, C., Ellis, L. M., Deavers, M. T., & Sood, A. K. (2011). A novel platform for detection of CK+ and CK- CTCs. *Cancer Discovery*, *1*(7), 580–586. doi:10.1158/2159-8290.CD-11-0215 PMID:22180853

Prasad Rao, J., & Geckeler, K. E. (2011). Polymer nanoparticles: Preparation techniques and size-control parameters. *Progress in Polymer Science*, *36*(7), 887–913. doi:10.1016/j.progpolymsci.2011.01.001

Primeau, A. J., Rendon, A., Hedley, D., Lilge, L., & Tannock, I. F. (2005). The distribution of the anti-cancer drug Doxorubicin in relation to blood vessels in solid tumors. *Clinical Cancer Research*, *11*(24), 8782–8788. doi:10.1158/1078-0432.CCR-05-1664 PMID:16361566

Raavé, R., van Kuppevelt, T. H., & Daamen, W. F. (2018). Chemotherapeutic drug delivery by tumoral extracellular matrix targeting. *Journal of Controlled Release*, *274*, 1–8. doi:10.1016/j.jconrel.2018.01.029 PMID:29382546

Ramacharyulu, P. V., Muhammad, R., Kumar, J. P., & Prasad, G. K. (2015). Iron phthalocyanine modified mesoporous titania nanoparticles for photocatalytic activity and CO 2 capture applications. *Physical Chemistry Chemical Physics*, *17*(39), 26456–26462. doi:10.1039/C5CP03576G PMID:26393761

Rawal, I., & Kaur, A. (2013). Synthesis of mesoporous polypyrrole nanowires/nanoparticles for ammonia gas sensing application. *Sensors and Actuators. A, Physical*, *203*, 92–102. doi:10.1016/j.sna.2013.08.023

Richard, P. F. (1959). There's plenty of room at the bottom: An invitation to enter a new field of physics. *Annual Meeting of the American Physical Society*.

Saeed, K., & Khan, I. (2016). Preparation and characterization of single-walled carbon nanotube/nylon 6, 6 nanocomposites. *Instrumentation Science & Technology*, *44*(4), 435–444. doi:10.1080/10739149.2015.1127256

Schiro, P. G., Zhao, M., Kuo, J. S., Koehler, K. M., Sabath, D. E., & Chiu, D. T. (2012). Sensitive and High-Throughput Isolation of Rare Cells from Peripheral Blood with Ensemble-Decision Aliquot Ranking. *Angewandte Chemie International Edition*, *51*(19), 4618–4622. doi:10.1002/anie.201108695 PMID:22359315

Shaalan, M., Saleh, M., El-Mahdy, M., & El-Matbouli, M. (2016). Recent progress in applications of nanoparticles in fish medicine: A review. *Nanomedicine; Nanotechnology, Biology, and Medicine*, *12*(3), 701–710. doi:10.1016/j.nano.2015.11.005 PMID:26656532

Shin, W. K., Cho, J., Kannan, A. G., Lee, Y. S., & Kim, D. W. (2016). Cross-linked composite gel polymer electrolyte using mesoporous methacrylate-functionalized SiO2 nanoparticles for lithium-ion polymer batteries. *Scientific Reports*, *18*(6), 26332. doi:10.1038rep26332 PMID:27189842

Stewart, M. E., Anderton, C. R., Thompson, L. B., Maria, J., Gray, S. K., Rogers, J. A., & Nuzzo, R. G. (2008). Nanostructured Plasmonic Sensors. *Chemical Reviews*, *108*(2), 494–521. doi:10.1021/cr068126n PMID:18229956

Sun, S., & Murray, C. B. (1999). Synthesis of monodisperse cobalt nanocrystals and their assembly into magnetic superlattices. *Journal of Applied Physics*, *85*(8), 4325–4330. doi:10.1063/1.370357

Svobodova, Z., Kucerova, J., Autebert, J., Horak, D., Bruckova, L., Viovy, J.-L., & Bilkova, Z. (2014). Application of an improved magnetic immunosorbent in an Ephesia chip designed for circulating tumor cell capture. *Electrophoresis*, *35*(2-3), 323–329. doi:10.1002/elps.201300196 PMID:23868447

Teo, B. K., & Sun, X. H. (2007). Silicon-based low-dimensional nanomaterials and nanodevices. *Chemical Reviews*, *107*(5), 1454–1532. doi:10.1021/cr030187n PMID:17488056

Thomas, M., & Klibanov, A. M. (2003). Conjugation to gold nanoparticles enhances polyethylenimine's transfer of plasmid DNA into mammalian cells. *Proceedings of the National Academy of Sciences of the United States of America*, *100*(16), 9138–9143. doi:10.1073/pnas.1233634100 PMID:12886020

Thomas, S., Kumar Mishra, P., & Talegaonkar, S. (2015). Ceramic nanoparticles: Fabrication methods and applications in drug delivery. *Current Pharmaceutical Design*, *21*(42), 6165–6188. doi:10.2174/1381612821666151027153246 PMID:26503144

Tiwari, J. N., Tiwari, R. N., & Kim, K. S. (2012). Progress in materials science three-dimensional nanostructured materials for advanced electrochemical energy devices. *Progress in Materials Science*, *57*(4), 724–803. doi:10.1016/j.pmatsci.2011.08.003

Waite, C.L., & Roth, C.M. (2012). Nanoscale drug delivery systems for enhanced drug penetration into solid tumors: current progress and opportunities. *Critical Reviews™ in Biomedical Engineering, 40*(1), 21-41.

Wallwiener, M., Riethdorf, S., Hartkopf, A. D., Modugno, C., Nees, J., Madhavan, D., Sprick, M. R., Schott, S., Domschke, C., Baccelli, I., Schönfisch, B., Burwinkel, B., Marmé, F., Heil, J., Sohn, C., Pantel, K., Trumpp, A., & Schneeweiss, A. (2014). Serial enumeration of circulating tumor cells predicts treatment response and prognosis in metastatic breast cancer: A prospective study in 393 patients. *BMC Cancer*, *14*(1), 512. doi:10.1186/1471-2407-14-512 PMID:25015676

Wang, S., Wang, H., Jiao, J., Chen, K. J., Owens, G. E., Kamei, K., Sun, J., Sherman, D. J., Behrenbruch, C. P., Wu, H., & Tseng, H.-R. (2009). Three-dimensional nanostructured substrates toward efficient capture of circulating tumor cells. *Angewandte Chemie International Edition in English*, *48*(47), 8970–8973. doi:10.1002/anie.200901668 PMID:19847834

Yamauchi, Y., & Kuroda, K. (2008). Rational design of mesoporous metals and related nanomaterials by a soft-template approach. *Chemistry, an Asian Journal*, *3*(4), 664–676. doi:10.1002/asia.200700350 PMID:18327878

Yoon, H. J., Kozminsky, M., & Nagrath, S. (2014). Emerging role of nanomaterials in circulating tumor cell isolation and analysis. *ACS Nano*, *8*(3), 1995–2017. doi:10.1021/nn5004277 PMID:24601556

Yu, M., Bardia, A., Aceto, N., Bersani, F., Madden, M. W., Donaldson, M. C., Desai, R., Zhu, H., Comaills, V., Zheng, Z., Wittner, B. S., Stojanov, P., Brachtel, E., Sgroi, D., Kapur, R., Shioda, T., Ting, D. T., Ramaswamy, S., Getz, G., ... Haber, D. A. (2014). Cancer therapy. Ex vivo culture of circulating breast tumor cells for individualized testing of drug susceptibility. *Science*, *345*(6193), 216–220. doi:10.1126cience.1253533 PMID:25013076

Yu, M., Ting, D. T., Stott, S. L., Wittner, B. S., Ozsolak, F., Paul, S., Ciciliano, J. C., Smas, M. E., Winokur, D., Gilman, A. J., Ulman, M. J., Xega, K., Contino, G., Alagesan, B., Brannigan, B. W., Milos, P. M., Ryan, D. P., Sequist, L. V., Bardeesy, N., ... Haber, D. A. (2012). RNA sequencing of pancreatic circulating tumour cells implicates WNT signalling in metastasis. *Nature*, *487*(7408), 510–513. doi:10.1038/nature11217 PMID:22763454

Zhang, X., Young, M. A., Lyandres, O., & Van Duyne, R. P. (2005). Rapid detection of an anthrax biomarker by surface-enhanced Raman spectroscopy. *Journal of the American Chemical Society*, *127*(12), 4484–4489. doi:10.1021/ja043623b PMID:15783231

Chapter 13
Emerging Nano-Based Drug Delivery Approach for Cancer Therapeutics

Muthupandian Saravanan
https://orcid.org/0000-0002-1480-3555
Mekelle University, Ethiopia & Saveetha Dental College, Saveetha Institute of Medical and Technical Sciences (SIMATS), India

S. Poornima
K. S. Rangasamy College of Technology, India

V. Karthik
K. S. Rangasamy College of Technology, India

A. Vigneshwaran
https://orcid.org/0000-0003-1898-3044
Vellore Institute of Technology, India

S. Manikandan
Saveetha School of Engineering, India

Subbaiya Ramasamy
The Copperbelt University, Zambia

R. Balachandar
Aarupadai Veedu Institute of Technology, India

P. Prakash
Thiagarajar College, India

Karthikeyan Mahendhran
Thiagarajar College, India

Murugappan Ramanathan
Thiagarajar College, India

ABSTRACT

Cancer remains the important cause for the mortality worldwide. In recent days, different biomedical strategies are in the limelight for the control of mortality due to cancer. As a promising strategy in cancer treatment, nano-formulations with advantages in drug delivery, health, and pharmacy are booming. The conventional nanocarrier continues, however, to suffer from low drug load efficiency, possible toxicity, unknown metabolism, and other uncertainties. To overcome these problems, carrier-free nanodrugs with desirable bioactivity were developed quickly and attracted considerable attention. In the meantime, the nanoarchitecture developed by a simple "natural" method with a multifunctional self-life has major advantages in multi-drug resistant synergistic cancer therapies and inhibition. Until now, the carrier-free nanoparticles for tumour therapy, phototherapy, chemotherapy, diagnostics, and synergistic therapy have made significant progress. In this review, the authors provide an integrated and detailed review of recent literature on nanodrug delivery systems composed of several active agents.

DOI: 10.4018/978-1-7998-5049-6.ch013

INTRODUCTION

As several researchers have identified that cancer is the leading cause of mortality worldwide and finding a cure remains the big task for many researchers (Akshatha Prasanna *et al* 2018). Researchers identified that cancer is a major public health problem and becomes the second foremost cause of death worldwide (Kun Yu *et al* 2020). Recent views of the young and growing scientists are having the enough knowledge and moreover they are having the interest on Targeted cancer therapy. Because it can differentiate the small differences between normal and cancer cells. Targeted therapies are usually more effective than other predictable treatments and display lesser unwanted contrary effects. Since the non-specific and systemic drug delivery leads to rapid removal of drug, administration of the highest tolerable dose of the drug is needed which is not reasonable and usually exhibits high toxicity (Behdokht Bahrami *et al* 2017). Although different types of drug delivery methods are present the Doxorubicin (DOX) is also one of the most frequently used anticancer drugs in chemotherapy and drug delivery systems (Chang-Seuk Lee *et al* 2017). Advances in tumor biology can be successful once the drug delivery and therapy can be identified because as we know that cancer continues to be a prevalent and lethal disease (Dandan Wu *et al* 2017). One effective plan is engaged by nanoparticulate drug delivery systems to delivery therapeutic agents into cancer cells (Kai Dong *et al* 2018). The important mode of the carcinogenesis arises is due to the critical roles played by the chromosomal rearrangements and gene mutations (Mokwena Mpho Gift *et al* 2018).

Recent Trends in the Formulation and Delivery of Nanocarriers to Cancer Cells

Rapid development of nanotechnology plays an important role in combating cancer with multiple ways including diagnosis, prevention and cure. Further, limitations in the conventional treatment options have driven the need for alternative options in the treatment of cancer patients globally. Especially nanocarriers are receiving considerable attention in recent days. This smart drug delivery unit are highly specific in the controlled release of drugs to the target sites at low concentration and reduces the side effects. Different nanocarrier systems such as metal nanoparticles, nanobiopolymers, liposomes, oligomers, micelles are investigated for their possible application in the smart delivery of anticancer drugs to the cancer targets. Further, influence of different characterstics features affecting efficacy of nanocarriers was the matter of concern in recent times by the scientific community to overcome the adverse reaction and to ensure the precise delivery of compounds to the malignant site (Giodini et al., 2017). However beyond their remarkable targeted drug delivery properties, factors such as biocompaitability and toxicity of nanocarrier vechicles made them to be still in infancy in commercial applications. In order to overcome these limitations cell membrane coated nanocarriers has been proposed as an viable alternative for targeted drug delivery system for cancer theranostics (Bose et al., 2018). These cell membrane coated nanocarriers specially targets the homologous cancer cells by recognizing cancer cell adhesion molecules and prevents cancer progression. Faraj et al. (2016) fabricated a conjugated paclitaxel and salinomycin drug encapsulated SWCNTs nanocarriers for the treatment of cancer stem cells. The findings the study offers a potential approach for the improved diagnosis and treatment of breast cancer and stem cell populations. Vinothinin et al. (2019) investigated the anticancer potential of folic acid conjugated grapheme oxide nanocarrier for the targeted delivery of paclitaxel in breast cancer cells. The paclitaxel loaded folic acid conjugated grapheme oxide nanocarrier promotes apopotopsis by increasing the activity of caspase 8 and caspase 3 activities. Similar grapheme based nanocarrier was developed (Gupta et al., 2020) for the site specific delivery of gallic acid against human lung cancer cells using 2D reduced grapheme oxide (rGO) nanocarriers. These biotinylated rGO system showed enhanced cellular internalization efficiency, high physicochemical stability for 45 days and sustainable drug release at low pH. Simultaneous targeted drug delivery and real time imaging of cancer cells is an important step in treatment and diagnosis of cancer. Liu et al. (2018) developed a dual-targeting DNA tetrahedron nanocarrier for breast cancer cell imaging and drug delivery. Findings of the study proposed a novel strategic design for the dual targeted drug loaded nanocarriers for enhancing the real time effect of chemotherapy. Nanotechnology based bitherapy has evolved as a significant strategy in combating the drug resistance of cancer therapy. Experimental studies on nano based biotherapy have proven in surmounting the delayed multidrug drug resistance and patient compliance. Leithy et al. (2019) highlighted the impact of tamoxifen citrate/ Coenzyme Q10 based smart nanaocarriers biotherapy in the treatment of breast cancer.

Nanocarrier Design and Synthesis

Nanocarriers can be used for particluar receptor-mediated activities, to magnify the anticancer effect of its loaded drug at a cellular level. The advent of smart nanoparticles (NPs) as a drug-delivery system, which incorporates passive, active and stimuli-responsive targeting, has enabled controlled drug release in response to particular physiological triggers at the precise target site at the suitable time (Hyun *et al.*, 2019).

Figure 1. Nano Drug Delivery System

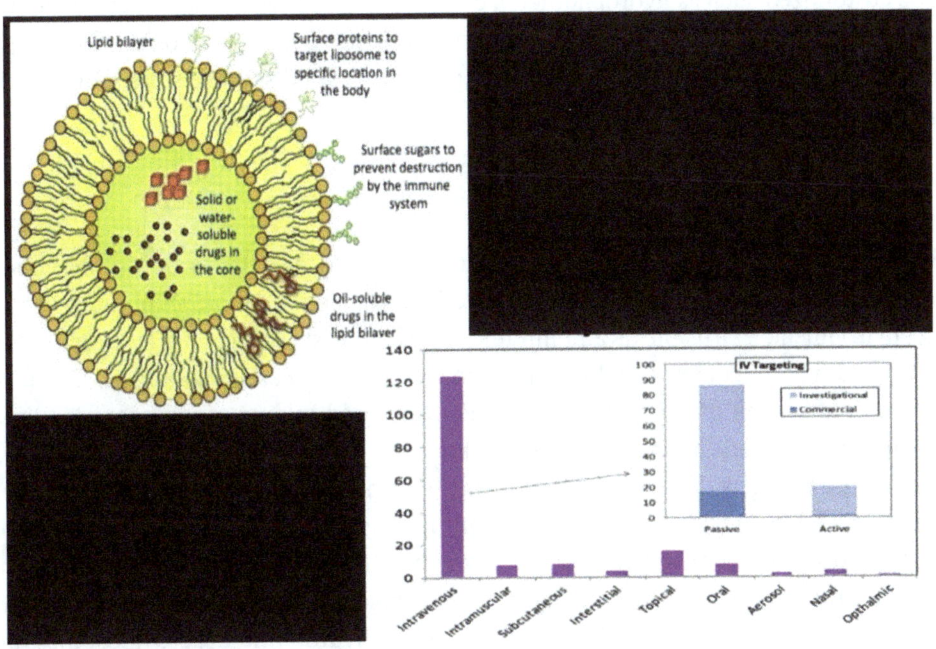

Protein-Drug Conjugated Nanocarriers

A protein-drug conjugate is typically a protein conjugated to the drug by using peptides or covalent binding. The drug gets released from the nanocarrier once it reaches the target cell, as the linker gets degraded once it arrives into the cell. But in case of protein-drug conjugates there is a possibility of premature release of drug, as the linker is easily degraded by the enzymes such as protease and other redox- altering agents (Sun *et al.,* 2014). The net charge of the whole protein-drug conjugate can play a major role in targeting also. The charge of the carrier can be modified using ligands. A positive charge molecule can effectively affect the cancer while a neutral charged molecule can easily diffuse in the cancer cell (Stylianopoulos *et al.,* 2010). Recently antibody protein conjugated to a nanocarrier has also been used to improve targeted treatment.

Polymeric Nanocarriers

Polymeric nanocarriers are synthetic nanocarriers, where a polymer is used to create a carrier and the desired drug (protein/ chemical molecule) is covalently bound to the exterior covalently or entrapped in a matrix within the polymer. Based on the origin of the polymer, it can be classified as synthetic and natural polymeric nanocarriers. Albumin and chitosan is some of the few naturally occurring nanocarriers. PEG (polyethylene glycol), and PGA (Poly-glutamic Acid) are examples of synthetic polymers (Bhatt *et al.,* 2003). These polymers can be used in gene therapy to carry the desired nucleic acid to the cells and as such. But, in case of treatment of targeted cells using a specific drug, synthetic polymer has advantages over natural polymers. In case of synthetic polymers, the desired characteristics of the carrier can be customized such as hydrophobicity, charge of the carrier, biodegradability and molecular weight of the carrier, while in natural polymeric carrier it is not possible. Though size, shape, success rate of synthesis/ fabrication and yield of nanocarriers at the desired size are some of the limitations of synthetic nanocarriers (Sun *et al.,* 2014).

Liposomal Nanocarriers

Liposomal nanocarriers, are lipid bilayer nanocarriers, created by distribution of amphiphilic molecule in a hydrophilic environment. Liposomes are created by sonication of amphiphilic molecule in any hydrophilic environment, extrusion of amphiphilic molecule distributed in a hydrophilic system via membranes to obtain homogenous sized liposomes or by reverse phase evaporation. The formed liposome mimics a lipid bilayer of cell. Upon interaction, the lipid bilayer of nanocarrier fuses with the lipid bilayer of the cell, resulting in release of the drug into the targeted cell. In some cases, the lipid bilayer of the nanocarrier gets degraded in the cell due to its harsh environment. Advantage of liposome is they are biodegradable, and successful delivery of the drug to cells. Thermo-sensitive liposome can be used to release drugs in the target cells, by directing high intensity microwaves or ultrasound, at the site of target cell (Yatvin *et al.,*1978). But, major disadvantage is that, the liposomes require a guidance or external force to drive them to the target cell. In case of hydrophilic drugs, liposomes are the best nanocarriers. But to carry amphiphilic or hydrophobic drugs, the drug must coated encapsulated in a polymeric substance for the liposome to carry it, and in such cases, the circulation time and half-life of the drug is highly increased (Sun *et al.,* 2014).

Hydrogels

Hydrogels are 3D network for cross linked water soluble polymers, that have the capability to retain fluid in large quantities. Hydrogels are positively charged in nature, thus can interact diffuse the drug payload freely in negatively charged cells as cancer cells. But hydrogels may aggregate to other negatively charged components such as serum proteins resulting less circulation time. By controlling the cross-linking matrix, one can increase the porosity of the hydrogel, which control drug loading volume (Hoare *et al.,*2008).

Metallic Nanocarriers

Inorganic metal nanoparticles, such as gold can be used as carriers of drugs or contrasting agent, which might act as a carrier in case of cancer theranostics. Gold nanoparticles, have high penetrating and diffusion properties. They have optical and thermal properties too. The drug can be conjugated onto the metallic nanoparticle and be guided to the targeted cells, via thermal or ultrasonic waves. In such a case the carriers are called as nanomotors. Similarly, ferric nanoparticles can also be used in drug delivery to cancer cells. The ferric nanoparticles can be guided to the target cell via magnetization and magnetic waves (Sun *et al.,*2014). Silver nanoparticles can also be used as nanocarrier for cancer treatment. Silver nanoparticles react to the acidic environment around a cancer cells, creating a reactive oxygen species, which helps in killing the cancer cells by apoptosis. Silver nanoparticle has also been observed to possess anti-angiogenic effects, thus inhibiting the blood vessel formation in cancer cells (Almajhdi *et al.,* 2014).

Though there are many nanoparticle/ Nano-sized molecule, which can be used as a nanocarrier for cancer treatment, only the above nanocarriers have been discussed because they possess the property to enter the cell, and diffuse into the cell environment. They also possess the ability to target and home on cancer cells based on the characteristics of the cancer cells.

Therapeutic Nanoparticles

Novel drug delivery methods are constantly explored by the pharmaceutical companies in order to deliver the drug at the target site in more efficient manner. The Nanoparticle mediated drug delivery in the field of biomedicine is acquiring an immense attention in the past decade as various methods are blooming out day by day. The main objective in using nanoparticle mediated drug delivery method is that the nano particles showcase better solubility for the drugs and chemical compounds which generally don't dissolve with the existing carriers.

Protein Based Nano Particles

The nanoparticles that are derived from proteins are very flexible that it can undergo easy modifications when they are employed as a carrier molecule. Albumin is a biodegradable, non-toxic protein compound which has various binding sites for several drug molecules. The drug molecule is made to bound to the albumin nanoparticle due to the presence of amino group and a carboxylic group by covalent derivatization. The nanoparticles are obtained by various processes viz., desolvation, thermal gelation and emulsification which has a greater possibility of covalent derivatization with ligand molecule (Elzobhby et al., 2012). The milk protein casein was used to encapsulate the anticancer drug flutamide, which is an effective anticancer drug for androgen dependent prostate cancer with very poor solubility and bioavailability.

Genipin, found in ancient Chinese medicine was used a cross linking agent to bind the nanoparticle with flutamide. The sustained release of the drug compound was achieved by the incorporation of the complex into the micellar core (Elgindy et al., 2014). The colloidal gels made of gelatin nanospheres were compared with the gelatin microspheres for the release of bone morphogenetic protein – 2(BMP-2), in which the results showcased that the nanospheres are way superior than the microspheres in terms of gel elasticity, injectability and other release kinetic parameters (Wang et al., 2016).

Polymer Based Nano Particle

The usage of polymeric substances has been extensively used in the pharmaceutical field. The polymers which are non-toxic to the body such as Poly (lactic-co-glycolic acid) PLGA and chitosan have been approved by the Food and Drug Administration (FDA) and European Medicine Agency (EMA) (Wang et al., 2016) The PLGA based nano particles are prepared by various techniques like single or double solvent evaporation method, Emulsification Solvent Diffusion (ESD) method, Emulsification reverse salting out method and Nanoprecipitation method. Among the various forms PLGA nanoparticles, only microspheres and microparticles are prevalently used. The PLGA nanoparticles provide better solubility and surface modification properties for the effective loading of drug compound. The anti-cancer drugs such as Paclitaxel, Vincristine Sulphate, Etoposide etc., are delivered using PLGA based nanoparticles (Tabatabaei et al., 2014). Using PLGA-PEG nanoparticles, lowering the expression levels of the target gene from the cancer tissue is made possible (Prabaharan, 2015). Similar to the PLGA nanoparticles, the chitosan nanoparticles also provide greater efficiency in drug delivery process. They can exhibit the property of passive drug distribution which improves the effect of therapy. They have various reactive functional groups which can be modified to load the ligand of interest (Dobson, 2008).

Magnetic Nanoparticle Based Drug Delivery

The magnetic nanoparticle-based drug delivery functions in such a way that the drug molecule is linked to the nanoparticles and they are injected into the body or they are made to reach near the target site with the help of catheter. They can also be assisted with the help of external magnetic field using rare earth magnets. The drugs which have been delivered using magnetic nanoparticles are epirubicine and doxorubicin. They are made to form a complex on the basis of electrostatic interactions due to the presence of phosphate group on the carrier surface and the amino group in the drug molecule. The entire process can be monitored through Magnetic Resonance Imaging (MRI). Although it is not effective in the drug delivery process, as it can interact with very less number of compounds, with continuous research and development, it can be made into an effective method (Ruiz et al., 2011). The iron-oxide nanoparticles loaded with fluorescein was studied for their effective drug delivery capacity *in vitro*, where it showcased the potential 'on-off' mechanism along with the capacity of varying the temperature around the cancer site which could serve in the thermo chemotherapy process.

Metallic and Silica Nanoparticles

The Gold nanoparticles can be effectively used for drug delivery applications as it is easy to synthesize gold nanoparticles in various size ranging from 1-100nm and in different shapes to conjugate the drug molecule. The main disadvantage in the application of gold nanoparticles is that they are not easily biodegradable and hence surface modification is necessary to make them a better drug carrier (Klebowski et al., 2018). The surface modification of the gold nanoparticles can be achieved by the usage of Polyethylene glycol (PEG) which acts a spacer (Kong et al., 2017). The Phyto fabricated silver nanoparticles were also used for treating ailments related to hepatocellular sites (Singh et al., 2018). The Palladium nano particle cages were designed for encapsulating the anticancer drug cisplatin which was proven to cytotoxic against the human cancer cell lines (Schmidt et al., 2014). The application of metal-based nanoparticle hybrids has greatly decreased the cancer drug resistance (Farooq et al., 2018).

The mesoporous silica nanoparticles are very flexible drug carriers in which the pore size is tunable by changing the parameters. Several antibiotics have been successfully encapsulated and delivered using silica nanoparticles (Kwon et al., 2013)

THE MECHANISMS OF NANOPARTICLE INTERNALIZATION INTO CANCER CELLS

Mechanism of Internalization of CPP

The mechanism of entry of CPP into the cell is still obscure till date, but there are three possible mechanisms by which a peptide can enter into a cell. They are,

Direct penetration,
Endocytosis pathway, and
Translocation through the formation of a transitory structure.

Direct Penetration

Direct penetration is an energy independent process. There are multiple proposed models for CPP entry into cells by direct penetration. But the first step in all models is that the positively charged CPP interacts with the negatively charged cell membrane, destabilizing the cell membrane by electrostatic interaction (Herce *et al.,* 2007). One of the proposed and highly agreed model direct penetration is transient pore formation. Here after the electrostatic interaction, the residual arginine interacts with the destabilized cell membrane nucleating a pore which helps the diffusion of the CPP into the cell (HU *et al.,* 2015).

Endocytosis Pathway

Endocytosis is an energy consuming pathway. Endocytosis is a method in which the cell membrane folds inside to bring the external particles into the cell and release it. Phagocytosis, pinocytosis and receptor-mediated endocytosis are the types of endocytosis. In phagocytosis the cell membrane takes larger size molecules, while in pinocytosis, only fluids and solutes are taken in (Ter-avetisyan *et al.,* 2009). In case of receptor-mediated endocytosis, a receptor helps in gaining entry into the cell for the CPP. The CPP will bind to its appropriate receptor on the cell membrane, and the receptor transports the CPP to a clathrin coated pit (Mc Mohan and Boucrot, 2011). Then the pit engulfs the CPP, forms a vesicle and enters the cell. Once it enters the cell and comes in contact with the endosome, the vesicle opens up and lets the CPP in endosome and the clathrin again resurfaces back on the cell membrane for the next transport (Richard *et al.,* 2005). Another example of receptor mediated endocytosis is caveolin mediated CPP internalization. The only difference is Caveolin coated vesicle has Smaller diameter than clathrin.

Figure 2. Direct penetration by CPP into the cell by transient pore formation

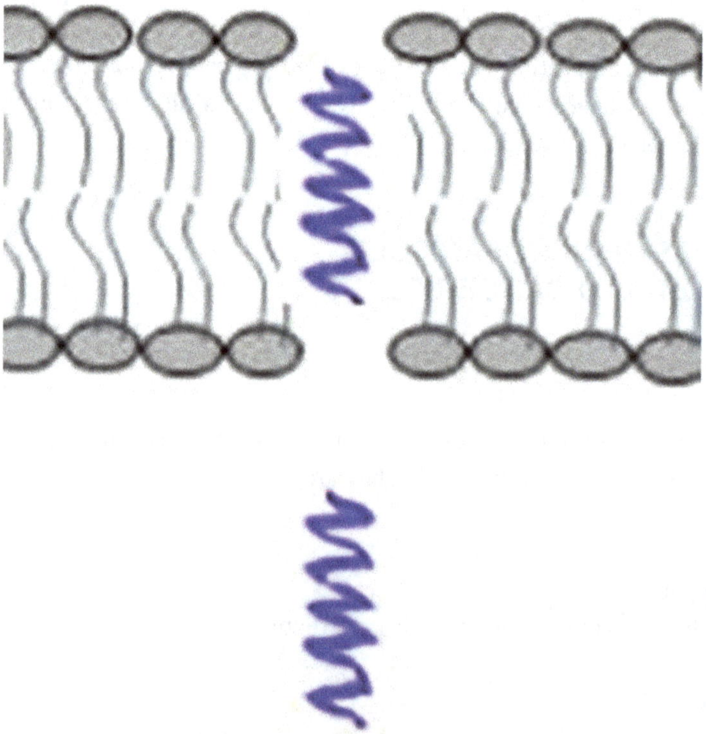

Translocation Through the Formation of a Transitory Structure

In this method the CPP's electrostatic interaction along with the interaction hydrophobic tryptophan, disrupts the lipid bilayer structure and forms a inverted micelle layer to transport the CPP inside the cell. This method is only compatible for small hydrophilic peptides, and is not suitable for highly cationic CPPs.

Conjugation of CPPs with Drug

The CPP that is to be used for drug transport must be nontoxic to the biological activity. But in some cases conjugating drug with vectors can lead to the loss of biological activity of the drug. This can be overcome by using linkers between the CPP and the cargo drug. Linkers connect the drug cargo and CPP and gets degraded or dissociated once it reaches the target site. The selected linker must be nontoxic, and have cleavable property. In cases, where the drug cargo is chemical compound disulfide bonds acts as linkers. But in case of protein as a drug cargo, linking becomes a problem as chemistry of proteins is highly complex. Any linking in a wrong manner can result in deactivation of the protein. Thus, in such cases the target protein is generated in tandem with the CPP to act as a transport into cell with the help of recombinant protein synthesis (Jamal *et al.,* 2004). The CPPs must be also proetected form protease to successfully transport the cargo drug into the cell. For this the L-amino acids are replaced with D-amino acid to escape degradation (Pujals *et al.,* 2008).

Figure 3. Receptor dependent endocytosis and receptor independent endocytosis

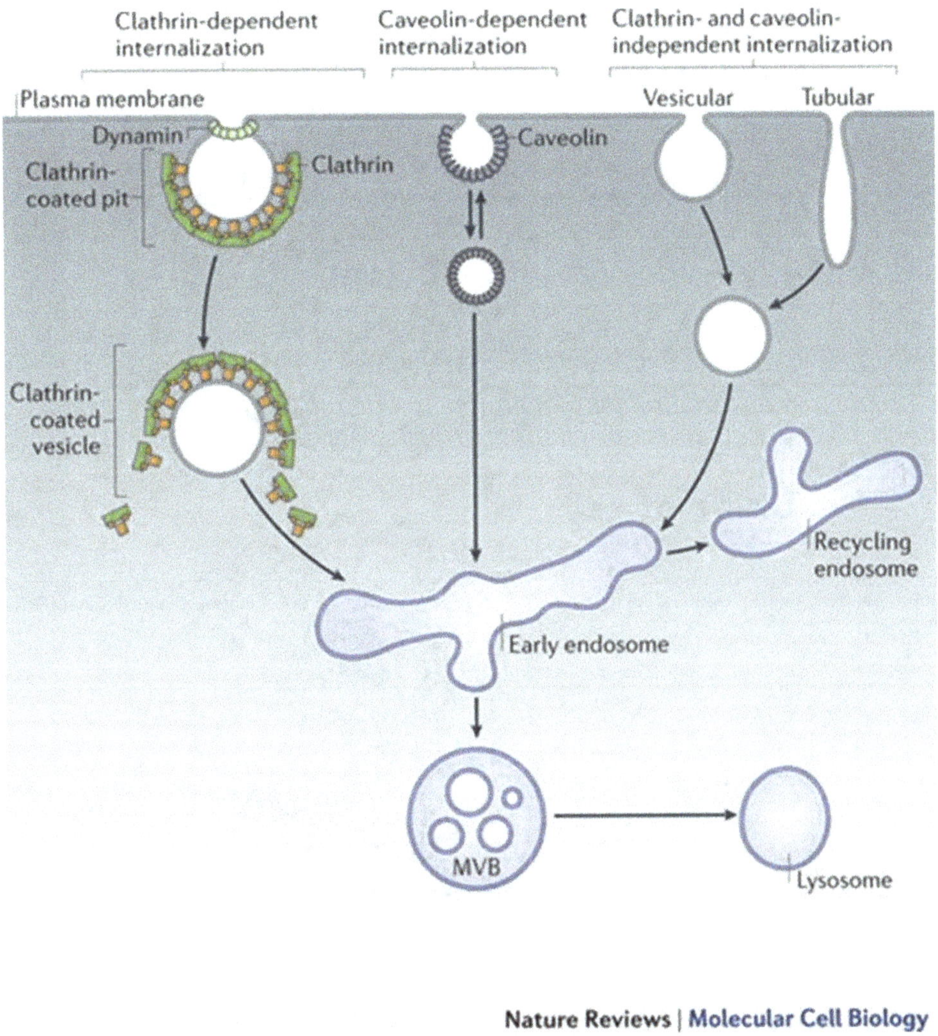

CHARACTERISTICS OF NANOCARRIERS

Physio-Chemical Properties of the Nanocarriers

The nanocarrier that is used for cancer treatment must have a specific size, shape, and surface characteristics to treat specific cancer cells. The size of nanoparticle plays a major role, as smaller size nanocarriers can easily distribute in the cell, but there is a risk of extravasation of the nanocarriers to the normal cells, other than the target cells. In case of larger size nanocarriers, there is no risk of extravasation, but the distribution via blood can be affected due to the size. The shape of the nanocarriers too, play a role in distribution in the blood, as the shape can have an impact on fluid dynamics. The surface of the nanocarriers can also be modified to gain entry into cells and for targeted cell homing (Stylianopoulos et al., 2010).

Figure 4. Translocation of CPP through the formation of transitory structure

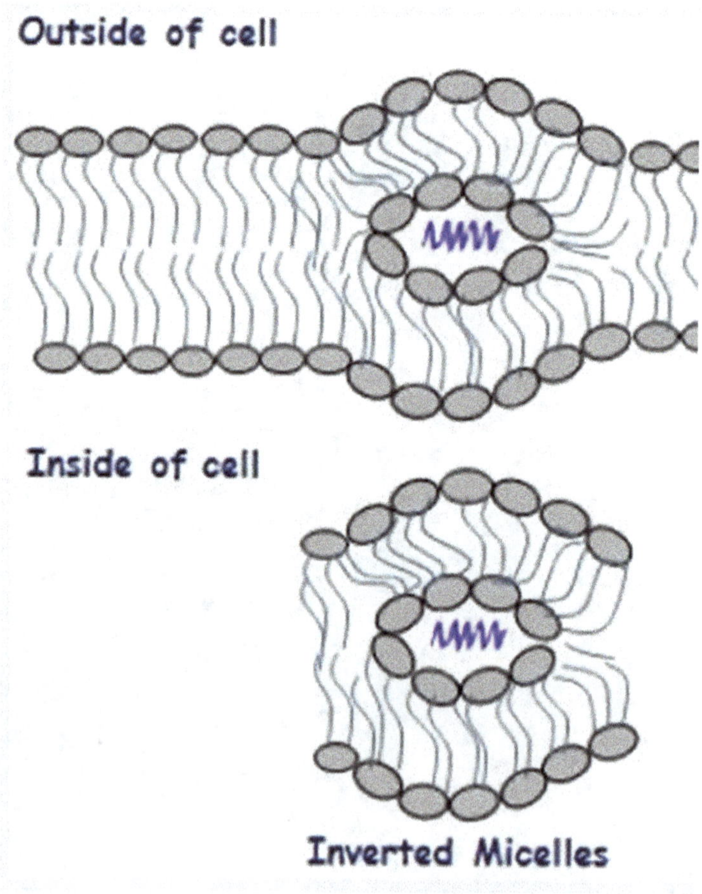

Solubility, Degradation and Clearance from the Biological System

The drug to be delivered must have good solubility, for it to show effect once the nanocarrier reaches the target site. Hydrophilic nanocarriers can be used to increase the solubility of the drug and increase its effect. Coating of the nanocarriers with water soluble, non-ionic components can also help in solubility (Wicki et al., 2015). Generally, in human body, hydrophobic molecules are considered as foreign particle and are engulfed by mononuclear phagocytic system. Thus opsonization of hydrophobic molecule can decrease the chance for the drug to reach the cancer cells. By using hydrophilic nanocarriers their hydrophobicity can be masked and be taken to the target cells. For example, PEG (Poly-Ethylene Glycol) coated nanocarriers can mask the hydrophobicity of the payload drug and help in transport and release of the drug successfully. Coating with a hydrophilic material not only helps in masking, the hydrophobicity of the molecule, but also increases the solubility, clearance and half life time of the drug in the system (Locatelli et al., 2012).

Homing of Targeted Cells by Nanocarriers

The nanaocarrier without homing property, will interact with all the cells instead of interacting with the target cells alone. A passive or active targeting is required to guide the nanocarriers to the target cells. In active targeting, a ligand can be conjugated onto the surface of the nanocarriers. Ligands having high specificity to the receptors, and characteristics cancer cells. Thus it does not invade normal cells (Cho et al.,2008). In a case of passive targeting, enhanced permeability and retention property can facilitate the accumulation of nanoparticles in the blood vessel of the cancer cells, without any requirement of conjugates for targeting. But in passive targeting the nanocarriers can also enter the normal functioning cells, liver and spleen (Wicki et al., 2015).

Drug Delivery Systems

Drug delivery systems can be deals with the different aspects as the Local Drug Delivery systems for postoperative cancer treatment provide separate merits it includes the enhanced dosage of targeted site and also the reduced side effects towards the organs and tissues (Lin *et al.,* 2019).

Cell-Penetrating Peptides

Cell penetrating peptides are a sequence of amino acids which are about 5-30 amino acids in length, that can penetrate into the cell by crossing the biological membranes. These positively charged peptides have high chance of cell penetration with low or no toxicity. Thus these Cell-penetrating peptides (CPP) or also known as Protein transduction domains are gaining popularity in research as a drug carrier to deliver the drug to the required site for treatment (Hossein *et al.,* 2018).

In 1991, Joliot and his team made a breakthrough discovery that the Antennapedia homeodomain protein, a drosophila transcription factor was able to enter nerve cells and control the neural morphogenesis genes (Joliot *et al.,* 1991). After this discovery, Derossi and colleagues discovered the first CPP by identifying that the third helix of the Antennapedia homeodomain protein, 'Penetratin', was the minimal peptide sequence necessary for cell entry across the barrier (Derossi *et al.,* 1994).

Classification of CPPs

After the identification of penetratin, the research on CPP increased and currently there are about 1700 CPP's tat have been identified. Based on the origin, CPP's can be classified as Protein derived, chimeric and synthetic. CPP's that are derived from naturally available proteins come under Protein-derived CPP. Example of a protein derived CPP is penetratin, which is derived from Antennapedia homeodomain protein. Chimeric CPP's are those derived by combining two different peptides. Example of a chimeric CPP is Transportan, which is derived from galanin and mastoparan. Polyarginine is an example of a synthetic peptide (Bechara *et al., 2013).*

CPP's can also be classified based on their pysiochemical property as Cationic, amphipathic and hydrophobic. Cationic CPP's are positively charged as they contain, polyarginine in their sequence. TAT, transcriptional activator protein in HIV-1, is an example of cationic CPP, which involves arginine and lysine residues (Kim *et al.,* 2018). CPP with lysine residues which create a negative charge while arginine creates positive charge, have high degree of amphicity, hence classified as amphipathic CPP (Huang *et al.,* 2018). Transportan is an amphipathic CPP. CPPs which have non-polar motifs only are called as hydrophobic CPPs.

CPP exhibited the ability to penetrate into the cell, but it does not specificity for penetration and penetrates into all cells. We need CPP's that can only penetrate cancer cells. Thus biomapping of phage carrying specific peptide is done to find which peptide can home onto cancer cell, so that specific peptide can be used as a CPP to deliver drug to cancer cells. In this methods, a target cancer cell type is exposed to phages carrying peptides of different length. Based on the phages that can attach to or internalize into the target cancer cell, the specific type of CPP to home the target cell can be identified. The strength of this method is that cell-specific peptides can be isolated without the prerequisite of knowing a surface biomarker (Shukla *et al., 2005).*

Using the same technique, Zhou *et al.* (2017), discovered the novel CPP, MT23 with mouse melanoma cell specificity. MT23 can only enter B16 melanoma cancer cells and MT23-apoptin can significantly inhibit tumor growth and induce the cell apoptosis in B16 tumor-bearing mice.

The preference of some peptides for cancer cells has been discovered more randomly. For instance, BR2, which was initially designed as a shorter derivative of the anti-tumoral peptide buforin IIb, showed unexpected tumor-homing ability through interaction with ganglioside, via lipid-mediated micropino-cytosis (Lim *et al.,* 2013).

Hypoxia-inducible factors (HIFs) are the main factors for cell response to hypoxia, by which they promote cancer cell survival and progression. Karagiota and co-workers designed TAT-EDT peptides to target cancer cells under hypoxia in hepatocarcinoma models. Their data showed that the CPPs showed cytotoxic effect towards those cells only and not on normal cells (Karagiota *et al.,* 2019).

Thus CPPs can be used to designing cancer homing drug delivery system, and passive drug delivery system. The only drawback is the CPP should be shield from degradation due to external factors, which can be achieved my chemical configuration.

LDL-LDL RECEPTOR

LDL in Cancer Theranostics

Lipoproteins are lipid-protein complexes, which exists either as a soluble complex or insoluble complex in the human body. Out of all the lipoproteins, those found in blood plasma are intensively studied for cancer treatment. Each lipoprotein in plasma has an apolipoprotein that helps in transporting fat molecules such as cholesterol, triglycerides, phospholipids and fat soluble vitamins to liver, intestine and peripheral tissues through the circulating blood plasma

Low density lipoproteins (LDL) are a natural nanoparticles of the size ranging between 20-25 nm. The LDL has a apolipoprotein B-100, that helps in transporting cholesterol across extrahepatic tissues. This B-100 is specifically recognized by LDL receptor on cell membrane, which gives entry into the cell. This receptor mediated entry is facilitated by endocytosis. LDL is preferred over other synthetic drug delivery system, as it is biodegradable and biocompatible. Once inside the cell, it is degraded to its biological units such as amino acids, triglycerides and cholesterol.

Thus LDL being a natural nanoparticle, it can be reconstituted to carry the required drug into the cell. There are three methods used to reconstitute the LDL with the treatment drug required to treat cancer, Conjugation of drug to the apolipoprotein11, Insertion of drug into phospholipid bilayer and Loading into the hydrophobic core.

Figure 5. LDL composition

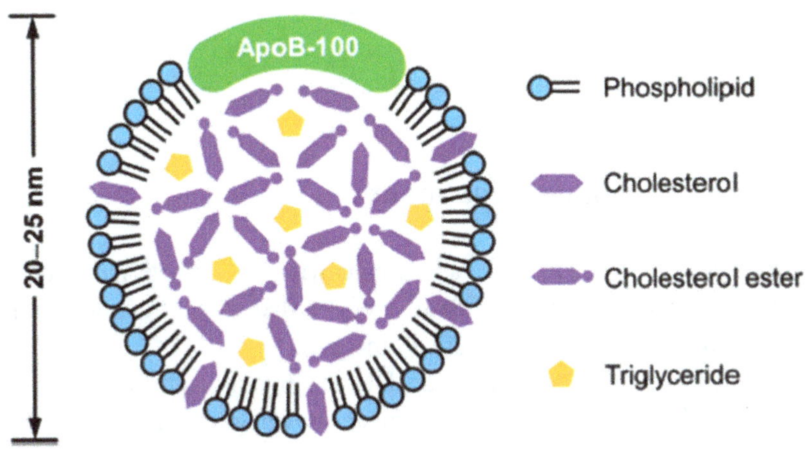

Conjugation of Drug into Apolipoprotein

In **conjugation of drug to apolipoprotein**, the drug is coupled covalently to the apolipoprotein through specific amino acids such as lysine, arginine, tyrosine and cysteine. Using this method, the cancer cells have been located using contrast imaging. But this method is not preferred as, any modification done to the crucial amino acid in the apolipoprotein can result in the inactivation of the B-100 in LDL (Glickson *et al.*, 2009). Thus phospholipid can be used as a site to non-covalently insert the drug rather than apolipoprotein. **Insertion of the drug in the phospholipid bilayer**, is far much easier as all that the drug need is an amphiphilic structure with a hydrophobic tail to insert itself into the phospholipid layer and a hydrophilic head which can interact with the aqueous external environment. But the disadvantage in this method is, that the interaction between drug and the LDL is by Vander Waals forces and because this weak force the drug can be dissociated from the LDL during transfer. The other method to reconstitute LDL to carry the drug is by **loading the drug into hydrophobic core**. The lipid core is extracted using non-polar solvents and the core is loaded with drug lipid conjugate. A hydrophobic core can carry ~1200 cholesterol esters. Thus this method is better suited as it can carry huge drug payload without dissociating from the LDL.

Figure 6. LDL drug conjugation (A) Drug covalently bound to apolipoprotein (B) Drug insertion in phospholipid layer (C) Drug loading into hydrophobic core (Chunlei and Younan, 2017).

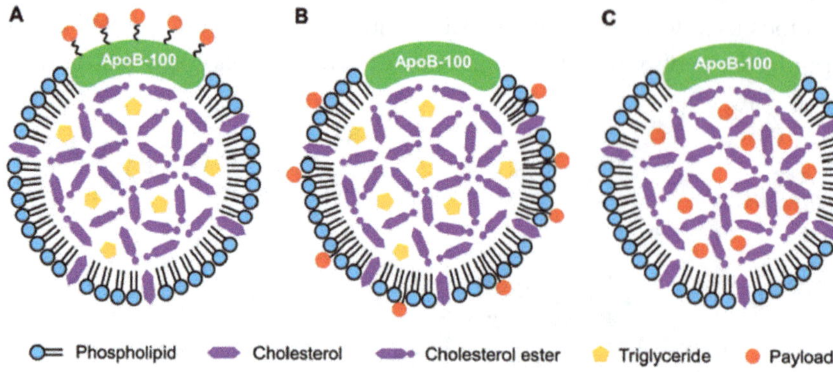

LDL as Drug Carrier for Cancer Treatment.

Chu and co-worker tested a DOX loaded LDL as a drug carrier, on a cancer induced mice. DOX loaded in LDL showed improved and better anti-proliferative effect, than direct treatment with DOX. LDL can also be loaded with contrasting agent for cancer cell imaging.

Versluis and team used an Iodine-125 labelled LDL, for imaging and found that LDL accumulated more in tumour tissues Of B-16 melanoma bearing mice. The imaging study proves that LDL can reach cancer cells (Versluis et al., 1996).

Hill and co-workers used 2-oleoylglycerol 1,3-bis(iodopanoate), a poly-iodinated triglyceride, to reconstruct a LDL and incubated it with HepG2 cell lines and found that, the CT imaging has been improved because of using LDL. In recent years, exogenous X-ray contrast agents have been introduced into LDL to facilitate tumor-targeted imaging by computed tomography (CT) (Hill et al., 2010).

When too many LDL gets oxidized they trigger the activity of macrophages. When the macrophages uptake and clear up the oxidized LDL, the macrophages are transformed into foam cells and triggers atherosclerotic plaques (The build-up of fats, cholesterol and other substances in and on the artery walls). To prevent the aggravation of atherosclerotic plaques, anti-inflammatory drugs can be used. Thus oxidized LDL has high affinity towards the atherosclerosis plaques and this property can be exploited for targeted drug delivery. Tauchi and coworkers demonstrated that oxidized LDL conjugated with a prodrug (dexamethasone palmitate, DP) had inhibitory effect on the formation of foam cells (Yoshihiko *et al.,*2000).

Naturally obtained LDL are very few in numbers and are easily compromised by pathogens. Thus LMPs (LDL- mimetic particles) were prepared from triglycerides, cholesterol, phospholipids and apoliporotein peptides. Without a moiety to gain access into the cell it's very difficult to deliver drug. Also, the fabrication of LMPs is very complex and is complicated without a moiety for entry (Thaxton *et al.*, 2016).

Assessment of Nanodoping Process and their Possible Cancer Motives

Nanomaterials has evolved as potential and inevitable tool for the diagnosis and treatment of cancer. The selective accretion of nanoparticles in tumor tissues and significant physicochemical properties of nanoparticles is the reason behind the success of nanomaterials in cancer theranostics. In last decade, cancer theranostics research has focused in the development of novel devices with combined potential diagnosis and drug delivery of anticancer compounds in cancer cells. Attia et al. (2018) developed a simple and effective nano optical sensor film with binuclear Pt-2-aminobenzimidazole-Bipyridine for the early diagnosis of liver cancer. Rare earth elements doped nano particles has escalated as a novel class of luminescent materials in cancer diagnosis and treatment (Hong et al., 2019). The unique features of rare earth doped upconversion nanoparticles such as deep tissue penetrability, excellent biocompaitability, water solubility, effective luminescent property and low signal to noise ratio had made them as an outstanding materials in cancer cell imaging and chemotherapy. Coa et al. (2011) developed a folic acid coupled LaF3:Yb, Er/Tm upconversion nanoarticles for bioimaging of tumour cells. Qian et al. (2010) assessed the anti tumour efficiency of Gold decorated NaYF4:Yb,Er/NaYF4/silica upconversion nanoparticles for photothermal destruction of neuroblastoma cells. Riberio et al. (2020) evaluated the potential of iron doped hydroxyappatite nanoparticles as drug free therapeutic agent and tumour cell imaging. The results of the study highlight the importance of intrinsic nano thernostic system with dual functionality. Zhang et al. (2019). demonstrated the effective practice of cancer cell membrane coated rare earth doped nanoparticles for the tumour surgery. In recent years carbon based photo luminescent nanomaterials had attracted significant attention in cancer abetment due to the chemical inertness, water dispersibility, cytotoxixity and stable photoluminescence (Sharma et al., 2018). Especially semi carbon dots had transformed as unique nanoparticles in bioimaging because of their hydrophilicity, cytocompaitability and low photobleaching (Ganguly et al., 2019). Application of rhodamine functionalized graphene quantum dots for the cancer stem cells was assessed by Guo et al. (2015). Das et al. (2019) fabricated a heteroatom doped carbon based naoprobe for targeted cancer drug delivery and cell labelling. The report of the study substantiated a trustworthy strategy for the imaging of cancer cells. In recent years microRNAs are emerged as authentic candidates in biomarkers for the rapid diagnosis of cancer. MicroRNAs play a decisive role in cell cycle process and apoptosis (Medina and Slack, 2008). A strong relationship between microRNAs expression pattern and cancer progression has been reported in 11 different cancer cells including glioblastoma, neck, prostate, colorectal, pancreas, lung and neuroblastoma. (Lawrie et al., 2008; Fabris et al., (2016). Sabahi et al. (2019) developed as carbon based electrochemical nano-genosensor for the early and selective detection of miR-21 biomarker in prostate cancer cells. The fabricated nanosenser exhibited excellent selectivity in discriminating cancer and normal cells in human serum samples. Similarly Gold nanoparticle based electrochemical DNA nano biosensor for the miR-21 was developed by Mandli et al. (2017).

Nano Drugs Groups used in Cancer Diagnosis and their Present Clinical State

In recent decades, immense efforts have been devoted for the development of effective protocols for cancer therapy and diagnosis. The specific features of tumor cells like over activated oncogenic proteins, abnormal active efflux pumps, low pH, modified redox potential and hypoxia conditions are responsible for the deactivation of anticancer drugs and development of chemotherapeutic resistance of cancer cells. The above complex behaviours of tumor cells has driven the need for the emergence of modern nanotheranostic approaches like tumor activated prodrug therapy, magnetothermal therapy, electrothermal therapy etc., Application of nano drugs have become a research hotspot in cancer diagnosis due to their less toxicity, precise targeting efficiency and enhanced bioavailability. Redox sensitive janus prodrug based on camptothecin and gemcitabine was developed by Xu et al. (2018). Kokkinos et al. (2020) investigated the possible applications of RNAi based nanodrugs for the treatment of pancreatic cancer.

Engineered nanoparticles combined with polymer matrix acts as scaffold materials for cancer cell imaging theranostic purposes (Iyer et al., 2006). Inorganic nanoparticles play an important role in thermbal ablation of tumors and imaging applications (Huang et al., 2011). Recent literatures have highlighted promising simultaneous application of nanoparticles in cancer imaging and chemotherapy (Nabil et al., 2019). A pH sensitive doxorubicin loaded pegylated gadolinium oxide nano particle was fabricated by Kumar et al. (2017). The fabricated nanocomplex facilitated the both therapeutic efficiency and optical as well as magnetic resonance imaging applications. Wang et al. (2017) developed docetaxel combined lipid polymer hybrid nanoparticles as imaging agent for the treatment of prostate cancer. Curcumin is an important anticancer natural product that has been extensively investigated for its antitumour behavior against multiple cancer types. Theranostic effect of curcumin in combination with anticancer drug paclitaxel was assessed by Nguyen et al. (2016) The findings of the study highlighted the possible application of fluorescent behavior of curcumin in monitoring the drug delivery and distribution in MCF7 skin cancer cell lines. Alberti et al. (2017) developed a PLA/PGA polymer nanoparticle based delivery of boron curcumin complex with simultaneous imaging of ovarian cancer cells.

CONCLUSION AND FUTURE PROSPECTS

The successful development of biomedical nanotechnology techniques nanodrug delivery and Nano theragnostic has proven to be an efficient strategy in the simultaneous online monitoring and drug delivery approaches for the treatment and early diagnosis of cancer . Several nanomaterials including metal nanoparticles, polymer nano carriers and liposomes have shown to be a potent agents for the construction of advanced nano carrier systems with diverse therapeutic applications. Interesting properties of nano drug delivery agents such as target specificity, controlled release of drugs, bio compatibility and theranostic properties has approved the usage of nanomaterials in clinical chemotherapy for tumor cells. Besides the development of technically inspired sophisticated nanosystems for cancer therapy, the mass production of these nanocarrier systems will bring the successful implementation of combined drug delivery and diagnosis approach form concept to clinic. Effective monitoring of these combined nanocarrier systems in pre and post clinical models will reflect the possible implementation in cancer treatment. Further, extensive toxicological investigation of nanocarrier systems will ultimately bring this technology as a successful concept in future.

REFERENCES

Al Faraj, A., Shaik, A. S., Ratemi, E., & Halwani, R. (2016). Combination of drug-conjugated SWCNT nanocarriers for efficient therapy of cancer stem cells in a breast cancer animal model. *Journal of Controlled Release*, *225*, 240–251. doi:10.1016/j.jconrel.2016.01.053 PMID:26827662

Alberti, D., Protti, N., Franck, M., Stefania, R., Bortolussi, S., Altieri, S., Deagostino, A., Aime, S., & Geninatti Crich, S. (2017). Theranostic Nanoparticles Loaded with Imaging Probes and Rubrocurcumin for Combined Cancer Therapy by Folate Receptor Targeting. *ChemMedChem*, *12*(7), 502–509. doi:10.1002/cmdc.201700039 PMID:28217982

Almajhdi, F., Fouad, H., Khalil, K., Awad, H., Mohamed, S., Elsarnagawy, T., & ... . (2014). In-vitro anticancer and antimicrobial activities of PLGA/ silver nanofiber composites prepared by electrospinning. *Journal of Materials Science*, *25*(4), 1045–1053. PMID:24375170

Attia, M. S., Youssef, A. O., Khan, Z. A., & Abou-Omar, M. N. (2018). Alpha fetoprotein assessment by using a nano optical sensor thin film binuclear Pt-2-aminobenzimidazole-Bipyridine for early diagnosis of liver cancer. *Talanta*, *186*, 36–43. doi:10.1016/j.talanta.2018.04.043 PMID:29784373

Bahramia, B., Hojjat-Farsangi, M., Mohammadi, H., Anvari, E., Ghalamfarsa, G., Yousefi, M., & Jadidi-Niaragh, F. (2017). Nanoparticles and targeted drug delivery in cancer therapy. *Immunology Letters*, *190*, 64–83. doi:10.1016/j.imlet.2017.07.015 PMID:28760499

Bechara, C., & Sagan, S. (2013). Cell-penetrating peptides: 20 years later, where do we stand? *FEBS Letters*, *587*(12), 1693–1702. doi:10.1016/j.febslet.2013.04.031 PMID:23669356

Bhatt, R. D. E., Vries, P., & Tulinsky, J. (2003). Synthesis and in vivo antitumor activity of poly (L-glutamic acid) conjugates of 20 (S)-camptothecin. *Journal of Medicinal Chemistry*, *46*(1), 190–193. doi:10.1021/jm020022r PMID:12502373

Bose, R. J., Paulmurugan, R., Moon, J., Lee, S. H., & Park, H. (2018). Cell membrane-coated nanocarriers: The emerging targeted delivery system for cancer theranostics. *Drug Discovery Today*, *23*(4), 891–899. doi:10.1016/j.drudis.2018.02.001 PMID:29426004

Bu, Junjie, Wanga, Ruan, Chen, Gunadhi, Bell, & Gu. (2019). Advances in drug delivery for post-surgical cancer treatment. *Biomaterials, 219,* 119-182. doi:10.1016/j.biomaterials.2019.04.027 PMID:31415954

Cao, T., Yang, Y., Gao, Y., Zhou, J., Li, Z., & Li, F. (2011). High-quality water-soluble and surface-functionalized upconversion nanocrystals as luminescent probes for bioimaging. *Biomaterials*, *32*(11), 2959–2968. doi:10.1016/j.biomaterials.2010.12.050 PMID:21262531

Cho, K., Wang, X., Nie, S., Chen, Z. G., & Shin, D. M. (2008). Therapeutic nanoparticles for drug delivery in cancer. *Clinical Cancer Research*, *14*(5), 1310–1316. doi:10.1158/1078-0432.CCR-07-1441 PMID:18316549

Das, P, Ganguly, S., Agarwal, T., Maity, P., Ghosh, S., Choudry, S., Gangopadhyay, S., Maiti, T.K., Dhara, S., Banerjee, S., & Das, N.C. (2019). Heteroatom doped blue luminescent carbon dots as a nano-probe for targeted cell labelling and anticancer drug delivery vehicle. *Material Chemistry and Physics, 237.*

Derakhshankhah, H., & Jafari, S. (2018). Cell penetrating peptides: A concise review with emphasis on biomedical applications. *Biomedicine and Pharmacotherapy*, *108*, 1090–1096. doi:10.1016/j.biopha.2018.09.097 PMID:30372809

Derossi, D., Joliot, A. H., Chassaing, G., & Prochiantz, A. (1994). The third helix of the Antennapedia homeodomain translocates through biological membranes. *The Journal of Biological Chemistry*, *269*, 10444–10450. PMID:8144628

Dobson. (2008). Magnetic nanoparticles for gene and drug delivery. *International Journal of Nanomedicine*, *3*(2), 169-180.

Dong, K., Zhang, Y., Zhang, L., Wang, Z., Ren, J., & Qu, X. (2019, March). Facile preparation of Metal–Organic Frameworks-based Hydrophobic Anticancer Drug Delivery Nanoplatform for targeted and enhanced cancer treatment, (2018). *Talanta*, *194*, 703–708. Advance online publication. doi:10.1016/j.talanta.2018.10.101 PMID:30609594

El-Leithy, E. S., Hassan, S. A., & Abdel-Rashid, R. S. (2019). Tamoxifen citrate/Coenzyme Q10 as smart nanocarriers Bitherapy for Breast Cancer: Cytotoxicity, genetoxicity, and antioxidant activity. *Journal of Drug Delivery Science and Technology*, *51*, 36–44. doi:10.1016/j.jddst.2019.02.010

Elgindy, N., Samy, W., & Elzoghby, A. (2014). Casein-based micelles: A novel vector for delivery of the poorly soluble anticancer drug, flutamide. *Therapeutic Delivery*, *5*(1), 7–9. doi:10.4155/tde.13.122 PMID:24341810

Elzoghby, A., Samy, W., & Elgindy, N. (2012). Albumin-based nanoparticles as potential controlled release drug delivery systems. *Journal of Controlled Release*, *157*(2), 168–182. doi:10.1016/j.jconrel.2011.07.031 PMID:21839127

Fabris, L., Ceder, Y., Chinnaiyan, A. M., Jenster, G. W., Sorensen, K. D., Tomlins, S., Visakorpi, T., & Calin, G. A. (2016). The potential of microRNAs as prostate cancer biomarkers. *European Urology*, *70*(2), 312–322. PubMed

Farooq, M., Novosad, V., Rozhkova, E., Wali, H., Ali, A., Fateh, A., Neogi, P. B., Neogi, A., & Wang, Z. (2018). Gold Nanoparticles-enabled Efficient Dual Delivery of Anticancer Therapeutics to HeLa Cells. *Scientific Reports*, *8*(1), 2907. doi:10.103841598-018-21331-y PMID:29440698

Ganguly, S., Das, P., Banerjee, S., & Das, N.C. (2019). Advancement in science and technology of carbon dot-polymer hybrid composites: a review. *Functional Composites and Structures*.

Gifta, M. M., Anna, K. C., Ivana, M.-T., & Heidi, A. (2010). A Review of Nanoparticle Photosensitizer Drug Delivery Uptake Systems for Photodynamic Treatment of Lung Cancer. (2018). *Photodiagnosis and Photodynamic Therapy*. Advance online publication. doi:10.1016/j.pdpdt.2018.03.006

Giodini, L., Re, F. L., Campagnol, D., Marangon, E., Posocco, B., Dreussi, E., & Toffoli, G. (2017). Nanocarriers in cancer clinical practice: A pharmacokinetic issue. *Nanomedicine; Nanotechnology, Biology, and Medicine*, *13*(2), 583–599. doi:10.1016/j.nano.2016.07.012 PMID:27520727

Glickson, J. D. (2009). *Lipoprotein Nanoplatform for Targeted Delivery of Diagnostic and Therapeutic Agents* (Vol. 645). Advances in Experimental Medicine and Biology. doi:10.1007/978-0-387-85998-9_35

Guo, R., Zhou, S., Li, Y., Li, X., Fan, L., & Voelcker, N. H. (2015). Rhodamine-Functionalized Graphene Quantum Dots for Detection of Fe3+ in Cancer Stem Cells. *ACS Applied Materials & Interfaces*, *7*(43), 23958–23966. doi:10.1021/acsami.5b06523 PMID:26317667

Gupta, N., Bhagat, S., Singh, M., Jangid, A. K., Bansal, V., Singh, S., Pooja, D., & Kulhari, H. (2020). Site-specific delivery of a natural chemotherapeutic agent to human lung cancer cells using biotinylated 2D rGO nanocarriers. *Materials Science and Engineering C*, *112*, 110884. doi:10.1016/j.msec.2020.110884 PMID:32409041

Han, H. J., Ekweremadu, C., & Patel, N. (2019). Advanced drug delivery system with nanomaterials for personalised medicine to treat breast cancer. *Journal of Drug Delivery Science and Technology*, *52*, 1051–1060. doi:10.1016/j.jddst.2019.05.024

Herce, H. D., & Garcia, A. E. (2007). Cell penetrating peptides: How do they do it? *Journal of Biological Physics*, *33*(5–6), 345–356. doi:10.100710867-008-9074-3 PMID:19669523

Hill, M. L., Corbin, I. R., Levitin, R. B., Cao, W., Mainprize, J. G., Yaffe, M. J., & Zheng, G. (2010). In-vitro assessment of poly-iodinated triglyceride reconstituted Low-density lipoprotein: Initial step towards CT molecular imaging. *Academic Radiology*, *17*(11), 1359–1365. doi:10.1016/j.acra.2010.06.006 PMID:20719547

Hoare, T. R., & Kohane, D. S. (2008). Hydrogels in drug delivery: Progress and challenges. *Polymer*, *49*(8), 1993–2007. doi:10.1016/j.polymer.2008.01.027

Hong, E., Liu, L., Bai, L., Xia, C., Gao, L., Zhang, L., & Wang, B. (2019). Control synthesis, subtle surface modification of rare-earth-doped upconversion nanoparticles and their applications in cancer diagnosis and treatment. *Materials Science and Engineering*, *105*, 110097. doi:10.1016/j.msec.2019.110097 PMID:31546381

Hu, Y., Sinha, S. K., & Patel, S. (2015). Investigating Hydrophilic Pores in Model Lipid Bilayers Using Molecular Simulations: Correlating Bilayer Properties with Pore-Formation Thermodynamics. *Langmuir*, *31*(24), 6615–6631. doi:10.1021/la504049q PMID:25614183

Huang, H. C., Barua, S., Sharma, G., Dey, S. K., & Rege, K. (2011). Inorganic nanoparticles for cancer imaging and therapy. *Journal of Controlled Release*, *155*(3), 344–357. doi:10.1016/j.jconrel.2011.06.004 PMID:21723891

Huang, Y. W., & Lee, H. J. (2018). Cell-penetrating peptides for medical theranostics and targeted drug delivery, peptide applications in biomedicine. In *Biotechnology and Bioengineering* (pp. 359–370). Elsevier.

Iyer, A. K., Khaled, G., Fang, J., & Maeda, H. (2006). Exploiting the enhanced permeability and retention effect for tumor targeting. *Drug Discovery Today*, *11*(17-18), 812–818. doi:10.1016/j.drudis.2006.07.005 PMID:16935749

Joliot, A., Pernelle, C., Deagostini-Bazin, H., & Prochiantz, A. (1991). Antennapedia homeobox peptide regulates neural morphogenesis. *Proceedings of the National Academy of Sciences of the United States of America*, *88*(5), 1864–1868. doi:10.1073/pnas.88.5.1864 PMID:1672046

Karagiota, A., Kourti, M., Simos, G., & Mylonis, I. (2019). HIF-1-derived cell-penetrating peptides inhibit ERK-dependent activation of HIF-1 and trigger apoptosis of cancer cells under hypoxia. *Cellular and Molecular Life Sciences*, *76*(4), 809–825. doi:10.100700018-018-2985-7 PMID:30535970

Kim, H., Kitamatsu, M., & Ohtsuki, T. (2018). Enhanced intracellular peptide delivery by multivalent cell-penetrating peptide with bioreducible linkage. *Bioorganic & Medicinal Chemistry Letters*, *28*(3), 378–381. doi:10.1016/j.bmcl.2017.12.035 PMID:29275934

Klębowski, B., Depciuch, J., Parlińska-Wojtan, M., & Baran, J. (2018). Applications of Noble Metal-Based Nanoparticles in Medicine. *International Journal of Molecular Sciences*, *19*(12), 4031. doi:10.3390/ijms19124031 PMID:30551592

Kokkinos, J., Ignacio, R. M. C., Sharbeen, G., Boyer, C., Gonzales-Aloy, E., Goldstein, D., Mccaroll, J. A., & Philips, P. A. (2020). Targeting the Undruggable in Pancreatic Cancer Using Nano-Based Gene Silencing Drugs. *Biomaterials*, *240*, 119742. doi:10.1016/j.biomaterials.2019.119742 PMID:32088410

Kong, F.-Y., Zhang, J.-W., Li, R.-F., Wang, Z.-X., Wang, W.-J., & Wang, W. (2017). Unique Roles of Gold Nanoparticles in Drug Delivery, Targeting and Imaging Applications. *Molecules (Basel, Switzerland)*, *22*(9), 1445. doi:10.3390/molecules22091445 PMID:28858253

Kumar, S., Meena, V. K., Hazari, P. P., & Sharma, R. K. (2017). PEG coated and doxorubicin loaded multimodal Gadolinium oxide nanoparticles for simultaneous drug delivery and imaging applications. *International Journal of Pharmaceutics*, *527*(1-2), 142–150. doi:10.1016/j.ijpharm.2017.05.027 PMID:28506803

Kwon, S., Singh, R., Perez, R., Abou Neel, E., Kim, H., & Chrzanowski, W. (2013). Silica-based mesoporous nanoparticles for controlled drug delivery. *Journal of Tissue Engineering*, *4*, 4. doi:10.1177/2041731413503357 PMID:24020012

Lawrie, C., Gal, S., Dunlop, H., Pushkaran, B., Liggins, A., Pulford, K., Banham, A. H., Pezzella, F., Boultwood, J., Wainscoat, J. S., Hatton, C. S., & Harris, A. L. (2008). Detection of elevated levels of tumour-associated microRNAs in serum of patients with diffuse large B-cell lymphoma. *British Journal of Haematology*, *141*, 672–675. PubMed

Lee, C.-S., Kim, H., Yu, J., Su, H. Y., Ban, S., Oh, S., Jeong, D., Im, J., Baek, M. J., & Kim, T. H. (2017). Doxorubicin-loaded oligonucleotide conjugated gold nanoparticles: A promising in vivo drug delivery system for colorectal cancer therapy. *European Journal of Medicinal Chemistry*, *142*, 416–423. Advance online publication. doi:10.1016/j.ejmech.2017.08.063 PMID:28870452

Lim, K. J., Sung, B. H., Shin, J. R., Lee, Y. W., Kim, D. J., Yang, K. S., & Kim, S. C. (2013). A cancer specific cell-penetrating peptide, BR2, for the efficient delivery of an scFv into cancer cells. *PLoS One*, *8*(6), e66084. doi:10.1371/journal.pone.0066084 PMID:23776609

Liu, X., Wu, L., Wang, L., & Jiang, W. (2018). A dual-targeting DNA tetrahedron nanocarrier for breast cancer cell imaging and drug delivery. *Talanta*, *179*, 356–363. doi:10.1016/j.talanta.2017.11.034 PMID:29310244

Locatelli, E., & Franchini, M. C. (2012). Biodegradable PLGA-b-PEG polymeric nanoparticles: Synthesis, properties, and nanomedical applications as drug delivery system. *Journal of Nanoparticle Research*, *14*(12), 1. doi:10.100711051-012-1316-4

Mandli, J., Mohammadi, H., & Amine, A. (2017). Electrochemical DNA sandwich biosensor based on enzyme amplified microRNA-21 detection and gold nanoparticles. *Bioelectrochemistry (Amsterdam, Netherlands)*, *116*, 17–23. PubMed

McMahon, H., & Boucrot, E. (2011). Molecular mechanism and physiological functions of clathrin-mediated endocytosis. *Nature Reviews. Molecular Cell Biology*, *12*(8), 517–533. doi:10.1038/nrm3151 PMID:21779028

Medina, P.P., & Slack, F.J. (2008). MicroRNAs and cancer: an overview. *Cell Cycle, 7*, 2485-2492.

Nabil, G., Bhise, K., Sau, S., Atef, M., El-Banna, H. A., & Iyer, A. K. (2019). Nanoengineered delivery systems for cancer imaging and therapy: Recent advances, future directions and patent evaluation. *Drug Discovery Today*, *24*(2), 462–491. doi:10.1016/j.drudis.2018.08.009 PMID:30121330

Nguyen, H.N., Ha,, P.T., Sao Nguyen,, A., Nguyen,, D.T., Do,, H.D., Thi,, Q.N., & Thi,, M.N.H. (2016). Curcumin as fluorescent probe for directly monitoring in vitro uptake of curcumin combined paclitaxel loaded PLA-TPGS nanoparticles. *Advances in Natural Sciences: Nanoscience and Nanotechnology*, *7*(2), 02500.

Prabaharan, M. (2015). Chitosan-based nanoparticles for tumor-targeted drug delivery. *International Journal of Biological Macromolecules*, *72*, 1313–1322. doi:10.1016/j.ijbiomac.2014.10.052 PMID:25450550

Prasanna, A., Pooja, R., Suchithra, V., Ravikumar, A., Kumar Gupta, P., & Niranjan, V. (2018). Akhil Ravikumar, Praveen Kumar Gupta, Vidya Niranjan. Smart Drug Delivery Systems for Cancer Treatment Using Nanomaterials. *Materials Today: Proceedings*, *5*(10), 21047–21054. doi:10.1016/j.matpr.2018.06.498

Pujals, S., & Giralt, E. (2008). Proline-rich, amphipathic cell-penetrating peptides. *Advanced Drug Delivery Reviews*, *60*(4-5), 473–484. doi:10.1016/j.addr.2007.09.012 PMID:18187229

Qian, L. P., Zhou, L. H., Too, H. P., & Chow, G. M. (2010). Gold decorated NaYF4:Yb,Er/NaYF4/silica (core/shell/shell) upconversion nanoparticles for photothermal destruction of BE(2)-C neuroblastoma cells. *Journal of Nanoparticle Research*, *13*(2), 499–510. doi:10.100711051-010-0080-6

Ribeiro, T. P., Monteiro, F. J., & Laranjeira, M. S. (2020). Duality of iron (III) doped nano hydroxyapatite in triple negative breast cancer monitoring and as a drug-free therapeutic agent. *Ceramics International*, *46*(10), 16590–16597. doi:10.1016/j.ceramint.2020.03.231

Richard, J. P., Melikov, K., Brooks, H., Prevot, P., Lebleu, B., & Chernomordik, L. V. (2005). Cellular uptake of unconjugated TAT peptide involves clathrin-dependent endocytosis and heparan sulfate receptors. *The Journal of Biological Chemistry*, *280*(15), 15300–15306. doi:10.1074/jbc.M401604200 PMID:15687490

Ruiz-Hernández, E., Baeza, A., & Vallet-Regí, M. (2011). Smart Drug Delivery through DNA/Magnetic Nanoparticle Gates. *ACS Nano*, *5*(2), 1259–1266. doi:10.1021/nn1029229 PMID:21250653

Sabahi, A., Salahandish, R., Ghaffarinejad, A., & Omidinia, E. (2019). Electrochemical nano-genosensor for highly sensitive detection of miR-21 biomarker based on SWCNT-grafted dendritic Au nanostructure for early detection of prostate cancer. *Talanta*. PMID:31892044

Schmidt, A., Molano, V., Hollering, M., Pöthig, A., Casini, A., & Kühn, F. (2016). Evaluation of New Palladium Cages as Potential Delivery Systems for the Anticancer Drug Cisplatin. *Chemistry (Weinheim an der Bergstrasse, Germany)*, *22*(7), 2253–2256. doi:10.1002/chem.201504930 PMID:26756963

Sharma, S., Umar, A., Sood, S., Mehta, S. K., & Kansal, S. K. (2018). Photoluminescent C-dots: An overview on the recent development in the synthesis, physiochemical properties and potential applications. *Journal of Alloys and Compounds*, *748*, 818–853. doi:10.1016/j.jallcom.2018.03.001

Shukla, G. S., & Krag, D. N. (2005). Phage display selection for cell-specific ligands: Development of a screening procedure suitable for small tumor specimens. *Journal of Drug Targeting*, *13*(1), 7–18. doi:10.1080/10611860400020464 PMID:15848950

Singh, A., Dar, M., Joshi, B., Sharma, B., Shrivastava, S., & Shukla, S. (2018). Phytofabrication of Silver nanoparticles: Novel Drug to overcome hepatocellular ailments. *Toxicology Reports*, *5*(5), 333–342. doi:10.1016/j.toxrep.2018.02.013 PMID:29854602

Stylianopoulos, T., Poh, M.-Z., Insin, N., Bawendi, M. G., Fukumura, D., Munn, L. L., & Jain, R. K. (2010). Diffusion of particles in the extracellular matrix: The effect of repulsive electrostatic interactions. *Biophysical Journal*, *99*(5), 1342–1349. doi:10.1016/j.bpj.2010.06.016 PMID:20816045

Stylianopoulos, T., Poh, M.-Z., Insin, N., Bawendi, M. G., Fukumura, D., Munn, L. L., & Jain, R. K. (2010). Diffusion of particles in the extracellular matrix: The effect of repulsive electrostatic interactions. *Biophysical Journal*, *99*(5), 1342–1349. doi:10.1016/j.bpj.2010.06.016 PMID:20816045

Sun, T., Zhang, Y. S., Pang, B., Hyun, D. C., Yang, M., & Xia, Y. (2014). Engineered nanoparticles for drug delivery in cancer therapy. *Angewandte Chemie International Edition*, *53*(46), 12320–12364. doi:10.1002/anie.201403036 PMID:25294565

Tabatabaei Mirakabad, F., Nejati-Koshki, K., Akbarzadeh, A., Yamchi, M., Milani, M., Zarghami, N., Zeighamian, V., Rahimzadeh, A., Alimohammadi, S., Hanifehpour, Y., & Joo, S. W. (2014). PLGA-Based Nanoparticles as Cancer Drug Delivery Systems. *Asian Pacific Journal of Cancer Prevention*, *15*(2), 517–535. doi:10.7314/APJCP.2014.15.2.517 PMID:24568455

Temsamani & Vidal. (2004). The use of cell penetrating peptides for drug delivery. *Research Focus DDT*, *9*, 1012-1019.

Ter-Avetisyan, G., Tünnemann, G., Nowak, D., Nitschke, M., Herrmann, A., Drab, M., & Cardoso, M. C. (2009). Cell entry of arginine-rich peptides is independent of endocytosis. *The Journal of Biological Chemistry*, *284*(6), 3370–3378. doi:10.1074/jbc.M805550200 PMID:19047062

Thaxton, C.S., Rink, J.S., Naha, P.C., & Cormode, D.P. (2016). Lipoproteins and lipoprotein mimetics for imaging and drug delivery. *Adv Drug Deliv Rev.*, *15*(106), 116-131.

Versluis, A., van Geel, P., Oppelaar, H., van Berkel, T. J. C., & Bijsterbosch, M. K. (1996). Receptor-mediated uptake of low-density lipoprotein by B16 melanoma cells *in vitro* and *in vivo* in mice. *British Journal of Cancer*, *74*(4), 525–532. doi:10.1038/bjc.1996.396 PMID:8761365

Vinothini, K., Rajendran, N. K., Ramu, A., Elumalai, N., & Rajan, M. (2019). Folate receptor targeted delivery of paclitaxel to breast cancer cells via folic acid conjugated graphene oxide grafted methyl acrylate nanocarrier. *Biomedicine and Pharmacotherapy*, *110*, 906–917. doi:10.1016/j.biopha.2018.12.008 PMID:30572195

Wang, H., Boerman, O., Sariibrahimoglu, K., Li, Y., Jansen, J., & Leeuwenburgh, S. (2012). Comparison of micro- vs. nanostructured colloidal gelatin gels for sustained delivery of osteogenic proteins: Bone morphogenetic protein-2 and alkaline phosphatase. *Biomaterials*, *33*(33), 8695–8703. doi:10.1016/j.biomaterials.2012.08.024 PMID:22922022

Wang, Q., Alshaker, H., Böhler, T., Srivats, S., Chao, Y., Cooper, C., & Pchejetski, D. (2017). Core shell lipid-polymer hybrid nanoparticles with combined docetaxel and molecular targeted therapy for the treatment of metastatic prostate cancer. *Scientific Reports*, *7*(7), 5901. doi:10.103841598-017-06142-x PMID:28724986

Wang, Y., Li, P., Truong-Dinh Tran, T., Zhang, J., & Kong, L. (2016). Manufacturing Techniques and Surface Engineering of Polymer Based Nanoparticles for Targeted Drug Delivery to Cancer. *Nanomaterials (Basel, Switzerland)*, *6*(2), 26. doi:10.3390/nano6020026 PMID:28344283

Wicki, A., Witzigmann, D., Balasubramanian, V., & Huwyler, J. (2015). Nanomedicine in cancer therapy: Challenges, opportunities, and clinical applications. *Journal of Controlled Release*, *200*, 138–157. doi:10.1016/j.jconrel.2014.12.030 PMID:25545217

Wu, D., Wang, L., Li, W., Xu, X., & Jiang, W. (2017, November). DNA nanostructure-based drug delivery nanosystems in cancer therapy. (2017). *International Journal of Pharmaceutics*, *533*(1), 169–178. Advance online publication. doi:10.1016/j.ijpharm.2017.09.032 PMID:28923770

Xu, Y., Huang, Y., Zhang, X., Lu, W., Yu, J., & Liu, S. (2018). Carrier-free Janus nano-prodrug based on camptothecin and gemcitabine: Reduction-triggered drug release and synergistic in vitro antiproliferative effect in multiple cancer cells. *International Journal of Pharmaceutics*, *550*(1-2), 45–56. doi:10.1016/j.ijpharm.2018.08.041 PMID:30138703

Yatvin, M. B., Weinstein, J. N., Dennis, W. H., & Blumenthal, R. (1978). Design of liposomes for enhanced local release of drugs by hyperthermia. *Science*, *202*(4374), 1290–1293. doi:10.1126cience.364652 PMID:364652

Yoshihiko, T., Iichi, Z., Miki, Y., Sumio, C., Juichi, S., Keiji, I., & Kazuhiro, M. (2000). Inhibitory Effect of Dexamethasone Palmitate-Low Density Lipoprotein Complex on Low Density Lipoprotein-Induced Macrophage Foam Cell Formation. *Biological & Pharmaceutical Bulletin*, *23*(4), 466–471. doi:10.1248/bpb.23.466 PMID:10784429

Yu, K., Liu, M., Dai, H., & Huang, X. (2020). Targeted drug delivery systems for bladder cancer therapy. *Journal of Drug Delivery Science and Technology*, *56*, 101535. Advance online publication. doi:10.1016/j.jddst.2020.101535

Zhang, X., He, S., Ding, B., Qu, C., Zhang, Q., Chen, H., Sun, Y., Fang, H., Long, Y., Zhang, R., Lan, X., & Cheng, Z. (2019). Cheng, Z. Cancer Cell Membrane-Coated Rare Earth Doped Nanoparticles for Tumor Surgery Navigation in NIR-II Imaging Window. *Chemical Engineering Journal*, *385*, 123959. doi:10.1016/j.cej.2019.123959

Zhu, C., & Xia, Y. (2017). Chunlei zhu and Younan Xia, "Biomimetics: Reconstitution of low-density lipoprotein for targeted drug delivery and related theranostic applications. *Chemical Society Reviews*, *46*(24), 7668–7682. doi:10.1039/C7CS00492C PMID:29104991

Chapter 14
Immune-Targeted Nanomedicine

Andrew Pradeep
The American College, India

R. Sethu Nagarajan
The American College, India

H. Fazil
The American College, India

ABSTRACT

Immunotherapy has become a preferable candidate for many diseases in recent days. The infusion or administration of immune complexes or components to elicit the own immune response against a particular disease by attracting the antigen presenting cells against the disease causing organism and eliciting the T-cell mediated killing and further activating cell mediated immunity based on the processed surface antigens underlies the basic concept behind the immunotherapy. Immunotherapy can be applied for all course of diseases even in the treatment of cancer. The limitation in using immunotherapy is that it needs a proper delivery vehicle to reach the diseased spot to shows its pharmacokinetic property. In case of cancer, the immune components administered itself are not able to pertain and penetrate the solid tumor mass. Nanoparticles are small-sized particles which are generally specific in action used in the field of medicine. Nanoparticles aid in targeted drug delivery to the specific spots and immune targeting of nanoparticles is due to its enhanced permeability and retention (EPR).

INTRODUCTION

In this modern scenario of scientific and technological development, there is an increased number of challenges for therapeutics design. However, the currently existing chemotherapeutics can do its role efficiently but there is a voidness over a period of time in treating several diseases especially in the field of Oncology. In general, the drugs are categorized into two categories-Broad spectrum drug and Narrow

DOI: 10.4018/978-1-7998-5049-6.ch014

spectrum drug. The broad spectrum of drug act as a therapeutics for about two or more complications on other hand narrow spectrum drug is administered in targeting limited target. In some cases it showed adverse effect and some major concerns are incompatibility to host and non-specificity of the drug to the site of action. In cancer treatment, Apart from chemotherapy or radiotherapy or surgical procedure, there is a need of combinational therapy to achieve the maximum therapeutic success. Generally the Immunotherapy is prescribed as a combination therapy with the traditional therapeutics with minimal side-effects. Immunotherapy is a form of biological therapy in which immune cells of the host is targeted and primed to fight against the antigens. The importance of immunotherapy has been acknowledged by James P. Allison (Nobel Prize for physiology or medicine 2018) for the discovery of CTLA-4 (cytotoxic T-lymphocyte-associated protein) and PD-1 / PD-L1 (Programmed cell death protein 1 / programmed cell death protein ligand 1) to Tasuku Honjo (Altmann DM, 2018). Immunotherapeutics can be a drug or a vaccine and depending upon the types of disease, site of action and need of the patients. Though it paves the positive hope in the new medical era, it cannot be used as an independent therapeutics. The major reason behind the failure of Immunotherapy was that it cannot able to reach the target sit and it lacks a proper delivery channel. To overcome these issues, it can be engineered with nanoparticles.

Over the few decades Nanotechnology has become a revolutionary field through an integrated approach. In accordance to the National Nanotechnology Initiative (NNI, US), nanotechnology implies the involvement of both science and engineering techniques in designing, synthesis, characterization, and application of various materials and devices on the nanoscale range typically ranges between 1–100 nanometers. The implication of nanotechnology in the development of newer drugs as a potent therapeutic or prophylactic agent is generally termed as Nanomedicine. On the other hand Nanomedicine can also be defined as the application of nanoparticles for purpose of diagnosis, treatment monitoring, control and prevention of diseases (Tinkle, S.*et al*, 2014).

The nanoparticles have several capability to achieve a greater adhesion to the target site and provide the required immune response. This greater immune targeting of nanoparticles are due to the Enhanced Permeability and Retention (EPR) effect and several unique properties. Apart from this, nanoparticles also act as an adjuvant thereby increasing the immune response to modulate the host anticancerous immunity in the treatment of cancer and will provoke new avenues for developing novel therapeutics. In this chapter, we discuss the salient features of nanoparticles and its need in targeted immunotherapy in treating several disease especially cancer.

CANCER- AN OVERVIEW

Cancer is one of Non-Communicable disease and remains as the leading death causing one globally, with an estimated 21 million by 2030 (D. Hanahan, R.A. Weinberg, 2011). Several reasons like altered mutation pattern, glycosylation level, methylation rate, addition or deletion of chromosome lead to the onset of cancer. The varied glycosylation is the major reason for tumor heterogeneity (S. S. Pinho & C. A. Reis, 2015) whereas the methylation rate desire the tumor suppressor gene silencing and oncogene activation activity (Sharma. S, *et al.*, 2009). Apart from this several other physical and chemical agents are also responsible for the onset of tumor development which is generally termed as carcinogen.

The onset of cancer is generally termed as benign tumor (neoplasm state) or primary tumor which is generally localized. Then reaches the Malignancy (injurious state), it leaves the localized niche and gets traversed to other site of body via lymphatic system or blood circulation to form its secondary niche.

In case if it remains in the blood circulation (Systemic), then it is termed as metastatic stage (Lodish. H. *et al.,* 2000) Nowadays, there are many examples of chemotherapeutic agents that have been used to control different cancers in clinics such as doxorubicin, paclitaxel, gemcitabine, and cisplatin (Chabner B. A., *et al.,* 2005)Apart from this, radiotherapy can also be widely used to treat cancer, however most of them are associated with varied side-effects. The success rate of the treatment majorly depends on the stage of cancer and host immune responses

CANCER AND HOST IMMUNOLOGY

Generally the immune system comprises of Innate and Adaptive immune responses. The innate immunity can also be described as primary immune response which act as a baseline of defense and the adaptive immunity can be defined as the secondary immune response which consists of T-cells and B-cells mediated immune responses and memory responses. The immune system is composed of several types of soluble bioactive molecules like Cytokines, proteins and so on to eliminate the non-host foreign particles from the host. This can be achieved through several mechanisms.

The tumor cells generally mimics the host immune response thereby it can achieve the needful vascularization and nutrition for development. The tumor cells generally undergoes several mechanism to evade the host immune responses against them. These mechanisms includes the defect in the effector arm of immune cells like CD8+ cytotoxic T-lymphocytes and natural killer cells; Abnormalities in the regulation of immune responses mediated by the regulatory T-cells, T-helper cells and dendritic cells (potent Antigen presenting cells) (Ossendorp F. *et al.,*2000). The immune-surveillance property of the host immune system and the evasion of tumor cells from the immune system generally leads to the Immuno-editing concept (Dunn, G. P., *et al.*, 2002) which can be achieved by enhancing or modulating the anti-tumor immune responses of the host. Our Immune system can be primed to rule out the cancer, One such notable example is altered or reduced expression of the cell surface marker of Major Histocompatibility (MHC) class I in cancer cells which leads to the reduced recognition by immune cells like Natural killer (NK) cells which thereby activating receptors present on NK cell surface such as NKG2D which bind to surface glycoproteins known as MICA/B that may be present on the tumors and induce cytolytic granule mediated apoptosis (Waldhauer, I., & Steinle, A., 2008).

REASON FOR A SEARCH OF ALTERNATIVE

Traditionally, Chemotherapy or Radiotherapy are believed to treat the cancer cells by direct killing. The failure of chemotherapy is majorly because of different extents of multidrug resistance (MDR) development which results with approximately 90% of cancer patient's death. MDR occurs when tumor cells develop resistance to unrelated classes of chemotherapeutic agents both structurally and functionally leading to drug unresponsiveness and/or drug efflux from cancer cells causes an major obstacle during the treatment (Saggar, J. K., et al., 2013)(Xue, X., & Liang, X. J. 2012). There are several reasons behind the development of multidrug resistance in cancerous cell and few of them are shown in Figure 1. To overcome this issue now a days, personalized medicine (Tailor made medicine) is preferred, in which the drugs are designed based on the individual host and nature of cancer respectively. But the concept of Personalized medicine is highly not feasible and expensive. In order to reach the affordable broad range outcome, the medical professionals seeks the implication of several new forms of therapeutics with reduced side-effects.

ROLE OF IMMUNOTHERAPY IN TREATING CANCER

In immunotherapy, the active components are primed to induce or elicit the host immune responses against the cancer cells. There are about five different classes of Immunotherapeutics and they are shown in the Figure2. The importance of immunotherapy has been acknowledged by James P. Allison (Nobel Prize for physiology or medicine 2018) for the discovery of CTLA-4 (cytotoxic T-lymphocyte-associated protein) and PD-1 / PD-L1 (Programmed cell death protein 1 / programmed cell death protein ligand 1) to Tasuku Honjo (Altmann DM, 2018). Zitvogel with her co-worker described the anti-neoplastic effect of chemotherapies and this leads to activation of innate immune system through immune stimulatory signals via pattern recognition receptors such as toll-like receptor 4 (TLR4) (Tang, J., et al., 2018). It is generally used as a combination therapy with the chemotherapeutic drugs and/or Radiation therapy.

In case of radiation therapy, unless infecting the tumor cell, it also kills the immunogenic cells thereby leading s to the depletion of the immune cells. Combining radiotherapy with immune-stimulatory PD-1 / PD-L1 (Programmed Death Ligand 1) blockade gives an appreciable results in Pre-clinical trials which highlights the synergistic potential of this combination while applying it as a tropical agent at the site of radiotherapy and it crosses the Phase I and II clinical trial (Park, S. S. *et al.,* 2015) (Hammerich, L., *et al.,* 2019).

Generally in most of the cases, the monoclonal antibody can be used in the immunotherapy which can act upon the cancerous cells. It can be a Murine(made from mouse proteins and the names of the treatments end in –omab), Chimeric (proteins are a combination of part mouse and part human and the names of the treatments end in -ximab) or Humanized (made from small parts of mouse proteins attached to human proteins and the names of the treatments end in -zumab). They trigger the immune system to fight against the tumor cells and leads to the antibody-dependent cell-mediated cytotoxicity (ADCC). In some cases they directly made to target the ligands or receptor in specific to the tumor cells and thereby control the cell proliferation and metastasis (Weiner, L. M. *et al.*, 2009). Some FDA approved Immunotherapeutics are listed in Table 1.

Though immunotherapy is a preferable one, it also have some challenges to overcome which include, the efficacy of the drug is often unpredictable and identification of targetable tumor-specific antigens (TSAs) which is also called neoantigens, differential drug permeability rate and increased chance for activation of autoimmune profile because of the increased release of drug at the site of action (Ventola, C. L. 2017) that can be overcome by implication of Nanotechnology in immunotherapy.

NANO IN THE FIELD OF THERAPEUTICS- AN IDYLLIC APPROACH

Several areas in the field of science had benefited significantly from the introduction of nanotechnology. The application of nanotechnology in the field of medicine is termed Nanomedicine. The use of the nanoparticles in various field is majorly because of it affordable size (1 to 100nm).In the field of pharmaceuticals, the manufacturing of nano-materials involves two different approaches: top down process and bottom down process. The top down process involves the breakdown of large materials into a smaller one by means of mechanical or chemical energy. On the other hand, the bottom down process starts with atomic or molecular size particle allowing the precursor particles to increase in size through chemical reaction (Boverhof, D. R. *et al.*, 2015).

There are different types of Nanoparticles which include Lipid based nanoparticles, Nanocapsules, Polymeric nanoparticles, Metal based nanoparticles (Silver nanoparticles, Zinc nanoparticles, Gold nanoparticles, Copper nanoparticles), Dendrimers (Dong, X., & Mumper, R. J. 2010) which was also shown in the Figure3.

The formulation is generally based on the conjugating drug and site of action. The key challenge in the nano-based drug development are the scale-up and reproducibility. The classical manufacturing processes does not makes the three dimensional medicines in the nanometer scale. It requires different methods as mentioned above, compromise top-down and bottom-down approaches, which include different types of steps, like homogenization, sonication, milling, emulsification, and the use of organic solvents. In a small-scale, it is easy to control and optimize the formulation, but in large scale production a small change in quantification or error will leads to the drastic change in physiochemical properties or can even modify the pharmacokinetics properties (Agrahari, V., & Hiremath, P. 2017).

CHARACTERISTIC FEATURES OF NANOPARTICLES

The nanoparticles possess several potent characteristic features to emerge as an eminent candidate in the field of medicine and it is generally used as a delivery vehicle to reach the appropriate site of action. The major advantages of using nanoparticles in the field of medicine are:

- Nanomaterials are functionalized with unique shape and size, which enabling them to penetrate and target specific site.
- Nanoparticles possess a special surface modification property which made them to overcome solubility and stability issues during localization of drug in the targeted area
- Nanoparticles possess a novel physical property which can be utilized for tracing the drug activity.
- Nanoparticles are composed of large surface area, so that it can provide a large payload which holds a drug and other adjuvants needs to be encapsulated.
- Nanoparticle has the sustained release property ultimately results in reaching the optimum concentration in a short time and maintained even at low drug concentration so that at once when the drug is delivered and recognized by a receptor, thereby prevent the damage to normal cells at the site of action (S. E. McNeil, 2009; Wang, R. *et al.*,2013)
- Nanoparticles can prevent the degradation of the conjugated drug.
- Nanoparticles can also have the property to alter the pharmacokinetic and distribution profile of the drug in the tissue and increase the intracellular efflux in cancer cells based on the surface permeability which cannot be easily achieved by normal drug (Elzoghby, A. O. *et al.*, 2012).

They are designed as active and passive targeting particles for cancer to deliver and elevate the intracellular anticancer activity also enhance the absorption through the epithelial diffusion. Enhancing the permeability and retention (EPR) effects of anticancer drugs is considered as passive targeting of Nanoparticles towards the tumors. However, actively targeted nanoparticles can be made based on tumor microenvironment and targeting surface ligand to the tumor cells (Bi, Y. et al., 2016). Enhanced Permeability and Retention effect will makes sure that the drugs are delivered at the appropriate site without damaging the other healthy cells/tissues and confirms highest accuracy with increased efficacy at the target site. Because of the unique inherent property even against the solid tumors, the nanoparticle is considered as an excellent tumor-targeting vehicle and this can makes the accumulation of nanoparticles with drug at the preferable tumor site. In addition, it allows targeting the tumor site that is directly connected to the main blood circulation (Schroeder, A. *et al.*, 2012).

Another main properties of the nanoparticles in the field of medicine are bio-compatibility and biodegradable which is considered as an important parameter in the field of pharmacology. The term biodegradability represents the biodegradable nature of the nanoparticles in the human body. Even if the drug cause some inflammation or irritation, the immune system will return to the regular function after elimination unless stay forever in the body and change the normal function of the immune system (Müller, R. H. *et al.*, 2011). And it should be compatible to the host otherwise it will activate the immune responses thereby leadings to the poor response of the drug.

TARGETED THERAPY USING NANOPARTICLES

Liposome Based Drugs

Liposome generally enhances both the pharmacokinetics and pharmacodynamics of the drugs thereby increasing the chance for survival of the patients. Liposomes composed of two layers of phospholipids with both hydrophilic and hydrophobic ends so that we can incorporate both types of drugs with aqueous core. It follows the colloidal drug delivery systems. The hydrophobic drugs are generally dissolved in the lipid layers, on other hand the hydrophilic drugs are loaded in the aqueous core region. There are several types of drugs made based on the Liposome fusion and some of them are as follow:

- Doxil liposomes composed of single lipid bilayer membrane with a Hydrogenated Soy Phosphatidyl Choline (HSPC) and cholesterol with doxorubicin encapsulated in the internal compartment along with the 2000 Da segments of Poly-Ethylene Glycol (PEG) grafting to the liposome surface. As the doxorubicin is a small molecular component, the PEG can be used for sustain release of drug and provides stability to the active component. In addition, the PEG functionalization makes the drug to cross the reticulo-endothelial system and helps to reach the tumor microenvironment with reduced cardio-toxicity and demonstrated the surface modification (Gabizon, A. A. 2001) (Gabizon, A, *et al.*, 2003). Now the Doxil is administered for second-line treatment of AIDS-related Kaposi's sarcoma, recurrent ovarian cancer, metastatic breast cancer, and multiple myeloma in combination with Bortezomib (Barenholz, Y. C. 2012).
- DaunoXome liposome is made up of Daunorubicin which consists of a bilayer membrane of Distearoyl Phosphatidylcholine and cholesterol at a 2:1 ratio. The citrate salt of Daunorubicin is encapsulated within the inner aqueous core of the vesicle and demonstrated the high stability with minimal leakage. It follows the passive-targeting mechanism of tumor sites and its size, net neutral charge, incorporation of cholesterol and a lipid molecule with a high phase-transition temperature and circulation half-life of 2–4 hours made it as an effective therapeutic tool (Forssen, E. A., & Ross, M. E. 1994)(Allen, T. M., & Martin, F. J. 2004). In US, it is generally used as a first line of choice in HIV based Kaposi's sarcoma treatment (Galen, U. S. 2011).

- Other drug includes AmBisome (amphotericin B), Visudyne (verteporfin) which can directly inhibit the P-gp efflux or bypassing P-gp through an endocytosis pathway thereby aiding the drug to reach the target site.

Apart from this there are several drugs are under clinical phase trials which can serve as an effective therapeutic tool in future.

Nanocapsules

Nanocapsules are comprised of polymeric membrane with hydrophilic/lipophilic surfactants that cover the liquid core where the hydrophobic drugs are loaded. The space allocated for loading the drug generally comprises large surface area and it have the sustained release ability. Nanocapsules generally overcome P-gp-mediated drug resistance thereby it can enhance the delivery of the active components to reach the site of action and to exhibit its efficacy. These kind of formulation is generally prescribed in the immune-based chemotherapy cases to recover the patients with minimal side-effects (Ganassin, R., *et al.*, 2018).

Apart from this, there are also several forms of Nano-based drugs which have the application of nanoparticle in drug formulation or nanoparticles used as conjugates. Some immunotherapy includes stimulating immune cells through cancer antigen vaccination, such as interleukin-2 (IL-2) and interferon-α (IFNα), or using dendritic cells for therapy enhancement. Some of the examples for nanobased immunotherapeutic drugs are:

- Tecemotide, for example, for non-small-lung cancer treatment in clinical trial III, is a transmembrane glycoprotein Mucin 1 (MUC1) antigen conjugated with liposome to induce immune response against cancer cells.
- CYT-6091 is an another example of a combination between tumor necrosis factor (TNF) antigen and colloid gold nanoparticle for the treatment of late-stage solid tumorsand in accordance to US National Library of Medicine 2012 record, it is in clinical trial I

Apart from these several drugs are in clinical trial phase. For instance, Dendrimer based hydrogels are used for Intra-ocular drug delivery in order to achieve the increased bioavailability and sustained release. In case of hydrogel formulation, the nanoparticles like Alginate, Xanthan gum, Cellulose, polymeric micelles are used in targeted drug delivery system.

FUTURE PROSPECTS

Even in this pandemic Covid-19 situation, many immune based drugs are under clinical study and one among them is Sepsivac. Sepsivac is composed of heat-killed *Mycobacterium W* (M_w) generally administered for Gram negative bacterial septicemia patients in order to minimize the activation of autoimmune system by means of increased cytokine production. In case of Corona infected patients, we can made a hypothesis that the virus enter the lung through ACE2 receptor thereby it multiplies and affects the alveolar cells. During this situation, there is an increased synthesis of cytokine in response to the viral load which have the more chance for activation of Autoimmune profile i.e. host immune cells itself affect the normal body cells. The increased synthesis of cytokine is generally termed as Cytokine storm as like in the septicemia patients and it is the major lethal factor behind the mortality. Sepsivac have the ability to reduce the Th2 response and thereby increasing the Th1 and TLR2 agonist which can prevent the host cell damage and viral destruction. As it was a less efficient drug, it can be combined with some nanoparticles to enhance its immunogenic property and make it as an effective therapeutic tool. We can adopt liposome fusion technique in which Sepsivac can be loaded along with the commercial antibiotic in the core region and made to reach the lung epithelial cell by means of aerosol spray. After it reaches the site of action, it is made to release the active components in a sustain manner so that there is a prolonged efficacy of the drug can be achieved with minimal side-effect. Likewise, the immune targeted nano-medicine can be used for treating different diseases and the nanoparticles will aid to overcome several obstacles in the field of medicine.

CONCLUSION

In the current scenario, the implication of immune based therapeutics paves a significant role over the medical field. Though it have some demerits, the Immuno-targeted nanomedicine holds a distinct role in the Pharmaceutical and its associated medical field. Further research need to be carried out towards this objective in order to develop several effective therapeutic tools for the future welfare.

ACKNOWLEDGMENT

We thank Mrs. P. Margaret Sangeetha, Assistant Professor, Department of Microbiology, The American College, Madurai. for her timely support during this chapter drafting and validation. We also thank reviewers for their valuable comments and guidance.

REFERENCES

Agrahari, V., & Hiremath, P. (2017). *Challenges associated and approaches for successful translation of nanomedicines into commercial products*. Academic Press.

Allen, T. M., & Martin, F. J. (2004). Advantages of liposomal delivery systems for anthracyclines. *Seminars in Oncology*, *31*, 5–15. doi:10.1053/j.seminoncol.2004.08.001 PMID:15717735

Altmann, D. M. (2018). A Nobel prize-worthy pursuit: Cancer immunology and harnessing immunity to tumour neoantigens. *Immunology*, *155*(3), 283–284. doi:10.1111/imm.13008 PMID:30320408

Barenholz, Y. C. (2012). Doxil®—the first FDA-approved nano-drug: Lessons learned. *Journal of Controlled Release*, *160*(2), 117–134. doi:10.1016/j.jconrel.2012.03.020 PMID:22484195

Bi, Y., Hao, F., Yan, G., Teng, L. J., Lee, R., & Xie, J. (2016). Actively targeted nanoparticles for drug delivery to tumor. *Current Drug Metabolism*, *17*(8), 763–782. doi:10.2174/13892002176661606191 91853 PMID:27335116

Boverhof, D. R., Bramante, C. M., Butala, J. H., Clancy, S. F., Lafranconi, M., West, J., & Gordon, S. C. (2015). Comparative assessment of nanomaterial definitions and safety evaluation considerations. *Regulatory Toxicology and Pharmacology*, *73*(1), 137–150. doi:10.1016/j.yrtph.2015.06.001 PMID:26111608

Chabner, B. A., & Roberts, T. G. Jr. (2005). Chemotherapy and the war on cancer. *Nature Reviews. Cancer*, *5*(1), 65–72. doi:10.1038/nrc1529 PMID:15630416

Connors, J. M., Jurczak, W., Straus, D. J., Ansell, S. M., Kim, W. S., Gallamini, A., Younes, A., Alekseev, S., Illés, Á., Picardi, M., Lech-Maranda, E., Oki, Y., Feldman, T., Smolewski, P., Savage, K. J., Bartlett, N. L., Walewski, J., Chen, R., Ramchandren, R., ... Radford, J. (2018). Brentuximab vedotin with chemotherapy for stage III or IV Hodgkin's lymphoma. *The New England Journal of Medicine*, *378*(4), 331–344. doi:10.1056/NEJMoa1708984 PMID:29224502

Dong, X., & Mumper, R. J. (2010). Nanomedicinal strategies to treat multidrug-resistant tumors: Current progress. *Nanomedicine (London)*, *5*(4), 597–615. doi:10.2217/nnm.10.35 PMID:20528455

Dunn, G. P., Bruce, A. T., Ikeda, H., Old, L. J., & Schreiber, R. D. (2002). Cancer immunoediting: From immunosurveillance to tumor escape. *Nature Immunology*, *3*(11), 991–998. doi:10.1038/ni1102-991 PMID:12407406

Elzoghby, A. O., Samy, W. M., & Elgindy, N. A. (2012). Albumin-based nanoparticles as potential controlled release drug delivery systems. *Journal of Controlled Release*, *157*(2), 168–182. doi:10.1016/j.jconrel.2011.07.031 PMID:21839127

Fan, A. C., & Leppert, J. T. (2019). Postmarketing Analysis of Sipuleucel-T—The Importance of Real-World Data. *JAMA Network Open*, *2*(8), e199233–e199233. doi:10.1001/jamanetworkopen.2019.9233 PMID:31411706

Fellner, C. (2012). Ipilimumab (yervoy) prolongs survival in advanced melanoma: Serious side effects and a hefty price tag may limit its use. *P&T*, *37*(9), 503–530. PMID:23066344

Forssen, E. A., & Ross, M. E. (1994). Daunoxome® treatment of solid tumors: Preclinical and clinical investigations. *Journal of Liposome Research*, *4*(1), 481–512. doi:10.3109/08982109409037058

Gabizon, A., Shmeeda, H., & Barenholz, Y. (2003). Pharmacokinetics of pegylated liposomal doxorubicin. *Clinical Pharmacokinetics*, *42*(5), 419–436. doi:10.2165/00003088-200342050-00002 PMID:12739982

Gabizon, A. A. (2001). Pegylated liposomal doxorubicin: Metamorphosis of an old drug into a new form of chemotherapy. *Cancer Investigation*, *19*(4), 424–436. doi:10.1081/CNV-100103136 PMID:11405181

Galen, U. S. (2011). *DaunoXome: daunorubicin citrate liposome injection*. Academic Press.

Ganassin, R., Horst, F. H., Camargo, N. S., Chaves, S. B., Morais, P. C., Mosiniewicz-Szablewska, E. & Muehlmann, L. A. (2018). Selol nanocapsules with a poly (methyl vinyl ether-co-maleic anhydride) shell conjugated to doxorubicin for combinatorial chemotherapy against murine breast adenocarcinoma in vivo. *Artificial Cells, Nanomedicine, and Biotechnology, 46*(sup2), 1046-1052.

Hammerich, L., Marron, T. U., Upadhyay, R., Svensson-Arvelund, J., Dhainaut, M., Hussein, S., & Salazar, A. M. (2019). Systemic clinical tumor regressions and potentiation of PD1 blockade with in situ vaccination. *Nature Medicine*, *25*(5), 814–824. doi:10.103841591-019-0410-x PMID:30962585

Hanahan, D., & Weinberg, R. A. (2011). Hallmarks of cancer: The next generation. *Cell, 144*(5), 646–674. doi:10.1016/j.cell.2011.02.013 PMID:21376230

Jain, M. D., Bachmeier, C. A., Phuoc, V. H., & Chavez, J. C. (2018). Axicabtagene ciloleucel (KTE-C19), an anti-CD19 CAR T therapy for the treatment of relapsed/refractory aggressive B-cell non-Hodgkin's lymphoma. *Ther Clin Risk Manag., 14*, 1007-1017. doi:10.2147/TCRM.S145039 PMID:29910620

Lodish, H., Berk, A., Zipursky, S. L., Matsudaira, P., Baltimore, D., & Darnell, J. (2000). *"Tumor cells and the onset of cancer," in Molecular Cell Biology* (4th ed.). W. H. Freeman.

McNeil, S. E. (2009). Nanoparticle therapeutics: A personal perspective. *Wiley Interdisciplinary Reviews, 1*(3), 264–271. PMID:20049796

Müller, R. H., Gohla, S., & Keck, C. M. (2011). State of the art of nanocrystals–special features, production, nanotoxicology aspects and intracellular delivery. *European Journal of Pharmaceutics and Biopharmaceutics, 78*(1), 1–9. doi:10.1016/j.ejpb.2011.01.007 PMID:21266197

National Science and Technology Council Committee on Technology. (2005). *The National Nanotechnology Initiative: Research and Development Leading to a Revolution in Technology and Industry*. Office of Sciences and Technology Policy.

Ossendorp, F., Toes, R. E., Offringa, R., van der Burg, S. H., & Melief, C. J. M. (2000). Importance of CD4(+) T helper cell responses in tumor immunity. *Immunology Letters, 74*(1), 75–79. doi:10.1016/S0165-2478(00)00252-2 PMID:10996631

Park, S. S., Dong, H., Liu, X., Harrington, S. M., Krco, C. J., Grams, M. P., & Kwon, E. D. (2015). PD-1 restrains radiotherapy-induced abscopal effect. *Cancer Immunology Research, 3*(6), 610–619. doi:10.1158/2326-6066.CIR-14-0138 PMID:25701325

Pinho, S. S., & Reis, C. A. (2015). Glycosylation in cancer: Mechanisms and clinical implications. *Nature Reviews. Cancer, 15*(9), 540–555. doi:10.1038/nrc3982 PMID:26289314

Raman, S. S., Hecht, J. R., & Chan, E. (2019). Talimogene laherparepvec: Review of its mechanism of action and clinical efficacy and safety. *Immunotherapy, 11*(8), 705–723. doi:10.2217/imt-2019-0033 PMID:31045464

Ruck, T., Bittner, S., Wiendl, H., & Meuth, S. G. (2015). Alemtuzumab in Multiple Sclerosis: Mechanism of Action and Beyond. *International Journal of Molecular Sciences, 16*(7), 16414-16439. doi:10.3390/ijms160716414

Saggar, J. K., Yu, M., Tan, Q., & Tannock, I. F. (2013). The tumor microenvironment and strategies to improve drug distribution. *Frontiers in Oncology, 3*, 154. doi:10.3389/fonc.2013.00154 PMID:23772420

Schroeder, A., Heller, D. A., Winslow, M. M., Dahlman, J. E., Pratt, G. W., Langer, R., & Anderson, D. G. (2012). Treating metastatic cancer with nanotechnology. *Nature Reviews. Cancer, 12*(1), 39–50. doi:10.1038/nrc3180 PMID:22193407

Sharma, S., Kelly, T. K., & Jones, P. A. (2009). Epigenetics in cancer. *Carcinogenesis, 31*(1), 27–36. doi:10.1093/carcin/bgp220 PMID:19752007

Stauber, D. J., Debler, E. W., Horton, P. A., Smith, K. A., & Wilson, I. A. (2006). Crystal structure of the IL-2 signaling complex: Paradigm for a heterotrimeric cytokine receptor. *Proceedings of the National Academy of Sciences of the United States of America*, *103*(8), 2788–2793. doi:10.1073/pnas.0511161103 PMID:16477002

Tang, J., Shalabi, A., & Hubbard-Lucey, V. M. (2018). Comprehensive analysis of the clinical immuno-oncology landscape. *Annals of Oncology: Official Journal of the European Society for Medical Oncology*, *29*(1), 84–91. doi:10.1093/annonc/mdx755 PMID:29228097

Tinkle, S., McNeil, S. E., Mühlebach, S., Bawa, R., Borchard, G., Barenholz, Y., Tamarkin, L., & Desai, N. (2014). Nanomedicines: Addressing the scientific and regulatory gap. *Annals of the New York Academy of Sciences*, *1313*(1), 35–56. doi:10.1111/nyas.12403 PMID:24673240

US National Library of Medicine. (2012). https://clinicaltrials.gov/ct2/show/NCT00356980?term

US National Library of Medicine. (2016). https://clinicaltrials.gov/ct2/show/NCT02191878?term

Ventola, C. L. (2017). Cancer immunotherapy, part 3: Challenges and future trends. *P&T*, *42*(8), 514. PMID:28781505

Waldhauer, I., & Steinle, A. (2008). NK cells and cancer immunosurveillance. *Oncogene*, *27*(45), 5932–5943. doi:10.1038/onc.2008.267 PMID:18836474

Wang, R., Billone, P. S., & Mullett, W. M. (2013). Nanomedicine in action: An overview of cancer nanomedicine on the market and in clinical trials. *Journal of Nanomaterials*, *2013*, 2013. doi:10.1155/2013/629681

Weiner, L. M., Dhodapkar, M. V., & Ferrone, S. (2009). Monoclonal antibodies for cancer immunotherapy. *Lancet*, *373*(9668), 1033–1040. doi:10.1016/S0140-6736(09)60251-8 PMID:19304016

Xue, X., & Liang, X. J. (2012). Overcoming drug efflux-based multidrug resistance in cancer with nanotechnology. *Chinese Journal of Cancer*, *31*(2), 100–109. doi:10.5732/cjc.011.10326 PMID:22237039

APPENDIX

Table 1. List of FDA Approved Immunotherapeutics

Drug	Class of Therapeutics	Description	Reference
Axicabtagene ciloleucel	CAR T Cell Therapy	CD19-directed genetically modified autologous T cell immunotherapy, expresses a new gene that targets and kills the lymphoma cells which is again infused	Jain MD *et al.*,2018
Ipilimumab	Check-point inhibitor	Humanized Monoclonal Ab binds CTLA-4 and preventing T-cell inhibition signal pathway	Fellner C. 2012
Aldesleukin	Cytokine mediated Immunomodulators	It is a lymphokine binds to the IL-2 receptor which leads enhanced production of Lymphocyte mitogenesis	Stauber D. *et al.*,2006
Alemtuzumab	Targetted Monoclonal Antibodies	Humanized Monoclonal antibody which binds to the CD52 antigens of B-Lymphocytes and T-Lymphocytes and further leads to ADCC	Ruck, T *et al.*, 2015
Brentuximab vedotin	Antibody-Drug conjugate	Composed of Human-murine IgG to target CD30, MMAE (Monomethyl auristatin E) to disrupt the Microtubules and a protease susceptible linker to link the both	Connors, J. *et al.*, 2018
Sipuleucel-T	Therapeutic vaccine	Autologous cellular immunotherapeutic vaccine which targets the Prostate Specific Cancer (PSA) by inducing the immune response	Fan, A. C., & Leppert, J. T., 2019
T-VEC	Oncolytic Viral Therapy	Talimogene laherparepvec is a genetically modified Herpes Simplex virus which is made to proliferate with the cancerous cells and release the immune stimulatory proteins, thereby elicit the anti-cancer activity	Raman, S. S, *et al.*, 2019

Chapter 15
Current Therapies and Future Prospects:
Peptide-Conjugated Nanoparticle in Targeted Drug Delivery for Cancer Treatment

Muddukrishnaiah K.
Anna University, Tiruchirappalli, India

ABSTRACT

Due to drug resistance, lack of cancer cell selectivity, and solubility, conventional cancer treatments lose their therapeutic uses, and as such, new therapeutic agents need to be developed. Nanomaterials and peptides are increasingly being used in the fields of cancer diagnosis, biomarker discovery due to their therapeutic values and novel way of targeting and curing the disease. Synergism among the peptide-conjugated nanoparticles is an exhilarating group of materials, not only sharing the benefits of conventional nanomedicine, but also possessing the unique properties of excellent biocompatibility, biodegradability, versatile sensitivity, specific biological purpose, and synthetic feasibility. These virtues inspired by the scientists and have taken advantage in the peptide-conjugated nano drugs for the accurate delivery of drugs reliably to the site of the lesion. This chapter offers a summary of emerging technologies that have recently been developed in the broad field of peptide-conjugated nanoparticles and offers guidance for targeted drug delivery and cancer therapy.

INTRODUCTION

Cancer is among the world's most significant diseases and is expected to result in 1,688,780 new diagnoses and 600,920 cancer deaths in 2017. The number of cancer patient cases is predicted to increase by almost seventy per cent in the next twenty years. Cancer treatment is a significant part of cancer diagnosis, along with chemotherapy and radiotherapy. (MDR) Multidrug resistance, the key pathway by

DOI: 10.4018/978-1-7998-5049-6.ch015

which most cancers develop resistance to cancer treatment, is one of the biggest obstacles to successful clinical care for various forms of cancer. Cancer treatment is the systemic technique most widely used to inhibit the growth of cancer cells, disease development and metastases. Anticancer medications, however, not only destroy cancer cells that are proliferating, and attack healthy cells that cause adverse effects. Antitumor drug carriers are also desperately required to preserve or enhance the effectiveness of chemotherapy and at the same time, the seriousness of symptoms and complications. When a drug, carriers for cancer therapy, nanomaterials are increasingly being used. In the fields of cancer diagnosis and biomarker discovery, nanomaterials also appeal to scientists. In preclinical and clinical trials, multiple antitumor nanodrugs are currently being tested and show promise in therapeutic and other environments. Nanoparticles, which be capable of being tailored to various biological activities and can be used in a verity of conditions, provide a safe and effective way to deliver chemotherapy (Langer, 1990; LaVan et al., 2003; Shi et al., 2011).

Nanomaterials as drug carriers have several advantages. Nanocarriers, i) Increase water solubility and preserve blood dissolved drugs and increase the pharmacokinetic and pharmacological of drugs; ii) Monitor drug delivery in a tissue- or cell-specific manner, thereby reducing drug accumulation in the chest, liver, spleen and other non-target organs and increasing therapeutic effectiveness; and iii) Develop a combination of imaging and therapeutic agents to track therapeutic efficacy in real-time.

CANCER MULTIDRUG RESISTANCE

Multidrug Resistance Factors (MDRF)

The resistance of multiple medications, the key process through which cancer cells develop chemotherapy resistance, relics a significant barrier to effective chemotherapy. It affects patients to buy through many cancers of the blood and tumours, as well as cancers of the cervical, ovary, renal, lung, liver, and intestinal cancers. Substantial efforts were made in the previous two decades to elucidate the cancer cycle of MDR. Many important factors responsible for mediation are briefly discussed in this analysis.

ATP-Binding Cassette Transporters (ABC Transports)

Some of the methods which play a crucial role in cancer patients are to inhibit the intracellular aggregation of anticancer drugs by releasing protein molecules that move drugs out from the cells. Such carriers operate on cell organelles and prevent anticancer drugs from reaching their cellular targets. Most of these molecules belong to the mammalian adenosine triphosphate (ATP) carrier family, a vast proportion of biologically complex transmembrane proteins shared with the plasma cell membrane. In human beings, depending on their sequence similarity and domain arrangement, 48 types of ATP-binding cassette transporters have been identified and classified into seven separate subfamilies (ABCA-G). Upon drug internalisation via the plasma membrane, transporters recognise active molecules use ATP hydrolysis energy to extract biomolecules from the cells resulting in poor bioavailability and potentially contributing to drug tolerance in cancer cells.

NANOPARTICLES

The term nanoparticle refers to materials ranging from 1000 nm to 1 nm in size of particles. At least in one aspect, nanoparticles used mainly as drug carriers molecules may being as large as 100 nm (Barry & Vertegel, 2013). The nanoparticle is a synthesis of different compounds lipids, metals, natural and synthetic polymers. Since cells absorb nanoparticles effectively than larger macromolecules, they are an excellent material for a delivery system. Over the past 50 years, several nanodrugs have been produced (figure 1).

Figure 1. Timeline of nanomedicine growth. (Source: Zhen LI et al. 2017.)

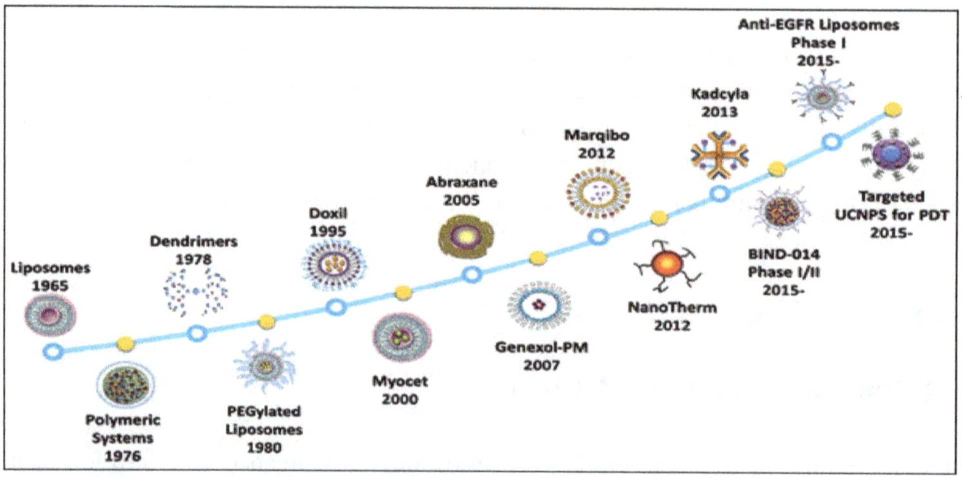

NDDs (Nano drug delivery system) is a fast developing and ever-growing area of active disease treatment and diagnoses, such as cancer, inflammation, neurological disease, and cardiovascular disease. Compared to small molecular weight compounds such as Taxotere and Cisplatin, Nano drug delivery system has benefits of enhanced circulation of blood and improved tumour tissue diffusion strength as well as the controllable release of cancer medicine into the cytoplasm. The traditional carriers for drug delivery are liposomes, nanoparticles and micelles (Chen et al., 2017; Huang et al., 2014; Li & Wang, 2018; Mahmoudi et al., 2017; Ruber Perez et al., 2015; Torchilin, 2005; Wang et al., 2017).

PROPERTIES OF NANOCARRIERS

Physico-Chemical Characteristics of Nanocarriers

Nanoparticles for cancer research may exist tailored to tailor different tumours in volume, shape, and surface characteristics. The shape is efficient for the transfer of nanocarriers across the bloodstream and eventual distribution to the cancer cells. Although nanoparticles in the leaky capillaries of tumours may accumulate more quickly than larger ones, they can also migrate into healthy tissues [Bregoli L, Movia D et al. 2016]. Larger nanoparticles, on the other hand, can not extravasate as quickly and therefore have a highly variable distribution in the bloodstream. Optimising the size of nanoparticles will help improve the precise absorption of tumour tissue [Truong NP, Whittaker MR et al. 2015]. Nanocarriers ' charge can also affect their blood stability and distribution. Previously, positively charge nanomaterials had been exposed to target cancer vessels most effectively; however, the neutral charge control after efflux allowed nanoparticles to migrate more rapidly to the cancer cells—[Stylianopoulos T, Poh M-Z, Insin N et al. 2010].

Figure 2. A schematic diagram demonstrating passive and active nanocarrier delivery to the target cell with a drug. (Source: Agnieszka Pudlarz et al. 2018.)

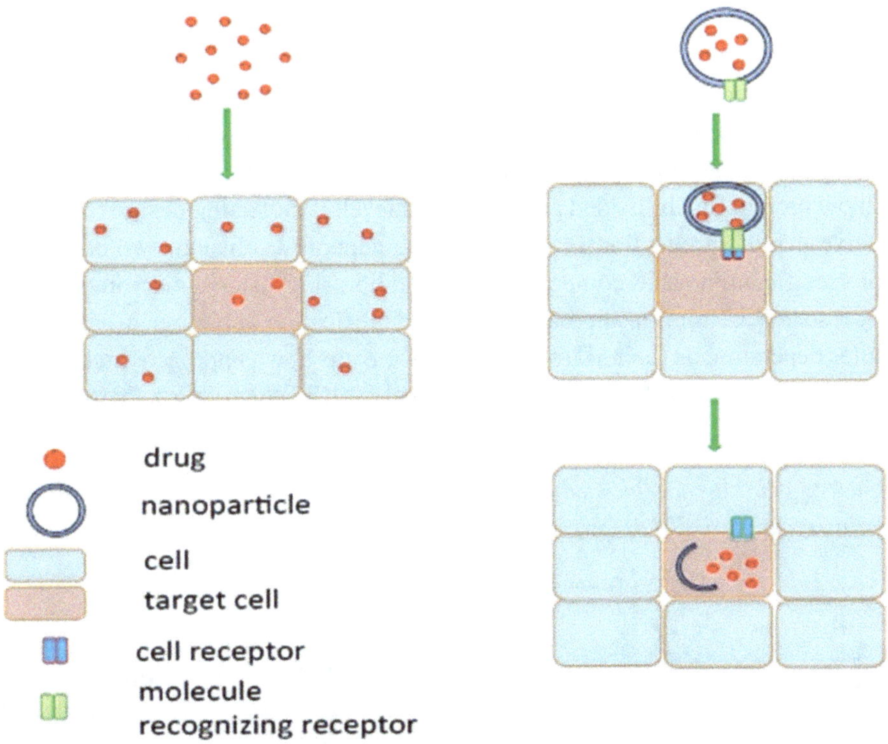

Solubility Degradation and Elimination

Medication with low aqueous solubility can be extracted from the blood after entrance into the cancer cells. Using lipophilic nanomaterials to encapsulate certain medications will increase their solubility, thereby enhancing in vivo bioavailability and allowing better effective delivery. Hydrophilic and non-ionic polymer coating nanoparticles with polyethylene glycol have been shown to improve the solubility and stability of nanoparticles. Once polyethylene glycol is removed, it should not interfere with the function of charged molecules such as RNA and DNA.

PEPTIDES

Biological and chemical peptides are compounds that have a variety of biological roles. These behave as signalling, substrates and antagonists of proteins, anticancer, biology regulators, etc. Peptides thus play a significant role as a medicinal and diagnosis component in biotechnological applications (Sahoo & Labhasetwar, 2003). Advantages depend on a manufacturing method, including biocompatibility, low price, controllable bioactivity, a range of chemicals, and precise targeting. In addition, they can be conveniently synthesised, for example, by means of solid-phase peptide processes, where the amino acid sequence be able to be accurately chosen at the cellular stage by changing the specific units. Data, peptides perform a significant function as the sequence of amino acids sequence continues to drive or tends to affect the molecular self-assembly by incorporating weak non-covalent bonds, electrostatic attractions, hydrogen bonds, hydrophobic interactions, and van der Waals connections. (Tesauro et al., 2019). In fact, peptides play a bioactive function as particular receptors, such as those overexpressed by defective cells, are known to be entirely or partially recognised by the sequence (Vanhee et al., 2011).

Peptides contain drawn an enormous arrangement of notice in biological applications as an innovative drug that can show protein flexibility and provide a high level of molecular design functionality. Present artificial bioactive peptide production strategies can be generally split into two categories. I) Design and analysis of specific amino acid composition peptide libraries inside a macromolecular topography (peptide library research, bottom-up approach). ii) The identification of biological active sequences from protein molecules depending on their 3D (structural design, top-down approach) architectures;(figure 3) (Brack & Orgel, 1975; Marasco et al., 2008; Ryvkin et al., 2018; Wang & Yu, 2018).

Figure 3. Bioactive peptides for biomedical application and their combination with nanoparticles. (Source: Woo-jin Jeong et al. 2018)

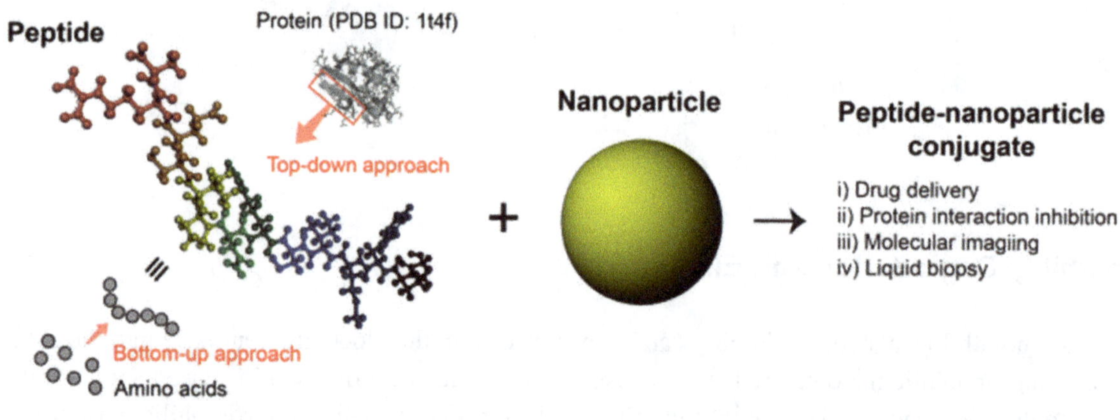

The identification of many protein receptors, peptides tumour-related and proteins is usual to produce a new generation of more effective and tailored anticancer medicines in the future, capturing a significant proportion of the demand for therapeutic cancer (figure 4).

The "biological molecules" therapeutic choice for cancer include the use of proteins, monoclonal antibodies and peptides. mAbs (Monoclonal antibodies) and large protein ligands have two main drawbacks relative to peptides: poor dissemination to tumours due to their large scale and dose-limiting sensitivity to the liver and bone marrow due to non-specific penetration in the reticuloendothelial system. Therefore, the use of such biomolecules was restricted to either the vascular target of endothelial or haematological malignancies on the luminal side of the tumour vessel.

Figure 4. Peptides can be used as an anticancer agent, as a drug carrier for cytotoxic products, as a vaccine, as well as hormones and as a carrier for radionuclides. (Source: Jyothi Thundimadathil 2012).

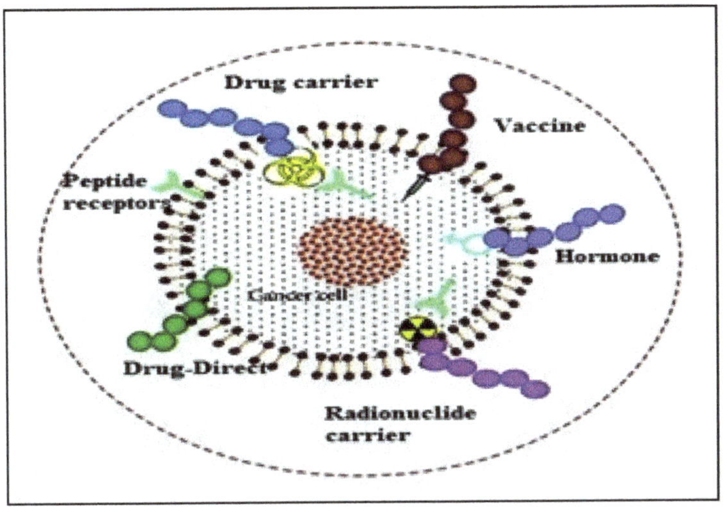

Table 1. Peptides in cancer treatment

Peptide	Indications
Agonists	
Buserelin	Prostate cancer
Gonadorelin	Cystic ovarian disease
Goserelin	hypothalamic-pituitary-gonadal (HPG) function
Histrelin	Prostate cancer;
Leuprolide	Prostate cancer;
Nafarelin	Treat symptoms of endometriosis
Triptorelin	Prostate cancer
Antagonists	
Abarelix	Prostate cancer
Cetrorelix	Prostate cancer
Degarelix	Prostate cancer
Ganirelix	Fertility treatment

1. PEPTIDE NANOPARTICLES

Peptide Self-Assembled Nanostructures

Peptides can accumulate in a variety of nanomaterials, like nanotubes, nanofibers, nanospheres plus nanovesicles, assisted with their processes and self-assembly environment. Various forms in addition to configurations of peptides can be self-assembled into nanostructures, such as cyclic and linear peptides, amphiphilic peptides, and a-helical peptides.

Figure 5. In supramolecular structures, different classes of peptides can be structured to tackle the self-assembling phenomenon. (Diego Tesauro et al. 2019)

a-Helical and *b*-Sheet Peptides

The essential feature in the development and production of peptide-based nanoparticles is the arrangement of the peptide backbone in the secondary structures of either helical or -sheet. This is a result of the correlations in the peptide form of the H-bonding mechanism in the amide and carbonyl groups. The strands then transform into a self-assembled arrangement that can be reconfigured in parallel or anti-parallel frames, based on the orientation of the peptide sequences. Typically, the peptide is meant to have residues of repetitive amino acids and distinct lipophilic and lipophobic regions. As a result, the lipophilic moiety inside the self-assembled nanomaterial could be shielded when the hydrophobic portion could be more exposed to the solvent system (Wagner et al., 2005). β-sheets and α-helices are shaped by solitary peptide chains in which mechanism of amide backbone are hydrogen bond formed intramolecularly. This structure contributes to the surfaces of each helix of side chains of amino acids.

Standard α-helical peptides with 2, 5 helices are combined and shaped in nanofibers (Moutevelis & Woolfson, 2009; Potekhin et al., 2001). These can also be self-assembled into nanofibers by helical coiled-coil configurations provided these have at least 30 amino acid residues (Tesauro et al., 2019).

Non-covalent reactions that often expose the key position in the overall structure are also accompanied by peptide self-assembly processes of this reason, in particular, non-covalent connections will be taken into account when developing self-assembled peptide nanoparticles of drug delivery. Indeed, non-covalent correlations should be implemented sensibly in the techniques. For example, pH values, temperature distribution, and solvent polarity, such non-covalent relations are easily unacceptable to the surrounding environment. Nevertheless, pH values are essential in peptide residues that are stronger in charged amino acids such as Glu, Asp, Lys, His, and Arg, as mentioned above. As a function of level, such peptides be able to exhibit positively or negatively shell loads. Instead, according to the pH values, these peptides be able to be self-assembled into various nanoformulations (Veiga et al., 2012).

1.1. Linear Peptides

Literature data showed which brief (under 6 residues) and insanely-short (known as the dipeptides and tripeptides) planar peptides are capable of being self-assembled into several different nanomaterials. This especially important dimension enables the production and purification steps to be reduced and the expense of the manufacturing process to be reduced (Reches & Gazit, 2003).

Among the most proof obligations of linear self-assembly, proteins is the extremely short homodipeptide Phe-Phe (FF), defined by Gazit and colleagues (in 2003) as the smallest distance of the aggregated A-amyloid peptides (A 1-40 and A 1-42). The new FF-based peptide materials are just a few examples of nanotubes, nanowires, nanofibrils, spherical vesicles and organogels. These materials have electrical (Adler-Abramovich et al., 2010), mechanical properties (Vasilev et al., 2016), electrochemical properties (Xie et al., 2016), optical properties (photoluminescence and optical waveguide properties). Hybrid hydrogel encapsulating docetaxel was also prepared using the Fmoc-FF peptide and the calcium-ion alginate co-assembly. The synergistic influence of these two elements, the final substance both in the water and the phosphate-buffered, demonstrated a very much excellent strength than the individual components. Regulated active molecule release was accomplished via adjusting the concentration ratio of peptide-polysaccharide (Aviv et al., 2018). Fmoc-FF dipeptide was also used to provide natural polymers such as hyaluronic acid (HA), a significant constituent of the matrix components, mechanical stiffness and stability. Fmoc-FF / HA composite hydrogels showed prolonged release of curcumin, a lipophilic polyphenol with antioxidant, anti-inflammatory and anticancer activity. In addition, a clear connection among the implementation of curcumin released and the concentration of Fmoc-FF peptide within the hydrogel matrix was found in this analysis (De Santis et al., 1974).

Cyclic Peptides

In 1974, theoretical research proposed the potential arrangement in a hollow tubular structure of a cyclic peptide (Ghadiri et al., 1993). Twenty years later, by stacking cyclic peptides with alternating amino acids, D and L, Ghadiri, and co-authors resolved the first crystalline nanotube structure (Chapman et al., 2012). On the outside region, the peptide side chains are formed. A result of structural changes L and D amino acids, they were organised in the typical nanotube structures. The nanotubes are self-assembled and stabilised by hydrogen bonds between amide groups of the cyclic backbone. In addition to alternating D-and L-type amino acids, some cyclic peptide sequences can be self-assembled by alternating between α-and β-amino acids, β-amino acids, and δ-amino acids by molecular stacking and H-bonds between backbones (Fernandez-Lopez et al., 2001; Hartgerink et al., 1996; Ishihara & Kimura, 2010; Wang et al., 2014). The cavity size depends on the cyclic peptide length, from 2 to 13 Å, which rises from a tetramer to a dodecamer. This side-chain charging parameter is essential for biotechnology applications. By tailoring the cyclic peptide's chemical structure, supramolecular self-assembled architectures can be used to meet application requirements, including stimulus-responsive nanomaterials for ant cancer agents, ion channelling, ion sensing, and gene delivery (Deehan et al., 2015; Jeong et al., 2013). There was a high drug encapsulation ratio of polyethyline glycol aggregates loaded with doxorubicin. The polyethyline glycol modified doxorubicin loaded CPNT bundles showed higher cytotoxicity, increased doxorubicin absorption, and altered intracellular distribution of doxorubicin in vitro breast cancer cells MCF-7/ADR. Twenty years later, by stacking cyclic peptides with alternating amino acids D and L (Hamley, 2011).

Amphiphilic Peptides

Biology has chosen hydrophobic and hydrophilic molecules to create a life by allowing them to universalise aspects of the program. Membranes will then hold biomolecules to make it possible for macromolecules to travel. Imitating architecture, hydrophilic and hydrophobic peptides are self-assembled into various nanostructures, including organelles, microemulsions, nanoparticles and nanotubes, thereby performing a major part in the growth of biotechnological nanostructures (Rizvi & Saleh, 2018; Versluis et al., 2010). The simplest self-aggregating peptides consist of small or extended homo chains of hydrophobic amino acids, such as Val, Ala, Gly and Phe, followed via one or two electrostatic residues, and these molecules have separate parts of hydrophobic and hydrophilic.

THE PEPTIDE–NP CONJUGATION

Nanoparticles have novel physical and chemical properties, separate from traditional polymer composites. Their mega-small scale and large surface-to-volume proportion are useful in the development of nanostructures which can interact in such a specific way through a range of nano-sized molecules(Habibi et al., 2016). The majority clear-cut approach to the development of peptide-based nanomaterials is self-assembly (Jeong et al., 2018; Kitov & Bundle, 2003). Peptide – nanoparticles conjugation provides improved manage over the structural characteristics of nanomaterials, allowing for easy adjustment of the overall form, size and volume of the conjugate by manufacturing NP scaffolds modified to the future applications. It's multiagency with the purpose of the PNCs will contain. Many biological system interactions are based on non-covalent interactions such as hydrogen, ionic bonds, van der waals, stacking forces (π–π), and lipophilic relationships. Although single bonds remain comparatively small, their competitive activity is increasingly contingent on the number of different binding pairs for steady affinity kinetics focused mostly on aggregate binding force principle. (Gargano et al., 2001; Hong et al., 2007).

TARGETED DRUG DELIVERY

Molecularly located within proteins and small molecular compounds, peptides in pharmacological studies can effectively leverage the morphological and chemical compensation of the two primary materials. As mentioned above, a quantity of peptides, along with nanoparticles, have proven their shows potential drug delivery capability, inhibition of pathogenic biological activity, molecular imaging, and liquid biopsy. Medical translation of nanocomposites appears challenging for the following reasons, increases the significant. First, the activity of the PNC was not completely understood under physiological circumstances, such as blood and intracellular space. Second, also on nanomaterial surfaces, peptides are often vulnerable to enzymatic degradation, needing extra defence techniques to preserve their work instead of expanding the structural level and physicochemical difficulty of the conjugates. Second, the possible immunogenicity of designed PNCs, which is a significant barrier to *in vivo* and clinical application, should be discussed. Finally, covalent binding with nanoparticles or other similar groups also marks in the impairment of the biomedical roles of peptide molecules. In answering these issues, it is likely that PNC systems will include a new polymeric material that could load the void in existing biotechnology fields, like as drug utilisation targets, the battle against multidrug-resistant pathogens, and separation. PNCs are likely to turn into a novel medium that can be used in conventional diagnostic and remedial systems, especially in tandem with rapid advancements in nanotechnology.

REFERENCES

Adler-Abramovich, L., Kol, N., Yanai, I., Barlam, D., Shneck, R. Z., Gazit, E., & Rousso, I. (2010). Self-assembled organic nanostructures with metallic-like stiffness. *Angewandte Chemie International Edition*, *49*(51), 9939–9942. doi:10.1002/anie.201002037 PMID:20878815

Aviv, M., Halperin-Sternfeld, M., Grigoriants, I., Buzhansky, L., Mironi-Harpaz, I., Seliktar, D., Einav, S., Nevo, Z., & Adler-Abramovich, L. (2018). Improving the Mechanical Rigidity of Hyaluronic Acid by Integration of a Supramolecular Peptide Matrix. *ACS Applied Materials & Interfaces*, *10*(49), 41883–41891. doi:10.1021/acsami.8b08423 PMID:30211538

Barry, J. N., & Vertegel, A. A. (2013). Nanomaterials for Protein Mediated Therapy and Delivery. *Nano LIFE*, *3*(4), 1–17. doi:10.1142/S1793984413430010 PMID:25414730

Brack, A., & Orgel, L. E. (1975). ß structures of alternating polypeptides and their possible prebiotic significance. *Nature*, *256*(5516), 383–387. doi:10.1038/256383a0 PMID:238134

Chapman, R., Danial, M., Koh, M. L., Jolliffe, K. A., & Perrier, S. (2012). Design and properties of functional nanotubes from the self-assembly of cyclic peptide templates. *Chemical Society Reviews*, *41*(18), 6023–6041. doi:10.1039/c2cs35172b PMID:22875035

Chen, H., Zhang, W., Zhu, G., Xie, J., & Chen, X. (2017). Rethinking cancer nanotheranostics. *Nature Reviews. Materials*, *2*(7), 1704. doi:10.1038/natrevmats.2017.24 PMID:29075517

De Santis, P., Morosetti, S., & Rizzo, R. (1974). Conformational Analysis of Regular Enantiomeric Sequences. *Macromolecules*, *7*(1), 52–58. doi:10.1021/ma60037a011

Deehan, M., Garces, S., Kramer, D., Baker, M. P., Rat, D., Roettger, Y., & Kromminga, A. (2015). Managing unwanted immunogenicity of biologicals. *Autoimmunity Reviews*, *14*(7), 569–574. doi:10.1016/j.autrev.2015.02.007 PMID:25742758

Fernandez-Lopez, S., Kim, H. S., Choi, E. C., Delgado, M., Granja, J. R., Khasanov, A., Kraehenbuehl, K., Long, G., Weinberger, D. A., Wilcoxen, K. M., & Ghadiri, M. R. (2001). Antibacterial agents based on the cyclic D,L-a-peptide architecture. *Nature*, *412*(6845), 452–455. doi:10.1038/35086601 PMID:11473322

Gargano, J. M., Ngo, T., Kim, J. Y., Acheson, D. W. K., & Lees, W. J. (2001). Multivalent inhibition of AB (5) toxins. *Journal of the American Chemical Society*, *123*(51), 12909–12910. doi:10.1021/ja016305a PMID:11749553

Ghadiri, M. R., Granja, J. R., Milligan, R. A., McRee, D. E., & Khazanovich, N. (1993). Self-assembling organic nanotubes based on a cyclic peptide architecture. *Nature*, *366*(6453), 324–327. doi:10.1038/366324a0 PMID:8247126

Habibi, N., Kamaly, N., Memic, A., & Shafiee, H. (2016). Self-assembled peptide based nanostructures: Smart nanomaterials toward targeted drug delivery. *Nano Today*, *11*(1), 41–60. doi:10.1016/j.nantod.2016.02.004 PMID:27103939

Hamley, I. W. (2011). Self-assembly of amphiphilic peptides. *Soft Matter*, *7*(9), 4122–4138. doi:10.1039/c0sm01218a

Hartgerink, J. D., Granja, J. R., Milligan, R. A., & Ghadiri, M. R. (1996). Peptide-amphiphile nanofibers: A versatile scaffold for the preparation of self-assembling materials. *Journal of the American Chemical Society*, *118*, 43–50. doi:10.1021/ja953070s

Hong, S., Leroueil, P. R., Majoros, I. J., Orr, B. G., Baker, J. R. Jr, & Banaszak Holl, M. M. (2007). M.M. Banaszak Holl, The binding avidity of a nanoparticle-based multivalent targeted drug delivery platform. *Chemistry & Biology*, *14*(1), 107–115. doi:10.1016/j.chembiol.2006.11.015 PMID:17254956

Huang, F., Wang, J., Qu, A., Shen, L., Liu, J., Liu, J., Zhang, Z., An, Y., & Shi, L. (2014). Y. an and L. Shi, Maintenance of amyloid beta peptide homeostasis by artificial chaperones based on mixed-shell polymeric micelles. *Angewandte Chemie International Edition*, *53*(34), 8985–8990. doi:10.1002/anie.201400735 PMID:24985739

Ishihara, Y., & Kimura, S. (2010). Nanofiber formation of amphiphilic cyclic tri-b-peptide. *Journal of Peptide Science*, *16*(2), 110–114. doi:10.1002/psc.1206 PMID:20063334

Jeong, W. J., Kwon, S. H., & Lim, Y. B. (2018). Modular self-assembling peptide platform with a tunable thermo responsiveness via a single amino acid substitution. *Advanced Functional Materials*, *28*, 35.

Jeong, W. J., Lee, M. S., & Lim, Y. B. (2013). Helix stabilised, thermostable, and protease-resistant self-assembled peptide nanostructures as potential inhibitors of protein–protein interactions. *Biomacromol*, *14*(8), 2684–2689. doi:10.1021/bm400532y PMID:23844716

Kitov, P. I., & Bundle, D. R. (2003). on the nature of the multivalency effect: A thermodynamic model. *Journal of the American Chemical Society*, *125*(52), 16271–16284. doi:10.1021/ja038223n PMID:14692768

Langer, R. (1990). New methods of drug delivery. *Science*, *249*(4976), 1527–1533. doi:10.1126cience.2218494 PMID:2218494

LaVan, D. A., McGuire, T., & Langer, R. (2003). Small-scale systems for in vivo drug delivery. *Nature Biotechnology*, *21*(10), 1184–1191. doi:10.1038/nbt876 PMID:14520404

Li, L. L., & Wang, H. (2018). Infection-targeted bactericidal nanoparticles. *Nature Biomedical Engineering*, *2*(2), 56–57. doi:10.103841551-018-0199-9 PMID:31015624

Mahmoudi, M., Yu, M., Serpooshan, V., Wu, J. C., Langer, R., Lee, R. T., Karp, J. M., & Farokhzad, O. C. (2017). Multiscale technologies for treatment of ischemic cardiomyopathy. *Nature Nanotechnology*, *12*(9), 845–855. doi:10.1038/nnano.2017.167 PMID:28875984

Marasco, D., Perretta, G., Sabatella, M., & Ruvo, M. (2008). Past and future perspectives of synthetic peptide libraries. *Current Protein & Peptide Science*, *9*(5), 447–467. doi:10.2174/138920308785915209 PMID:18855697

Moutevelis, E., & Woolfson, D. N. (2009). A Periodic Table of Coiled-Coil Protein Structures. *Journal of Molecular Biology*, *385*(3), 726–732. doi:10.1016/j.jmb.2008.11.028 PMID:19059267

Potekhin, S. A., Melnik, T. N., Popov, V., Lanina, N. F., Vazina, A. A., Rigler, P., Verdini, A. S., Corradin, G., & Kajava, A. V. (2001). De novo design of fibrils made of short helical coiled coil peptides. *Chemistry & Biology*, *8*(11), 1025–1032. doi:10.1016/S1074-5521(01)00073-4 PMID:11731294

Reches, M., & Gazit, E. (2003). Casting metal nanowires within discrete self-assembled peptide nanotubes. *Science*, *300*(5619), 625–627. doi:10.1126cience.1082387 PMID:12714741

Rizvi, S. A. A., & Saleh, A. M. (2018). Applications of nanoparticle systems in drug delivery technology. *Saudi Pharmaceutical Journal*, *26*(1), 64–70. doi:10.1016/j.jsps.2017.10.012 PMID:29379334

Ruber Perez, C. M., Stephanopoulos, N., Sur, S., Lee, S. S., Newcomb, C., & Stupp, S. I. (2015). The Powerful Functions of Peptide-Based Bioactive Matrices for Regenerative Medicine. *Annals of Biomedical Engineering*, *43*(3), 501–514. doi:10.100710439-014-1166-6 PMID:25366903

Ryvkin, H., Ashkenazy, H., Weiss-Ottolenghi, Y., Piller, C., Pupko, T., & Gershoni, J. M. (2018). Phage display peptide libraries: Deviations from randomness and correctives. *Nucleic Acids Research*, *46*(9), e52. doi:10.1093/nar/gky077 PMID:29420788

Sahoo, S. K., & Labhasetwar, V. (2003). Nanotech approaches to drug delivery and imaging. *Drug Discovery Today*, *8*(24), 1112–1120. doi:10.1016/S1359-6446(03)02903-9 PMID:14678737

Shi, J., Xiao, Z., Kamaly, N., & Farokhzad, O. C. (2011). Self-assembled targeted nanoparticles: Evolution of technologies and bench to bedside translation. *Accounts of Chemical Research, 44*(10), 1123–1134. doi:10.1021/ar200054n PMID:21692448

Siegel, R. L., Miller, K. D., Fedewa, S. A., Ahnen, D. J., Meester, R. G., Barzi, A., & Jemal, A. (2017). Colorectal cancer statistics. *CA: a Cancer Journal for Clinicians, 67*(3), 177–193. doi:10.3322/caac.21395 PMID:28248415

Tesauro, D., Accardo, A., Diaferia, C., Milano, V., Guillon, J., Ronga, L., & Rossi, F. (2019). Peptide-Based Drug-Delivery Systems in Biotechnological Applications: Recent Advances and Perspectives. *Molecules (Basel, Switzerland), 24*(2), 351. doi:10.3390/molecules24020351 PMID:30669445

Torchilin, V. P. (2005). Recent advances with liposomes as pharmaceutical carriers. *Nature Reviews. Drug Discovery, 4*(2), 145–160. doi:10.1038/nrd1632 PMID:15688077

Vanhee, van der Sloot, Verschueren, Serrano, Rousseau, & Schymkowitz. (2011). Computational design of peptide ligands. *Trends Biotechnol., 29*(5), 231-239.

Vasilev, S., Zelenovskiy, P., Vasileva, D., Nuraeva, A., Shur, V. Y., & Kholkin, A. L. (2016). ShurYa, V.; Kholkin, A.L. Piezoelectric properties of diphenylalanine microtubes prepared from the solution. *Journal of Physics and Chemistry of Solids, 93*, 68–72. doi:10.1016/j.jpcs.2016.02.002

Veiga, A. S., Sinthuvanich, C., Gaspar, D., Franquelim, H. G., Castanho, M. A. R. B., & Schneider, J. P. (2012). Arginine-rich self-assembling peptides as potent antibacterial gels. *Biomaterials, 33*(35), 8907–8916. doi:10.1016/j.biomaterials.2012.08.046 PMID:22995710

Versluis, F., Marsden, H. R., & Kros, A. (2010). Power struggles in peptide-amphiphile nanostructures. *Chemical Society Reviews, 39*(9), 3434–3444. doi:10.1039/b919446k PMID:20644886

Wagner, D. E., Philips, C. L., Ali, W. M., Nybakken, G. E., Crawford, E. D., Schwab, A. D., Smith, W. F., & Fairman, R. (2005). Toward the development of peptide nanofilaments and nanopores as smart materials. *Proceedings of the National Academy of Sciences of the United States of America, 102*(36), 12656–12661. doi:10.1073/pnas.0505871102 PMID:16129839

Wang, M., Du, L., Wu, X., Xiong, S., & Chu, P. K. (2011). Charged Diphenylalanine Nanotubes and Controlled Hierarchical Self-Assembly. *ACS Nano, 5*(6), 4448–4454. doi:10.1021/nn2016524 PMID:21591732

Wang, S. H., & Yu, J. (2018). Structure-based design for binding peptides in anticancer therapy. *Biomaterials, 156*, 1–15. doi:10.1016/j.biomaterials.2017.11.024 PMID:29182932

Wang, Y., Yi, S., Sun, L., Huang, Y., Lenaghan, S. C., & Zhang, M. (2014). Doxorubicin-loaded cyclic peptide nanotube bundles overcome chemoresistance in breast cancer cells. *Journal of Biomedical Nanotechnology, 10*(3), 445–45. doi:10.1166/jbn.2014.1724 PMID:24730240

Wang, Y. F., Liu, L., Xue, X., & Liang, X. J. (2017). Nanoparticle-based drug delivery systems: What can they really do in vivo? *F1000 Research, 6*, 681. doi:10.12688/f1000research.9690.1 PMID:28620465

Xie, Y., Zhao, J., Huang, R., Qi, W., Wang, Y., Su, R., & He, Z. (2016). Calcium-Ion-Triggered Co-assembly of Peptide and Polysaccharide into a Hybrid Hydrogel for Drug Delivery. *Nanoscale Research Letters, 11*(1), 184. doi:10.118611671-016-1415-8 PMID:27067732

Chapter 16
Smart Theranostic Applications of Metal Nanoparticles Against Cancer

Malathi Balasubramaniyan
Bannari Amman Institute of Technology, India

Abdul Azeez Nazeer
Bannari Amman Institute of Technology, India

Vimalraj Vinayagam
Bannari Amman Institute of Technology, India

Sudarshana Deepa Vijaykumar
National Institute of Technology, Tadepalligudem, India

ABSTRACT

Metals such as silver, gold, and copper were used in ancient times for their medicinal properties. When these metals are converted to nanoparticles, they show unique and advanced physicochemical and biological properties due to their enhanced surface to volume ratio. Hence, these properties are utilized by researchers to develop highly specific diagnostic tools as well as a therapeutic agent against cancer. Cancer is a complex disease-causing desolation and death. Early detection and treatment is the only way to evade mortality. This chapter focuses on metal nanoparticles used as a theranostic agent against cancer. It summarizes the synthesis methodology along with their advantages, drawbacks and characterizations. Their recent application in diagnosing and treating cancer has also been highlighted.

DOI: 10.4018/978-1-7998-5049-6.ch016

INTRODUCTION

Cancer is a debilitating disease standing as one of the major reason for the high rate of mortality in the world annually. The current lifestyle: choices of food, pollution, high usage of chemicals in agriculture, aquaculture, cosmetics, and construction, exposure to radiation etc., are the leading causes for triggering cancer. These triggering agent cause hindrance in normal cellular functions by altering cellular signals, replication cycle, and apoptosis hence making cancer difficult to diagnose and treat (J. Ferlay et al., 2019; Jacques Ferlay et al., 2015).

importance of Early Diagnosis and Treatment

Cancer can slowly spread to other parts of the body, making it excruciating to cure. The current diagnostic for cancers involves pathological analysis using radiological and histopathological examinations. These techniques do not detect cancer in their primary stage hence making them highly difficult to cure. The chemotherapy, immunotherapy, tumour treating field and surgery are few treatment techniques currently being followed. These treatments are linked with the extreme level of drawbacks such as side effects of non-specific drug distribution, cytotoxicity, a low percentage of patients response and development of drug resistance (Mun, Babiker, Weinberg, Kirson, & Von Hoff, 2018). The morbidity and mortality rate caused due to cancer is very high due to delayed diagnosis and side effects of current treatments. Hence, the development of early detection and treatment system against cancer has become highly crucial.

Nanotechnology Against Cancer

A cancer therapy should be smart in nature, delivering minimal drug at the specific site to obtain maximum outcome and reduced side effects. The heterogenicity of cancer and the person it infects makes the eradication highly difficult but not impossible. One of the most upcoming and promising solutions is tailor-made diagnostic as well as therapeutic nanomaterials. These nanoparticles with dual properties as also know as “theranostic nanoparticles or nano theranostics”.

The term “theranostics” was first coined in 2002 by Funkhouser to define the applications that utilizes both diagnostic and therapeutic process (Funkhouser, 2002). Understanding the microenvironment of the tumour and its properties and the cellular phenotype is important to achieve the theranostic property of NPs. Theranostics is advantageous because of its dual ability to both diagnose and treat the tumour (Kelkar & Reineke, 2011). The two halves of theranostics are diagnosis and therapy.

These theranostic nanoparticles are used for imaging, detection of cancer cell, the release of the drug and monitoring drug distribution (Chen, Zhang, Zhu, Xie, & Chen, 2017; Harmsen, Wall, Huang, & Kircher, 2017). Recent research developments have led to the development of novel platforms in nanotechnology for smart particles such as metal nanoparticles, quantum dot therapy, magnetic nanoparticles, liposomes, nanotubes and many more. This chapter focuses mainly on selected metal nanoparticles and their application in the identifying and treating cancer.

SYNTHESIS OF NANOPARTICLES

T nanoparticles can be synthesized using a top-down or bottom-up approach. In the top-down approach, the bulk materials are carved down to form the desired shape, while in the bottom-up approach, the particles are rearranged in the molecular level to form the nanoparticles. The synthesis methodology of metal nanoparticles can be broadly classified into three types: physical, chemical and biological techniques. In all the techniques, the metal nanoparticles are synthesized from their respective precursor salt(Y. Li et al., 2018). The reducing agent in the case of bottom-up breaks down the precursor leading to the formation of metal ions followed by growth together to form nanoparticle of defined shape and size based on the environment.

Physical Method

In the physical method the top-down approach is followed by applying mechanical force, different energy radiation for ablation, leaching to form nanoparticles. It involves techniques such as laser ablation, arc plasma, high energy ball milling, ultra-sonication, physical vapour deposition, electrospraying, sonication, irradiation and microwave-assisted synthesis. The product obtained from this technology is highly pure and is of industrial grade. However, the drawbacks include surface irregularity and high energy consumption (Siddiqi, Husen, & Rao, 2018; W. Zhang, Qiao, & Chen, 2007)

Chemical Method

The chemical method of synthesis can be classified under the bottom-up approach. They can be classified further into three type-chemical assisted, electrochemical and pyrolysis (W. Zhang et al., 2007). This synthesis technique utilizes a chemical reducing agent such as sodium hydroxide, reducing sugars, trisodium citrate, N, N-dimethyl- formamide, polyol, ascorbic acid and poly(ethylene glycol) (PEG). The common process of synthesis involves and organic or inorganic precursor solution of the metal. To this solution, the reducing agent is added of defined concentration with or without the catalytic activity of temperature. Followed by the addition of chemical stabilizing agents such as surfactants, ligands or polymers to prevent agglomeration. The major advantage of this technique is that nanoparticles can be produced at a faster rate. Also, the desired size, shape and surface property can be fabricated by altering the reaction parameters. The major disadvantage of this technique is the toxicity associated with these particles to humans as well as the environment, hence making them worthless for medical applications (Deepak et al., 2019; Y. Li et al., 2018).

Biological Method

This process can also be classified under the bottom-up approach. Biological methods utilize the metabolites and enzymes produced by plants and micro-organism as reducing and capping agent thereby making the newly formed nanoparticles non-toxic, cheap and environmentally friendly. In the microbial method, the precursor is added either to the culture directly or to the buffer containing the bacterial or fungal cells and incubated before retrieving the nanoparticles. Though this method is cheap, the interference of the media components or loss of metabolites during isolation of cell can hinder the formation of metal nanoparticles and making the purification process difficult (Panda & Deepa, 2011; Santhoshkumar, Agarwal, Menon, Rajeshkumar, & Venkat Kumar, 2019). The phytosynthesis process utilizes the metabolites such as flavonoids, alkaloids, tannins, terpenoids, saponins, phenols and proteins act as both bioreductant and capping agent. The production of nanoparticles here is easier and inexpensive compared to microbial synthesis as there is no media required nor the tedious purification. The Phyto-extracts are usually prepared by using the washed and chopped plants parts such as the flower, leaves, stem, fruits or roots either by boiling or grinding in deionized water. The interaction of this extract with the precursor leads to the formation of the metal nanoparticles. The reaction time can vary from hours to days depending on the plants' nature. The major disadvantage of this method is the low yield and the long duration required to synthesis nanoparticles compared to the chemical method (Das & Chatterjee, 2019; Deepak et al., 2019).

Characterization of Nanoparticles

The analysis of any newly formed nanoparticle is highly essential as it gives an enhanced insight into the size, shape, size-distribution and physicochemical properties. The purity, as well as the compounds responsible for the reduction, can also be identified.

MACROSCOPIC/VISUAL OBSERVATIONS

The fundamental and effortless characterization is by visually observing the change in the color. The reduction in size to nano-range, makes nanoparticles excellent absorbers by increasing the occurrence of electronic transitions. The collective excitation near the surface is called surface plasmon resonance (SPR). The quantum confinement lets the nanomaterials have a SPR spectrum shift giving them their unique size based color (Campion & Kambhampati, 1998).

Particle Size and Zeta Potential Analyzer

This characterization technique quantifies both size distribution as well as charge of particles (Krishnamoorthy & Jayalakshmi, 2012a; Sadhasivam, Dey, Francis, & Devasena, 2015). Zeta potential measures the effective electric charge of nanoparticles on the surface and for its stability in aqueous suspensions(Rajeshkumar & Bharath, 2017).

UV-Visible Spectroscopy

UV-Visible Spectroscopy is next basic methodology to assess the formation of nanoparticles, which is swift and comparatively economical (Behera, 2012; Paul Das, Rebecca Livingstone, Veluswamy, & Das, 2018). In nanoparticles, the valence and conduction band are close leading to the free movement of electrons. The translation leads to reduction of band gap between valence and conduction band thereby increasing the electronic transition which leads to broadening of plasmon resonance causing occurrence of SPR. The SPR forms peak from 200-800 nm signifying the formation of nanoparticles. The shift in the absorption peak towards higher wavelength signifies increase in the size of nanoparticles. The intensity and bandwidth of the peak illustrate the variance in the shape and size distribution of particles. When the SPR bands are sharp, they indicate monodispersity, whereas the broad width of resonance band indicates aggregation or poly dispersed and aggregated structures (Akinsiku et al., 2018; Campion & Kambhampati, 1998)

Electron Microscopy and Energy Dispersive X-Ray Analysis (EDAX)

The electron microscope helps to visualize and confirm the nanoparticle morphology. Both Transmission electron microscopy (TEM) and Scanning electron microscopy (SEM) determines the shape, size and dispersity of the synthesized particles. The analyses of size distribution using these methodologies are not entirely reliable as the average size of a particle in the image of a single section may not be representative for the whole sample distribution understudy. To overcome this issue, images of different sections can also be taken to analyze the particle size dispersion. The obtained results when cross analyzed with PSA and XRD (X-ray diffractometry) results can authenticate the monodispersity or polydispersity of the particle. Electron microscopes utilizes a high energy electron beam to scan/penetrate the surface of the particles and the scattered or reflected or transmitted electrons provides high-resolution characteristic features(Buhr et al., 2009). SEM coupled with EDAX can aid in micro-elemental analysis of the particles which indicates the presence of elements in weight and atomic percentage. This technique validates the purity of the metal in the synthesized sample (Anandalakshmi, Venugobal, & Ramasamy, 2016; Elsayed, Othman, Hassan, & Elshafei, 2018).

X-ray Diffractometry (XRD)

The amorphous or crystalline structure of the nanoparticles can be evaluated using X-ray diffraction (XRD) techniques. The plant extract is amorphous in nature and gives broad peak. Literature shows that SNP with Braggs angle between 35° to 80° representing (111) (200) (220) (311) planes indicate that the nanoparticle has face centered cubic (FCC) crystals structures (Ma et al. 2001, Doddapaneni et al 2018; Rasheed et al 2018). The peak values can be used to obtain the size of the crystals using Debye-Scherrer's formula. The equations is as given below,

$$D = \frac{K\lambda}{\beta \cos\theta}$$

Where D is the size of the particle, k is a shape factor, λ is the incident wavelength, β is the full width at half maximum of the diffraction peak, and θ is the Bragg's angle of peak. The diffraction angle value is considered as 2θ.

FOURIER-TRANSFORM INFRARED SPECTROSCOPY (FTIR)

FTIR is a simple and effective technique in investigation of bio-molecules involved in the convertion of metal-precursor to metal nanoparticle or metal oxide nanoparticle (Y. Zhang, Kohler, & Zhang, 2002). It is used to find out the chemicals and bio-molecules involved in the nanoparticles synthesis(Baraton, 2002). This analysis also identifies the phyto-molecules responsible for reduction of the metal. The FTIR is recorded in from 4000–400 cm^{-1} wavelength. The analysis peaks of reducing agent, precursor and purified nanoparticles are compared to understand the process involved. The peaks at 3307–3325 cm^{-1} region attributes the characteristic stretching of OH molecule of anti-oxidant compounds from the biological source indicating the activity of the antioxidants in bottom-up synthesis process (Arumai Selvan, Mahendiran, Senthil Kumar, & Kalilur Rahiman, 2018). The absorption band around 2912–2929 cm^{-1} corresponds to stretching of alkenes C-H group. The sharp peak in the 1630–1634 cm^{-1} is responsible for the stretching of C=O in amide I groups of protein. These peaks when remain unaltered in the nanoparticle ir spectrum, indicates the unaffected secondary nature of the protein. The band from 1533–1574 cm^{-1} indicates stretching of N-H of amide II. The – NO3 groups gives a peak around 1389 cm^{-1} for nitrate salts. (Huang et al., 2007). The C-N group stretching can be found at 1248–1260 cm^{-1}. flavanoids (J. Huang et al., 2007). These are the basic functional groups that are spotted in the biosynthetic metallic nanoparticles.

EFFECT OF PROCESS PARAMETERS

Concentration of Precursor

Precursor is the metallic salt that provides the metallic core for the nanoparticle. The concentration of the precursors and the corresponding bio extract concentration cast a huge impact in the size and yield of the nanoparticles (Krishnamoorthy & Jayalakshmi, 2012a). With the increasing concentration of the substrate, nanoparticles are formed more quickly (Akinsiku et al., 2018; Krishnamoorthy & Jayalakshmi, 2012b; Verma & Mehata, 2016).

pH of the Solution

The pH of the reaction mixture also plays an important role in synthesis process. It affects the size, shape and colloidal nature of nanoparticles and has the ability to alter the biomolecules, which might affect their capping and stabilizing abilities. The quantitative tests have proved that nanoparticles are formed rapidly at neutral pH. The enhanced formation can be attributed to ionization of phenol group in the phyto-extract As the pH is lowered to acidic environment, there are repulsive forces between the metal anions leading to aggregation of particles rather than nucleation forming larger particles. Increase in pH, increases nucleation thereby enhancing synthesis of nanoparticles with smaller size (Muthu & Priya, 2017; Rasheed et al., 2018)

Temperature

When the temperature of the mixture is increased, it speeds up the reaction rate for reduction of the precursor salt. Also, the enhanced temperature increases the kinetic energy of the reduced metal ions. This leads to faster nucleation and utilization of the ions, thereby controlling the particle growth leading to monodispersed tiny particles. Thus the size of the particles decreases with increased temperature (Verma & Mehata, 2016).

APPLICATIONS OF METAL NANOPARTICLES

Applications of Biosynthetic Silver Nanoparticles

The silver nanoparticles (SNPs) have important applications such as antibacterial DNA sequencing and others (Desai, Mankad, Gupta, & Jha, 2012). Recently, the nano-silver has gained overwhelming recognition towards the theranostic applications due to their physicochemical, electro-magnetic and optical properties (Narayanan & Sakthivel, 2010). Applications of SNPs are dependent on their size, shape and stability. They show potential applications in catalysis, optics, electronics and as biomedical agents. Among various nanoparticles, SNPs have been used preferably in diverse biomedical applications as antibacterial, antifungal, antiviral, and anti-inflammatory agents (Ahn, Eom, Yang, Meyer, & Choi, 2014). These particles control the spreading of cancer by by hindering the ATP synthesis and alter the essential metabolic pathways thereby causing cell death (Paul Das et al., 2018). These green-synthesized particles have a wide applications in various medical fields, such as anti-proliferative agents against cancerous cells in humans (Rasheed, Bilal, Iqbal, & Li, 2017).

In the toxicity studies, biosynthetic SNPs produced insignificant/no toxicity on normal human cells but was able to inhibit the survival of various cancer cells such as MCF-7 human breast cancer cell line, B16 mouse melanoma cell line and A549 human lung cancer cell line (Mukherjee et al., 2014). SNPs can also be synthesized using biological source to act as fluorescent materials. The biomolecules present in the surface of the biogenic SNPs could also enhance the fluorescence greatly and thus can be used for diagnostic application. (Chowdhury et al., 2007; Peyser, Vinson, Bartko, & Dickson, 2001; Treguer et al., 2005; Z. Wang & Rothberg, 2005) Thus the biogenic SNPs has been reported to be the effective theranostic tool.

Applications of Biosynthetic Gold Nanoparticles

Gold in its bulk stage is a non-reactive metal, but on convertion to nanoparticle their properties change completely due to the enhanced surface to volume ratio thereby enabling quantum confinement (Jain, Lee, El-sayed, & El-sayed, 2006). However, the changes in the properties do not affect the stability of the metals at the nanoscale and thus the colloidal gold nanoparticles (AuNPs) are one of the most stable nanoparticles preferred for theranostic applications.

The unique optical property of AuNPs corresponds to the surface plasmon resonance (SPR) and it can be readily used to enhance optical imaging based on their absorption, scattering, fluorescence, Raman scattering, etc. When the lights of certain wavelengths are illuminated on the AuNPs, the surface electrons begin to oscillate in resonance with the frequency of light due to the interaction between the electromagnetic field of light and the conduction electrons on the AuNP surface (X. Huang & El-Sayed, 2010). Exploiting these plasmonic properties of AuNPS, there has been extensive research for the advancement of optical imaging. The contrast between the signals scattering from AuNPs and background from cells and tissues observed through the dark field (DF) microscopy reports the presence of AuNPs within biological systems. The AuNPs produce stronger signals than the cells and this difference can be analysed for imaging of cells.

The photothermal imaging is also achieved as the light absorbed by AuNPs generates heat. These properties of AuNPs can be used to detect the cancer cells inside the body and further the same can be used to treat the cancer cells (Jeyarani et al., 2020). The supremacy of AuNPs in treating cancer has been reported in multiple works of literature (Kajani, Bordbar, Zarkesh Esfahani, & Razmjou, 2016; Rajeshkumar, 2016). The AuNPs were found to restrict cell migration and kill the tumour cells through caspase -3 & 9 induced apoptosis. Research on anti-cancer activity by the biosynthesized gold nanoparticle revealed that the AuNP modified the morphology of the tumour cells and stopped the cell proliferation(Kajani et al., 2016; Liu, Jiang, Nam, Moon, & Kim, 2018).

Applications of Biosynthetic Copper Nanoparticles

Copper sulfide nanoparticles (CuSNPs) are the most efficient and cost-effective Photo Thermal Therapy (PTT) and Photo Dynamic Therapy (PDT) (Feng et al., 2016; L. Li et al., 2017). CuSNPs are biocompatible and has low or no toxicity (Poudel et al., 2019). The CuSNPs absorb near-infrared (NIR) better than AuNPs. The better absorption by CuSNPs is attributed to the d-d electron transition which is not influenced by the dielectric constant of the environment of copper ions whereas the AuNPs absorb light by surface plasmon resonance and thus diminishes over a period of exposure to laser light (L. Wang, 2016; Zha et al., 2013).

The homogenous size-controlled PEGylated CuSNPs synthesized in an aqueous system forms stable particles. The efficiency of contrast imaging of the nano-platforms is inversely proportional to the particle size in HeLa cells (Ding et al., 2015). Near-infrared absorbing CuS NPs are conjugated to a black hole quencher 3 (BHQ3) via a Matrix metallo-proteinases (MMPs) -cleavable peptide linker, which will absorb light at 630 nm and 930 nm. Bond cleavage with MMPs would release the BHQ3 and retain the CuSNPs inside the tumour for the theranostic action. In the spectral imaging of the cleaved tumour, the peak at 680 nm is abated rapidly, with the intensifying 930 nm peak indicating the MMP activity inside the tumour cell.

In MRI scanning, CuSNP is ligated with magnetic materials such as manganese for better image resolution and accuracy. Liu et al. (2015) have treated the human breast cancer cells (MDA-MB-231) for imaging using PEGylated CuS NP ligated with manganese (II) chelate (MPG NPs) prepared by a facile hydrothermal method (Ding et al., 2015). The CuSNP aided in the visualization of cells *in vivo* followed by photothermal therapy of only cancer cells. Therefore, CuS NPs functionalized with other metals and ligands hold considerable promise as a theranostic nanomedicine for MR imaging in cancer. Zhang et al. (2013) have developed functionalized CuS with gadolinium- diethylenetriamine penta acetic acid (DTPA) as a means for MR imaging of HeLa cells (Zha et al., 2013). These results indicate that the CuSNP is a hopeful contender as a contrast agent in MR imaging for cancers.

The CuSNPs are used in the treatment of tumours by acting as the nanocarriers to transport chemotherapeutic agents into tumour tissues, and enhancers for chemotherapy, immunotherapy and radiotherapy (Feng et al., 2016; Han, Zhang, Chen, Shu, & Wang, 2016). In the photothermal therapy (PTT), the CuS NPs are photosensitizers which are shot into the tumour cells and irradiated with Infra-Red lights (IR). The CuS NPs inside the tumour cells convert the optical energy into thermal energy and burn the cells through hyperthermia. The hyperthermia further increases regional blood flow, cellular metabolism, and membrane permeability, which is also beneficial for cellular uptake when chemophototherapy is the therapeutic design (Poudel et al., 2019).

The high photostability of CuSNPs makes it an effective photothermal agent in photodynamic therapy (PDT). The hollow CuS NPs have a high definite surface area, abundant mesoporous pores, and better biocompatibility, and better clearance from the body which makes them one of the ideal agents for the targeted delivery of the possibly cytotoxic drugs into cancer cells (Han et al., 2016). The photosensitizer

(PS) CuS NPs are administered into targeted pathological cells and the cells are irradiated with light of wavelength corresponding to the PS. The irradiated PS excites and generates the cytotoxic reactive oxygen species (ROS) such as singlet molecular oxygen, hydroxyl radicals, and/or superoxide anions, and cytotoxicity through oxidative stress of pathological cells, and further damage the lipids, proteins, or nucleic acids, which are important cellular macromolecules. The photodynamic activity is based on inducing series of consecutive biochemical and morphological reactions i.e. photooxidative reactions (Abrahamse, Kruger, Kadanyo, & Mishra, 2017)

CONCLUSION

Despite the advancement in science & technology, cancer treatment is still a faraway goal. This chapter described the importance of nanotechnology in cancer detection and therapy. The various synthesis and characterization techniques used for metal nanoparticle production are discussed. The progress of silver, gold and copper nanoparticles in the field of cancer theranostics have also been highlighted. These nanoparticles overcome the limitation of the existing technique thereby providing a step ahead in cancer therapy.

ACKNOWLEDGMENT

This research received no specific grant from any funding agency in the public, commercial, or not-for-profit sectors.

REFERENCES

Abrahamse, H., Kruger, C. A., Kadanyo, S., & Mishra, A. (2017). Nanoparticles for Advanced Photodynamic Therapy of Cancer. *Photomedicine and Laser Surgery, 35*(11), 581–588. doi:10.1089/pho.2017.4308 PMID:28937916

Ahn, J. M., Eom, H. J., Yang, X., Meyer, J. N., & Choi, J. (2014). Comparative toxicity of silver nanoparticles on oxidative stress and DNA damage in the nematode, Caenorhabditis elegans. *Chemosphere, 108*, 343–352. doi:10.1016/j.chemosphere.2014.01.078 PMID:24726479

Akinsiku, A. A., Dare, E. O., Ajanaku, K. O., Ajani, O. O., Olugbuyiro, J. A. O., Siyanbola, T. O., … Emetere, M. E. (2018). Modeling and synthesis of ag and ag/Ni allied bimetallic nanoparticles by green method: Optical and biological properties. *International Journal of Biomaterials, 2018*(3). doi:10.1155/2018/9658080

Anandalakshmi, K., Venugobal, J., & Ramasamy, V. (2016). Characterization of silver nanoparticles by green synthesis method using Pedalium murex leaf extract and their antibacterial activity. *Applied Nanoscience (Switzerland), 6*(3), 399–408. doi:10.100713204-015-0449-z

Arumai Selvan, D., Mahendiran, D., Senthil Kumar, R., & Kalilur Rahiman, A. (2018). Garlic, green tea and turmeric extracts-mediated green synthesis of silver nanoparticles: Phytochemical, antioxidant and in vitro cytotoxicity studies. *Journal of Photochemistry and Photobiology B: Biology, 180*(2017), 243–252. doi:10.1016/j.jphotobiol.2018.02.014

Baraton, M.-I. (2002). Surface Analysis of Semiconducting Nanoparticles by FTIR Spectroscopy. *Nanocrystalline Metals and Oxides*, *7*(1), 165–187. doi:10.1007/0-306-47609-6_6

Behera, S. (2012). UV-Visible Spectrophotometric Method Development and Validation of Assay of Paracetamol Tablet Formulation. *Journal of Analytical & Bioanalytical Techniques*, *03*(06), 2175–2187. doi:10.4172/2155-9872.1000151

Buhr, E., Senftleben, N., Klein, T., Bergmann, D., Gnieser, D., Frase, C. G., & Bosse, H. (2009). Characterization of nanoparticles by scanning electron microscopy in transmission mode. *Measurement Science & Technology*, *20*(8), 084025. Advance online publication. doi:10.1088/0957-0233/20/8/084025

Campion, A., & Kambhampati, P. (1998). Surface-Enhanced raman scattering. *Chemical Society Reviews*, *27*(4), 241–250. doi:10.1039/a827241z

Chen, H., Zhang, W., Zhu, G., Xie, J., & Chen, X. (2017). Rethinking cancer nanotheranostics. *Nature Reviews. Materials*, *2*(May), 17024. Advance online publication. doi:10.1038/natrevmats.2017.24 PMID:29075517

Chowdhury, M. H., Gray, S. K., Pond, J., Geddes, C. D., Aslan, K., & Lakowicz, J. R. (2007). Computational study of fluorescence scattering by silver nanoparticles. *Journal of the Optical Society of America. B, Optical Physics*, *24*(9), 2259. doi:10.1364/JOSAB.24.002259 PMID:19777118

Das, M., & Chatterjee, S. (2019). Green synthesis of metal/metal oxide nanoparticles toward biomedical applications: Boon or bane. In *Green Synthesis* (pp. 265–301). Characterization and Applications of Nanoparticles. doi:10.1016/B978-0-08-102579-6.00011-3

Deepak, P., Amutha, V., Kamaraj, C., Balasubramani, G., Aiswarya, D., & Perumal, P. (2019). Chemical and green synthesis of nanoparticles and their efficacy on cancer cells. In *Green Synthesis* (pp. 369–387). Characterization and Applications of Nanoparticles. doi:10.1016/B978-0-08-102579-6.00016-2

Desai, R., Mankad, V., Gupta, S., & Jha, P. (2012). Size Distribution of Silver Nanoparticles: UV-Visible Spectroscopic Assessment. *Nanoscience and Nanotechnology Letters*, *4*(1), 30–34. doi:10.1166/nnl.2012.1278

Ding, K., Zeng, J., Jing, L., Qiao, R., Liu, C., Jiao, M., Li, Z., & Gao, M. (2015). Aqueous synthesis of PEGylated copper sulfide nanoparticles for photoacoustic imaging of tumors. *Nanoscale*, *7*(25), 11075–11081. doi:10.1039/C5NR02180D PMID:26055816

Elsayed, M. A., Othman, A. M., Hassan, M. M., & Elshafei, A. M. (2018). Optimization of silver nanoparticles biosynthesis mediated by Aspergillus niger NRC1731 through application of statistical methods: enhancement and characterization. *Biotech, 8*(3). doi:10.100713205-018-1158-6

Feng, Q., Zhang, Y., Zhang, W., Shan, X., Yuan, Y., Zhang, H., Hou, L., & Zhang, Z. (2016). Tumor-targeted and multi-stimuli responsive drug delivery system for near-infrared light induced chemo-phototherapy and photoacoustic tomography. *Acta Biomaterialia*, *38*, 129–142. doi:10.1016/j.actbio.2016.04.024 PMID:27090593

Ferlay, J., Colombet, M., Soerjomataram, I., Mathers, C., Parkin, D. M., Piñeros, M., Znaor, A., & Bray, F. (2019). Estimating the global cancer incidence and mortality in 2018: GLOBOCAN sources and methods. *International Journal of Cancer*, *144*(8), 1941–1953. doi:10.1002/ijc.31937 PMID:30350310

Ferlay, J., Soerjomataram, I., Dikshit, R., Eser, S., Mathers, C., Rebelo, M., Parkin, D. M., Forman, D., & Bray, F. (2015). Cancer incidence and mortality worldwide: Sources, methods and major patterns in GLOBOCAN 2012. *International Journal of Cancer*, *136*(5), E359–E386. doi:10.1002/ijc.29210 PMID:25220842

Han, L., Zhang, Y., Chen, X. W., Shu, Y., & Wang, J. H. (2016). Protein-modified hollow copper sulfide nanoparticles carrying indocyanine green for photothermal and photodynamic therapy. *Journal of Materials Chemistry. B, Materials for Biology and Medicine*, *4*(1), 105–112. doi:10.1039/C5TB02002F PMID:32262813

Harmsen, S., Wall, M. A., Huang, R., & Kircher, M. F. (2017). Cancer imaging using surface-enhanced resonance Raman scattering nanoparticles. *Nature Protocols*, *12*(7), 1400–1414. doi:10.1038/nprot.2017.031 PMID:28686581

Huang, J., Li, Q., Sun, D., Lu, Y., Su, Y., Yang, X., Wang, H., Wang, Y., Shao, W., He, N., Hong, J., & Chen, C. (2007). Biosynthesis of silver and gold nanoparticles by novel sundried Cinnamomum camphora leaf. *Nanotechnology*, *18*(10), 105104. Advance online publication. doi:10.1088/0957-4484/18/10/105104

Huang, X., & El-Sayed, M. A. (2010). Gold nanoparticles: Optical properties and implementations in cancer diagnosis and photothermal therapy. *Journal of Advanced Research*, *1*(1), 13–28. doi:10.1016/j.jare.2010.02.002

Jain, P. K., Lee, K. S., El-sayed, I. H., & El-sayed, M. A. (2006). Calculated Absorption and Scattering Properties of Gold Nanoparticles of Different Size, Shape, and Composition: Applications in Biological Imaging and Biomedicine. *Journal of Photochemistry and Photobiology. B, Biology*, *110*, 7238–7248. PMID:16599493

Jeyarani, S., Vinita, N. M., Puja, P., Senthamilselvi, S., Devan, U., Velangani, A. J., Biruntha, M., Pugazhendhi, A., & Kumar, P. (2020). Biomimetic gold nanoparticles for its cytotoxicity and biocompatibility evidenced by fluorescence-based assays in cancer (MDA-MB-231) and non-cancerous (HEK-293) cells. *Journal of Photochemistry and Photobiology. B, Biology*, *202*, 111715. doi:10.1016/j.jphotobiol.2019.111715 PMID:31790882

Kajani, A. A., Bordbar, A. K., Zarkesh Esfahani, S. H., & Razmjou, A. (2016). Gold nanoparticles as potent anticancer agent: Green synthesis, characterization, and: In vitro study. *RSC Advances*, *6*(68), 63973–63983. doi:10.1039/C6RA09050H

Kelkar, S. S., & Reineke, T. M. (2011). Theranostics: Combining imaging and therapy. *Bioconjugate Chemistry*, *22*(10), 1879–1903. doi:10.1021/bc200151q PMID:21830812

Krishnamoorthy, P., & Jayalakshmi, T. (2012a). Preparation, characterization and synthesis of silver nanoparticles by using phyllanthusniruri for the antimicrobial activity and cytotoxic effects. *Journal of Chemical and Pharmaceutical Research*, *4*(11), 4783–4794.

Krishnamoorthy, P., & Jayalakshmi, T. (2012b). Preparation, characterization and synthesis of silver nanoparticles by using phyllanthusniruri for the antimicrobial activity and cytotoxic effects. *Journal of Chemical and Pharmaceutical Research*, *4*(11), 4783–4794.

Li, L., Rashidi, L. H., Yao, M., Ma, L., Chen, L., Zhang, J., Zhang, Y., & Chen, W. (2017). CuS nanoagents for photodynamic and photothermal therapies: Phenomena and possible mechanisms. *Photodiagnosis and Photodynamic Therapy*, *19*, 5–14. doi:10.1016/j.pdpdt.2017.04.001 PMID:28389371

Li, Y., Chang, Y., Lian, X., Zhou, L., Yu, Z., Wang, H., & An, F. (2018). Silver nanoparticles for enhanced cancer theranostics: In vitro and in vivo perspectives. *Journal of Biomedical Nanotechnology*, *14*(9), 1515–1542. doi:10.1166/jbn.2018.2614 PMID:29958548

Liu, Z., Jiang, W., Nam, J., Moon, J. J., & Kim, B. Y. S. (2018). Immunomodulating Nanomedicine for Cancer Therapy. *Nano Letters*, *18*(11), 6655–6659. doi:10.1021/acs.nanolett.8b02340 PMID:30185039

Mukherjee, S., Chowdhury, D., Kotcherlakota, R., Patra, S., Vinothkumar, B., Bhadra, M. P., ... Patra, C. R. (2014). Potential theranostics application of bio-synthesized silver nanoparticles (4-in-1 system). *Theranostics*, *4*(3), 316–335. doi:10.7150/thno.7819 PMID:24505239

Mun, E. J., Babiker, H. M., Weinberg, U., Kirson, E. D., & Von Hoff, D. D. (2018). Tumor-treating fields: A fourth modality in cancer treatment. *Clinical Cancer Research*, *24*(2), 266–275. doi:10.1158/1078-0432.CCR-17-1117 PMID:28765323

Muthu, K., & Priya, S. (2017). Green synthesis, characterization and catalytic activity of silver nanoparticles using Cassia auriculata flower extract separated fraction. *Spectrochimica Acta. Part A: Molecular and Biomolecular Spectroscopy*, *179*, 66–72. doi:10.1016/j.saa.2017.02.024 PMID:28219038

Narayanan, K. B., & Sakthivel, N. (2010). Biological synthesis of metal nanoparticles by microbes. *Advances in Colloid and Interface Science*, *156*(1–2), 1–13. doi:10.1016/j.cis.2010.02.001 PMID:20181326

Panda, T., & Deepa, K. (2011). Biosynthesis of gold nanoparticles. *Journal of Nanoscience and Nanotechnology*, *11*(12), 10279–10294. doi:10.1166/jnn.2011.5021 PMID:22408900

Paul Das, M., Rebecca Livingstone, J., Veluswamy, P., & Das, J. (2018). Exploration of Wedelia chinensis leaf-assisted silver nanoparticles for antioxidant, antibacterial and in vitro cytotoxic applications. *Yao Wu Shi Pin Fen Xi*, *26*(2), 917–925. doi:10.1016/j.jfda.2017.07.014 PMID:29567263

Peyser, L. A., Vinson, A. E., Bartko, A. P., & Dickson, R. M. (2001). Photoactivated fluorescence from individual silver nanoclusters. *Science*, *291*(5501), 103–106. doi:10.1126cience.291.5501.103 PMID:11141556

Poudel, K., Gautam, M., Jin, S. G., Choi, H. G., Yong, C. S., & Kim, J. O. (2019). Copper sulfide: An emerging adaptable nanoplatform in cancer theranostics. *International Journal of Pharmaceutics*, *562*(March), 135–150. doi:10.1016/j.ijpharm.2019.03.043 PMID:30904728

Rajeshkumar, S. (2016). Anticancer activity of eco-friendly gold nanoparticles against lung and liver cancer cells. *Journal of Genetic Engineering and Biotechnology*, *14*(1), 195–202. doi:10.1016/j.jgeb.2016.05.007 PMID:30647615

Rajeshkumar, S., & Bharath, L. V. (2017). Mechanism of plant-mediated synthesis of silver nanoparticles – A review on biomolecules involved, characterisation and antibacterial activity. *Chemico-Biological Interactions*, *273*, 219–227. doi:10.1016/j.cbi.2017.06.019 PMID:28647323

Rasheed, T., Bilal, M., Iqbal, H. M. N., & Li, C. (2017). Green biosynthesis of silver nanoparticles using leaves extract of Artemisia vulgaris and their potential biomedical applications. *Colloids and Surfaces. B, Biointerfaces*, *158*, 408–415. doi:10.1016/j.colsurfb.2017.07.020 PMID:28719862

Rasheed, T., Bilal, M., Li, C., Nabeel, F., Khalid, M., & Iqbal, H. M. N. (2018). Catalytic potential of biosynthesized silver nanoparticles using Convolvulus arvensis extract for the degradation of environmental pollutants. *Journal of Photochemistry and Photobiology B: Biology, 181*(2017), 44–52. doi:10.1016/j.jphotobiol.2018.02.024

Sadhasivam, L., Dey, N., Francis, A. P., & Devasena, T. (2015). Innovare Academic Sciences. *Transdermal Patches of Chitosan Nanoparticles for Insulin Delivery, 7*(5).

Santhoshkumar, J., Agarwal, H., Menon, S., Rajeshkumar, S., & Venkat Kumar, S. (2019). A biological synthesis of copper nanoparticles and its potential applications. In *Green Synthesis* (pp. 199–221). Characterization and Applications of Nanoparticles. doi:10.1016/B978-0-08-102579-6.00009-5

Siddiqi, K. S., Husen, A., & Rao, R. A. K. (2018). A review on biosynthesis of silver nanoparticles and their biocidal properties. *Journal of Nanobiotechnology*, *16*(1), 14. Advance online publication. doi:10.118612951-018-0334-5 PMID:29452593

Treguer, M., Rocco, F., Lelong, G., Le Nestour, A., Cardinal, T., Maali, A., & Lounis, B. (2005). Fluorescent silver oligomeric clusters and colloidal particles. *Solid State Sciences*, *7*(7), 812–818. doi:10.1016/j.solidstatesciences.2005.01.017

Verma, A., & Mehata, M. S. (2016). Controllable synthesis of silver nanoparticles using Neem leaves and their antimicrobial activity. *Journal of Radiation Research and Applied Sciences*, *9*(1), 109–115. doi:10.1016/j.jrras.2015.11.001

Wang, J., Hsu, S.-W., Tao, A. R., & Jokerst, J. V. (2017). Copper sulfide nanodisk as photoacoustic contrast agent for ovarian tumor detection. *Photons Plus Ultrasound: Imaging and Sensing*, *2017*(10064), 100642D. doi:10.1117/12.2251184

Wang, L. (2016). Synthetic methods of CuS nanoparticles and their applications for imaging and cancer therapy. *RSC Advances*, *6*(86), 82596–82615. doi:10.1039/C6RA18355G

Wang, Z., & Rothberg, L. J. (2005). Origins of blinking in single-molecule raman spectroscopy. *The Journal of Physical Chemistry B*, *109*(8), 3387–3391. doi:10.1021/jp0460947 PMID:16851369

Zha, Z., Zhang, S., Deng, Z., Li, Y., Li, C., & Dai, Z. (2013). Enzyme-responsive copper sulphide nanoparticles for combined photoacoustic imaging, tumor-selective chemotherapy and photothermal therapy. *Chemical Communications*, *49*(33), 3455–3457. doi:10.1039/c3cc40608c PMID:23507786

Zhang, W., Qiao, X., & Chen, J. (2007). Synthesis of silver nanoparticles-Effects of concerned parameters in water/oil microemulsion. *Materials Science and Engineering B. Solid-State Materials for Advanced Technology*, *142*(1), 1–15. doi:10.1016/j.mseb.2007.06.014

Zhang, Y., Kohler, N., & Zhang, M. (2002). Surface modification of superparamagnetic magnetite nanoparticles and their intracellular uptake. *Biomaterials*, *23*(7), 1553–1561. doi:10.1016/S0142-9612(01)00267-8 PMID:11922461

ADDITIONAL READING

Abrahamse, H., Kruger, C. A., Kadanyo, S., & Mishra, A. (2017). Nanoparticles for Advanced Photodynamic Therapy of Cancer. *Photomedicine and Laser Surgery*, *35*(11), 581–588. doi:10.1089/pho.2017.4308 PMID:28937916

Ahn, J. M., Eom, H. J., Yang, X., Meyer, J. N., & Choi, J. (2014). Comparative toxicity of silver nanoparticles on oxidative stress and DNA damage in the nematode, Caenorhabditis elegans. *Chemosphere*, *108*, 343–352. doi:10.1016/j.chemosphere.2014.01.078 PMID:24726479

Akinsiku, A. A., Dare, E. O., Ajanaku, K. O., Ajani, O. O., Olugbuyiro, J. A. O., Siyanbola, T. O., … Emetere, M. E. (2018). Modeling and synthesis of ag and ag/Ni allied bimetallic nanoparticles by green method: Optical and biological properties. *International Journal of Biomaterials, 2018*(3). doi:10.1155/2018/9658080

Anandalakshmi, K., Venugobal, J., & Ramasamy, V. (2016). Characterization of silver nanoparticles by green synthesis method using Pedalium murex leaf extract and their antibacterial activity. *Applied Nanoscience (Switzerland)*, *6*(3), 399–408. doi:10.100713204-015-0449-z

Arumai Selvan, D., Mahendiran, D., Senthil Kumar, R., & Kalilur Rahiman, A. (2018). Garlic, green tea and turmeric extracts-mediated green synthesis of silver nanoparticles: Phytochemical, antioxidant and in vitro cytotoxicity studies. *Journal of Photochemistry and Photobiology B: Biology, 180*(2017), 243–252. doi:10.1016/j.jphotobiol.2018.02.014

Baraton, M.-I. (2002). Surface Analysis of Semiconducting Nanoparticles by FTIR Spectroscopy. *Nanocrystalline Metals and Oxides*, *7*(1), 165–187. doi:10.1007/0-306-47609-6_6

Behera, S. (2012). UV-Visible Spectrophotometric Method Development and Validation of Assay of Paracetamol Tablet Formulation. *Journal of Analytical & Bioanalytical Techniques*, *03*(06), 2175–2187. doi:10.4172/2155-9872.1000151

Buhr, E., Senftleben, N., Klein, T., Bergmann, D., Gnieser, D., Frase, C. G., & Bosse, H. (2009). Characterization of nanoparticles by scanning electron microscopy in transmission mode. *Measurement Science & Technology*, *20*(8), 084025. Advance online publication. doi:10.1088/0957-0233/20/8/084025

Campion, A., & Kambhampati, P. (1998). Surface-Enhanced raman scattering. *Chemical Society Reviews*, *27*(4), 241–250. doi:10.1039/a827241z

Chen, H., Zhang, W., Zhu, G., Xie, J., & Chen, X. (2017). Rethinking cancer nanotheranostics. *Nature Reviews. Materials*, *2*(May), 17024. Advance online publication. doi:10.1038/natrevmats.2017.24 PMID:29075517

Chowdhury, M. H., Gray, S. K., Pond, J., Geddes, C. D., Aslan, K., & Lakowicz, J. R. (2007). Computational study of fluorescence scattering by silver nanoparticles. *Journal of the Optical Society of America. B, Optical Physics*, *24*(9), 2259. doi:10.1364/JOSAB.24.002259 PMID:19777118

Das, M., & Chatterjee, S. (2019). Green synthesis of metal/metal oxide nanoparticles toward biomedical applications: Boon or bane. In *Green Synthesis* (pp. 265–301). Characterization and Applications of Nanoparticles., doi:10.1016/B978-0-08-102579-6.00011-3

Deepak, P., Amutha, V., Kamaraj, C., Balasubramani, G., Aiswarya, D., & Perumal, P. (2019). Chemical and green synthesis of nanoparticles and their efficacy on cancer cells. In *Green Synthesis* (pp. 369–387). Characterization and Applications of Nanoparticles., doi:10.1016/B978-0-08-102579-6.00016-2

Desai, R., Mankad, V., Gupta, S., & Jha, P. (2012). Size Distribution of Silver Nanoparticles: UV-Visible Spectroscopic Assessment. *Nanoscience and Nanotechnology Letters*, *4*(1), 30–34. doi:10.1166/nnl.2012.1278

Ding, K., Zeng, J., Jing, L., Qiao, R., Liu, C., Jiao, M., Li, Z., & Gao, M. (2015). Aqueous synthesis of PEGylated copper sulfide nanoparticles for photoacoustic imaging of tumors. *Nanoscale*, *7*(25), 11075–11081. doi:10.1039/C5NR02180D PMID:26055816

Elsayed, M. A., Othman, A. M., Hassan, M. M., & Elshafei, A. M. (2018). Optimization of silver nanoparticles biosynthesis mediated by Aspergillus niger NRC1731 through application of statistical methods: enhancement and characterization. *3 Biotech*, *8*(3). doi:10.100713205-018-1158-6

Feng, Q., Zhang, Y., Zhang, W., Shan, X., Yuan, Y., Zhang, H., Hou, L., & Zhang, Z. (2016). Tumor-targeted and multi-stimuli responsive drug delivery system for near-infrared light induced chemo-phototherapy and photoacoustic tomography. *Acta Biomaterialia*, *38*, 129–142. doi:10.1016/j.actbio.2016.04.024 PMID:27090593

Ferlay, J., Colombet, M., Soerjomataram, I., Mathers, C., Parkin, D. M., Piñeros, M., Znaor, A., & Bray, F. (2019). Estimating the global cancer incidence and mortality in 2018: GLOBOCAN sources and methods. *International Journal of Cancer*, *144*(8), 1941–1953. doi:10.1002/ijc.31937 PMID:30350310

Ferlay, J., Soerjomataram, I., Dikshit, R., Eser, S., Mathers, C., Rebelo, M., Parkin, D. M., Forman, D., & Bray, F. (2015). Cancer incidence and mortality worldwide: Sources, methods and major patterns in GLOBOCAN 2012. *International Journal of Cancer*, *136*(5), E359–E386. doi:10.1002/ijc.29210 PMID:25220842

Han, L., Zhang, Y., Chen, X. W., Shu, Y., & Wang, J. H. (2016). Protein-modified hollow copper sulfide nanoparticles carrying indocyanine green for photothermal and photodynamic therapy. *Journal of Materials Chemistry. B, Materials for Biology and Medicine*, *4*(1), 105–112. doi:10.1039/C5TB02002F PMID:32262813

Harmsen, S., Wall, M. A., Huang, R., & Kircher, M. F. (2017). Cancer imaging using surface-enhanced resonance Raman scattering nanoparticles. *Nature Protocols*, *12*(7), 1400–1414. doi:10.1038/nprot.2017.031 PMID:28686581

Huang, J., Li, Q., Sun, D., Lu, Y., Su, Y., Yang, X., Wang, H., Wang, Y., Shao, W., He, N., Hong, J., & Chen, C. (2007). Biosynthesis of silver and gold nanoparticles by novel sundried Cinnamomum camphora leaf. *Nanotechnology*, *18*(10), 105104. Advance online publication. doi:10.1088/0957-4484/18/10/105104

Huang, X., & El-Sayed, M. A. (2010). Gold nanoparticles: Optical properties and implementations in cancer diagnosis and photothermal therapy. *Journal of Advanced Research*, *1*(1), 13–28. doi:10.1016/j.jare.2010.02.002

Jain, P. K., Lee, K. S., El-sayed, I. H., & El-sayed, M. A. (2006). Calculated Absorption and Scattering Properties of Gold Nanoparticles of Different Size, Shape, and Composition: Applications in Biological Imaging and Biomedicine. *Journal of Photochemistry and Photobiology. B, Biology*, *110*, 7238–7248. PMID:16599493

Jeyarani, S., Vinita, N. M., Puja, P., Senthamilselvi, S., Devan, U., Velangani, A. J., Biruntha, M., Pugazhendhi, A., & Kumar, P. (2020). Biomimetic gold nanoparticles for its cytotoxicity and biocompatibility evidenced by fluorescence-based assays in cancer (MDA-MB-231) and non-cancerous (HEK-293) cells. *Journal of Photochemistry and Photobiology. B, Biology*, *202*, 111715. doi:10.1016/j.jphotobiol.2019.111715 PMID:31790882

Kajani, A. A., Bordbar, A. K., Zarkesh Esfahani, S. H., & Razmjou, A. (2016). Gold nanoparticles as potent anticancer agent: Green synthesis, characterization, and: In vitro study. *RSC Advances*, *6*(68), 63973–63983. doi:10.1039/C6RA09050H

Kelkar, S. S., & Reineke, T. M. (2011). Theranostics: Combining imaging and therapy. *Bioconjugate Chemistry*, *22*(10), 1879–1903. doi:10.1021/bc200151q PMID:21830812

Krishnamoorthy, P., & Jayalakshmi, T. (2012a). Preparation, characterization and synthesis of silver nanoparticles by using phyllanthusniruri for the antimicrobial activity and cytotoxic effects. *Journal of Chemical and Pharmaceutical Research*, *4*(11), 4783–4794.

Krishnamoorthy, P., & Jayalakshmi, T. (2012b). Preparation, characterization and synthesis of silver nanoparticles by using phyllanthusniruri for the antimicrobial activity and cytotoxic effects. *Journal of Chemical and Pharmaceutical Research*, *4*(11), 4783–4794.

Li, L., Rashidi, L. H., Yao, M., Ma, L., Chen, L., Zhang, J., Zhang, Y., & Chen, W. (2017). CuS nanoagents for photodynamic and photothermal therapies: Phenomena and possible mechanisms. *Photodiagnosis and Photodynamic Therapy*, *19*, 5–14. doi:10.1016/j.pdpdt.2017.04.001 PMID:28389371

Li, Y., Chang, Y., Lian, X., Zhou, L., Yu, Z., Wang, H., & An, F. (2018). Silver nanoparticles for enhanced cancer theranostics: In vitro and in vivo perspectives. *Journal of Biomedical Nanotechnology*, *14*(9), 1515–1542. doi:10.1166/jbn.2018.2614 PMID:29958548

Liu, Z., Jiang, W., Nam, J., Moon, J. J., & Kim, B. Y. S. (2018). Immunomodulating Nanomedicine for Cancer Therapy. *Nano Letters*, *18*(11), 6655–6659. doi:10.1021/acs.nanolett.8b02340 PMID:30185039

Mukherjee, S., Chowdhury, D., Kotcherlakota, R., Patra, S., Vinothkumar, B., Bhadra, M. P., ... Patra, C. R. (2014). Potential theranostics application of bio-synthesized silver nanoparticles (4-in-1 system). *Theranostics*, *4*(3), 316–335. doi:10.7150/thno.7819 PMID:24505239

Mun, E. J., Babiker, H. M., Weinberg, U., Kirson, E. D., & Von Hoff, D. D. (2018). Tumor-treating fields: A fourth modality in cancer treatment. *Clinical Cancer Research*, *24*(2), 266–275. doi:10.1158/1078-0432.CCR-17-1117 PMID:28765323

Muthu, K., & Priya, S. (2017). Green synthesis, characterization and catalytic activity of silver nanoparticles using Cassia auriculata flower extract separated fraction. *Spectrochimica Acta. Part A: Molecular and Biomolecular Spectroscopy*, *179*, 66–72. doi:10.1016/j.saa.2017.02.024 PMID:28219038

Narayanan, K. B., & Sakthivel, N. (2010). Biological synthesis of metal nanoparticles by microbes. *Advances in Colloid and Interface Science*, *156*(1–2), 1–13. doi:10.1016/j.cis.2010.02.001 PMID:20181326

Panda, T., & Deepa, K. (2011). Biosynthesis of gold nanoparticles. *Journal of Nanoscience and Nanotechnology*, *11*(12), 10279–10294. doi:10.1166/jnn.2011.5021 PMID:22408900

Paul Das, M., Rebecca Livingstone, J., Veluswamy, P., & Das, J. (2018). Exploration of Wedelia chinensis leaf-assisted silver nanoparticles for antioxidant, antibacterial and in vitro cytotoxic applications. *Yao Wu Shi Pin Fen Xi*, *26*(2), 917–925. doi:10.1016/j.jfda.2017.07.014 PMID:29567263

Peyser, L. A., Vinson, A. E., Bartko, A. P., & Dickson, R. M. (2001). Photoactivated fluorescence from individual silver nanoclusters. *Science*, *291*(5501), 103–106. doi:10.1126cience.291.5501.103 PMID:11141556

Poudel, K., Gautam, M., Jin, S. G., Choi, H. G., Yong, C. S., & Kim, J. O. (2019). Copper sulfide: An emerging adaptable nanoplatform in cancer theranostics. *International Journal of Pharmaceutics*, *562*(March), 135–150. doi:10.1016/j.ijpharm.2019.03.043 PMID:30904728

Rajeshkumar, S. (2016). Anticancer activity of eco-friendly gold nanoparticles against lung and liver cancer cells. *Journal of Genetic Engineering and Biotechnology*, *14*(1), 195–202. doi:10.1016/j.jgeb.2016.05.007 PMID:30647615

Rajeshkumar, S., & Bharath, L. V. (2017). Mechanism of plant-mediated synthesis of silver nanoparticles – A review on biomolecules involved, characterisation and antibacterial activity. *Chemico-Biological Interactions*, *273*, 219–227. doi:10.1016/j.cbi.2017.06.019 PMID:28647323

Rasheed, T., Bilal, M., Iqbal, H. M. N., & Li, C. (2017). Green biosynthesis of silver nanoparticles using leaves extract of Artemisia vulgaris and their potential biomedical applications. *Colloids and Surfaces. B, Biointerfaces*, *158*, 408–415. doi:10.1016/j.colsurfb.2017.07.020 PMID:28719862

Rasheed, T., Bilal, M., Li, C., Nabeel, F., Khalid, M., & Iqbal, H. M. N. (2018). Catalytic potential of bio-synthesized silver nanoparticles using Convolvulus arvensis extract for the degradation of environmental pollutants. *Journal of Photochemistry and Photobiology B: Biology*, *181*(2017), 44–52. doi:10.1016/j.jphotobiol.2018.02.024

Santhoshkumar, J., Agarwal, H., Menon, S., Rajeshkumar, S., & Venkat Kumar, S. (2019). A biological synthesis of copper nanoparticles and its potential applications. In *Green Synthesis* (pp. 199–221). Characterization and Applications of Nanoparticles., doi:10.1016/B978-0-08-102579-6.00009-5

Siddiqi, K. S., Husen, A., & Rao, R. A. K. (2018). A review on biosynthesis of silver nanoparticles and their biocidal properties. *Journal of Nanobiotechnology*, *16*(1), 14. Advance online publication. doi:10.118612951-018-0334-5 PMID:29452593

Treguer, M., Rocco, F., Lelong, G., Le Nestour, A., Cardinal, T., Maali, A., & Lounis, B. (2005). Fluorescent silver oligomeric clusters and colloidal particles. *Solid State Sciences*, *7*(7), 812–818. doi:10.1016/j.solidstatesciences.2005.01.017

Verma, A., & Mehata, M. S. (2016). Controllable synthesis of silver nanoparticles using Neem leaves and their antimicrobial activity. *Journal of Radiation Research and Applied Sciences*, *9*(1), 109–115. doi:10.1016/j.jrras.2015.11.001

Wang, J., Hsu, S.-W., Tao, A. R., & Jokerst, J. V. (2017). Copper sulfide nanodisk as photoacoustic contrast agent for ovarian tumor detection. *Photons Plus Ultrasound: Imaging and Sensing*, *2017*(10064), 100642D. doi:10.1117/12.2251184

Wang, L. (2016). Synthetic methods of CuS nanoparticles and their applications for imaging and cancer therapy. *RSC Advances*, *6*(86), 82596–82615. doi:10.1039/C6RA18355G

Wang, Z., & Rothberg, L. J. (2005). Origins of blinking in single-molecule raman spectroscopy. *The Journal of Physical Chemistry B*, *109*(8), 3387–3391. doi:10.1021/jp0460947 PMID:16851369

Zha, Z., Zhang, S., Deng, Z., Li, Y., Li, C., & Dai, Z. (2013). Enzyme-responsive copper sulphide nanoparticles for combined photoacoustic imaging, tumor-selective chemotherapy and photothermal therapy. *Chemical Communications*, *49*(33), 3455–3457. doi:10.1039/c3cc40608c PMID:23507786

Zhang, W., Qiao, X., & Chen, J. (2007). Synthesis of silver nanoparticles-Effects of concerned parameters in water/oil microemulsion. *Materials Science and Engineering B. Solid-State Materials for Advanced Technology*, *142*(1), 1–15. doi:10.1016/j.mseb.2007.06.014

Zhang, Y., Kohler, N., & Zhang, M. (2002). Surface modification of superparamagnetic magnetite nanoparticles and their intracellular uptake. *Biomaterials*, *23*(7), 1553–1561. doi:10.1016/S0142-9612(01)00267-8 PMID:11922461

Chapter 17
Magnetic Nanoparticles: Role in Next Generation Nanomedicine

Subbiah Latha
Anna University, Tiruchirappalli, India

Palanisamy Selvamani
Anna University, Tiruchirappalli, India

Suresh Babu Palanisamy
The Copperbelt University, Zambia

Deepak B. Thimiri Govindaraj
Council for Scientific and Industrial Research, South Africa

Prabha Thangavelu
Nandha College of Pharmacy, India

ABSTRACT

The magnetic nanoparticles are said to be a class of nanoparticles or nanomaterials that can be manipulated by the help of externally applied magnetic field. These magnetic nanoparticles constitute materials such as nickel, cobalt, iron, and their derivatives. These are normally smaller than 1 µm in diameter possess wide range of properties and attractive characteristics suitable for biomedical such as used as hyperthermia, enhancing magnetic resonance imaging (MRI) data, supplementing tissue engineering efforts, and improving the target-based drug delivery and many other technological applications. In the field of cancer research, the role of nanoparticles and nanotechnology-based methods and novel strategies have been increasing swiftly for cancer identification and cancer therapy. The iron oxide (Fe_3O_4, γ-Fe_2O_3) nanoparticles (NPs) are widely used for the drug delivery, magnetic nanoparticle-enhanced hyperthermia, and also as MRI contrast agents due to its biocompatibility, low toxicity, etc. lead to the growth of novel biopharmaceutical technologies.

DOI: 10.4018/978-1-7998-5049-6.ch017

INTRODUCTION

Nowadays, Magnetic nanoparticles, playing an important role in various fields especially in the medical field to detect the presence of antibiotics from a diverse sources like human fluid, water, food, and beverages and which may cause antibiotic resistance to the prescribed antibiotic treatment for some infections. However, this magnetic nanoparticle in various analytical filed like in biosensor has been used for the detection of the antibiotic (Cristea et al., 2017). The recent development of nanotechnology is synthesizing the biocompatible and functionalized magnetic nanoparticles that have various applications in biomedicine. Most often, the ferrites' magnetic nanoparticle has been focused significantly owing to their specific properties like its superparamagnetic, non-toxic, smaller in size, etc. (Doan, 2009). Moreover, there is an emergence to search for nanomaterials that could meet present biomedical needs. The MNPs could exhibit higher magnetic moments and surface area to volume ratios, which could make them more efficient for the treatment of cancer and as well for their target-based drug delivery. Besides, it acts as a contrast media for MRI and could produce the sensitivity to biosensors and as well as the diagnostic tools. The recent improvement in nanotechnology trends has to lead to the realization of the next generation magnetic nanoparticles that is suitable for biomedical applications (Farzin et al., 2020).

Figure 1. Role of Magnetic Nanoparticles in Next Generation Nanomedicine

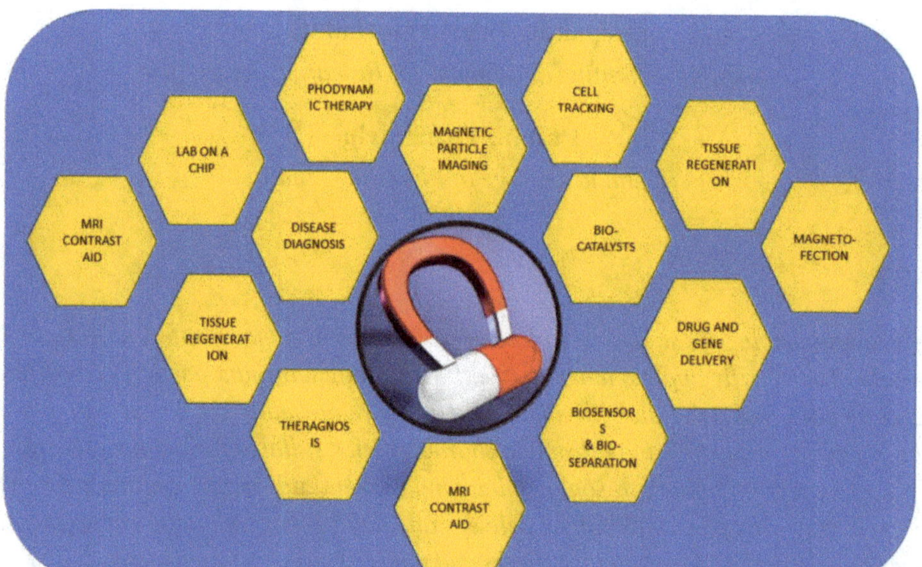

Anti-Microbial Activity

The magnetic nanoparticles can penetrate over the bacterial surface leads to cellular changes and exerts microbicidal or microbiostatic actions. Hence, its application is popular for the control of infectious diseases. Magnetic nanoparticles coated with chitosan loaded with ampicillin showed significant activity against both gram-positive, gram-negative bacteria, *Mycobacterium tuberculosis* and *Candida albicans*. The synthesized nanocomposites exhibited antibacterial and antifungal properties, as well as antimycobacterial effects. (Hussein-Al-Ali et al., 2014). However, a study by Layla et al. (Alumutairi et al., 2020) demonstrated the feasibility for the use of magnetic nanoparticle/alternating magnetic field mild hyperthermia as a minimally invasive strategy to treat *S. aureus* biofilm infections, which has the potential to work synergistically with standard antibiotic treatment, In addition, these nanoconjugates aids host innate immune response of macrophages for the clearance of intracellular bacteria. Parasmani et al., reported the preparation of recyclable magnetic iron oxide nanoparticles grafted with charged cobaltocenium-containing metallopolymers by surface-initiated reversible addition-fragmentation chain transfer (RAFT) polymerization. The magnetic nanoparticles loaded with β-Lactam antibiotic penicillin-G as conjugate showed good bactericidal efficiency against both Gram-positive and Gram-negative pathogens, which facilitated the revitalization of the penicillin.(Pageni et al., 2018)

The MNPs coated with lipoamino acid (Size: 7 nm) were loaded with the antibiotics viz. Ampicillin and Gentamycin reported promising for combination therapy when evaluated against *S.aureus* and *E.coli* pathogens. The lipoamino acid-coated magnetic nanoparticles showed the promising anti-bacterial activity and anti-biofilm activity of ampicillin and gentamicin (Gholami et al., 2018). To treat drug resistant fungal pathogens, magnetic nanoparticles loaded with anti-fungal agents that could promote the apoptosis of fungal cells was proposed and validated. An increased anti-microbial activity was observed with vancomycin loaded magnetic nanoparticles against resistant bacterial strain (Niemirowicz & Bucki, 2017). Similarly, Jie et al. (Li, 2019) reported that methicillin-resistant *S.aureus* biofilm was mechanically damaged by magnetic nanoparticles in association with magnetic hyperthermia to an extent of five log10 reductions in biofilm bacteria after the treatment with 30 mg/ml of 11 nm magnetic nanoparticles using the magnetic hyperthermia. Ampicillin loaded magnetic nanoparticles (20 µl) were added to the bacterial strains and incubated overnight showed 95% inhibition (Binandeh, 2018).

Novel antibiotic coated with iron oxide nanoparticles were synthesized to treat bacterial infections. These magnetic nanoparticles have been prepared by using the oleic acid as the first anchor layer and then the bioactive components such as rifampicin, a broad-spectrum antibiotic agent was coated as a second layer. These double-layered magnetic nanoparticles with a mean particle size of 11.9 nm was used for the treatment of bacterial infections (Ciurlicaa et al., 2010). The drug-loaded magnetic nanoparticles could be locally accumulated with an external magnetic field that offers selective therapy, leads to the avoidance of systemic and related adverse effects. The intravenous administration of magnetic nanoparticles conjugated with rhodamine B isothiocyanate and PEG to the mice previously implanted with the magnetic field such as ferritic steel and Ti90Al6V4 could offer the local accumulation of the conjugated antibiotic (Janßen et al., 2020). The magnetic nanoparticles coupled with PBP10 peptide and human cathelicidin-derived LL-37 peptide along rhodamine B results in the significant bactericidal activity against *S.aureus*, *P.aeruginosa* and fungicidal activity against the fungal strain *C.albicans*, which is perhaps due to the action of binding of phosphoinositide with the domain gelsolin (Bucki et al., 2019; Wnorowska et al., 2020). MNPs assisted bacterial detection based on polymerase chain reaction (PCR), colorimetric detection, fluorescent detection, and surface-enhanced Raman detection techniques (Yuan et al., 2018b) have been reported. Previous studies reported that MNPs with superparamagnetism could shorten the longitudinal or transverse relaxation time of water protons nearby, thus MNPs-based contrast agents have been used to enhance the signal of the abnormal anatomy for the detection and monitoring of infectious diseases (Lefevre et al., 2011; Chen et al., 2016a). Particularly, MRI has been explored to noninvasively track bacteria and monitor antibiotic therapy of bacterial infection by using MNPs-based contrast agents (Lefevre et al., 2011; Hoerr et al., 2013). Hence, the magnetic nanoparticles possess the ability to inhibit the function of efflux drug transporters, promote apoptosis of microbes and reduce the formation of biofilm seems much promising for the anti-microbial therapy and also for infection imaging and prognosis of antimicrobial therapy.

Clinical Hyperthermia

Magnetic hyperthermia is thermal therapy to treat particularly the cancerous cell using magnetic nanoparticles. Heating of a magnetic nanomaterial in an alternating magnetic field can occur by four mechanisms: (1) dielectric losses in a material of low electrical conductivity; (2) eddy currents losses in a material of high electrical conductivity; (3) frictional heating due to the physical rotation of an anisotropic magnetic particle and; (4) hysteresis losses in a magnetic material. When the tumour cell is exposed to a relatively high temperature (113°F), the cancer cells could be killed selectively without affecting the normal cells. Furthermore, this technique also shrinks the size of the cancer cell by the induction of heat. Various types of superparamagnetic and ferromagnetic nanoparticles grafted with target-specific agents could be delivered and targeted to the specific site and under an applied magnetic field elicit hyperthermia and can be used as adjuvant therapy in radiation and as well chemotherapy for treating the cancer cell. (Hildebrandt et al., 2002; Reddy et al., 2012; Van der Zee, 2002; Wust et al., 2002). Hyperthermia induced by magnetic cationic liposomes can reduce the tumor growth when combined with TNF-α gene delivery (Ito et al., 2001). MNPs loaded with epirubicine was the first magnetic carrier used to target and destruct the solid cancer cells through magnetic hyperthermia (Liibbe & Bergemann, 1997). Similarly, hyperthermia induced by MNPs coated with amino silane groups was proven beneficial to curtail tumour cell growth (Jordan et al., 1997; Laurent & Mahmoudi, 2011; Mura et al., 2013; Reddy et al., 2012).

Chronic Kidney Disease (CKD)

Chronic kidney disease is a fatal disease, results in chronic anaemia and treated by giving surplus iron supplements. The magnetic nanoparticles ($MnFe_2O_4$) was used as a tracer to detect the rats' kidney perfusion profile. The animals underwent a single injection of doxorubicin to induce a progressive kidney disease and the kidney perfusion and washout profiles of infected animals have been differentiated with the healthy animals by using the magnetic nanoparticle techniques. Due to inefficient uptake of iron supplements through oral and parental routes. Alternatively, Ferumoxytol, a formulation with MNPs, coated over a carbohydrate shell of poly glucose sorbitol carboxymethyl ether (Stoumpos et al., 2019) as IV injection for the CKD treatment. Ferumoxytol is an alternative to gadolinium-based compounds as a vascular contrast agent for magnetic resonance angiography (MRA), particularly for patients with chronic kidney disease (CKD).

Cancer Therapy

The magnetic nanoparticles (MNPs) are magnetic nanomaterials with small particle size ranging in nanodimensions, larger surface area, and superparamagnetic property with an excellent magnetic response to the applied magnetic field without any magnetic remanence. Such MNPs have a prominent role in cancer diagnosis and target drug delivery and treatment (Wu & Huang, 2017).

Widder et al to reported that cancer cells were targeted albumin coated doxorubicin conjugated with magnetic nanoparticles locally injected and reported its applications in MRI and anticancer activity. But, when given intra-arterially an enhanced effect of drug was observed (Widder et al., 1981; Widder et al., 1979). Doxorubicin loaded within the core of Iron-gold Nanoparticle (Fe@Au core) also showed reliable anticancer activity (Kayal & Ramanujan, 2010). The PEGylated cubic Fe_3O_4 NPs loaded with doxorubicin when targeted to the mouse skin fibrosarcoma (weHI-164) cells showed promising anti-cancer activity (Dutta et al., 2018).

SPIONs decorated with MethoxyPeG-poly (d,l-lactide-co-glycolide)(MPeG-PLGA) and loaded with paclitaxel was magnetically targeted in HeLa, NIH3T3 cells and A549 cells respectively showed significant efficacy at a much lower dose (Lu et al., 2016; Ray Chowdhuri et al., 2016). Paclitaxel conjugated magnetic nanoparticles was magnetically targeted and showed profound effect in prostate cancer (Chorny et al., 2010).

Daunorubicin loaded with MNPs as polymeric microcontainers released the drug in a controlled fashion through a pH-sensitive controlled release mechanism showed significant activity (Chatzipavlidis et al., 2011). MNPs loaded with daunorubicin was magnetically targeted and evaluated in human chronic myelogenous leukemia cells (K562 cells) showed significant elevation in anticancer activity (Wang et al., 2011a).

PLA multiwalled carbon nanotubes based iron oxide composite nanofibers loaded with daunorubicin was magnetically targeted in leukemia cancer cells for their therapeutic action (Hosseini et al., 2016) and the methoxy PEG-lysine-oleic acid-modified Iron Oxide nanoparticles are loaded with Paclitaxel has used to target the carcinomic cells in liver and kidney and showed their anticancer activity (Tian et al., 2017) significantly. Whereas, the magnetic iron oxide nanoparticles duly loaded with Daunorubicine and 5-bromotetrandrin was used to target the Human leukemia K562/A02 cells for combination chemotherapy (Wang et al., 2011b). Hydrophilic MNPs coated with nonionic surfactant pluronic F127–polyethylene-polypropylene oxide polymeric carrier was used to target the LNCaP and PC-3 prostate cancer cell lines promisingly (Nizamov et al., 2018).

In addition to this, the anticancer drug doxirubicin has its wide range of applications as a targeted therapy for treating the caner through its magnetic nanoparticle formulation. There are many researchers has done the work on this drug and reported fruitfully for treating and target-specific delivery of this drug, some of the studies include, the

Fe_3O_4-inositol hexaphosphate functionalized magnetic nanocarrier loaded with Doxorubicin was targeted on MG63 cell lines (Xu et al., 2017), core-shell mesoporous silica and SPIONs entrapped within the liposomes were loaded with doxorubicin have shown improved efficiency against U87 cells (Sharifabad et al., 2016), the dextran-coated SPIONs loaded with doxorubicin was magnetically targeted to treat the pancreatic carcinoma cell lines of human (MIA PaCa-2) (Arachchige et al., 2017), Fe_3O_4 nanoparticles coated with polyvinyl alcohol as hydrogel was used as a magnetic carrier for delivery of doxorubicin in cancer treatment (Abasian et al., 2017), and the iron oxide NPs loaded with doxorubicin calcium phosphate or arginyl glycyl aspartic acid was used to target and treat T24 bladder cancer cells for therapeutic activity (Wang et al., 2017) respectively.

Similarly, the iron oxide nanoparticles with folic acid conjugate have been used for the targeted hyperthermia treatment of lymph node metastases of prostate cancer along with MRI detection (Bonvin et al., 2017), the magnetic nanoparticles loaded with the 4-epidoxorubicine have initially used to target solid tumor cells for destruction (Lubbe et al., 1996).

The Fe_3O_4-NH_2, an amine group immobilized iron oxides, attached on the surface of poly[(acrylic acid)-block-(N-isopropylacrylamide)-block-(acrylic acid)], a self-assembled triblock copolymer, which was used for targeted drug delivery and for inducing hyperthermia in MCF-7 cells (Kuo et al., 2016). Whereas, the cobalt ferrite ($CoFe_2O_4$) particles covalently bonded with a photosensitizer was targeted in

PC-3 cells to confirm the photodynamic anticancer activity (Choi et al., 2016). Iron oxide nanoparticles incorporated with dopamine-bovine serum albumin/hyaluronic acid further loaded with paclitaxel and demonstrated for targeted drug delivery and imaging for the diagnosis of cancer cells (Vismara et al., 2017).

The doxorubicin is one of the most reported drug to demonstrate magnetically targeted drug delivery to the target site. The PEGylated polyethylene bis(carboxymethyl ether) iron oxide nanoparticles loaded with doxorubicin (Ji et al., 2018), the starch/ZnO-coated iron oxide nanoparticles loaded with curcumin and its analogues was evaluated on Human lymphocytes, HepG2, and MCF-7 cell lines (Saikia et al., 2017) and reported for its anticancer potency. The dendritic-Fe_3O_4 supramolecular NPs were loaded with doxorubicin and the tumor regression in the spheroid murine melanoma model was reported (Nigam & Bahadur, 2018). Similarly, dendritic-Fe_3O_4 supramolecular NPs were encapsulated into the microvascular vesicles to deliver the drug cargos to the carcinomic cells by direct targeting had shown a profound anticancer activity than the conventional delivery of the doxorubicin (Hua et al., 2010).

There are many research studies have been done by various groups in establishing the role of magnetic nanoparticles for treating cancer and other biomedical conditions. The bovine serum albumin-coated superparamagnetic γ-Fe_2O_3 nanoparticles was used to target the U25I of the brain (Wang, Tu, Tian et al, 2016), the Fe_2O_3 magnetic nanoparticles were loaded with 5-fluorouracil was evaluated in 7,901 cell lines had shown the enhanced anticancer activity on the tested cells (Shen, Ma, Yu et al, 2016). Likewise, the polylactic-co-glycolic acid (PLGA) nanoparticles with or without the iron oxide nanoparticles loaded with 5-fluorouracil and the relative efficacy and the role of MNPs in cancer therapy was reported (Eynali et al., 2017). The magnetic field-sensitive methylcellulose and PCL gel based magnetic nanocomposite loaded with 5-fluorouracil showed controlled drug release when targeted against cancer cells (Nikjoo & Aroguz, 2016).

MNPs mediated drug delivery systems are capable of simultaneous drug delivery as well as diagnosis using Magnetic Resonance Imaging defined as theragnostics. SPION's loaded with doxorubicin reported to possess theragnostic property (Souto & Ali, 2012; Jain et al., 2008; Yang et al., 2007). PLGA-coated magnetite-graphene oxide loaded with 5-fluorouracil system was proven as an efficient theragnostic aid under infrared laser in rabbits (Mohammadi Gazestani et al., 2018).

Magnetic drug targeting (MDT) is a familiar approach for trusted therapeutic activity because this approach gives site-specific action and drug release could be controlled and magnetically triggered. Epirubicin loaded ferrofluid delivered on the surface of the site of action gives reliable cancer activity when the magnets were maintained at a distance of 0.5 cm from the tumour surface (L¨ubbe et al., 2001). Albumin/lauric acid hybrid decorated SPIONs loaded with mitoxantrone upon exposure with Jurkat cells for *in-vitro* evaluation had shown potent anticancer activity (Matthew, 2018; Zaloga et al., 2014). SPION's loaded with mitoxantrone had efficiently targeted the infected malignant tumor cells locally and have been used successfully for the treatment of squamous cell carcinoma without any systemic toxicity (Alexiou et al., 2000; Souto & Ali, 2012). In another study, the SPIONs coated with lactide-co-glycolide ethylene oxide fumarate is loaded with paclitaxel has been used in the treatment of cancer which has a benefit of decreasing undesired effects on the host and increases circulation time of the magnetic nanoparticles (Souto & Ali, 2012; Bharde et al., 2008) and the SPIONs coated with unsaturated aliphatic polyesters such as Poly (ethylene glycol) co- fumarate vectors filled with anticancer drug tamoxifen had shown enhanced delivery of drug without burst drug release effect and particle digestion. It is reported that, this type of novel delivery vectors is useful in delivering the viral gene and drugs (Mahmoudi et al., 2009). Similarly, genistein a polysaccharide used in the treatment of cancer was covalently attached to the SPIONs coated with carboxymethyl chitosan was targeted to tumor cells and observed to have a reliable cancer cell apoptosis (Si et al., 2010).

The Fe_3O_4-ZnO hybrid nanoparticles loaded with folic acid are used for theragnostic purposes of human epithelial colorectal adenocarcinoma and are also used in the photodynamic therapy (Patel et al., 2017). The iron oxide nanoparticles doped hyperbranched polyglycerol conjugated with folic acid is used to target MCF-7 cells (Mostaghasi et al., 2018). SPIONs functionalized with (3-aminopropyl) triethoxysilane and polyethylene glycol and conjugated with folic acid targeted the U87 (Akal et al., 2016a) cancer cells.

Magnetic nanoparticles prepared by the agglomeration of ZnO and Fe_2O_4 loaded with doxorubicin as a controlled release formulation had shown excellent destruction of malignant tumour cells specifically targeted through the use of alternating magnetic field. (Lee et al., 2013; Polo et al., 2018) The doxorubicin-loaded to the PEGylated iron oxide nano cubes was successfully employed in the destruction of the tumour microenvironment in epidermal carcinoma upon intratumoral injection embodied with Magnetic Fluid Hyperthermia (Kolosnjaj-Tabi et al., 2014; Polo et al., 2018).

Though there are various binding methods used to load the drugs to the magnetic nanoparticles, conjugation of drugs through ionic linkage has got many advantages like multiple drug loading and to tailor the drug release profile governance. Metaxantrone, a positively charged drug was coupled with negatively charged phosphate moeity through ionic linkages and loaded to SPIONs was used in permanent tumor destruction (Alexiou et al., 2000; Amanda et al., 2016). Similarly, epirubicin was also ionically linked to anionic iron oxide nanoparticles through its amino sugar and used for targeting the cancer cells (Amanda et al., 2016; Bergemann et al., 1999; Douziech-Eyrolles, 2007). Methotrexate was covalently conjugated to the surface of SPIONs by amidation reaction with the -COOH groups on the drug and amino ends in vector surface and was employed for the cancer cell targeted chemotherapy (Amanda et al., 2016; Kohler et al., 2005). Likewise, a photosensitizer (laser-assisted) and a tumour remission drug chlorine e6 (Ce6) was covalently bonded to the surface of the SPIONs and used for antitumor activity (Amanda et al., 2016; Huang et al., 2011).

The ultra-magnetic liposomes loaded with drug 5, 10, 15, 20-tetra (m-hydroxyphenyl) chlorine (m-THPC) is used in the cancer therapy which is achieved by combined photosensitivity and magnetic hyperthermia principles of the magnetic nanoparticles (Di Corato et al., 2015; Kuznetsov et al., 2001). The anticancer drugs and proteins such as doxorubicin, tissue-plasminogen activator (t-PA), disulfonated tetra phenyl chlorine (TPCS2a) and 5, 10, 15, 20-tetra(m-hydroxyphenyl) chlorine (mTHPC) were loaded with human macrophages along with the magnetic nanoparticles which has been used to treat cancer by enhanced magnetic targeting strategy.

The antitumor activity with the gold-iron oxide (Au-Fe_3O_4) magnetic nanoparticles along with polycationic polyethylenimine (PEI) polymer was loaded with Notch3 chimera has shown potential anticancer activity. Furthermore, it possess the capability and ideal qualities of a gene delivery vector (Boussif et al., 1995; Chen et al., 2017; Shaw et al., 2013). Gold in nanometric dimensions, can behave like magnetic nanoparticles (Grèget et al., n.d.) and the delivery of antibiotics such as ampicillin, streptomycin, and kanamycin could be achieved when these drug molecules are conjugated with Gold Nanoparticles and reported to possess significant activity with a lower dose when compared to conventional drug delivery methods (Burygin et al., 2009; Geller et al., 1986; Gu et al., 2003; Pissuwan et al., 2011; Rosemary et al., 2006; Saha et al., 2007). Methotrexate and folic acid is loaded to the gold nanoparticles reported to shown seven-fold higher cytotoxic activity than the free methotrexate molecule was observed in the lewis lung carcinoma cells (Chen et al., 2007; Pissuwan et al., 2011).

PLGA coated with iron oxide acts as a magnetic carrier loaded with quercetin was used to treat the lung cancer shown effective targeting capacity and used as an alternate for aerosols (Verma et al., 2013). A single dose of magnetically triggered drug carriers containing doxorubicin was injected intra arterially and tested in animals to treat cancer and found as a good vector of drug carriers and no toxicity at a low dose level (Goodwin et al., 2001). The magnetic nanoparticle coupled with doxorubicin hydrochloride was injected as infusion, which is used to magnetically target the hepatocellular carcinoma (Dobson, 2006) and the SPIONs loaded with doxorubicin was targeted to U251 cell lines showed good therapeutic activity (Wang, Chang, Zhang et al, 2016).

In another study, the MNPs coupled with Lipofectamine 2000/Cationic lipid 67 (GL67)/plasmid DNA (pDNA) liposome complex is used for effective viral gene transfer in airway endothelium (Xenariou et al., 2006). Whereas, the iron oxide nanoparticles functionalized with folic acid and rhodamine was used in a cell proliferation assay on HeLa cells upon targeting (Ancira-Cortez et al., 2017). When the hyaluronic acid coated iron oxide nanoparticles was used to target the macrophages cell lines THP-1for immune activation which reveals that this is a potential tool to deliver drugs (Kamat et al., 2010) and peptides by targeting HEK293 and A549 (Kumar et al., 2007). The magnetic nano-biohybrids comprising silica, magnetic nanoparticles, and enzyme capsulated polymer was prepared and loaded with a prodrug indole-3-acetic acid demonstrated potent anticancer activity (Correa et al., 2018).

Enhanced drug delivery in nanoscale drug targeting into the cancerous cell could be achieved by sonoporation through formation of small pores in cell membranes by using the ultrasound for the transformation of nucleic acid materials/drugs to target the cells. Pifithrin-μ loaded in PEG-coated MNPs was delivered to the site of action (Mérida et al., 2018). Likewise, in another study, the zinc ferrite magnetic nanoparticles were prepared by the introduction of zinc ions into ferrite lattice had shown cytotoxic effect on cells through hyperthermia and photothermia with zinc ferrite MNPs (Kasparis et al., 2018).

Chitosan, poly(1→4)-2-amino-2-deoxy-d-glucan, a partially acetylated glucosamine biopolymer, which is present in the cell wall of fungi *Mucorales*, is a polyamino saccharide with many significant biological and chemical properties and therefore used in many types of applications such as, waste water treatment, chromatographic support, enzyme immobilization, drug-delivery systems. (Doan, 2009) The pH-sensitive chitosan complexed with amine-terminated poly-N-isopropyl acrylamide magnetic nanoparticles tagged with folic acid and florescent agent was used as a reliable carrier to load curcumin analog which has produced intended cell apoptosis in cancer lines. (Yadavalli et al., 2015) Chitosan-co-PEG/poly(glycerol sebacate)-co-PEG-coated iron oxide acts as an extraordinary carrier for 5-Fluorouracil to treat carcinomic cells and also in HT29 cell lines (Naghizadeh et al., 2018) and as well the magnetic iron oxide mesoporous silica nanocomposites (m-MCM-41) was loaded with 5-fluorouracil and used in the treatment of cancer and proved to be an excellent drug carrier (Egodawatte et al., 2017).

Breast Cancer

Globally, the second leading cause of death in women community is due to breast cancer. Increased rate of metastasis together with drug resistance with conventional drug therapy, makes a difficult situation to treat breast cancer fruitfully, ultimately leads to morbidity and mortality. However, the various conventional treatment options such as chemotherapy, surgery, radiotherapy, etc. suffers various limitations such as high cost, non-specific toxicity, genetic mutation of the normal cells, etc. Many researchers and scientists have been focusing on the early diagnosis of breast tumours so that, they could treat them in the early stage itself. Considering the need to develop a novel treatment regimen that could specifically target and treat the cancer cells without affecting or with minimally toxic to the normal cells. The iron oxide magnetic nanoparticles are widely used nowadays both *in-vivo* and *in-vitro* to specifically target cancer cells. MNPs are conjugated with drugs, antibodies, proteins, enzymes, nucleotides, dyes, biomarkers, isotopes etc to deliver them to the targeted cancer cells utilizing an applied magnetic field enhances drug efficacy and reduce drug toxicity. Owing to the superparamagnetic, biocompatible, and biodegradable properties of the iron oxide magnetic nanoparticles, it has been widely used in the breast cancer diagnosis, treatment, and therapy monitoring (Shakil et al., 2019). The anti-mitotic drug paclitaxel isolated from the Pacific Yew tree, approved by US FDA for the therapy of various types of cancers like ovarian, breast, lung, head, and neck cancer, and Kaposi's sarcoma. The superparamagnetic iron oxide magnetic nanoparticles loaded with paclitaxel potentiates its activity in insensitive and resistant breast adenocarcinoma MCF-7 cells (Rivera-Rodriguez et al., 2018). MNPs coated with chitosan and folic acid-conjugated chitosan is used to target on MCF-7 offers theragnostic applications (Zarrin et al., 2016), MNPs conjugated with chitosan-graft-poly(2-dimethylaminoethyl methacrylate) together with folic acid targeted in the following cell lines L929 (mouse fibroblast), MCF-7 (human breast cancer), and MDA-MB-231 (human breast cancer) showed significant therapeutic efficacy (Anirudhan & Christa Binusreejayan, 2018). MNPs coated with luteinising hormone releasing hormone (LHRH) shown reliable therapeutic effect in breast cancer cells and lung metastasis cell (Zhou et al., 2006). Similarly, the core-shell mesoporous silica-coated MNPs loaded with doxorubicin as liposomes (Sharifabad et al., 2016), MNPs loaded with Doxorubicin for dual targeting (Varshosaz et al., 2016), the azobis[N-(2-carboxyethyl)–2-methylpropionamidine iron oxide nanoparticles loaded with doxorubicin were used to target MCF-7 breast cancer cells proven beneficial (Chen et al., 2016). In a recent study, magnetic resonance imaging (MRI)-guided chemo-photothermal therapy of human breast cancer xenograft in nude mice was demonstrated using a novel core/shell structure of Fe_3O_4@MoS_2 nanocubes via the integration of Molybdenum disulfide MoS_2 film onto iron oxide nanocubes through a facile hydrothermal method (Li et al., 2020; Xie et al., 2018).

Adimoolam et al., reported chitosan-coated iron-oxide nanoparticles with pH-sensitive glutaraldehyde linker for pH responsive delivery of doxorubicin (pH range of 4.4–6.4) for specific delivery of doxorubicin in the intracellular components of human breast cancer (MCF-7) and ovarian cancer (SK-OV-3) cell lines (Adimoolam et al., 2018). β-cyclodextrin–PEG–polyethyleneimine-coated iron-oxide nanoparticles loaded with 5-fluorouracil proven for its promising therapeutic applications upon evaluation in L929 and MCF-7 cells (Prabha & Raj, 2016). the iron oxide NPs/chitosan, folic acid loaded with poorly soluble quercetin reported for its theragnostic applications in PC-3 and MCF-7 cells (Nandagopal et al., 2016). Steroidal anti-estrogen drugs loaded in liposomes containing iron oxide nanoparticles used to target and deliver the drug at the breast cancer (Plassat et al., 2011).

Cervical Cancer

Cervical cancer is one of the most common cancers causes fatality among women in most of the developing countries. The oncoproteins of human papillomavirus, an elevated levels of estrogen, smoking, multiparity, and other factors are the main culprit cause of cervical cancer. Nanomedicine through use of quantum dots, nanoparticles, nanowires, and magnetic nanoparticles offers remarkable improvements in the therapeutic strategies to target and treat cervical cancer devoid of side effects associated with conventional drug therapy (Hira & Reena, 2018). The folate conjugated cobalt ferrite nanoparticles coated with PEG verified in HSF 1184 (human skin fibroblast cells) and HeLa (human cervical cancer cell) displayed potential theragnostic uses (Nasiri & Hassanzadeh-Tabrizi, 2018). Poly(N-isopropylacrylamide) hydrogel incorporated with MNPs conjugated with folic acid exhibited enhanced uptake of the MNPs to the cervical cancer cells (HeLa) and synergism (Kim et al., 2017). MNPs coated with polyethylene glycol containing folic acid targeted towards Human cervix epitheloid carcinoma - HeLa cells, human breast carcinoma cells - MCF-7 and MDA-MB435 shown profound penetration and amplified activity (Allard-Vannier et al., 2017). The iron oxide nanoparticles encapsulated within the inner core loaded with dual anticancer drugs curcumin and doxorubicin for delivery in the targeted site (HeLa cells) for augumented anticancer activity (Wang, Jing, Zhang et al, 2016). The cucurbit (7) uril-modified iron-oxide nanoparticles are used in the delivery of doxorubicin to HeLa cells, A2780 and Dox resistant A2780 cells and the results reported seems encouraging (Benyettou et al., 2017).

Doxorubicin is one of the most potent chemotherapeutic agents widely used for treating diverse types of cancers. However, its clinical use is very much limited owing to its side effects and drug resistance. Consequently, many new nanocarriers including magnetic nanoparticles, are extensively used to reduce the side effects of anticancer chemotherapy, which includes, the surface-modified multifunctional $ZnFe_2O_4$ nanoparticles loaded with Daunorubicine is targeted towards HeLa cells for the treatment of cervical cancer (Maiti et al., 2016). The chitosan-functionalized iron-oxide nanoparticles loaded with doxorubicin was targeted towards ovarian cancer cell (SK-Ov-3) for targeted therapy (Adimoolam et al., 2018). Similarly, Genipin-cross-linked iron (III) oxide/polyetherimide nanoparticles are loaded with doxorubicin is used in cancer therapy targeting HeLa cells (Long et al., 2017), iron-oxide NPs coated with poly(N-isopropyl acrylamide) is loaded with doxorubicin and targeted towards cervical cancer cells for theranostic applications (Yar et al., 2018), and the folic acid conjugated PEG-coated MNPs are loaded with doxorubicin and targeted towards HeLa cells to bring efficient anticancer activity in minimal dose (Erdem et al., 2017).

Folic acid receptor mediated endocytosis depends solely upon the cell specific folate receptors available on the surface of the cell membrane which only attach to the specific components found in the extracellular space. Cancer cells, unlike normal human cells, overexpress folate receptors which have a very high affinity for folic acid. The folic acid is a short-chain smaller in size and thus it promotes the internalization of the magnetic nanoparticles conjugated with folic acid and conjugated drug cargos. Folic acid is desirable due to its stability, inexpensive, non-immunogenic, and high affinity to the cell surface receptors. Therefore, the attachment of the folic acid ligand to superparamagnetic iron oxide nanoparticles is an important step for the folic acid receptor-mediated targeted drug delivery of iron oxide nanoparticles (Li et al., 2016). There are many research reports based on this concept viz., folic acid mediated iron oxide (Fe_3O_4) MNPs are used to target human serous ovarian cell line (Skov-3) for theragnostic purposes (Zhang et al., 2016), folic acid conjugated polyglycerol-coated iron oxide nanoparticles for improving radiosensitivity to target on human cervical cancer (Fakhimikabir et al., 2018), the magnetite at silicon dioxide nanostructures are loaded with folic acid was used in the treatment of human cervical cancer by the aid of hyperthermia (Bettini et al., 2017). Similarly, the polyacrylic acid-functionalized Fe_3O_4 MNPs were conjugated with folic acid through peptide bonding nanocarriers are magnetically driven to target against HepG2 to get rid of cancerous cells in the liver (Gupta et al., 2016). The carboxylated quercetin coupled to superparamagnetic iron oxide MNPs modified by (3-aminopropyl) triethoxysilane, folic acid, and polyethylene glycol conjugates were efficiently targeted in MCF-7, HeLa and A549 cancer cells (Akal et al., 2016b). The magnetic mesoporous silica NPs, copper nanocrystals, photothermal agent was functionalized together with polyethylene glycol was further conjugated with folic acid and targeted against cell lines such as HeLa for theragnostic purposes (Gao et al., 2016). Branched polyethyleneimine (bPEI) which has weak blue luminescence (that can be used in diagnostic purposes) was extensively used in gene delivery. This luminescent blue bPEI-SPIONs obtained after oxidizing amine groups in bPEI and reducing the Fe^{3+} ions of SPIONs was used in the transport of oligonucleotide, PIC to HeLa cells, etc (Acar et al., 2018). Considerably, the iron oxide mesoporous silica nano carriers were magnetically driven to target against HepG2 to get rid of cancerous cells in the liver (Luo et al., 2016).

Gastric Cancer

Gastric cancer is one of the most morbid cancers and it is very difficult to cure. The present prognosis of gastric cancer is very poor with a 5-year survival rate is less than 24%. Conventional treatment including surgery, radiation and chemotherapy are very effective only in early stages and in *in-situ* gastric cancer. However, gastric cancer in its advanced stages do not respond to these types of treatments. Moreover, resistance to chemotherapy-induced apoptosis is one of the major cause for the failure of conventional therapies of gastric cancer and early-stage detection of gastric cancer remains as a great challenge to the clinicians (Cui et al., 2015). Use of MNPs seems to be a promising strategy to treat gastric cancer through drug targeting specifically to the cancerous site improves the therapeutic efficiency than the conventional therapeutic modalities. The iron oxide magnetic nanoparticles coupled with the epidermal growth factor (EGF) has been used in treating the colon cancer and breast cancer (Creixell et al., 2010), the murine colon carcinoma cells are treated with Epirubicin 5TR1 aptamer SPIONs which targets C26 cells has been reported to have the reliable therapeutic activity in the treatment of colon cancer (Jalalian et al., 2013). The superparamagnetic iron oxide nanoparticles (Fe_3O_4) encapsulated with 5-flurouracil is used in targeting cancer cells (colon cancer) for therapy (Shen, Gao, Yu et al, 2016), the polycaprolactone/chitosan-coated superparamagnetic iron oxide nanographene oxide has been used to treat CT29 colon cancer cells with combination of radiofrequency core loaded with 5-flurouacil (Asadi et al., 2018), the aminosaline magnetic nanoparticles with folic acid is targeted towards collateral cancer which is tested with cancer xenografts of animal model (Niemirowicz et al., 2017) and the polyethylenimine superparamagnetic iron oxide nanoparticles has been used in the PD-L1 siRNA delivery for gastric cancer and MRI (Luo et al., 2017).

Liver Cancer

Hepatocellular carcinoma, is another fatal cancer occurs in liver and one of the most common cause of death. MNPs have been used in recent days for the diagnosis and treatment of liver cancer reported as a promising avenue through magnetic hyperthermia and magnetic drug targeting. (Ungureanu et al., 2016). The sodium alginate-polyvinyl alcohol-bovine serum albumin-coated Fe_3O_4 NPs loaded with Doxorubicin have been used in the magnetically guided therapy of liver cancer by targeting HepG2 and L-02 cell lines (Prabha & Raj, 2017), the magnetic mesoporous silica Nanoparticles conjugated with folic acid has been used as a potential carrier for controlled release to treat cancerous HeLa cells (Varshosaz et al., 2016), the PeG carboxyl–poly(ε-caprolactone)-modified magnetic nanoparticle loaded with paclitaxel is targeted towards Mouse H22 hepatocarcinoma cells (H22) and hepatoma carcinoma cell (Hep G2) for anticancer activity (Li et al., 2017), the SPIONs loaded with Doxorubicin is targeted towards LO2, L929, and HepG2 cell lines to bring about the required therapeutic activity (Komarala et al., 2016; Peng et al., 2016) and the Fe_3O_4 NP-capped mesoporous silica nanocarriers are magnetically driven to target against HepG2 to get rid of cancerous cells in the liver (Wu et al., 2016). These research reports further support the importance of the magnetic nanoparticle as targeting therapy of liver cancer cells.

Lung Cancer

Another leading cause of death among cancer types in the world is lung cancer, specifically non-small cell lung cancer. Non-small cell lung cancer is present in advanced stages in most of the patients diagnosed with lung cancer owing to its asymptomatic nature. Hence there is an urgent need for early diagnosis. In addition, the poor response of the currently available treatment options for this lung cancer urged to develop the alternative therapy for treating this type of cancer. Clinical use of MNPs for cancer therapy, especially magnetic hyperthermia is one of the trending strategies for the tumour ablation owing to improved drug solubility, bioavailability, increased site-specific targeting, reduced systemic toxicity, and reduced multidrug resistance (Menon et al., 2017).

Owing to above said reason, the magnetic aerosol drug targeting method was reported to be useful for the efficient non-invasive delivery of drugs to lung tumours (especially located on the lower wall of airway branches). The introduction of a permanent magnet next to the tumor increased the fraction of particle deposition from 49% to 68% (Manshadi et al., 2019) and the TAT peptide-conjugated magnetic PLA-PEG Nanocapsules loaded with paclitaxel was delivered to the human lung adenocarcinoma epithelial cells A549 showed better efficiency.

Arthritis

The chronic inflammatory autoimmune disease, Rheumatoid Arthritis (RA), has a complex pathology which is characterized by the inflammation in joints, pannus formation, bones and cartilage destruction, the devastation of the synovium often associated with severe pain, stiffness, and work disability. The traditional way of the treatment of RA has a short biological half-life, poor bioavailability, and the need for a higher dose. Therefore, drug targeting at the RA joints has been introduced to improve the therapeutic efficacy using magnetic nanoparticles (Pirmardvand Chegini et al., 2018).

Rheumatoid arthritis can also be treated with a technology called magnetically targeted chemo-photo thermal treatment where the drug methotrexate (MTX) is loaded with poly(lactic-co-glycolic acid, PLGA) gold/iron core-shell nanoparticles conjugated along with arginine-glycine-aspartic acid reported to show improved therapeutic efficacy (Kim et al., 2015). The polyethyleneimine (PEI) coated SPIONs is loaded with siRNA which acts against the IL-2/-15 receptor β chain is delivered intravenously and targeted towards the inflamed site with a magnetic field applied externally has shown an improved therapeutic effect (Duan et al., 2014). Similarly, the SPIONs tagged with corticosteroid drug dexamethasone to the inflamed site by intra-atrial route produce enhanced therapeutic activity this type of delivery has been prescribed for arthritis and osteoarthritis (Butoescu et al., 2009). Diclofenac sodium loaded on to the magnetic iron core and targeted to the arthritic site brings about the therapeutic response with a smaller dose compared to conventional drug delivery (Arias et al., 2009). SPIONs are made targeted towards the inflamed joints and have shown a very good therapeutic effect (Zhang et al., 2018). The methotrexate magnetic microcapsules for the treatment of RA has been a promising agent in bringing the therapeutic effect and show a sustained release of drugs according to the Hixson-Crowell model release kinetics (Chakkarapani et al., 2015).

Theragnostic Applications

MNPs were used for simultaneous diagnosis and therapy as theragnostic agents. MNPs used as diagnostic aid and utilised widely in magnetic resonance imaging and magnetic particle imaging. Similarly, the therapeutic applications of MNPs are as therapeutic supplements in anemia and in advanced cancer treatment including magnetic drug targeting and hyperthermia. Based on their magnetic properties, MNPs were mainly used for theragnostic purposes such as image-guided drug delivery, microbubble mediated opening of the blood-brain barrier, and theragnostic tissue engineering (Dadfar et al., 2019).

The polyethylene glycol coated magnetic nanoparticles were used to destroy the cancerous cells through hyperthermia induction and simultaneous diagnosis using magnetic particle imaging. This gives an insight of the therapy and feedback of the live situation in tumor cells (Chiu-Lam et al., 2018), the magnetic mesoporous silica with doped carbon dots was loaded with 5-fluorouracil was used in theragnostic applications and targeting the overexpressed HePG2 cancer cells (Das et al., 2018), the peptide H7K(R2)2-modified magnetic liposomes loaded with Paclitaxel was used for anticancer activity (Zheng et al., 2018).

Temperature-sensitive liposomes (TSL) was magnetized by loading magnetic particles in a bilayer gap and the core is loaded with the drug. Experimental results have shown more efficiency than any other conventional route of delivery (Kostevšek et al., 2018). The antibiotic-resistant bacterial microorganisms were killed by targeted delivery of magnetic nanoparticles which are functionalized by specific glycoconjugate to adhere to *Escherichia coli* strain (EC K99) and induced by GM3-MNPs. Thus the resulting MNPs shown successful destruction of the bacteria by magnetically mediated energy delivery (MagMED), which is an innovative path to destroy microbes other than conventional antibiotics (Fellows, Raval, Murbach et al, 2018).

However, the cancerous cell overexpressing HER2 was targeted by the MNPs bounded with proteins of bacterial origin shown reliable possibilities to act as a theragnostic tool demonstrated by MPQ-cytometry method (Shipunoval et al., 2018), the MNPs loaded with the heparin was used to target and treat neointimal hyperplasia (vascular tissue proliferation) which gets accumulated in the vascular stent placed patients (Fellows, Ghobrial, Mappus et al, 2018). The thermosensitive magneto liposomes containing recombinant tissue-type plasminogen activators were targeted to obtain thermal thrombolysis and have produced a profound effect (Liu et al., 2018). Similarly, the human serum albumin coated MNPs had a high affinity to get attached with doxorubicin was used in the delivery of anticancer drug and photosensitizer such as Bacteriochlorine-A was used to image the targeted site by using MRI or *in-vivo* fluorescence or to induce the photoinduced cytotoxicity (Abakumov et al., 2018). Together, these theragnostic applications of the MNPs highlight the versatility for use in the future medical practice.

Tissue Regeneration

Tissue and organ transplantation pose many pitfalls like less availability of donors, need for immunosuppression, as well as poor success rate. there is an urgent demand in tissue engineering and regenerative medicine (TERM). However, nanoparticle production and the demand for control over the nanoscale system significantly have an impact on the TERM. MNPs could provide high control over the properties of scaffolds such as tuning the mechanical strength and providing magnetically triggered controlled release of bioactive agents. However, the magnetic labelling of the cells using MNPs allows magnetic manipulation of cells and also able to control the cellular functions. Moreover, for the cellular manipulation, by applying the magnetic force, the functional magnetite nanoparticles have been developed and applied to the tissue engineering processes, which are designated as magnetic force-based tissue engineering (Mag-TE) (Ito & Kamihira, 2011). Tissue scaffolding is a general technique used for tissue regeneration and delivery of drugs in the affected region. Poly aspartic acid is used to prepare a matrix with fibres ranging in nanoscale and loaded with MNPs proved to be a promising agent for tissue engineering and lesion generations (Veres et al., 2018). The magnetic nanoparticles embedded in the polymeric biocompatible fibres was used to treat the spinal injury by producing the above scaffold into an injectable formulation. The ingested MNPs align to form a callus when magnetic field is applied externally allows tissue regeneration in the lesion cavity (Dappric et al., 2018). 3D-printed scaffold prepared with iron oxide nanoparticles doped with starch and polycaprolactone blend incubated with human adipose stem cells (HSACs) promotes TGF-β signalling and generates tendon tissues (Gonçalves et al., 2018). Likewise, the iron oxide nanoparticles were used for cell tagging, sorting, tracking, grafting and targeted *in-vivo* delivery in stem cell replacement therapy (Bulte et al., 2001). Similarly, the magnetic nanoparticles were also used to construct and harvest multilayered keratinocyte sheet-like 3-D structures. These self-assembled magnetic nanowire arrays was reported for tissue engineering applications (Ito et al., 2005).

Biomolecular Application

The magnetic nanoparticle-biomolecule hybrid systems were used to recognize the pairs of DNA-DNA, DNA-protein, protein-protein, and protein in organics and in several nanobiotechnology areas, viz. molecular biology, diagnostics, medical treatment, and environmental separation, etc. The MNPs are a kind of solid platform, provide additional information on allocating the strand of interest to the desired location by applying an external magnetic field without direct contact. The advantage of the MNPs-DNA hybrid methods facilitates the sequence-specific DNA separation and as well as in some extremely challenging conditions, such as capture and enrichment of multiple low-abundant DNA targets from the huge genomic background (Shen et al., 2018). The magnetic intercellular stealth nanoparticles produced from ferritin based nanobiomaterials are employed in the activation and probing of GTPase and to carry out similar biomolecular reactions that can be used to bring up much molecular therapy (Kappen & Budke-Gieseking, 2018).

Hypertension

The burden of heart disease, stroke, kidney failure, and premature death and disability is the characterization of Hypertension (high blood pressure), which is a global public health issue at present. The renin-angiotensin-aldosterone system (RAAS) is a coordinated hormonal cascade that governs cardiovascular, renal, and adrenal functions by regulating fluid and electrolyte balance along with arterial blood pressure. The drugs were delivered through oral route by using magnetic nanoparticles that consist of a magnetic material and an active chemical is an interesting technology that allows controlling the particle aggregation, site specific drug release, magnetically triggered drug release etc. The anti-hypertension drug aliskiren was loaded in the poly (D, L-lactide) was observed to be a promising anti-hypertensive drug formulation which decreased blood pressure of the host with higher efficiency than normal (Antal & Kubovcikova, 2015).

Alzheimer's Disease

The formulation of drugs and innovative methods for an early diagnosis plays an important role in the development of theragnostic towards degenerative neurological diseases. Alzheimer's disease, a common form of dementia, is being diagnosed by a non-invasive clinical technique like magnetic resonance imaging for the early diagnosis of AD. However, the recent approach is emerging in nanotechnology is a promising tool for diagnosis and therapy of AD where the role of superparamagnetic iron oxide nanoparticles is predominant (Amiri et al., 2013). Dextran-coated iron oxide magnetic nanoparticle loaded with osmotin was delivered to the brain with the aid of magnetic field to cross the blood-brain barrier (BBB). MNPs delivery across the BBB has proven that the above technique has the potential to serve as a novel drug delivery tool (Amin, 2017). The curcumin conjugated with polyvinylpyrrolidone and polyethylene glycol-polylactic acid block copolymer by antisolvent precipitation produces stable and biocompatible polymer which is further coated with MNPs were used in the diagnosis and imaging of Alzheimer's disease (Cheng et al., 2015). The chelating drug desferrioxamine conjugated with the iron oxide nanoparticles is also reported for use for Alzheimer's disease treatment to cross the blood-brain barrier for good therapeutic efficacy (i.e., improved efficacy with minimal side effects) (Liu et al., 1998). Similarly, the advanced nano-sized iron detection techniques in neuronal tissues, such as superconducting quantum interference magnetometry, have found to be used as a novel diagnostic technique to classify iron deposits and to grade neuroferritinopathy in Alzheimer's patient's brains. MNPs can be also used as an iron chelator, hence plays a significant role as a therapeutic agent for neurodegenerative diseases (Doraiswamy & Finefrock, 2004; Hautot, 2003; Hautot et al., 2007; Liu et al., 2006).

Antibody Delivery

The antibody-conjugated nanoparticles have received more attention nowadays in bioseparation and clinical diagnostics. They were able to detect and isolate the various biomolecules and cells offers cost-effectiveness due to specific bioconjugation with the reasonable antigen capturing capacity. The antibodies conjugated to the magnetic nanoparticles were guided magnetically and making them bind specifically to the target and also for the controlled drug release at the target receptor site (Hossein & Haghighia, 2020). There are various obstacles in delivering antibodies to a human host by using nanosystems as a vector due to its large size and ability to disrupt the nanocarrier's penetration through barriers (Artemov et al., 2003a; Artemov et al., 2003b). Herceptin monoclonal antibody (mAb) is coupled with dimercaptosuccinic acid-coated magnetic nanoparticles as vector against cancer cell NIH3T6.7 and the HER2/neu marker expression showed profound activity (Huh et al., 2005; Sun et al., 2008). Likewise, the human epidermal growth factor 2 (HER2) has significant implications in human breast cancer where its antibody (HER2 antibody) was conjugated with glycerol mono-oleate coated Superparamagnetic Iron Oxide Nanoparticles (SPIONs) and delivered to target cells. The human breast carcinoma cell lines had shown enhanced uptake of the prepared MNPs (Dilnawaz et al., 2010; Weber & Falkenhagen, 1997).

Retinoblastoma

The children are more prompt to have a Retinoblastoma, which is one of the most common malignant intraocular tumors type. Magnetic nanoparticles of various materials can enhance the intraocular drug delivery to the most restricted posterior segment of the eye and increased the intravitreal half-life of the chemotherapy agents. The Y79 retinoblastoma cells (carcinomic cells in the eye) can be destructed by targeting dextran-coated iron oxide nanoparticles (DCIONs) by controlled management of magnetic hyperthermia which activated cell apoptosis and significant efficacy was reported (Demirci et al., 2019). Considerably, the iron oxide nanoparticles were stabilized by oleic acid doped with Sulforhodamine B (or Kiton Red 620) dye will act as a good contrast agent and observations have shown cytosolic absorption of such MNPs and their imaging capacity were found to be tremendous without any toxicity (Jaidev et al., 2014).

Future Prospects:

The development of nanotechnology and nanoscience and the recent research on magnetic nanoparticles have made reasonable progress in the biomedical field. Moreover, the iron oxide nanoparticles along with their magnetic properties and contrast have already been used successfully in the areas of biomedicine, including MRI scanning for the detection of diseases in the brain, cardiovascular system, liver, blood vessels, and other vital organs, etc. More recently the iron oxide nanoparticles based nanohybrids have placed a prominent role in hyperthermia treatment for cancer cells. Besides, to defeat the drawbacks in cancer and multi-drug resistant diseases, the magnetic nanoparticle based strategies has been playing an important role as a targeting aid, diagnosis, chemo, and phototherapy agents.

In future, the magnetic nanoparticles can hold a big assurance in biomedical applications in the area of diagnosis, early detection of disease, tissue imaging, drug and gene delivery to the target, etc. The incorporation of magnetic nanoparticles, especially iron oxide nanoparticles, could broaden the application horizon of the theragnostics, which combines therapeutics with diagnostics. In near future, the multifunctional iron oxide magnetic nanoparticles would be an attractive material for biomedical applications and may help to modify the usual business strategies of pharmaceutical industries.

CONCLUSION

In the present scenario, an extensive fundamental research work on MNPs, had shown the underlying potential of MNPs for translational biomedical applications. Facile synthesis of MNPs and its development as novel formulations found potential applications as MRI contrast agents, magnetic hyperthermia, magnetically targeted drug delivery, magnetic particle imaging etc. It is evidenced from the published literatures in the past decades that there is a wide expansion of MNPs research and their applications in numerous biomedical conditions. However the toxic potential of MNPs cannot be overlooked. Further studies are essentially required to identify a safe dose and formulation of MNPs with optimal such as the development of particles with enhanced magnetic properties and polymeric coating, etc. Similarly, the surface functionalization of MNPs with polymers, biomolecules, and ligands is very much crucial in order to impart biological recognition and interaction skills. However, at present, for stimulation and modelling of the magnetic nanoparticles, micromagnetic tools have opened a new platform of research that helps to predict the properties of biosensors and their applications which are more cost-effective, error-free, and therefore saves time. It is rational to expect that the future progress in magnetism-based nanomedicine lies in the development of the latest technological advances in the fields of micro- and nano-fabrication. Also, regulatory guidelines for the clinical use of magnetic nanoparticles and their formulations needs to be evolved. Hence, it is envisaged that the magnetic nanoparticles will be the "materials of the future", which could significantly contribute not limited to nanomedicine and beyond.

REFERENCES

Abakumov, M., Majouga, A., Kabanov, A., & Chekhonin, V. (2018). Protein coated magnetic nanoparticles for cancer treatment and diagnostics. *12th International conference on Scientific and clinical applications of Magnetic carriers.*

Abasian, M., Hooshangi, V., & Moghadam, P. N. (2017). Synthesis of polyvinyl alcohol hydrogel grafted by modified Fe3 O4 nanoparticles: Characterization and doxorubicin delivery studies. *Iranian Polymer Journal, 26*(5), 313–322. doi:10.100713726-017-0521-5

Acar, Unal, Khodadust, Durmusoglu, Erdem, Yagci, Cleva, Yang, & Yurtsever. (2018). Discovery of Self-luminescent Polyethyleneimine Coated SPIONs and Their Theranostic Applications. *12th International conference on Scientific and clinical applications of Magnetic carrier.*

Adimoolam, M. G., Amreddy, N., Nalam, M. R., & Sunkara, M. V. (2018). A simple approach to design chitosan functionalized Fe3O4 nanoparticles for pH responsive delivery of doxorubicin for cancer therapy. *Journal of Magnetism and Magnetic Materials, 448*, 199–207. doi:10.1016/j.jmmm.2017.09.018

Akal, Z. Ü., Alpsoy, L., & Baykal, A. (2016a). Superparamagnetic iron oxide conjugated with folic acid and carboxylated quercetin for chemotherapy applications. *Ceramics International, 42*(7), 9065–9072. doi:10.1016/j.ceramint.2016.02.166

Akal, Z. Ü., Alpsoy, L., & Baykal, A. (2016b). Biomedical applications of SPION@ APTES@PEG-folic acid@carboxylated quercetin nanodrug on various cancer cells. *Applied Surface Science, 378*, 572–581. doi:10.1016/j.apsusc.2016.03.217

Alexiou, C., Arnold, W., Klein, R. J., Parak, F. G., Hulin, P., Bergemann, C., & … . (2000). Locoregional cancer treatment with magnetic drug targeting. *Cancer Research, 60*(23), 6641–6648. PMID:11118047

Ali, J. (2012). *Polymeric Nanoparticles, Magnetic Nanoparticles and Quantum Dots: Current and Future Perspectives* (E. B. Souto, Ed.). Patenting Nanomedicines, Springer-Verlag Berlin Heidelberg.

Allard-Vannier, E., Hervé-Aubert, K., Kaaki, K., Blondy, T., Shebanova, A., Shaitan, K. V., Ignatova, A. A., Saboungi, M.-L., Feofanov, A. V., & Chourpa, I. (2017). Folic acid-capped PEGylated magnetic nanoparticles enter cancer cells mostly via clathrin-dependent endocytosis. *Biochimica et Biophysica Acta. G, General Subjects, 1861*(6), 1578–1586. doi:10.1016/j.bbagen.2016.11.045 PMID:27919801

Alumutairi, L., Yu, B., Filka, M., Nayfach, J., & Kim, M.-H. (2020). Mild magnetic nanoparticle hyperthermia enhances the susceptibility of *Staphylococcus aureus* biofilm to antibiotics. *International Journal of Hyperthermia, 37*(1), 66–75. doi:10.1080/02656736.2019.1707886 PMID:31964196

Amanda, Silva, Espinosa, Kolosnjaj-Tabi, Wilhelm, & Gazeau. (2016). *Medical Applications of Iron Oxide Nanoparticles. In Iron Oxides: From Nature to Applications.* Wiley-VCH Verlag GmbH & Co. KGaA.

Amin. (2017). Osmotin-loaded Magnetic Nanoparticles with an Electromagnetic Guidance for the Treatment of Alzheimer's disease. *Nanoscale*.

Amiri, H., Saeidi, K., Borhani, P., Manafirad, A., Ghavami, M., & Zerbi, V. (2013). Alzheimer's Disease: Pathophysiology and Applications of Magnetic Nanoparticles as MRI Theranostic Agents. *ACS Chemical Neuroscience*, *4*(11), 1417–1429. Advance online publication. doi:10.1021/cn4001582 PMID:24024702

Ancira-Cortez, A., Morales-Avila, E., Ocampo-García, B. E., González-Romero, C., Medina, L. A., López-Téllez, G., & Cuevas-Yáñez, E. (2017). Preparation and characterization of a tumor-targeting dual-image system based on iron oxide nanoparticles functionalized with folic acid and rhodamine. *Journal of Nanomaterials*, *2017*(1), 1–11. doi:10.1155/2017/5184167

Anirudhan, T. S., & Christa Binusreejayan, J. (2018). pH and magnetic field sensitive folic acid conjugated protein–polyelectrolyte complex for the controlled and targeted delivery of 5-fluorouracil. *Journal of Industrial and Engineering Chemistry*, *57*, 199–207. doi:10.1016/j.jiec.2017.08.024

Antal, I., & Kubovcikova, M. (2015). Magnetic poly(D,L-lactide) nanoparticles loaded with aliskiren: A promising tool for hypertension treatment. *Journal of Magnetism and Magnetic Materials*, *380*, 280–284. doi:10.1016/j.jmmm.2014.10.089

Arachchige, M. P., Laha, S. S., Naik, A. R., Lewis, K. T., Naik, R., & Jena, B. P. (2017). Functionalized nanoparticles enable tracking the rapid entry and release of doxorubicin in human pancreatic cancer cells. *Micron (Oxford, England)*, *92*, 25–31. doi:10.1016/j.micron.2016.10.005 PMID:27846432

Arias, J. L., López-Viota, M., López-Viota, J., & Delgado, Á. V. (2009). Development of iron/ethylcellulose (core/shell) nanoparticles loaded with diclofenac sodium for arthritis treatment. *International Journal of Pharmaceutics*, *382*(1-2), 270–276. doi:10.1016/j.ijpharm.2009.08.019 PMID:19712736

Artemov, D., Mori, N., Okollie, B., & Bhujwalla, Z. M. (2003a). MR molecular imaging of the Her-2/ neu receptor in breast cancer cells using targeted iron oxide nanoparticles. *Magnetic Resonance in Medicine*, *49*(3), 403–408. doi:10.1002/mrm.10406 PMID:12594741

Artemov, D., Mori, N., Ravi, R., & Bhujwalla, Z. M. (2003b). Magnetic resonance molecular imaging of the HER-2/neu receptor. *Cancer Research*, *63*, 2723–2727. PMID:12782573

Asadi, L., Shirvalilou, S., Khoee, S., & Khoei, S. (2018). Cytotoxic effect of 5-fluorouracil-loaded polymer-coated magnetite nanographene oxide combined with radiofrequency. *Anti-cancer Agents in Medicinal Chemistry*, *18*(8), 1148–1155. doi:10.2174/1871520618666180404151218 PMID:29623854

Benyettou, F., Fahs, H., Elkharrag, R., Bilbeisi, R. A., Asma, B., Rezgui, R., Motte, L., Magzoub, M., Brandel, J., Olsen, J.-C., Piano, F., Gunsalus, K. C., Platas-Iglesias, C., & Trabolsi, A. (2017). Selective growth inhibition of cancer cells with doxorubicin-loaded CB (Gholami et al., 2018)-modified iron-oxide nanoparticles. *RSC Advances*, *7*(38), 23827–23834. doi:10.1039/C7RA02693E

Bergemann, C., Müller-Schulte, D., Oster, J., à Brassard, L., & Lübbe, A. S. (1999). Magnetic ion-exchange nano- and microparticles for medical, biochemical and molecular biological applications. *Journal of Magnetism and Magnetic Materials*, *194*(1-3), 45–52. doi:10.1016/S0304-8853(98)00554-X

Bettini, S., Giancane, G., Pagano, R., Bonfrate, V., Salvatore, L., Madaghiele, M., Buccolieri, A., Manno, D., Serra, A., Maruccio, G., Monteduro, A. G., Syrgiannis, Z., Valli, L., & Prato, M. (2017). A simple approach to synthetize folic acid decorated magnetite@SiO2 nanostructures for hyperthermia applications. *Journal of Materials Chemistry. B, Materials for Biology and Medicine*, *5*(36), 7547–7556. doi:10.1039/C7TB02002C PMID:32264230

Bharde, A. A., Parikh, R. Y., Baidakova, M., Jouen, S., Hannoyer, B., Enoki, T., Prasad, B. L. V., Shouche, Y. S., Ogale, S., & Sastry, M. (2008). Bacteria-mediated precursor-dependent biosynthesis of superparamagnetic iron oxide and iron sulfide nanoparticles. *Langmuir*, *24*(11), 5787–5794. doi:10.1021/la704019p PMID:18454562

Binandeh, M. (2018). *Application of Magnetic Nanoparticles with structure of Core-Shell, as bactericidal. J Immunology and Inflammation Diseases Therapy*. doi:10.31579/2637-8876/011

Bonvin, D., Bastiaansen, J. A. M., Stuber, M., Hofmann, H., Mionić Ebersold, M., & Ebersold, M. M. (2017). Folic acid on iron oxide nanoparticles: Platform with high potential for simultaneous targeting, MRI detection and hyperthermia treatment of lymph node metastases of prostate cancer. *Dalton Transactions (Cambridge, England)*, *46*(37), 12692–12704. doi:10.1039/C7DT02139A PMID:28914298

Boussif, O., Lezoualc'h, F., Zanta, M. A., Mergny, M. D., Scherman, D., Demeneix, B., & Behr, J. P. (1995). A versatile vector for gene and oligonucleotide transfer into cells in culture and in vivo: Polyethylenimine. *Proceedings of the National Academy of Sciences of the United States of America*, *92*(16), 7297–7301. doi:10.1073/pnas.92.16.7297 PMID:7638184

Bucki, R., Niemirowicz-Laskowska, K., Deptuła, P., Wilczewska, A. Z., Misiak, P., Durnaś, B., Fiedoruk, K., Piktel, E., Mystkowska, J., & Janmey, P. A. (2019). Susceptibility of microbial cells to the modified PIP_2-binding sequence of gelsolin anchored on the surface of magnetic nanoparticles. *Journal of Nanobiotechnology*, *17*(1), 81. doi:10.118612951-019-0511-1 PMID:31286976

Bulte, J. W., Douglas, T., Witwer, B., Zhang, S.-C., Strable, E., Lewis, B. K., Zywicke, H., Miller, B., van Gelderen, P., Moskowitz, B. M., Duncan, I. D., & Frank, J. A. (2001). Magnetodendrimers allow endosomal magnetic labeling and in vivo tracking of stem cells. *Nature Biotechnology*, *19*(12), 1141–1147. doi:10.1038/nbt1201-1141 PMID:11731783

Burygin, G. L., Khlebtsov, B. N., Shantrokha, A. N., Dykman, L. A., Bogatyrev, V. A., & Khlebtsov, N. G. (2009). On the enhanced antibacteria activity of antibiotics mixed with gold nanoparticles. *Nanoscale Research Letters*, *4*(8), 794–801. doi:10.100711671-009-9316-8 PMID:20596384

Butoescu, N., Seemayer, C. A., Foti, M., Jordan, O., & Doelker, E. (2009). Dexamethasone-containing PLGA superparamagnetic microparticles as carriers for the local treatment of arthritis. *Biomaterials*, *30*(9), 1772–1780. doi:10.1016/j.biomaterials.2008.12.017 PMID:19135244

Chakkarapani, P., Subbiah, L., Palanisamy, S., Bibiana, A., Ahrentorp, F., Jonasson, C., & Johansson, C. (2015). Encapsulation of methotrexate loaded magnetic microcapsules for magnetic drug targeting and controlled drug release. *Journal of Magnetism and Magnetic Materials*, *380*, 280–294. doi:10.1016/j.jmmm.2014.11.006

Chatzipavlidis, A., Bilalis, P., Efthimiadou, E. K., Boukos, N., & Kordas, G. C. (2011). Sacrificial template-directed fabrication of superparamagnetic polymer microcontainers for pH-activated controlled release of daunorubicin. *Langmuir, 27*(13), 8478–8485. doi:10.1021/la201240h PMID:21648457

Chen, L., Wu, L., Liu, F., Qi, X., Ge, Y., & Shen, S. (2016). Azo-functionalized Fe3 O4 nanoparticles: A near-infrared light triggered drug delivery system for combined therapy of cancer with low toxicity. *Journal of Materials Chemistry, 4*(21), 3660–3669. doi:10.1039/C5TB02704G PMID:32263305

Chen, Y., Xu, M., Guo, Y., Tu, K., Wu, W., Wang, J., Tong, X., Wu, W., Qi, L., & Shi, D. (2017). Targeted chimera delivery to ovarian cancer cells by heterogeneous gold magnetic nanoparticle. *Nanotechnology, 28*(2), 025101. doi:10.1088/0957-4484/28/2/025101 PMID:27906685

Chen, Y.-H., Tsai, C.-Y., Huang, P.-Y., Chang, M.-Y., Cheng, P.-C., Chou, C.-H., Chen, D.-H., Wang, C.-R., Shiau, A.-L., & Wu, C.-L. (2007). Methotrexate conjugated to gold nanoparticles inhibits tumor growth in a syngeneic lungtumor model. *Molecular Pharmaceutics, 4*(5), 713–722. doi:10.1021/mp060132k PMID:17708653

Cheng, K. K., Chan, P. S., Fan, S., Kwan, S. M., Yeung, K. L., & Wang, Y.-X. J. (2015). Curcumin-conjugated magnetic nanoparticles for detecting amyloid plaques in Alzheimer's disease mice using magnetic resonance imaging (MRI). Biomaterials, 44, 155-172.

Chiu-Lam, A., Dhavalikar, R., Tay, Z. W., Chandrasekharan, P., Conolly, S. M., & Rinaldi, C. (2018). Spatially Controlledand Image Guided Nanoscale Thermal Therapy. *12th International conference on Scientific and clinical applications of Magnetic carriers.*

Choi, K.-H., Nam, K., Malkinski, L., Choi, E., Jung, J.-S., & Park, B. (2016). Sizedependent photodynamic anticancer activity of biocompatible multifunctional magnetic submicron particles in prostate cancer cells. *Molecules (Basel, Switzerland), 21*(9), 1187. doi:10.3390/molecules21091187

Chorny, M., Fishbein, I., Yellen, B. B., Alferiev, I. S., Bakay, M., Ganta, S., Adamo, R., Amiji, M., Friedman, G., & Levy, R. J. (2010). Targeting stents with local delivery of paclitaxel-loaded magnetic nanoparticles using uniform Welds. *Proceedings of the National Academy of Sciences of the United States of America, 107*(18), 8346–8351. doi:10.1073/pnas.0909506107 PMID:20404175

Ciurlicaa, Nadejdea, Creangaa, Carlescua, & Badescub. (2010). *Antibiotic Coated Magnetite Nanoparticles For Biological Applications.* Olomouc.

Correa, Sierra, Asín, Gutiérrez, de la Fuente, Betancor, & Grazú. (2018). Nanobiohybrids for the thermoactivation of a therapeutic enzyme by magnetic hyperthermia. *12th International Conference on Scientific and Clinical Applications of Magnetic Carriers.*

Creixell, M., Herrera, A. P., & Ayala, V. (2010). Preparation of epidermal growth factor (EGF) conjugated iron oxide nanoparti- cles and their internalization into colon cancer cells. *Journal of Magnetism and Magnetic Materials, 322*, 2244–2250. doi:10.1016/j.jmmm.2010.02.019

Cristea, C., Tertis, M., & Galatus, R. (2017). Magnetic Nanoparticles for Antibiotics Detection. *Nanomaterials (Basel, Switzerland), 7*(6), 119. doi:10.3390/nano7060119 PMID:28538684

Cui, D., Zhang, C., Liu, B., Shu, Y., Du, T., Shu, D., Wang, K., Dai, F., Liu, Y., Li, C., Pan, F., Yang, Y., Ni, J., Li, H., Brand-Saberi, B., & Guo, P. (2015). Regression of Gastric Cancer by Systemic Injection of RNA Nanoparticles Carrying both Ligand and siRNA. *Scientific Reports*, *5*(1), 10726. doi:10.1038rep10726 PMID:26137913

Dadfar, S. M., Roemhild, K., Drude, N. I., von Stillfried, S., Knüchel, R., Kiessling, F., & Lammers, T. (2019, January 1). Iron oxide nanoparticles: Diagnostic, therapeutic and theranostic applications. *Advanced Drug Delivery Reviews*, *138*, 302–325. doi:10.1016/j.addr.2019.01.005 PMID:30639256

Dappric, J., Johnson, C., Gilbert, R., McKay, R., & Dresdner, K. (2018). Superparamagnetic Particle Scaffold for Regenerating Damaged Neural Tissue. *12th International conference on Scientific and clinical applications of Magnetic carriers.*

Das, R. K., Pramanik, A., Majhi, M., & Mohapatra, S. (2018). Magnetic mesoporous silica gated with doped carbon dot for site-specific drug delivery, fluorescence, and MR imaging. *Langmuir*, *34*(18), 5253–5262. doi:10.1021/acs.langmuir.7b04268 PMID:29634272

Demirci, H., Slimani, N., Pawar, M., Kumon, R. E., Vaishnava, P., & Besirli, C. G. (2019). Magnetic Hyperthermia in Y79 Retinoblastoma and ARPE-19 Retinal Epithelial Cells: Tumor Selective Apoptotic Activity of Iron Oxide Nanoparticle. *Translational Vision Science & Technology*, *8*(5), 18. doi:10.1167/tvst.8.5.18 PMID:31602343

Di Corato, R., Béalle, G., Kolosnjaj-Tabi, J., Espinosa, A., Clément, O., Silva, A. K. A., Ménager, C., & Wilhelm, C. (2015). Combining magnetic hyperthermia and photody- namic therapy for tumor ablation with photoresponsive magnetic liposomes. *ACS Nano*, *9*(3), 2904–2916. doi:10.1021/nn506949t PMID:25695371

Dilnawaz, F., Singh, A., Mohanty, C., & Sahoo, S. K. (2010). Dual drug loaded superparamagnetic iron oxide nanoparticles for targeted cancer therapy. *Biomaterials*, *31*(13), 3694–3706. doi:10.1016/j.biomaterials.2010.01.057 PMID:20144478

Doan, T. K. D. (2009). Article. *Journal of Physics: Conference Series*, *187*, 012036. doi:10.1088/1742-6596/187/1/012036

Doan, T. K. D. (2009). Article. *Journal of Physics: Conference Series*, *187*, 012036. doi:10.1088/1742-6596/187/1/012036

Dobson. (2006). Macro and Nanoparticle based targeting of drug and gene delivery. *Nanomedicine.*

Doraiswamy, P. M., & Finefrock, A. E. (2004). Metals in our minds: Therapeutic implications for neurodegenerative disorders. *Lancet Neurology*, *3*(7), 431–434. doi:10.1016/S1474-4422(04)00809-9 PMID:15207800

Douziech-Eyrolles, L. (2007). Nanovectors for anticancer agents based on superparamagnetic iron oxide nanoparticles. *International Journal of Nanomedicine*, *2*, 541–550. PMID:18203422

Duan, J., Dong, J., Zhang, T., Su, Z., Ding, J., Zhang, Y., & Mao, X. (2014). Polyethyleneimine-functionalized iron oxide nanoparticles for systemic siRNA delivery in experimental arthritis. *Nanomedicine (London)*, *9*(6), 789–801. doi:10.2217/nnm.13.217 PMID:24392891

Dutta, B., Shetake, N. G., Gawali, S. L., Barick, B. K., Barick, K. C., Babu, P. D., Pandey, B. N., Priyadarsini, K. I., & Hassan, P. A. (2018). PEG mediated shape-selective synthesis of cubic Fe3O4 nanoparticles for cancer therapeutics. *Journal of Alloys and Compounds*, *737*, 347–355. doi:10.1016/j.jallcom.2017.12.028

Egodawatte, S., Dominguez, S. Jr, & Larsen, S. C. (2017). Solvent effects in the development of a drug delivery system for 5-fluorouracil using magnetic mesoporous silica nanoparticles. *Microporous and Mesoporous Materials*, *237*, 108–116. doi:10.1016/j.micromeso.2016.09.024

Erdem, M., Yalcin, S., & Gunduz, U. (2017). Folic acid-conjugated polyethylene glycol-coated magnetic nanoparticles for doxorubicin delivery in cancer chemotherapy: Preparation, characterization and cytotoxicity on HeLa cell line. *Human and Experimental Toxicology*, *36*(8), 833–845. doi:10.1177/0960327116672910 PMID:27758842

Eynali, S., Khoei, S., Khoee, S., & Esmaelbeygi, E. (2017). Evaluation of the cytotoxic effects of hyperthermia and 5-fluorouracil-loaded magnetic nanoparticles on human colon cancer cell line HT-29. *International Journal of Hyperthermia*, *33*(3), 327–335. doi:10.1080/02656736.2016.1243260 PMID:27701929

Fakhimikabir, H., Tavakoli, M. B., Zarrabi, A., Amouheidari, A., & Rahgozar, S. (2018). The role of folic acid-conjugated polyglycerol coated iron oxide nanoparticles on radiosensitivity with clinical electron beam (6 MeV) on human cervical carcinoma cell line: In vitro study. *Journal of Photochemistry and Photobiology. B, Biology*, *182*, 71–76. doi:10.1016/j.jphotobiol.2018.03.023 PMID:29626804

Farzin, A., Etesami, S. A., Quint, J., Memic, A., & Tamayol, A. (2020). Magnetic Nanoparticles in Cancer Therapy and Diagnosis. *Advanced Healthcare Materials*, *9*(9), 1901058. doi:10.1002/adhm.201901058 PMID:32196144

Fellows, Ghobrial, Mappus, Hargett, Bolding, Dean, & Mefford. (2018). Heparin Coated Magnetic Nanoparticles for Treatment of In-Stent Restenosis and Prevention of Late Stent Thrombosis. *12th International conference on Scientific and clinical applications of Magnetic carriers.*

Fellows, B. D., Raval, Y. S., Murbach, J., Cordeau, Y., Tzeng, T.-R. J., & Thompson Mefford, O. (2018). Glycoconjugate-Functionalized Magnetic Nanoparticles: A Tool for Selective Killing of Targeted Bacteria Via Magnetically Mediated Energy Delivery. *12th International conference on Scientific and clinical applications of Magnetic carriers.*

Gao, Z., Liu, X., Deng, G., Zhou, F., Zhang, L., Wang, Q., & Lu, J. (2016). Fe3 O4 @mSiO2 -FA-CuS-PEG nanocomposites for magnetic resonance imaging and targeted chemophotothermal synergistic therapy of cancer cells. *Dalton Transactions (Cambridge, England)*, *45*(34), 13456–13465. doi:10.1039/C6DT01714B PMID:27493065

Geller, R. J., Chevalier, R. L., & Spyker, D. A. (1986). Acute amoxicillin nephrotoxicity following an overdose. *Clinical Toxicology*, *24*(2), 175–182. PMID:3712526

Gholami, A., Ebrahiminezhad, A., Abootalebi, N., & Ghasemi, Y. (2018). Synergistic Evaluation of Functionalized Magnetic Nanoparticles and Antibiotics Against Staphylococcus aureus and Escherichia coli. *Pharmaceutical Nanotechnology*, *6*(4), 276–286. doi:10.2174/2211738506666181031143048 PMID:30381089

Gonçalves, A. I., Rodrigues, M. T., Reis, R. L., Gomes, M. E., & El Haj, A. J. (2018). Magnetic actuated SPCL scaffolds dopped with iron oxide magnetic nanoparticles as mechano-instructive platforms for tendon tissue engineering. *12th International conference on Scientific and clinical applications of Magnetic carriers.*

Goodwin, S. C., Bittner, C. A., Peterson, C. L., & Wong, G. (2001). Single-dose toxicity study of hepatic intra- arterial infusion of doxorubicin coupled to a novel magnetically targeted drug carrier. *Toxicological Sciences, 60*(1), 177–183. doi:10.1093/toxsci/60.1.177 PMID:11222884

Grèget, Nealon, Vileno, Turek, Mèny, Derory, Voirin, Rivière, Rogalev, Wilhelm, Joly, Knafo, Ballon, Terazzi, Kappler, Donnio, & Gallani. (n.d.). *Magnetic Properties of Gold Nanoparticles: A Room-Temperature Quantum Effect.* . doi:10.1002/cphc.201200394

Gu, H., Ho, P. L., Tong, L., Wang, L., & Xu, B. (2003). Presenting vancomycin on nanoparticles to enhance antimicrobial activities. *Nano Letters, 3*(9), 1261–1263. doi:10.1021/nl034396z

Gupta, J., Mohapatra, J., Bhargava, P., & Bahadur, D. (2016). A pH-responsive folate conjugated magnetic nanoparticle for targeted chemo-thermal therapy and MRI diagnosis. *Dalton Transactions (Cambridge, England), 45*(6), 2454–2461. doi:10.1039/C5DT04135J PMID:26685824

Hautot, D. (2003). Preliminary evaluation of nanoscale biogenic magnetite in Alzheimer's disease brain tissue. *Proceedings. Biological Sciences, 270*(Suppl 1), S62–S64. PMID:12952638

Hautot, D., Pankhurst, Q. A., Morris, C. M., Curtis, A., Burn, J., & Dobson, J. (2007). Preliminary observation of elevated levels of nanocrystalline iron oxide in the basal ganglia of neuroferritinopathy patients. *Biochimica et Biophysica Acta, 1772*(1), 21–25. doi:10.1016/j.bbadis.2006.09.011 PMID:17097860

Hildebrandt, B., Wust, P., & Ahlers, O. (2002). The cellular and molecular basis of hyperthermia. *Critical Reviews in Oncology/Hematology, 43*(1), 33–56. doi:10.1016/S1040-8428(01)00179-2 PMID:12098606

Hira, I., & Reena, V. (2018). Inspection and Remedy of Cervical Cancer Using Nanoparticles. *Nanoscience & Nanotechnology-Asia, 8*(2), 186–192. doi:10.2174/2210681207666170323155831

Hossein, A., & Haghighia, M. T. K. Z. F. F. F. (2020, April). Effects of different quantities of antibody conjugated with magnetic nanoparticles on cell separation efficiency. *Heliyon, 6*(4), e03677. doi:10.1016/j.heliyon.2020.e03677 PMID:32280795

Hosseini, L., Mahboobnia, K., & Irani, M. (2016). Fabrication of PLA/MWCNT/ Fe3 O4 composite nanofibers for leukemia cancer cells. *Int J Polym Mater Polymer Biomater., 65*(4), 176–182. doi:10.1080/00914037.2015.1074912

Hua, M. Y., Yang, H. W., Chuang, C. K., Tsai, R. Y., Chen, W. J., Chuang, K. L., Chang, Y. H., Chuang, H. C., & Pang, S. T. (2010). Magnetic-nanoparticlemodiWed paclitaxel for targeted therapy for prostate cancer. *Biomaterials, 31*(28), 7355–7363. doi:10.1016/j.biomaterials.2010.05.061 PMID:20609471

Huang, P., Li, Z., Lin, J., Yang, D., Gao, G., Xu, C., Bao, L., Zhang, C., Wang, K., & Song, H. (2011). Photosensitizer-conjugated magnetic nanoparticles for in vivo simultaneous magnetofluores- cent imaging and targeting therapy. *Biomaterials, 32*(13), 3447–3458. doi:10.1016/j.biomaterials.2011.01.032 PMID:21303717

Huh, Y. M., Jun, Y. W., Song, H. T., Kim, S., Choi, J. S., Lee, J. H., Yoon, S., Kim, K. S., Shin, J. S., Suh, J. S., & Cheon, J. (2005). In vivo magnetic resonance detection of cancer by using multifunctional magnetic nanocrystals. *Journal of the American Chemical Society, 127*(35), 12387–12391. doi:10.1021/ja052337c PMID:16131220

Hussein-Al-Ali, S. H., El Zowalaty, M. E., Hussein, M. Z., & Benjamin, M. (2014). Synthesis, characterization, and antimicrobial activity of an ampicillin-conjugated magnetic nanoantibiotic for medical applications. *International Journal of Nanomedicine, 9*, 3801–3814. doi:10.2147/IJN.S61143 PMID:25143729

Ito, A., & Kamihira, M. (2011). Tissue engineering using magnetite nanoparticles. *Progress in Molecular Biology and Translational Science, 104*, 355–395. doi:10.1016/B978-0-12-416020-0.00009-7 PMID:22093224

Ito, A., Shinkai, M., Honda, H., & Kobayashi, T. (2001). Heat-inducible TNF-alpha gene therapy combined with hyperthermia using magnetic nanoparticles as a novel tumor-targeted therapy. *Cancer Gene Therapy, 8*(9), 649–654. doi:10.1038j.cgt.7700357 PMID:11593333

Ito, A., Shinkai, M., Honda, H., & Kobayashi, T. (2005). Medical application of functionalized magnetic nanoparticles. *Journal of Bioscience and Bioengineering, 100*(1), 1–11. doi:10.1263/jbb.100.1 PMID:16233845

Jaidev, L. R., Bhavsar, D. V., Sharma, U., Jagannathan, N. R., Krishnan, U. M., & Sethuraman, S. (2014). Engineered multifunctional nanomaterials for multimodal imaging of retinoblastoma cellsin vitro. *Journal of Biomaterials Science. Polymer Edition, 25*(11), 1093–1109. doi:10.1080/09205063.2014.917040 PMID:24911385

Jain, T. K., Richey, J., Strand, M., Leslie-Pelecky, D. L., Flask, C. A., & Labhasetwar, V. (2008). Magnetic nanoparticles with dual functional properties: Drug delivery and magnetic resonance imaging. *Biomaterials, 29*(29), 4012–4021. doi:10.1016/j.biomaterials.2008.07.004 PMID:18649936

Jalalian, S. H., Taghdisi, S. M., Hamedani, N. S., Kalat, S. A. M., Lavaee, P., ZandKarimi, M., Ghows, N., Jaafari, M.R., Naghibi, S., & Danesh, N.M. (2013). *EpirubicinLoaded Super Paramagnetic Iron Oxide Nanoparticle? Aptamer Bioconjugate for Combined Colon Cancer*. Academic Press.

Janßen, H. C., Angrisani, N., Kalies, S., Hansmann, F., Kietzmann, M., Warwas, D. P., Behrens, P., & Reifenrath, J. (2020). Biodistribution, biocompatibility and targeted accumulation of magnetic nanoporous silica nanoparticles as drug carrier in orthopedics. *Journal of Nanobiotechnology, 18*(1), 14. doi:10.118612951-020-0578-8 PMID:31941495

Ji, F., Zhang, K., Li, J., Gu, Y., Zhao, J., & Zhang, J. (2018). A dual pH/magnetic responsive nanocarrier based on PEGylated Fe3 O4 nanoparticles for doxorubicin delivery. *Journal of Nanoscience and Nanotechnology, 18*(7), 4464–4470. doi:10.1166/jnn.2018.15275 PMID:29442620

Jordan, A., Scholz, R., Wust, P., Fahling, H., Krause, J., Wlodarczyk, W., Sander, B., Vogl, T., & Felix, R. (1997). Effects of Magnetic Fluid Hyperthermia (MFH) on C3H Mammary Carcinoma in vivo. *International Journal of Hyperthermia, 13*(6), 587–605. doi:10.3109/02656739709023559 PMID:9421741

Kamat, M., El-Boubbou, K., Zhu, D. C., Lansdell, T., Lu, X., Li, W., & Huang, X. (2010). Hyaluronic Acid Immobilized Magnetic Nanoparticles for Active Targeting and Imaging of Macrophages. *Bioconjugate Chemistry, 21*(11), 2128–2135. doi:10.1021/bc100354m PMID:20977242

Kappen, M., & Budke-Gieseking, A. (2018). MagnetogeneticActivationof Small GTPaseswithMagIcS-NanoparticlesInside Living Cells. *12th International conference on Scientific and clinical applications of Magnetic carriers.*

Kasparis, G., Wang, L., Plan, A., LaGrow, A., Sergides, A., Wilhelm, C., & Nguyen, T. K. T. (2018). Synthesis and Application of Zinc Ferrite Nanoparticles as Agents for Cancer Thermotherapy. *12th International conference on Scientific and clinical applications of Magnetic carriers.*

Kayal, & Ramanujan. (2010). Anti-Cancer Drug Loaded Iron–Gold Core–Shell Nanoparticles (Fe@Au) for Magnetic Drug Targeting. *Journal of Nanoscience and Nanotechnology.* PMID:21133071

Kim, Lee, Park, HeeMun, Park, & Yoo. (2015). Drug-loaded gold/iron/gold plasmonic nanoparticles for magnetic targeted chemo photothermal treatment of rheumatoid arthritis. *J. Biomaterials.*

Kim, H., Jo, A., Baek, S., Lim, D., Park, S.-Y., Cho, S. K., Chung, J. W., & Yoon, J. (2017). Synergistically enhanced selective intracellular uptake of anticancer drug carrier comprising folic acidconjugated hydrogels containing magnetite nanoparticles. *Scientific Reports, 7*(1), 41090. doi:10.1038rep41090 PMID:28106163

Kohler, N., Sun, C., Wang, J., & Zhang, M. (2005). Methotrexate- modified superparamagnetic nanoparticles and their intracellular uptake into human cancer cells. *Langmuir, 21*(19), 8858–8864. doi:10.1021/la0503451 PMID:16142971

Kolosnjaj-Tabi, J., Di Corato, R., Lartigue, L., Marangon, I., Guardia, P., Silva, A. K. A., Luciani, N., Clement, O., Flaud, P., Singh, J. V., Decuzzi, P., Pellegrino, T., Wilhelm, C., & Gazeau, F. (2014). Heat-generating iron oxide nanocubes: Subtle "Destructurators" of the tumoral microenvironment. *ACS Nano, 8*(5), 4268–4283. doi:10.1021/nn405356r PMID:24738788

Komarala, E. P., Nigam, S., Aslam, M., & Bahadur, D. (2016). In-vitro evaluation of layered double hydroxide–Fe3 O4 magnetic nanohybrids for thermochemotherapy. *New Journal of Chemistry, 40*(1), 423–433. doi:10.1039/C5NJ01701G

Kostevšek, N., Monaco, I., & Al-Jamalc, W. (2018). Magneto-liposomes as Thernostic agents. *12th International conference on Scientific and clinical applications of Magnetic carriers.*

Kumar, A., Sahoo, B., Montpetit, A., Behera, S., Lockey, R. F., & Mohapatra, S. S. (2007). Development of hyaluronic acid–Fe2O3 hybrid magnetic nanoparticles for targeted delivery of peptides. *Nanomedicine; Nanotechnology, Biology, and Medicine, 3*(2), 132–137. doi:10.1016/j.nano.2007.03.001 PMID:17572355

Kuo, C.-Y., Liu, T.-Y., Hardiansyah, A., & Chiu, W.-Y. (2016). Magnetically polymeric nanocarriers for targeting delivery of curcumin and hyperthermia treatments toward cancer cells. *Journal of Polymer Science. Part A, Polymer Chemistry, 54*(17), 2706–2713. doi:10.1002/pola.28150

Kuznetsov, A. A., Filippov, V. I., Alyautdin, R. N., Torshina, N. L., & Kuznetsov, O. A. (2001). Application of magnetic liposomes for magnetically guided transport of muscle relaxants and anti-cancer photodynamic drugs. *Journal of Magnetism and Magnetic Materials*, *225*(1-2), 95–100. doi:10.1016/S0304-8853(00)01235-X

Laurent, S., & Mahmoudi, M. (2011). Superparamagnetic iron oxide nanoparticles: Promises for diagnosis and treatment of cancer. *International Journal of Molecular Epidemiology and Genetics*, *2*, 367–390. PMID:22199999

Lee, J. H., Chen, K. J., Noh, S. H., Garcia, M. A., Wang, H., Lin, W. Y., Jeong, H., Kong, B. J., Stout, D. B., Cheon, J., & Tseng, H. R. (2013). On-demand drug release system for in vivo cancer treatment through self- assembled magnetic nanoparticles. *Angewandte Chemie International Edition in English*, *52*(16), 4384–4388. doi:10.1002/anie.201207721 PMID:23519915

Li, J. (2019). A new tool to attack biofilms: Driving magnetic iron-oxide nanoparticles to disrupt the atrix. *Nanoscale*, *11*, 6905–6915. doi:10.1039/C8NR09802F PMID:30912773

Li, L., Gao, F., Jiang, W., Wu, X., Cai, Y., Tang, J., Gao, X., & Gao, F. (2016). Folic acid-conjugated superparamagnetic iron oxide nanoparticles for tumor-targeting MR imaging. *Drug Delivery*, *23*(5), 1726–1733. doi:10.3109/10717544.2015.1006404 PMID:25715808

Li, S., Gao, Y., Ding, Y., Xu, A., & Tan, H. (2020). Supramolecular nano drug delivery systems mediated via host-guest chemistry of cucurbit[n]uril (n = 6 and 7). *Chinese Chemical Letters*. Advance online publication. doi:10.1016/j.cclet.2020.04.049

Li, X., Yang, Y., Jia, Y., Pu, X., Yang, T., Wang, Y., Ma, X., Chen, Q., Sun, M., Wei, D., Kuang, Y., Li, Y., & Liu, Y. (2017). Enhanced tumor targeting effects of a novel paclitaxel-loaded polymer: PEG-PCCL-modified magnetic iron oxide nanoparticles. *Drug Delivery*, *24*(1), 1284–1294. doi:10.1080/10717544.2017.1373167 PMID:28891337

Liibbe, A. S., & Bergemann, C. (1997). Scientific and Clinical Applications of Magnetic Carriers. Plenum Press.

Liu, Men, Perry, & Smith. (1998). *Nanoparticle and Iron Chelators as a Potential Novel Alzheimer Therapy*. Humana Press.

Liu, C.-H., Hsu, H.-L., Chen, J.-P., & Ma, Y.-H. (2018). Intravenous Delivery of Tissue Plasminogen Activator by Thermosensitive Magneto-liposomes for Target Thrombolysis. *12th International conference on Scientific and clinical applications of Magnetic carriers*.

Liu, G., Men, P., Harris, P. L. R., Rolston, R. K., Perry, G., & Smith, M. A. (2006). Nanoparticle iron chelators: A new therapeutic approach in Alzheimer disease and other neurologic disorders associated with trace metal imbalance. *Neuroscience Letters*, *406*(3), 189–193. doi:10.1016/j.neulet.2006.07.020 PMID:16919875

Long, R., Liu, Y., Wu, S., Wang, S., & Wang, P. (2017). Co-delivery of doxorubicin and small interfering RNA with genipin-cross-linked iron (III) oxide-polyetherimide nanoparticles for combined cancer therapy. *Nanomaterials Nanotechnology.*, *7*, 184798041770409. doi:10.1177/1847980417704094

Lu, B., Huang, X., Mo, J., & Zhao, W. (2016). Drug delivery using nanoparticles for cancer stem-like cell targeting. *Frontiers in Pharmacology*, *7*(065103), 84. doi:10.3389/fphar.2016.00084 PMID:27148051

L¨ubbe, A. S., Alexiou, C., & Bergemann, C. (2001). Clinical applications of magnetic drug targeting. *The Journal of Surgical Research*, *95*(2), 200–206. doi:10.1006/jsre.2000.6030 PMID:11162046

Lubbe, A. S., Bergemann, C., Huhnt, W., Fricke, T., & Riess, H. (1996). Preclinical experiences with magnetic drug targeting: Tolerance and efficacy. *Cancer Research*, *56*, 4694–4701. PMID:8840986

Luo, X., Peng, X., Hou, J., Wu, S., Shen, J., & Wang, L. (2017). Folic acid-functionalized polyethylenimine superparamagnetic iron oxide nanoparticles as theranostic agents for magnetic resonance imaging and PD-L1 siRNA delivery for gastric cancer. *International Journal of Nanomedicine*, *12*, 5331–5343. doi:10.2147/IJN.S137245 PMID:28794626

Luo, X., Wang, Y., Lin, H., & Qu, F. (2016). DOX-Fe3 O4 @mSiO2 -PO-FA nanocomposite for synergistic chemo- and photothermal therapy. *RSC Advances*, *6*(113), 112232–112240. doi:10.1039/C6RA23292B

Mahmoudi, M., Shokrgozar, M. A., Simchi, A., Imani, M., Milani, A. S., Stroeve, P., Vali, H., Hafeli, U. O., & Bonakdar, S. (2009). Multiphysics flow modeling and in vitro toxicity of iron oxide nanoparticles coated with poly(vinyl alcohol). *The Journal of Physical Chemistry C*, *113*(6), 2322–2331. doi:10.1021/jp809453v

Maiti, D., Saha, A., & Devi, P. S. (2016). Surface modified multifunctional ZnFe2O4 nanoparticles for hydrophobic and hydrophilic anti-cancer drug molecule loading. *Physical Chemistry Chemical Physics*, *18*(3), 1439–1450. doi:10.1039/C5CP05840F PMID:26524183

Manshadi, M. K. D., Saadat, M., Mohammadi, M., Kamali, R., Shamsi, M., Naseh, M., & Sanati-Nezhad, A. (2019). Magnetic aerosol drug targeting in lung cancer therapy using permanent magnet. *Drug Delivery*, *26*(1), 120–128. doi:10.1080/10717544.2018.1561765 PMID:30798633

Matthew, D. (2018). *Externally Induced Drug Release Systems with Magnetic Nanoparticle Carriers: An Emerging Field in Nanomedicine*. Adv. Therap.

Menon, J.U., Kuriakose, A., & Iyer, R. (2017). Dual-Drug Containing Core-Shell Nanoparticles for Lung Cancer Therapy. *Sci Rep.*, *7*(1), 13249. . doi:10.103841598-017-13320-4

Mérida, F., Juan, E., & Torres-Lugo, M. (2018). Ultrasonic Potentiation of Magnetic Fluid Hyperthermia/ Pifithrin-μ Combined Therapies in Ovarian Cancer Cells. *12th International conference on Scientific and clinical applications of Magnetic carriers.*

Mohammadi Gazestani, A., Khoei, S., Khoee, S., Emamgholizadeh Minaei, S., & Motevalian, M. (2018). In vivo evaluation of the combination effect of near-infrared laser and 5-fluorouracil-loaded PLGA-coated magnetite nanographene oxide. *Artificial Cells, Nanomedicine, and Biotechnology*, *10*(sup2), 1–9. doi:10.1080/21691401.2018.1450265 PMID:29543075

Mostaghasi, E., Zarepour, A., & Zarrabi, A. (2018). Folic acid armed Fe3O4-HPG nanoparticles as a safe nano vehicle for biomedical theranostics. *Journal of the Taiwan Institute of Chemical Engineers*, *82*, 33–41. doi:10.1016/j.jtice.2017.11.004

Mura, S., Nicolas, J., & Couvreur, P. (2013). Stimuli-responsivenanocarriers for drug delivery. *Nature Materials*, *12*(11), 991–1003. doi:10.1038/nmat3776 PMID:24150417

Naghizadeh, S., Hassanzadeh Nemati, N., Hassani Najafabadi, A., Niknejad, H., & Khani, M.-M. (2018). Controlled release of fluorouracil (5-FU) from chitosan-co-poly(ethylene glycol)/poly(glycerol sebacate)-copoly(ethylene glycol)-coated iron oxide. *Int J Polymer Mater Polymer Biomater.*, *67*(4), 212–220. doi:10.1080/00914037.2017.1320657

Nandagopal, G. D., Periyathambi, P., & Sastry, T. P. (2016). Fabrication of a multifunctional nanocomposite containing INP coated with chitosan coupled with folic acid and loaded with quercetin in diagnosis and its possible treatment of cancer. *RSC Advances*, *6*(101), 99514–99523. doi:10.1039/C6RA19901A

Nasiri, M., & Hassanzadeh-Tabrizi, S. A. (2018). Synthesis and characterization of folate-decorated cobalt ferrite nanoparticles coated with poly(ethylene glycol) for biomedical applications. *Journal of the Chinese Chemical Society (Taipei)*, *65*(2), 231–242. doi:10.1002/jccs.201700271

Niemirowicz, K., & Bucki, R. (2017). Enhancing the fungicidal activity of antibiotics: Are magnetic nanoparticles the key? *Nanomedicine (London)*, *12*(15), 1747–1749. doi:10.2217/nnm-2017-0051 PMID:28703013

Niemirowicz, K., Car, H., Sadowska, A., Wątek, M., Kręowski, R., Cechowska-Pasko, M., Wilczewska, A. Z., Mystkowska, J., Kasacka, I., Torres, A., & Bucki, R. (2017). Pharmacokinetics and anticancer activity of folic acid-functionalized magnetic nanoparticles. *Journal of Biomedical Nanotechnology*, *13*(6), 665–677. doi:10.1166/jbn.2017.2363

Nigam, S., & Bahadur, D. (2018). Doxorubicin-loaded dendritic-Fe3 O4 supramolecular nanoparticles for magnetic drug targeting and tumor regression in spheroid murine melanoma model. *Nanomedicine; Nanotechnology, Biology, and Medicine*, *14*(3), 759–768. doi:10.1016/j.nano.2018.01.005 PMID:29339187

Nikjoo, D., & Aroguz, A. Z. (2016). Magnetic field responsive methylcellulosepolycaprolactone nanocomposite gels for targeted and controlled release of 5-fluorouracil. *International Journal of Polymeric Materials*, *65*(8), 421–432. doi:10.1080/00914037.2015.1129954

Nizamov, T. R., Garanina, A. S., Grebennikov, I. S., Zhironkina, O. A., Strelkova, O. S., Alieva, I. B., Kireev, I. I., Abakumov, M. A., Savchenko, A. G., & Majouga, A. G. (2018). Effect of iron oxide nanoparticle shape on doxorubicin drug delivery toward LNCaP and PC-3 cell lines. *BioNanoScience*, *8*(1), 394–406. doi:10.100712668-018-0502-y

Pageni, P., Yang, P., Bam, M., Zhu, T., Chen, Y. P., Decho, A. W., Nagarkatti, M., & Tang, C. (2018). Recyclable magnetic nanoparticles grafted with antimicrobial metallopolymer-antibiotic bioconjugates. *Biomaterials*, *178*(September), 363–372. doi:10.1016/j.biomaterials.2018.05.007 PMID:29759729

Patel, K., Raj, B. S., Chen, Y., & Lou, X. (2017). Novel folic acid conjugated Fe3 O4 - ZnO hybrid nanoparticles for targeted photodynamic therapy. *Colloids and Surfaces. B, Biointerfaces*, *150*, 317–325. doi:10.1016/j.colsurfb.2016.10.045 PMID:27810128

Peng, N., Wu, B., Wang, L., He, W., Ai, Z., Zhang, X., Wang, Y., Fan, L., & Ye, Q. (2016). High drug loading and pH-responsive targeted nanocarriers from alginate-modified SPIONs for anti-tumor chemotherapy. *Biomaterials Science*, *4*(12), 1802–1813. doi:10.1039/C6BM00504G PMID:27792228

Pirmardvand Chegini, S., Varshosaz, J., & Taymouri, S. (2018). Recent approaches for targeted drug delivery in rheumatoid arthritis diagnosis and treatment. *Artificial Cells, Nanomedicine, and Biotechnology, 46*(sup2), 502–514. doi:10.1080/21691401.2018.1460373 PMID:29661045

Pissuwan, D., Niidome, T., & Cortie, M. B. (2011). The forthcoming applications of gold nanoparticles in drug and gene delivery systems. *Journal of Controlled Release, 149*(1), 65–71. doi:10.1016/j.jconrel.2009.12.006 PMID:20004222

Plassat, V., Wilhelm, C., Marsaud, V., Ménager, C., Gazeau, F., Renoir, J.-M., & Lesieur, S. (2011). Anti-estrogen- loaded superparamagnetic liposomes for intracellular magnetic targeting and treatment of breast cancer tumors. *Advanced Functional Materials, 21*(1), 83–92. doi:10.1002/adfm.201001450

Polo, E., del Pino, P., Pardo, A., Taboada, P., & Pelaz, B. (2018). Magnetic Nanoparticles for Cancer Therapy and Bioimaging. In G. Gonçalves & G. Tobias (Eds.), Nanooncology, Nanomedicine and Nanotoxicology. Springer International Publishing AG. doi:10.1007/978-3-319-89878-0_7

Prabha, G., & Raj, V. (2016). Formation and characterization of β-cyclodextrin (β-CD)–polyethyleneglycol (PEG)–polyethyleneimine (PEI) coated Fe3O4 nanoparticles for loading and releasing 5-Fluorouracil drug. *Biomedicine and Pharmacotherapy, 80*, 173–182. doi:10.1016/j.biopha.2016.03.015 PMID:27133054

Prabha, G., & Raj, V. (2017). Sodium alginate–polyvinyl alcohol–bovin serum albumin coated Fe3 O4 nanoparticles as anticancer drug delivery vehicle: Doxorubicin loading and in vitro release study and cytotoxicity to HepG2 and L02 cells. *Materials Science and Engineering, 79*, 410–422. doi:10.1016/j.msec.2017.04.075 PMID:28629035

Ray Chowdhuri, A., Bhattacharya, D., & Sahu, S. K. (2016). Magnetic nanoscale metal organic frameworks for potential targeted anticancer drug delivery, imaging and as an MRI contrast agent. *Dalton Transactions (Cambridge, England), 45*(7), 2963–2973. doi:10.1039/C5DT03736K PMID:26754449

Reddy, L. H., Arias, J. L., Nicolas, J., & Couvreur, P. (2012). Magnetic Nanoparticles: Design and Characterization, Toxicity and Biocompatibility, Pharmaceutical and Biomedical Applications. *Chemical Reviews, 112*(11), 5818–5878. doi:10.1021/cr300068p PMID:23043508

Rivera-Rodriguez, A., Chiu-Lam, A., Morozov, V.M., Ishov, A.M., & Rinaldi, C. (2018). Magnetic nanoparticle hyperthermia potentiates paclitaxel activity in sensitive and resistant breast cancer cells. *Int J Nanomedicine, 13*, 4771-4779. . doi:10.2147/IJN.S171130

Rosemary, M. J., MacLaren, I., & Pradeep, T. (2006). Investigations of the antibacterial properties of ciprofloxacin@SiO2. *Langmuir, 22*(24), 10125–10129. doi:10.1021/la061411h PMID:17107009

Saha, B., Bhattacharya, J., Mukherjee, A., Ghosh, A. K., Santra, C. R., Dasgupta, A. K., & Karmakar, P. (2007). In vitro structural and functional evaluation of gold nanoparticles conjugated antibiotics. *Nanoscale Research Letters, 2*(12), 614–622. doi:10.100711671-007-9104-2

Saikia, C., Das, M. K., Ramteke, A., & Maji, T. K. (2017). Evaluation of folic acid tagged aminated starch/ZnO coated iron oxide nanoparticles as targeted curcumin delivery system. *Carbohydrate Polymers, 157*, 391–399. doi:10.1016/j.carbpol.2016.09.087 PMID:27987943

Shakil, M. S., Hasan, M. A., & Sarker, S. R. (2019). Iron Oxide Nanoparticles for Breast Cancer Theranostics. *Current Drug Metabolism*, *20*(6), 446–456. doi:10.2174/1389200220666181122105043 PMID:30465497

Sharifabad, M. E., Mercer, T., & Sen, T. (2016). Drug-loaded liposome-capped mesoporous core–shell magnetic nanoparticles for cellular toxicity study. *Nanomedicine (London)*, *11*, 2757–2767. doi:10.2217/nnm-2016-0248 PMID:27759497

Shaw, D., Clamp, A., & Jayson, G. C. (2013). Angiogenesis as a target for the treatment of ovarian cancer. *Current Opinion in Oncology*, *25*(5), 558–565. doi:10.1097/CCO.0b013e328363e0da PMID:23942301

Shen, B., Ma, Y., Yu, S., & Ji, C. (2016). Smart multifunctional magnetic nanoparticle-based drug delivery system for cancer thermochemotherapy and intracellular imaging. *ACS Applied Materials & Interfaces*, *8*(37), 24502–24508. doi:10.1021/acsami.6b09772 PMID:27573061

Shen, B. B., Gao, X. C., Yu, S. Y., Ma, Y., & Ji, C. H. (2016). Fabrication and potential application of a di-functional magnetic system: Magnetic hyperthermia therapy and drug delivery. *CrystEngComm*, *18*(7), 1133–1138. doi:10.1039/C5CE02267C

Shen, W., Cetinel, S., & Montemagno, C. (2018). Application of biomolecular recognition via magnetic nanoparticle in nanobiotechnology. *Journal of Nanoparticle Research*, *20*(5), 130. doi:10.100711051-018-4232-4

Shipunoval, Stremovskiy, Nikitin, Care, Sunna, Zvyagin, & Deyev. (2018). A novel method of nanoparticle surface biomodification for the development of effective agents for cancer theranostics. *12th International conference on Scientific and clinical applications of Magnetic carriers*.

Si, H. Y., Li, D. P., Wang, T. M., Zhang, H. L., Ren, F. Y., Xu, Z. G., & Zhao, Y. Y. (2010). Improving the anti-tumor effect of genistein with a biocompatible superpara- magnetic drug delivery system. *Journal of Nanoscience and Nanotechnology*, *10*(4), 2325–2331. doi:10.1166/jnn.2010.1913 PMID:20355429

Stoumpos, S., Hennessy, M., Vesey, A. T., Radjenovic, A., Kasthuri, R., Kingsmore, D. B., Mark, P. B., & Roditi, G. (2019). Ferumoxytol magnetic resonance angiography: A dose-finding study in patients with chronic kidney disease. *European Radiology*, *29*(7), 3543–3552. doi:10.100700330-019-06137-4 PMID:30919067

Sun, C., Lee, J., & Zhang, M. (2008). Magnetic nanoparticles in MR imaging and drug delivery. *Advanced Drug Delivery Reviews*, *60*(11), 1252–1265. doi:10.1016/j.addr.2008.03.018 PMID:18558452

Tian, J., Yan, C., Liu, K., Tao, J., Guo, Z., Liu, J., Zhang, Y., Xiong, F., & Gu, N. (2017). Paclitaxel-loaded magnetic nanoparticles: Synthesis, characterization, and application in targeting. *Journal of Pharmaceutical Sciences*, *106*(8), 2115–2122. doi:10.1016/j.xphs.2017.04.023 PMID:28456728

Ungureanu, B. S., Teodorescu, C. M., & Săftoiu, A. (2016, September). Magnetic Nanoparticles for Hepatocellular Carcinoma Diagnosis and Therapy. *Journal of Gastrointestinal and Liver Diseases; JGLD*, *25*(3), 375–383. doi:10.15403/jgld.2014.1121.253.nan PMID:27689203

Van der Zee, J. (2002). Heating the patient: A promising approach? *Annals of Oncology: Official Journal of the European Society for Medical Oncology, 13*(8), 1173–1184. doi:10.1093/annonc/mdf280 PMID:12181239

Varshosaz, J., Hassanzadeh, F., Aliabadi, H.S., Khoraskani, F.R., Mirian, M., & Behdadfar, B. (2016). Targeted delivery of doxorubicin to breast cancer cells by magnetic LHRH chitosan bioconjugated nanoparticles. *Int J Biol Macromol., 93*(Pt A), 1192–1205.

Veres, T., Molnár, K., Nesztor, D., Tombácz, E., Voniatis, C., Fehér, D., Weber, Gy., Zrinyi, M., & Jedlovszky-Hajdu, A. (2018). Magnetic gelfiber based scaffoldfortheranosticapplication. *12th International conference on Scientific and clinical applications of Magnetic carriers.*

Verma, Crosbie-Staunton, Amrosatti, Gallaagher, Rayn, Doody, McAtamney, MacLoughlin, Galvin, Burke, Volkov, & Gun'ko. (2013). Magnetic core shell nanoparticle for Drug delivery by nebulization. *Journal of Nanobiotechnology.*

Vismara, E., Bongio, C., Coletti, A., Edelman, R., Serafini, A., Mauri, M., Simonutti, R., Bertini, S., Urso, E., Assaraf, Y., & Livney, Y. (2017). Albumin and hyaluronic acid-coated superparamagnetic iron oxide nanoparticles loaded with paclitaxel for biomedical applications. *Molecules (Basel, Switzerland), 22*(7), 1030. doi:10.3390/molecules22071030 PMID:28640222

Wang, J., Chen, B., & Chen, J. (2011a). Synthesis and antitumor efficacy of daunorubicin-loaded magnetic nanoparticles. *International Journal of Nanomedicine, 6*, 203. PMID:21445276

Wang, J., Chen, B., & Cheng, J. (2011b). Apoptotic mechanism of human leukemia K562/A02 cells induced by magnetic iron oxide nanoparticles coloaded with daunorubicin and 5-bromotetrandrin. *International Journal of Nanomedicine, 6*, 1027. PMID:21720514

Wang, X., Chang, Y., Zhang, D., Tian, B., Yang, Y., & Wei, F. (2016). Transferrinconjugated drug/dye-co-encapsulated magnetic nanocarriers for active-targeting fluorescent/magnetic resonance imaging and antitumor effects in human brain tumor cells. *RSC Advances, 6*(107), 105661–105675. doi:10.1039/C6RA20903C

Wang, X., Jing, X., Zhang, X., Liu, Q., Liu, J., Song, D., Wang, J., & Liu, L. (2016). A versatile platform of magnetic microspheres loaded with dual-anticancer drugs for drug release. *Materials Chemistry and Physics, 177*, 213–219. doi:10.1016/j.matchemphys.2016.04.021

Wang, X., Tu, M., Tian, B., Yi, Y., Wei, Z., & Wei, F. (2016). Synthesis of tumortargeted folate conjugated fluorescent magnetic albumin nanoparticles for enhanced intracellular dual-modal imaging into human brain tumor cells. *Analytical Biochemistry, 512*, 8–17. doi:10.1016/j.ab.2016.08.010 PMID:27523645

Wang, Y.-P., Liao, Y.-T., Liu, C.-H., Yu, J., Alamri, H. R., Alothman, Z. A., Hossain, M. S. A., Yamauchi, Y., & Wu, K. C.-W. (2017). Trifunctional Fe3 O4 /CaP/ Alginate core–shell–corona nanoparticles for magnetically guided, pH-responsive, and chemically targeted chemotherapy. *ACS Biomaterials Science & Engineering, 3*(10), 2366–2374. doi:10.1021/acsbiomaterials.7b00230 PMID:30854421

Weber, C., & Falkenhagen, D. (1997). Specific blood purification by means of antibodyconjugated magnetic microspheres. In *Scientific and Clinical Applications of Magnetic Carriers* (pp. 371–378). Plenum Press. doi:10.1007/978-1-4757-6482-6_28

Widder, K. J., Morris, R. M., Howard, D. P., & Senyei, A. E. (1981). Tumor remission in Yoshida sarcoma-bearing rats by selective targeting of magnetic albu- min microspheres containing doxorubicin. *Proceedings of the National Academy of Sciences of the United States of America, 78*(1), 579–581. doi:10.1073/pnas.78.1.579 PMID:6941258

Widder, K. J., Senyei, A. E., & Ranney, D. F. (1979). Magneti- cally responsive microspheres and other carri- ers for the biophysical targeting of antitumor agents. *Advances in Pharmacology and Chemotherapy, 16*, 213–271. doi:10.1016/S1054-3589(08)60246-X PMID:382799

Wnorowska, U., Fiedoruk, K., Piktel, E., Prasad, S. V., Sulik, M., Janion, M., Daniluk, T., Savage, P. B., & Bucki, R. (2020). Nanoantibiotics containing membrane-active human cathelicidin LL-37 or synthetic ceragenins attached to the surface of magnetic nanoparticles as novel and innovative therapeutic tools: Current status and potential future applications. *Journal of Nanobiotechnology, 18*(1), 3. doi:10.118612951-019-0566-z PMID:31898542

Wu, L., Zhang, Z., Li, H., Wei, W., & Liu, S. (2016). Fe3 O4 -capped mesoporous silica foam for pH-responsive drug delivery. *Journal of Nanoscience and Nanotechnology, 16*(7), 6781–6787. doi:10.1166/jnn.2016.11385

Wu, M., & Huang, S. (2017). Magnetic nanoparticles in cancer diagnosis, drug delivery and treatment. *Molecular and Clinical Oncology, 7*(5), 738–746. doi:10.3892/mco.2017.1399 PMID:29075487

Wust, P., Hildebrandt, B., Sreenivasa, G., Rau, B., Gellermann, J., Riess, H., Felix, R., & Schlag, P. M. (2002). Hyperthermia in combined treatment of cancer. *The Lancet. Oncology, 3*(8), 487–497. doi:10.1016/S1470-2045(02)00818-5 PMID:12147435

Xenariou, S., Griesenbach, U., Ferrari, S., Dean, P., Scheule, R. K., Cheng, S. H., Geddes, D. M., Plank, C., & Alton, E. W. F. W. (2006). Using mag- netic forces to enhance non-viral gene transfer to airway epi- thelium in vivo. *Gene Therapy, 13*(21), 1545–1552. doi:10.1038j.gt.3302803 PMID:16738690

Xie, W., Gao, Q., Wang, D., Guo, Z., Gao, F., Wang, X., Cai, Q., Feng, S., Fan, H., Sun, X., & Zhao, L. (2018). Doxorubicin-loaded Fe3O4@MoS2- PEG-2DG nanocubes as a theranostic platform for magnetic resonance imaging-guided chemo-photothermal therapy of breast cancer. *Nano Research, 11*(5), 2470–2487. doi:10.100712274-017-1871-1

Xu, D., Xu, T., Guo, X., Liu, Q., Liu, J., Lv, W., Jing, X., Zhang, H., & Wang, J. (2017). Effect of the synthesis method on the performance of Fe3 O4 inositol hexaphosphate as a drug delivery vehicle for combination therapeutics with doxorubicin. *New Journal of Chemistry, 41*(13), 5305–5312. doi:10.1039/C7NJ00599G

Yadavalli, T., Ramasamy, S., Chandrasekaran, G., Michael, I., Therese, H. A., & Chennakesavulu, R. (2015). Dual responsive PNIPAM–chitosan targeted magnetic nanopolymers for targeted drug delivery. *Journal of Magnetism and Magnetic Materials, 380*, 315–320. doi:10.1016/j.jmmm.2014.09.035

Yang, J., Lee, C.-H., Ko, H.-J., Suh, J.-S., Yoon, H.-G., Lee, K., Huh, Y.-M., & Haam, S. (2007). Multifunctional magneto-polymeric nanohybrids for targeted detection and synergistic therapeutic effects on breast cancer. *Angewandte Chemie International Edition in English, 46*(46), 8836–8839. doi:10.1002/anie.200703554 PMID:17943947

Yar, Y., Khodadust, R., Akkoc, Y., Utkur, M., Saritas, E. U., Gozuacik, D., & Yagci Acar, H. (2018). Development of tailored SPIONPNIPAM nanoparticles by ATRP for dually responsive doxorubicin delivery and MR imaging. *Journal of Materials Chemistry*, *6*(2), 289–300. doi:10.1039/C7TB00646B PMID:32254171

Zaloga, J., Janko, C., Nowak, J., Matuszak, J., Knaup, S., Eberbeck, D., Tietze, R., Unterweger, H., Friedrich, R. P., Duerr, S., Heimke-Brinck, R., Baum, E., Cicha, I., D¨orje, F., Odenbach, S., Lyer, S., Lee, G., & Alexiou, C. (2014). Development of a lauric acid/albumin hybrid iron oxide nanoparticle system with improved biocompatibility. *International Journal of Nanomedicine*, *9*, 4847. doi:10.2147/IJN.S68539 PMID:25364244

Zarrin, A., Sadighian, S., Rostamizadeh, K., Firuzi, O., Hamidi, M., Mohammadi-Samani, S., & Miri, R. (2016). Design, preparation, and in vitro characterization of a trimodally-targeted nanomagnetic onco-theranostic system for cancer diagnosis and therapy. *International Journal of Pharmaceutics*, *500*(1–2), 62–76. doi:10.1016/j.ijpharm.2015.12.051 PMID:26721723

Zhang, H., Li, J., Hu, Y., Shen, M., Shi, X., & Zhang, G. (2016). Folic acid-targeted iron oxide nanoparticles as contrast agents for magnetic resonance imaging of human ovarian cancer. *Journal of Ovarian Research*, *9*(1), 19. doi:10.118613048-016-0230-2 PMID:27025582

Zhang, S., Wu, L., Cao, J., Wang, K., Ge, Y., Ma, W., & Shen, S. (2018). Effect of magnetic nanoparticles size on rheumatoid arthritis targeting and photothermal therapy. *Colloids and Surfaces. B, Biointerfaces*, *170*, 224–232. doi:10.1016/j.colsurfb.2018.06.016 PMID:29929166

Zheng, X. C., Ren, W., Zhang, S., Zhong, T., Duan, X., Yin, Y.-F., Xu, M.-Q., Hao, Y.-L., Li, Z.-T., Li, H., Liu, M., Li, Z.-Y., & Zhang, X. (2018). The theranostic efficiency of tumor-specific, pH-responsive, peptide-modified, liposome-containing paclitaxel and superparamagnetic iron oxide nanoparticles. *International Journal of Nanomedicine*, *13*, 1495–1504. doi:10.2147/IJN.S157082 PMID:29559778

Zhou, J., Leuschner, C., Kumar, C., Hormes, J. F., & Soboyejo, W. O. (2006). Sub-cellular accumulation of magnetic nanoparticles in breast tumors and metastases. *Biomaterials*, *27*(9), 2001–2008. doi:10.1016/j.biomaterials.2005.10.013 PMID:16280161

Chapter 18

Copper Nanoparticles:
Easily Accessible Nano-Weapons Against the Escalating Antibiotic Resistance

Ubaid Rasool
B. S. Abdur Rahman Crescent Institute of Science and Technology, India

Davoodbasha Mubarak Ali
B. S. Abdur Rahman Crescent Institute of Science and Technology, India

Hemalatha S.
B. S. Abdur Rahman Crescent Institute of Science and Technology, India

ABSTRACT

Nanoparticles are also proving a great alternative to antibiotics in combating the deadly bacterial infections. Copper nanoparticles, in particular, are being utilized as antimicrobial agents as well as their interaction with other particles helps to improve the antimicrobial efficacy. Compared to other metallic nanoparticles, copper nanoparticles are highly reactive, and their small size allows them to be used in different areas. When exposed to air, copper nanoparticles form copper oxide, and to prevent this oxidation, silica and carbon are utilized to coat copper nanoparticles. Metallic copper nanoparticles have been used in place of silver nanoparticles and other noble metals as anti-infective agents and also utilized in in water purification.

SUMMARY

Nanotechnology is an ancient conception which was originally coined by Richard Feynman in 1959 while lecturing the topic "There's plenty of room at the bottom" at the California Institute of Technology. The name nanotechnology originated from a Greek word "dwarf" meaning small and is often defined as a collective art of science and technology of small things. At nanoscale dimensions, there is a behavioural change in the particles and they function exceptionally well. For example, gold at nanoscale dimensions

DOI: 10.4018/978-1-7998-5049-6.ch018

turns out to be a good catalyst for fuel cells and semiconductors. Nanotechnology manipulates things at their molecular or atomic level with a positive attempt of creating unique materials with huge potential. Nanotechnology also has an enormous potential in creating smaller, inexpensive and faster devices with improved performance at nanoscale dimensions. It changes the properties of a material in such a way that the material behaves more efficiently at a nanoscale dimension compared to the same material at a large scale. Most of the vital metabolic processes in living beings occur at the nanoscale range and one such example is haemoglobin, which helps to carry oxygen across the bloodstream is approximately 5 nanometers in diameter.

In recent years, nanoparticles are getting more attention because of their diverse applications in the biomedical field including targeted drug delivery, hyperthermia, photoablation therapy, bioimaging and biosensors. Nanoparticles are also proving a great alternative to antibiotics in combating the deadly bacterial infections. Copper nanoparticles, in particular, are being utilized as antimicrobial agents as well as their interaction with other particles helps to improve the antimicrobial efficacy. Compared to other metallic nanoparticles, copper nanoparticles are highly reactive and their small size allow them to be used in different areas. When exposed to air, copper nanoparticles form copper oxide and to prevent this oxidation, silica and carbon are utilized to coat copper nanoparticles. Metallic copper nanoparticles have been used in place of silver nanoparticles and other noble metals as anti-infective agents and also utilized in in water purification.

DRUG RESISTANCE AND MICROBIAL PATHOGENESIS

Drug Resistance

With an unprecedented rise of antibiotic-resistant pathogenic microbes, a serious concern has arisen regarding the treatment of the infections due to the hindrance posed by drug-resistant microbes. It is generally believed that more than 70% of the bacterial related infections display resistance towards more than one antibiotic that are commonly utilized to treat the infection(Preethi et.al., 2019). Multidrug-resistant microorganisms have become a serious concern and a major health issue because of the complications they cause in treating infections. For example, Gram-negative *Pseudomonas aeruginosa* which is a ubiquitous microorganism and an opportunistic pathogen affects the patients with overlying conditions including AIDS, cancer, urinary tract infections (UTI) or those patients who are immunocompromised with cytotoxic drugs. The ability of this organism to form biofilm and synthesize the toxic compound called pyocyanin enhances its ability towards more antibiotic resistance (Cabot et al., 2016; Girard & Bloemberg, 2008).

Two other Gram-negative bacteria that have developed antibiotic resistance are *E. coli and K. pneumoniae.* These two bacteria are capable of synthesizing ESBLs which has resulted in the development of resistance towards all the beta-lactam antibiotics except carbapenems, cephamycins and clavulanic acid. Beta-lactamases are the enzymes with a potential to hydrolyze the beta-lactam ring that is present in beta-lactam antibiotics leading to antibiotic resistance. These bacteria are a reason for various diseases including diarrhoea, septicaemia, pneumonia and urinary tract infections. These bacteria also have the ability to produce biofilms which makes them more resistant towards antibiotics. Bacterial biofilms are responsible for more than 60% of the infections among the biofilm producing bacteria and those infections are generally chronic which pose a hindrance in treating those infections. Until now, there has not been any drug discovered that specifically targets biofilms, hence there is an urgent necessity to find an alternative way of breaking this antibiotic resistance (Shaikh et al., 2015; Ansari et al., 2014).

Multidrug-resistant Gram-positive bacteria are also involved in causing infections and these bacteria are generally found in the gastrointestinal tract of human as well as mucous membranes such as mouth, vagina and skin. Gram-positive bacteria including Staphylococci and Streptococci are the main reason for worldwide death and a major reason for nosocomial meningitis which can have serious clinical implications. These bacteria are predominantly found in hospital settings (Appelbaum & Jacobs, 2005; Palabiyikoglu, 2003). *Staphylococcus aureus* is among those bacteria which is responsible for skin infections as well as the infections of the other organs. This organism is predominantly found in hospital healthcare workers and is able to enter the human body through skin injuries and mucous membranes. Gram-positive Enterococci are the main reason for bloodstream infections in hospitalized patients. Gram-positive *Enterococcus faecalis* is a dominant bacterium and the main reason for Enterococcal infections in bedridden patients and can cause life-threatening infections including endocarditis, bladder, prostate and epididymal infections (Archer, 1998).

Apart from pathogenic bacteria, fungi are also responsible for causing various infections in human beings. Fungi are generally present on the skin and mucous membranes including mouth, rectum and vagina and they can travel through the bloodstream affecting throat, intestine and heart valves. Fungi, as well as yeasts, have been reported to cause various infections in hospitalized and particularly immunocompromised patients (Casadevall, 2002). One such example of human pathogenic fungi is yeast which is a single-celled organism. Among yeasts, *Candida albicans* is the most common and pathogenic. This organism is believed to be the most pathogenic and persistent Candida species isolated clinically. *Candida albicans* is usually found in intestine, vagina and skin and this organism plays a major role in causing Candidiasis and Candidemia. *Candida albicans* has the ability to develop resistance against antimycotic drugs such as fluconazole which is generally used in treating Candidiasis. The ability of Candida species to grow on human mucous membranes is due to different virulence factors and the synthesis of enzymes including phospholipases, proteases and haemolysis (Pfaller & Wenzel, 1992; Deorukhkar et al., 2014).

With an unprecedented spread of antibiotic resistance among bacteria and fungi, the treatment of various life-threatening infections has become very difficult. Hence, there is a need for finding an alternative approach which can help in reducing the antibiotic resistance.

MICROBIAL PATHOGENESIS

A pathogen is referred to a microorganism with an ability to cause a disease in any animal, plant or an insect. The term pathogenicity refers to the disease-causing capability of any microorganism inside a host. Microbes are able to express their pathogenicity through their virulence, a term which signifies the degree of pathogenicity of the microbe. Some of the important microbes and their pathogenicity is briefly described below.

Pathogenicity of ESBL Producing Bacteria

ESBL production by different bacteria has resulted in an unprecedented increase in resistance towards antibiotics. Beta-lactamases are defined as those bacterial enzymes which have the capability to hydrolyze the beta-lactam ring that is usually found in all the beta-lactam antibiotics which eventually leads to antibiotic resistance. ESBLs constitute a diverse and challenging group of enzymes that are being looked upon closely by clinical laboratories. Presently, ESBLs are one of the diverse group of enzymes and are most often the variants of broad-spectrum TEM and SHV beta-lactamases. These enzymes have the ability to hydrolyze most of the penicillins, in addition to cephalosporins which confers resistance towards these agents. Only those ESBLs which are able to hydrolyze the substrates (penicillins and cephalosporins or monobactams) at measurable rates pose a challenge in the clinical practice (Eliopoulos & Bush, 2001). The ability of bacteria to produce ESBLs led to resistance towards third-generation cephalosporins. ESBL producing bacteria are linked to high morbidity and mortality rates, particularly among those patients which are under intensive care. Hence, precise laboratory detection is vital in order to avoid any failures with antimicrobial therapies (Singh et al., 2015). ESBLs are encoded by three major genes including blaTEM, blaSHV and blaCTX-M (Jemima & Verghese, 2008) and are being recognized at a fast pace. These bacteria are on a rise since the past few years and are a cause of many diseases such as diarrhoea, septicaemia, pneumonia and urinary tract infections (Schauñer et al., 2018). ESBLs are either plasmid or chromosomally mediated but the plasmid-mediated ESBLs are predominantly found among Enterobacteriaceae. These enzymes are widely spread throughout the world and furthermore, *E. coli* and *K. pneumoniae* contain a significant percentage of ESBLs (Sah et.al., 2019a and 2019b). A continuous modification in ESBLs leads to a change in the amino acid sequence present near the active site of these beta-lactamases. There are 400 different variants of ESBLs that have been identified so far and all those variants are classified into different groups (Gholipour, 2014).

ESBL production has become a major hurdle in treating the infections caused by Gram-negative bacteria. These bacteria are spreading at a very fast pace and are a matter of concern due to their adverse effects on human health. The most important reason behind the increased resistance of ESBLs towards antibiotics is that these enzymes undergo continuous mutations which result in an expansion of substrate profile. Gene encoding CTX-M-15 which is a major factor for ESBL production is prevalent worldwide and is more prevalent in South India (Bali et al., 2010; Sah and Hemalatha, 2015).

Pathogenicity of an Opportunistic Pathogen P. Aeruginosa

With the discovery of the phenomenon of bacterial QS that is employed by the bacteria in order to regulate their pathogenic and virulence mechanisms has attracted the researchers in developing various novel therapeutic agents. QS is a system of communication developed by several bacteria in order to coordinate group behaviours. QS controls a series of extracellular virulence factors and biofilm formation in numerous pathogenic bacteria including *P. aeruginosa* (Lee & Zhang, 2015). *P. aeruginosa* is a ubiquitous and an opportunistic pathogen targeting the patients with overlying conditions including AIDS, cancer, patients immunocompromised by cytotoxic drugs, cystic fibrosis, blood, skin and genito-urinary infections. This organism is a major and a severe cause of nosocomial infections. *P. aeruginosa* has the capability to synthesize various extracellular compounds including elastase, alkaline protease, rhamnolipids, pyocyanin, exotoxin A, and hydrogen cyanide (Cabot et al., 2016; Girard & Bloemberg, 2008). All these compounds are crucial for *P. aeruginosa* to regulate the virulence (Shanmugasundaram et.al., 2019). QS system in *P. aeruginosa* controls the production of these extracellular factors together with the biofilm formation. QS also acts as a defence system in protecting *P. aeruginosa*. A decrease in the production of virulence factors through inactivation of the QS system can provide many ways to fight the infection caused by drug-resistant bacteria (Haussler & Fuqua, 2013). From the past 15 years, a deep research is being going on in order to find different QS inhibitors. Some of those inhibitors possess diverse structures such as furanone derivatives, different heterocycles, drugs such as macrolide and non-macrolide antibiotics, salicylic acid etc (Galloway et al., 2011). Apart from these inhibitors, some natural substances such as plant extracts have been reported to possess a great capability in inhibiting QS systems of Gram-negative bacteria (Koh et al., 2013).

Pathogenicity of Enterococcus Species

The past few decades have experienced the emergence of multi-drug resistant enterococci which is a leading cause of various drug resistant infections and have become a serious health-related issues including UTI, endocarditis and bacteraemia (Jayaprakash et.al 2016b). The virulence factors are responsible for antibiotic resistance, toxin production, biofilm formation, etc. The antibiotic therapy of such infections is difficult since enterococci display resistance to several drugs. Enterococci may also develop or acquire resistance to other antibacterial agents including chloramphenicol, ciprofloxacin, erythromycin, tetracycline, trimethoprim and vancomycin. The resistance displayed by Enterococci against antibiotics poses a threat to the effectiveness of new antibiotics. New approaches including active or passive immunization and development of novel molecules and phyto compounds (Sathishkumar et.al., 2018; Tahira et.al., 2017) may need to be taken into consideration (Ike, 2017; Lebreton et al., 2014).

Pathogenicity of Streptococcus Species

The genus Streptococcus includes numerous pathogenic species which cause diseases in human and animals and have evolved characteristic repositories of protein and non-protein toxins. *S. pyogenes* is an important species of this genus which is a gram-positive extracellular bacterial pathogen. Group A streptococci (GAS) are a cause of various pyogenic infections ranging from mild to severe, involving the mucous membranes, tonsils, skin, and deeper tissues, which may lead to a fatal outcome due to the production of exotoxin B (SpeB) (Krzyściak et al., 2013; Cole et al., 2008). *S. pyogenes* pathogenesis is facilitated by a massive repository of extracellular virulence factors. The persistence of the organism at the early stage of infection is mediated by a series of antiphagocytic molecules (Medina et al., 2003).

Pathogenicity of Salmonella Typhimurium

In human beings, the infections caused by *Salmonella* strains depend on certain factors which include the type of strain involved as well as the physiological condition of the host. Salmonella strains mostly affect the children and aged people. In comparison to healthy individuals, immunocompromised patients are more sensitive towards *Salmonella* infections. (Jayaprakash et.al. 2016a). Most of the *Salmonella* strains are pathogenic due to their potential of invasion, replication and survival inside the human host which leads to severe consequences. The remarkable feature displayed by Salmonella while invading a non-phagocytic human host cell includes the induction of its own phagocytosis in order to take a control of the host cell machinery. The persistence of Salmonella within the host is the main factor for its virulence and those strains without this capability are non-virulent. *S. typhi* and *S. paratyphi* A, B and C infections lead to enteric fever (Connor & Schwartz, 2005; Eng et al., 2015). Salmonella infections are a serious threat all over the world and since these strains have developed resistance towards the first line of antibiotics, new antimicrobial agents of choice have been introduced which include fluoroquinolones and extended-spectrum cephalosporins and green extracts and compounds (Ubaid et. al., 2018b).

These new antimicrobials are utilized in treating the infections due to the multidrug-resistant strains of Salmonella (Eng et al., 2015; Sood et al., 1999).

Pathogenicity of Proteus Mirabilis

Proteus mirabilis is a rod-shaped Gram-negative bacterium mostly known for its swarming motility and urease activity. This organism is a frequent source of catheter-associated urinary tract infections (CAUTI) that are often polymicrobial. *P. mirabilis* resides in a diverse range of habitats including soil, water sources, and sewage, but this organism is mostly found as a commensal of the gastrointestinal tract of human and animals. This bacterium is capable of causing a wide range of infections in human including those of wounds, eye, gastrointestinal tract, and the urinary tract, but it is mostly noted for infections of the catheterized urinary tract, known as CAUTI. *P. mirabilis* associated UTIs and CAUTIs are usually complicated resulting in urolithiasis and permanent renal damage. This bacterium is an agent of catheter biofilm formation, rapidly fouling the surface of a newly inserted urinary catheter. In this process, surface organelles such as fimbriae and other adhesins appear to play an important role. Urease enzyme also contributes dramatically to this process (Armbruster et al., 2018; Schaffer & Pearson, 2015). Currently, there are no licensed vaccines available for this organism, and moreover, multidrug-resistant isolates of *P. mirabilis* are increasing at a fast pace. Therefore, efforts to generate effective vaccines or therapeutic treatments are required.

Pathogenicity of MRSA

MRSA is the primary cause of hospital and community-associated infections and this organism is resistant towards the whole class of β-lactam antibiotics, including methicillin and penicillin. Hospital-associated MRSA strains are generally multi-drug resistant, hence leaving the treatment options on the use of less efficient drugs for example vancomycin. MRSA, similar to several other *S. aureus* strains, synthesize virulence factors including toxins, adhesion proteins, immune evasion and other virulence determinants. In *S. aureus* pathogenesis, surface proteins have several important roles. These proteins play a role in the cell wall metabolism of this bacterium apart from facilitating the binding to host tissue, internalization and immune evasion. These proteins also help in the aggregation and the biofilm formation of this bacterium (Raygada & Levine, 2009; Boswihi & Udo, 2018).

Pathogenicity of a Fungal Pathogen Candida Albicans

Candida species are the most predominant in causing several health-related issues including serious infections and they pose a serious threat in clinical practice. The rate of fungal infections is dependent on the immunological condition of host, application of broad-spectrum antibiotics, transplants, continued use of intravascular and urethral catheters, use of corticosteroids and parenteral nutrition (François et al., 2013). *Candida albicans* is considered to be the most pathogenic Candida species and so far, this species remains the most persistent Candida species that is isolated clinically (Borman et al., 2016). Different species of Candida cause candidemia (Silva et al., 2012) and produce harmful enzymes to thrive on human mucous membranes (Deorukhkar et al., 2014).

C. albicans is also able to produce biofilm and the biofilm formation by *C. albicans* is not a simple gathering of cells, but rather highly structured microbial communities, which is being assumed to carry out different functions such as to ease the process of the influx of nutrients and waste disposal (Ubaid et.al., 2019). A complex three-dimensional structure is formed by mature *C. albicans* during biofilm formation (Uppuluri et al., 2009).

Applications of Nanoparticles

Nanoparticles are gradually gaining a considerable interest owing to their enormous applications in different fields. Researchers have started using nanoparticles in drug and gene delivery systems, antibacterial agents, biosensors, cancer therapies, catalysts, detection of proteins and much more. Extensive research has been published regarding the applications of nanoparticles. Few examples describing these applications are given below.

Drug Delivery

The significant factor which is required in developing a novel drug delivery system and to attain the highest therapeutic effect is the precise and secure targeted drug delivery. The smaller size of nanoparticles allows them to avoid different cellular barriers. Nanocarriers enhance the diffusion of hydrophobic compounds making them appropriate for parenteral administration. Nanocarriers also enhance the stability of various therapeutic agents including peptides and oligonucleotides (Li et al., 2011). Magnetosomes from *Magnetospirillum gryphiswaldense* have been investigated for *in vitro* toxicity on mouse ðbroblasts and have been found non-toxic to mouse fibroblasts (Xiang et al., 2007). Bacterial magnetic particles have been investigated for their impact on mouse immune response and have shown no toxic effects and hence can be the emerging candidates for novel drug or gene carriers for tumour therapies (Meng et al., 2010). Low cardiac toxicity of doxorubicin (DOX) loaded bacterial magnetosomes has been reported together with their ability to inhibit the tumour growth (Sun et al., 2007). The *in vitro* and *in vivo* delivery of β-galactosidase plasmids utilizing polyethylenimine (PEI)-associated magnetotactic bacterial nanoparticles (MTB-PEI-NP) has been reported (Xie et al., 2009). Gold nanoparticles can be readily altered with different ligands and functional groups and display low toxicity, high surface area and stability. Hence, gold nanoparticles can be emerging candidates for novel drug deliveries. Gold nanoparticles display numerous biomedical applications, however the reports on the use of biosynthesized gold nanoparticles as gene delivery agents are less (Li et al., 2011).

Silver nanoparticles have been widely utilized for their novel therapeutic applications and moreover, they also display antibacterial, antifungal, antiviral and ant-inñammatory properties. Different concentrations of silver nanoparticles synthesized from *B. licheniformis* showed anti-angiogenic activity and reduced the cell survival (Kalishwaralal et al., 2009).

Copper nanoparticles also have the potential drug loading and efficient photoluminescence capability which makes them as potential carriers for drug deliveries. Copper nanoparticles can act as the carriers of anticancer drugs where the degradation of DNA through the action of copper nanomaterials by means of singlet oxygen production can take place. Chemical modifications of copper nanoparticles to produce active molecules can make them suitable for interacting with more macromolecules (Salavati-niasari et al., 2008).

Nanoparticle-mediated delivery of drugs to specific target sites is being anticipated to cut down the utilization of anticancer drugs and with improved specificity, greater efficiency as well as lesser toxicity. In the next few years, nanotechnology will rise to a whole new level with growing therapeutic and diagnostic applications (Mandal, 2017).

Antimicrobial Activity

With an unprecedented rise of antibiotic-resistant pathogenic microbes, the treatment of the infections has become difficult due to the hindrance posed by drug-resistant microbes (Ubaid et.al., 2018a). More than 70% of the bacterial infections display resistance towards any one of the antibiotics that are commonly used for treating the infection (Prestinaci et al., 2015). For centuries, metals including silver, copper, gold, titanium and zinc are being utilized since they are known to possess antimicrobial properties (Tahira et.al., 2018). In recent times, nanotechnology has provided enormous opportunities in different areas of science and technology. Metal nanoparticles including silver, copper, gold, titanium, magnesium have been recognized for their antimicrobial activity(Shariq et.al., 2019) The larger surface/volume ratio and the lesser size of nanoparticles compared to their respective metals leads to a change in their properties including heat treatment, mass transfer, the rate of dissolution and catalytic potential (Ravishankar & Jamuna, 2011).

In accordance with the literature, silver nanoparticles represent the most prevalent inorganic nanoparticles utilized as antimicrobial agents. Silver nanoparticles act on drug-resistant pathogens and they are believed to damage the outer bacterial membrane (Rohini et.al., 2019) Researchers also believe that silver nanoparticles are responsible for inducing pits and gaps in the bacterial membranes leading to cell fragmentation. Silver nanoparticles and their antimicrobial efficiency upon size reduction have been studied on *Bipolarissor sorokiniana* and *Magnaporthe grisea*. Silver nanoparticles have also been utilized to control the growth of soilborne fungi which often do not produce spores. Silver nanoparticles were found very active in controlling the growth of spore-forming fungal plant pathogens and may possibly display less toxicity compared to synthetic fungicides (Ravishankar & Jamuna, 2011; Jo et al., 2009). Silver nanoparticles have been tested for their antimicrobial effect against 4 different foodborne pathogens viz. *Escherichia coli, Listeria monocytogenes, Salmonella typhimurium and Vibrio parahaemolyticus* and it has been stated that silver nanoparticles can act as alternative disinfectants and cleansing agents of equipment and surfaces in food-related settings (Zarei et al., 2014; Ranjani et.al., 2019b).

Gold nanoparticles are considered extremely valuable for their antibacterial properties since they are less toxic, highly functional, have polyvalent effects, display photothermal activity and are easily detected. Most of the antibiotics and antibacterial nanomaterials generally act through the production of ROS in causing cell death, however, gold nanoparticles do not act through the ROS-related mechanism. The antibacterial effect of gold nanoparticles has been attributed to two factors, viz. (1) attachment to the bacterial membrane and a modification in membrane potential together with a reduction in the ATP level (2) inhibition in tRNA-ribosome binding (Zawrah & Sherein, 2011). The antimicrobial efficiency of gold nanoparticles was demonstrated in *E. coli, S. aureus, P. aeruginosa, Micrococcus luteus, Aspergillus fumigates and Aspergillus niger* (Selvaraj & Alagar, 2007).

Copper is an easily accessible metal and an essential trace element in a majority of living organisms. Copper particles in nanoscale dimensions are utilized in gas sensors, high-temperature superconductors, solar cells and wood preservatives. Copper in its metallic form has been utilized as an antibacterial agent since decades (Shirin et al., 2018). The antimicrobial activities of silver and copper nanoparticles on *B. subtilis, S. aureus* and *E. coli* showed that copper nanoparticles are more effective in controlling the growth of *B. subtilis* when compared to silver nanoparticles and the reason suggested is the higher affinity of copper nanoparticles to surface amines and carboxyl groups. Silver nanoparticles, on the other hand, were more effective on *E. coli* and *S. aureus* compared to copper nanoparticles (Ruparelia et al., 2008; Tahira et.al., 2019a, b and c). An effort to enhance the antimicrobial properties by grafting copper nanoparticles on the surface of multiwall carbon nanotubes (MWCNT) has been put forth and it has been stated that carbon nanotubes increased the surface area of copper nanoparticles which was the reason for higher activity of Cu-MWCNT system compared to pure copper nanoparticles and MWCNT on *E. coli*. The efficiency (% kill) of Cu-MWCNT system was found higher (75% $\pm$ 0.8) compared to pure copper nanoparticles (52% $\pm$ 1.8) in *E. coli*. The possible antibacterial mechanism of Cu-MWCNT system could be the release of copper ions which after entering the bacterial cell resulted in the disruption of biochemical processes (Raja et al., 2010). Chitosan-stabilized copper nanoparticles were found to be effective against bacteria and yeast (Usman et al., 2013).

Owing to the enormous development in the field of nanotechnology and the wide range of applications that nanoparticles display, there needs to be a check on the toxicity that nanoparticles can display. The main toxic effects of nanoparticles that are of concern include neurological and respiratory damage, circulatory damage apart from various other toxic effects. While as numerous nanoparticles appear non-toxic yet some of them display toxicity which needs to be studied. The antimicrobial effect of nanoparticles and the abolition of the bacterial infections could be a valuable development in the health sector (Wang et al., 2017; Slavin et al., 2017).

Biosensors

For constructing a biosensor, the two important properties of nanoparticles including surface area and free energy are important in the immobilization of biomolecules as these properties help in maintaining the strong interactions. The coating of biomolecules on larger materials often leads to denaturation and also damages their bioactivity but the adsorption of the same biomolecules on nanoparticle surfaces helps in retaining their bioactivity owing to the biocompatibility of nanoparticles. Meanwhile, a majority of nanoparticles carry charges which helps them to adsorb different biomolecules through electrostatic interactions. The immobilization of proteins by gold nanoparticles has been reported as a result of covalent bonding. Compared to other nanoparticles, gold nanoparticles are possibly the most commonly used nanoparticles for immobilization of proteins (Xiliang et al., 2011). In the early 90s, gold nanoparticles have been utilized for immobilizing different types of enzymes and in the development different enzyme electrodes which retained the enzymatic activity (Ahirwal & Chanchal, 2009). A successful immobilization of horseradish peroxidase on gold nanoparticles has been reported. The smaller sized nanoparticles were observed to be more appropriate for enzyme immobilization (Xiao et al., 1999). A reagentless amperometric immunosensor with long-term stability by immobilizing an antibody to control Japanese encephalitis by immobilizing a related antibody on gold nanoparticles were developed (Zhuo et al., 2005).

Nanoparticles as Catalysts

The large surface area and the specific characteristics of nanoparticles have allowed them to be broadly used as reductants and/or catalysts in order to enhance the reaction rates (Sergio & García, 2016). For instance, the catalytic function and the good dispersing ability of magnetic nanoparticles have resulted in their utilization as reaction rate enhancers in microbiological processes. Fe3O4 nanoparticles together with the coated microbial cells of *Pseudomonas delaðeldii* have been utilized to perform the desulfurization of dibenzothiophene and no reduction in the desulfurization efficiency of *P. delaðeldii* was observed (Shan et al., 2005).

Nanopharmaceuticals

Nanopharmaceuticals can be utilized in detecting diseases at a very early phase and the diagnostic applications could build upon conventional procedures using nanoparticles. Nanopharmaceuticals is a developing field which involves the nanoscale functioning of the drug particles or therapeutic delivery systems. The delivery of an appropriate dose of a specific active agent to a specific disease site is still challenging in the pharmaceutical industry. The huge potential of nanopharmaceuticals can tackle this failure as it involves the site-specific targeting of active agents. Nanopharmaceuticals can also reduce the toxic side effects of the system and in that way can result in better patient compliance. There is an enormous pressure faced by pharmaceutical industries to deliver high quality and successful products and hence, nanotechnology can provide a boost in enhancing the formulation and targeted discovery of drugs (Ranjani et.al., 2019A) Nanopharmaceuticals have enormous potential in the production of cost-effective drugs and the success rate in research and development and thus can reduce the time for both drug discovery and diagnostics (Volkmar et al., 2014).

ANTIMICROBIAL ACTIVITY OF COPPER NANOPARTICLES AND THEIR MECHANISM OF ACTION ON GRAM-POSITIVE AND GRAM-NEGATIVE BACTERIAL CELLS

Metallic copper has been in use since decades because of its antimicrobial properties. Since copper is a low cost and a readily available material, hence the synthesis of copper nanoparticles is economic and beneficial. Copper nanoparticles are active against Enterobacteriaceae and *P. aeruginosa* in addition to the drug-resistant *C. albicans*. These microorganisms cause several grave infections which can lead to death. Since, drug-resistant pathogenic microbes are continuously increasing at a fast pace, there is a need to find an alternative so that the problem of increasing antibiotic resistance comes to an end. The activity of copper nanoparticles against various pathogenic microorganisms can make them an alternative to various present-day antibiotics to which the microorganisms display resistance. Copper nanoparticles can be commercialized to replace the less effective antibiotics and this process requires a proper modification because copper nanoparticles can also display toxicity which needs to be minimized.

Activity of Copper Nanoparticles Based on the Bacterial Cell Wall Structure and Composition

Gram-negative bacterial cell wall, such as *E. coli* are enclosed by a layer of lipopolysaccharides and peptidoglycans and this cell wall arrangement may assist in the penetration of nanoparticles into the cell. Gram-positive bacteria (such as *Staphylococcus aureus*) on the other side possess a much thicker peptidoglycan layer (Figure 1). The interaction between copper nanoparticles and the cell wall proves more harmful to Gram-negative bacteria as they lack a thick peptidoglycan layer. One more possible reason for the penetration of copper nanoparticles into the Gram-negative bacterial cell is that the lipopolysaccharide molecules present in the cell wall of Gram-negative bacteria carry a negative charge and most of the nanoparticles (including copper nanoparticles) release positive ions which leads to an increased uptake of ions causing intracellular damage (Figure 2) (Slavin et al., 2017; Tahira et.al. 2018a).

Gram-positive and Gram-negative bacterial cell walls are negatively charged which enhance the interaction of copper nanoparticles with bacterial cells (Magnusson & Bayer, 1982). Studies carried out on Gram-negative bacteria such as *Salmonella typhimurium* have revealed the mosaic pattern of anionic surfaces rather than continuous layers which eventually leads to higher interactions of nanoparticles at these positions (Sonohara et al., 1995).

Figure 1. Bacterial cell wall structure and composition

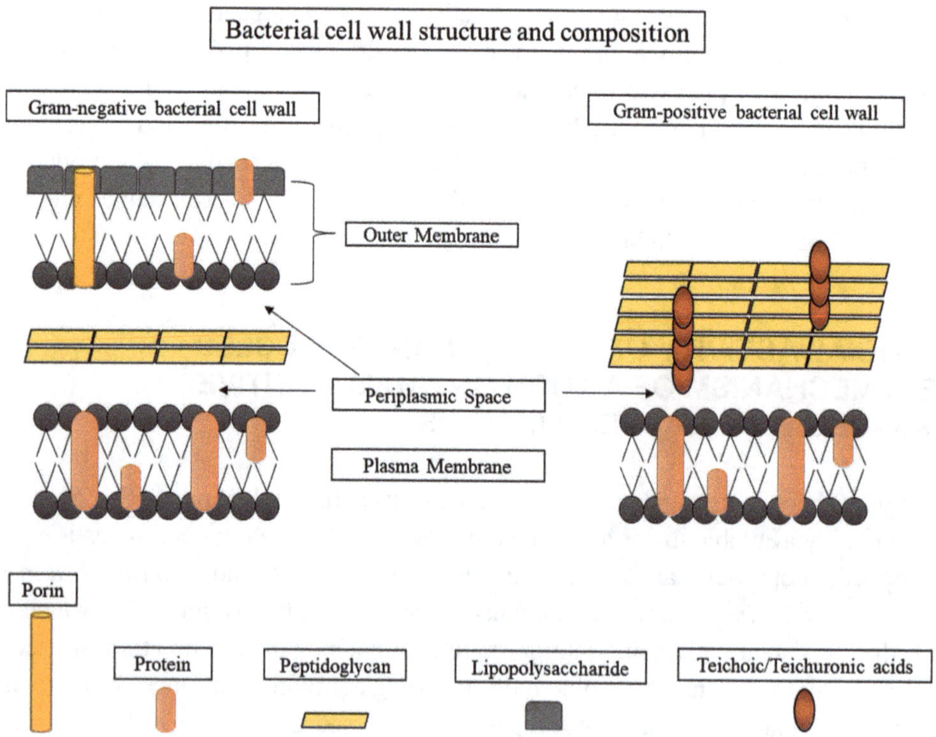

Copper Nanoparticles as Antibacterial Agents

A study carried out by Ruparelia and co-workers involved the utilization of silver and copper nanoparticles against *E. coli, B. subtilis and S. aureus* (Ruparelia et al., 2008; Ubaid and Hemalatha, 2017) and it was observed that copper nanoparticles displayed more efficiency against *B. subtilis* compared to silver nanoparticles (silver nanoparticles displayed more antimicrobial effect against *E. coli and S. aureus* compared to copper nanoparticles). The reason suggested for this higher efficiency of copper nanoparticles is the higher affinity of the copper nanoparticles to surface amines and carboxyl groups of *B. subtilis.*

Raffi and co-workers studied the feasibility of the use of copper nanoparticles against *E. coli* in both the solid and liquid media. Upon observing under scanning electron (SEM), pits and cavities were observed on the bacterial cell wall along with morphological changes. The results also demonstrated the concentration dependent effect of copper nanoparticles on *E. coli* (Raffi et al., 2010).

Figure 2. The electrostatic interaction between copper nanoparticles and the bacterial cell wall

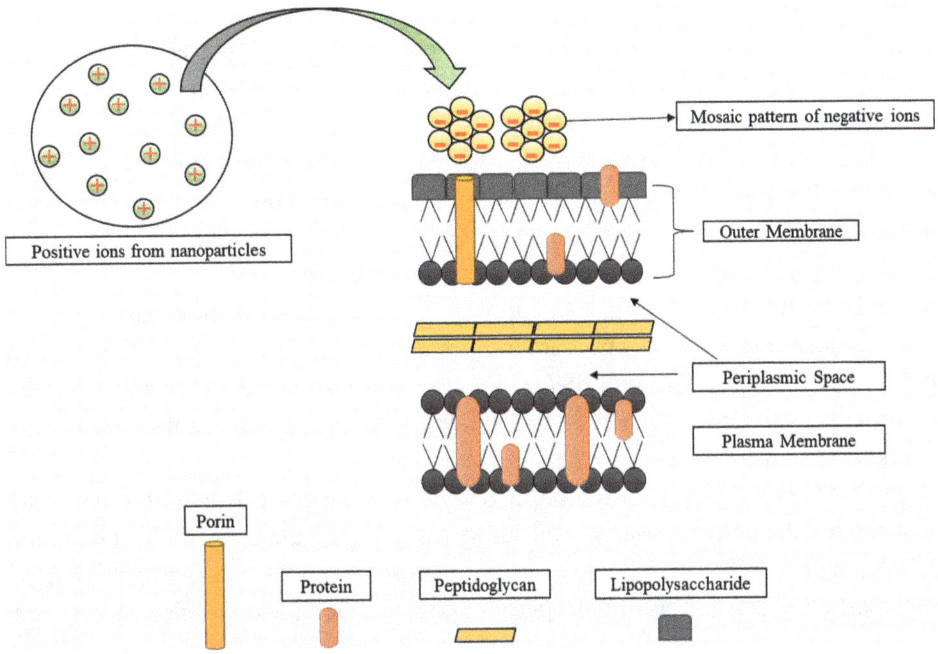

Theivasanthi and Alagar studied the comparative effects of copper nanoparticles synthesized using electrolysis technique and chemical reduction process. It was observed that copper nanoparticles synthesized through electrolysis technique were more effective against *E. coli* compared to those synthesized through a chemical reduction process (Theivasanthi & Alagar, 2011). The preparation of copper nanoparticles using electrical power resulted in the modification of the antibacterial effects.

Chatterjee and co-workers presented a method involving the reduction of CuCl2 using gelatin as a stabilizer for synthesis of copper nanoparticles (Chatterjee et al., 2012). When *E. coli* was treated with those nanoparticles, it resulted in the formation of filamentous cells with an average filament size. These nanoparticles showed higher efficiency against multi-drug resistant *E. coli* besides Gram-positive *B. subtilis and S. aureus.*

An interesting study carried out on the antibacterial effect of copper oxide nanoparticles on *Legionella pneumophila* (Lu et al., 2013) revealed a significant effect of nanoparticles on the expression of genes required for metabolism, transcription, translation, DNA replication and repair, virulence, and unknown/hypothetical proteins.

Thekkae Padil and co-workers synthesized highly stable copper oxide nanoparticles using gum karaya, a naturally occurring polysaccharide in plants through green method (Padil & Cernik, 2013). Small sized copper oxide nanoparticles displayed higher antibacterial effect. The authors also pointed out the efficiency of copper oxide nanoparticles in wound dressing, bed lining, active cotton bandages, and medical and food industries.

A thermal decomposition method adopted by Das and co-workers to synthesize copper oxide nanoparticles showed positive antioxidant and antibacterial effects (Das et al., 2013). Compared to other metal nanoparticles, these nanoparticles displayed higher free radical scavenging activity as well as showed concentration-dependent antibacterial effect against *E. coli* and *P. aeruginosa*.

Usman and co-workers utilized a simple and a cost-effective approach in synthesizing chitosan-stabilized copper nanoparticles which were active against Gram-positive, Gram-negative as well as the yeast *Candida albicans*. These chitosan-stabilized copper nanoparticles displayed more antibacterial effect on Gram-negative bacteria (*P. aeruginosa*) compared to Gram-positive bacteria (Usman et al., 2013).

A novel green approach using ginger (*Zingiber officinale*) extract in synthesizing copper nanoparticles was put forward (Subhankari & Nayak, 2013). The synthesized nanoparticles were more efficient in controlling *E. coli* compared to copper sulphate solution and pure ginger extract. The researchers also stated that this green method can be beneficial in the purification of water, management of the air quality and antibacterial packaging since this method utilized non-toxic materials.

A green chemistry approach in synthesizing highly stable copper oxide nanoparticles from the aqueous extract of *Acalypha indica* leaf was put forward (Sivaraj et al., 2014). The synthesized nanoparticles were effective in controlling the growth of *E. coli, Pseudomonas fluorescens* and *Candida albicans*. Furthermore, the synthesized nanoparticles also displayed cytotoxic activity against MCF-7 breast cancer cell lines through MTT assay.

A comparative study on the biofilm-inhibiting activity of copper oxide and iron oxide nanoparticles against multi-drug resistant Uropathogens was put forward (Agarwala et al., 2014) and it was observed that copper oxide nanoparticles were more effective in controlling the biofilm formation compared to iron oxide nanoparticles. The antibiofilm property of copper oxide nanoparticles was found to be dose-dependent.

Parikh and co-workers applied the green nanotechnology approach in synthesizing copper nanoparticles using Datura Metal leaf extract (Parikh et al., 2014). The proposed method has been suggested as an efficient, easy, economical and eco-friendly method for nanoparticle synthesis. The synthesized nanoparticles displayed antibacterial activity against *E. coli, Bacillus megaterium, and B. subtilis.*

Tomasz and co-workers developed copper nanostructures with high antibacterial effects against Gram positive bacteria such as MRSA strains (Kruk et al., 2015). The antibacterial effects were found to be even higher compared to silver nanoparticles. The copper nanoparticles displayed a higher antibacterial efficiency against Gram positive bacteria such as clinical methicillin resistant *S. aureus* strains and were also active against *Candida* species.

CONCLUSION

There is a growing scientific evidence confirming the antimicrobial properties of metallic nanoparticles against Gram-positive and Gram-negative bacteria besides fungi. According to the literature, copper nanoparticles have been found to be more effective against disease causing bacteria compared to other metallic nanoparticles. In recent years, copper nanoparticles are gaining a substantial attention because of their antibacterial effects against a vast range of microorganisms. This emphasizes the capability of copper nanoparticles in nano-scale dimensions as the efficient antimicrobial agents in biomedical field. As of now, copper nanoparticles are being successfully utilized in *in-vitro* studies where they are showing positive results but, in order to utilize copper nanoparticles in clinical settings, they need to be tested under in-vivo conditions using animal models. We are hoping that in the near future, copper nanoparticles can probably be used as the efficient antibacterial agents to fight the disease-causing pathogenic microorganisms and to break the growing antibiotic resistance.

REFERENCES

Agarwala, M., Choudhury, B., & Yadav, R. (2014). Comparative study of antibiofilm activity of copper oxide and iron oxide nanoparticles against multidrug resistant biofilm forming uropathogens. *Indian Journal of Microbiology*, *54*(3), 365–368. doi:10.1007/s12088-014-0462-z PubMed

Ahirwal, G. K., & Chanchal, K. M. (2009). Direct Electrochemistry of Horseradish Peroxidase-Gold Nanoparticles Conjugate. *Sensors (Basel)*, *9*(2), 881–894. doi:10.3390/s90200881 PubMed

Ansari, M. A., Khan, H. M., Khan, A. A., Cameotra, S. S., & Pal, R. (2014). Antibioÿlm efÿcacy of silver nanoparticles against bioÿlm of extended spectrum b-lactamase isolates of Escherichia coli and Klebsiella pneumoniae. Applied Nanoscience, 4(7), 859–868. doi:10.1007/s13204-013-0266-1

Appelbaum, P. C., & Jacobs, M. R. (2005). Recently approved and investigational antibiotics for treatment of severe infections caused by Gram-positive bacteria. *Current Opinion in Microbiology*, *8*(5), 510–517. doi:10.1016/j.mib.2005.07.001 PubMed

Archer, G. L. (1998). Staphylococcus aureus: A well-armed pathogen. *Clinical Infectious Diseases*, *26*(5), 1179–1181. doi:10.1086/520289 PubMed

Armbruster, C. E., Mobley, H. L. T., & Pearson, M. M. (2018). Pathogenesis of Proteus mirabilis Infection. *EcoSal Plus*, *8*(1), ESP-0009–ESP-2017. doi:10.1128/ecosalplus.ESP-0009-2017 PubMed

Bali, E. B., Açık, L., & Sultan, N. (2010). Phenotypic and molecular characterization of SHV, TEM, CTX-M and extended-spectrum β-lactamase produced by Escherichia coli, Acinetobacter baumannii and Klebsiella isolates in a Turkish hospital. *African Journal of Microbiological Research*, *4*, 650–654.

Borman, A. M., Szekely, A., & Johnson, E. M. (2016). Comparative Pathogenicity of United Kingdom Isolates of the Emerging Pathogen Candida auris and Other Key Pathogenic Candida Species. Clinical Science and Epidemiology, 1(4), e00189–e16. doi:10.1128/mSphere.00189-16 PubMed

Boswihi, S. S., & Udo, E. E. (2018). Methicillin-resistant Staphylococcus aureus: An update on the epidemiology, treatment options and infection control. Current Medicine Research and Practice, 8(1), 18–24. doi:10.1016/j.cmrp.2018.01.001

Cabot, G., Zamorano, L., Moyà, B., Juan, C., Navas, A., Blázquez, J., & Oliver, A. (2016). Evolution of Pseudomonas aeruginosa antimicrobial resistance and ÿtness under low and high mutation rates. *Antimicrobial Agents and Chemotherapy*, *60*(3), 1767–1778. doi:10.1128/AAC.02676-15 PubMed

Cabot, G., Zamorano, L., Moyà, B., Juan, C., Navas, A., Blázquez, J., & Oliver, A. (2016). Evolution of Pseudomonas aeruginosa antimicrobial resistance and ÿtness under low and high mutation rates. *Antimicrobial Agents and Chemotherapy*, *60*(3), 1767–1778. doi:10.1128/AAC.02676-15 PubMed

Casadevall, A., Feldmesser, M., & Pirofski, L. (2002). Induced humoral immunity and vaccination against major human fungal pathogens. *Current Opinion in Microbiology*, *5*(4), 386–391. doi:10.1016/S1369-5274(02)00337-5 PubMed

Chatterjee, A. K., Sarkar, R. K., Chattopadhyay, A. P., Aich, P., Chakraborty, R., & Basu, T. (2012). A simple robust method for synthesis of metallic copper nanoparticles of high antibacterial potency against E. coli. *Nanotechnology*, *23*(8), 085103. doi:10.1088/0957-4484/23/8/085103 PubMed

Cole, J. N., Henningham, A., Gillen, C. M., Ramachandran, V., & Walker, M. J. (2008). Human pathogenic streptococcal proteomics and vaccine development. *Proteomics. Clinical Applications*, *2*(3), 387–410. doi:10.1002/prca.200780048 PubMed

Connor, B. A., & Schwartz, E. (2005). Typhoid and paratyphoid fever in travellers. *The Lancet Infectious Diseases*, *5*(10), 623–628. doi:10.1016/S1473-3099(05)70239-5 PubMed

Das, D., Nath, B. C., Phukon, P., & Dolui, S. K. (2013). Synthesis and evaluation of antioxidant and antibacterial behavior of CuO nanoparticles. *Colloids and Surfaces. B, Biointerfaces*, *101*, 430–433. doi:10.1016/j.colsurfb.2012.07.002 PubMed

Deorukhkar, S. C., Saini, S., & Mathew, S. (2014). Virulence Factors Contributing to Pathogenicity of Candida tropicalis and Its Antifungal Susceptibility Profile. *International Journal of Microbiology*, *2014*, 1–6. doi:10.1155/2014/456878 PubMed

Deorukhkar, S. C., Saini, S., & Mathew, S. (2014). Virulence Factors Contributing to Pathogenicity of Candida tropicalis and Its Antifungal Susceptibility Profile. *International Journal of Microbiology*, *2014*, 1–6. doi:10.1155/2014/456878 PubMed

Eliopoulos, G. M., & Bush, K. (2001). New beta-lactamases in gram-negative bacteria: Diversity and impact on the selection of antimicrobial therapy. *Clinical Infectious Diseases*, *32*(7), 1085–1089. doi:10.1086/319610 PubMed

Eng, S.-K., Pusparajah, P., Mutalib, N.-S., Ser, H.-L., Chan, K.-G., & Lee, L. H. (2015). Salmonella: A review on pathogenesis, epidemiology and antibiotic resistance. *Frontiers in Life Science*, *8*(3), 284–293. doi:10.1080/21553769.2015.1051243

François, L. M., Wilson, D., & Hube, B. (2013). Candida albicans pathogenicity mechanisms. *Virulence*, *4*(2), 119–128. doi:10.4161/viru.22913 PubMed

Galloway, W. R., Hodgkinson, J. T., Bowden, S. D., Welch, M., & Spring, D. R. (2011). Quorum sensing in gram-negative bacteria: Small molecule modulation of AHL and AI-2 quorum sensing pathways. *Chemical Reviews*, *111*(1), 28–67. doi:10.1021/cr100109t PubMed

Gholipour, A., Soleimani, N., Shokri, D., Mobasherizadeh, S., Kardi, M., & Baradaran, A. (2014). Phenotypic and Molecular Characterization of Extended-Spectrum β-Lactamase Produced by Escherichia coli, and Klebsiella pneumoniae Isolates in an Educational Hospital. *Jundishapur Journal of Microbiology*, *7*(8), e11758. doi:10.5812/jjm.11758 PubMed

Girard, G., & Bloemberg, G. V. (2008). Central role of quorum sensing in regulating the production of pathogenicity factor in Pseudomonas aeruginosa. *Future Microbiology*, *3*(1), 97–106. doi:10.2217/17460913.3.1.97 PubMed

Girard, G., & Bloemberg, G. V. (2008). Central role of quorum sensing in regulating the production of pathogenicity factor in Pseudomonas aeruginosa. *Future Microbiology*, *3*(1), 97–106. doi:10.2217/17460913.3.1.97 PubMed

Haussler, S., & Fuqua, C. (2013). Bioÿlms 2012: New discoveries and signiÿcant wrinkles in a dynamic ÿeld. *Journal of Bacteriology*, *195*(13), 2947–2958. doi:10.1128/JB.00239-13 PubMed

Ike, Y. (2017). Pathogenicity of Enterococci. *Japanese Journal of Bacteriology*, *72*(2), 189–211. doi:10.3412/jsb.72.189 PubMed

Jayaprakash, R., Sah, S. k., Hemalatha, S., & Easwaramoorthy, D. (2016a). Synthesis, Characterization, QSAR, Docking, Antibacterial Activity, and Brine Shrimp Lethal Studies on L-Phenylalanine Schiff Bases. Asian J. Pharma. clin. *Res.*, *9*, 3.

Jayaprakash, R., Sah, S. k., Hemalatha, S., & Easwaramoorthy, D. (2016b). QSAR, Brine Shrimp Lethal Assay and antimicrobial studies on Synthesized L-Tryptophan-2,4-dihydroxy benzaldehyde Schiff Base. *Int. J. Chem. Tech*, *9*, 48–54.

Jemima, S. A., & Verghese, S. (2008). Multiplex PCR for blaCTX-M & blaSHV in the extended spectrum beta lactamase (ESBL) producing Gram-negative isolates. *The Indian Journal of Medical Research*, *128*, 313–317. PubMed

Jo, Y.-K., Jung, G., & Kim, B. H. (2009). Antifungal Activity of Silver Ions and Nanoparticles on Phytopathogenic Fungi. *Plant Disease*, *93*(10), 1037–1042. doi:10.1094/PDIS-93-10-1037 PubMed

Kalishwaralal, K., Banumathi, E., Pandian, S. R. K., Deepak, V., Muniyandi, J., Eom, S. H., & Gurunathan, S. (2009). Silver nanoparticles inhibit VEGF induced cell proliferation and migration in bovine retinal endothelial cells. *Colloids and Surfaces. B, Biointerfaces*, *73*(1), 51–57. doi:10.1016/j.colsurfb.2009.04.025 PubMed

Koh, C. L., Sam, C. K., Yin, W. F., Tan, L. Y., Krishnan, T., Chong, Y. M., & Chan, K. G. (2013). Plant-derived natural products as sources of anti-quorum sensing compounds. *Sensors (Basel)*, *13*(5), 6217–6228. doi:10.3390/s130506217 PubMed

Kruk, T., Szczepanowicz, K., Stefańska, J., Socha, R. P., & Warszyński, P. (2015). Synthesis and antimicrobial activity of monodisperse copper nanoparticles. *Colloids and Surfaces. B, Biointerfaces*, *128*, 17–22. doi:10.1016/j.colsurfb.2015.02.009 PubMed

Krzyściak, W., Pluskwa, K. K., Jurczak, A., & Kościelniak, D. (2013). The pathogenicity of the Streptococcus genus. *European Journal of Clinical Microbiology & Infectious Diseases*, *32*(11), 1361–1376. doi:10.1007/s10096-013-1914-9 PubMed

Lebreton, F., Willems, R. J. L., & Gilmore, M. S. (2014). Enterococcus Diversity, Origins in Nature, and Gut Colonization. In M. S. Gilmore, D. B. Clewell, Y. Ike, & N. Shankar (Eds.), *Enterococci: From Commensals to Leading Causes of Drug Resistant Infection* (pp. 5–65). Massachusetts Eye and Ear Infirmary.

Lee, J., & Zhang, L. (2015). The hierarchy quorum sensing network in Pseudomonas aeruginosa. *Protein & Cell*, *6*(1), 26–41. doi:10.1007/s13238-014-0100-x PubMed

Li, X., Xu, H., Chen, Z.-S., & Chen, G. (2011). Biosynthesis of Nanoparticles by Microorganisms and Their Applications. Journal of Nanomaterials, 2011, 270974. doi:10.1155/2011/270974

Lu, J., Struewing, I., Buse, H. Y., Kou, J., Shuman, H. A., Faucher, S. P., & Ashbolt, N. J. (2013). Legionella pneumophila transcriptional response following exposure to CuO nanoparticles. *Applied and Environmental Microbiology*, *79*(8), 2713–2720. doi:10.1128/AEM.03462-12 PubMed

Magnusson, K.-E., & Bayer, M. E. (1982). Anionic sites on the envelope of Salmonella typhimurium mapped with cationized ferritin. *Cell Biophysics*, *4*(2-3), 163–175. doi:10.1007/BF02918311 PubMed

Mandal, A. K. (2017). Copper Nanomaterials as Drug Delivery System against Infectious Agents and Cancerous Cells. JALSI, 15(4), 1–8. doi:10.9734/JALSI/2017/38444

Medina, E., Goldmann, O., Toppel, A. W., & Chhatwal, G. S. (2003). Survival of Streptococcus pyogenes within host phagocytic cells: A pathogenic mechanism for persistence and systemic invasion. *The Journal of Infectious Diseases*, *187*(4), 597–603. doi:10.1086/373998 PubMed

Meng, C., Tian, J., Li, Y., & Zheng, S. (2010). Inñuence of native bacterial magnetic particles on mouse immune response. Wei Sheng Wu Xue Bao, 50, 817–821. PubMed

Padil, V. V. T., & Cernik, M. (2013). Green synthesis of copper oxide nanoparticles using gum karaya as a biotemplate and their antibacterial application. *International Journal of Nanomedicine*, *8*, 889–898. PubMed

Palabiyikoglu, I. (2003). Pathogenesis of intensive care infections. *Turkish Journal of Intensive Care Medicine*, *3*, 81–101.

Parikh, P., Zala, D., & Makwana, B. A. (2014). Biosynthesis of Copper Nanoparticles and Their Antimicrobial Activity. OAlib, 01(01), 1–15. doi:10.4236/oalib.preprints.1200067

Pfaller, M. A., & Wenzel, R. (1992). Impact of the changing epidemiology of fungal infections in the 1990s. *European Journal of Clinical Microbiology & Infectious Diseases*, *11*(4), 287–291. doi:10.1007/BF01962067 PubMed

Preethi, B., Sha, S. K., Jayaprakash, R., Kutti Rani, S., & Hemalatha, S. (2019). Synthesis, characterization and biological studies on 4-bromo-2-{(z)-[(furan-2-ylmethyl) imino]methyl}phenol praseodymium complex. *Rasayan Journal of Chemistry*, *12*(3), 1455–1462. doi:10.31788/RJC.2019.1235296

Prestinaci, F., Patrizio, P., & Annalisa, P. (2015). Antimicrobial Resistance: A Global Multifaceted Phenomenon. *Pathogens and Global Health*, *109*(7), 309–318. doi:10.1179/2047773215Y.0000000030 PubMed

Raffi, M., Mehrwan, S., Bhatti, T. M., Akhter, J. I., Hameed, A., Yawar, W., & Hasan, M. M. (2010). Investigations into the antibacterial behavior of copper nanoparticles against Escherichia coli. *Annals of Microbiology*, *60*(1), 75–80. doi:10.1007/s13213-010-0015-6

Raja, M., Shanmugharaj, A. M., & Hun, R. S. (2010). An efficient growth of silver and copper nanoparticles on multiwalled carbon nanotube with enhanced antimicrobial activity. *Journal of Biomedical Materials Research*, *96B*, 119–126. PubMed

Ranjani, S., Shariq Ahmed, M., Ruckmani, K., & Hemalatha, S. (2019b). Green nanocolloids control Multi drug resistant pathogenic bacteria. Journal of Cluster Science. Advance online publication. doi:10.1007/s10876-019-01694-6

Ranjani, S., Tamanna, K., & Hemalatha, S. (2019a) Triphala green nano colloids: Synthesis, characterization and screening biomarkers. J. App. Nano Sci.

Ravishankar, R. V., & Jamuna, B. A. (2011). Nanoparticles and their potential application as antimicrobials. Science against microbial pathogens: communicating current research and technological advances, 197-209.

Raygada, J. L., & Levine, D. P. (2009). Methicillin-Resistant Staphylococcus aureus: A Growing Risk in the Hospital and in the Community. *American Health & Drug Benefits*, *2*, 86–95. PubMed

Rohini, B., Tahira, A., Waseem, M., Khan, J., Kashif, M., & Hemalatha, S. (2019). AgNPs from Nigella sativa control breast cancer: An In vitro study. *Journal of Environmental Pathology, Toxicology and Oncology*, *38*(2), 185–194. doi:10.1615/JEnvironPatholToxicolOncol.2019027318 PubMed

Ruparelia, J. P., Chatterjee, A. K., Duttagupta, S. P., & Mukherji, S. (2008). Strain specificity in antimicrobial activity of silver and copper nanoparticles. *Acta Biomaterialia*, *4*(3), 707–716. doi:10.1016/j.actbio.2007.11.006 PubMed

Sah, S. K., & Hemalatha, S. (2015). Extended spectrum Beta lactamase (ESBL) Mechanism of antibiotic resistance and Epidemiology. *International Journal of Pharm Tech Research*, *7*, 303–309.

Sah, S. K., Rasool, U., & Hemalatha, S. (2019a) Andrographis paniculata extract inhibit growth, biofilm formation in multidrug resistant strains of Klebsiella pneumonia. J. Trad. Compl. Med.

Sah, S. K., Rasool, U., & Hemalatha, S. (2019b). Efficacy of Andrographis paniculata against AmpC producing multi drug resistant E. coli Biocat. *Agricultural Biotechnology*, *21*, 101139. doi:10.1016/j.bcab.2019.101139

Salavati, M., Davar, F., & Mir, N. (2008). Synthesis and characterization of metallic copper nanoparticles via thermal decomposition. *Polyhedron*, *27*(17), 3514–3518. doi:10.1016/j.poly.2008.08.020

Sathish Kumar, S., Fazeela, S. M., Sah, S. K., Kulandaisamy, S., Mahasampath Gowri, S., Hemalatha, S., & Balasubramanian, K. K. (2018). Synthesis, characterization, in vitro, and in silico studies of 4-(20-hydroxybenzoyl) and 4-(20-hydroxynaphthoyl)- thiabenzene-1-methyl-1-oxides. *Synthetic Communications*, *48*(5), 553–560. doi:10.1080/00397911.2017.1411950

Schaffer, J. N., & Pearson, M. M. (2015). Proteus mirabilis and Urinary Tract Infections. *Microbiology Spectrum*, *3*(5), UTI-0017–UTI-2013. doi:10.1128/microbiolspec.UTI-0017-2013 PubMed

Schauñer, K., Nowak, K., Düx, A., Semmler, T., Villa, L., Kourouma, L., Bangoura, K., Wieler, L. H., Leendertz, F. H., & Guenther, S. (2018). Clinically Relevant ESBL-Producing K. pneumoniae ST307 and E. coli ST38 in an Urban West African Rat Population. Frontiers in Microbiology, 9, 150. doi:10.3389/fmicb.2018.00150 PubMed

Selvaraj, V., & Alagar, M. (2007). Analytical detection and biological assay of antileukemic drug 5-fluorouracil using gold nanoparticles as probe. *International Journal of Pharmaceutics*, *337*(1-2), 275–281. doi:10.1016/j.ijpharm.2006.12.027 PubMed

Sergio, N., & García, H. (2016). Nanoparticles for Catalysis. *Ed. Thomas Nann. Nanomaterials*, *6*, 123.

Shaikh, S., Fatima, J., Shakil, S., Rizvi, S. M. D., & Kamal, M. A. (2015). Antibiotic resistance and extended spectrum beta-lactamases: Types, epidemiology and treatment. *Saudi Journal of Biological Sciences*, *22*(1), 90–101. doi:10.1016/j.sjbs.2014.08.002 PubMed

Shan, G., Xing, J., Zhang, H., & Liu, H. (2005). Biodesulfurization of Dibenzothiophene by Microbial Cells Coated with Magnetite Nanoparticles. *Applied and Environmental Microbiology*, *71*(8), 4497–4502. doi:10.1128/AEM.71.8.4497-4502.2005 PubMed

Shanmugasundaram, P., Saroj, S. K., Ubaid, R., Mahasampath Gowri, S., Easwaramoorthy, D., & Hemalatha, S. (2019). *Novel curcumin analogs act as antagonists to control nosocomial infection causing Pseudomonas aeruginosa*. Biocat. Agri. Biotech., doi:10.1016/j.bcab.2019

Shariq, A. M., Ranjani, S., Tahira, A., Waseem, M., Khan, J., Kashif, M., & Hemalatha, S. (2019). Biogenic AgNps synthesized via endophytic bacteria and its biological applications. *Environmental Science and Pollution Research International*, *26*(26), 26939–26946. Advance online publication. doi:10.1007/s11356-019-05869-6 PubMed

Shirin, M., Asghar, E., & Somayeh, H.-N. (2018). Copper Nanoparticles as Antibacterial Agents. *Journal of Molecular Pharmaceutics & Organic Process Research*, *6*, 140.

Silva, S., Negri, M., Henriques, M., Oliveira, R., Williams, D. W., & Azeredo, J. (2012). Candida glabrata, Candida parapsilosis and Candida tropicalis: Biology, epidemiology, pathogenicity and antifungal resistance. *FEMS Microbiology Reviews*, *36*(2), 288–305. doi:10.1111/j.1574-6976.2011.00278.x PubMed

Singh, V. K., Tuladhar, R., & Chaudhary, M. K. (2015). Beta Lactamase Producing Escherichia coli, Klebsiella pneumoniae and Methicillin Resistant Staphylococcus aureus among Uropathogens. *Nepal Journal of Science and Technology*, *16*(1), 105–112. doi:10.3126/njst.v16i1.14364

Sivaraj, R., Rahman, P. K., Rajiv, P., Narendhran, S., & Venckatesh, R. (2014). Biosynthesis and characterization of Acalypha indica mediated copper oxide nanoparticles and evaluation of its antimicrobial and anticancer activity. *Spectrochimica Acta. Part A: Molecular and Biomolecular Spectroscopy*, *129*, 255–258. doi:10.1016/j.saa.2014.03.027 PubMed

Slavin, Y. N., Asnis, J., Häfeli, U. O., & Bach, H. (2017). Metal nanoparticles: Understanding the mechanisms behind antibacterial activity. *Journal of Nanobiotechnology*, *15*(1), 65. doi:10.1186/s12951-017-0308-z PubMed

Sonohara, R., Muramatsu, N., Ohshima, H., & Kondo, T. (1995). Difference in surface properties between Escherichia coli and Staphylococcus aureus as revealed by electrophoretic mobility measurements. *Biophysical Chemistry*, *55*(3), 273–277. doi:10.1016/0301-4622(95)00004-H PubMed

Sood, S., Kapil, A., Das, B., Jain, Y., & Kabra, S. K. (1999). Re-emergence of chloramphenicol-sensitive Salmonella typhi. *Lancet*, *353*(9160), 1241–1242. doi:10.1016/S0140-6736(99)00637-6 PubMed

Subhankari, I., & Nayak, P. (2013). Antimicrobial Activity of Copper Nanoparticles Synthesised by Ginger (Zingiber officinale) Extract. *World J Nano Sci Technol*, *2*, 10–13.

Sun, J. B., Duan, J. H., Dai, S. L., Ren, J., Zhang, Y. D., Tian, J. S., & Li, Y. (2007). In vitro and in vivo antitumor effects of doxorubicin loaded with bacterial magnetosomes (DBMs) on H22 cells: The magnetic bionanoparticles as drug carriers. *Cancer Letters*, *258*(1), 109–117. doi:10.1016/j.canlet.2007.08.018 PubMed

Tahira, A., & Hemalatha, S. (2019b). Mycosilver Nanoparticles: Synthesis, characterization and screening the efficacy against plant pathogenic fungi. *BioNanoScience.*

Tahira, A., Khan, M. S., & Hemalatha, S. (2017). Extraction of Flavonoids from various Parts of Couroupita guianensis and Its Efficacy Against Pathogenic Bacteria. *Asian Journal of Pharmaceutical and Clinical Research*, *10*, 4.

Tahira, A., Khan, M. S., & Hemalatha, S. (2018a). Novel Silver Nanoparticles Synthesized from Anthers of Couroupita guianensis Abul. Control Growth and Biofilm Formation in Human Pathogenic Bacteria. *Nano Biomedicine and Engineering*, *10*, 3.

Tahira, A., Khan, M. S., & Hemalatha, S. (2018b). A facile and rapid method for green synthesis of Silver Myco nanoparticles from endophytic fungi. *International Journal of Nanodimension*, *9*, 435–441.

Tahira, A., Priya, S., Sah, S. K., Khan, M. S., & Hemalatha, S. (2019c). Ta-AgNps are potential antimicrobial resistance breakers. J. Nano. Str.

Tahira, A., Vabeiryureilai, M., & Senthil Kumar, N. (2019a). Fungal-mediated synthesis of pharmaceutically active silver nanoparticles and anticancer property against A549 cells through apoptosis Environ. Sci. Pol. Res., 26(13), 13649–13657. doi:10.1007/s11356-019-04718-w

Theivasanthi, T., & Alagar, M. (2011). Studies of Copper Nanoparticles Effects on Micro-organisms. *Annals of Biological Research*, *2*, 368–373.

Ubaid, R., & Hemalatha, S. (2017). Marine endophytic actinomycetes assisted synthesis of copper nanoparticles (CuNPs): Characterization and antibacterial efficacy against human pathogens. *Materials Letters*, *194*, 176–180. doi:10.1016/j.matlet.2017.02.055

Ubaid, R., & Hemalatha, S. (2019). Effect of biosynthesized copper nanoparticles (Cunps) on growth and biofilm formation in fluconazole resistant Candida albicans. *Journal of Microbiology, Biotechnology and Food Sciences*, *9*, 21–24.

Ubaid, R., Priya, S., Afsana, P., Sah, S. K., & Hemalatha, S. (2018b). Efficacy of Andrographis paniculata against extended spectrum β-lactamase (ESBL) producing E. coli BMC comp. *Alt. Med.*, *18*, 244.

Ubaid, R., Sah, S. K., & Hemalatha, S. (2018a). Growth inhibitory effect of oven dried copper nanoparticles (cunps) on drug resistant clinical isolates Ira. *Journal of Marine Science and Engineering*, *15*, 12–20.

Uppuluri, P., Pierce, C. G., & López-Ribot, J. L. (2009). Candida albicans Biofilm Formation and its Clinical Consequences. *Future Microbiology*, *4*(10), 1235–1237. doi:10.2217/fmb.09.85 PubMed

Usman, M. S., El Zowalaty, M. E., Shameli, K., Zainuddin, N., Salama, M., & Ibrahim, N. A. (2013). Synthesis, characterization, and antimicrobial properties of copper nanoparticles. *International Journal of Nanomedicine*, *8*, 4467–4479. PubMed

Volkmar, W., Pettinger, T. K., & Murdock, N. (2014). Nanopharmaceuticals (part 1): Products on the Market. *International Journal of Nanomedicine*, *9*, 4357–4373. PubMed

Wang, L., Hu, C., & Shao, L. (2017). The antimicrobial activity of nanoparticles: Present situation and prospects for the future. *International Journal of Nanomedicine*, *12*, 1227–1249. doi:10.2147/IJN.S121956 PubMed

Xiang, L., Wei, J., Jianbo, S., Guili, W., Feng, G., & Ying, L. (2007). Puriÿed and sterilized magnetosomes from Magnetospirillum gryphiswaldense MSR-1 were not toxic to mouse ÿbroblasts in vitro. *Letters in Applied Microbiology*, *45*(1), 75–81. doi:10.1111/j.1472-765X.2007.02143.x PubMed

Xiao, Y., Ju, H.-X., & Chen, H.-Y. (1999). Hydrogen peroxide sensor based on horseradish peroxidase-labeled Au colloids immobilized on gold electrode surface by cysteamine monolayer. *Analytica Chimica Acta*, *391*(1), 73–82. doi:10.1016/S0003-2670(99)00196-8

Xie, J., Chen, K., & Chen, X. (2009). Production, modiÿcation and bio-applications of magnetic nanoparticles gestated by magnetotactic bacteria. *Nano Research*, *2*(4), 261–278. doi:10.1007/s12274-009-9025-8 PubMed

Xiliang, L., Aoife, M., Anthony, J. K., & Malcolm, S. (2011). Nanoparticles in Electrochemical Biosensors. In H. S. Nalwa (Ed.), *Encyclopedia of Nanoscience and Nanotechnology* (pp. 473–501). Academic Press.

Zarei, M., Jamnejad, A., & Elahe, K. (2014). Antibacterial Effect of Silver Nanoparticles Against Four Foodborne Pathogens. *Jundishapur Journal of Microbiology*, *7*(1), e8720. doi:10.5812/jjm.8720 PubMed

Zawrah, M. F., & Sherein, I. A. E.-M. (2011). Antimicrobial Activities of Gold Nanoparticles against Major Foodborne Pathogens. *Life Science Journal*, *8*, 37–44.

Zhuo, Y., Yuan, R., Chai, Y., Tang, D., Zhang, Y., Wang, N., Li, X., & Zhu, Q. (2005). A reagentless amperometric immunosensor based on gold nanoparticles/thionine/Nafion-membrane-modified gold electrode for determination of α-1-fetoprotein. *Electrochemistry Communications*, *7*(4), 355–360. doi:10.1016/j.elecom.2005.02.001

Chapter 19
Toxicity and Risk Assessment of Nanomaterials

Gulzar Ahmed Rather
https://orcid.org/0000-0001-7513-8109
Sathyabama Institute of Science and Technology (Deemed), Chennai, India

Mir Zahoor Gul
https://orcid.org/0000-0002-1088-4255
Department of Biochemistry, Osmania University, India

Muzafar Riyaz
https://orcid.org/0000-0001-9372-681X
Entomology Research Institute, Loyola College, Chennai, India

Arghya Chakravorty
https://orcid.org/0000-0002-7522-8304
Vellore Institute of Technology, India

Mohd Hashim Khan
B. S. Abdur Rahman Crescent Institute of Science and Technology, India

Anima Nanda
Sathyabama Institute of Science and Technology (Deemed), Chennai, India

Mohd Yasin Bhat
Department of Plant Sciences, School of Life Sciences, University of Hyderabad, India

ABSTRACT

The development and advancement of nanotechnology has led to widespread use of nanomaterials (NMs) in a wide variety of fields, including the environment, agriculture, biomedicine, industry, and human health. Nanomaterials may interfere with biological systems due to their smaller size and increased surface area ratios in terms of volume which may raise unanticipated toxicological apprehensions. The minute nature of the particles may cross biological margins and induce toxicity in vital organs like the human brain and placenta. Moreover, environmental exposures to NMs are inevitable, which have become crucial parts of our daily routine lives and consequently, search for the nanotoxicity is gaining attention to lessen or get rid of the toxic impacts of the NMs. The safety concerns and risk assessment of NMs have been raised significantly in research community circles, industries, and regulatory bodies. The information in the concerned section, obtained from current literature and toxicological effects of NMs research are assorted.

DOI: 10.4018/978-1-7998-5049-6.ch019

INTRODUCTION

Every technology has two sides like that of a coin, one is the substantial beneficial roles and on the other hand, it gives birth to a few limitations as well as negatives (Khan et al., 2020). The nanoscience and nanotechnology deal with nanometers (nm) and the size of typical potent nanomaterials (NMs) is 100nm (Chakravorty et al., 2020a). Nanomaterials of all 0D, 1D, 2D, and 3D dimensions are mainly two types, i.e. carbon and metal NMs (Chakravorty et al., 2020a). In last few decades several scientific reports have been reported that all kinds of NMs as potent antimicrobial agents (Chakravorty et al., 2020b), drug delivery agent (Raghavan et al., 2020; Deb et al., 2017), anticancer agent (Deb et al., 2018), cosmetics (Ahmad et al., 2019). The same material has been also enhanced textile industry as processing agent (Singh et al., 2020), agriculture industry as a biofertilizer (Boutchuen et al., 2019), the electrochemical industry mainly for batter energy storage device (Velappan et al., 2020); this material dramatically changed the current landscape of phytoremediation (Das et al., 2015; Das et al., 2013) and so many scenarios just because of their unique physical and chemical nature (Paul et al., 2020). However, there is a huge chance to get exposed to the producers, environment, workers and consumers during the production and preparation of NM (Savolainen et al., 2010; Valsami et al., 2015). While NM is increasingly work-related, as seen as indirect means by the range of nano-enabled products available in the market, as stated. In 2004, Germany registered 20,000 to 114,000 workers in 450 different nanotechnology companies (Pietroiusti et al., 2018). Everywhere the results of these kinds of exposure cannot be evaluated or predicted with the help of the current knowledge, because of fewer identification characteristics on the molecular level and hence any biological system may undergo disturbance. To overcome the problem some novel attributes need to be evolved to predict nanoscale materials (Kinaret et al., 2017). Across the globe several scientific projects are underway to develop such kind of predictions tools (Pietroiusti et al., 2018).

Though it has been reported that everywhere the toxicity of nanobiomaterials (NBMs) are not size dependent, sometimes larger NMs with similar chemical properties (Catalon et al., 2016; Rossi et al., 2010; Palomäkiet al., 2011) may refer to the same and the target specificity on the biological system can be also a reason (Kreyling et al., et al., 2009). In this case, a significant problem is the biocorona shaped to envelop each nanoparticle and a bigger particle until it enters the biological system. However, biocorona formation is expected to have a greater effect on the properties of nanoparticles, among others owing to the relatively larger increase in particle dimensions relative to larger particles. Therefore, the lipid and protein coating mounted on the surface of specific nanoparticles significantly influences measurement, dramatically affecting cellular and molecular targets that particles can meet (Monopoli et al., 2012; Walkey et al., 2014; Miclăuş et al., 2016).Another big obstacle in the evaluation of NBM risks for laboratory higher animals, humans (for clinical trial purpose) and environmental ecosystems is that in a particulate state, NBM 's activity varies significantly from that of typical soluble materials and have an impact on not only the kinetics of NBM in biological settings but also in the certain potential risk (Savolainen et al., 2013).Nanomaterials' properties pose major difficulties in determining danger through the lungs, but also by other exposure paths. Nevertheless, in the lungs, where nanoparticles enter the body more easily, they are often quickly filled by biomolecules making kinetic actions and effects more difficult to determine. The airways' unique features, therefore, add uncertainty to NBM's inhalation path hazard evaluation. Generally, determining NBM consequences is difficult as correlations with adverse effects with NBM properties (physiochemical) are not well known. Oral exposure to NBM might not be essential for human health; however, the significance of this problem is expected to expand as the use of NBM, particularly in food products, becomes more common. Reliable risk and hazard evaluation approaches can be rapidly established in response to potential demands (Pietroiusti et al., 2018). The well over 1,600 "nano-enabled" goods in industry, all the employees needed to do so (Pietroiusti et al., 2018) and, according to new projections, 6 million employees are theoretically subjected to NMs by 2020 (Roco et al., 2011). As the toxicity of NMs different metal, metal oxide and carbon materials can cause different pulmonary ailments including inflammations and fibrosis, granuloma (Gosens et al., 2016; Guo et al., 2019; Mohanta et al., 2019). Specifically, carbon nanotubes can lead to deposition of collagens in the subpleural region of lungs, which may result in chronic pulmonary disorders (Sun et al., 2017; Türk et al., 2018; Ding et al., 2019).

In this regard, workplace-related concerns should primarily be discussed in this analysis as the most important sensitivity to nanomaterials exists in workplaces. Often, customer and environmental sensitivity is even lesser.

BASIS FOR TOXICITY OF NANOMATERIALS

Several physicochemical properties act as intermediaries of NMs toxicity. The foremost and significant physical and chemical properties which determine NMs toxicity are size, shape, chemical composition and concentration, surface chemistry and charge, agglomeration/aggregation, medium and storage time (Fig 1). In the following section, the above-mentioned factors are broadly described.

Figure 1. Physicochemical factors determining NP Toxicity

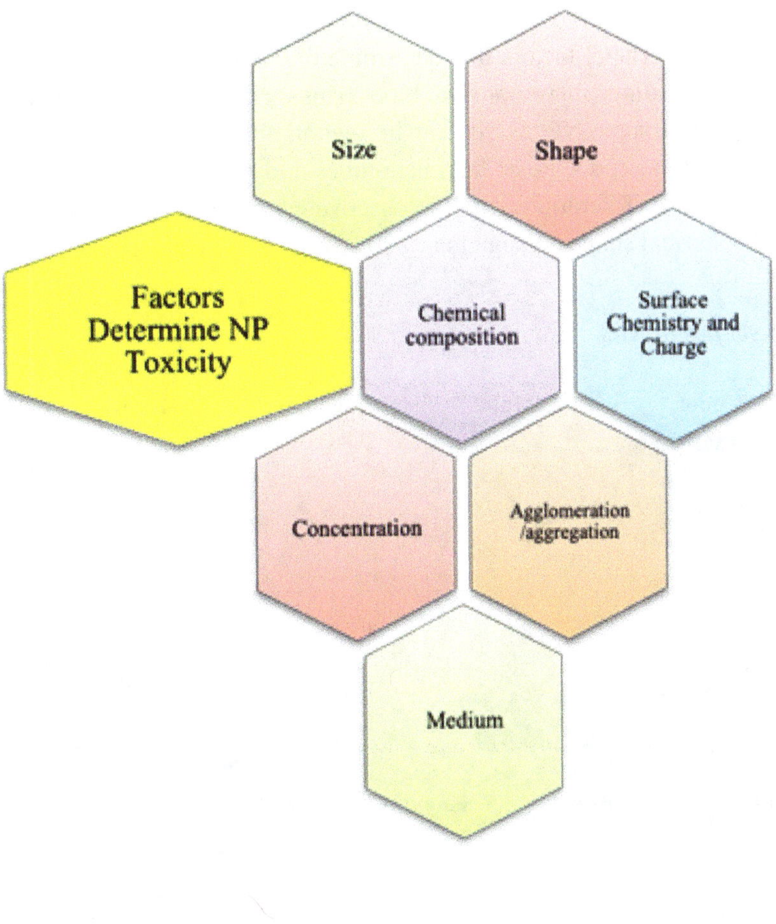

SIZE

Size plays a crucial role in determining the toxicity of nanoparticles (NPs). The smaller NPs can get an easy passage through the physiological barriers due to higher uptake than large NPs and can spread through all organs where damage is certain by increased cytotoxicity (Nabeshi et al., 2010). After getting the passage inside the body through the cell membrane, the NPs with minute size can penetrate through different organs and interact with different cellular organelles like mitochondria, nucleus followed by a surge in toxicity and these minute sized NPs can also avoid excretion by accumulating themselves in organs like liver and spleen (Saifi et al., 2018). The NPs with sizes smaller than 100 nm can also dodge the defense mechanism of the body because smaller NPs are difficult to be phagocytized compared to the larger NPs (Blandino et al., 2012). The decrease in size of a nanoparticle up to nanoscale alters the physical, chemical, and structural properties, which fluctuates the reactivity of NPs results in the amelioration of their toxicity potential. The experiments on animal models have revealed that smaller NPs are the root cause of inflammatory responses in rat lungs in comparison to that of large NPs (Oberdorster et al., 1994) and often causes more oxidation and DNA damage due to large surface area of a given volume of smaller NPs than an equal volume of larger NPs (Buzea et al., 2007). Many *in vivo* studies have also analyzed that with the course of intravenous inoculation, NPs smaller than 10 nm can rapidly spread across all the organs and tissues as compared to the NPs with size ranging (from 50-250 nm) which can accumulate in liver, spleen, and blood (Jong et al., 2008).

SHAPE

The NP shape is one of the crucial factors that determine the toxicity of NPs. The NMs are coming in different forms (Fig.2) including fibers, spheres, tubes, rings, and planes, rods, cubes, helices (Sharifi et al., 2012). The shape-dependent NP toxicity can be seen in different forms like NPs having spherical shape are more susceptible to endocytosis than nanotubes and nanofibers (Champion and Mitragotri, 2006). Studies in human HaCaT keratinocytes have discovered that nanorods are having more cytotoxicity than gold NPs of spherical shape (Wang et al., 2008).

Figure 2. Different shapes of NPs.

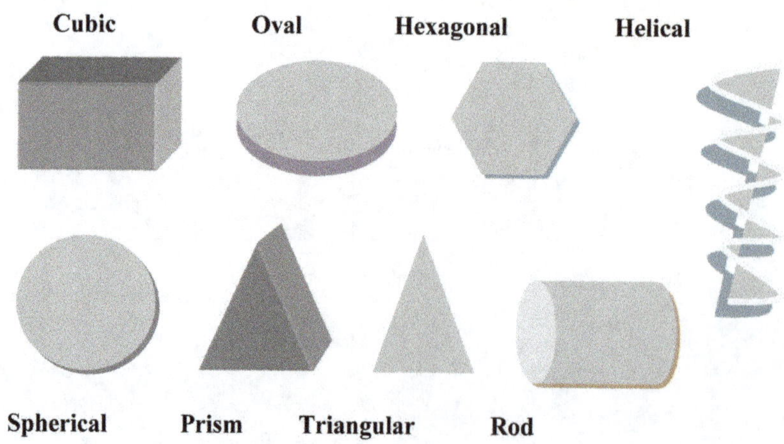

The toxicity potential of single-walled carbon nanotubes (SWCNTs) and multi-walled carbon nanotubes (MWCNTs) have been extensively studied and it was observed SWCNTs are more impactful in obstructing calcium channels particularly in comparison to circular fullerenes. (Park et al., 2003). Comparing both the CNTs it was found that shape had a strong influence in the cytotoxicity of these NPs (Grabinski et al., 2010; Zhang et al., 2015). The CNTs have diameter ranges between 0.4 nm and 100 nm and lengths may vary between some nanometers to centimeters (Dai, 2002). The CNTs tend to aggregate due to their hydrophobic nature that can be very detrimental to cells in a culture medium (Riviere et al, 2005; Cui et al., 2005) and were found to be toxic at adequately high doses of 400 µg/ml on human T cells followed by cell death (Bottini et al., 2006).

SURFACE CHEMISTRY AND CHARGE

The surface chemistry of NMs is one of the major factors that can influence a biological system in response to the contact of NMs and have a large control over the diffusion of NPs inside the body, yet they can be mostly compatible with the immune system (Dobrovolskaia and McNeil, 2007). *In vitro* studies have suggested that a large surface area of NMs leads to an increase in reactivity (Roduner, 2006) and the toxicity of a NMs can be severely affected its surface area. A dose-dependent increase in oxidation and damage to DNA can be seen in the NPs with higher surface area (Donaldson and Stone, 2003).

With the interactions in the biological systems, the surface charge of NPs is one of the key factors in developing the toxicity (Schaeublin et al., 2011; Badawy et al., 2011). The surface charge can transform the shape and size of a NMs by influencing the response of an organism through aggregate or agglomerate development (Hoshino et al., 2004). A study has observed that the surface charge of a NMs can revamp the porousness of the blood-brain barrier (Lockman et al., 2004). In case of positively and negativity charged Gold NPs, the positive NPs are more toxic as absorbed by the cells in great proportions than the negative ones (Hühn et al., 2013) as the positively charged NPs have an improved proportion for opsonization (Alexis et al., 2008).

CHEMICAL COMPOSITION AND CONCENTRATION

Through various studies and experiments, the toxicity of NPs was sturdily based on the physicochemical factors including shape, size, surface chemistry and charge. Chemical composition and concentration are some of the factors that determine the toxicity of a NP. In an experiment on mouse fibroblasts, the effect of 20-nm silicon dioxide (SiO2) and zinc oxide (ZnO) NPs were taken into consideration for comparison between the effects of these NPs. The ZnO NPs showed oxidative stress and SiO2 NPs changed the DNA structure (Yang et al., 2009). The chemical composition is certainly a key factor that determines the toxicity of NMs. The environment conditions such as pH or ionic strength can be a cause of NP degradation. While interacting with the cells, some metal ions can escape from the NP core results in the origin of NP toxicity. Some metal ions can cause high toxicity as most metal ions damaging the cells are already toxic and even in higher concentrations, many beneficial metal ions can even be toxic (Sukhanova et al., 2018). However, at different concentrations, the toxic effect of NPs may vary. As analyzed by many studies, several materials were found to be harmless in comparison with other materials. The variations stood in the aggregation properties of NPs in air and water, followed by a characteristic variation among inhalation, instillation or in vitro experiments (Buzea et al., 2007) Many studies have proved that the higher concentration of NPs would promote particle aggregation which can have toxic effects compared to lower concentrations (Churg et al., 1998; Takenaka et al., 2001; Gurr et al., 2005).

AGGLOMERATION/AGGREGATION

The NPs in small concentrations are having a tendency to spread in the circulatory system and other organs while in case of NPs with higher concentrations may lead to the formation of large aggregates which can cause a lung injury due to the development of higher accumulation (Takenaka et al., 2001) which can take months or years to get eliminated based on the type of NP. However, some NPs might never be eliminated results in negative health effects due to extensive residence times of NPs within the lungs, circulatory system, and organs (Blandino et al., 2012). The size and shape of NPs will change due to agglomeration, therefore, compromise the separation efficiency. Due to the higher surface energy, NPs aggregate easily followed by extreme influence on their toxicity profile including CNTs and metal oxide (Wick et al., 2007; Limbach et al., 2005). Conversely, the toxicity of metal-based NMs is due to the metal ions associated with them (Sabella et al., 2014).

MEDIUM AND STORAGE TIME

The NPs in a medium for stable delivery is very important for their biological dispersal and succeeding action. In vivo exposure of NPs has been found to affect the physiological solutions due to agglomeration. The particle dispersion can also be affected by the medium conditions such as ionic strength and pH (Sharifi et al., 2012). A greater size can be seen in particles of TiO2, ZnO, or carbon black in PBS as compared to that in Water. A study suggested that the behavior of NP systems depends on the medium. The solitary or aggregated NMs may have diverse biological impacts and the solvent media may affect the dermal uptake of these NMs. Media dispersion or wetting agents can also hurt the NMs toxicity. (Sager et al., 2007; Colvin, 2003).

NANOMATERIALS AND THE ENVIRONMENT:

Nanotechnology's amazing potential and future benefits often pose serious risks for the environment and human health. The ubiquitous use of NMs has eventually resulted in their entry into the environment, either as the initial NMs (as manufactured) or more likely as the degradations of societal nano-enabled products, unfortunately, render them potential toxic substances. Those very NMs are theoretically highly non-degradable and can persist in the environment for a longer period. Consequently, they could pose new threats to health and the environment that were not experienced before.

EFFECTS ON AQUATIC ENVIRONMENT:

The water ecosystems are contaminated year after year by a plethora of xenobiotics by different anthropogenic activities. In recent times, the aquatic environment has become more vulnerable to contamination with NMs. The NMs enter the water sources in a variety of ways, including direct entry, atmospheric precipitation, aquifers transfer, and soil runoff. These NMs are highly toxic to aquatic life such as fish, daphnia and even to single-celled organisms (Nowack and Bucheli 2007). Due to their continuous interaction with some of these substances through gill penetration, swallowing, dermal contactor cellular adsorption, etc. these water species are too frequently subjected to NMs toxicity than their terrestrial counterparts. Accumulated NMs in aquatic species pass via the food supply chain and impact all the trophic levels in every ecosystem (Klaine et al., 2008). The physiochemical properties like morphology, aggregation, surface charge, reactivity, etc. and the features of extracellular as well as intracellular environments (lipid adsorption or protein models) can result in a particular position of NMs in the target site and that may establish the toxicity of those NMs (Yue et al. 2017; Selck et al. 2016). Due to the wide surface area, such fine particles may generate reactive oxygen species (ROS) which are quite detrimental and cause significant damage to membranes, proteins, and DNA (Reijnders 2006; Brown et al. 2001). The ROS like $^{-}O_2$, HO_2, $^{\cdot}OH$, and H_2O_2) have been reported to be a well-established nano-toxicity phenomenon whose increased production may result to lipid peroxidation, cytoskeletal disorganization, mitochondrial degradation, protein oxidation, and DNA damage in the tissues of an organism (Yue et al. 2017; Chatel et al., 2018; Fu, et al., 2014; Vale et al., 2016; Rocha et al., 2015). Several other nanoparticles can lead to the production of ROS that further deteriorate the fragile equilibrium between the ability of both the biological system to generate and inactivation of ROS (Arora et al., 2012). Most of the NMs like ZnO, CuO, MgO, TiO_2, Ag, carbon nanotubes, quantum dots, and fullerenes have been recognized to have deleterious effects on aquatic life, even though the frequency and intensity depend on several attributes such as type of NMs, size, charge and exposed species. The silver nanomaterials can actively interact with DNA, triggering crosslinks between DNA-protein and DNA-DNA (Klingelfus et al. 2017). These Ag NMs have also been reported to cause substantial changes in antioxidant markers viz., enhanced lipid peroxidation and thereby, substantially decreased expression of CAT, SOD and GSH (Taju et al. 2014). Further, short-term exposure to CuO NMs in fish resulted in distorted morphology and cellular damage (Gupta et al. 2016). The CuO- NMs have been found to amass in mussel gills and trigger severe effects on biochemical, organelles, cells and tissues resulting in overproduction of ROS and therefore, disrupt internal homeostasis (Gupta et al. 2016). The NMs of cerium dioxide (CeO_2) have been demonstrated to trigger alterations in lysosome membrane stability, phagocytic potential and extracellular reactive oxygen species in *Mytilus galloprovincialis* hemocytes. These NMs often contribute to the downregulation of the immune system as illustrated by the failure of lysosome membrane stability and a significant decline in phagocytosis ability. The widely used carbon-based NMs, such as nC60 also lead to oxidative stress and lipid peroxidation in the brain of fish and significantly reduces glutathione (GSH) enzymes (Oberdorster 2004). Other findings indicated that the exposure to graphene oxide (GO) induced greater levels of malondialdehyde, SOD, CAT, and decreased levels of GST, all highlighting oxidative stress (Chen et al. 2016). Because of their photocatalytic activity, the NMs are ready to interact with many other materials that can even create new noxious substances that indirectly lead to environmental hazards. Carbon nanotubes along with other NMs trigger markers of oxidative stress and thereby induce damage to the DNA. Furthermore, CNTs may bind to macromolecules such as proteins and results in remnant metabolite formation. The exposure of fish to metal ions may alter their physiological roles which include acid-base balancing, respiratory activity, variations in osmoregulation, epithelial edema and mucus secretions. Histological alterations in the epithelium of gills result in the blood loss circulation with the ensuing collapse of cardiovascular and eventually fatality (Shaw and Handy 2011; BrazMota et al. 2018; Shaw and Handy 2011; Souza et al. 2018).

EFFECTS ON SOIL ENVIRONMENT:

As a result of anthropogenic practices like continual release through water and soil decontamination methods, use in agricultural production (pesticides/fertilizers) and accidental releases through air-water, sewage, and sludges, NMs gain entry and therefore accumulate in the soil. They can be highly resistant to degradation and can stay long in the soil (Tourinho et al. 2012; Maurer-Jones et al. 2013; Shandilya et al. 2015; Coll et al. 2016). The concentration of NMs in the soil is higher than in air or water (Klaine et al. 2008). Considering the availability of NMs in soils, the assessment of their effects on soil biodiversity is imperative (Bondarenko et al. 2013). The existence of these NMs discharged into the soil is greatly affected by chemical and physical attributes of NMs and their ability to interact with natural colloidal components and soil characteristics (Boxall et al. 2007). Copious NMs aggregates may be immobilized in smaller pores by staining, filtration, and sedimentation processes. On the other hand, NMs can move farther than their bulkier counterparts (surface runoff) as their unique smaller design (size and shape) enables them to freely migrate between soil particles across the pre spaces.

The main soil characteristics like the presence of organic matter, morphology, pH, etc affect the microbial population of soil, hence play a fundamental role in the toxic implications of NMs (Fierer and Jackson 2006; Simonin and Richaume 2015). The toxicity of these NMs is adversely affecting the plants and microbial communities (bacteria and fungi). Research findings have shown that the NMs are toxic to certain species of bacteria including *Bacillus subtilis, Vibrio fischeri* and *Escherichia coli* (Jiling et al., 2016). In these bacterial species, the toxicity is stimulated by different mechanisms like membrane dysfunctionality, ROS and photocatalytic oxidation associated with a surface coating. NMs of CuO can also destabilize the membrane of *E. coli* by K^+ leakage (Zhao et al., 2013). The plate counts and enzymatic activities of Azobacter, P and K solubilizing bacteria were severely affected by CeO_2 and ZnO(Chai et al., 2015). Nanomaterials such as $TiO_2$and CuO greatly reduces the frequency of functional soil bacteria and enzymatic activity, impacting their population dynamics in inundated agricultural soils (Xu et al., 2015). The toxicity of these NMs may not only directly impact crop quality as well the yield, but also the microbial diversity and their environment (in the endosphere and rhizosphere). Another negative impact on soils induced by NMs is the effect on the rate of self-cleansing and nutrient levels which is the baseline for controlling plant nutrient processes and increasing fertility of the soil (Janvier et al. 2007; Suresh et al. 2013). The earlier studies showed the build-up of toxicity on earthworms due to different nanotubes (Peterson et al. 2008). Other NMs such as $Al_2O_{3,}$ CeO_2, TiO_2 and ZnO are also found to exert adverse effects on the rate of reproduction, growth, altered gene expression and mortality in soil nematodes like *Eisenia fetida* (earthworms) and *Caenorhabditis elegans*(Hu et al., 2010; Lahive et al. 2014; Roh et al., 2009; Lahive et al. 2014). The impact of NMs on the physiological and morphological nature of plants evaluated by different toxicity tests is apparent from numerous several studies (Ma et al., 2015; Faisal et al., 2016). Metal-based NMs were shown to influence the variation in chlorophyll content several crop plants, such as wheat, tomato, green peas (chlorophyll content is^{-}) whereas beans, cucumber, mustard(where chlorophyll content is) (Rastogi et al., 2017). Many studies have shown lipid peroxidation, genotoxic effects, DNA damage upon exposure with NMs (ZnO, NiO, nAg, CeO_2, TiO_2) on crop plants like cucumber, tomato, rice, maize, onions, broad beans, etc. (Pokhrel& Dubey, 2013; Zhao et al., 2013; Patlolla et al., 2012; Faisal et al., 2013; Dimpka et al., 2012; Rico et al., 2013;).

EFFECT OF NANOPARTICLES AND HUMAN HEALTH

Nanomaterials have the same dimensions as protein-like biological molecules. At present, exposure to NMs is growing, and due to uncertainties over their toxicity, serious issues have emerged that these substances pose new risks to human health (Handy and shaw, 2007). The toxic effects of the chemicals/ materials depend on their ability to enter the body, or rather tissues or cells with the human body. All the materials including NMs have three different ways of getting into the human body:

EXPOSURE THROUGH SKIN

The importance of subcutaneous exposure to toxic substances continuously rises (Mackevica and Hansen, 2016). The exposure of human skin to NMs can happen through deliberate as well as unintended ways. Intentional exposures of NMs may occur from the application of cosmetics such as lotions, creams, etc. that contain ZnO and TiO_2 coated nanoparticles. Unintentional exposure occurs directly during the process of synthesis, emission, and recycling NM-based products, environmental pollution, occupational activities (Oberdörster et al., 2005, Zimmer et al., 2002). There are two potential mechanisms of NM adsorptions into the skin: trans-epidermal intracellular pathway or hair follicle diffusion and sweat ducts intercellular trans-epidermal pathway or diffusion through the hair follicle, and sweat ducts (Monteiro-riviere and Inman, 2006). Human skin is an excellent deterrent to NMs or other hazardous materials, but, hair follicles and sweat glands render it vulnerable (Teow et al. 2011). The NMs are typically visualized least via viable skin, as more penetration occurs in hair follicles when the protective layer of the skin is damaged or removed (Mavon et al. 2007). The TiO_2 surface coating can indirectly affect the skin thereby allowing NMs to penetrate. The antimicrobial potential of Ag NPs has made these NMs most widely used in cosmetic products (Miethling-graff et al. 2014). The toxicity of AgNPs is triggered by the initiation of oxidative stress associated with reduced viability, loss of mitochondrial function and apoptosis and cell death (Foldbjerg et al. 2009). Nanomaterials have also been shown to trigger injury-responses that lead to inflammation inside the skin. These materials can denature proteins, unmask epitopes, and even damage DNA and cells (Barlow et al. 2005). NMs administered to the skin can cause photocatalytic activity in the subcutaneous layers, resulting in the generation of ROS, degrading DNA, and upsetting normal cellular functions (Crosera et al., 2009). Quantum dots (QDs) applied on the skin can migrate to local lymph nodes, probably via Langerhans cells and skin macrophages, therefore, increasing possible immunomodulation concerns (Kim et al., 2004). It has been further reported that the NMs such as QDs, SWCNTs, MWCNTs with titanium surface coating have deleterious consequences on epidermal keratinocytes and fibroblasts and are capable to alter their gene as well as protein expression (Haliullin et al., 2015).

EXPOSURE THROUGH DIGESTIVE SYSTEM

Inhalation of NMs is a significant mode of exposure to the human body (Foldbjerg and Autrup, 2013). Nanomaterials that are used directly or indirectly in packaged foods and pharmaceuticals are orally consumed and absorbed by the gastrointestinal tract (GI), through which they gain entry into the lymphatic tissues (Teow et al. 2011). Based on their nature, these NMs are either eliminated from body if unstable, or clumped by physical and chemical changes, resulting in GI tract obstruction and subsequently death (Wang et al., 2006). Once internalized, these NMs interfere with different cellular organs and macromolecules, leading to genotoxic and cell inhibitory and effects which ultimately result in death due to membrane lipid peroxidation, energy depletion, organelle destruction and changes in interconnected signaling pathways (Jin et al., 2007; Bhabra et al., 2009; Pereira et al., 2011). These effects are the result of oxidative stress caused by a disequilibrium among ROS and antioxidant mechanisms.It has been reported that NMs interfere with proteins, either by chaperone-like interaction or by altering protein structure, results in protein misfolding (Wagner et al., 2010). The toxicity caused by the ingested TiO_2 results in impairment of the cellular membrane of the digestive gland through oxidative stress (Valant et al., 2012). Upon entering hepatic circulation, NMs behave differently. They can be hepatoxic or detritus in the biliary system which can block hepatic ducts or cause progressive fibrosis. Furthermore, this can affect the gastrointestinal lining's perviousness and therefore, can lead to ulcers, epithelium weakening, metaplasia or dysplasia, nutrient malabsorption, or extreme causes chronic bleeding.

EXPOSURE THROUGH THE RESPIRATORY TRACT

The respiratory tract is a relevant target for NMs exposure both from an occupational and consumer perspective. The pathogenic implications and pathology of inhaled NMs are attracting the interest of researches over the years. Accumulation of NMs in the pulmonary system varies greatly depending on the granulometry of NMs and their airborne performance (Ma et al. 2015). The particle granulometry has a major influence on the respiratory deposition location (Ma et al., 2015). In many NMs synthesis methods, granulometry can also differ considerably corresponding stages of synthesis. Such variations in the distribution of nanoparticles in the lungs might have significant implications for the health impacts of NMs inhaled and the process of removal involved (Zhang et al. 2005; Ma et al., 2015). Diffusion plays a prominent role in the penetration of NMs deeper into the respiratory system as they hit with air molecules due to displacement (Borm et al. 2006). In case of inhalation, the toxic effects of CNTs are more fatal, give rise to severe inflammatory response as compared to oral or dermal exposures (Foldvari and Bagonluri 2008). Once inhaled, these CNTs encounter proteins and other cell components and retain their structure or metabolize according to their physicochemical attributes, and then translocate through blood circulation to different body parts. (Pichardo et al. 2012; Sharifi et al. 2012). Also, NMs could even induce nasal pathology. In such cases, nasal epithelial cells are hurt, and mucous membranes start rupturing which leads to reduced odor sensation and nasal humidification. The intensity of effects becomes more pronounced as the scale of NMs decreases. Since the particles are so small, eliminating these particles gets difficult and eventually eroding the nasal air passage defense. Due to the sensitive nature of NMs, they trigger damage to epithelial cells and macrophages lead to inflammation of lungs (Donaldson and MacNee, 1998).

RISK ASSESSMENT OF NANOMATERIALS

Nanotechnology as a cutting-edge technology for the production and execution of new NMs and has prompted the need for a lot of risk assessment. Much attention has been paid in recent years to how certain objects can be measured alongside each other. It is believed that a multifold boom in particular research publications and patents follow the growth of nanotechnology. The current application of classical and modern NMs is explored in the sense of regulatory requirements and standardization of chemicals, and other consumer products. Nano based industries specifically emphasize the best problem- solving capability and cost-effective theme, rather restricting the introduction of novel products for the need of time. However, nano-based consumer goods are highly demanded in the medical field for health and strength, home and garden, followed by food and beverage and coatings of articles (Vance et al. 2015). Because of their modest size and large volume area, the physicochemical characteristics of NMs attract further attention owing to their prospect in the biological applications. These NMs are have considered having tremendous biomedical potential as novel drug transporters with increased cellular engagement and better permeation ability through biological membranes. Nevertheless, these innovative features NMs, elevate the concept of toxicity as a side effect for human life and their concerned properties. Nowadays, a plethora of reports on NMs is directed towards adjusting the toxic effect of the NMs to explore the biological side effects to make them compatible and suitable for use in daily routine life. Setting a high standard of characterization are the foundation elements of NM risk assessment. Reporting of continuing debates in the field of scientific communities and the relevant information of the given concept will provide an accurate basis for the risk assessment of impending toxicities.

BASIC PRINCIPLES OF RISK ASSESSMENT

Risk assessment is a process or activity ineffective administration of workplace health and safety to evaluate the exposure, hazard, and dose-response data to distinguish risk in human resources. Risk estimations deliver information and data to maintain the establishment of occupational exposure limits, which should be conducted particularly when changes in tools, drafts or procedures occur in a work area.

Fundamentally, the process of risk assessment involves the six steps as follows:

Step 1: Identification of hazardous material (source of risk or danger)
Step 2: Analysis of consequences (property damage, potential injury)
Step 3: Assessment of risk (probability/chance, the magnitude of risk and frequency)
Step 4: Determination of the mode of action (mitigating risk or completely removal)
Step 5: Application of controls (remodel, removal, new techniques)
Step 6: Appraisal of Controls (audit or proportional evaluation)

The comprehensive bookmarks of risk assessment paces (NRC, 2009) include the following:

1. PROBLEM FORMULATION

It is a preliminary assessment of the essence of the threat, the exposure management alternatives and the data required to decide among several viable options.

2. RISK ASSESSMENT

Risk evaluation makes it more effective to define, at the beginning of the process, the most important knowledge necessary for decision-making in which the subdivisions are classified as:

1. **Hazard assessment:** Hazard evaluation is an assessment of the extent and magnitude of biological impacts usually associated with toxicology studies.
2. **Exposure assessment:** Exposure assessment involves measuring or estimating the exposures of workers, by the task or by complete shift.
3. **Dose-response assessment:** Dose-response assessment provides information on the approximate dosage or model (e.g. in animal studies) and the biological responses considered significant to human health.
4. **Risk characterization:** Risk characterization sums up the findings from exposure, hazard and dose-response evaluations to provide figures or data for proper decision-making and risk communication in risk management.

RISK MANAGEMENT AND RISK COMMUNICATION

Risk management focuses mainly on identifying and implementing appropriate risk control mechanisms and reducing public exposure to hazardous substances (Marchant et al., 2008). Risk communication is an open method of sharing thoughts or information and a continuing part of the exercise of risk analysis. The kind of communication apprises all the stakeholders of the practice at every step of the risk assessment which helps them to understand the risk.

RISK ASSESSMENT PERSPECTIVE

Life Cycle of Nanomaterials

It is known that during their life cycle, the physicochemical properties of NMs that are applicable for their possible risk are subject to change (Grieger et al., 2012). The life cycle of all existing materials like chemicals, NMs are being produced, used, and disposed of are termed as cycled but assessing the risk of certain chemicals is a complex route. Many NMs are used in numerous products to strengthen or make them more effective to the target but the life cycle consideration even more challenging to destroy or complete removal of the part of NMs. For instant, some NMs are incorporated into a milieu of plastic which in turn may impact the possibility of exposure and therefore the potential risk. Some NMs end up in the environment depending on their structure and properties by dissolving, transformation, combining with the relevant components, and therefore ultimately can sediment or be transported further away from the healed portion. A risk assessment of NMs throughout its life cycle should also be able to distinguish between risks to different populations such as workers and consumers. The risk assessment of NMs covers an estimation of their indirect exposure through the environment such as from groundwater or crop production following uptake of substances through the soil.

INHALATION OF NANOMATERIALS

The most pertinent route of exposure to NMs is breathing them in through respiration in animal bodies. Sometimes NMs can reach deeper and deeper via the lungs, where uptake into the bloodstream is more likely and particles may form a lining on lungs fluid or they may cross the pulmonary epithelium to reach the underlying tissues or may flow into the systematic circulation. Whether sometimes when the NMs are inhaled, they are entrapped into the mucous cells and effective clearance processes may be started by the macrophage in the deep lungs. If the particles are taken up, the process of translocation started to transport the particles into the other organs of the body through the blood flow. The translocation of the particle depends on their physical properties and may influence by the size, volume, or any surface treatments made with the particles. Some NMs also irritate at their point of entry by inhalation. Irritation is a sign that the lungs' defense system works, resulting in activation of the body's natural clearing process, causing the lining of the lung to create phlegm that helps push particles up and out of the lungs. The inhalation of nanoparticles through air pollution pointed out a link for illnesses such as chronic bronchitis, asthma, emphysema, lung cancer, cardiovascular problems, and heart associated diseases (Landsiedel et al., 2014). Primarily lung diseases (e.g. pneumoconiosis) are an especially important issue to consider when assessing risks for workers such as miners, bakers and diesel exhaust fume inhalers which may cause acute and chronic health issues.

DERMAL PENETRATION

The absorption of the NMs through the skin is also possible when it comes to the existing contact of dermal pores. Due to smaller size, they may penetrate the skin in a greater range than larger particles. Ultimately the size of the nanoparticles at most priorities for absorption but some other parameters such as the composition, solubility and surface chemistry are also considered in dermal toxicity causing factors (Landsiedel et al., 2014).

ENVIRONMENTAL EFFECTS

Nanomaterials can affect organisms in the surroundings and their physical compartments such as ground, atmosphere, soil surface, sediments, and groundwater. The assessment of the environmental impact of NMs is challenging because nanoscale particles can interact with the surrounding compartments (Taghavi et al., 2013). Both the direct and indirect effects have to be considered to assess the true environmental impact of NMs. Indirect effects are mainly related to the transformation and interactions of nanoparticles with the environment, and further to the transformed state of organisms. It is difficult to differentiate the indirect effects of the nanoparticles from direct effects. Indirect organism toxicity means modifications of the particles which do not occur in their original makeup. Nanomaterials can be dispersed in the aqueous medium, degraded into another type of particle with texture differences or attached to other substances creating difficulties in quantifying or measuring the relevant concentrations of NMs to which organisms are exposed in the environment. Due to this phenomenon, the transformed NMs may have a more or less severe impact than the original NMs by changing their existence in a natural environment.

SUMMARY AND CONCLUSION

In this chapter, we tried to lay out an encyclopedic perspective of the toxic nature of NMs along with several other details. There are several research works present which give a crystal-clear view of NMs cytotoxicity and its collision in the existence of human. The physicochemical characteristic is the main responsible factor for the toxic nature of NMs. The figure of toxicity will change if there is any remodeling found in physicochemical characteristics. There is shortfall of interconnection in various nanomaterials between *in vivo, in vitro,* and *in-silico* statistics. A database of toxicity is required due to the increase in the toxic nature of NMs in the recent era. A combination of different toxicity-indicating techniques is required to decrease the toxic nature of NMs and it is also utilized to assume toxicity of newly synthesized NMs.

REFERENCES

Ahmad, I., Ahmad, A., Iftekhar, S., Khalid, S., Aftab, A., & Raza, S. A. (2019). Role of Nanoparticles in Cosmetics Industries. In *Biological Synthesis of Nanoparticles and Their Applications* (pp. 173–204). CRC Press. doi:10.1201/9780429265235-14

Alexis, F., Pridgen, E., Molnar, L. K., & Farokhzad, O. C. (2008). Factors affecting the clearance and biodistribution of polymeric nanoparticles. *Molecular Pharmaceutics*, *5*(4), 505–515. doi:10.1021/mp800051m PMID:18672949

Arora, S., Rajwade, J. M., & Paknikar, K. M. (2012). Nanotoxicology and in vitro studies: The need of the hour. *Toxicology and Applied Pharmacology*, *258*(2), 151–165. doi:10.1016/j.taap.2011.11.010 PMID:22178382

Barlow, P. G., Donaldson, K., MacCallum, J., Clouter, A., & Stone, V. (2005). Serum exposed tonanoparticle carbon black displays increased potential to induce macrophage migration. *Toxicology Letters*, *155*(3), 397–401. doi:10.1016/j.toxlet.2004.11.006 PMID:15649623

Bhabra, G., Sood, A., Fisher, B., Cartwright, L., Saunders, M., Evans, W. H., Surprenant, A., Lopez-Castejon, G., Mann, S., Davis, S. A., Hails, L. A., Ingham, E., Verkade, P., Lane, J., Heesom, K., Newson, R., & Case, C. P. (2009). Nanoparticles can cause DNA damage across a cellular barrier. *Nature Nanotechnology*, *4*(12), 876–883. doi:10.1038/nnano.2009.313 PMID:19893513

Bondarenko, O., Juganson, K., Ivask, A., Kasemets, K., Mortimer, M., & Kahru, A. (2013). Toxicity of Ag, CuO and ZnO nanoparticles to selected environmentally relevant test organisms and mammalian cells in vitro: A critical review. *Archives of Toxicology*, *87*(7), 1181–1200. doi:10.100700204-013-1079-4 PMID:23728526

Borm, P. J., Robbins, D., Haubold, S., Kuhlbusch, T., Fissan, H., Donaldson, K., Schins, R., Stone, V., Kreyling, W., Lademann, J., Krutmann, J., Warheit, D., & Oberdorster, E. (2006). The potential risks of nanomaterials: A review carried out for ECETOC. *Particle and Fibre Toxicology*, *3*(1), 11. doi:10.1186/1743-8977-3-11 PMID:16907977

Bottini, M., Bruckner, S., Nika, K., Bottini, N., Bellucci, S., Magrini, A., & Mustelin, T. (2006). Multi-walled carbon nanotubes induce T lymphocyte apoptosis. *Toxicology Letters*, *160*(2), 121–126. doi:10.1016/j.toxlet.2005.06.020 PMID:16125885

Boutchuen, A., Zimmerman, D., Aich, N., Masud, A. M., Arabshahi, A., & Palchoudhury, S. (2019). Increased plant growth with hematite nanoparticle fertilizer drop and determining nanoparticle uptake in plants using multimodal approach. *Journal of Nanomaterials*, *2019*, 2019. doi:10.1155/2019/6890572

Boxall, A. B., Tiede, K., & Chaudhry, Q. (2007). Engineered nanomaterials in soils and water: How do they behave and could they pose a risk to human health? *Nanomedicine (London)*, *2*(6), 919–927. doi:10.2217/17435889.2.6.919 PMID:18095854

Braz-Mota, S., Campos, D. F., MacCormack, T. J., Duarte, R. M., Val, A. L., & Almeida-Val, V. M. F. (2018). Mechanisms of toxic action of copper and copper nanoparticles in two Amazon fish species: Dwarf cichlid (Apistogramma agassizii) and cardinal tetra (Paracheirodonaxeirodi). *The Science of the Total Environment*, *630*, 1168–1180. doi:10.1016/j.scitotenv.2018.02.216 PMID:29554738

Brown, D. M., Wilson, M. R., MacNee, W., Stone, V., & Donaldson, K. (2001). Size-dependent proinflammatory effects of ultrafine polystyrene particles: A role for surface area and oxidative stress in the enhanced activity of ultrafines. *Toxicology and Applied Pharmacology*, *175*(3), 191–199. doi:10.1006/taap.2001.9240 PMID:11559017

Buzea, C., Pacheco, I. I., & Robbie, K. (2007). Nanomaterials and nanoparticles: Sources and toxicity. *Biointerphases*, *2*(4), MR17–MR71. doi:10.1116/1.2815690 PMID:20419892

Chai, H., Yao, J., Sun, J., Zhang, C., Liu, W., Zhu, M., & Ceccanti, B. (2015). The effect of metal oxide nanoparticles on functional bacteria and metabolic profiles in agricultural soil. *Bulletin of Environmental Contamination and Toxicology*, *94*(4), 490–495. doi:10.100700128-015-1485-9 PMID:25636440

Chakravorty, A., Biswas, B., Sana, S. S., Rayan, R. A., Lala, N. L., & Ramakrishna, S. (2020). A review on toxicity of turmeric derived Nano-Formulates against bacterial and fungal cells with special emphasis on electrospunnanofibers. *Materials Today: Proceedings*. Advance online publication. doi:10.1016/j.matpr.2020.05.820

Chakravorty, A., Rather, G. A., Ali, A., Bhat, B. A., Sana, S. S., Abhishek, N., & Nanda, A. (2020). *Nano Approach: Indian Spices as Antimicrobial Agents. In Ethnopharmacological Investigation of Indian Spices*. IGI Global.

Champion, J. A., & Mitragotri, S. (2006). Role of target geometry in phagocytosis. *Proceedings of the National Academy of Sciences of the United States of America, 103*(13), 4930–4934. doi:10.1073/pnas.0600997103 PMID:16549762

Chatel, A., Lievre, C., Barrick, A., Bruneau, M., & Mouneyrac, C. (2018). Transcriptomic approach: A promising tool for rapid screening nanomaterial-mediated toxicity in the marine bivalve Mytilus edulis – application to copper oxide nanoparticles. *Comparative Biochemistry and Physiology. Toxicology & Pharmacology: CBP, 205*, 26–33. doi:10.1016/j.cbpc.2018.01.003 PMID:29382575

Chen, M., Yin, J., Liang, Y., Yuan, S., Wang, F., Song, M., & Wang, H. (2016). Oxidative stress and immunotoxicity induced by graphene oxide in zebrafish. *Aquatic Toxicology (Amsterdam, Netherlands), 174*, 54–60. doi:10.1016/j.aquatox.2016.02.015 PMID:26921726

Churg, A., Stevens, B., & Wright, J. L. (1998). Comparison of the uptake of fine and ultrafine TiO2 in a tracheal explant system. *American Journal of Physiology. Lung Cellular and Molecular Physiology, 274*(1), L81–L86. doi:10.1152/ajplung.1998.274.1.L81 PMID:9458804

Coll, C., Notter, D., Gottschalk, F., Sun, T., Som, C., & Nowack, B. (2016). Probabilistic environmental risk assessment of five nanomaterials (nano-TiO_2, nano-Ag, nano-ZnO, CNT, and fullerenes). *Nanotoxicology, 10*(4), 4. doi:10.3109/17435390.2015.1073812 PMID:26554717

Colvin, V. L. (2003). The potential environmental impact of engineered nanomaterials. *Nature Biotechnology, 21*(10), 1166–1170. doi:10.1038/nbt875 PMID:14520401

Crosera, M., Bovenzi, M., Maina, G., Adami, G., Zanette, C., Florio, C., & Larese, F.F. (2009). Nanoparticle dermal absorption and toxicity: a review of literature. *International Archives of Occupational and Environmental Health, 82*, 1043–1055.

Cui, D., Tian, F., Ozkan, C. S., Wang, M., & Gao, H. (2005). Effect of single wall carbon nanotubes on human HEK293 cells. *Toxicology Letters, 155*(1), 73–85. doi:10.1016/j.toxlet.2004.08.015 PMID:15585362

Dai, H. (2002). Carbon nanotubes: Opportunities and challenges. *Surface Science, 500*(1-3), 218–241. doi:10.1016/S0039-6028(01)01558-8

Das, D., Vimala, R., & Das, N. (2013). Column Studies on Removal of Ag (I) from Electroplating Wastewater by MacrofungusPleurotus platypus: Use of Modelling and Response Surface Methodology. *Nature Environment and Pollution Technology, 12*(2), 273.

Das, D., Vimala, R., & Das, N. (2015). Removal of Ag (I) and Zn (II) ions from single and binary solution using sulfonated form of gum arabic-powdered mushroom composite hollow semispheres: Equilibrium, kinetic, thermodynamic and ex-situ studies. *Ecological Engineering, 75*, 116–122. doi:10.1016/j.ecoleng.2014.11.037

De Jong, W. H., Hagens, W. I., Krystek, P., Burger, M. C., Sips, A. J., & Geertsma, R. E. (2008). Particle size-dependent organ distribution of gold nanoparticles after intravenous administration. *Biomaterials, 29*(12), 1912–1919. doi:10.1016/j.biomaterials.2007.12.037 PMID:18242692

Deb, A., & Vimala, R. (2017). Graphene Mediated Drug Delivery-A Boon to Cancer Therapy. *Research Journal of Pharmacy and Technology*, *10*(5), 1571–1576. doi:10.5958/0974-360X.2017.00276.1

Deb, A., & Vimala, R. (2018). Camptothecin loaded graphene oxide nanoparticle functionalized with polyethylene glycol and folic acid for anticancer drug delivery. *Journal of Drug Delivery Science and Technology*, *43*, 333–342. doi:10.1016/j.jddst.2017.10.025

Dimkpa, C. O., McLean, J. E., Latta, D. E., Manangón, E., Britt, D. W., Johnson, W. P., & Anderson, A. J. (2012). CuO and ZnO nanoparticles: Phytotoxicity, metal speciation, and induction of oxidative stress in sand-grown wheat. *Journal of Nanoparticle Research*, *14*(9), 1–15. doi:10.100711051-012-1125-9

Ding, C., Du, J., Cao, Y., Yue, C., & Cheng, B. (2019). Effects of the aspect ratio of multi-walled carbon nanotubes on the structure and properties of regenerated collagen fibers. *International Journal of Biological Macromolecules*, *126*, 595–602. doi:10.1016/j.ijbiomac.2018.12.144 PMID:30576737

Dobrovolskaia, M. A., & McNeil, S. E. (2007). Immunological properties of engineered nanomaterials. *Nature Nanotechnology*, *2*(8), 469–478. doi:10.1038/nnano.2007.223 PMID:18654343

Donaldson, K., Li, X., & MacNee, W. (1998). Ultrafine (nanometre) particlemediated lung injury. *Journal of Aerosol Science*, *29*(5-6), 553–560. doi:10.1016/S0021-8502(97)00464-3

Donaldson, K., & Stone, V. (2003). Current hypotheses on the mechanisms of toxicity of ultrafine particles. *Annalidell'Istitutosuperiore di sanitã*, *39*(3), 405-410.

El Badawy, A. M., Silva, R. G., Morris, B., Scheckel, K. G., Suidan, M. T., & Tolaymat, T. M. (2011). Surface charge-dependent toxicity of silver nanoparticles. *Environmental Science & Technology*, *45*(1), 283–287. doi:10.1021/es1034188 PMID:21133412

Faisal, M., Saquib, Q., Alatar, A. A., Al-Khedhairy, A. A., Hegazy, A. K., & Musarrat, J. (2013). Phytotoxic hazards of NiO-nanoparticles in tomato: A study on mechanism of cell death. *Journal of Hazardous Materials*, *250–251*(April), 318–332. doi:10.1016/j.jhazmat.2013.01.063 PMID:23474406

Fierer, N., & Jackson, R. B. (2006). The diversity and biogeography of soil bacterial communities. *Proceedings of the National Academy of Sciences of the United States of America*, *103*(3), 626–631. doi:10.1073/pnas.0507535103 PMID:16407148

Foldbjerg, R., & Autrup, H. (2013). Article. *Mech Silver Nanoparticle Toxicity*, *1*, 5–15.

Foldbjerg, R., Olesen, P., Hougaard, M., Dang, D. A., Hoffmann, H. J., & Autrup, H. (2009). PVP-coated silver nanoparticles and silver ions induce reactive oxygen species, apoptosis and necrosis in THP-1 monocytes. *Toxicology Letters*, *190*(2), 156–162. doi:10.1016/j.toxlet.2009.07.009 PMID:19607894

Foldvari, M., & Bagonluri, M. (2008). Carbon nanotubes as functional excipients for nanomedicines: II. Drug delivery and biocompatibilityissues. *Nanomedicine (London)*, *4*(3), 183–200. doi:10.1016/j.nano.2008.04.003 PMID:18550450

Fu, P. P., Xia, Q., Hwang, H. M., Ray, P. C., & Yu, H. (2014). Mechanisms of nanotoxicity: Generation of reactive oxygen species. *Yao Wu Shi Pin Fen Xi*, *22*(1), 64–75. PMID:24673904

Gosens, I., Cassee, F. R., Zanella, M., Manodori, L., Brunelli, A., Costa, A. L., & Stone, V. (2016). Organ burden and pulmonary toxicity of nano-sized copper (II) oxide particles after short-term inhalation exposure. *Nanotoxicology*, *10*(8), 1084–1095. doi:10.3109/17435390.2016.1172678 PMID:27132941

Grabinski, C., Hussain, S., Lafdi, K., Braydich-Stolle, L., & Schlager, J. (2007). Effect of particle dimension on biocompatibility of carbon nanomaterials. *Carbon*, *45*(14), 2828–2835. doi:10.1016/j.carbon.2007.08.039

Grieger, K. D., Linkov, I., Hansen, S. F., & Baun, A. (2012). Environmental risk analysis for nanomaterials: Review and evaluation of frameworks. *Nanotoxicology*, *6*(2), 196–212. doi:10.3109/17435390.2011.569095 PMID:21486187

Guo, C., Robertson, S., Weber, R. J., Buckley, A., Warren, J., Hodgson, A., ... Macchiarulo, S. (2019). Pulmonary toxicity of inhaled nano-sized cerium oxide aerosols in Sprague–Dawley rats. *Nanotoxicology*, *13*(6), 733–750. doi:10.1080/17435390.2018.1554751 PMID:30704321

Gupta, Y. R., Sellegounder, D., Kannan, M., Deepa, S., Senthilkumaran, B., & Basavaraju, Y. (2016). Effect of copper nanoparticles exposure in the physiology of the common carp (Cyprinus carpio): Biochemical, histological and proteomic approaches. *Aquac Fish*, *1*, 15–23. doi:10.1016/j.aaf.2016.09.003

Gurr, J. R., Wang, A. S., Chen, C. H., & Jan, K. Y. (2005). Ultrafine titanium dioxide particles in the absence of photoactivation can induce oxidative damage to human bronchial epithelial cells. *Toxicology*, *213*(1-2), 66–73. doi:10.1016/j.tox.2005.05.007 PMID:15970370

Haliullin, T. O., Zalyalov, R. R., Shvedova, A. A., & Tkachov, A. G. (2015). Hygienic evaluation of multilayercarbon nanotubes. *Meditsina Truda i Promyshlennaia Ekologiia*, *7*, 37–42. PMID:26470480

Handy, R., & Shaw, B. (2007). Toxic effects of nanoparticles and nanomaterials:Implications for public health, risk assessment and the public perception of nano-technology. *Health Risk & Society*, *9*(2), 125–144. doi:10.1080/13698570701306807

Hoshino, A., Fujioka, K., Oku, T., Suga, M., Sasaki, Y. F., Ohta, T., Yasuhara, M., Suzuki, K., & Yamamoto, K. (2004). Physicochemical properties and cellular toxicity of nanocrystal quantum dots depend on their surface modification. *Nano Letters*, *4*(11), 2163–2169. doi:10.1021/nl048715d

Hu, C. W., Li, M., Cui, Y. B., Li, D. S., Chen, J., & Yang, L. Y. (2010). Toxicological effects ofTiO 2 and ZnO nanoparticles in soil on earthworm Eisenia fetida. *Soil Biology & Biochemistry*, *42*(4), 586–591. doi:10.1016/j.soilbio.2009.12.007

Hühn, D., Kantner, K., Geidel, C., Brandholt, S., De Cock, I., Soenen, S. J., ... Nienhaus, G. U. (2013). Polymer-coated nanoparticles interacting with proteins and cells: Focusing on the sign of the net charge. *ACS Nano*, *7*(4), 3253–3263. doi:10.1021/nn3059295 PMID:23566380

Janvier, C., Villeneuve, F., Alabouvette, C., Edel-Hermann, V., Mateille, T., & Steinberg, C. (2007). Soil health through soil disease suppression: Which strategy from descriptors to indicators? *Soil Biology & Biochemistry*, *39*(1), 1–23. doi:10.1016/j.soilbio.2006.07.001

Jiling, C., Youzhi, F., Xiangui, L., & Junhua, W. (2016). Arbuscular mycorrhizal fungi alleviate the negative effects of iron oxide nanoparticles on bacterial community in rhizospheric soils. *Frontiers in Environmental Science*, *4*, 10.

Jin, Y., Kannan, S., Wu, M., & Zhao, J.X. (2007). Toxicity of luminescent silica nanoparticles to living cells. *Chemical Research in Toxicology, 20*, 1126–1133.

Khan, M. G., Yezdani, U., Chakravorty, A., &Shukla, T. (2020). Efforts and Challenges paved by India to confront of Corona Virus (COVID-19). *Bangladesh Journal of Medical Science*, 88-S.

Kim, S., Lim, Y. T., Soltesz, E. G., De Grand, A. M., Lee, J., Nakayama, A., Parker, J. A., Mihaljevic, T., Laurence, R. G., Dor, D. M., Cohn, L. H., Bawendi, M. G., & Frangioni, J. V. (2004). Near-infrared fluorescent type ii quantum dots for sentinel lymph node mapping. *Nature Biotechnology*, *22*(1), 93–97. doi:10.1038/nbt920

Kinaret, P., Ilves, M., Fortino, V., Rydman, E., Karisola, P., Lähde, A., …Alenius, H. (2017). *Inhalation and oropharyngeal aspiration exposure to rod-like carbon*. Academic Press.

Kinaret, P., Ilves, M., Fortino, V., Rydman, E., Karisola, P., Lähde, A., Koivisto, J., Jokiniemi, J., Wolff, H., Savolainen, K., Greco, D., & Alenius, H. (2017, January 24). nanotubes induce similar airway inflammation and biological responses in mouse lungs. *ACS Nano*, *11*(1), 291–303. doi:10.1021/acsnano.6b05652 PMID:28045493

Klaine, S. J., Alvarez, P. J., Batley, G. E., Fernandes, T. F., Handy, R. D., Lyon, D. Y., Mahendra, S., McLaughlin, M. J., & Lead, J. R. (2008). Nanomaterials in the environment: Behavior, fate, bioavailability, and effects. *Environmental Toxicology and Chemistry*, *27*(9), 1825–1851. doi:10.1897/08-090.1 PMID:19086204

Klingelfus, T., Lirola, J. R., Oya Silva, L. F., Disner, G. R., Vicentini, M., Nadaline, M. J. B., Robles, J. C. Z., Trein, L. M., Voigt, C. L., Silva de Assis, H. C., Mela, M., Leme, D. M., & Cestari, M. M. (2017). Acute and longterm effects of trophic exposure to silver nanospheres in the central nervous system of a neotropical fish Hoplias intermedius. *Neurotoxicology*, *63*, 146–154. doi:10.1016/j.neuro.2017.10.003 PMID:29031576

Kreyling, W. G., Semmler-Behnke, M., Seitz, J., Scymczak, W., Wenk, A., Mayer, P., Takenaka, S., & Oberdörster, G. (2009). Size dependence of the translocation of inhaled iridium and carbon nanoparticle aggregates from the lung of rats to the blood and secondary target organs. *Inhalation Toxicology*, *21*(sup1, Suppl 1), 55–60. doi:10.1080/08958370902942517 PMID:19558234

Lahive, E., Jurkschat, K., Shaw, B. J., Handy, R. D., Spurgeon, D. J., & Svendsen, C. (2014). Toxicity of cerium oxide nanoparticles to the earthworm Eisenia fetida: Subtle effects. *Environmental Chemistry*, *11*(3), 268. doi:10.1071/EN14028

Landsiedel, R., Sauer, U. G., Ma-Hock, L., Schnekenburger, J., & Wiemann, M. (2014). Pulmonary toxicity of nanomaterials: A critical comparison of published *in vitro* assays and *in vivo* inhalation or instillatiogn studies. *Nanomedicine (London)*, *9*(16), 2557–2585. doi:10.2217/nnm.14.149 PMID:25490426

Limbach, L. K., Li, Y., Grass, R. N., Brunner, T. J., Hintermann, M. A., Muller, M., Gunther, D., & Stark, W. J. (2005). Oxide nanoparticle uptake in human lung fibroblasts: Effects of particle size, agglomeration, and diffusion at low concentrations. *Environmental Science & Technology*, *39*(23), 9370–9376. doi:10.1021/es051043o PMID:16382966

Lockman, P. R., Koziara, J. M., Mumper, R. J., & Allen, D. D. (2004). Nanoparticle surface charges alter blood–brain barrier integrity and permeability. *Journal of Drug Targeting*, *12*(9-10), 635–641. doi:10.1080/10611860400015936 PMID:15621689

Ma, C., White, J. C., Dhankher, O. P., & Xing, B. (2015). Metal-based nanotoxicity and detoxification pathways in higher plants. *Environmental Science & Technology*, *49*(12), 7109–7122. doi:10.1021/acs.est.5b00685 PMID:25974388

Ma, J., Mercer, R. R., Barger, M., Schwegler-Berry, D., Cohen, J. M., Demokritou, P., & Castranova, V. (2015). Effects of amorphous silica coating on cerium oxide nanoparticles induced pulmonaryresponses. *Toxicology and Applied Pharmacology*, *288*(1), 63–73. doi:10.1016/j.taap.2015.07.012 PMID:26210349

Mackevica, A., & Foss Hansen, S. (2016). Release of nanomaterials from solid nano-composites and consumer exposure assessment—A forward-looking review. *Nanotoxicology*, *10*(6), 641–653. doi:10.3109/17435390.2015.1132346 PMID:26667577

Marchant, G. E., Sylvester, D. J., & Abbott, K. W. (2008). Risk management principles for nanotechnology. *NanoEthics*, *2*(1), 43–60. doi:10.100711569-008-0028-9

Maurer-Jones, M. A., Ian, L., Gunsolus, I. L., Murphy, C. J., & Haynes, C. L. (2013). Toxicity of engineered nanoparticles in the environment. *Analytical Chemistry*, *85*(6), 3036–3049. doi:10.1021/ac303636s PMID:23427995

Mavon, A., Miquel, C., Lejeune, O., Payre, B., & Moretto, P. (2007). In vitro percutaneous absorption and in vivo stratum corneum distribution of an organic and a mineral sunscreen. *Skin Pharmacology and Physiology*, *20*(1), 10–20. doi:10.1159/000096167 PMID:17035717

Miclăuş, T., Beer, C., Chevallier, J., Scavenius, C., Bochenkov, V. E., Enghild, J. J., & Sutherland, D. S. (2016). Dynamic protein coronas revealed as a modulator of silver nanoparticle sulphidation in vitro. *Nature Communications*, *7*(1), 1–10. doi:10.1038/ncomms11770

Miethling-graff, R., Rumpker, R., Richter, M., Verano-Braga, T., Kjeldsen, F., Brewer, J., Hoyland, J., Rubahn, H.-G., & Erdmann, H. (2014). Toxicology in vitro exposure to silver nanoparticles induces size- and dose-dependent oxidative stress and cytotoxicity in human colon carcinoma cells. *Toxicology In Vitro*, *28*(7), 1280–1289. doi:10.1016/j.tiv.2014.06.005 PMID:24997297

Mohanta, D., Patnaik, S., Sood, S., & Das, N. (2019). Carbon nanotubes: Evaluation of toxicity at biointerfaces. *Journal of Pharmaceutical Analysis*, *9*(5), 293–300. doi:10.1016/j.jpha.2019.04.003 PMID:31929938

Monopoli, M. P., Aberg, C., Salvati, A., & Dawson, K. (2012). A biomolecular coronas provide the biological identity of nanosized materials. *Nature Nanotechnology*, *7*(12), 779–786. doi:10.1038/nnano.2012.207 PMID:23212421

Monteiro-riviere, N. A., & Inman, A. O. (2006). Challenges for assessing carbon nanomaterial toxicity to the skin. *Carbon N Y*, *44*(6), 1070–1078. doi:10.1016/j.carbon.2005.11.004

Monteiro-Riviere, N. A., Nemanich, R. J., Inman, A. O., Wang, Y. Y., & Riviere, J. E. (2005). Multi-walled carbon nanotube interactions with human epidermal keratinocytes. *Toxicology Letters*, *155*(3), 377–384. doi:10.1016/j.toxlet.2004.11.004 PMID:15649621

Nabeshi, H., Yoshikawa, T., Matsuyama, K., Nakazato, Y., Arimori, A., Isobe, M., ... Yamashita, T. (2010). Size-dependent cytotoxic effects of amorphous silica nanoparticles on Langerhans cells. *Die Pharmazie-An International Journal of Pharmaceutical Sciences*, *65*(3), 199–201. PMID:20383940

Nowack, B., & Bucheli, T. D. (2007). Occurrence, behavior and effects of nanoparticles in the environment. *Environmental Pollution*, *150*(1), 5–22. doi:10.1016/j.envpol.2007.06.006 PMID:17658673

NRC. (2009). *Science and Decisions: Advancing Risk Assessment. Committee on Improving Risk Analysis Approaches Used by the U.S. EPA, Board on Environmental Studies and Toxicology, Division on Earth and Life Studies, National Research Council of the National Academies*. The National Academies Press.

Oberdorster, E. (2004). Manufactured nanomaterials (fullerenes, C60) induce oxidative stress in the brain of juvenile largemouth bass. *Environmental Health Perspectives*, *112*(10), 1058–1062. doi:10.1289/ehp.7021 PMID:15238277

Oberdörster, G., Ferin, J., & Lehnert, B. E. (1994). Correlation between particle size, in vivo particle persistence, and lung injury. *Environmental Health Perspectives*, *102*(suppl 5), 173–179. PMID:7882925

Oberdörster, G., Maynard, A., Donaldson, K., Castranova, V., Fitzpatrick, J., Ausman, K., & Olin, S. (2005). Principles for characterizing the potential human health effectsfrom exposure to nanomaterials: Elements of a screening strategy. *Particle and Fibre Toxicology*, *2*(1), 1.

Pacheco-Blandino, I., Vanner, R., & Buzea, C. (2012). Toxicity of nanoparticles. In *Toxicity of building materials* (pp. 427–475). Woodhead Publishing. doi:10.1533/9780857096357.427

Palomäki, J., Välimäki, E., Sund, J., Vippola, M., Clausen, P. A., Jensen, K. A., Savolainen, K., Matikainen, S., & Alenius, H. (2011, September 27). NLRP3 inflammasome through a similar mechanism. *ACS Nano*, *5*(9), 6861–6870. doi:10.1021/nn200595c PMID:21800904

Park, K. H., Chhowalla, M., Iqbal, Z., & Sesti, F. (2003). Single-walled carbon nanotubes are a new class of ion channel blockers. *The Journal of Biological Chemistry*, *278*(50), 50212–50216. doi:10.1074/jbc.M310216200 PMID:14522977

Patlolla, A. K., Berry, A., May, L., & Tchounwou, P. B. (2012). Genotoxicity of silver nanoparticles in Viciafaba: A pilot study on the environmental monitoring of nanoparticles. *International Journal of Environmental Research and Public Health*, *9*(5), 1649–1662. doi:10.3390/ijerph9051649 PMID:22754463

Paul, A. M., Sajeev, A., Nivetha, R., Gothandapani, K., Bhardwaj, P., Govardhan, K., ... Grace, A. N. (2020). Cuprous oxide (Cu2O)/graphitic carbon nitride (g-C3N4) nanocomposites for electrocatalytic hydrogen evolution reaction. *Diamond and Related Materials*, *107*, 107899. doi:10.1016/j.diamond.2020.107899

Pereira, R., Rocha-Santos, T. A. P., Antunes, F. E., Rasteiro, M. G., Ribeiro, R., Goncalves, F., Soares, A. M. V. M., & Lopes, I. (2011). Screening evaluation of the ecotoxicity and genotoxicity of soils contaminated with organic and inorganic nanoparticles: The role of ageing. *Journal of Hazardous Materials*, *194*, 345–354. doi:10.1016/j.jhazmat.2011.07.112 PMID:21871729

Petersen, E. J., Huang, Q., & Weber, W. J. Jr. (2008). Bioaccumulation of radio-labeled carbon nanotubes by Eisenia foetida. *Environmental Science & Technology*, *42*(8), 3090–3095. doi:10.1021/es071366f PMID:18497171

Pichardo, S., Gutiérrez-Praena, D., Puerto, M., Sánchez, E., Grilo, A., Cameán, A. M., & Jos, Á. (2012). Oxidative stressresponses to carboxylic acid functionalized single wall carbon nano-tubes on the human intestinal cell line Caco-2. *Toxicology In Vitro*, *26*(5), 672–677. doi:10.1016/j.tiv.2012.03.007 PMID:22449549

Pietroiusti, A., Stockmann-Juvala, H., Lucaroni, F., & Savolainen, K. (2018). Nanomaterial exposure, toxicity, and impact on human health. *Wiley Interdisciplinary Reviews. Nanomedicine and Nanobiotechnology*, *10*(5), e1513. doi:10.1002/wnan.1513 PMID:29473695

Pokhrel, L. R., & Dubey, B. (2013). Evaluation of developmental responses of two crop plants exposed to silver and zinc oxide nanoparticles. *The Science of the Total Environment*, *452*, 321–332. doi:10.1016/j.scitotenv.2013.02.059 PMID:23532040

Raghavan, V., Deb, A., & Grace, A. N. (2020). *Honokiol-Camptothecin loaded graphene oxide nanoparticle towards combinatorial anti-cancer drug delivery*. IET Nanobiotechnology.

Rastogi, A., Zivcak, M., Sytar, O., Kalaji, H. M., He, X., Mbarki, S., & Brestic, M. (2017). Impact of Metal and Metal Oxide Nanoparticles on Plant: A Critical Review. *Frontiers in Chemistry*, *5*, 78. doi:10.3389/fchem.2017.00078 PMID:29075626

Reijnders, L. (2006). Cleaner nanotechnology and hazard reduction of manufactured nanoparticles. *Journal of Cleaner Production*, *14*(2), 124–133. doi:10.1016/j.jclepro.2005.03.018

Rico, C. M., Hong, J., Morales, M. I., Zhao, L., Barrios, A. C., Zhang, J.-Y., Peralta-Videa, J. R., & Gardea-Torresdey, J. L. (2013). Effect of cerium oxide nanoparticles on rice: A study involving the antioxidant defense system and in vivo fluorescence imaging. *Environmental Science & Technology*, *47*(11), 5635–5642. doi:10.1021/es401032m PMID:23662857

Rocha, T. L., Gomes, T., Sousa, V. S., Mestre, N. C., & Bebianno, M. J. (2015). Ecotoxicological impact of engineered nanomaterials in bivalve molluscs: An overview. *Marine Environmental Research*, *111*, 74–88. doi:10.1016/j.marenvres.2015.06.013 PMID:26152602

Roco, M. C. (2011). The long view of nanotechnology development: The National Nanotechnology Initiative at 10 years. *Journal of Nanoparticle Research*, 13.

Roduner, E. (2006). Size matters: Why nanomaterials are different. *Chemical Society Reviews*, *35*(7), 583–592. doi:10.1039/b502142c PMID:16791330

Roh, J. Y., Sim, S. J., Yi, J., Park, K., Chung, K. H., Ryu, D. Y., & Choi, J. (2009). Ecotoxicity of silver nanoparticles on the soil nematode Caenorhabditis elegans using functional ecotoxicogenomics. *Environmental Science & Technology*, *43*(10), 3933–3940. doi:10.1021/es803477u PMID:19544910

Rossi, E. M., Pylkkänen, L., Koivisto, A. J., Nykäsenoja, H., Wolff, H., Savolainen, K., & Alenius, H. (2010). Inhalation exposure to nanosized and fine TiO2 particles inhibits features of allergic asthma in a murine model. *Particle and Fibre Toxicology*, *7*(1), 35. doi:10.1186/1743-8977-7-35 PMID:21108815

Sabella, S., Carney, R. P., Brunetti, V., Malvindi, M. A., Al-Juffali, N., Vecchio, G., Janes, S. M., Bakr, O. M., Cingolani, R., Stellacci, F., & Pompa, P. P. (2014). A general mechanism for intracellular toxicity of metal-containing nanoparticles. *Nanoscale*, *6*(12), 7052–7061. doi:10.1039/c4nr01234h PMID:24842463

Sager, T. M., Porter, D. W., Robinson, V. A., Lindsley, W. G., Schwegler-Berry, D. E., & Castranova, V. (2007). Improved method to disperse nanoparticles for in vitro and in vivo investigation of toxicity. *Nanotoxicology*, *1*(2), 118–129. doi:10.1080/17435390701381596

Saifi, M. A., Khurana, A., & Godugu, C. (2018). Nanotoxicology: toxicity and risk assessment of nanomaterials. In Nanomaterials in chromatography (pp. 437-465). Elsevier.

Savolainen, K., & Alenius, H. (2013). Disseminating widely. *Nature Nanotechnology*, *8*(2), 72. doi:10.1038/nnano.2013.3 PMID:23380925

Savolainen, K., Alenius, H., Norppa, H., Pylkkänen, L., Tuomi, T., & Kasper, G. (2010). Risk assessment of engineered nanomaterials and nanotechnologies—A review. *Toxicology, 269*, 92–104.

Savolainen, K., & Norppa, H. (2016). In vitro and in vivo genotoxic effects of straight versus tangled multi-walled carbon nanotubes. *Nanotoxicology*, *10*(6). PMID:26674712

Schaeublin, N. M., Braydich-Stolle, L. K., Schrand, A. M., Miller, J. M., Hutchison, J., Schlager, J. J., & Hussain, S. M. (2011). Surface charge of gold nanoparticles mediates mechanism of toxicity. *Nanoscale*, *3*(2), 410–420. doi:10.1039/c0nr00478b PMID:21229159

Selck, H., Handy, R. D., Fernandes, T. F., Klaine, S. J., & Petersen, E. J. (2016). Nanomaterials in the aquatic environment: A European Union-United States perspective on the status of ecotoxicity testing, research priorities, and challenges ahead. *Environmental Toxicology and Chemistry*, *35*(5), 1055–1067. doi:10.1002/etc.3385 PMID:27089437

Shandilya, N., Le, B. O., Bressot, C., & Morgeneyer, M. (2015). Emission of titanium dioxide nanoparticles from building materials to the environment by wear and weather. *Environmental Science & Technology*, *49*(4), 2163–2170. doi:10.1021/es504710p PMID:25590625

Sharifi, S., Behzadi, S., Laurent, S., Forrest, M. L., Stroeve, P., & Mahmoudi, M. (2012). Toxicity of nanomaterials. *Chemical Society Reviews*, *41*(6), 2323–2343. doi:10.1039/C1CS15188F PMID:22170510

Shaw, B. J., & Handy, R. D. (2011). Physiological effects of nanoparticles on fish: A comparison of nano - metals versus metal ions. *Environment International*, *37*(6), 1083–1097. doi:10.1016/j.envint.2011.03.009 PMID:21474182

Simonin, M., & Richaume, A. (2015). Impact of engineered nanoparticles on the activity, abundance, and diversity of soil microbial communities: A review. *Environmental Science and Pollution Research International*, *22*(18), 13710–13723. doi:10.100711356-015-4171-x PMID:25647498

Singh, V., & Srivastava, V. C. (2020). Self-engineered iron oxide nanoparticle incorporated on mesoporousbiochar derived from textile mill sludge for the removal of an emerging pharmaceutical pollutant. *Environmental Pollution*, *259*, 113822. doi:10.1016/j.envpol.2019.113822 PMID:31887588

Souza, J. P., Venturini, F. P., Santos, F., & Zucolotto, V. (2018). Chronic toxicity in Ceriodaphniadubia induced by graphene oxide. *Chemosphere*, *190*, 218–224. doi:10.1016/j.chemosphere.2017.10.018 PMID:28992473

Sukhanova, A., Bozrova, S., Sokolov, P., Berestovoy, M., Karaulov, A., & Nabiev, I. (2018). Dependence of nanoparticle toxicity on their physical and chemical properties. *Nanoscale Research Letters*, *13*(1), 44. doi:10.118611671-018-2457-x PMID:29417375

Sun, H., Zhou, J., Huang, Z., Qu, L., Lin, N., Liang, C., Dai, R., Tang, L., & Tian, F. (2017). Carbon nanotube-incorporated collagen hydrogels improve cell alignment and the performance of cardiac constructs. *International Journal of Nanomedicine*, *12*, 3109–3120. doi:10.2147/IJN.S128030 PMID:28450785

Suresh, A. K., Pelletier, D. A., & Doktycz, M. J. (2013). Relating nanomaterial properties and microbial toxicity. *Nano*, *5*, 463–474. PMID:23203029

Taghavi, S. M., Momenpour, M., Azarian, M., Ahmadian, M., Souri, F., Taghavi, S. A., Sadeghain, M., & Karchani, M. (2013). Effects of Nanoparticles on the Environment and Outdoor Workplaces. *Electronic Physician*, *5*(4), 706–712. PMID:26120406

Taju, G., Majeed, S. A., Nambi, K. S. N., & Hameed, A. S. S. (2014). In vitro assay for the toxicity of silver nanoparticles using heart and gill cell lines of Catlacatla and gill cell line of Labeorohita. *Comparative Biochemistry and Physiology. Toxicology & Pharmacology: CBP*, *161*, 41–52. doi:10.1016/j.cbpc.2014.01.007 PMID:24524868

Takenaka, S., Karg, E., Roth, C., Schulz, H., Ziesenis, A., Heinzmann, U., ... Heyder, J. (2001). Pulmonary and systemic distribution of inhaled ultrafine silver particles in rats. *Environmental Health Perspectives*, *109*(suppl 4), 547–551. PMID:11544161

Teow, Y., Asharani, P. V., Hande, M. P., & Valiyaveettil, S. (2011). Health impact and safety of engineered nanomaterials. *Chemical Communications (Cambridge)*, *47*(25), 7025–7038. doi:10.1039/c0cc05271j PMID:21479319

Tourinho, P. S., van Gestel, C. A., Lofts, S., Svendsen, C., Soares, A. M., & Loureiro, S. (2012). Metal-based nanoparticles in soil: Fate, behavior, and effects on soil invertebrates. *Environmental Toxicology and Chemistry*, *31*(8), 1679–1692. doi:10.1002/etc.1880 PMID:22573562

Türk, S., Altınsoy, I., Efe, G. Ç., İpek, M., Özacar, M., & Bindal, C. (2018). 3D porous collagen/functionalized multiwalled carbon nanotube/chitosan/hydroxyapatite composite scaffolds for bone tissue engineering. *Materials Science and Engineering C*, *92*, 757–768. doi:10.1016/j.msec.2018.07.020 PMID:30184804

Valant, J., Drobne, D., & Novak, S. (2012). Effect of ingested titanium dioxidenanoparticles on the digestive gland cell membrane of terrestrialisopods. *Chemosphere, 87*(1), 19–25. doi:10.1016/j.chemosphere.2011.11.047 PMID:22189379

Vale, G., Mehennaoui, K., Cambier, S., Libralato, G., Jomini, S., & Domingos, R. F. (2016). Manufactured nanoparticles in the aquatic environment-biochemical responses on freshwater organisms: A critical overview. *Aquatic Toxicology (Amsterdam, Netherlands), 170*, 162–174. doi:10.1016/j.aquatox.2015.11.019 PMID:26655660

Valsami-Jones, E., & Lynch, I. (2015). NANOSAFETY. How safe are nanomaterials? *Science, 350*(6259), 388–389. doi:10.1126cience.aad0768 PMID:26494749

Vance, M. E., Kuiken, T., Vejerano, E. P., McGinnis, S. P., Hochella, M. F. Jr, Rejeski, D., & Hull, M. S. (2015). Nanotechnology in the real world: Redeveloping the nanomaterial consumer products inventory. *Beilstein Journal of Nanotechnology, 6*, 1769–1780. doi:10.3762/bjnano.6.181 PMID:26425429

Velappan, S., Nivedhita, P., Vimala, R., & Raja, S. (2020). Role of nanotitania on the thermomechanical properties of silicon carbide refractories. *Ceramics International, 46*(16), 25921–25926. doi:10.1016/j.ceramint.2020.07.077

Wagner, S. C., Roskamp, M., Pallerla, M., Araghi, R. R., Schlecht, S., & Koksch, B. (2010). pallerla M, Araghi RR, Schlecht S, koksch B. Nanoparticle-induced folding and fibril formation of coiled coil-based model peptides. *Small, 6*(12), 1321–1328. doi:10.1002mll.200902067 PMID:20517875

Walkey, C. D., Olsen, J. B., Song, F., Liu, R., Guo, H., Olsen, D. W. H., Cohen, Y., Emili, A., & Chan, W. C. (2014). Protein corona fingerprinting predicts the cellular interaction of gold and silver nanoparticles. *ACS Nano, 8*(3), 2439–2455. doi:10.1021/nn406018q PMID:24517450

Wang, S., Lu, W., Tovmachenko, O., Rai, U. S., Yu, H., & Ray, P. C. (2008). Challenge in understanding size and shape dependent toxicity of gold nanomaterials in humanskin keratinocytes. *Chemical Physics Letters, 463*(1-3), 145–149. doi:10.1016/j.cplett.2008.08.039 PMID:24068836

Wick, P., Manser, P., Limbach, L. K., Dettlaff-Weglikowska, U., Krumeich, F., Roth, S., Stark, W., & Bruinink, A. (2007). The degree and kind of agglomeration affect carbon nanotube cytotoxicity. *Toxicology Letters, 168*(2), 121–131. doi:10.1016/j.toxlet.2006.08.019 PMID:17169512

Xu, C., Peng, C., Sun, L., Zhang, S., Huang, H., Chen, Y., & Shi, J. (2015). Distinctive effects of TiO 2 and CuO nanoparticles on soil microbes and their community structures in flooded paddy soil. *Soil Biology & Biochemistry, 86*, 24–33. doi:10.1016/j.soilbio.2015.03.011

Yang, H., Liu, C., Yang, D., Zhang, H., & Xi, Z. (2009). Comparative study of cytotoxicity, oxidative stress and genotoxicity induced by four typical nanomaterials: The role of particle size, shape and composition. *Journal of Applied Toxicology, 29*(1), 69–78. doi:10.1002/jat.1385 PMID:18756589

Yue, Y., Li, X., Sigg, L., Suter, M. J., Pillai, S., Behra, R., & Schirmer, K. (2017). Interaction of silver nanoparticles with algae and fish cells: A side by side comparison. *Journal of Nanobiotechnology, 15*(1), 16. doi:10.118612951-017-0254-9 PMID:28245850

Zhang, X., Hu, S., Wang, M., Yu, J., Khan, Q., Shang, J., & Ba, L. (2015). Continuous graphene and carbon nanotube based high flexible and transparent pressure sensor arrays. *Nanotechnology*, *26*(11), 115501. doi:10.1088/0957-4484/26/11/115501 PMID:25712583

Zhang, Z., Kleinstreuer, C., Donohue, J. F., & Kim, C. S. (2005). Comparison of micro- and nano-size particledepositions in a human upper airway model. *Journal of Aerosol Science*, *36*(2), 123–129. doi:10.1016/j.jaerosci.2004.08.006

Zhao, L., Sun, Y., Hernandez-Viezcas, J. A., Servin, A. D., Hong, J., Niu, G., & Gardea-Torresdey, J. L. (2013). Influence of CeO2 andZnO nanoparticles on cucumber phys-iological markers and bioaccumulation of Ce and Zn: A life cycle study. *Journal ofAgricultural and Food Chemistry*, *61*(49), 11945–11951. doi:10.1021/jf404328e PMID:24245665

Zimmer, A. T., Baron, P. A., & Biswas, P. (2002). The influence of operating parameters on number-weighted aerosol size distribution generated from a gas metal arc welding process. *Journal of Aerosol Science*, *33*(3), 519–531. doi:10.1016/S0021-8502(01)00189-6

Compilation of References

A Brief Overview of Classes of Antibiotics – Compound Interest. (n.d.). Available at:https://www.compoundchem.com/2014/09/08/antibiotics/

Aadil, K. R., Nathani, A., Sharma, C. S., Lenka, N., & Gupta, P. (2018). Fabrication of biocompatible alginate-polyvinyl (alcohol) nanofibers scaffold. *Materials Technology*, *33*(8), 507–512. doi:10.1080/10667857.2018.1473234

Abakumov, M., Majouga, A., Kabanov, A., & Chekhonin, V. (2018). Protein coated magnetic nanoparticles for cancer treatment and diagnostics. *12th International conference on Scientific and clinical applications of Magnetic carriers.*

Abasian, M., Hooshangi, V., & Moghadam, P. N. (2017). Synthesis of polyvinyl alcohol hydrogel grafted by modified Fe3 O4 nanoparticles: Characterization and doxorubicin delivery studies. *Iranian Polymer Journal*, *26*(5), 313–322. doi:10.100713726-017-0521-5

Abdelmohsen, L. K. E. A., Peng, F., Tu, Y., & Wilson, D. A. (2014). Micro- and nano-motors for biomedical applications. *Journal of Materials Chemistry. B, Materials for Biology and Medicine*, *2*(17), 2395–2408. doi:10.1039/C3TB21451F PMID:32261411

Abdul-Aziz, M., Lipman, J., Mouton, J., Hope, W., & Roberts, J. (2015). Applying Pharmacokinetic / Pharmacodynamic Principles in Critically Ill Patients: Optimizing Efficacy and Reducing Resistance Development. *Seminars in Respiratory and Critical Care Medicine*, *36*(01), 136–153. doi:10.1055-0034-1398490 PMID:25643277

Abedon, S. T., Thomas-Abedon, C., Thomas, A., & Mazure, H. (2011). Bacteriophage prehistory: Is or is not Hankin, 1896, a phage reference? *Bacteriophage*, *1*(3), 174–178. doi:10.4161/bact.1.3.16591 PMID:22164351

Abid, J. P., Frigoli, M., Pansu, R., Szeftel, J., Zyss, J., Larpent, C., & Brasselet, S. (2011). Light-driven directed motion of azobenzene-coated polymer nanoparticles in an aqueous medium. *Langmuir*, *27*(13), 7967–7971. doi:10.1021/la200682p PMID:21651298

Abinaya, M., Rekha, R., Sivakumar, S., Govindarajan, M., Alharbi, N. S., Kadaikunnan, S., Khaled, J. M., Alobaidi, A. S., Al-Anbr, M. N., & Vaseeharan, B. (2019). Novel and Facile Synthesis of Sea Anemone Adhesive Protein-Coated ZnO Nanoparticles: Antioxidant, Antibiofilm, and Mosquito Larvicidal Activity Against Aedes aegypti. *Journal of Cluster Science*, *30*(6), 1393–1402. doi:10.100710876-019-01581-0

Abinaya, M., Vaseeharan, B., Divya, M., Sharmili, A., Govindarajan, M., Alharbi, N. S., Kadaikunnan, S., Khaled, J. M., & Benelli, G. (2018). Bacterial exopolysaccharide (EPS)-coated ZnO nanoparticles showed high antibiofilm activity and larvicidal toxicity against malaria and Zika virus vectors. *Journal of Trace Elements in Medicine and Biology*, *45*, 93–103. doi:10.1016/j.jtemb.2017.10.002 PMID:29173489

Abraham, E. P., & Chain, E. (1940). An enzyme from bacteria able to destroy penicillin. *Nature*, *146*(3713), 837. doi:10.1038/146837a0

Abrahamse, H., Kruger, C. A., Kadanyo, S., & Mishra, A. (2017). Nanoparticles for Advanced Photodynamic Therapy of Cancer. *Photomedicine and Laser Surgery*, *35*(11), 581–588. doi:10.1089/pho.2017.4308 PMID:28937916

Abral, H., Kadriadi, Mahardika, M., Handayani, D., Sugiarti, E., & Muslimin, A. N. (2019). Characterization of disintegrated bacterial cellulose nanofibers/PVA bionanocomposites prepared via ultrasonication. *International Journal of Biological Macromolecules*, *135*, 591–599. doi:10.1016/j.ijbiomac.2019.05.178 PMID:31132440

Abreu, A. S., Castanheira, E. M., Queiroz, M. J. R., Ferreira, P. M., Vale-Silva, L. A., & Pinto, E. (2011). Nanoliposomes for encapsulation and delivery of the potential antitumoral methyl 6-methoxy-3-(4-methoxyphenyl)-1H-indole-2-carboxylate. *Nanoscale Research Letters*, *6*(1), 482. doi:10.1186/1556-276X-6-482 PMID:21812989

Acar, Unal, Khodadust, Durmusoglu, Erdem, Yagci, Cleva, Yang, & Yurtsever. (2018). Discovery of Self-luminescent Polyethyleneimine Coated SPIONs and Their Theranostic Applications. *12th International conference on Scientific and clinical applications of Magnetic carrier.*

Acevedo, F., Villegas, P., Urtuvia, V., Hermosilia, J., Navia, R., & Seeger, M. (2018). Bacterial polyhydroxybutyrate for electrospun fiber production. *International Journal of Biological Macromolecules*, *106*, 692–697. doi:10.1016/j.ijbiomac.2017.08.066 PMID:28823514

Adams, A. A., Okagbare, P. I., Feng, J., Hupert, M. L., Patterson, D., Gottert, J., McCarley, R. L., Nikitopoulos, D., Murphy, M. C., & Soper, S. A. (2008). Highly efficient circulating tumor cell isolation from whole blood and label-free enumeration using polymer-based microfluidics with an integrated conductivity sensor. *Journal of the American Chemical Society*, *130*(27), 8633–8641. doi:10.1021/ja8015022 PMID:18557614

Adeli, H., Khorasani, T., & Parvazinia, M. (2019). Development of antimicrobial and antioxidant electrospun soluble potato starch nanofibers loaded with carvacrol. *International Journal of Biological Macromolecules*, *1*(122), 238–254. doi:10.1016/j.ijbiomac.2018.10.115 PMID:30342125

Adhikari, U., Ghosh, A., & Chandra, G. (2013). Nano particles of herbal origin: A recent eco-friend trend in mosquito control. *Asian Pacific Journal of Tropical Disease*, *3*(2), 167–168. doi:10.1016/S2222-1808(13)60065-1

Adimoolam, M. G., Amreddy, N., Nalam, M. R., & Sunkara, M. V. (2018). A simple approach to design chitosan functionalized Fe3O4 nanoparticles for pH responsive delivery of doxorubicin for cancer therapy. *Journal of Magnetism and Magnetic Materials*, *448*, 199–207. doi:10.1016/j.jmmm.2017.09.018

Adler-Abramovich, L., Kol, N., Yanai, I., Barlam, D., Shneck, R. Z., Gazit, E., & Rousso, I. (2010). Self-assembled organic nanostructures with metallic-like stiffness. *Angewandte Chemie International Edition*, *49*(51), 9939–9942. doi:10.1002/anie.201002037 PMID:20878815

Agarwala, M., Choudhury, B., & Yadav, R. N. S. (2014). Comparative study of antibiofilm activity of copper oxide and iron oxide nanoparticles against multidrug resistant biofilm forming uropathogens. *Indian Journal of Microbiology*, *54*(3), 365–368. doi:10.100712088-014-0462-z PMID:24891746

Agarwal, P., Mehta, A., Kachhwaha, S., & Kothari, S. L. (2013). Green Synthesis of Silver Nanoparticles and Their Activity Against *Mycobacterium tuberculosis*. *Advanced Science, Engineering and Medicine*, *5*(7), 709–714.

Agemy, L., Friedmann-Morvinski, D., Kotamraju, V. R., Roth, L., Sugahara, K. N., Girard, O. M., Mattrey, R. F., Verma, I. M., & Ruoslahti, E. (2014). Targeted nanoparticle enhanced proapoptotic peptide as potential therapy for glioblastoma. *Proceedings of the National Academy of Sciences of the United States of America*, *108*(42), 17450–17455. doi:10.1073/pnas.1114518108 PMID:21969599

Agrahari, V., & Hiremath, P. (2017). *Challenges associated and approaches for successful translation of nanomedicines into commercial products*. Academic Press.

Ahirwal, G. K., & Chanchal, K. M. (2009). Direct Electrochemistry of Horseradish Peroxidase-Gold Nanoparticles Conjugate. *Sensors (Basel)*, *9*(2), 881–894. doi:10.3390/s90200881 PubMed

Ahmad, I., Ahmad, A., Iftekhar, S., Khalid, S., Aftab, A., & Raza, S. A. (2019). Role of Nanoparticles in Cosmetics Industries. In *Biological Synthesis of Nanoparticles and Their Applications* (pp. 173–204). CRC Press. doi:10.1201/9780429265235-14

Ahmad, N., Amin, M. C. I. M., Mahali, S. M., Ismail, I., & Chuang, V. T. G. (2014). Biocompatible and mucoadhesive bacterial cellulose-g-poly (acrylic acid) hydrogels fororal protein delivery. *Molecular Pharmaceutics*, *11*(11), 4130–4142. doi:10.1021/mp5003015 PMID:25252107

Ahmad, Z., Shah, A., Siddiq, M., & Kraatz, H. B. (2014). Polymeric micelles as drug delivery vehicles. *RSC Advances*, *4*(33), 17028–17038. doi:10.1039/C3RA47370H

Ahmed, T., & Mukhtar, M. (2018). *Climate Change and mosquitoes: need and ways to diminish mosquito related diseases in Pakistan*. Academic Press. Technology Times, 24 May 2018. https://www.technologytimes.pk/2018/05/24/climate-change-mosquitoes/

Ahmed, A. D., Mohammed, K. H., Lee, S. B., Kim, K., Saha, S. K., ... Gwang-Mo, Y. (2017). T.he rol.e of reactive oxygen species (ROS) in the biological activities of metallic nanoparticles. *International Journal of Molecular Sciences*, *18*(1), 120. doi:10.3390/ijms18010120

Ahmed, E. M. (2015). Hydrogel: Preparation, characterization, and applications: A review. *Journal of Advanced Research*, *6*(2), 105–121. doi:10.1016/j.jare.2013.07.006 PMID:25750745

Ahmed, M., Smith, D. M., Hamouda, T., Rangel-Moreno, J., Fattom, A., & Khader, S. A. (2017). A novel nanoemulsion vaccine induces mucosal Interleukin-17 responses and confers protection upon Mycobacterium tuberculosis challenge in mice. *Vaccine*, *35*(37), 4983–4989. doi:10.1016/j.vaccine.2017.07.073 PMID:28774560

Ahmed, T., Hyder, M. Z., Liaqat, I., & Scholz, M. (2019). Climatic Conditions: Conventional and Nanotechnology-Based Methods for the Control of Mosquito Vectors Causing Human Health Issues. *International Journal of Environmental Research and Public Health*, *16*(17), 3165. doi:10.3390/ijerph16173165 PMID:31480254

Ahn, J. M., Eom, H. J., Yang, X., Meyer, J. N., & Choi, J. (2014). Comparative toxicity of silver nanoparticles on oxidative stress and DNA damage in the nematode, Caenorhabditis elegans. *Chemosphere*, *108*, 343–352. doi:10.1016/j.chemosphere.2014.01.078 PMID:24726479

Aitken, R., Ks, C., & Tran, L. (2004). Nanoparticle: An Occupational Hygiene Review. *Institute of Occupational Medicine for the Health and Safety Executive*, 1-102. doi:10.1055os-SD-013-01214

Ajayan, P. M. (2004). Nanotechnology: How does a nanofibre grow? *Nature*, *427*(6973), 402–403. doi:10.1038/427402a PMID:14749810

Akal, Z. Ü., Alpsoy, L., & Baykal, A. (2016a). Superparamagnetic iron oxide conjugated with folic acid and carboxylated quercetin for chemotherapy applications. *Ceramics International*, *42*(7), 9065–9072. doi:10.1016/j.ceramint.2016.02.166

Akal, Z. Ü., Alpsoy, L., & Baykal, A. (2016b). Biomedical applications of SPION@APTES@PEG-folic acid@carboxylated quercetin nanodrug on various cancer cells. *Applied Surface Science*, *378*, 572–581. doi:10.1016/j.apsusc.2016.03.217

Akbari, B., Tavandashti, M. P., & Zandrahimi, M. (2011). Particle size characterization of nanoparticles—A practicalapproach. *Iranian Journal of Materials Science and Engineering*, *8*, 48–56.

Akbari, V., Abedi, D., Pardakhty, A., & Sadeghi-Aliabadi, H. (2013). Ciprofloxacin nano niosomes for targeting intracellular infections: An *in vitro* evaluation. *Journal of Nanoparticle Research*, *15*(4), 1–14. doi:10.100711051-013-1556-y

Akbarzadeh, A., Samiei, M., & Davaran, S. (2012). Magnetic nanoparticles: Preparation, physical properties, and applications in biomedicine. *Nanoscale Research Letters*, *7*(1), 144. doi:10.1186/1556-276X-7-144 PMID:22348683

Akinsiku, A. A., Dare, E. O., Ajanaku, K. O., Ajani, O. O., Olugbuyiro, J. A. O., Siyanbola, T. O., … Emetere, M. E. (2018). Modeling and synthesis of ag and ag/Ni allied bimetallic nanoparticles by green method: Optical and biological properties. *International Journal of Biomaterials, 2018*(3). doi:10.1155/2018/9658080

Al Faraj, A., Shaik, A. S., Ratemi, E., & Halwani, R. (2016). Combination of drug-conjugated SWCNT nanocarriers for efficient therapy of cancer stem cells in a breast cancer animal model. *Journal of Controlled Release*, *225*, 240–251. doi:10.1016/j.jconrel.2016.01.053 PMID:26827662

Alam, T., Khan, R. A. A., Ali, A., Sher, H., Ullah, Z., & Ali, M. (2019). Biogenic synthesis of iron oxide nanoparticles via Skimmia laureola and their antibacterial efficacy against bacterial wilt pathogen Ralstonia solanacearum. *Materials Science and Engineering C*, *98*, 101–108. doi:10.1016/j.msec.2018.12.117 PMID:30812984

Alan, H. F., Neil, A., Keith, R. M., & Andrew, P. W. (2017). Drug resistance in eukaryotic microorganisms. *Nature Microbiology*, *1*(7), 16092. doi:10.1038/nmicrobiol.2016.92 PMID:27572976

Alavi, M., & Rai, M. (2019). Recent advances in antibacterial applications of metal nanoparticles (MNPs) and metal nanocomposites (MNCs) against multi-drug-resistant (MDR) bacteria. *Expert Review of Anti-Infective Therapy*, *17*(6), 419–428. doi:10.1080/14787210.2019.1614914 PMID:31046483

Alberti, D., Protti, N., Franck, M., Stefania, R., Bortolussi, S., Altieri, S., Deagostino, A., Aime, S., & Geninatti Crich, S. (2017). Theranostic Nanoparticles Loaded with Imaging Probes and Rubrocurcumin for Combined Cancer Therapy by Folate Receptor Targeting. *ChemMedChem*, *12*(7), 502–509. doi:10.1002/cmdc.201700039 PMID:28217982

Alcalde-Rico, M., Hernando-Amado, S., Blanco, P., Martínez, J. L. (2016). Multidrug Efflux Pumps at the Crossroad between Antibiotic Resistance and Bacterial Virulence. *Frontiers in Microbiology, 21*(7), 1483.

Alexander, A. (2012). Self-propelled nanotubes. *ACS Nano*, *6*(2), 1751–1756. doi:10.1021/nn204762w PMID:22233271

Alexander, B. D., Johnson, M. D., Pfeiffer, C. D., Jiménez-Ortigosa, C., Catania, J., Booker, R., Castanheira, M., Messer, S. A., Perlin, D. S., & Pfaller, M. A. (2013). Increasing echinocandin resistance in Candida glabrata: Clinical failure correlates with presence of FKS mutations and elevated minimum inhibitory concentrations. *Clinical Infectious Diseases*, *56*(12), 1724–1732. doi:10.1093/cid/cit136 PMID:23487382

Alexander, J. W. (2009). History of the medical use of silver. *Surgical Infections*, *10*(3), 289–292. doi:10.1089ur.2008.9941 PMID:19566416

Alexiou, C., Arnold, W., Klein, R. J., Parak, F. G., Hulin, P., Bergemann, C., & … . (2000). Locoregional cancer treatment with magnetic drug targeting. *Cancer Research*, *60*(23), 6641–6648. PMID:11118047

Alexis, F., Pridgen, E., Molnar, L. K., & Farokhzad, O. C. (2008). Factors affecting the clearance and biodistribution of polymeric nanoparticles. *Molecular Pharmaceutics*, *5*(4), 505–515. doi:10.1021/mp800051m PMID:18672949

Alex, S., & Tiwari, A. (2015). Functionalized gold nanoparticles: Synthesis, properties and applications—A review. *Journal of Nanoscience and Nanotechnology*, *15*(3), 1869–1894. doi:10.1166/jnn.2015.9718 PMID:26413604

Ali, J. (2012). *Polymeric Nanoparticles, Magnetic Nanoparticles and Quantum Dots: Current and Future Perspectives* (E. B. Souto, Ed.). Patenting Nanomedicines, Springer-Verlag Berlin Heidelberg.

Alix Panabieres, C., & Pantel, K. (2014). Challenges in circulating tumour cell research. *Nature Reviews. Cancer*, *14*(9), 623–631. doi:10.1038/nrc3820 PMID:25154812

Allahverdiyev, A. M., Kon, K. V., Abamor, E. S., Bagirova, M., & Rafailovich, M. (2011). Coping with antibiotic resistance: Combining nanoparticles with antibiotics and other antimicrobial agents. *Expert Review of Anti-Infective Therapy*, *9*(11), 1035–1052. doi:10.1586/eri.11.121 PMID:22029522

Allaker, R. P., & Ren, G. (2008). Potential impact of nanotechnology on the control of infectious disease. *Transactions of the Royal Society of Tropical Medicine and Hygiene*, *102*(1), 1–2. doi:10.1016/j.trstmh.2007.07.003 PMID:17706258

Allard-Vannier, E., Hervé-Aubert, K., Kaaki, K., Blondy, T., Shebanova, A., Shaitan, K. V., Ignatova, A. A., Saboungi, M.-L., Feofanov, A. V., & Chourpa, I. (2017). Folic acid-capped PEGylated magnetic nanoparticles enter cancer cells mostly via clathrin-dependent endocytosis. *Biochimica et Biophysica Acta. G, General Subjects*, *1861*(6), 1578–1586. doi:10.1016/j.bbagen.2016.11.045 PMID:27919801

Allen, T. M., & Martin, F. J. (2004). Advantages of liposomal delivery systems for anthracyclines. *Seminars in Oncology*, *31*, 5–15. doi:10.1053/j.seminoncol.2004.08.001 PMID:15717735

Almajhdi, F., Fouad, H., Khalil, K., Awad, H., Mohamed, S., Elsarnagawy, T., & ... . (2014). In-vitro anticancer and antimicrobial activities of PLGA/ silver nanofiber composites prepared by electrospinning. *Journal of Materials Science*, *25*(4), 1045–1053. PMID:24375170

Almeida Da Silva, P. E., & Palomino, J. C. (2011). Molecular basis and mechanisms of drug resistance in Mycobacterium tuberculosis: Classical and new drugs. *The Journal of Antimicrobial Chemotherapy*, *66*(7), 1417–1430. doi:10.1093/jac/dkr173 PMID:21558086

Aloush, V., Navon-Venezia, S., Seigman-Igra, Y., Cabili, S., & Carmeli, Y. (2006). Multidrug-resistant Pseudomonas aeruginosa: Risk factors and clinical impact. *Antimicrobial Agents and Chemotherapy*, *50*(1), 43–48. doi:10.1128/AAC.50.1.43-48.2006 PMID:16377665

Al-Shabib, N. A., Husain, F. M., Ahmed, F., Khan, R. A., Khan, M. S., Ansari, F. A., ... Khan, J. M. (2018). Low temperature synthesis of superparamagnetic iron oxide (Fe3O4) nanoparticles and their ROS mediated inhibition of biofilm formed by food-associated bacteria. *Frontiers in Microbiology*, *9*, 2567. doi:10.3389/fmicb.2018.02567 PMID:30464757

Altmann, D. M. (2018). A Nobel prize-worthy pursuit: Cancer immunology and harnessing immunity to tumour neoantigens. *Immunology*, *155*(3), 283–284. doi:10.1111/imm.13008 PMID:30320408

Alumutairi, L., Yu, B., Filka, M., Nayfach, J., & Kim, M.-H. (2020). Mild magnetic nanoparticle hyperthermia enhances the susceptibility of *Staphylococcus aureus* biofilm to antibiotics. *International Journal of Hyperthermia*, *37*(1), 66–75. doi:10.1080/02656736.2019.1707886 PMID:31964196

Álvarez-Paino, M., Muñoz-Bonilla, A., & Fernández-García, M. (2017). Antimicrobial polymers in the nano-world. *Nanomaterials (Basel, Switzerland)*, *7*(2), 48. doi:10.3390/nano7020048 PMID:28336882

Amanda, Silva, Espinosa, Kolosnjaj-Tabi, Wilhelm, & Gazeau. (2016). *Medical Applications of Iron Oxide Nanoparticles. In Iron Oxides: From Nature to Applications*. Wiley-VCH Verlag GmbH & Co. KGaA.

Amber, M. (2014). Delivery of antiviral small interfering RNA with gold nanoparticles inhibits dengue virus infection in vitro. *The Journal of General Virology*, *95*(8), 1712–1722. doi:10.1099/vir.0.066084-0 PMID:24828333

Amerasan, D., Nataraj, T., Murugan, K., Panneerselvam, C., Madhiyazhagan, P., Nicoletti, M., & Benelli, G. (2016). Myco-synthesis of silver nanoparticles using Metarhizium anisopliae against the rural malaria vector Anopheles culicifacies Giles (Diptera: Culicidae). *Journal of Pest Science*, *89*(1), 249–256. doi:10.100710340-015-0675-x PMID:28217039

Amin. (2017). Osmotin-loaded Magnetic Nanoparticles with an Electromagnetic Guidance for the Treatment of Alzheimer's disease. *Nanoscale*.

Amini, F., Semnani, D., Karbasi, S., & Banitaba, S. N. (2018). A novel bilayer drug-loaded wound dressing of PVDF and PHB/Chitosan nanofibers applicable for post-surgical ulcers. *International Journal of Polymeric Materials and Polymeric Biomaterials*, *68*(13), 772–777. doi:10.1080/00914037.2018.1506982

Amin, M. C. I. M., Ahmad, N., Halib, N., & Ahmad, I. (2012). Synthesis and characterization of thermo-and pH-responsive bacterial cellulose/acrylic acid hydrogels for drug delivery. *Carbohydrate Polymers*, *88*(2), 465–473. doi:10.1016/j.carbpol.2011.12.022

Amiri, H., Saeidi, K., Borhani, P., Manafirad, A., Ghavami, M., & Zerbi, V. (2013). Alzheimer's Disease: Pathophysiology and Applications of Magnetic Nanoparticles as MRI Theranostic Agents. *ACS Chemical Neuroscience*, *4*(11), 1417–1429. Advance online publication. doi:10.1021/cn4001582 PMID:24024702

Anandalakshmi, K., Venugobal, J., & Ramasamy, V. (2016). Characterization of silver nanoparticles by green synthesis method using Pedalium murex leaf extract and their antibacterial activity. *Applied Nanoscience (Switzerland)*, *6*(3), 399–408. doi:10.100713204-015-0449-z

Anchordoquy, T. J., Barenholz, Y., Boraschi, D., Chorny, M., Decuzzi, P., Dobrovolskaia, M. A., ... Godin, B. (2017). *Mechanisms and barriers in cancer nanomedicine: addressing challenges, looking for solutions*. Academic Press.

Ancira-Cortez, A., Morales-Avila, E., Ocampo-García, B. E., González-Romero, C., Medina, L. A., López-Téllez, G., & Cuevas-Yáñez, E. (2017). Preparation and characterization of a tumor-targeting dual-image system based on iron oxide nanoparticles functionalized with folic acid and rhodamine. *Journal of Nanomaterials*, *2017*(1), 1–11. doi:10.1155/2017/5184167

Anderson, D. M. W., & Bridgeman, M. M. E. (1985). The composition of the proteinaceous polysaccharides exuded by astragalus microcephalus, A. Gummifer and A. Kurdicus - The sources of turkish gum tragacanth. *Phytochemistry*, *24*(10), 2301–2304. doi:10.1016/S0031-9422(00)83031-9

Andersson, D. I., & Hughes, D. (2010). Antibiotic resistance and its cost: Is it possible to reverse resistance? *Nature Reviews. Microbiology*, *8*(4), 260–271. doi:10.1038/nrmicro2319 PMID:20208551

Anghel, I., Grumezescu, A. M., Andronescu, E., Anghel, A. G., Ficai, A., Saviuc, C., Grumezescu, V., Vasile, B. S., & Chifiriuc, M. C. (2012). Magnetite nanoparticles for functionalized textile dressing to prevent fungal biofilms development. *Nanoscale Research Letters*, *7*(1), 7. doi:10.1186/1556-276X-7-501 PMID:22950367

Anirudhan, T. S., & Christa Binusreejayan, J. (2018). pH and magnetic field sensitive folic acid conjugated protein–polyelectrolyte complex for the controlled and targeted delivery of 5-fluorouracil. *Journal of Industrial and Engineering Chemistry*, *57*, 199–207. doi:10.1016/j.jiec.2017.08.024

Ansari, M. A., Khan, H. M., Khan, A. A., Cameotra, S. S., & Pal, R. (2014). Antibioÿlm efÿcacy of silver nanoparticles against bioÿlm of extended spectrum b-lactamase isolates of Escherichia coli and Klebsiella pneumoniae. Applied Nanoscience, 4(7), 859–868. doi:10.1007/s13204-013-0266-1

Ansari, M. A., Khan, H. M., Alzohairy, M. A., Jalal, M., Ali, S. G., Pal, R., & Musarrat, J. (2015). Green synthesis of Al_2O_3 nanoparticles and their bactericidal potential against clinical isolates of multi-drug resistant *Pseudomonas aeruginosa*. *World Journal of Microbiology & Biotechnology*, *31*(1), 153–164. doi:10.100711274-014-1757-2 PMID:25304025

Ansari, M. A., Khan, H. M., Khan, A. A., Cameotra, S. S., Saquib, Q., & Musarrat, J. (2014). Interaction of Al2O3 nanoparticles with Escherichia coli and their cell envelope biomolecules. *Journal of Applied Microbiology*, *116*(4), 772–783. doi:10.1111/jam.12423 PMID:24354999

Anselmo, A. C., & Mitragotri, S. (2016). Nanoparticles in the clinic. *Bioengineering & Translational Medicine*, *1*(1), 10–29. doi:10.1002/btm2.10003 PMID:29313004

Antal, I., & Kubovcikova, M. (2015). Magnetic poly(D,L-lactide) nanoparticles loaded with aliskiren: A promising tool for hypertension treatment. *Journal of Magnetism and Magnetic Materials*, *380*, 280–284. doi:10.1016/j.jmmm.2014.10.089

Antonelli, G., & Turriziani, O. (2012). Antiviral therapy: Old and current issues. *International Journal of Antimicrobial Agents*, *40*(2), 95–102. doi:10.1016/j.ijantimicag.2012.04.005 PMID:22727532

Appelbaum, P. (2006). The emergence of vancomycin-intermediate and vancomycin-resistant *Staphylococcus aureus*. *Clinical Microbiology and Infection*, *12*, 16–23. doi:10.1111/j.1469-0691.2006.01344.x PMID:16445720

Appelbaum, P. C., & Jacobs, M. R. (2005). Recently approved and investigational antibiotics for treatment of severe infections caused by Gram-positive bacteria. *Current Opinion in Microbiology*, *8*(5), 510–517. doi:10.1016/j.mib.2005.07.001 PubMed

Aqel, M., El Nour, R.A., Ammar, A., & Warthan, A. (2012). Carbon nanotubes, science and technology part (I) structure, synthesis and characterisation. *Arabian Journal of Chemistry*, *5*(1), 1-23.

Arachchige, M. P., Laha, S. S., Naik, A. R., Lewis, K. T., Naik, R., & Jena, B. P. (2017). Functionalized nanoparticles enable tracking the rapid entry and release of doxorubicin in human pancreatic cancer cells. *Micron (Oxford, England)*, *92*, 25–31. doi:10.1016/j.micron.2016.10.005 PMID:27846432

Arakha, M., Pal, S., Samantarrai, D., Panigrahi, T. K., Mallick, B. C., Pramanik, K., Mallick, B., & Jha, S. (2015). Antimicrobial activity of iron oxide nanoparticle upon modulation of nanoparticle-bacteria interface. *Scientific Reports*, *5*(1), 14813. doi:10.1038rep14813 PMID:26437582

Aravindan, V., Sundaramurthy, J., Suresh Kumar, P., Lee, Y. S., Ramakrishna, S., & Madhavi, E. (2015). Electrospun nanofibers:a prospective electroactive material for constructing high performance Li-ion batteries. *Chemical Communications*, *51*(12), 2225–2234. doi:10.1039/C4CC07824A PMID:25493289

Archer, G. L. (1998). Staphylococcus aureus: A well-armed pathogen. *Clinical Infectious Diseases*, *26*(5), 1179–1181. doi:10.1086/520289 PubMed

Arciola, C. R., Campoccia, D., Speziale, P., Montanaro, L., & Costerton, J. W. (2012). Biofilm formation in *Staphylococcus* implant infections. A review of molecular mechanisms and implications for biofilm-resistant materials. *Biomaterials*, *33*(26), 5967–5982. doi:10.1016/j.biomaterials.2012.05.031 PMID:22695065

Argyo, C., Weiss, V., Bräuchle, C., & Bein, T. (2014). Multifunctional mesoporous silica nanoparticles as a universal platform for drug delivery. *Chemistry of Materials*, *26*(1), 435–451. doi:10.1021/cm402592t

Arias, J. L., López-Viota, M., López-Viota, J., & Delgado, Á. V. (2009). Development of iron/ethylcellulose (core/shell) nanoparticles loaded with diclofenac sodium for arthritis treatment. *International Journal of Pharmaceutics*, *382*(1-2), 270–276. doi:10.1016/j.ijpharm.2009.08.019 PMID:19712736

Arias, L. S., Pessan, J. P., Vieira, A. P. M., Lima, T. M. T. D., Delbem, A. C. B., & Monteiro, D. R. (2018). Iron oxide nanoparticles for biomedical applications: A perspective on synthesis, drugs, antimicrobial activity, and toxicity. *Antibiotics (Basel, Switzerland)*, *7*(2), 46. doi:10.3390/antibiotics7020046 PMID:29890753

Arlov, O., Aachmann, F. L., Sundan, A., Espevik, T., & Skjåk-Bræk, G. (2014). Heparin-like properties of sulfated alginates with defined sequences and sulfation degrees. *Biomacromolecules*, *15*(7), 2744–2750. doi:10.1021/bm500602w PMID:24844124

Armbruster, C. E., Mobley, H. L. T., & Pearson, M. M. (2018). Pathogenesis of Proteus mirabilis Infection. *EcoSal Plus*, *8*(1), ESP-0009–ESP-2017. doi:10.1128/ecosalplus.ESP-0009-2017 PubMed

Armendariz, V., Herrera, I., Jose-yacaman, M., Troiani, H., & Santiago, P. (2004). Size controlled gold nanoparticle formation by Avena sativa biomass: Use of plants in nanobiotechnology. *Journal of Nanoparticle Research*, *6*(4), 377–382. doi:10.100711051-004-0741-4

Arokiyaraj, S., Saravanan, M., Prakash, N. U., Arasu, M. V., Vijayakumar, B., & Vincent, S. (2013). Enhanced antibacterial activity of iron oxide magnetic nanoparticles treated with Argemone mexicana L. leaf extract: An in vitro study. *Materials Research Bulletin*, *8*(9), 3323–3327. doi:10.1016/j.materresbull.2013.05.059

Arora, S., Jain, J., Rajwade, J. M., & Paknikar, K. M. (2009). Interactions of silver nanoparticles with primary mouse fibroblasts and liver cells. *Toxicology and Applied Pharmacology*, *236*(3), 310–318. doi:10.1016/j.taap.2009.02.020 PMID:19269301

Arora, S., Rajwade, J. M., & Paknikar, K. M. (2012). Nanotoxicology and in vitro studies: The need of the hour. *Toxicology and Applied Pharmacology*, *258*(2), 151–165. doi:10.1016/j.taap.2011.11.010 PMID:22178382

Artemov, D., Mori, N., Okollie, B., & Bhujwalla, Z. M. (2003a). MR molecular imaging of the Her-2/ neu receptor in breast cancer cells using targeted iron oxide nanoparticles. *Magnetic Resonance in Medicine*, *49*(3), 403–408. doi:10.1002/mrm.10406 PMID:12594741

Artemov, D., Mori, N., Ravi, R., & Bhujwalla, Z. M. (2003b). Magnetic resonance molecular imaging of the HER-2/ neu receptor. *Cancer Research*, *63*, 2723–2727. PMID:12782573

Arumai Selvan, D., Mahendiran, D., Senthil Kumar, R., & Kalilur Rahiman, A. (2018). Garlic, green tea and turmeric extracts-mediated green synthesis of silver nanoparticles: Phytochemical, antioxidant and in vitro cytotoxicity studies. *Journal of Photochemistry and Photobiology B:. Biology*, *180*(2017), 243–252. doi:10.1016/j.jphotobiol.2018.02.014

Arya, A., Gupta, K., Chundawat, T. S., & Vaya, D. (2018). Biogenic synthesis of copper and silver nanoparticles using green alga Botryococcus braunii and its antimicrobial activity. *Bioinorganic Chemistry and Applications*, *2018*, 2018. doi:10.1155/2018/7879403 PMID:30420873

Asadi, L., Shirvalilou, S., Khoee, S., & Khoei, S. (2018). Cytotoxic effect of 5-fluorouracil-loaded polymer-coated magnetite nanographene oxide combined with radiofrequency. *Anti-cancer Agents in Medicinal Chemistry*, *18*(8), 1148–1155. doi:10.2174/1871520618666180404151218 PMID:29623854

AshaRani, P. V., Low Kah Mun, G., Hande, M. P., & Valiyaveettil, S. (2009). Cytotoxicity and genotoxicity of siver nanoparticles in human cells. *ACS Nano*, *3*(2), 279–290. doi:10.1021/nn800596w PMID:19236062

Assa, F., Jafarizadeh-Malmiri, H., Ajamein, H., Anarjan, N., Vaghari, H., Sayyar, Z., & Berenjian, A. (2016). A biotechnological perspective on the application of iron oxide nanoparticles. *Nano Research*, *9*(8), 2203–2225. doi:10.100712274-016-1131-9

Astefanei, A., Núñez, O., & Galceran, M. T. (2015). Characterisation and determination of fullerenes: A critical review. *Analytica Chimica Acta*, *882*, 1–21. doi:10.1016/j.aca.2015.03.025 PMID:26043086

ATSDR (Agency for toxic substances and Disease Registry). (1990). *Toxicological profile for Silver*. Prepared by Clement international corporation, under Contract 205-88-0608. U.S. public Health Service. ATSDR/TP-90-24.

Atterbury, R. J. (2009). Bacteriophage biocontrol in animals and meat products. *Microbial Biotechnology*, *2*(6), 601–612. doi:10.1111/j.1751-7915.2009.00089.x PMID:21255295

Attia, M. S., Youssef, A. O., Khan, Z. A., & Abou-Omar, M. N. (2018). Alpha fetoprotein assessment by using a nano optical sensor thin film binuclear Pt-2-aminobenzimidazole-Bipyridine for early diagnosis of liver cancer. *Talanta*, *186*, 36–43. doi:10.1016/j.talanta.2018.04.043 PMID:29784373

Aviv, M., Halperin-Sternfeld, M., Grigoriants, I., Buzhansky, L., Mironi-Harpaz, I., Seliktar, D., Einav, S., Nevo, Z., & Adler-Abramovich, L. (2018). Improving the Mechanical Rigidity of Hyaluronic Acid by Integration of a Supramolecular Peptide Matrix. *ACS Applied Materials & Interfaces*, *10*(49), 41883–41891. doi:10.1021/acsami.8b08423 PMID:30211538

Awad, H. M., Kamal, Y. S., Aziz, R., Sarmidi, M. R., & El-Enshasy, H. A. (2012). Antibiotics as microbial secondary metabolites: Production and application. *Jurnal Teknologi*, *59*, 101–111. doi:10.11113/jt.v59.1593

Azam, A., Ahmed, A. S., Oves, M., Khan, M. S., Habib, S. S., & Memic, A. (2012). Antimicrobial activity of metal oxide nanoparticles against Gram-positive and Gram-negative bacteria: A comparative study. *International Journal of Nanomedicine*, *7*, 6003. doi:10.2147/IJN.S35347 PMID:23233805

Azarniya, M., Tamjid, E., Eslahi, N., & Simchi, A. (2019). Modification of bacterial cellulose/keratin nanofibrous mats by a tragacanth gum-conjugated hydrogel for wound healing. *International Journal of Biological Macromolecules*, *134*, 280–289. doi:10.1016/j.ijbiomac.2019.05.023 PMID:31071405

Aznar, R., Barahona, F., Geiss, O., Ponti, J., Luis, T. J., & Moreno, J. B. (2017). Quantification and size characterization of silver nanoparticles in environmental aquous samples and consumer products by single particle ICPMS. *Talanta*, *175*, 200–208. doi:10.1016/j.talanta.2017.07.048 PMID:28841979

Badri Narayanan, K., & Sakthivel, N. (2018). Coriander leaf mediated biosynthesis of gold nanoparticles. *Materials Letters*, *62*(30), 4588–4590. doi:10.1016/j.matlet.2008.08.044

Bae, Y. H., & Park, K. (2011). Targeted drug delivery to tumors: Myths, reality and possibility. *Journal of Controlled Release*, *153*(3), 198–205. doi:10.1016/j.jconrel.2011.06.001 PMID:21663778

Baeza, A., & Vallet-Regí, M. (2018). Nanomotors for Nucleic Acid, Proteins, Pollutants and Cells Detection. *International Journal of Molecular Sciences*, *19*(6), 1579. doi:10.3390/ijms19061579 PMID:29799489

Bahramia, B., Hojjat-Farsangi, M., Mohammadi, H., Anvari, E., Ghalamfarsa, G., Yousefi, M., & Jadidi-Niaragh, F. (2017). Nanoparticles and targeted drug delivery in cancer therapy. *Immunology Letters*, *190*, 64–83. doi:10.1016/j.imlet.2017.07.015 PMID:28760499

Bai, J., Kim, Y. T., Ryu, S., & Lee, J. H. (2016). Biocontrol and Rapid Detection of Food-Borne Pathogens Using Bacteriophages and Endolysins. *Frontiers in Microbiology*, *7*, 474. doi:10.3389/fmicb.2016.00474 PMID:27092128

Bailey, V. A., Mackey, L. N., & Trokhan, P. D. (2010). *Starch fiber.* US Patent, 7,704328.

Baker, K. R., Jana, B., Hansen, A. M., Vissing, K. J., Nielsen, H. M., Franzyk, H., & Guardabassi, L. (2019). Repurposing azithromycin and rifampicin against Gram-negative pathogens by combination with peptide potentiators. *International Journal of Antimicrobial Agents*, *53*(6), 868–872. doi:10.1016/j.ijantimicag.2018.10.025 PMID:30447380

Baker, S., Pasha, A., & Satish, S. (2017). Biogenic nanoparticles bearing antibacterial activity and their synergistic effect with broad spectrum antibiotics: Emerging strategy to combat drug resistant pathogens. *Saudi Pharmaceutical Journal*, *25*(1), 44–51. doi:10.1016/j.jsps.2015.06.011 PMID:28223861

Baker, S., Rakshith, D., & Kavitha, K. S. (2013). Plants: Emerging as nanofactories towards facile route in synthesis of nanoparticles. *BioImpacts*, *3*, 111–117. PMID:24163802

Balaghi, S., Mohammadifar, M. A., Zargaraan, A., Gavlighi, H. A., & Mohammadi, M. (2011). Compositional analysis and rheological characterization of gum tragacanth exudates from six species of Iranian Astragalus. *Food Hydrocolloids*, *25*(7), 1775–1784. doi:10.1016/j.foodhyd.2011.04.003

Balakrishnan, S., Srinivasan, M., & Mohanraj, J. (2016). Biosynthesis of silver nanoparticles from mangrove plant (Avicennia marina) extract and their potential mosquito larvicidal property. *Journal of Parasitic Diseases : Official Organ of the Indian Society for Parasitology*, *40*(3), 991–996. doi:10.100712639-014-0621-5 PMID:27605825

Balasubramanian, S., Kagan, D., Hu, C.M.J., Campuzano, S., Castañon, M.J.L., Dae, N.L., Zimmerman,Y.K.M., Zhang, L., & Wang, J. (2011). *Micromachine-Enabled Capture and Isolation of Cancer Cells in Complex Media*. doi:.1002/anie.20110011

Balasubramanyam, A., Sailaja, N., Mahboob, M., Rahman, M. F., Hussain, S. M., & Grover, P. (2010). In vitro mutagenicity assessment ofaluminium oxide nanomaterials using the Salmonella/microsome assay. *Toxicology In Vitro*, *24*(6), 1871–1876. doi:10.1016/j.tiv.2010.07.004 PMID:20624453

Bali, E. B., Açık, L., & Sultan, N. (2010). Phenotypic and molecular characterization of SHV, TEM, CTX-M and extended-spectrum β-lactamase produced by Escherichia coli, Acinetobacter baumannii and Klebsiella isolates in a Turkish hospital. *African Journal of Microbiological Research*, *4*, 650–654.

Balsells, E., Shi, T., Leese, C., Lyell, I., Burrows, J., Wiuff, C., Campbell, H., Kyaw, M. H., & Nair, H. (2019). Global burden of Clostridium difficile infections: A systematic review and meta-analysis. *Journal of Global Health*, *9*(1), 010407. doi:10.7189/jogh.09.010407 PMID:30603078

Balzarini, J., Ford, A., Maguire, N. M., John, J., Das, K., & Arnold, E. (n.d.). Alpha-carboxynucleoside phosphonates: direct-acting inhibitors of viral DNA polymerases. *Future Medicinal Chemistry, 11*(2). doi:10.4155/fmc-2018-0324

Bandyopadhyay, S., Jose, R. P., & Jorge, L. G. (2013). Advanced analytical techniques for the measurement of nanomaterials in food and agricultural samples: A review. *Environmental Engineering Science*, *30*(3), 118–125. doi:10.1089/ees.2012.0325 PMID:23483065

Banik, B. L., Fattahi, P., & Brown, J. L. (2016). Polymeric nanoparticles: The future of nanomedicine. *Wiley Interdisciplinary Reviews. Nanomedicine and Nanobiotechnology*, *8*(2), 271–299. doi:10.1002/wnan.1364 PMID:26314803

Baptista, P. V., McCusker, M. P., Carvalho, A., Ferreira, D. A., Mohan, N. M., Martins, M., & Fernandes, A. R. (2018). Nano-strategies to fight multidrug resistant bacteria—"A Battle of the Titans. *Frontiers in Microbiology*, *9*, 1441. doi:10.3389/fmicb.2018.01441 PMID:30013539

Barabadi, H., Alizadeh, Z., Rahimi, M. T., Barac, A., Maraolo, A. E., Robertson, L. J., Masjedi, A., Shahrivar, F., & Ahmadpour, E. (2019). Nanobiotechnology as an emerging approach to combat malaria: A systematic review. *Nanomedicine; Nanotechnology, Biology, and Medicine*, *18*, 221–233. doi:10.1016/j.nano.2019.02.017 PMID:30904586

Baraton, M.-I. (2002). Surface Analysis of Semiconducting Nanoparticles by FTIR Spectroscopy. *Nanocrystalline Metals and Oxides*, *7*(1), 165–187. doi:10.1007/0-306-47609-6_6

Barber, M. (1961). Methicillin-resistant staphylococci. *Journal of Clinical Pathology*, *14*(4), 385–393. doi:10.1136/jcp.14.4.385 PMID:13686776

Barenholz, Y. C. (2012). Doxil®—the first FDA-approved nano-drug: Lessons learned. *Journal of Controlled Release*, *160*(2), 117–134. doi:10.1016/j.jconrel.2012.03.020 PMID:22484195

Barhoum, A., Pal, K., Rahierd, H., Uludage, H., Kim, I. S., & Bechelanya, M. (2019). Nanofibers as new-generation materials: From spinning and nano-spinning fabrication techniques to emerging applications. *Applied Materials Today*, *17*, 1–35. doi:10.1016/j.apmt.2019.06.015

Barlow, P. G., Donaldson, K., MacCallum, J., Clouter, A., & Stone, V. (2005). Serum exposed tonanoparticle carbon black displays increased potential to induce macrophage migration. *Toxicology Letters*, *155*(3), 397–401. doi:10.1016/j.toxlet.2004.11.006 PMID:15649623

Barrak, H., Saied, T., Chevallier, P., Laroche, G., M'nif, A., & Hamzaoui, A. H. (2019). Synthesis, characterization, and functionalization of ZnO nanoparticles by N-(trimethoxysilylpropyl) ethylenediamine triacetic acid (TMSEDTA). *Arabian Journal of Chemistry*, *12*(8), 4340–4347. doi:10.1016/j.arabjc.2016.04.019

Barrangou, R., Fremaux, C., Deveau, H., Richards, M., Boyaval, P., Moineau, S., Romero, D. A., & Horvath, P. (2007). CRISPR provides acquired resistance against viruses in prokaryotes. *Science*, *315*(5819), 1709–1712. doi:10.1126cience.1138140 PMID:17379808

Barry, J. N., & Vertegel, A. A. (2013). Nanomaterials for Protein Mediated Therapy and Delivery. *Nano LIFE*, *3*(4), 1–17. doi:10.1142/S1793984413430010 PMID:25414730

Basha, R. Y., TS, S. K., & Doble, M. (. (2019). Dual delivery of tuberculosis drugs via cyclodextrin conjugated curdlan nanoparticles to infected macrophages. *Carbohydrate Polymers*, *218*, 53–62. doi:10.1016/j.carbpol.2019.04.056 PMID:31221343

Bate, R., Jin, G. Z., Mathur, A., & Attaran, A. (2016). Poor Quality Drugs and Global Trade: A Pilot Study. *American Journal of Health Economics*, *2*(3), 373–398. doi:10.1162/AJHE_a_00049

Baym, M., Stone, L. K., & Kishony, R. (2016). Multidrug evolutionary strategies to reverse antibiotic resistance. *Science*, *351*(6268), 3292. doi:10.1126cience.aad3292 PMID:26722002

Bazrgari, F., Khameneh, B., Fazly Bazzaz, B. S., Mahmoudi, A., & Malaekeh-Nikouei, B. (2020). Effect of the nanoliposomal formulations of rifampin and N-acetyl cysteine on staphylococcus epidermidis biofilm. *Nanomedicine Journal*, *7*(2), 131–137.

Bechara, C., & Sagan, S. (2013). Cell-penetrating peptides: 20 years later, where do we stand? *FEBS Letters*, *587*(12), 1693–1702. doi:10.1016/j.febslet.2013.04.031 PMID:23669356

Becker, N., Petric, D., Zgomba, M., Boase, C., Madon, M., Dahl, C., & Kaiser, A. (2010). *Mosquitoes and their control*. Springer Science & Business Media. doi:10.1007/978-3-540-92874-4

Behera, S. (2012). UV-Visible Spectrophotometric Method Development and Validation of Assay of Paracetamol Tablet Formulation. *Journal of Analytical & Bioanalytical Techniques*, *03*(06), 2175–2187. doi:10.4172/2155-9872.1000151

Behera, S. S., Patra, J. K., Pramanik, K., Panda, N., & Thatoi, H. (2012). Characterization and Evaluation of antibacterial activities of chemically synthesized Iron Oxide nanoparticles. *World J. Nano Sci. Eng.*, *2*(04), 196–200. doi:10.4236/wjnse.2012.24026

Benelli, G. (2015b). Plant-borne ovicides in the fight against mosquito vectors of medical and veterinary importance: A systematic review. *Parasitology Research*, *114*(9), 3201–3212. doi:10.100700436-015-4656-z PMID:26239801

Benelli, G. (2016). Green synthesized nanoparticles in the fight against mosquito-borne diseases and cancer—A brief review. *Enzyme and Microbial Technology*, *95*, 58–68. doi:10.1016/j.enzmictec.2016.08.022 PMID:27866627

Benelli, G. (2018). Mode of action of nanoparticles against insects. *Environmental Science and Pollution Research International*, *25*(13), 12329–12341. doi:10.100711356-018-1850-4 PMID:29611126

Benelli, G., Govindarajan, M., Kadaikunnan, S., & Alharbi, N. S. (2017). What Kind of Reducing Botanical? High Mosquitocidal Efficacy of a Silver Nanocomposite Synthesized Using a Leaf Aqueous Extract of Fumaria indica. *Journal of Cluster Science*, *28*(1), 637–643. doi:10.100710876-017-1159-3

Bengtsson-Palme, J., Kristiansson, E., & Larsson, D. G. J. (2018). Environmental factors influencing the development and spread of antibiotic resistance. *FEMS Microbiology Reviews*, *42*(1). Advance online publication. doi:10.1093/femsre/fux053 PMID:29069382

Bennett, K.M., Jo, J., Cabral, H., Bakalova, R., & Aoki, I. (2014). MR imaging techniques for nano-pathophysiology and theranostics. *Adv. Drug Deliv*, 75-94.

Benoit, D. A., & Michel, B. T. (2007). ROS as signaling molecules: Mechanisms that generate specificity in ROS homeostasis. *Nature Reviews. Molecular Cell Biology*, *8*(10), 813–824. doi:10.1038/nrm2256 PMID:17848967

Benyettou, F., Fahs, H., Elkharrag, R., Bilbeisi, R. A., Asma, B., Rezgui, R., Motte, L., Magzoub, M., Brandel, J., Olsen, J.-C., Piano, F., Gunsalus, K. C., Platas-Iglesias, C., & Trabolsi, A. (2017). Selective growth inhibition of cancer cells with doxorubicin-loaded CB (Gholami et al., 2018)-modified iron-oxide nanoparticles. *RSC Advances*, *7*(38), 23827–23834. doi:10.1039/C7RA02693E

Bergemann, C., Müller-Schulte, D., Oster, J., à Brassard, L., & Lübbe, A. S. (1999). Magnetic ion-exchange nano- and microparticles for medical, biochemical and molecular biological applications. *Journal of Magnetism and Magnetic Materials*, *194*(1-3), 45–52. doi:10.1016/S0304-8853(98)00554-X

Bergmiller, T., Andersson, A. M. C., Tomasek, K., Balleza, E., Kiviet, D. J., Hauschild, R., & Guet, C. C. (2017). Biased partitioning of the multidrug efflux pump AcrAB-TolC underlies long-lived phenotypic heterogeneity. *Science*, *356*(6335), 311–315. doi:10.1126cience.aaf4762 PMID:28428424

Bernardi, P., Scorrano, L., Colonna, R., Petronilli, V., & Di Lisa, F. (1999). Mitochondria and cell death. *European Journal of Biochemistry*, *264*(3), 687–701. doi:10.1046/j.1432-1327.1999.00725.x PMID:10491114

Bertrand, N., & Leroux, J. C. (2012). The journey of a drug-carrier in the body: An anatomo-physiological perspective. *Journal of Controlled Release*, *161*(2), 152–163. doi:10.1016/j.jconrel.2011.09.098 PMID:22001607

Besinis, A., de Peralta, T., & Handy, R. D. (2014). The antibacterial effects of silver, titanium dioxide and silica dioxide nanoparticles compared to the dental disinfectant chlorhexidine on Streptococcus mutans using a suite of bioassays. *Nanotoxicology*, 8. PMID:23092443

Bettini, S., Giancane, G., Pagano, R., Bonfrate, V., Salvatore, L., Madaghiele, M., Buccolieri, A., Manno, D., Serra, A., Maruccio, G., Monteduro, A. G., Syrgiannis, Z., Valli, L., & Prato, M. (2017). A simple approach to synthetize folic acid decorated magnetite@SiO2 nanostructures for hyperthermia applications. *Journal of Materials Chemistry. B, Materials for Biology and Medicine*, *5*(36), 7547–7556. doi:10.1039/C7TB02002C PMID:32264230

Beveridge, T. J., & Murray, R. G. (1980). Sites of metal deposition in the cell wall of *Bacillus subtilis*. *Journal of Bacteriology*, *141*(2), 876–887. PMID:6767692

Beyth, N., Houri-Haddad, Y., Domb, A., Khan, W., & Hazan, R. (2015). Alternative antimicrobial approach: nano-antimicrobial materials. *Evid. Based Complement. Altern. Med.*, 1–16.

Beyth, N., Houri-Haddad, Y., Domb, A., Khan, W., & Hazan, R. (2015). Alternative antimicrobial approach: Nano-antimicrobial materials. *Evidence-Based Complementary and Alternative Medicine*, *20*, 151–169. PMID:25861355

Bhabra, G., Sood, A., Fisher, B., Cartwright, L., Saunders, M., Evans, W. H., Surprenant, A., Lopez-Castejon, G., Mann, S., Davis, S. A., Hails, L. A., Ingham, E., Verkade, P., Lane, J., Heesom, K., Newson, R., & Case, C. P. (2009). Nanoparticles can cause DNA damage across a cellular barrier. *Nature Nanotechnology*, *4*(12), 876–883. doi:10.1038/nnano.2009.313 PMID:19893513

Bharde, A. A., Parikh, R. Y., Baidakova, M., Jouen, S., Hannoyer, B., Enoki, T., Prasad, B. L. V., Shouche, Y. S., Ogale, S., & Sastry, M. (2008). Bacteria-mediated precursor-dependent biosynthesis of superparamagnetic iron oxide and iron sulfide nanoparticles. *Langmuir, 24*(11), 5787–5794. doi:10.1021/la704019p PMID:18454562

Bhardwaj, A., Kumar, L., Narang, R. K., & Murthy, R. S. R. (2013). Development and characterization of ligand-appended liposomes for multiple drug therapy for pulmonary tuberculosis. *Artificial Cells, Nanomedicine, and Biotechnology, 41*(1), 52–59. doi:10.3109/10731199.2012.702316 PMID:22889361

Bhardwaj, A., Mehta, S., Yadav, S., Singh, S. K., Grobler, A., Goyal, A. K., & Mehta, A. (2016). Pulmonary delivery of antitubercular drugs using spray-dried lipid–polymer hybrid nanoparticles. *Artificial Cells, Nanomedicine, and Biotechnology, 44*(6), 1544–1555. doi:10.3109/21691401.2015.1062389 PMID:26178768

Bhargava, A., Jain, N., Khan, M. A., Pareek, V., Dilip, R. V., & Panwar, J. (2016). Utilizing metal tolerance potential of soil fungus for efficient synthesis of gold nanoparticles with superior catalytic activity for degradation of rhodamine B. *Journal of Environmental Management, 1*(183), 22–32. doi:10.1016/j.jenvman.2016.08.021 PMID:27567934

Bharti, C., Nagaich, U., Pal, A. K., & Gulati, N. (2015). Mesoporous silica nanoparticles in target drug delivery system: A review. *International Journal of Pharmaceutical Investigation, 5*(3), 124. doi:10.4103/2230-973X.160844 PMID:26258053

Bhatia, S. (2016). Nanoparticles types, classification, characterization, fabrication methods and drug delivery applications. In *Natural polymer drug delivery systems* (pp. 33–93). Springer. doi:10.1007/978-3-319-41129-3_2

Bhattacharjee, A., Kumar, K., Arora, A., & Katti, D. S. (2016). Fabrication and characterization of Pluronic modified poly(hydroxybutyrate) fibers for potential wound dressing applications. *Materials Science and Engineering C, 63*, 266–273. doi:10.1016/j.msec.2016.02.074 PMID:27040219

Bhattarai, N., Li, Z., Edmondson, D., & Zhang, M. (2006). Alginate- based Nanofibrous Scaffolds: Structural, mechanical, and Biological Properties. *Advanced Materials, 18*(11), 1463–1467. doi:10.1002/adma.200502537

Bhatt, R. D. E., Vries, P., & Tulinsky, J. (2003). Synthesis and in vivo antitumor activity of poly (L-glutamic acid) conjugates of 20 (S)-camptothecin. *Journal of Medicinal Chemistry, 46*(1), 190–193. doi:10.1021/jm020022r PMID:12502373

Bhumkar, D. R., Joshi, H. M., Sastry, M., & Pokharkar, V. B. (2007). Chitosan reduced gold nanoparticles as novel carriers for transmucosal delivery of insulin. *Pharmaceutical Research, 24*(8), 1415–1426. doi:10.100711095-007-9257-9 PMID:17380266

Biehl, P., Luhe, M. V., Dutz, S., & Schacher, F. H. (2018). Synthesis, characterization and applications of magnetic nanoparticles featuring polyzwitterionic coatings. *Polymers, 10*(1), 1–28. doi:10.3390/polym10010091 PMID:30966126

Binandeh, M. (2018). *Application of Magnetic Nanoparticles with structure of Core-Shell, as bactericidal. J Immunology and Inflammation Diseases Therapy*. doi:10.31579/2637-8876/011

Biswas, A., Haldane, A., Arnold, E., & Levy, R. (2019). Epistasis and entrenchment of drug resistance in HIV-1 subtype B. *eLife, 50*(24), 1-30.

Bi, Y., Hao, F., Yan, G., Teng, L. J., Lee, R., & Xie, J. (2016). Actively targeted nanoparticles for drug delivery to tumor. *Current Drug Metabolism, 17*(8), 763–782. doi:10.2174/1389200217666160619191853 PMID:27335116

Björnmalm, M., Faria, M., & Caruso, F. (2016). Increasing the impact of materials in and beyond bio-nano science. *Journal of the American Chemical Society, 138*(41), 13449–13456. doi:10.1021/jacs.6b08673 PMID:27672703

Blair, J. M. A., Bavro, V. N., Ricci, V., Modi, N., Cacciotto, P., Kleinekathöfer, U., & Piddock, L. J. V. (2015). AcrB drug-binding pocket substitution confers clinically relevant resistance and altered substrate specificity. *Proceedings of the National Academy of Sciences of the United States of America*, *112*(11), 3511–3516. doi:10.1073/pnas.1419939112 PMID:25737552

Blair, J. M. A., Webber, M. A., Baylay, A. J., Ogbolu, D. O., & Piddock, L. J. V. (2015). Molecular mechanisms of antibiotic resistance. *Nature Reviews. Microbiology*, *13*(1), 42–51. doi:10.1038/nrmicro3380 PMID:25435309

Blaiszik, B. J., Kramer, S. L. B., Olugebefola, S. C., Moore, J. S., Sottos, N. R., & White, S. R. (2010). Self- healing polymers and composites. *Annual Review of Materials Research*, *40*(1), 179–211. doi:10.1146/annurev-matsci-070909-104532

Blanco, E., Shen, H., & Ferrari, M. (2015). Principles of nanoparticle design for overcoming biological barriers to drug delivery. *Nature Biotechnology*, *33*(9), 941–951. doi:10.1038/nbt.3330 PMID:26348965

Blecher, K., Nasir, A., & Friedman, A. (2011). The growing role of nanotechnology in combating infectious disease. *Virulence*, *2*(5), 395–401. doi:10.4161/viru.2.5.17035 PMID:21921677

Bloom, G., Merrett, G. B., Wilkinson, A., Lin, V., & Paulin, S. (2017). Antimicrobial resistance and universal health coverage. *BMJ Global Health*, *2*(4), e000518. doi:10.1136/bmjgh-2017-000518 PMID:29225955

Bogireddy, N. K. R., Pal, U., Gomez, L. M., & Agarwal, V. (2018). Size controlled green synthesis of gold nanoparticles using Coffea arabica seed extract and their catalytic performance in 4-nitrophenol reduction. *RSC Advances*, *8*(44), 24819–24826. doi:10.1039/C8RA04332A

Bondarenko, O., Juganson, K., Ivask, A., Kasemets, K., Mortimer, M., & Kahru, A. (2013). Toxicity of Ag, CuO and ZnO nanoparticles to selected environmentally relevant test organisms and mammalian cells in vitro: A critical review. *Archives of Toxicology*, *87*(7), 1181–1200. doi:10.100700204-013-1079-4 PMID:23728526

Bonvin, D., Bastiaansen, J. A. M., Stuber, M., Hofmann, H., Mionić Ebersold, M., & Ebersold, M. M. (2017). Folic acid on iron oxide nanoparticles: Platform with high potential for simultaneous targeting, MRI detection and hyperthermia treatment of lymph node metastases of prostate cancer. *Dalton Transactions (Cambridge, England)*, *46*(37), 12692–12704. doi:10.1039/C7DT02139A PMID:28914298

Borges, M. C., Castro, L. A., & Fonseca, B. A. L. (2013). Chloroquine use improves dengue-related symptoms. *Memorias do Instituto Oswaldo Cruz*, *108*(5), 596–599. doi:10.1590/S0074-02762013000500010 PMID:23903975

Borman, A. M., Szekely, A., & Johnson, E. M. (2016). Comparative Pathogenicity of United Kingdom Isolates of the Emerging Pathogen Candida auris and Other Key Pathogenic Candida Species. Clinical Science and Epidemiology, 1(4), e00189–e16. doi:10.1128/mSphere.00189-16 PubMed

Borm, P. J., Robbins, D., Haubold, S., Kuhlbusch, T., Fissan, H., Donaldson, K., Schins, R., Stone, V., Kreyling, W., Lademann, J., Krutmann, J., Warheit, D., & Oberdorster, E. (2006). The potential risks of nanomaterials: A review carried out for ECETOC. *Particle and Fibre Toxicology*, *3*(1), 11. doi:10.1186/1743-8977-3-11 PMID:16907977

Bose, R. J., Paulmurugan, R., Moon, J., Lee, S. H., & Park, H. (2018). Cell membrane-coated nanocarriers: The emerging targeted delivery system for cancer theranostics. *Drug Discovery Today*, *23*(4), 891–899. doi:10.1016/j.drudis.2018.02.001 PMID:29426004

Boswihi, S. S., & Udo, E. E. (2018). Methicillin-resistant Staphylococcus aureus: An update on the epidemiology, treatment options and infection control. Current Medicine Research and Practice, 8(1), 18–24. doi:10.1016/j.cmrp.2018.01.001

Bottagisio, M., Lovati, A. B., Galbusera, F., Drago, L., & Banfi, G. (2019). A precautionary approach to guide the use of transition metal-based nanotechnology to prevent orthopedic infections. *Materials (Basel), 12*(2), 314. doi:10.3390/ma12020314 PMID:30669523

Bottini, M., Bruckner, S., Nika, K., Bottini, N., Bellucci, S., Magrini, A., & Mustelin, T. (2006). Multi-walled carbon nanotubes induce T lymphocyte apoptosis. *Toxicology Letters, 160*(2), 121–126. doi:10.1016/j.toxlet.2005.06.020 PMID:16125885

Boussif, O., Lezoualc'h, F., Zanta, M. A., Mergny, M. D., Scherman, D., Demeneix, B., & Behr, J. P. (1995). A versatile vector for gene and oligonucleotide transfer into cells in culture and in vivo: Polyethylenimine. *Proceedings of the National Academy of Sciences of the United States of America, 92*(16), 7297–7301. doi:10.1073/pnas.92.16.7297 PMID:7638184

Boutchuen, A., Zimmerman, D., Aich, N., Masud, A. M., Arabshahi, A., & Palchoudhury, S. (2019). Increased plant growth with hematite nanoparticle fertilizer drop and determining nanoparticle uptake in plants using multimodal approach. *Journal of Nanomaterials, 2019*, 2019. doi:10.1155/2019/6890572

Boverhof, D. R., Bramante, C. M., Butala, J. H., Clancy, S. F., Lafranconi, M., West, J., & Gordon, S. C. (2015). Comparative assessment of nanomaterial definitions and safety evaluation considerations. *Regulatory Toxicology and Pharmacology, 73*(1), 137–150. doi:10.1016/j.yrtph.2015.06.001 PMID:26111608

Boxall, A. B., Tiede, K., & Chaudhry, Q. (2007). Engineered nanomaterials in soils and water: How do they behave and could they pose a risk to human health? *Nanomedicine (London), 2*(6), 919–927. doi:10.2217/17435889.2.6.919 PMID:18095854

Bozzuto, G., & Molinari, A. (2015). Liposomes as nanomedical devices. *International Journal of Nanomedicine, 10*, 975. doi:10.2147/IJN.S68861 PMID:25678787

Brack, A., & Orgel, L. E. (1975). ß structures of alternating polypeptides and their possible prebiotic significance. *Nature, 256*(5516), 383–387. doi:10.1038/256383a0 PMID:238134

Brandelli, A. (2012). Nanostructures as promising tools for delivery of antimicrobial peptides. *Mini-Reviews in Medicinal Chemistry, 12*(8), 731–741. doi:10.2174/138955712801264774 PMID:22512554

Braun, L. B., Hessberger, T., Pütz, E., Müller, C., Giesselmann, F., Serra, C. A., & Zentel, R. (2018). Actuating thermo- and photo-responsive tubes from liquid crystalline elastomers. *Journal of Materials Chemistry. C, Materials for Optical and Electronic Devices, 6*(34), 9093–9101. doi:10.1039/C8TC02873G

Brayner, R., Ferarri-Iliou, R., Brivois, N., Djediat, S., Benedetti, M. F., & Fiévet, F. (2016). Toxicological impact studies based on Escherichia coli bacteria in ultrafine ZnO nanoparticles colloidal medium. *Nano Letters, 6*(4), 866–870. doi:10.1021/nl052326h PMID:16608300

Braz-Mota, S., Campos, D. F., MacCormack, T. J., Duarte, R. M., Val, A. L., & Almeida-Val, V. M. F. (2018). Mechanisms of toxic action of copper and copper nanoparticles in two Amazon fish species: Dwarf cichlid (Apistogramma agassizii) and cardinal tetra (Paracheirodonaxeirodi). *The Science of the Total Environment, 630*, 1168–1180. doi:10.1016/j.scitotenv.2018.02.216 PMID:29554738

Brice-Profeta, S., Arrio, M. A., Tronc, E., Menguy, N., Letard, I., Cartier dit Moulin, C., Noguès, M., Chanéac, C., Jolivet, J.-P., & Sainctavit, P. (2005). Magnetic order in γ-Fe2O3 nanoparticles: A XMCD study. *Journal of Magnetism and Magnetic Materials, 288*, 354–365. doi:10.1016/j.jmmm.2004.09.120

Briers, Y., & Lavigne, R. (2015). Breaking barriers: Expansion of the use of endolysins as novel antibacterials against Gram-negative bacteria. *Future Microbiology, 10*(3), 377–390. doi:10.2217/fmb.15.8 PMID:25812461

Briers, Y., Walmagh, M., Van Puyenbroeck, V., Cornelissen, A., Cenens, W., Aertsen, A., Oliveira, H., Azeredo, J., Verween, G., Pirnay, J.-P., Miller, S., Volckaert, G., & Lavigne, R. (2014). Engineered endolysin-based "Artilysins" to combat multidrug-resistant gram-negative pathogens. *mBio*, *5*(4), e01379–e01314. doi:10.1128/mBio.01379-14 PMID:24987094

Briuglia, M. L., Rotella, C., McFarlane, A., & Lamprou, D. A. (2015). Influence of cholesterol on liposome stability and on in vitro drug release. *Drug Delivery and Translational Research*, *5*(3), 231–242. doi:10.100713346-015-0220-8 PMID:25787731

Brossier, F., Veziris, N., Truffot-Pernot, C., Jarlier, V., & Sougakoff, W. (2011). Molecular investigation of resistance to the antituberculous drug ethionamide in multidrug-resistant clinical isolates of Mycobacterium tuberculosis. *Antimicrobial Agents and Chemotherapy*, *55*(1), 355–360. doi:10.1128/AAC.01030-10 PMID:20974869

Brown, C. L., Bushell, G., Whitehouse, M. W., Agrawal, D., Tupe, S., Paknikar, K., & Tiekink, E. R. (2007). Nanogoldpharmaceutics. *Gold Bulletin*, *40*(3), 245–250. doi:10.1007/BF03215588

Brown, D. M., Wilson, M. R., MacNee, W., Stone, V., & Donaldson, K. (2001). Size-dependent proinflammatory effects of ultrafine polystyrene particles: A role for surface area and oxidative stress in the enhanced activity of ultrafines. *Toxicology and Applied Pharmacology*, *175*(3), 191–199. doi:10.1006/taap.2001.9240 PMID:11559017

Bucki, R., Niemirowicz-Laskowska, K., Deptuła, P., Wilczewska, A. Z., Misiak, P., Durnaś, B., Fiedoruk, K., Piktel, E., Mystkowska, J., & Janmey, P. A. (2019). Susceptibility of microbial cells to the modified PIP 2-binding sequence of gelsolin anchored on the surface of magnetic nanoparticles. *Journal of Nanobiotechnology*, *17*(1), 81. doi:10.118612951-019-0511-1 PMID:31286976

Buhr, E., Senftleben, N., Klein, T., Bergmann, D., Gnieser, D., Frase, C. G., & Bosse, H. (2009). Characterization of nanoparticles by scanning electron microscopy in transmission mode. *Measurement Science & Technology*, *20*(8), 084025. Advance online publication. doi:10.1088/0957-0233/20/8/084025

Bu, Junjie, Wanga, Ruan, Chen, Gunadhi, Bell, & Gu. (2019). Advances in drug delivery for post-surgical cancer treatment. *Biomaterials, 219,* 119-182. doi:10.1016/j.biomaterials.2019.04.027 PMID:31415954

Bull, M. (2008). 基因的改变NIH Public Access. *Bone*, *23*, 1–7.

Bulte, J. W., Douglas, T., Witwer, B., Zhang, S.-C., Strable, E., Lewis, B. K., Zywicke, H., Miller, B., van Gelderen, P., Moskowitz, B. M., Duncan, I. D., & Frank, J. A. (2001). Magnetodendrimers allow endosomal magnetic labeling and in vivo tracking of stem cells. *Nature Biotechnology*, *19*(12), 1141–1147. doi:10.1038/nbt1201-1141 PMID:11731783

Bundschuh, T., Knopp, R., & Kim, J. I. (2001). Laser-induced breakdown detection (LIBD) of aquatic colloids with different laser systems. *Colloids and Surfaces. A, Physicochemical and Engineering Aspects*, *177*(1), 47–55. doi:10.1016/S0927-7757(99)00497-5

Burygin, G. L., Khlebtsov, B. N., Shantrokha, A. N., Dykman, L. A., Bogatyrev, V. A., & Khlebtsov, N. G. (2009). On the enhanced antibacteria activity of antibiotics mixed with gold nanoparticles. *Nanoscale Research Letters*, *4*(8), 794–801. doi:10.100711671-009-9316-8 PMID:20596384

Butoescu, N., Seemayer, C. A., Foti, M., Jordan, O., & Doelker, E. (2009). Dexamethasone-containing PLGA superparamagnetic microparticles as carriers for the local treatment of arthritis. *Biomaterials*, *30*(9), 1772–1780. doi:10.1016/j.biomaterials.2008.12.017 PMID:19135244

Cabello, F. C. (2006). Heavy use of prophylactic antibiotics in aquaculture: A growing problem for human and animal health and for the environment. *Environmental Microbiology*, *8*(7), 1137–1144. doi:10.1111/j.1462-2920.2006.01054.x PMID:16817922

Cabot, G., Zamorano, L., Moyà, B., Juan, C., Navas, A., Blázquez, J., & Oliver, A. (2016). Evolution of Pseudomonas aeruginosa antimicrobial resistance and ÿtness under low and high mutation rates. *Antimicrobial Agents and Chemotherapy*, *60*(3), 1767–1778. doi:10.1128/AAC.02676-15 PubMed

Cai, W., Gao, T., Hong, H., & Sun, J. (2008). Applications of gold nanoparticles in cancer nanotechnology. *Nanotechnology, Science and Applications*, *1*, 17–32. doi:10.2147/NSA.S3788 PMID:24198458

Cai, X., Luo, Y., Zhang, W., Du, D., & Lin, Y. (2016). pH-Sensitive ZnO quantum dots–doxorubicin nanoparticles for lung cancer targeted drug delivery. *ACS Applied Materials & Interfaces*, *8*(34), 22442–22450. doi:10.1021/acsami.6b04933 PMID:27463610

Calvo-Marzal, P., Sattayasamitsathit, S., Balasubramanian, S., Windmiller, J. R., Dao, C., & Wang, J. (2010). Propulsion of nanowire diodes. *Chemical Communications*, *46*(10), 1623. doi:10.1039/b925568k PMID:20177595

Cameron, J. M., Wales, D. J., & Newton, G. N. (2018). Shining a light on the photo-sensitisation of organic-inorganic hybrid polyoxometalates. *Dalton Transactions (Cambridge, England)*, *47*(15), 5120–5136. doi:10.1039/C8DT00400E PMID:29517788

Campbell, S., & K, S.-F. (2019). *Antiparasitic Drugs*. StatPearls Publishing.

Campion, A., & Kambhampati, P. (1998). Surface-Enhanced raman scattering. *Chemical Society Reviews*, *27*(4), 241–250. doi:10.1039/a827241z

Campuzano, U., Orozco, J., Kagan, D., Guix, M., Gao, W., Sattayasamitsathit, S., Jonathan, C. C., Merkoçi, A., & Wang, J. (2012). Bacterial isolation by lectin-modified microengines. *Nano Letters*, *12*(1), 396–401. doi:10.1021/nl203717q PMID:22136558

Cantas, L., Shah, S. Q., Cavaco, L. M., Manaia, C. M., Walsh, F., Popowska, M., Garelick, H., Bürgmann, H., & Sorum, H. (2013). A brief multi-disciplinary review on antimicrobial resistance in medicine and its linkage to the global environmental microbiota. *Frontiers in Microbiology*, *4*, 96. doi:10.3389/fmicb.2013.00096 PMID:23675371

Cao, C. Y., Chen, Y., Wu, F. Z., Deng, Y., & Liang, G. L. (2011). Caspase-3 controlled assembly of nanoparticles for fluorescence turn on. *Chemical Communications*, *47*(37), 10320–10322. doi:10.1039/c1cc14112k PMID:21853204

Cao, T., Yang, Y., Gao, Y., Zhou, J., Li, Z., & Li, F. (2011). High-quality water-soluble and surface-functionalized upconversion nanocrystals as luminescent probes for bioimaging. *Biomaterials*, *32*(11), 2959–2968. doi:10.1016/j.biomaterials.2010.12.050 PMID:21262531

Carattoli, A. (2001). Importance of integrons in the diffusion of resistance. *Veterinary Research*, *32*(3-4), 243–259. doi:10.1051/vetres:2001122 PMID:11432416

Carling, P. C., & Bartley, J. M. (2010). Evaluating hygienic cleaning in health care settings: What you do not know can harm your patients. *American Journal of Infection Control*, *38*(5, Suppl 1), S41–S50. doi:10.1016/j.ajic.2010.03.004 PMID:20569855

Carloni, E., Andreoni, F., Omiccioli, E., Villa, L., Magnani, M., & Carattoli, A. (2017). Comparative analysis of the standard PCR-Based Replicon Typing (PBRT) with the commercial PBRT-KIT. *Plasmid*, *90*, 10–14. doi:10.1016/j.plasmid.2017.01.005 PMID:28137396

Carlson, C., Hussain, S. M., Schrand, A. M. K., Braydich-Stolle, L., Hess, K. L., Jones, R. L., & Schlager, J. J. (2008). Unique cellular interaction of silver nanoparticles: Size-dependent generation of reactive oxygen species. *The Journal of Physical Chemistry B*, *112*(43), 13608–13619. doi:10.1021/jp712087m PMID:18831567

Casadevall, A., Feldmesser, M., & Pirofski, L. (2002). Induced humoral immunity and vaccination against major human fungal pathogens. *Current Opinion in Microbiology*, *5*(4), 386–391. doi:10.1016/S1369-5274(02)00337-5 PubMed

Casewell, M., Friis, C., Marco, E., McMullin, P., & Phillips, I. (2003). The European ban on growth-promoting antibiotics and emerging consequences for human and animal health. *The Journal of Antimicrobial Chemotherapy*, *52*(2), 159–161. doi:10.1093/jac/dkg313 PMID:12837737

Cassidy, P. J., & Radda, G. K. (2005). Molecular imaging perspectives. *Journal of the Royal Society, Interface*, *2*(3), 133–144. doi:10.1098/rsif.2005.0040 PMID:16849174

Castellano, J. J., Shafii, S. M., Ko, F., Donate, G., Wright, T. E., Mannari, R. J., & Robson, M. C. (2007). Comparative evaluation of silver-containing antimicrobial dressings and drugs. *International Wound Journal*, *4*(2), 114–122. doi:10.1111/j.1742-481X.2007.00316.x PMID:17651227

Catalao, M. J., Gil, F., Moniz-Pereira, J., & Pimentel, M. (2011). Functional analysis of the holin-like proteins of mycobacteriophage Ms6. *Journal of Bacteriology*, *193*(11), 2793–2803. doi:10.1128/JB.01519-10 PMID:21441511

Cattoir, V., & Felden, B. (2009). Future antibacterial strategies: From basic concepts to clinical challenges. *The Journal of Infectious Diseases*, *220*(3), 350–360. doi:10.1093/infdis/jiz134 PMID:30893436

Chabner, B. A., & Roberts, T. G. Jr. (2005). Chemotherapy and the war on cancer. *Nature Reviews. Cancer*, *5*(1), 65–72. doi:10.1038/nrc1529 PMID:15630416

Chae, T., Yang, H., Leung, V., Ko, F., & Troczynski, T. (2013). Novel biomimetic hydroxyapatite/alginate nanocomposite fibrous scaffolds. *Journal of Materials Science. Materials in Medicine*, *24*(8), 1885–1894. doi:10.100710856-013-4957-7 PMID:23695359

Chae, W. P., Xing, Z. C., Kim, Y. J., Sang, H. S., Huh, M. W., & Kang, I. K. (2011). Fabrication and biocompatibility of rutin-containing PHBV nanofibrous scaffolds. *Polymer Korea*, *35*(3), 210–215. doi:10.7317/pk.2011.35.3.210

Chai, H., Yao, J., Sun, J., Zhang, C., Liu, W., Zhu, M., & Ceccanti, B. (2015). The effect of metal oxide nanoparticles on functional bacteria and metabolic profiles in agricultural soil. *Bulletin of Environmental Contamination and Toxicology*, *94*(4), 490–495. doi:10.100700128-015-1485-9 PMID:25636440

Cha, K. E., & Myung, H. (2007). Cytotoxic effects of nanoparticles assessed *in vitro* and *in vivo*. *Journal of Microbiology and Biotechnology*, *17*, 1573–1578. PMID:18062241

Chakkarapani, P., Subbiah, L., Palanisamy, S., Bibiana, A., Ahrentorp, F., Jonasson, C., & Johansson, C. (2015). Encapsulation of methotrexate loaded magnetic microcapsules for magnetic drug targeting and controlled drug release. *Journal of Magnetism and Magnetic Materials*, *380*, 280–294. doi:10.1016/j.jmmm.2014.11.006

Chakravorty, A., Biswas, B., Sana, S. S., Rayan, R. A., Lala, N. L., & Ramakrishna, S. (2020). A review on toxicity of turmeric derived Nano-Formulates against bacterial and fungal cells with special emphasis on electrospunnanofibers. *Materials Today: Proceedings*. Advance online publication. doi:10.1016/j.matpr.2020.05.820

Chakravorty, A., Rather, G. A., Ali, A., Bhat, B. A., Sana, S. S., Abhishek, N., & Nanda, A. (2020). *Nano Approach: Indian Spices as Antimicrobial Agents. In Ethnopharmacological Investigation of Indian Spices*. IGI Global.

Chambers, H. F., & DeLeo, F. R. (2009). Waves of resistance: *Staphylococcus aureus* in the antibiotic era. *Nature Reviews. Microbiology*, *7*(9), 629–641. doi:10.1038/nrmicro2200 PMID:19680247

Champion, J. A., & Mitragotri, S. (2006). Role of target geometry in phagocytosis. *Proceedings of the National Academy of Sciences of the United States of America*, *103*(13), 4930–4934. doi:10.1073/pnas.0600997103 PMID:16549762

Chamundeeswari, M., Sobhana, S. S., Jacob, J. P., Kumar, M. G., Devi, M. P., Sastry, T. P., & Mandal, A. B. (2010). Preparation, characterization and evaluation of a biopolymeric gold nanocomposite with antimicrobial activity. *Biotechnology and Applied Biochemistry*, *55*(1), 29–35. doi:10.1042/BA20090198 PMID:19929854

Chandran, S. P., Chaudhary, M., Pasricha, R., Ahmad, A., & Sastry, M. (2006). Synthesis of gold nanotriangles and silver nanoparticles using Aloe vera plant extract. *Biotechnology Progress*, *22*(2), 577–583. doi:10.1021/bp0501423 PMID:16599579

Chapman, R., Danial, M., Koh, M. L., Jolliffe, K. A., & Perrier, S. (2012). Design and properties of functional nanotubes from the self-assembly of cyclic peptide templates. *Chemical Society Reviews*, *41*(18), 6023–6041. doi:10.1039/c2cs35172b PMID:22875035

Chapman, S., Dobrovolskaia, M., Farahani, K., Goodwin, A., Joshi, A., Lee, H., ... Singh, R. (2013). Nanoparticles for cancer imaging: The good, the bad, and the promise. *Nano Today*, *8*(5), 454–460. doi:10.1016/j.nantod.2013.06.001 PMID:25419228

Chatel, A., Lievre, C., Barrick, A., Bruneau, M., & Mouneyrac, C. (2018). Transcriptomic approach: A promising tool for rapid screening nanomaterial-mediated toxicity in the marine bivalve Mytilus edulis – application to copper oxide nanoparticles. *Comparative Biochemistry and Physiology. Toxicology & Pharmacology: CBP*, *205*, 26–33. doi:10.1016/j.cbpc.2018.01.003 PMID:29382575

Chatterjee, A. K., Sarkar, R. K., Chattopadhyay, A. P., Aich, P., Chakraborty, R., & Basu, T. (2012). A simple robust method for synthesis of metallic copper nanoparticles of high antibacterial potency against E. coli. *Nanotechnology*, *23*(8), 85–103. doi:10.1088/0957-4484/23/8/085103 PMID:22293320

Chatzipavlidis, A., Bilalis, P., Efthimiadou, E. K., Boukos, N., & Kordas, G. C. (2011). Sacrificial template-directed fabrication of superparamagnetic polymer microcontainers for pH-activated controlled release of daunorubicin. *Langmuir*, *27*(13), 8478–8485. doi:10.1021/la201240h PMID:21648457

Chaudhary, J., Tailor, G., Yadav, B. L., & Michael, O. (2019). Synthesis and biological function of Nickel and Copper nanoparticles. *Heliyon*, *5*(6), e01878. doi:10.1016/j.heliyon.2019.e01878 PMID:31198877

Chaudhry, Q., Scotter, M., Blackburn, J., Ross, B., Boxall, A., Castle, L., Aitken, R., & Watkins, R. (2008). Applications and implications of nanotechnologies for the food sector. *Food Additives and Contaminants*, *25*(3), 241–244. doi:10.1080/02652030701744538 PMID:18311618

Chauhan, A., Zubair, S., Tufail, S., Sherwani, A., Sajid, M., Raman, S. C., Azam, A., & Owais, M. (2011). Fungus-mediated biological synthesis of gold nanoparticles: Potential in detection of liver cancer. *International Journal of Nanomedicine*, *6*, 2305–2319. PMID:22072868

Chawla, P. R., Bajaj, I. B., Survase, S. A., & Singhal, R. S. (2009). Microbial cellulose: Fermentative production and applications. *Food Technology and Biotechnology*, *47*(2), 107–124.

Chen, C. H., & Lu, T. K. (2020). Development and challenges of antimicrobial peptides for therapeutic applications. *Antibiotics (Basel, Switzerland)*, *9*(1), 24. doi:10.3390/antibiotics9010024 PMID:31941022

Chen, C. R., Mou, F. Z., Xu, L. L., Wang, S. F., Guan, J. G., Feng, Z. P., Wang, Q. W., Kong, L., Li, W., Wang, J., & Zhang, Q. J. (2017). Light-steered isotropic semiconductor micromotors. *Advanced Materials*, *29*(3), 1603374. doi:10.1002/adma.201603374 PMID:27748536

Chen, C., Li, D., Yano, H., & Abe, K. (2019a). Bioinspired hydrogels: Quinone crosslinking reaction for chitin nanofibers with enhanced mechanical strength via surface deacetylation. *Carbohydrate Polymers*, *207*, 411–417. doi:10.1016/j.carbpol.2018.12.007 PMID:30600023

Cheng, K. K., Chan, P. S., Fan, S., Kwan, S. M., Yeung, K. L., & Wang, Y.-X. J. (2015). Curcumin-conjugated magnetic nanoparticles for detecting amyloid plaques in Alzheimer's disease mice using magnetic resonance imaging (MRI). Biomaterials, 44, 155-172.

Chen, G., Wu, C., Weng, W., Wu, D., & Yan, W. (2003). Preparation of polystyrene graphite nanosheet composites. *Polymer*, *44*(6), 1781–1784. doi:10.1016/S0032-3861(03)00050-8

Cheng, Y. T., Yang, C. C., & Shyur, L. F. (2016). Phytomedicine-Modulating oxidative stress and the tumor microenvironment for cancer therapy. *Pharmacological Research*, *114*, 128–143. doi:10.1016/j.phrs.2016.10.022 PMID:27794498

Chen, H., Zhang, W., Zhu, G., Xie, J., & Chen, X. (2017). Rethinking cancer nanotheranostics. *Nature Reviews. Materials*, *2*(7), 1704. doi:10.1038/natrevmats.2017.24 PMID:29075517

Chen, J., Liu, J., Li, J., Xu, L., & Qiao, Y. (2017). One-pot synthesis of nitrogen and sulphur co-doped carbon dots and its application for sensor and multi-color cellular imaging. *Journal of Colloid and Interface Science*, *485*, 167–174. doi:10.1016/j.jcis.2016.09.040 PMID:27662029

Chen, L., Wu, L., Liu, F., Qi, X., Ge, Y., & Shen, S. (2016). Azo-functionalized Fe3 O4 nanoparticles: A near-infrared light triggered drug delivery system for combined therapy of cancer with low toxicity. *Journal of Materials Chemistry*, *4*(21), 3660–3669. doi:10.1039/C5TB02704G PMID:32263305

Chen, M., Yin, J., Liang, Y., Yuan, S., Wang, F., Song, M., & Wang, H. (2016). Oxidative stress and immunotoxicity induced by graphene oxide in zebrafish. *Aquatic Toxicology (Amsterdam, Netherlands)*, *174*, 54–60. doi:10.1016/j.aquatox.2016.02.015 PMID:26921726

Chen, P., Liu, L., Pan, J., Mei, J., Li, C., & Zheng, Y. (2019). Biomimetic composite scaffold of hydroxyapatite/gelatinchitosan core-shell nanofibers for bone tissue engineering. *Materials Science and Engineering C*, *97*, 325–335. doi:10.1016/j.msec.2018.12.027 PMID:30678918

Chen, W. J., Tsai, P. J., & Chen, Y. C. (2008). Functional Fe3O4/TiO2 core/shell magnetic nanoparticles as photokilling agents for pathogenic bacteria. *Small*, *4*(4), 485–491. doi:10.1002mll.200701164 PMID:18348230

Chen, W., Wang, C., Gao, Y., Wu, Y., Wu, G., Shi, X., Du, Y., & Deng, H. (2019b). Incorporating Chitin Derived Glucosamine Sulfate into Nanofibers via Coaxial electrospinning for Cartilage Regeneration. *Carbohydrate Polymers*, *229*, 115544. doi:10.1016/j.carbpol.2019.115544 PMID:31826435

Chen, X., Wong, R., Khalidov, I., Wang, A. Y., Leelawattanachai, J., Wang, Y., & Jin, M. M. (2011). Inflamed leukocyte-mimetic nanoparticles for molecular imaging of inflammation. *Biomaterials*, *32*(30), 7651–7661. doi:10.1016/j.biomaterials.2011.06.030 PMID:21783245

Chen, Y.-H., Tsai, C.-Y., Huang, P.-Y., Chang, M.-Y., Cheng, P.-C., Chou, C.-H., Chen, D.-H., Wang, C.-R., Shiau, A.-L., & Wu, C.-L. (2007). Methotrexate conjugated to gold nanoparticles inhibits tumor growth in a syngeneic lungtumor model. *Molecular Pharmaceutics*, *4*(5), 713–722. doi:10.1021/mp060132k PMID:17708653

Chen, Y., Xu, M., Guo, Y., Tu, K., Wu, W., Wang, J., Tong, X., Wu, W., Qi, L., & Shi, D. (2017). Targeted chimera delivery to ovarian cancer cells by heterogeneous gold magnetic nanoparticle. *Nanotechnology*, *28*(2), 025101. doi:10.1088/0957-4484/28/2/025101 PMID:27906685

Cheow, W. S., Chang, M. W., & Hadinoto, K. (2010). Antibacterial efficacy of inhalable levofloxacin-loaded polymeric nanoparticles against E. coli biofilm cells: The effect of antibiotic release profile. *Pharmaceutical Research*, *27*(8), 1597–1609. doi:10.100711095-010-0142-6 PMID:20407918

Cheung, R. C. F., Ng, T. B., Wong, J. H., & Chan, W. Y. (2015). Chitosan: An update on potential biomedical and pharmaceutical applications. *Marine Drugs, 13*(8), 5156–5186. doi:10.3390/md13085156 PMID:26287217

Chia, S. L., & Leong, D. T. (2016). Reducing ZnO nanoparticles toxicity through silica coating. *Heliyon, 2*(10), e00177. doi:10.1016/j.heliyon.2016.e00177 PMID:27812550

Chifiriuc, C., Grumezescu, V., Grumezescu, A. M., Saviuc, C., Lazˇar, V., & Andronescu, E. (2012). Hybrid magnetite nanoparticles/Rosmarinus officinalis essential oil nanobiosystem with antibiofilm activity. *Nanoscale Research Letters, 7*(1), 14. doi:10.1186/1556-276X-7-209 PMID:22490675

Ching, S. H., Bansal, N., & Bhandari, B. (2017). Alginate gel particles–A review of production techniques and physical properties. *Critical Reviews in Food Science and Nutrition, 57*(6), 1133–1152. doi:10.1080/10408398.2014.965773 PMID:25976619

Chiu-Lam, A., Dhavalikar, R., Tay, Z. W., Chandrasekharan, P., Conolly, S. M., & Rinaldi, C. (2018). Spatially Controlledand Image Guided Nanoscale Thermal Therapy. *12th International conference on Scientific and clinical applications of Magnetic carriers.*

Choi, K.-H., Nam, K., Malkinski, L., Choi, E., Jung, J.-S., & Park, B. (2016). Sizedependent photodynamic anticancer activity of biocompatible multifunctional magnetic submicron particles in prostate cancer cells. *Molecules (Basel, Switzerland), 21*(9), 1187. doi:10.3390/molecules21091187

Cho, K., Wang, X., Nie, S., Chen, Z. G., & Shin, D. M. (2008). Therapeutic nanoparticles for drug delivery in cancer. *Clinical Cancer Research, 14*(5), 1310–1316. doi:10.1158/1078-0432.CCR-07-1441 PMID:18316549

Chopra, I., & Roberts, M. (2001). Tetracycline antibiotics: mode of action, applications, molecular biology, and epidemiology of bacterial resistance. *Microbiol Mol Biol Rev, 65*(2), 232-260. doi:10.1128/MMBR.65.2.232-260.2001

Chopra, I., Hesse, L., & O'Neill, A. J. (2002). Exploiting current understanding of antibiotic action for discovery of new drugs. *Journal of Applied Microbiology, 92*, 4S–15S. doi:10.1046/j.1365-2672.92.5s1.13.x PMID:12000608

Chorny, M., Fishbein, I., Yellen, B. B., Alferiev, I. S., Bakay, M., Ganta, S., Adamo, R., Amiji, M., Friedman, G., & Levy, R. J. (2010). Targeting stents with local delivery of paclitaxel-loaded magnetic nanoparticles using uniform Welds. *Proceedings of the National Academy of Sciences of the United States of America, 107*(18), 8346–8351. doi:10.1073/pnas.0909506107 PMID:20404175

Chowdhury, M. H., Gray, S. K., Pond, J., Geddes, C. D., Aslan, K., & Lakowicz, J. R. (2007). Computational study of fluorescence scattering by silver nanoparticles. *Journal of the Optical Society of America. B, Optical Physics, 24*(9), 2259. doi:10.1364/JOSAB.24.002259 PMID:19777118

Chung, H. J., Castro, C. M., Im, H., Lee, H., & Weissleder, R. (2013). A magneto-DNA nanoparticle system for rapid detection and phenotyping of bacteria. *Nature Nanotechnology, 8*(5), 369–375. doi:10.1038/nnano.2013.70 PMID:23644570

Chung, M. H., Beck, I. A., Dross, S., Tapia, K., Kiarie, J. N., Richardson, B. A., Overbaugh, J., Sakr, S. R., John-Stewart, G. C., & Frenkel, L. M. (2014). Oligonucleotide ligation assay detects HIV drug resistance associated with virologic failure among antiretroviral-naive adults in Kenya. *Journal of Acquired Immune Deficiency Syndromes, 67*(3), 246–253. doi:10.1097/QAI.0000000000000312 PMID:25140907

Churg, A., Stevens, B., & Wright, J. L. (1998). Comparison of the uptake of fine and ultrafine TiO2 in a tracheal explant system. *American Journal of Physiology. Lung Cellular and Molecular Physiology, 274*(1), L81–L86. doi:10.1152/ajplung.1998.274.1.L81 PMID:9458804

Cinteza, L. O., Scomoroscenco, C., Voicu, S. N., Nistor, C. L., Nitu, S., Trica, B., Jecu, M.-L., & Petcu, C. (2018). Chitosa.n-stabilized Ag nanoparticles with superior biocompatibility and their synergistic antibacterial effect in mixtures with essential oils. *Nanomaterials (Basel, Switzerland)*, *8*(10), 826. doi:10.3390/nano8100826 PMID:30322127

Ciurlicaa, Nadejdea, Creangaa, Carlescua, & Badescub. (2010). *Antibiotic Coated Magnetite Nanoparticles For Biological Applications*. Olomouc.

Claeys, K. C., Heil, E. L., Pogue, J. M., Lephart, P. R., & Johnson, J. K. (2018). The Verigene dilemma: Gram-negative polymicrobial bloodstream infections and clinical decision making. *Diagnostic Microbiology and Infectious Disease*, *91*(2), 144–146. doi:10.1016/j.diagmicrobio.2018.01.012 PMID:29449043

Clarence, Y., & Geoffrey, S. (2015). Nanoparticles as potential new generation broad spectrum antimicrobial agents. *Daru: Journal of Faculty of Pharmacy, Tehran University of Medical Sciences*, *23*(1), 43. doi:10.118640199-015-0125-6 PMID:26329777

Clokie, M. R., Millard, A. D., Letarov, A. V., & Heaphy, S. (2011). Phages in nature. *Bacteriophage*, *1*(1), 31–45. doi:10.4161/bact.1.1.14942 PMID:21687533

Coccia, F., Tonucci, L., Bosco, D., Bressan, M., & Alessandro, N. (2012). One pot synthesis of lignin-stabilized platinum and palladium nanoparticles and their catalytic behaviors in oxidation and reduction reactions. *Green Chemistry*, *14*(4), 1073–1078. doi:10.1039/c2gc16524d

Coffey, B., Mills, S., Coffey, A., McAuliffe, O., & Ross, R. P. (2010). Phage and their lysins as biocontrol agents for food safety applications. *Annual Review of Food Science and Technology*, *1*(1), 449–468. doi:10.1146/annurev.food.102308.124046 PMID:22129344

Cohen, S. J., Punt, C. J. A., Iannotti, N., Saidman, B. H., Sabbath, K. D., Gabrail, N. Y., Picus, J., Morse, M., Mitchell, E., Miller, M. C., Doyle, G. V., Tissing, H., Terstappen, L. W. M. M., & Meropol, N. J. (2009). Relationship of Circulating Tumor Cells to Tumor Response, Progression-Free Survival, and Overall Survival in Patients With Metastatic Colorectal Cancer. *Journal of Clinical Oncology*, *26*(19), 3213–3221. doi:10.1200/JCO.2007.15.8923 PMID:18591556

Cole, J. N., Henningham, A., Gillen, C. M., Ramachandran, V., & Walker, M. J. (2008). Human pathogenic streptococcal proteomics and vaccine development. *Proteomics. Clinical Applications*, *2*(3), 387–410. doi:10.1002/prca.200780048 PubMed

Coll, C., Notter, D., Gottschalk, F., Sun, T., Som, C., & Nowack, B. (2016). Probabilistic environmental risk assessment of five nanomaterials (nano-TiO_2, nano-Ag, nano-ZnO, CNT, and fullerenes). *Nanotoxicology*, *10*(4), 4. doi:10.3109/17435390.2015.1073812 PMID:26554717

Colvin, V. L. (2003). The potential environmental impact of engineered nanomaterials. *Nature Biotechnology*, *21*(10), 1166–1170. doi:10.1038/nbt875 PMID:14520401

Comeau, A. M., Tetart, F., Trojet, S. N., Prere, M. F., & Krisch, H. M. (2007). Phage-Antibiotic Synergy (PAS): Beta-lactam and quinolone antibiotics stimulate virulent phage growth. *PLoS One*, *2*(8), e799. doi:10.1371/journal.pone.0000799 PMID:17726529

Connor, B. A., & Schwartz, E. (2005). Typhoid and paratyphoid fever in travellers. *The Lancet Infectious Diseases*, *5*(10), 623–628. doi:10.1016/S1473-3099(05)70239-5 PubMed

Connors, J. M., Jurczak, W., Straus, D. J., Ansell, S. M., Kim, W. S., Gallamini, A., Younes, A., Alekseev, S., Illés, Á., Picardi, M., Lech-Maranda, E., Oki, Y., Feldman, T., Smolewski, P., Savage, K. J., Bartlett, N. L., Walewski, J., Chen, R., Ramchandren, R., ... Radford, J. (2018). Brentuximab vedotin with chemotherapy for stage III or IV Hodgkin's lymphoma. *The New England Journal of Medicine*, *378*(4), 331–344. doi:10.1056/NEJMoa1708984 PMID:29224502

Cooley, J.F. (1900). *Improved methods of apparatus for electrically separating the relatively volatile liquid component from the component of relatively fixed substances of composite fluids*. United Kindom patent, 6385.

Correa, Sierra, Asín, Gutiérrez, de la Fuente, Betancor, & Grazú. (2018). Nanobiohybrids for the thermoactivation of a therapeutic enzyme by magnetic hyperthermia. *12th International Conference on Scientific and Clinical Applications of Magnetic Carriers*.

Corrêa, A. C., Pereira, P. R., & Paschoalin, V. M. (2019). Preparation and Characterization of Nanoliposomes for the Entrapment of Bioactive Hydrophilic Globular Proteins. *JoVE*, (150), e59900. doi:10.3791/59900 PMID:31524878

Costa, A., Sarmento, B., & Seabra, V. (2018). Mannose-functionalized solid lipid nanoparticles are effective in targeting alveolar macrophages. *European Journal of Pharmaceutical Sciences*, *114*, 103–113. doi:10.1016/j.ejps.2017.12.006 PMID:29229273

Costa, C. S., Ronconi, J. V. V., Daufenbach, J. F., Gonçalves, C. L., Rezin, G. T., Streck, E. L., & da Silva Paula, M. M. (2010). In vitro effects of silver nanoparticles on the mitochondrial respiratory chain. *Molecular and Cellular Biochemistry*, *342*(1-2), 51–56. PMID:20411305

Costa-Gouveia, J., Pancani, E., Jouny, S., Machelart, A., Delorme, V., Salzano, G., ... Flipo, M. (2017). Combination therapy for tuberculosis treatment: Pulmonary administration of ethionamide and booster co-loaded nanoparticles. *Scientific Reports*, *7*(1), 1–14. doi:10.103841598-017-05453-3 PMID:28710351

Couturier, M., Bex, F., Bergquist, P. L., & Maas, W. K. (1988). Identification and classification of bacterial plasmids. *Clinical Microbiology Reviews*, *52*(3), 375–395. doi:10.1128/MR.52.3.375-395.1988 PMID:3054468

Cree, I. A., & Charlton, P. (2017). Molecular chess? Hallmarks of anti-cancer drug resistance. *BMC Cancer*, *17*(1), 10. doi:10.118612885-016-2999-1 PMID:28056859

Creixell, M., Herrera, A. P., & Ayala, V. (2010). Preparation of epidermal growth factor (EGF) conjugated iron oxide nanoparti- cles and their internalization into colon cancer cells. *Journal of Magnetism and Magnetic Materials*, *322*, 2244–2250. doi:10.1016/j.jmmm.2010.02.019

Cristea, C., Tertis, M., & Galatus, R. (2017). Magnetic Nanoparticles for Antibiotics Detection. *Nanomaterials (Basel, Switzerland)*, *7*(6), 119. doi:10.3390/nano7060119 PMID:28538684

Cristina Buzea, I. I. P., & Robbie, K. (2007). Nanomaterials and nanoparticles: Sources and toxicity. *Biointerphases*, *2*(4), 17–71. doi:10.1116/1.2815690 PMID:20419892

Cristofanilli, M., Budd, G. T., Ellis, M. J., Stopeck, A., Matera, J., Miller, M. C., Reuben, J. M., Doyle, G. V., Allard, W. J., Terstappen, L. W. M. M., & Hayes, D. F. (2004). Circulating Tumor Cells, Disease Progression, and Survival in Metastatic Breast Cancer. *The New England Journal of Medicine*, *351*(8), 781–791. doi:10.1056/NEJMoa040766 PMID:15317891

Croci, R., Bottaro, E., Chan, K., Watanabe, S., Pezzullo, M., Mastrangelo, E., & Nastruzzi, C. (2016). Liposomal Systems as Nanocarriers for the Antiviral Agent Ivermectin. *International Journal of Biomaterials*, *2016*, 1–15. Advance online publication. doi:10.1155/2016/8043983 PMID:27242902

Croissant, J. G., Fatieiev, Y., & Khashab, N. M. (2017). Degradability and clearance of silicon, organosilica, silsesquioxane, silica mixed oxide, and mesoporous silica nanoparticles. *Advanced Materials*, *29*(9), 1604634. doi:10.1002/adma.201604634 PMID:28084658

Crosera, M., Bovenzi, M., Maina, G., Adami, G., Zanette, C., Florio, C., & Larese, F.F. (2009). Nanoparticle dermal absorption and toxicity: a review of literature. *International Archives of Occupational and Environmental Health, 82*, 1043–1055.

Cui, D., Tian, F., Ozkan, C. S., Wang, M., & Gao, H. (2005). Effect of single wall carbon nanotubes on human HEK293 cells. *Toxicology Letters*, *155*(1), 73–85. doi:10.1016/j.toxlet.2004.08.015 PMID:15585362

Cui, D., Zhang, C., Liu, B., Shu, Y., Du, T., Shu, D., Wang, K., Dai, F., Liu, Y., Li, C., Pan, F., Yang, Y., Ni, J., Li, H., Brand-Saberi, B., & Guo, P. (2015). Regression of Gastric Cancer by Systemic Injection of RNA Nanoparticles Carrying both Ligand and siRNA. *Scientific Reports*, *5*(1), 10726. doi:10.1038rep10726 PMID:26137913

Cui, J., Faria, M., Björnmalm, M., Ju, Y., Suma, T., Gunawan, S. T., Richardson, J. J., Heidari, H., Bals, S., Crampin, E. J., & Caruso, F. (2016). A framework to account for sedimentation and diffusion in particle–cell interactions. *Langmuir*, *32*(47), 12394–12402. doi:10.1021/acs.langmuir.6b01634 PMID:27384770

Curnis, F., Fiocchi, M., Sacchi, A., Gori, A., Gasparri, A., & Corti, A. (2016). NGR-tagged nano-gold: A new CD13-selective carrier for cytokine delivery to tumors. *Nano Research*, *9*(5), 1393–1408. doi:10.100712274-016-1035-8 PMID:27226823

Currie, C. J., Berni, E., Jenkins-Jones, S., Poole, C. D., Ouwens, M., Driessen, S., de Voogd, H., Butler, C. C., & Morgan, C. L. (2014). Antibiotic treatment failure in four common infections in UK primary care 1991-2012: Longitudinal analysis. *BMJ (Clinical Research Ed.)*, *349*(sep23 3), g5493. doi:10.1136/bmj.g5493 PMID:25249162

da Silva, P. B., de Freitas, E. S., Solcia, M. C., de Souza, P. C., da Silva, M. M., Batista, A. A., ... Chorilli, M. (2018). A Nanostructured Lipid System to Improve the Oral Bioavailability of Ruthenium (II) Complexes for the Treatment of Infections Caused by Mycobacterium tuberculosis. *Frontiers in Microbiology*, *9*, 2930. doi:10.3389/fmicb.2018.02930 PMID:30574128

Dadfar, S. M., Roemhild, K., Drude, N. I., von Stillfried, S., Knüchel, R., Kiessling, F., & Lammers, T. (2019, January 1). Iron oxide nanoparticles: Diagnostic, therapeutic and theranostic applications. *Advanced Drug Delivery Reviews*, *138*, 302–325. doi:10.1016/j.addr.2019.01.005 PMID:30639256

Daemi, H., Barikani, M., & Sardon, H. (2017). Transition-metal-free synthesis of supramolecular ionic alginate-based polyurethanes. *Carbohydrate Polymers*, *157*, 1949–1954. doi:10.1016/j.carbpol.2016.11.086 PMID:27987915

Daemi, H., Mashayekhi, M., & Modaress, M. P. (2018). Facile fabrication of sulfated alginate electrospunnanofibers. *Carbohydrate Polymers*, *198*, 481–485. doi:10.1016/j.carbpol.2018.06.105 PMID:30093025

Dai, B. H., Wang, J. Z., Xiong, Z., Zhan, X. J., Dai, W., Li, C. C., Feng, S. P., & Tang, J. Y. (2016). Programmable artificial phototactic microswimmer. *Nature Nanotechnology*, *11*(12), 1087–1092. doi:10.1038/nnano.2016.187 PMID:27749832

Dai, H. (2002). Carbon nanotubes: Opportunities and challenges. *Surface Science*, *500*(1-3), 218–241. doi:10.1016/S0039-6028(01)01558-8

Dai, L., Liu, J., Luo, Z., Li, M., & Cai, K. (2016). Tumor therapy: Targeted drug delivery systems. *Journal of Materials Chemistry. B, Materials for Biology and Medicine*, *4*(42), 6758–6772. doi:10.1039/C6TB01743F PMID:32263571

Dakal, T. C., Kumar, A., Majumdar, R. S., & Yadav, V. (2016). Mechanistic basis of antimicrobial actions of silver nanoparticles. *Frontiers in Microbiology*, *7*, 1831. doi:10.3389/fmicb.2016.01831 PMID:27899918

Dale-Skinner, J. W., & Bonev, B. B. (2009). Molecular Mechanisms of Antibiotic Resistance: The need for novel antimicrobial therapies. Wiley-VCH verlag GmbH & Co.

Danaei, M., Kalantari, M., Raji, M., Fekri, H. S., Saber, R., Asnani, G. P., ... Taheriazam, A. (2018). Probing nanoliposomes using single particle analytical techniques: Effect of excipients, solvents, phase transition and zeta potential. *Heliyon*, *4*(12), e01088. doi:10.1016/j.heliyon.2018.e01088 PMID:30603716

Danhier, F. (2016). To exploit the tumor microenvironment: Since the EPR effect fails in the clinic, what is the future of nanomedicine? *Journal of Controlled Release*, *244*, 108–121. doi:10.1016/j.jconrel.2016.11.015 PMID:27871992

Daniele, M. A., Adams, A. A., Naciri, J., North, S. H., & Ligler, F. S. (2014). Interpenetrating networks based on gelatin methacrylamide and PEG formed using concurrent thiol click chemistries for hydrogel tissue engineering scaffolds. *Biomaterials*, *35*(6), 1845–1856. doi:10.1016/j.biomaterials.2013.11.009 PMID:24314597

Daniel, K., Susana, C., & Shankar, B. (2011). Functionalized Micromachines for Selective and Rapid Isolation of Nucleic Acid Targets from Complex Samples. *Nano Letters*, *11*(5), 2083–2087. doi:10.1021/nl2005687 PMID:21491941

Daniel, M. C., & Astruc, D. (2004). Gold nanoparticles: Assembly, supramolecular chemistry, quantum-size-related properties, and applications toward biology, catalysis, and nanotechnology. *Chemical Reviews*, *104*(1), 293–346. doi:10.1021/cr030698+ PMID:14719978

Dantas, G., Sommer, M. O., Oluwasegun, R. D., & Church, G. M. (2008). Bacteria subsisting on antibiotics. *Science*, *320*(5872), 100–103. doi:10.1126cience.1155157 PMID:18388292

Dappric, J., Johnson, C., Gilbert, R., McKay, R., & Dresdner, K. (2018). Superparamagnetic Particle Scaffold for Regenerating Damaged Neural Tissue. *12th International conference on Scientific and clinical applications of Magnetic carriers.*

Dar, M. A., Ingle, A., & Rai, M. (2013). Enhanced antimicrobial activity of silver nanoparticles synthesized by cryphonectria sp. Evaluated singly and in combination with antibiotics. *Nanomedicine (London)*, *9*(1), 105–110. doi:10.1016/j.nano.2012.04.007 PMID:22633901

Darvishi, B., Manoochehri, S., Kamalinia, G., Samadi, N., Amini, M., Mostafavi, S. H., ... Dinarvand, R. (2015). Preparation and antibacterial activity evaluation of 18-β-glycyrrhetinic acid loaded PLGA nanoparticles. *IJPR*, *14*(2), 373. PMID:25901144

Das, P, Ganguly, S., Agarwal, T., Maity, P., Ghosh, S., Choudry, S., Gangopadhyay, S., Maiti, T.K., Dhara, S., Banerjee, S., & Das, N.C. (2019). Heteroatom doped blue luminescent carbon dots as a nano-probe for targeted cell labelling and anticancer drug delivery vehicle. *Material Chemistry and Physics, 237.*

Das, D., Nath, B. C., Phukon, P., & Dolui, S. K. (2013). Synthesis and evaluation of antioxidant and antibacterial behavior of CuO nanoparticles. *Colloids and Surfaces. B, Biointerfaces*, *101*, 430–433. doi:10.1016/j.colsurfb.2012.07.002 PubMed

Das, D., Vimala, R., & Das, N. (2013). Column Studies on Removal of Ag (I) from Electroplating Wastewater by MacrofungusPleurotus platypus: Use of Modelling and Response Surface Methodology. *Nature Environment and Pollution Technology*, *12*(2), 273.

Das, D., Vimala, R., & Das, N. (2015). Removal of Ag (I) and Zn (II) ions from single and binary solution using sulfonated form of gum arabic-powdered mushroom composite hollow semispheres: Equilibrium, kinetic, thermodynamic and ex-situ studies. *Ecological Engineering*, *75*, 116–122. doi:10.1016/j.ecoleng.2014.11.037

Das, K., & Roychoudhury, A. (2014). Reactive oxygen species (ROS) and response of antioxidants as ROS-scavengers during environmental stress plants. *Frontiers in Environmental Science*, *2*, 1–13. doi:10.3389/fenvs.2014.00053

Das, M., & Chatterjee, S. (2019). Green synthesis of metal/metal oxide nanoparticles toward biomedical applications: Boon or bane. In *Green Synthesis* (pp. 265–301). Characterization and Applications of Nanoparticles. doi:10.1016/B978-0-08-102579-6.00011-3

Das, R. K., Pramanik, A., Majhi, M., & Mohapatra, S. (2018). Magnetic mesoporous silica gated with doped carbon dot for site-specific drug delivery, fluorescence, and MR imaging. *Langmuir*, *34*(18), 5253–5262. doi:10.1021/acs.langmuir.7b04268 PMID:29634272

Davey, P., Sneddon, J., & Nathwani, D. (2017). Overview of strategies for overcoming the challenge of antimicrobial resistance. *Expert Review of Clinical Pharmacology, 3*(5), 667–686. doi:10.1586/ecp.10.46 PMID:22111749

De Jong, W. H., Hagens, W. I., Krystek, P., Burger, M. C., Sips, A. J., & Geertsma, R. E. (2008). Particle size-dependent organ distribution of gold nanoparticles after intravenous administration. *Biomaterials, 29*(12), 1912–1919. doi:10.1016/j.biomaterials.2007.12.037 PMID:18242692

de Kraker, M. E., Stewardson, A. J., & Harbarth, S. (2016). Will 10 Million People Die a Year due to Antimicrobial Resistance by 2050? *PLoS Medicine, 13*(11), e1002184. doi:10.1371/journal.pmed.1002184 PMID:27898664

De Matteis, L., Jary, D., Lucía, A., García-Embid, S., Serrano-Sevilla, I., Pérez, D., ... De la Fuente, J. M. (2018). New active formulations against M. tuberculosis: Bedaquiline encapsulation in lipid nanoparticles and chitosan nanocapsules. *Chemical Engineering Journal, 340*, 181–191. doi:10.1016/j.cej.2017.12.110

De Santis, P., Morosetti, S., & Rizzo, R. (1974). Conformational Analysis of Regular Enantiomeric Sequences. *Macromolecules, 7*(1), 52–58. doi:10.1021/ma60037a011

Deb, A., & Vimala, R. (2017). Graphene Mediated Drug Delivery-A Boon to Cancer Therapy. *Research Journal of Pharmacy and Technology, 10*(5), 1571–1576. doi:10.5958/0974-360X.2017.00276.1

Deb, A., & Vimala, R. (2018). Camptothecin loaded graphene oxide nanoparticle functionalized with polyethylene glycol and folic acid for anticancer drug delivery. *Journal of Drug Delivery Science and Technology, 43*, 333–342. doi:10.1016/j.jddst.2017.10.025

Decher, G. (1997). Fuzzy nanoassemblies: Toward layered polymeric multicomposites. *Science, 277*(5330), 1232–1237. doi:10.1126cience.277.5330.1232

Deck, D. H., & Winston, L. G. (2012). Sulfonamides, trimethhoprim and quinolones. In Basic and clinical pharmacology (12th ed.). McGraw-Hill.

Dedrick, R. M., Guerrero-Bustamante, C. A., Garlena, R. A., Russell, D. A., Ford, K., Harris, K., ... Hatfull, G. F. (2019). Engineered bacteriophages for treatment of a patient with a disseminated drug-resistant Mycobacterium abscessus. *Nature Medicine, 25*(5), 730–733. doi:10.103841591-019-0437-z PMID:31068712

Deehan, M., Garces, S., Kramer, D., Baker, M. P., Rat, D., Roettger, Y., & Kromminga, A. (2015). Managing unwanted immunogenicity of biologicals. *Autoimmunity Reviews, 14*(7), 569–574. doi:10.1016/j.autrev.2015.02.007 PMID:25742758

Deepak, P., Amutha, V., Kamaraj, C., Balasubramani, G., Aiswarya, D., & Perumal, P. (2019). Chemical and green synthesis of nanoparticles and their efficacy on cancer cells. In *Green Synthesis* (pp. 369–387). Characterization and Applications of Nanoparticles. doi:10.1016/B978-0-08-102579-6.00016-2

Dehcheshmeh, M. A., & Fathi, M. (2019). Production of core-shell nanofibers from zein and tragacanth for encapsulation of saffron extract. *International Journal of Biological Macromolecules, 1*(22), 272–279. doi:10.1016/j.ijbiomac.2018.10.176 PMID:30416096

Del Grosso, M., Camilli, R., Barbabella, G., Blackman Northwood, J., Farrell, D. J., & Pantosti, A. (2011). Genetic resistance elements carrying mef subclasses other than mef(A) in *Streptococcus pyogenes. Antimicrobial Agents and Chemotherapy, 55*(7), 3226–3230. doi:10.1128/AAC.01713-10 PMID:21502613

del Pozo, J. L., & Patel, R. (2007). The challenge of treating biofilm-associated bacterial infections. *Clinical Pharmacology and Therapeutics, 82*(2), 204–209. doi:10.1038j.clpt.6100247 PMID:17538551

Demirci, H., Slimani, N., Pawar, M., Kumon, R. E., Vaishnava, P., & Besirli, C. G. (2019). Magnetic Hyperthermia in Y79 Retinoblastoma and ARPE-19 Retinal Epithelial Cells: Tumor Selective Apoptotic Activity of Iron Oxide Nanoparticle. *Translational Vision Science & Technology*, *8*(5), 18. doi:10.1167/tvst.8.5.18 PMID:31602343

Denning, D. W., & Bromley, M. J. (2015). Infectious disease. How to bolster the antifungal pipeline. *Science*, *347*(6229), 1414–1416. doi:10.1126cience.aaa6097 PMID:25814567

Deorukhkar, S. C., Saini, S., & Mathew, S. (2014). Virulence Factors Contributing to Pathogenicity of Candida tropicalis and Its Antifungal Susceptibility Profile. *International Journal of Microbiology*, *2014*, 1–6. doi:10.1155/2014/456878 PubMed

Depardieu, F., Podglajen, I., Leclercq, R., Collatz, E., & Courvalin, P. (2007). Modes and Modulations of Antibiotic Resistance Gene Expression. *Clinical Microbiology Reviews*, *20*(1), 79–114. doi:10.1128/CMR.00015-06 PMID:17223624

DeQueiroz, G. A., & Day, D. F. (2007). Antimicrobial activity and effectiveness of a combination of sodium hypochlorite and hydrogen peroxide in killing and removing *Pseudomonas aeruginosa*biofilms from surfaces. *Journal of Applied Microbiology*, *103*(4), 794–802. doi:10.1111/j.1365-2672.2007.03299.x PMID:17897181

Derakhshankhah, H., & Jafari, S. (2018). Cell penetrating peptides: A concise review with emphasis on biomedical applications. *Biomedicine and Pharmacotherapy*, *108*, 1090–1096. doi:10.1016/j.biopha.2018.09.097 PMID:30372809

Derossi, D., Joliot, A. H., Chassaing, G., & Prochiantz, A. (1994). The third helix of the Antennapedia homeodomain translocates through biological membranes. *The Journal of Biological Chemistry*, *269*, 10444–10450. PMID:8144628

Desai, R., Mankad, V., Gupta, S., & Jha, P. (2012). Size Distribution of Silver Nanoparticles: UV-Visible Spectroscopic Assessment. *Nanoscience and Nanotechnology Letters*, *4*(1), 30–34. doi:10.1166/nnl.2012.1278

Di Corato, R., Béalle, G., Kolosnjaj-Tabi, J., Espinosa, A., Clément, O., Silva, A. K. A., Ménager, C., & Wilhelm, C. (2015). Combining magnetic hyperthermia and photody- namic therapy for tumor ablation with photoresponsive magnetic liposomes. *ACS Nano*, *9*(3), 2904–2916. doi:10.1021/nn506949t PMID:25695371

Diarra, M. S., Giguere, K., Malouin, F., Lefebvre, B., Bach, S., Delaquis, P., Aslam, M., Ziebell, K. A., & Roy, G. (2009). Genotype, serotype, and antibiotic resistance of sorbitol-negative Escherichia coli isolates from feedlot cattle. *Journal of Food Protection*, *72*(1), 28–36. doi:10.4315/0362-028X-72.1.28 PMID:19205460

Dicks, K. V., & Stout, J. E. (2019). Molecular diagnostics for Mycobacterium tuberculosis infection. *Annual Review of Medicine*, *70*(1), 77–90. doi:10.1146/annurev-med-040717-051502 PMID:30125128

Dieme, C., Bechah, Y., Socolovschi, C., Audoly, G., Berenger, J. M., Faye, O., Raoult, D., & Parola, P. (2015). Transmission potential of Rickettsia felis infection by Anopheles gambiae mosquitoes. *Proceedings of the National Academy of Sciences of the United States of America*, *112*(26), 8088–8093. doi:10.1073/pnas.1413835112 PMID:26056256

Dilnawaz, F., Singh, A., Mohanty, C., & Sahoo, S. K. (2010). Dual drug loaded superparamagnetic iron oxide nanoparticles for targeted cancer therapy. *Biomaterials*, *31*(13), 3694–3706. doi:10.1016/j.biomaterials.2010.01.057 PMID:20144478

Dimkpa, C. O., McLean, J. E., Latta, D. E., Manangón, E., Britt, D. W., Johnson, W. P., & Anderson, A. J. (2012). CuO and ZnO nanoparticles: Phytotoxicity, metal speciation, and induction of oxidative stress in sand-grown wheat. *Journal of Nanoparticle Research*, *14*(9), 1–15. doi:10.100711051-012-1125-9

Dinali, R., Ebrahiminezhad, A., Manley-Harris, M., Ghasemi, Y., & Berenjian, A. (2017). Iron oxide nanoparticles in modern microbiology and biotechnology. *Critical Reviews in Microbiology*, *43*(4), 493–507. doi:10.1080/1040841X.2016.1267708 PMID:28068855

Dinesh, D., Murugan, K., Madhiyazhagan, P., Panneerselvam, C., Kumar, P. M., Nicoletti, M., Jiang, W., Benelli, G., Chandramohan, B., & Suresh, U. (2015). Mosquitocidal and antibacterial activity of green-synthesized silver nanoparticles from Aloe vera extracts: Towards an effective tool against the malaria vector Anopheles stephensi? *Parasitology Research, 114*(4), 1519–1529. doi:10.100700436-015-4336-z PMID:25653031

Ding, B., Wang, X., Yu, J., & Wang, M. (2011). Polyamide 6 composite nano-fiber/net func- tionalized by polyethyleneimine on quartz crystal microbalance for highly sensitive formaldehyde sensors. *Journal of Materials Chemistry, 21*(34), 12784–12792. doi:10.1039/c1jm11847a

Ding, C., Du, J., Cao, Y., Yue, C., & Cheng, B. (2019). Effects of the aspect ratio of multi-walled carbon nanotubes on the structure and properties of regenerated collagen fibers. *International Journal of Biological Macromolecules, 126*, 595–602. doi:10.1016/j.ijbiomac.2018.12.144 PMID:30576737

Ding, H., Gangalum, P. R., Galstyan, A., Fox, I., Patil, R., Hubbard, P., Murali, R., Ljubimova, J. Y., & Holler, E. (2017). HER2-positive breast cancer targeting and treatment by a peptide-conjugated mini nanodrug. *Nanomedicine; Nanotechnology, Biology, and Medicine, 13*(2), 631–639. doi:10.1016/j.nano.2016.07.013 PMID:27520726

Ding, K., Zeng, J., Jing, L., Qiao, R., Liu, C., Jiao, M., Li, Z., & Gao, M. (2015). Aqueous synthesis of PEGylated copper sulfide nanoparticles for photoacoustic imaging of tumors. *Nanoscale, 7*(25), 11075–11081. doi:10.1039/C5NR02180D PMID:26055816

DMC. (2020). *Malaria control Program*. DMC.

Dmitriev, B. A., Toukach, F. V., Holst, O., Rietschel, E. T., & Ehlers, S. (2004). Tertiary structure of *Staphylococcus aureus* cell wall murein. *Journal of Bacteriology, 186*(21), 7141–7148. doi:10.1128/JB.186.21.7141-7148.2004 PMID:15489425

Doan, T. K. D. (2009). Article. *Journal of Physics: Conference Series, 187*, 012036. doi:10.1088/1742-6596/187/1/012036

Doan, T. V. P., & Olivier, J. C. (2009). Preparation of rifampicin-loaded PLGA microspheres for lung delivery as aerosol by premix membrane homogenization. *International Journal of Pharmaceutics, 382*(1-2), 61–66. doi:10.1016/j.ijpharm.2009.08.008 PMID:19682562

Dobrovolskaia, M. A., & McNeil, S. E. (2007). Immunological properties of engineered nanomaterials. *Nature Nanotechnology, 2*(8), 469–478. doi:10.1038/nnano.2007.223 PMID:18654343

Dobrucka, R., Kaczmarek, M., Łagiedo, M., Kielan, A., & Dlugaszewska, J. (2019). Evaluation of biologically synthesized Au-CuO and CuO-ZnO nanoparticles against glioma cells and microorganisms. *Saudi Pharmaceutical Journal, 27*(3), 373–383. doi:10.1016/j.jsps.2018.12.006 PMID:30976181

Dobson. (2006). Macro and Nanoparticle based targeting of drug and gene delivery. *Nanomedicine*.

Dobson. (2008). Magnetic nanoparticles for gene and drug delivery. *International Journal of Nanomedicine, 3*(2), 169-180.

Domingo-Calap, P., & Delgado-Martínez, J. (2018). Bacteriophages: Protagonists of a post-antibiotic era. *Antibiotics (Basel, Switzerland), 7*(3), 66. doi:10.3390/antibiotics7030066 PMID:30060506

Donaldson, K., & Stone, V. (2003). Current hypotheses on the mechanisms of toxicity of ultrafine particles. *Annalidell'Istitutosuperiore di sanitã, 39*(3), 405-410.

Donaldson, K., Li, X., & MacNee, W. (1998). Ultrafine (nanometre) particlemediated lung injury. *Journal of Aerosol Science, 29*(5-6), 553–560. doi:10.1016/S0021-8502(97)00464-3

Dong, J., Signo, K. S. L., Vanderlinder, E. M., Yost, C. K., & Dahms, T. E. S. (2011). Atomic force microscopy of a ctpA mutant in *Rhizobium leguminosarum* reveals surface defects linking CtpA function to biofilm formation. *Microbiology, 157*(11), 3049–3058. doi:10.1099/mic.0.051045-0 PMID:21852352

Dong, K., Zhang, Y., Zhang, L., Wang, Z., Ren, J., & Qu, X. (2019, March). Facile preparation of Metal–Organic Frameworks-based Hydrophobic Anticancer Drug Delivery Nanoplatform for targeted and enhanced cancer treatment, (2018). *Talanta, 194*, 703–708. Advance online publication. doi:10.1016/j.talanta.2018.10.101 PMID:30609594

Dong, X., & Mumper, R. J. (2010). Nanomedicinal strategies to treat multidrug-resistant tumors: Current progress. *Nanomedicine (London), 5*(4), 597–615. doi:10.2217/nnm.10.35 PMID:20528455

Dong, Z. Y., Narsing Rao, M. P., Xiao, M., Wang, H. F., Hozzein, W. N., Chen, W., & Li, W. J. (2017). Antibacterial activity of silver nanoparticles against Staphylococcus warneri synthesized using endophytic bacteria by photo-irradiation. *Frontiers in Microbiology, 8*, 1090. doi:10.3389/fmicb.2017.01090 PMID:28659903

Doraiswamy, P. M., & Finefrock, A. E. (2004). Metals in our minds: Therapeutic implications for neurodegenerative disorders. *Lancet Neurology, 3*(7), 431–434. doi:10.1016/S1474-4422(04)00809-9 PMID:15207800

Douziech-Eyrolles, L. (2007). Nanovectors for anticancer agents based on superparamagnetic iron oxide nanoparticles. *International Journal of Nanomedicine, 2*, 541–550. PMID:18203422

Dreaden, E. C., Alkilany, A. M., Huang, X., Murphy, C. J., & El-Sayed, M. A. (2012). The golden age: Gold nanoparticles for biomedicine. *Chemical Society Reviews, 41*(7), 2740–2779. doi:10.1039/C1CS15237H PMID:22109657

Drlica, K., Malik, M., Kernsm, R. J., & Zhaom, X. (2008). Quinolone-mediated bacterial death. *Antimicrobial Agents and Chemotherapy, 52*(2), 385–392. doi:10.1128/AAC.01617-06 PMID:17724149

Drulis-Kawa, Z., Majkowska-Skrobek, G., & Maciejewska, B. (2015). Bacteriophages and phage-derived proteins—Application approaches. *Current Medicinal Chemistry, 22*(14), 1757–1773. doi:10.2174/0929867322666150209152851 PMID:25666799

Duan, J., Dong, J., Zhang, T., Su, Z., Ding, J., Zhang, Y., & Mao, X. (2014). Polyethyleneimine-functionalized iron oxide nanoparticles for systemic siRNA delivery in experimental arthritis. *Nanomedicine (London), 9*(6), 789–801. doi:10.2217/nnm.13.217 PMID:24392891

Du, F., Zhang, L., Zhang, L., Zhang, M., Gong, A., Tan, Y., & Zou, S. (2017). Engineered gadolinium-doped carbon dots for magnetic resonance imaging-guided radiotherapy of tumors. *Biomaterials, 121*, 109–120. doi:10.1016/j.biomaterials.2016.07.008 PMID:28086179

Dunn, G. P., Bruce, A. T., Ikeda, H., Old, L. J., & Schreiber, R. D. (2002). Cancer immunoediting: From immunosurveillance to tumor escape. *Nature Immunology, 3*(11), 991–998. doi:10.1038/ni1102-991 PMID:12407406

Duraffour, S., Andrei, G., Topalis, D., Krečmerová, M., Crance, J. M., Garin, D., & Snoeck, R. (2012, July 01). Mutations Conferring Resistance to Viral DNA Polymerase Inhibitors in Camelpox Virus Give Different Drug-Susceptibility Profiles in Vaccinia Virus. *Journal of Virology, 86*(13), 7310–7325. Advance online publication. doi:10.1128/JVI.00355-12 PMID:22532673

Durán, N., Islan, G. A., Durán, M., & Castro, G. R. (2016). Nanobiotechnology Solutions against Aedes aegypti %J. *Journal of the Brazilian Chemical Society, 27*, 1139–1149. http://www.scielo.br/scielo.php?script=sci_arttext&pid=S0103-50532016000701139&nrm=iso

Durante-Mangoni, E., Andini, R., & Zampino, R. (2019). Management of carbapenem-resistant Enterobacteriaceae infections. *Clinical Microbiology and Infection, 25*(8), 943–950. doi:10.1016/j.cmi.2019.04.013 PMID:31004767

Dutta, B., Shetake, N. G., Gawali, S. L., Barick, B. K., Barick, K. C., Babu, P. D., Pandey, B. N., Priyadarsini, K. I., & Hassan, P. A. (2018). PEG mediated shape-selective synthesis of cubic Fe3O4 nanoparticles for cancer therapeutics. *Journal of Alloys and Compounds*, *737*, 347–355. doi:10.1016/j.jallcom.2017.12.028

ECDC. (2019a). *Communicable Disease Threats Report*. European Centre for Disease Prevention and Control Issue. www.ecdc.europa.eu/en/publications-data/communicable-disease-threats-report-15-21-december-2019-week-51

ECDC. (2019b). *Communicable Disease Threats Report, 15-21 December 2019, week 51*. ECDC.

Edgar, R., Friedman, N., Molshanski-Mor, S., & Qimron, U. (2012). Reversing bacterial resistance to antibiotics by phage-mediated delivery of dominant sensitive genes. *Applied and Environmental Microbiology*, *78*(3), 744–751. doi:10.1128/AEM.05741-11 PMID:22113912

Edson, J. A., & Kwon, Y. J. (2016). Design, challenge, and promise of stimuli-responsive nanoantibiotics. *Nano Convergence*, *3*(1), 26. doi:10.118640580-016-0085-7 PMID:28191436

Egodawatte, S., Dominguez, S. Jr, & Larsen, S. C. (2017). Solvent effects in the development of a drug delivery system for 5-fluorouracil using magnetic mesoporous silica nanoparticles. *Microporous and Mesoporous Materials*, *237*, 108–116. doi:10.1016/j.micromeso.2016.09.024

El Badawy, A. M., Silva, R. G., Morris, B., Scheckel, K. G., Suidan, M. T., & Tolaymat, T. M. (2011). Surface charge-dependent toxicity of silver nanoparticles. *Environmental Science & Technology*, *45*(1), 283–287. doi:10.1021/es1034188 PMID:21133412

El Haddad, L., Harb, C. P., Gebara, M. A., Stibich, M. A., & Chemaly, R. F. (2019). A systematic and critical review of bacteriophage therapy against multidrug-resistant ESKAPE organisms in humans. *Clinical Infectious Diseases*, *69*(1), 167–178. doi:10.1093/cid/ciy947 PMID:30395179

El Shafey, A. M. (2020). Green synthesis of metal and metal oxide nanoparticles from plant leaf extracts and their applications: A review. *Green Processing and Synthesis*, *9*(1), 304–339. doi:10.1515/gps-2020-0031

Eleraky, N. E., Allam, A., Hassan, S. B., & Omar, M. M. (2020). Nanomedicine fight against antibacterial resistance: An overview of the recent pharmaceutical innovations. *Pharmaceutics*, *12*(2), 142. doi:10.3390/pharmaceutics12020142 PMID:32046289

Elgindy, N., Samy, W., & Elzoghby, A. (2014). Casein-based micelles: A novel vector for delivery of the poorly soluble anticancer drug, flutamide. *Therapeutic Delivery*, *5*(1), 7–9. doi:10.4155/tde.13.122 PMID:24341810

Elie-Ali-Komi, D., & Hamblin, M. R. (2016). Chitin and Chitosan: Production and Application of Versatile Biomedical Nanomaterials. *International Journal of Advanced Research*, *4*(3), 411–427. PMID:27819009

Eliopoulos, G. M., & Bush, K. (2001). New beta-lactamases in gram-negative bacteria: Diversity and impact on the selection of antimicrobial therapy. *Clinical Infectious Diseases*, *32*(7), 1085–1089. doi:10.1086/319610 PubMed

El-Leithy, E. S., Hassan, S. A., & Abdel-Rashid, R. S. (2019). Tamoxifen citrate/Coenzyme Q10 as smart nanocarriers Bitherapy for Breast Cancer: Cytotoxicity, genetoxicity, and antioxidant activity. *Journal of Drug Delivery Science and Technology*, *51*, 36–44. doi:10.1016/j.jddst.2019.02.010

Elliott, J. A., Shibuta, Y., Amara, H., Bichara, C., & Neyts, E. C. (2013). Atomistic modelling of CVD synthesis of carbon nanotubes and graphene. *Nanoscale*, *5*(15), 6662–6676. doi:10.1039/c3nr01925j PMID:23774798

El-Ridy, M. S., Yehia, S. A., Kassem, M. A. E. M., Mostafa, D. M., Nasr, E. A., & Asfour, M. H. (2015). Niosomal encapsulation of ethambutol hydrochloride for increasing its efficacy and safety. *Drug Delivery*, *22*(1), 21–36. doi:10.3109/10717544.2013.868556 PMID:24359403

Elsaesser, A., & Howard, C. V. (2012). Toxicology of nanoparticles. *Advanced Drug Delivery Reviews*, *64*(2), 129–137. doi:10.1016/j.addr.2011.09.001 PMID:21925220

Elsayed, M. A., Othman, A. M., Hassan, M. M., & Elshafei, A. M. (2018). Optimization of silver nanoparticles biosynthesis mediated by Aspergillus niger NRC1731 through application of statistical methods: enhancement and characterization. *Biotech*, *8*(3). doi:10.100713205-018-1158-6

Elzoghby, A., Samy, W., & Elgindy, N. (2012). Albumin-based nanoparticles as potential controlled release drug delivery systems. *Journal of Controlled Release*, *157*(2), 168–182. doi:10.1016/j.jconrel.2011.07.031 PMID:21839127

Eng, S.-K., Pusparajah, P., Mutalib, N.-S., Ser, H.-L., Chan, K.-G., & Lee, L. H. (2015). Salmonella: A review on pathogenesis, epidemiology and antibiotic resistance. *Frontiers in Life Science*, *8*(3), 284–293. doi:10.1080/21553769.2015.1051243

Epaulard, O., Villier, C., Ravaud, P., Chosidow, O., Blanche, S., Mamzer-Bruneel, M.-F., Thiebaut, A., Leccia, M.-T., & Lortholary, O. (2013). A multistep voriconazole-related phototoxic pathway may lead to skin carcinoma: Results from a French nationwide study. *Clinical Infectious Diseases*, *57*(12), e182–e188. doi:10.1093/cid/cit600 PMID:24046296

Erdem, M., Yalcin, S., & Gunduz, U. (2017). Folic acid-conjugated polyethylene glycol-coated magnetic nanoparticles for doxorubicin delivery in cancer chemotherapy: Preparation, characterization and cytotoxicity on HeLa cell line. *Human and Experimental Toxicology*, *36*(8), 833–845. doi:10.1177/0960327116672910 PMID:27758842

Erman, A., Hergouth, V. K., Blango, M. G., Kos, M. K., Mulvey, M. A., & Veranic, P. (2017). Repeated treatments with chitosan in combination with antibiotics completely eradicate uropathogenic Escherichia coli from infected mouse urinary bladders. *The Journal of Infectious Diseases*, *216*, 375–381. doi:10.1093/infdis/jix023 PMID:28119486

Ershov, B. G., Abkhalimov, E. V., Solovov, R. D., & Roldughin, V. I. (2016). Gold nanoparticles in aqueous solutions: Influence of size and pH on hydrogen dissociative adsorption and Au (III) ion reduction. *Physical Chemistry Chemical Physics*, *18*(19), 13459–13466. doi:10.1039/C6CP01996J PMID:27125624

Esmaeili, F., Hosseini-Nasr, M., Rad-Malekshahi, M., Samadi, N., Atyabi, F., & Dinarvand, R. (2007). Preparation and antibacterial activity evaluation of rifampicin-loaded poly lactide-co-glycolide nanoparticles. *Nanomedicine; Nanotechnology, Biology, and Medicine*, *3*(2), 161–167. doi:10.1016/j.nano.2007.03.003 PMID:17468055

Esmaeillou, M., Zarrini, G., Ahangarzadeh, R. M., Shahbazi, M. J., & Bahadori, A. (2017). Vancomycin capped with silver nanoparticles as an antibacterial agent against multi-drug resistance bacteria. *Advanced Pharmaceutical Bulletin*, *7*(3), 479–483. doi:10.15171/apb.2017.058 PMID:29071232

Estelrich, J., Sánchez-Martín, M. J., & Busquets, M. A. (2015). Nanoparticles in magnetic resonance imaging: From simple to dual contrast agents. *International Journal of Nanomedicine*, *10*, 1727. PMID:25834422

Evans, J. S., Sun, Y., Senyuk, B., Keller, P., Pergamenshchik, V. M., Lee, T., & Smalyukh, I. I. (2013). Active shape-morphing elastomeric colloids in short-pitch cholesteric liquid crystals. *Physical Review Letters*, *110*(18), 187802. doi:10.1103/PhysRevLett.110.187802 PMID:23683245

Eynali, S., Khoei, S., Khoee, S., & Esmaelbeygi, E. (2017). Evaluation of the cytotoxic effects of hyperthermia and 5-fluorouracil-loaded magnetic nanoparticles on human colon cancer cell line HT-29. *International Journal of Hyperthermia*, *33*(3), 327–335. doi:10.1080/02656736.2016.1243260 PMID:27701929

Ezzati, J. N. D., Valizadeh, H., & Hamishehkar, H. (2015). Solid Lipid nanoparticles as Efficient Drug and Gene Delivery Systems: Recent Breakthroughs. *Advanced Pharmaceutical Bulletin*, *5*(2), 151–159. doi:10.15171/apb.2015.022 PMID:26236652

Fabricius, A. L., Duester, L., Meermann, B., & Ternes, T. (2014). ICP-MS based characterization of inorganic nanoparticles-sample preparation and off line fractionation strategies. *Analytical and Bioanalytical Chemistry*, *406*(2), 467–479. doi:10.100700216-013-7480-2 PMID:24292431

Fabris, L., Ceder, Y., Chinnaiyan, A. M., Jenster, G. W., Sorensen, K. D., Tomlins, S., Visakorpi, T., & Calin, G. A. (2016). The potential of microRNAs as prostate cancer biomarkers. *European Urology*, *70*(2), 312–322. PubMed

Faisal, M., Saquib, Q., Alatar, A. A., Al-Khedhairy, A. A., Hegazy, A. K., & Musarrat, J. (2013). Phytotoxic hazards of NiO-nanoparticles in tomato: A study on mechanism of cell death. *Journal of Hazardous Materials*, *250–251*(April), 318–332. doi:10.1016/j.jhazmat.2013.01.063 PMID:23474406

Fakhimikabir, H., Tavakoli, M. B., Zarrabi, A., Amouheidari, A., & Rahgozar, S. (2018). The role of folic acid-conjugated polyglycerol coated iron oxide nanoparticles on radiosensitivity with clinical electron beam (6 MeV) on human cervical carcinoma cell line: In vitro study. *Journal of Photochemistry and Photobiology. B, Biology*, *182*, 71–76. doi:10.1016/j.jphotobiol.2018.03.023 PMID:29626804

Fan, A. C., & Leppert, J. T. (2019). Postmarketing Analysis of Sipuleucel-T—The Importance of Real-World Data. *JAMA Network Open*, *2*(8), e199233–e199233. doi:10.1001/jamanetworkopen.2019.9233 PMID:31411706

Fang, D., Liu, Y., Jiang, S., Nie, J., & Ma, G. (2011). Effect of intermolecular interaction on electrospinning of sodium alginate. *Carbohydrate Polymers*, *85*(1), 276–279. doi:10.1016/j.carbpol.2011.01.054

Fan, Z., Chang, Y., Cui, C., Sun, L., Wang, D. H., Pan, Z., & Zhang, M. (2018). Near infrared fluorescent peptide nanoparticles for enhancing esophageal cancer therapeutic efficacy. *Nature Communications*, *9*(1), 1–11. doi:10.103841467-018-04763-y PMID:29973582

Farahani, M. S., Falahati, M., & Askari, S. (2017). Nano-toxicology of metal nanoparticles in stem cells (in vitro). *Journal of Nanomedicine & Nanotechnology*, *8*(6). Advance online publication. doi:10.4172/2157-7439-C1-055

Faraji, M., Yamini, Y., & Rezaee, M. (2010). Magnetic nanoparticles: Synthesis, stabilization, functionalization, characterization, and applications. *Journal of the Indian Chemical Society*, *7*(1), 1–37. doi:10.1007/BF03245856

Farooq, M., Novosad, V., Rozhkova, E., Wali, H., Ali, A., Fateh, A., Neogi, P. B., Neogi, A., & Wang, Z. (2018). Gold Nanoparticles-enabled Efficient Dual Delivery of Anticancer Therapeutics to HeLa Cells. *Scientific Reports*, *8*(1), 2907. doi:10.103841598-018-21331-y PMID:29440698

Farzin, A., Etesami, S. A., Quint, J., Memic, A., & Tamayol, A. (2020). Magnetic Nanoparticles in Cancer Therapy and Diagnosis. *Advanced Healthcare Materials*, *9*(9), 1901058. doi:10.1002/adhm.201901058 PMID:32196144

Fayaz, A. M., Balaji, K., Girilal, M., Yadav, R., Tech, M., Kalaichelvan, P. T., & Venketesan, R. (2010). Biogenic synthesis of silver nanoparticles and their synergistic effect with antibiotics: A study against gram-positive and gram-negative bacteria. *Nanomedicine (London)*, *6*(1), 103–109. doi:10.1016/j.nano.2009.04.006 PMID:19447203

Fearns, R., & Plemper, R. K. (2017, April 15). Polymerases Of Paramyxoviruses And Pneumoviruses. *Virus Research*, *234*, 87–102. doi:10.1016/j.virusres.2017.01.008 PMID:28104450

Feliu, N., Sun, X., Alvarez Puebla, R. A., & Parak, W. J. (2017). Quantitative particle–cell interaction: Some basic physicochemical pitfalls. *Langmuir*, *33*(27), 6639–6646. doi:10.1021/acs.langmuir.6b04629 PMID:28379704

Felix, L. C., Ortega, V. A., & Goss, G. G. (2017). Cellular uptake and intracellular localization of poly (acrylic acid) nanoparticles in a rainbow trout (Oncorhynchus mykiss) gill epithelial cell line RTgill-W1. *Aquatic Toxicology (Amsterdam, Netherlands)*, *192*, 58–68. doi:10.1016/j.aquatox.2017.09.008 PMID:28917946

Fellner, C. (2012). Ipilimumab (yervoy) prolongs survival in advanced melanoma: Serious side effects and a hefty price tag may limit its use. *P&T*, *37*(9), 503–530. PMID:23066344

Fellows, B. D., Raval, Y. S., Murbach, J., Cordeau, Y., Tzeng, T.-R. J., & Thompson Mefford, O. (2018). Glycoconjugate-Functionalized Magnetic Nanoparticles: A Tool for Selective Killing of Targeted Bacteria Via Magnetically Mediated Energy Delivery. *12th International conference on Scientific and clinical applications of Magnetic carriers.*

Fellows, Ghobrial, Mappus, Hargett, Bolding, Dean, & Mefford. (2018). Heparin Coated Magnetic Nanoparticles for Treatment of In-Stent Restenosis and Prevention of Late Stent Thrombosis. *12th International conference on Scientific and clinical applications of Magnetic carriers.*

Fenchel, T., Blackburn, H., & King, G. M. (2012). *Bacterial Biogeochemistry: The ecophysiology of mineral cycling* (3rd ed.). Academic Press/Elsevier.

Feng, C., Khulbe, K. C., & Matsuura, T. (2010). Recent progress in the preparation, characterization, and applications of nanofibers and nanofiber membranes via electrospinning/interfacial polymerization. *Journal of Applied Polymer Science*, *115*(2), 756–776. doi:10.1002/app.31059

Feng, L., Li, S., Li, H., Zhai, J., Song, Y., Jiang, L., & Zhu, D. (2002). Super- Hydrophobic Surface of Aligned Polyacrylonitrile Nanofibers. *Angewandte Chemie International Edition*, *41*(7), 1221–1223. doi:10.1002/1521-3773(20020402)41:7<1221::AID-ANIE1221>3.0.CO;2-G PMID:12491265

Feng, Q. L., Wu, J., Chen, G. Q., Cui, F. Z., Kim, T. N., & Kim, J. O. (2000). A mechanistic study of the antibacterial effect of silver ions on *Escherichia coli* and *Staphylococcus aureus*. *Journal of Biomedical Materials Research*, *52*(4), 662–668. doi:10.1002/1097-4636(20001215)52:4<662::AID-JBM10>3.0.CO;2-3 PMID:11033548

Feng, Q., Zhang, Y., Zhang, W., Shan, X., Yuan, Y., Zhang, H., Hou, L., & Zhang, Z. (2016). Tumor-targeted and multi-stimuli responsive drug delivery system for near-infrared light induced chemo-phototherapy and photoacoustic tomography. *Acta Biomaterialia*, *38*, 129–142. doi:10.1016/j.actbio.2016.04.024 PMID:27090593

Feng, Z. Q., Chu, X., Huang, N. P., Wang, T., Wang, Y., Shi, X., Ding, Y., & Gu, Z. Z. (2009). The effect of nanofibrous galactosylated chitosan scaffolds on the formation of rat primary hepatocyte aggregates and the maintenance of liver function. *Biomaterials*, *30*(14), 2753–2763. doi:10.1016/j.biomaterials.2009.01.053 PMID:19232710

Ferlay, J., Shin, H.R., Bray, F., Forman, D., Mathers, C., & Parkin, D.M. (2010). GLOBOCAN 2008; cancer incidence and mortality worldwide. *IArC CancerBase*, 10.

Ferlay, J., Colombet, M., Soerjomataram, I., Mathers, C., Parkin, D. M., Piñeros, M., Znaor, A., & Bray, F. (2019). Estimating the global cancer incidence and mortality in 2018: GLOBOCAN sources and methods. *International Journal of Cancer*, *144*(8), 1941–1953. doi:10.1002/ijc.31937 PMID:30350310

Ferlay, J., Soerjomataram, I., Dikshit, R., Eser, S., Mathers, C., Rebelo, M., Parkin, D. M., Forman, D., & Bray, F. (2015). Cancer incidence and mortality worldwide: Sources, methods and major patterns in GLOBOCAN 2012. *International Journal of Cancer*, *136*(5), E359–E386. doi:10.1002/ijc.29210 PMID:25220842

Fernandes, C., Suares, D., & Yergeri, M. C. (2018). Tumor microenvironment targeted nanotherapy. *Frontiers in Pharmacology*, *9*, 1230. doi:10.3389/fphar.2018.01230 PMID:30429787

Fernandes, J. G., Correia, D. M., Botelho, G., Padrão, J., Dourado, F., Ribeiro, C., Lanceros-Méndez, S., & Sencadas, V. (2014). PHB- PEO electrospun fiber membranes containing chlorhexidine for drug delivery applications. *Polymer Testing*, *34*(1), 64–71. doi:10.1016/j.polymertesting.2013.12.007

Fernández, L., Gutiérrez, D., García, P., & Rodríguez, A. (2019). The Perfect Bacteriophage for Therapeutic Applications-A Quick Guide. *Antibiotics (Basel, Switzerland)*, *8*(3), 126. Advance online publication. doi:10.3390/antibiotics8030126 PMID:31443585

Fernández, L., Gutiérrez, D., Rodríguez, A., & García, P. (2018). Application of Bacteriophages in the Agro-Food Sector: A Long Way Toward Approval. *Frontiers in Cellular and Infection Microbiology*, *8*, 296. doi:10.3389/fcimb.2018.00296 PMID:30186776

Fernandez-Lopez, S., Kim, H. S., Choi, E. C., Delgado, M., Granja, J. R., Khasanov, A., Kraehenbuehl, K., Long, G., Weinberger, D. A., Wilcoxen, K. M., & Ghadiri, M. R. (2001). Antibacterial agents based on the cyclic D,L-a-peptide architecture. *Nature*, *412*(6845), 452–455. doi:10.1038/35086601 PMID:11473322

Ferraris, S., & Padoan, G. (2018). *Preparation and characterization of Ti-6Al-4V surfaces functionalized with silver for antibacterial purposes: study of ionic and nanoparticles release mechanism.* Academic Press.

Ferraz-Carvalho, R. S., Pereira, M. A., Linhares, L. A., Lira-Nogueira, M. C., Cavalcanti, I. M., Santos-Magalhães, N. S., & Montenegro, L. M. (2016). Effects of the encapsulation of usnic acid into liposomes and interactions with antituberculous agents against multidrug-resistant tuberculosis clinical isolates. *Memorias do Instituto Oswaldo Cruz*, *111*(5), 330–334. doi:10.1590/0074-02760150454 PMID:27143488

Fierer, N., & Jackson, R. B. (2006). The diversity and biogeography of soil bacterial communities. *Proceedings of the National Academy of Sciences of the United States of America*, *103*(3), 626–631. doi:10.1073/pnas.0507535103 PMID:16407148

Fischbach, M. A., & Walsh, C. T. (2009). Antibiotics for emerging pathogens. *Science*, *325*(5944), 1089–1093. doi:10.1126cience.1176667 PMID:19713519

Fluit, A. C., & Schmitz, F. J. (2004). Resistance integrons and super-integrons. *Clinical Microbiology and Infection*, *10*(4), 272–288. doi:10.1111/j.1198-743X.2004.00858.x PMID:15059115

Foldbjerg, R., & Autrup, H. (2013). Article. *Mech Silver Nanoparticle Toxicity*, *1*, 5–15.

Foldbjerg, R., Dang, D. A., & Autrup, H. (2011). Cytotoxicity and genotoxicity of silver nanoparticles in the human lung cancer cell line, A549. *Archives of Toxicology*, *85*(7), 743–750. doi:10.100700204-010-0545-5 PMID:20428844

Foldbjerg, R., Olesen, P., Hougaard, M., Dang, D. A., Hoffmann, H. J., & Autrup, H. (2009). PVP-coated silver nanoparticles and silver ions induce reactive oxygen species, apoptosis and necrosis in THP-1 monocytes. *Toxicology Letters*, *190*(2), 156–162. doi:10.1016/j.toxlet.2009.07.009 PMID:19607894

Foldvari, M., & Bagonluri, M. (2008). Carbon nanotubes as functional excipients for nanomedicines: II. Drug delivery and biocompatibilityissues. *Nanomedicine (London)*, *4*(3), 183–200. doi:10.1016/j.nano.2008.04.003 PMID:18550450

Fonseca, L. M., Cruxen, C. E. D. S., Bruni, G. P., Fiorentini, A. M., Zayareze, E. D. R., & Lim, L. T. (2019). Development of antimicrobial and antioxidant electrospun soluble potato starch nanofibers loaded with carvacrol. *International Journal of Biological Macromolecules*, *15*(139), 1182–1190. doi:10.1016/j.ijbiomac.2019.08.096 PMID:31415859

Forbes, B. A., Sahm, D. F., & Weissfeld, A. S. (1998). *Bailey And Scott's Diagnostic Microbiology* (10th ed.). Mosby Inc.

Forssen, E. A., & Ross, M. E. (1994). Daunoxome® treatment of solid tumors: Preclinical and clinical investigations. *Journal of Liposome Research*, *4*(1), 481–512. doi:10.3109/08982109409037058

Fosgerau, K., & Hoffmann, T. (2015). Peptide therapeutics: Current status and future directions. *Drug Discovery Today*, *20*(1), 122–128. doi:10.1016/j.drudis.2014.10.003 PMID:25450771

Fournier-Bidoz, S., Arsenault, A. C., Manners, I., & Ozin, G. A. (2005). Synthetic self-propelled nanorotors. *Chemical Communications*, *441*(4), 441. doi:10.1039/b414896g PMID:15654363

Fraimow, H. S., & Tsigrelis, C. (2011). Antimicrobial resistance in the intensive care unit: Mechanisms, epidemiology, and management of specific resistant pathogens. *Critical Care Clinics*, *27*(1), 163–205. doi:10.1016/j.ccc.2010.11.002 PMID:21144992

Franci, G., Falanga, A., Galdiero, S., Palomba, L., Rai, M., Morelli, G., & Galdiero, M. (2015). Silver nanoparticles as potential antibacterial agents. *Molecules (Basel, Switzerland)*, *20*(5), 8856–8874. doi:10.3390/molecules20058856 PMID:25993417

François, L. M., Wilson, D., & Hube, B. (2013). Candida albicans pathogenicity mechanisms. *Virulence*, *4*(2), 119–128. doi:10.4161/viru.22913 PubMed

Francolini, I., & Piozzi, A. (2019). *Polymeric Systems as Antimicrobial or Antifouling Agents*. Academic Press.

Franco-Molina, M. A., Mendoza-Gamboa, E., Sierra-Rivera, C. A., Gómez-Flores, R. A., Zapata-Benavides, P., Castillo-Tello, P., & Rodríguez-Padilla, C. (2010). Antitumor activity of colloidal silver on MCF-7 human breast cancer cells. *Journal of Experimental & Clinical Cancer Research*, *29*(1), 148. doi:10.1186/1756-9966-29-148 PMID:21080962

Frost, L. S., & Koraimann, G. (2010). Regulation of bacterial conjugation: Balancing opportunity with adversity. *Future Microbiology*, *5*(7), 1057–1071. doi:10.2217/fmb.10.70 PMID:20632805

Fung, F., Wang, H. S., & Menon, S. (2018). Food safety in the 21st century. *Biomedical Journal*, *41*(2), 88–95. doi:10.1016/j.bj.2018.03.003 PMID:29866604

Fu, P. P., Xia, Q., Hwang, H. M., Ray, P. C., & Yu, H. (2014). Mechanisms of nanotoxicity: Generation of reactive oxygen species. *Yao Wu Shi Pin Fen Xi*, *22*(1), 64–75. PMID:24673904

Gabizon, A. A. (2001). Pegylated liposomal doxorubicin: Metamorphosis of an old drug into a new form of chemotherapy. *Cancer Investigation*, *19*(4), 424–436. doi:10.1081/CNV-100103136 PMID:11405181

Gabizon, A., Shmeeda, H., & Barenholz, Y. (2003). Pharmacokinetics of pegylated liposomal doxorubicin. *Clinical Pharmacokinetics*, *42*(5), 419–436. doi:10.2165/00003088-200342050-00002 PMID:12739982

Gad, S.C. (2017). Oral drug formulation development in pharmaceutical lead selection stage. *Oral Formulation Roadmap from Early Drug Discovery to Development*, 39-88.

Gaharwar, A. K., Peppas, N. A., & Khademhosseini, A. (2014). Nanocomposite hydrogels for biomedical applications. *Biotechnology and Bioengineering*, *111*(3), 441–453. doi:10.1002/bit.25160 PMID:24264728

Gahlawat, G., Shikha, S., Chaddha, B. S., Chaudhuri, S. R., Mayilraj, S., & Choudhury, A. R. (2016). Microbial glycolipoprotein-capped silver nanoparticles as emerging antibacterial agents against cholera. *Microbial Cell Factories*, *15*(1), 25. doi:10.118612934-016-0422-x PMID:26829922

Gaidukevich, S. K., Mikulovich, Y. L., Smirnova, T. G., Andreevskaya, S. N., Sorokoumova, G. M., Chernousova, L. N., Selishcheva, A. A., & Shvets, V. I. (2016). Antibacterial effects of liposomes containing phospholipid cardiolipin and fluoroquinolone levofloxacin on Mycobacterium tuberculosis with extensive drug resistance. *Bulletin of Experimental Biology and Medicine*, *160*(5), 675–678. doi:10.100710517-016-3247-z PMID:27021087

Gajbhiye, M., Kesharwani, J., Ingle, A., Gade, A., & Rai, M. (2009). Fungus-mediated synthesis of silver nanoparticles and their activity against pathogenic fungi in combination with fluconazole. *Nanomedicine (London)*, *5*(4), 382–386. doi:10.1016/j.nano.2009.06.005 PMID:19616127

Galagan, J. E. (2014). Genomic insights into tuberculosis. *Nature Reviews. Genetics*, *15*(5), 307–320. doi:10.1038/nrg3664 PMID:24662221

Galanzha, E. I., Shashkov, E., Sarimollaoglu, M., Beenken, K. E., ... Shirtliff, M. E. (2012). *In vivo* magnetic enrichment, photoacoustic diagnosis, and photothermal purging of infected blood using multifunctional gold and magnetic nanoparticles. *PLoS One*, *7*(9), e45557. doi:10.1371/journal.pone.0045557 PMID:23049814

Galen, U. S. (2011). *DaunoXome: daunorubicin citrate liposome injection*. Academic Press.

Galloway, W. R., Hodgkinson, J. T., Bowden, S. D., Welch, M., & Spring, D. R. (2011). Quorum sensing in gram-negative bacteria: Small molecule modulation of AHL and AI-2 quorum sensing pathways. *Chemical Reviews*, *111*(1), 28–67. doi:10.1021/cr100109t PubMed

Ganassin, R., Horst, F. H., Camargo, N. S., Chaves, S. B., Morais, P. C., Mosiniewicz-Szablewska, E. & Muehlmann, L. A. (2018). Selol nanocapsules with a poly (methyl vinyl ether-co-maleic anhydride) shell conjugated to doxorubicin for combinatorial chemotherapy against murine breast adenocarcinoma in vivo. *Artificial Cells, Nanomedicine, and Biotechnology, 46*(sup2), 1046-1052.

Gandhi, A. D., Murugan, K., Umamahesh, K., Babujanarthanam, R., Kavitha, P., & Selvi, A. (2019). Lichen Parmelia sulcata mediated synthesis of gold nanoparticles: An eco-friendly tool against Anopheles stephensi and Aedes aegypti. *Environmental Science and Pollution Research International*, *26*(23), 23886–23898. doi:10.100711356-019-05726-6 PMID:31218582

Ganesh, M., Hemalatha, P., Peng, M. M., & Jang, H. T. (2017). One pot synthesized Li, Zr doped porous silica nanoparticle for low temperature CO2 adsorption. *Arabian Journal of Chemistry*, *10*, S1501–S1505. doi:10.1016/j.arabjc.2013.04.031

Ganguly, S., Das, P., Banerjee, S., & Das, N.C. (2019). Advancement in science and technology of carbon dot-polymer hybrid composites: a review. *Functional Composites and Structures*.

Ganjian, H., Nikokar, A., Tieshayar, A., Mostafaei, N., Amirmozafari, S., & Kiani, S. (2012). Effects of salt stress on the antimicrobial drug resistance and protein profile of *Staphylococcus aureus*. *Jundishapur Journal of Microbiology*, *5*(28), 328–331.

Gan, P. P., Ng, S. H., Huang, Y., & Li, S. F. Y. (2012). Green synthesis of gold nanoparticles using palm oil mill effluent (POME): A low-cost and eco-friendly viable approach. *Bioresource Technology*, *113*, 132–135. doi:10.1016/j.biortech.2012.01.015 PMID:22297042

Gao, H., & He, Q. (2014). The interaction of nanoparticles with plasma proteins and the consequent influence on nanoparticles behavior. *Expert Opinion on Drug Delivery*, *11*(3), 409–420. doi:10.1517/17425247.2014.877442 PMID:24397260

Gao, J., Huang, X., Liu, H., Zan, F., & Ren, J. (2012). Colloidal stability of gold nanoparticles modified with thiol compounds: Bioconjugation and application in cancer cell imaging. *Langmuir*, *28*(9), 4464–4471. doi:10.1021/la204289k PMID:22276658

Gao, P., Xia, G., Bao, Z., Feng, C., Cheng, X., Kong, M., Liu, Y., & Chen, X. (2016). Chitosan based nanoparticles as protein carriers for efficient oral antigen delivery. *International Journal of Biological Macromolecules*, *91*, 716–723. doi:10.1016/j.ijbiomac.2016.06.015 PMID:27287772

Gao, W., Sattayasamitsathit, S., Manesh, K. M., Weihs, D., & Wang, J. (2010). Magnetically Powered Flexible Metal Nanowire Motors. *Journal of the American Chemical Society*, *132*(41), 14403–14405. doi:10.1021/ja1072349 PMID:20879711

Gao, W., Thamphiwatana, S., Angsantikul, P., & Zhang, L. (2014). Nanoparticle approaches against bacterial infections. *Wiley Interdisciplinary Reviews. Nanomedicine and Nanobiotechnology*, *6*(6), 532–547. doi:10.1002/wnan.1282 PMID:25044325

Gao, Y., Sarfraz, M. K., Clas, S. D., Roa, W., & Löbenberg, R. (2015). Hyaluronic acid-tocopherol succinate-based self-assembling micelles for targeted delivery of rifampicin to alveolar macrophages. *Journal of Biomedical Nanotechnology*, *11*(8), 1312–1329. doi:10.1166/jbn.2015.2091 PMID:26295135

Gao, Y., Wu, J., Ren, X., Tan, X., Hayat, T., Alsaedi, A., Cheng, C., & Chen, C. (2017). Impact of graphene oxide on the antibacterial activity of antibiotics against bacteria. *Environmental Science. Nano*, *4*(5), 1016–1024. doi:10.1039/C7EN00052A

Gao, Z., Liu, X., Deng, G., Zhou, F., Zhang, L., Wang, Q., & Lu, J. (2016). Fe3 O4 @mSiO2 -FA-CuS-PEG nanocomposites for magnetic resonance imaging and targeted chemophotothermal synergistic therapy of cancer cells. *Dalton Transactions (Cambridge, England)*, *45*(34), 13456–13465. doi:10.1039/C6DT01714B PMID:27493065

Garcia-Gradilla, V., Sattayasamitsathit, S., Soto, F., Kuralay, F., Yardımcı, C., Wiitala, D., & Wang, J. (2014). Ultrasound-Propelled Nanoporous Gold Wire for Efficient Drug Loading and Release. *Small*, n/a. Advance online publication. doi:10.1002mll.201401013 PMID:24995778

Gargano, J. M., Ngo, T., Kim, J. Y., Acheson, D. W. K., & Lees, W. J. (2001). Multivalent inhibition of AB (5) toxins. *Journal of the American Chemical Society*, *123*(51), 12909–12910. doi:10.1021/ja016305a PMID:11749553

Gatfield, J., Albrecht, I., Zanolari, B., Steinmetz, M. O., & Pieters, J. (2005). Association of the leukocyte plasma membrane with the actin cytoskeleton through coiled coil-mediated trimeric coronin 1 molecules. *Molecular Biology of the Cell*, *16*(6), 2786–2798. doi:10.1091/mbc.e05-01-0042 PMID:15800061

Ge, L., Li, Q., Wang, M., Ouyang, J., Li, X., & Xing, M. M. (2014). Nanosilver particles in medical applications: Synthesis, performance, and toxicity. *International Journal of Nanomedicine*, *9*, 2399. PMID:24876773

Geller, R. J., Chevalier, R. L., & Spyker, D. A. (1986). Acute amoxicillin nephrotoxicity following an overdose. *Clinical Toxicology*, *24*(2), 175–182. PMID:3712526

Gelmanova, I. Y., Keshavjee, S., Golubchikova, V. T., Berezina, V. I., Strelis, A. K., Yanova, G. V., ... Murray, M. (2007). Barriers to successful tuberculosis treatment in Tomsk, Russian Federation: Non-adherence, default and the acquisition of multidrug resistance. *Bulletin of the World Health Organization*, *85*(09), 703–711. doi:10.2471/BLT.06.038331 PMID:18026627

Geng, L., Wang, Z., Jia, X., Han, Q., Xiang, Z., Li, D., ... Hu, Z. (2016). HER2 targeting peptides screening and applications in tumor imaging and drug delivery. *Theranostics*, *6*(8), 1261–1273. doi:10.7150/thno.14302 PMID:27279916

Georghiou, S. B., Magana, M., Garfein, R. S., Catanzaro, D. G., Catanzaro, A., & Rodwell, T. C. (2012). Evaluation of genetic mutations associated with Mycobacterium tuberculosis resistance to amikacin, kanamycin and capreomycin: A systematic review. *PLoS One*, *7*(3), e33275. doi:10.1371/journal.pone.0033275 PMID:22479378

Gerencser, A. A., Chinopoulos, C., Birket, M. J., Jastroch, M., Vitelli, C., Nicholls, D. G., & Brand, M. D. (2012). Quantitative measurement of mitochondrial membrane potential in cultured cells: Calcium-induced de- and hyperpolarization of neuronal mitochondria. *The Journal of Physiology*, *590*(12), 2845–2871. doi:10.1113/jphysiol.2012.228387 PMID:22495585

Ghadiri, M. R., Granja, J. R., Milligan, R. A., McRee, D. E., & Khazanovich, N. (1993). Self-assembling organic nanotubes based on a cyclic peptide architecture. *Nature*, *366*(6453), 324–327. doi:10.1038/366324a0 PMID:8247126

Ghayempour, S., & Montazer, M. (2019). A novel controlled release system based on Tragacanth nanofibers loaded pepper mint oil. *Carbohydrate Polymers*, *205*, 589–595. doi:10.1016/j.carbpol.2018.10.078 PMID:30446145

Gholami, A., Ebrahiminezhad, A., Abootalebi, N., & Ghasemi, Y. (2018). Synergistic Evaluation of Functionalized Magnetic Nanoparticles and Antibiotics Against Staphylococcus aureus and Escherichia coli. *Pharmaceutical Nanotechnology*, *6*(4), 276–286. doi:10.2174/2211738506666181031143048 PMID:30381089

Gholipour, A., Soleimani, N., Shokri, D., Mobasherizadeh, S., Kardi, M., & Baradaran, A. (2014). Phenotypic and Molecular Characterization of Extended-Spectrum β-Lactamase Produced by Escherichia coli, and Klebsiella pneumoniae Isolates in an Educational Hospital. *Jundishapur Journal of Microbiology*, *7*(8), e11758. doi:10.5812/jjm.11758 PubMed

Ghorbani, H. R. (2014). A review of methods for synthesis of Al nanoparticles. *Oriental Journal of Chemistry*, *30*(4), 1941–1949. doi:10.13005/ojc/300456

Ghosh, A., Chowdhury, N., & Chandra, G. (2012). Plant extracts as potential mosquito larvicides. *The Indian Journal of Medical Research*, *135*(5), 581–598. PMID:22771587

Ghosh, S., Patil, S., Ahire, M., Kitture, R., Kale, S., ... Pardesi, K. (2012). Synthesis of silver nanoparticles using *Dioscorea bulbifera* tuber extract and evaluation of its synergistic potential in combination with antimicrobial agents. *International Journal of Nanomedicine*, *7*, 483–496. PMID:22334779

Gibbs, J., & Zhao, Y. (2011). Catalytic nanomotors: Fabrication, mechanism, and applications. *Frontiers of Materials Science*, *5*(1), 25–39. doi:10.100711706-011-0120-x

Gifta, M. M., Anna, K. C., Ivana, M.-T., & Heidi, A. (2010). A Review of Nanoparticle Photosensitizer Drug Delivery Uptake Systems for Photodynamic Treatment of Lung Cancer. (2018). *Photodiagnosis and Photodynamic Therapy*. Advance online publication. doi:10.1016/j.pdpdt.2018.03.006

Giguère, S. (2006). Antimicrobial Drug Action and Interaction: An Introduction. Antimicrobial therapy in Veterinary Medicine (4th ed.). Blackwell Publishing.

Gillespie, S. H. (2002). Evolution of drug resistance in Mycobacterium tuberculosis: Clinical and molecular perspective. *Antimicrobial Agents and Chemotherapy*, *46*(2), 267–274. doi:10.1128/AAC.46.2.267-274.2002 PMID:11796329

Gil-Tomas, J., Tubby, S., Parkin, I. P., Narband, N., Dekker, L., Nair, S. P., Wilson, M., & Street, C. (2007). Lethal photosensitisation of Staphylococcus aureus using a toluidine blue O-tiopronin-gold nanoparticles conjugate. *Journal of Materials Chemistry*, *17*(35), 3739–3746. doi:10.1039/b706615e

Giner-Casares, J. J., Henriksen-Lacey, M., Coronado-Puchau, M., & Liz-Marzán, L. M. (2016). Inorganic nanoparticles for biomedicine: Where materials scientists meet medical research. *Materials Today*, *19*(1), 19–28. doi:10.1016/j.mattod.2015.07.004

Giodini, L., Re, F. L., Campagnol, D., Marangon, E., Posocco, B., Dreussi, E., & Toffoli, G. (2017). Nanocarriers in cancer clinical practice: A pharmacokinetic issue. *Nanomedicine; Nanotechnology, Biology, and Medicine*, *13*(2), 583–599. doi:10.1016/j.nano.2016.07.012 PMID:27520727

Girard, G., & Bloemberg, G. V. (2008). Central role of quorum sensing in regulating the production of pathogenicity factor in Pseudomonas aeruginosa. *Future Microbiology*, *3*(1), 97–106. doi:10.2217/17460913.3.1.97 PubMed

Giri Dev, V. R., & Hemamalini, T. (2018). Porous electrospun starch rich polycaprolactone blend nanofibers for severe hemorrhage. *International Journal of Biological Macromolecules*, *118*, 1276–1283. doi:10.1016/j.ijbiomac.2018.06.163 PMID:29964114

Gleghorn, J. P., Pratt, E. D., Denning, D., Liu, H., Bander, N. H., Tagawa, S. T., Nanus, D. M., Giannakakou, P. A., & Kirby, B. J. (2010). Capture of circulating tumor cells from whole blood of prostate cancer patients using geometrically enhanced differential immunocapture (GEDI) and a prostate-specific antibody. *Lab on a Chip*, *10*(1), 27–29. doi:10.1039/B917959C PMID:20024046

Glickson, J. D. (2009). *Lipoprotein Nanoplatform for Targeted Delivery of Diagnostic and Therapeutic Agents* (Vol. 645). Advances in Experimental Medicine and Biology. doi:10.1007/978-0-387-85998-9_35

Golizadeh, M., Karimi, A., Gandomi-Ravandi, S., Vossoughi, M., Khafaji, M., Johateai, M. T., & Faghihi, F. (2019). Evaluation of cellular attachment and proliferation on different surface charged functional cellulose electrospun nanofibers. *Carbohydrate Polymers*, *207*, 796–805. doi:10.1016/j.carbpol.2018.12.028 PMID:30600068

Golkar, Z., Bagasra, O., & Pace, D. G. (2014). Bacteriophage therapy: A potential solution for the antibiotic resistance crisis. *Journal of Infection in Developing Countries*, *8*(2), 129–136. doi:10.3855/jidc.3573 PMID:24518621

Gonçalves, A. I., Rodrigues, M. T., Reis, R. L., Gomes, M. E., & El Haj, A. J. (2018). Magnetic actuated SPCL scaffolds dopped with iron oxide magnetic nanoparticles as mechano-instructive platforms for tendon tissue engineering. *12th International conference on Scientific and clinical applications of Magnetic carriers.*

González-Menéndez, E., Fernández, L., Gutiérrez, D., Pando, D., Martínez, B., Rodríguez, A., & García, P. (2018). Strategies to encapsulate the Staphylococcus aureus bacteriophage phiIPLA-RODI. *Viruses*, *10*(9), 495. doi:10.3390/v10090495 PMID:30217072

Goodsell, D. S. (2004). Bionanotechnology: Lessons from Nature (illustrated ed.). John Wiley & Sons.

Goodwin, S. C., Bittner, C. A., Peterson, C. L., & Wong, G. (2001). Single-dose toxicity study of hepatic intra- arterial infusion of doxorubicin coupled to a novel magnetically targeted drug carrier. *Toxicological Sciences*, *60*(1), 177–183. doi:10.1093/toxsci/60.1.177 PMID:11222884

Gootz, T. D. (1990). Discovery and development of new antimicrobial agents. *Clinical Microbiology Reviews*, *3*(1), 13–31. doi:10.1128/CMR.3.1.13 PMID:2404566

Gopidas, K. R., Whitesell, J. K., & Fox, M. A. (2003). Synthesis, characterization and catalytic activity of a palladium nanoparticle cored dendrimier. *Nano Letters*, *3*(12), 1757–1760. doi:10.1021/nl0348490

Gopinath, P. M., Narchonai, G., Dhanasekaran, D., Ranjani, A., & Thajuddin, N. (2015). Mycosynthesis, characterization and antibacterial properties of AgNPs against multidrug resistant (MDR) bacterial pathogens of female infertility cases. *Asian Journal of Pharmaceutical Sciences*, *10*(2), 138–145. doi:10.1016/j.ajps.2014.08.007

Gosens, I., Cassee, F. R., Zanella, M., Manodori, L., Brunelli, A., Costa, A. L., & Stone, V. (2016). Organ burden and pulmonary toxicity of nano-sized copper (II) oxide particles after short-term inhalation exposure. *Nanotoxicology*, *10*(8), 1084–1095. doi:10.3109/17435390.2016.1172678 PMID:27132941

Grabinski, C., Hussain, S., Lafdi, K., Braydich-Stolle, L., & Schlager, J. (2007). Effect of particle dimension on biocompatibility of carbon nanomaterials. *Carbon*, *45*(14), 2828–2835. doi:10.1016/j.carbon.2007.08.039

Gradmann, C. (2016). Re-Inventing Infectious Disease: Antibiotic Resistance and Drug Development at the Bayer Company 1945-80. *Medical History*, *60*(2), 155–180. doi:10.1017/mdh.2016.2 PMID:26971595

Grandgirard, D., Loeffler, J. M., Fischetti, V. A., & Leib, S. L. (2008). Phage lytic enzyme Cpl-1 for antibacterial therapy in experimental pneumococcal meningitis. *The Journal of Infectious Diseases*, *197*(11), 1519–1522. doi:10.1086/587942 PMID:18471063

Greenblatt, M. S., Bennet, W. P., Hollstein, M., & Harris, C. C. (1994). Mutations in the p53 tumor suppressor gene: Clues to cancer etiology and molecular pathogenesis. *Cancer Research*, *54*(18), 4855. PMID:8069852

Greene, C., Vadlamudi, G., Newton, D., Foxman, B., & Xi, C. (2016). The influence of biofilm formation and multidrug resistance on environmental survival of clinical and environmental isolates of *Acinetbacter baumannii*. *American Journal of Infection Control*, *44*(5), e65–e71. doi:10.1016/j.ajic.2015.12.012 PMID:26851196

Grieger, K. D., Linkov, I., Hansen, S. F., & Baun, A. (2012). Environmental risk analysis for nanomaterials: Review and evaluation of frameworks. *Nanotoxicology*, *6*(2), 196–212. doi:10.3109/17435390.2011.569095 PMID:21486187

Grigore, M. E., Biscu, E., Holban, A., Gestal, M., & Grumezescu, A. (2016). Methods of Synthesis, Properties and Biomedical Applications of CuO Nanoparticles. *Pharmaceuticals (Basel, Switzerland)*, *9*(4), 75. doi:10.3390/ph9040075 PMID:27916867

Grimsey, N. J., Lin, Y., Narala, R., Rada, C. C., Mejia-Pena, H., & Trejo, J. (2019). G protein–coupled receptors activate p38 MAPK via a non-canonical TAB1–TAB2–and TAB1–TAB3–dependent pathway in endothelial cells. *The Journal of Biological Chemistry*, *294*(15), 5867–5878. doi:10.1074/jbc.RA119.007495 PMID:30760523

Grumezescu, A. M., Gestal, M. C., Holban, A. M., Grumezescu, V., Vasile, B. Ş., Mogoantă, L., Iordache, F., Bleotu, C., & Mogoşanu, G. D. (2014). Biocompatible Fe3O4 increases the efficacy of amoxicillin delivery against Gram-positive and Gram-negative bacteria. *Molecules (Basel, Switzerland)*, *19*(4), 5013–5027. doi:10.3390/molecules19045013 PMID:24759068

Grèget, Nealon, Vileno, Turek, Mèny, Derory, Voirin, Rivière, Rogalev, Wilhelm, Joly, Knafo, Ballon, Terazzi, Kappler, Donnio, & Gallani. (n.d.). *Magnetic Properties of Gold Nanoparticles: A Room-Temperature Quantum Effect*. . doi:10.1002/cphc.201200394

Gubler, D. J. (2002). The Global Emergence/Resurgence of Arboviral Diseases As Public Health Problems. *Archives of Medical Research*, *33*(4), 330–342. doi:10.1016/S0188-4409(02)00378-8 PMID:12234522

Gu, F., Wang, W., Cai, Z., Xue, F., Jin, Y., & Zhu, J. Y. (2018). Water retention value for characterizing fibrillation degree of cellulosic fibers at micro and nanometer scales. *Cellulose (London, England)*, *25*(5), 2861–2871. doi:10.100710570-018-1765-8

Gu, H., Ho, P. L., Tong, E., Wang, L., & Xu, B. (2013). Presenting vancomycin on nanoparticles to enhance antimicrobial activities. *Nano Letters*, *3*(9), 1261–1263. doi:10.1021/nl034396z

Guidotti, G., Brambilla, L., & Rossi, D. (2017). Cell-penetrating peptides: From basic research to clinics. *Trends in Pharmacological Sciences*, *38*(4), 406–424. doi:10.1016/j.tips.2017.01.003 PMID:28209404

Guilhelmelli, F., Vilela, N., Albuquerque, P., Derengowski, L., Silva-Pereira, I., & Kyaw, C. (2013). Antibiotic development challenges: The various mechanisms of action of antimicrobial peptides and of bacterial resistance. *Frontiers in Microbiology*, *4*, 353. doi:10.3389/fmicb.2013.00353 PMID:24367355

Guix, M., Meyer, A. K., Koch, B., & Schmidt, O. G. (2016). Carbonate-based Janus micromotors moving in ultra-light acidic environment generated by HeLa cells in situ. *Scientific Reports*, *6*(1), 21701. Advance online publication. doi:10.1038rep21701 PMID:26905939

Gunes, O. C., Unalan, I., Cecen, B., Albayrak, A. Z., Havitciogluc, H., Ustun, O., & Ergur, B. U. (2018). Three-dimensional silk impregnated HAp/PHBV nanofibrous scaffolds for bone regeneration. *International Journal of Polymeric Materials and Polymeric Biomaterials*, *68*(5), 217–228. doi:10.1080/00914037.2018.1443928

Guo, C., Robertson, S., Weber, R. J., Buckley, A., Warren, J., Hodgson, A., ... Macchiarulo, S. (2019). Pulmonary toxicity of inhaled nano-sized cerium oxide aerosols in Sprague–Dawley rats. *Nanotoxicology*, *13*(6), 733–750. doi:10.1080/17435390.2018.1554751 PMID:30704321

Guo, R., Zhou, S., Li, Y., Li, X., Fan, L., & Voelcker, N. H. (2015). Rhodamine-Functionalized Graphene Quantum Dots for Detection of Fe3+ in Cancer Stem Cells. *ACS Applied Materials & Interfaces*, *7*(43), 23958–23966. doi:10.1021/acsami.5b06523 PMID:26317667

Gupta, A., Moyano, D. F., Parnsubsakul, A., Papadopoulos, A., Wang, L. S., Landis, R. F., Das, R., & Rotello, V. M. (2016). Ultrastable and biofunctionalizable gold nanoparticles. *ACS Applied Materials & Interfaces*, *8*(22), 14096–14101. doi:10.1021/acsami.6b02548 PMID:27191946

Gupta, A., Mumtaz, S., Li, C. H., Hussain, I., & Rotello, V. M. (2019). Combatting antibiotic-resistant bacteria using nanomaterials. *Chemical Society Reviews*, *48*(2), 415–427. doi:10.1039/C7CS00748E PMID:30462112

Gupta, J., Mohapatra, J., Bhargava, P., & Bahadur, D. (2016). A pH-responsive folate conjugated magnetic nanoparticle for targeted chemo-thermal therapy and MRI diagnosis. *Dalton Transactions (Cambridge, England)*, *45*(6), 2454–2461. doi:10.1039/C5DT04135J PMID:26685824

Gupta, M., Chashoo, G., Sharma, P. R., Saxena, A. K., Gupta, P. N., Agrawal, G. P., & Vyas, S. P. (2014). Dual targeted polymeric nanoparticles based on tumor endothelium and tumor cells for enhanced antitumor drug delivery. *Molecular Pharmaceutics*, *11*(3), 697–715. doi:10.1021/mp400404p PMID:24512060

Gupta, N., Bhagat, S., Singh, M., Jangid, A. K., Bansal, V., Singh, S., Pooja, D., & Kulhari, H. (2020). Site-specific delivery of a natural chemotherapeutic agent to human lung cancer cells using biotinylated 2D rGO nanocarriers. *Materials Science and Engineering C*, *112*, 110884. doi:10.1016/j.msec.2020.110884 PMID:32409041

Gupta, Y. R., Sellegounder, D., Kannan, M., Deepa, S., Senthilkumaran, B., & Basavaraju, Y. (2016). Effect of copper nanoparticles exposure in the physiology of the common carp (Cyprinus carpio): Biochemical, histological and proteomic approaches. *Aquac Fish*, *1*, 15–23. doi:10.1016/j.aaf.2016.09.003

Gurr, J. R., Wang, A. S., Chen, C. H., & Jan, K. Y. (2005). Ultrafine titanium dioxide particles in the absence of photoactivation can induce oxidative damage to human bronchial epithelial cells. *Toxicology*, *213*(1-2), 66–73. doi:10.1016/j.tox.2005.05.007 PMID:15970370

Habibi, N., Kamaly, N., Memic, A., & Shafiee, H. (2016). Self-assembled peptide based nanostructures: Smart nanomaterials toward targeted drug delivery. *Nano Today*, *11*(1), 41–60. doi:10.1016/j.nantod.2016.02.004 PMID:27103939

Haddaoui, M., Sola, C., Raouafi, N., & Korri-Youssoufi, H. (2019). E-DNA detection of rpoB gene resistance in Mycobacterium tuberculosis in real samples using Fe3O4/polypyrrole nanocomposite. *Biosensors & Bioelectronics*, *128*, 76–82. doi:10.1016/j.bios.2018.11.045 PMID:30640123

Hadisi, Z., Nourmohammadi, J., & Nassiri, S. M. (2018). The antibacterialand anti-inflammatory investigation of Lawsonia Inermis gelatin starch nanofibrous dressing in burn wound. *International Journal of Biological Macromolecules*, *107*, 2008–2019. doi:10.1016/j.ijbiomac.2017.10.061 PMID:29037870

Hadrup, N., Sharma, A. K., & Loeschner, K. (2018). Toxicity of silver ions, metallic silver, and silver nanoparticle materials after in vivo dermal and mucosal surface exposure: A review. *Regulatory Toxicology and Pharmacology*, *98*, 257–267. doi:10.1016/j.yrtph.2018.08.007 PMID:30125612

Hagens, W. I., Oomen, A. G., de Jong, W. H., Cassee, F. R., & Sips, A. J. (2007). What do we (need to) know about the kinetic properties of nanoparticles in the body? *Regulatory Toxicology and Pharmacology*, *49*(3), 217–229. doi:10.1016/j.yrtph.2007.07.006 PMID:17868963

Haider, A., & Kang, I. K. (2015). Preparation of silver nanoparticles and their industrial and biomedical applications: A comprehensive review. *Advances in Materials Science and Engineering*, *2015*, 2015. doi:10.1155/2015/165257

Hajiahmadi, F., Alikhani, M. Y., Shariatifar, H., Arabestani, M. R., & Ahmadvand, D. (2019). The bactericidal effect of lysostaphin coupled with liposomal vancomycin as a dual combating system applied directly on methicillin-resistant Staphylococcus aureus infected skin wounds in mice. *International Journal of Nanomedicine*, *14*, 5943–5955. doi:10.2147/IJN.S214521 PMID:31447553

Hajipour, M.J., & Fromm, K.M., AkbarAshkarran, A., & Mahmoudi, M. (2012). Antibacterial properties of nanoparticles. *Trends*, *30*(10), 499–511. PMID:22884769

Haliullin, T. O., Zalyalov, R. R., Shvedova, A. A., & Tkachov, A. G. (2015). Hygienic evaluation of multilayercarbon nanotubes. *Meditsina Truda i Promyshlennaia Ekologiia*, *7*, 37–42. PMID:26470480

Hamed, A., Osman, R., Al-Jamal, K. T., Holayel, S. M., & Geneidi, A. S. (2019). Enhanced antitubercular activity, alveolar deposition and macrophages uptake of mannosylated stable nanoliposomes. *Journal of Drug Delivery Science and Technology*, *51*, 513–523. doi:10.1016/j.jddst.2019.03.032

Hamidi, M., Azadi, A., & Rafiei, P. (2008). Hydrogel nanoparticles in drug delivery. *Advanced Drug Delivery Reviews*, *60*(15), 1638–1649. doi:10.1016/j.addr.2008.08.002 PMID:18840488

Hamilton, W. L., Amato, R., van der Pluijm, R. W., Jacob, C. G., Quang, H. H., Thuy-Nhien, N. T., Hien, T. T., Hongvanthong, B., Chindavongsa, K., Mayxay, M., Huy, R., Leang, R., Huch, C., Dysoley, L., Amaratunga, C., Suon, S., Fairhurst, R. M., Tripura, R., Peto, T. J., ... Miotto, O. (2019). Evolution and expansion of multidrug-resistant malaria in southeast Asia: A genomic epidemiology study. *The Lancet. Infectious Diseases*, *19*(9), 943–951. doi:10.1016/S1473-3099(19)30392-5 PMID:31345709

Hamley, I. W. (2011). Self-assembly of amphiphilic peptides. *Soft Matter*, *7*(9), 4122–4138. doi:10.1039/c0sm01218a

Hammerich, L., Marron, T. U., Upadhyay, R., Svensson-Arvelund, J., Dhainaut, M., Hussein, S., & Salazar, A. M. (2019). Systemic clinical tumor regressions and potentiation of PD1 blockade with in situ vaccination. *Nature Medicine*, *25*(5), 814–824. doi:10.103841591-019-0410-x PMID:30962585

Hamouda, R. A., Hussein, M. H., Abo-elmagd, R. A., & Bawazir, S. S. (2019). Synthesis and biological characterization of silver nanoparticles derived from the cyanobacterium Oscillatoria limnetica. *Scientific Reports*, *9*(1), 1–17. doi:10.103841598-019-49444-y PMID:31506473

Hanahan, D., & Weinberg, R. A. (2011). Hallmarks of cancer: The next generation. *Cell*, *144*(5), 646–674. doi:10.1016/j.cell.2011.02.013 PMID:21376230

Hancock, R. E., & Chapple, D. S. (1999). Peptide antibiotics. *Antimicrobial Agents and Chemotherapy*, *43*(6), 1317–1323. doi:10.1128/AAC.43.6.1317 PMID:10348745

Handy, R. D., Owen, R., & Valsami-Jones, E. (2008). The ecotoxicology of nanoparticles and nanomaterials: Current status, knowledge gaps, challenges, and future needs. *Ecotoxicology (London, England)*, *17*(5), 315–325. doi:10.100710646-008-0206-0 PMID:18408994

Handy, R., & Shaw, B. (2007). Toxic effects of nanoparticles and nanomaterials:Implications for public health, risk assessment and the public perception of nano-technology. *Health Risk & Society*, *9*(2), 125–144. doi:10.1080/13698570701306807

Han, H. J., Ekweremadu, C., & Patel, N. (2019). Advanced drug delivery system with nanomaterials for personalised medicine to treat breast cancer. *Journal of Drug Delivery Science and Technology*, *52*, 1051–1060. doi:10.1016/j.jddst.2019.05.024

Han, L., Zhang, Y., Chen, X. W., Shu, Y., & Wang, J. H. (2016). Protein-modified hollow copper sulfide nanoparticles carrying indocyanine green for photothermal and photodynamic therapy. *Journal of Materials Chemistry. B, Materials for Biology and Medicine*, *4*(1), 105–112. doi:10.1039/C5TB02002F PMID:32262813

Han, S. O., Son, W. K., Youk, J. H., & Park, W. H. (2008). Electrospinning of ultrafine cellulose fibers and fabrication of poly (butylene succinate) biocomposites reinforced by them. *Journal of Applied Polymer Science*, *107*(3), 1954–1959. doi:10.1002/app.26643

Harada, L. K., Silva, E. C., Campos, W. F., Del Fiol, F. S., Vila, M., Dąbrowska, K., Krylov, V. N., & Balcão, V. M. (2018). Biotechnological applications of bacteriophages: State of the art. *Microbiological Research*, *212*, 38–58. doi:10.1016/j.micres.2018.04.007 PMID:29853167

Harbach, R. E. (2013). *Mosquito taxonomic inventory*. Academic Press.

Harmsen, S., Wall, M. A., Huang, R., & Kircher, M. F. (2017). Cancer imaging using surface-enhanced resonance Raman scattering nanoparticles. *Nature Protocols*, *12*(7), 1400–1414. doi:10.1038/nprot.2017.031 PMID:28686581

Hartgerink, J. D., Granja, J. R., Milligan, R. A., & Ghadiri, M. R. (1996). Peptide-amphiphile nanofibers: A versatile scaffold for the preparation of self-assembling materials. *Journal of the American Chemical Society*, *118*, 43–50. doi:10.1021/ja953070s

Haussler, S., & Fuqua, C. (2013). Bioÿlms 2012: New discoveries and signiÿcant wrinkles in a dynamic ÿeld. *Journal of Bacteriology*, *195*(13), 2947–2958. doi:10.1128/JB.00239-13 PubMed

Hautot, D. (2003). Preliminary evaluation of nanoscale biogenic magnetite in Alzheimer's disease brain tissue. *Proceedings. Biological Sciences*, *270*(Suppl 1), S62–S64. PMID:12952638

Hautot, D., Pankhurst, Q. A., Morris, C. M., Curtis, A., Burn, J., & Dobson, J. (2007). Preliminary observation of elevated levels of nanocrystalline iron oxide in the basal ganglia of neuroferritinopathy patients. *Biochimica et Biophysica Acta*, *1772*(1), 21–25. doi:10.1016/j.bbadis.2006.09.011 PMID:17097860

Hazarika, S. N., Gupta, K., Shamin, K. N. A. M., Bhardwaj, P., Boruah, R., Yadav, K. K., Naglot, A., Deb, P., Mandal, M., Doley, R., Veer, V., Baruah, I., & Namsa, N. D. (2016). One-pot facile green synthesis of biocidal silver nanoparticles. *Materials Research Express*, *3*(7), 075401. doi:10.1088/2053-1591/3/7/075401

Hazbón, M. H., Brimacombe, M., Del Valle, M. B., Cavatore, M., Guerrero, M. I., Varma-Basil, M., ... León, C. I. (2006). Population genetics study of isoniazid resistance mutations and evolution of multidrug-resistant Mycobacterium tuberculosis. *Antimicrobial Agents and Chemotherapy*, *50*(8), 2640–2649. doi:10.1128/AAC.00112-06 PMID:16870753

Hegstad, K., Mikalsen, T., Coque, T. M., Werner, G., & Sundsfjord, A. (2010). Mobile genetic elements and their contribution to the emergence of antimicrobial resistant Enterococcus faecalis and Enterococcus faecium. *Clinical Microbiology and Infection*, *16*(6), 541–554. doi:10.1111/j.1469-0691.2010.03226.x PMID:20569265

Heinis, C., & Winter, G. (2015). Encoded libraries of chemically modified peptides. *Current Opinion in Chemical Biology*, *26*, 89–98. doi:10.1016/j.cbpa.2015.02.008 PMID:25768886

Hellberg, R. S., & Chu, E. (2016). Effects of climate change on the persistence and dispersal of foodborne bacterial pathogens in the outdoor environment: A review. *Critical Reviews in Microbiology*, *42*(4), 548–572. doi:10.3109/1040841X.2014.972335 PMID:25612827

Hemeg, H. A. (2017). Nanomaterials for alternative antibacterial therapy. *International Journal of Nanomedicine*, *12*, 8211–8225. doi:10.2147/IJN.S132163 PMID:29184409

Henostroza, M. A. B., Melo, K. J. C., Yukuyama, M. N., Löbenberg, R., & Bou-Chacra, N. A. (2020). Cationic rifampicin nanoemulsion for the treatment of ocular tuberculosis. *Colloids and Surfaces. A, Physicochemical and Engineering Aspects*, *597*, 124755. doi:10.1016/j.colsurfa.2020.124755

Herce, H. D., & Garcia, A. E. (2007). Cell penetrating peptides: How do they do it? *Journal of Biological Physics*, *33*(5–6), 345–356. doi:10.100710867-008-9074-3 PMID:19669523

Herce, H. D., Garcia, A. E., & Cardoso, M. C. (2014). Fundamental molecular mechanism for the cellular uptake of guanidinium-rich molecules. *Journal of the American Chemical Society*, *136*(50), 17459–17467. doi:10.1021/ja507790z PMID:25405895

He, S., Guo, Z., Zhang, Y., Zhang, S., Wang, J., & Gu, N. (2017). Biosynthesis of gold nanoparticles using the bacteria Rhodopseudomonas capsulata. *Materials Letters*, *61*(18), 3984–3987. doi:10.1016/j.matlet.2007.01.018

Hetrick, E. M., Shin, J. H., Paul, H. S., & Schoenfisch, M. H. (2009). Antibiofilm efficacy of nitric oxide-releasing silica nanoparticles. *Biomaterials*, *30*(14), 2782–2789. doi:10.1016/j.biomaterials.2009.01.052 PMID:19233464

He, W., Jia, H., Cai, J., Han, X., Zheng, Z., Wamer, W. G., & Yin, J. J. (2016). Production of reactive oxygen species and electrons from photoexcited ZnO and ZnS nanoparticles: A comparative study for unraveling their distinct photocatalytic activities. *The Journal of Physical Chemistry C*, *120*(6), 3187–3195. doi:10.1021/acs.jpcc.5b11456

Heydary, A. H., Karamian, E., Poorazizi, E., Khandan, A., & Heydaripour, J. (2015). A novel nanofiber of Iranian Gum Tragacanth-Polyvinyl Alcohol/ Nanoclay composite for wound healing applications. *Procedia Materials Science*, *11*, 176–182. doi:10.1016/j.mspro.2015.11.079

Hidayat, L. K., Hsu, D. I., Quist, R., Shriner, K. A., & Wong-Beringer, A. (2006). A. High-dose vancomycin therapy for methicillin-resistant Staphylococcus aureus infections: Efficacy and toxicity. *Archives of Internal Medicine*, *166*(19), 2138–2144. doi:10.1001/archinte.166.19.2138 PMID:17060545

Hildebrandt, B., Wust, P., & Ahlers, O. (2002). The cellular and molecular basis of hyperthermia. *Critical Reviews in Oncology/Hematology*, *43*(1), 33–56. doi:10.1016/S1040-8428(01)00179-2 PMID:12098606

Hill, S.R., & GK, S. (2019). *Antiparasitic Drugs*. StatPearls Publishing.

Hill, M. L., Corbin, I. R., Levitin, R. B., Cao, W., Mainprize, J. G., Yaffe, M. J., & Zheng, G. (2010). In-vitro assessment of poly-iodinated triglyceride reconstituted Low-density lipoprotein: Initial step towards CT molecular imaging. *Academic Radiology*, *17*(11), 1359–1365. doi:10.1016/j.acra.2010.06.006 PMID:20719547

Hill, W. R., & Pillsbury, D. M. (1939). *Argyria–The Pharmacology of Silver Baltimore*. Williams and Wilkins.

Hira, I., & Reena, V. (2018). Inspection and Remedy of Cervical Cancer Using Nanoparticles. *Nanoscience & Nanotechnology-Asia*, *8*(2), 186–192. doi:10.2174/2210681207666170323155831

Hisatomi, T., Kubota, J., & Domen, K. (2014). Recent advances in semiconductors for photocatalytic and photoelectrochemical water splitting. *Chemical Society Reviews*, *43*(22), 7520–7535. doi:10.1039/C3CS60378D PMID:24413305

Hivechi, A., Bahrami, S. H., & Siegel, R. A. (2018). Investigation of morphological, mechanical and biological properties of cellulose nanocrystal reinforced electrospun gelatin nanofibers. *International Journal of Biological Macromolecules*, *124*, 411–417. doi:10.1016/j.ijbiomac.2018.11.214 PMID:30476513

Hoare, T. R., & Kohane, D. S. (2008). Hydrogels in drug delivery: Progress and challenges. *Polymer*, *49*(8), 1993–2007. doi:10.1016/j.polymer.2008.01.027

Hoge, C. W., Gambel, J. M., Srijan, A., Pitarangsi, C., & Echeverria, P. (1998). Trends in antibiotic resistance among diarrheal pathogens isolated in Thailand over 15 years. *Clinical Infectious Diseases*, *26*(2), 341–345. doi:10.1086/516303 PMID:9502453

Høiby, N., Bjarnsholt, T., Givskov, M., Molin, S., & Ciofu, O. (2010). Antibiotic resistance of bacterial biofilms. *International Journal of Antimicrobial Agents*, *35*(4), 322–332. doi:10.1016/j.ijantimicag.2009.12.011 PMID:20149602

Hoiby, N., Ciofu, O., Johansen, H. K., Song, Z. J., Moser, C., Jensen, P. O., Molin, S., Givskov, M., Tolker-Nielsen, T., & Bjarnsholt, T. (2011). The clinical impact of bacterial biofilms. *International Journal of Oral Science*, *3*(2), 55–65. doi:10.4248/IJOS11026 PMID:21485309

Holban, A. M., Gestal, M. C., & Grumezescu, A. M. (2016). Control of biofilm-associated infections by signaling molecules and nanoparticles. *International Journal of Pharmaceutics*, *510*(2), 409–418. doi:10.1016/j.ijpharm.2016.02.044 PMID:26945736

Holmes, A. H., Moore, L. S., Sundsfjord, A., Steinbakk, M., Regmi, S., Karkey, A., Guerin, P. J., & Piddock, L. J. (2015). Understanding the mechanisms and drivers of antimicrobial resistance. *Lancet*, *387*(10014), 176–187. doi:10.1016/S0140-6736(15)00473-0 PMID:26603922

Holt, K. B., & Bard, A. J. (2005). Interaction of silver (I) ions with the respiratory chain of *Escherichia coli*: An electrochemical and scanning electrochemical microscopy study of the antimicrobial mechanism of micromolar Ag^+. *Biochemie.*, *44*(39), 13214–13223. doi:10.1021/bi0508542 PMID:16185089

Ho, M. H., Liao, M. H., Lin, Y. L., Lai, C. H., Lin, P. I., & Chen, R. M. (2014). Improving effects of chitosan nanofiber scaffolds on osteoblast proliferation and maturation. *International Journal of Nanomedicine*, *9*, 4293–4304. PMID:25246786

Homayoni, H., Ravandi, S. A. H., & Valizadeh, M. (2009). Influence of the Molecular Weight of Chitosan on the spinnability of chitosan/poly(vinyl alcohol) blend nanofibers. *Journal of Applied Polymer Science*, *113*(4), 2507–2513. doi:10.1002/app.30148

Honarvar, Z., Hadian, Z., & Mashayekh, M. (2016). Nanocomposites in food packaging applications and their risk assessment for health. *Electronic Physician*, *8*(6), 2531–2538. doi:10.19082/2531 PMID:27504168

Hong, E., Liu, L., Bai, L., Xia, C., Gao, L., Zhang, L., & Wang, B. (2019). Control synthesis, subtle surface modification of rare-earth-doped upconversion nanoparticles and their applications in cancer diagnosis and treatment. *Materials Science and Engineering*, *105*, 110097. doi:10.1016/j.msec.2019.110097 PMID:31546381

Hong, S., Leroueil, P. R., Majoros, I. J., Orr, B. G., Baker, J. R. Jr, & Banaszak Holl, M. M. (2007). M.M. Banaszak Holl, The binding avidity of a nanoparticle-based multivalent targeted drug delivery platform. *Chemistry & Biology*, *14*(1), 107–115. doi:10.1016/j.chembiol.2006.11.015 PMID:17254956

Hong, Y., Diaz, M., Córdova-Figueroa, U. M., & Sen, A. (2010). Light-Driven Titanium-Dioxide-Based Reversible Microfireworks and Micromotor/Micropump Systems. *Advanced Functional Materials*, *20*(10), 1568–1576. doi:10.1002/adfm.201000063

Hori, S.S., & Gambhir, S.S. (2011). Mathematical model identifies blood biomarker-based early cancer detection strategies and limitations. *Sci. Transl. Med.*, 109ra116–109ra116.

Hornish, R. E., & Kotarski, S. F. (2002). Cephalosporins in veterinary medicine - ceftiofur use in food animals. *Current Topics in Medicinal Chemistry*, *2*(7), 717–731. doi:10.2174/1568026023393679 PMID:12052187

Hoseinzadeh, E., Makhdoumi, P., Taha, P., Hossini, H., Stelling, J., Kamal, M. A., & Ashraf, G. M. (2017). A review on nano-antimicrobials: Metal nanoparticles, methods and mechanisms. *Current Drug Metabolism*, *18*, 120–128. doi:10.2174/1389200217666161201111146 PMID:27908256

Hoshino, A., Fujioka, K., Oku, T., Suga, M., Sasaki, Y. F., Ohta, T., Yasuhara, M., Suzuki, K., & Yamamoto, K. (2004). Physicochemical properties and cellular toxicity of nanocrystal quantum dots depend on their surface modification. *Nano Letters*, *4*(11), 2163–2169. doi:10.1021/nl048715d

Hossein, A., & Haghighia, M. T. K. Z. F. F. F. (2020, April). Effects of different quantities of antibody conjugated with magnetic nanoparticles on cell separation efficiency. *Heliyon*, *6*(4), e03677. doi:10.1016/j.heliyon.2020.e03677 PMID:32280795

Hosseini, L., Mahboobnia, K., & Irani, M. (2016). Fabrication of PLA/MWCNT/ Fe3 O4 composite nanofibers for leukemia cancer cells. *Int J Polym Mater Polymer Biomater.*, *65*(4), 176–182. doi:10.1080/00914037.2015.1074912

Hou, S., Zhao, H., Zhao, L., Shen, Q., Wei, K. S., Suh, D. Y., Nakao, A., Garcia, M. A., Song, M., Lee, T., Xiong, B., Luo, S.-C., Tseng, H.-R., & Yu, H. (2013). Capture and stimulated release of circulating tumor cells on polymer-grafted silicon nanostructures. *Advanced Materials*, *25*(11), 1547–1551. doi:10.1002/adma.201203185 PMID:23255101

Howes, P. D., Chandrawati, R., & Stevens, M. M. (2014). Colloidal nanoparticles as advanced biological sensors. *Science*, *346*(6205), 1247390. doi:10.1126cience.1247390 PMID:25278614

Howes, P. D., Rana, S., & Stevens, M. M. (2014). Plasmonic nanomaterials for biodiagnostics. *Chemical Society Reviews*, *43*(11), 3835–3853. doi:10.1039/C3CS60346F PMID:24323079

Hozyen, H. F., Ibrahim, E. S., Khairy, E. A., & El-Dek, S. I. (2019). Enhanced antibacterial activity of capped zinc oxide nanoparticles: A step towards the control of clinical bovine mastitis. *Veterinary World*, *12*(8), 1225–1232. doi:10.14202/vetworld.2019.1225-1232 PMID:31641301

Hrapovic, S., Liu, Y., Male, K. B., & Luong, J. H. T. (2004). Electrochemical biosensingplateform using platinum nanoparticles and carbon nanotubes. *Analytical Chemistry*, *76*(4), 1083–1088. doi:10.1021/ac035143t PMID:14961742

Hsieh, P. F., Lin, H. H., Lin, T. L., Chen, Y. Y., & Wang, J. T. (2017). Two T7-like bacteriophages, K5-2 and K5-4, each encodes two capsule depolymerases: Isolation and functional characterization. *Scientific Reports*, *7*(1), 1–13. doi:10.103841598-017-04644-2 PMID:28676686

Hua, M. Y., Yang, H. W., Chuang, C. K., Tsai, R. Y., Chen, W. J., Chuang, K. L., Chang, Y. H., Chuang, H. C., & Pang, S. T. (2010). Magnetic-nanoparticlemodiWed paclitaxel for targeted therapy for prostate cancer. *Biomaterials*, *31*(28), 7355–7363. doi:10.1016/j.biomaterials.2010.05.061 PMID:20609471

Huang, C. M., Chen, C. H., Pornpattananangkul, D., Zhang, L., Chan, M., Hsieh, M. F., & Zhang, L. (2011). Eradication of drug resistant *Staphylococcus aureus* by liposomal oleic acids. *Biomaterials*, *32*(1), 214–221. doi:10.1016/j.biomaterials.2010.08.076 PMID:20880576

Huang, C., Soenen, S. J., Gulck, E., Vanhamb, G., Rejman, J., Van Calenbergh, S., Vervaet, C., Coenye, T., Verstraelen, H., Temmerman, M., Demeester, J., & De Smwdt, C. (2012). Electrospun cellulose acetate phthalate fibers for semen induced anti-HIV vaginal drug delivery. *Biomaterials*, *3*(33), 962–969. doi:10.1016/j.biomaterials.2011.10.004 PMID:22018388

Huang, F., Wang, J., Qu, A., Shen, L., Liu, J., Liu, J., Zhang, Z., An, Y., & Shi, L. (2014). Y. an and L. Shi, Maintenance of amyloid beta peptide homeostasis by artificial chaperones based on mixed-shell polymeric micelles. *Angewandte Chemie International Edition*, *53*(34), 8985–8990. doi:10.1002/anie.201400735 PMID:24985739

Huang, H. C., Barua, S., Sharma, G., Dey, S. K., & Rege, K. (2011). Inorganic nanoparticles for cancer imaging and therapy. *Journal of Controlled Release*, *155*(3), 344–357. doi:10.1016/j.jconrel.2011.06.004 PMID:21723891

Huang, J., Li, Q., Sun, D., Lu, Y., Su, Y., Yang, X., Wang, H., Wang, Y., Shao, W., He, N., Hong, J., & Chen, C. (2007). Biosynthesis of silver and gold nanoparticles by novel sundried Cinnamomum camphora leaf. *Nanotechnology*, *18*(10), 105104. Advance online publication. doi:10.1088/0957-4484/18/10/105104

Huang, L., Li, D. Q., Lin, Y. J., Wei, M., Evans, D. G., & Duan, X. (2005). Controllable preparation of Nano-MgO and investigation of its bactericidal properties. *Journal of Inorganic Biochemistry*, *99*(5), 986–993. doi:10.1016/j.jinorgbio.2004.12.022 PMID:15833320

Huang, N., Chen, X., Zhu, X., Xu, M., & Liu, J. (2017). Ruthenium complexes/polypeptide self-assembled nanoparticles for identification of bacterial infection and targeted antibacterial research. *Biomaterials*, *141*, 296–313. doi:10.1016/j.biomaterials.2017.07.005 PMID:28709020

Huang, P., Li, Z., Lin, J., Yang, D., Gao, G., Xu, C., Bao, L., Zhang, C., Wang, K., & Song, H. (2011). Photosensitizer-conjugated magnetic nanoparticles for in vivo simultaneous magnetofluores- cent imaging and targeting therapy. *Biomaterials*, *32*(13), 3447–3458. doi:10.1016/j.biomaterials.2011.01.032 PMID:21303717

Huang, X., & El-Sayed, M. A. (2010). Gold nanoparticles: Optical properties and implementations in cancer diagnosis and photothermal therapy. *Journal of Advanced Research*, *1*(1), 13–28. doi:10.1016/j.jare.2010.02.002

Huang, Y. W., & Lee, H. J. (2018). Cell-penetrating peptides for medical theranostics and targeted drug delivery, peptide applications in biomedicine. In *Biotechnology and Bioengineering* (pp. 359–370). Elsevier.

Huang, Y., Li, X., Liao, Z., Zhang, G., Liu, Q., Tang, J., Peng, Y., Liu, X., & Luo, Q. (2007). A randomized comparative trial between Acticoat and SD-Ag in the treatment of residual burn wounds including safety analysis. *Burns*, *33*(2), 161–166. doi:10.1016/j.burns.2006.06.020 PMID:17175106

Huang, Z., Zheng, X., Yan, D., Yin, G., Liao, X., Kang, Y., Yao, Y., Huang, D., & Hao, B. (2008). Toxicological effect of ZnO nanoparticles based on bacteria. *Langmuir*, *24*(8), 4140–4144. doi:10.1021/la7035949 PMID:18341364

Hua, S., & Wu, S. Y. (2013). The use of lipid-based nanocarriers for targeted pain therapies. *Frontiers in Pharmacology*, *4*, 143. doi:10.3389/fphar.2013.00143 PMID:24319430

Hu, C. W., Li, M., Cui, Y. B., Li, D. S., Chen, J., & Yang, L. Y. (2010). Toxicological effects ofTiO 2 and ZnO nanoparticles in soil on earthworm Eisenia fetida. *Soil Biology & Biochemistry*, *42*(4), 586–591. doi:10.1016/j.soilbio.2009.12.007

Huh, A. J., & Kwon, Y. J. (2011). "Nanoantibiotics": A new paradigm for treating infectious diseases using nanomaterials in the antibiotics resistant era. *Journal of Controlled Release*, *156*(2), 128–145. doi:10.1016/j.jconrel.2011.07.002 PMID:21763369

Hühn, D., Kantner, K., Geidel, C., Brandholt, S., De Cock, I., Soenen, S. J., ... Nienhaus, G. U. (2013). Polymer-coated nanoparticles interacting with proteins and cells: Focusing on the sign of the net charge. *ACS Nano*, *7*(4), 3253–3263. doi:10.1021/nn3059295 PMID:23566380

Huh, Y. M., Jun, Y. W., Song, H. T., Kim, S., Choi, J. S., Lee, J. H., Yoon, S., Kim, K. S., Shin, J. S., Suh, J. S., & Cheon, J. (2005). In vivo magnetic resonance detection of cancer by using multifunctional magnetic nanocrystals. *Journal of the American Chemical Society*, *127*(35), 12387–12391. doi:10.1021/ja052337c PMID:16131220

Hui Nicole Chia Poh. (2005). *Nanomedicine and Cancer. Science and Technology Studies, Section 84405*. Retrieved August 14, 2020, from http://www.tahan.com/charlie/nanosociety/course201/nanos/NH.pdf

Hule, R. A., & Pochan, D. J. (2007). Polymer nanocomposites for biomedical applications. *MRS Bulletin, 32*(4), 354–358. doi:10.1557/mrs2007.235

Hunter, R., & Actor, J. (2019). The pathogenesis of post-primary tuberculosis. A game changer for vaccine development. *Tuberculosis (Edinburgh, Scotland), 116*, S114–S117. doi:10.1016/j.tube.2019.04.018 PMID:31076321

Hussain, S. M., Hess, K. L., Gearhart, J. M., Geiss, K. T., & Schlager, J. J. (2005). In vitro toxicity of nanoparticles in BRL 3A rat liver cells. *Toxicology In Vitro, 19*(7), 975–983. doi:10.1016/j.tiv.2005.06.034 PMID:16125895

Hussein-Al-Ali, S. H., El Zowalaty, M. E., Hussein, M. Z., & Benjamin, M. (2014). Synthesis, characterization, and antimicrobial activity of an ampicillin-conjugated magnetic nanoantibiotic for medical applications. *International Journal of Nanomedicine, 9*, 3801–3814. doi:10.2147/IJN.S61143 PMID:25143729

Hussein-Al-Ali, S. H., El Zowalaty, M. E., Hussein, M. Z., Ismail, M., & Webster, T. J. (2014). Synthesis, characterization, controlled release, and antibacterial studies of a novel streptomycin chitosan magnetic nanoantibiotic. *International Journal of Nanomedicine, 9*, 549. PMID:24549109

Hu, Y., & Catchmark, J. M. (2011). In vitro biodegradability and mechanical properties of bioabsorbable bacterial cellulose incorporating cellulases. *Acta Biomaterialia, 7*(7), 2835–2845. doi:10.1016/j.actbio.2011.03.028 PMID:21459165

Hu, Y., Liu, A., Vaudrey, J., Vaiciunaite, B., Moigboi, C., McTavish, S. M., Kearns, A., & Coates, A. (2015). Combinations of beta-lactam or aminoglycoside antibiotics with plectasin are synergistic against methicillin-sensitive and methicillin-resistant Staphylococcus aureus. *PLoS One, 10*(2), e0117664. doi:10.1371/journal.pone.0117664 PMID:25692771

Hu, Y., Sinha, S. K., & Patel, S. (2015). Investigating Hydrophilic Pores in Model Lipid Bilayers Using Molecular Simulations: Correlating Bilayer Properties with Pore-Formation Thermodynamics. *Langmuir, 31*(24), 6615–6631. doi:10.1021/la504049q PMID:25614183

Ibele, M., Mallouk, T., & Sen, A. (2009). Schooling Behavior of Light-Powered Autonomous Micromotors in Water. *Angewandte Chemie International Edition, 48*(18), 3308–3312. doi:10.1002/anie.200804704 PMID:19338004

Ibrahim, K.S. (2013). Carbon nanotubes-properties and applications: a review. *Carbon Letters, 14*(3), 131-144.

Ignatova, M., Starbova, K., Markova, N., Manolova, N., & Rashkov, I. (2006). Electrospunnano-fibre mats with antibacterial properties from quaternized chitosan and poly(vinyl alcohol). *Carbohydrate Research, 341*(12), 2098–2107. doi:10.1016/j.carres.2006.05.006 PMID:16750180

Ike, Y. (2017). Pathogenicity of Enterococci. *Japanese Journal of Bacteriology, 72*(2), 189–211. doi:10.3412/jsb.72.189 PubMed

Imran, M., Iqbal, J., & Mehmood, T. (2006). Synthesis, characterization and (*in vitro*) Screening of Amoxicillin and its complexes with Ag(I), Zn(II), Cu(II), Co(II), Ni(II). *Journal of Biosciences, 5*, 946–949.

Infection: Types, causes, and differences. (n.d.). Available at: https://www.medicalnewstoday.com/articles/196271.php#types

Ingale, A. G., & Chaudhari, A. N. (2013). Biogenic synthesis of nanoparticles and potential applications: An eco-friendly approach. *Journal of Nanomedicine & Nanotechnology, 4*, 2–4. doi:10.4172/2157-7439.1000165

Iqbal, N., Kadir, M. R. A., Mahmood, N. H., Salim, N., Froemming, G. R. A., Balaji, H. R., & Kamarul, T. (2014). Characterization, antibacterial and in vitro compatibility of zinc–silver doped hydroxyapatite nanoparticles prepared through microwave synthesis. *Ceramics International, 40*(3), 4507–4513. doi:10.1016/j.ceramint.2013.08.125

Iravani, S. (2011). Critical review – green synthesis of metal nanoparticles using plants. *Green Chemistry, 13*(10), 2638–2650. doi:10.1039/c1gc15386b

Ishihara, Y., & Kimura, S. (2010). Nanofiber formation of amphiphilic cyclic tri-b-peptide. *Journal of Peptide Science, 16*(2), 110–114. doi:10.1002/psc.1206 PMID:20063334

Ito, A., & Kamihira, M. (2011). Tissue engineering using magnetite nanoparticles. *Progress in Molecular Biology and Translational Science, 104*, 355–395. doi:10.1016/B978-0-12-416020-0.00009-7 PMID:22093224

Ito, A., Shinkai, M., Honda, H., & Kobayashi, T. (2001). Heat-inducible TNF-alpha gene therapy combined with hyperthermia using magnetic nanoparticles as a novel tumor-targeted therapy. *Cancer Gene Therapy, 8*(9), 649–654. doi:10.1038j.cgt.7700357 PMID:11593333

Ito, A., Shinkai, M., Honda, H., & Kobayashi, T. (2005). Medical application of functionalized magnetic nanoparticles. *Journal of Bioscience and Bioengineering, 100*(1), 1–11. doi:10.1263/jbb.100.1 PMID:16233845

Ivashchenko, O., Lewandowski, M., Peplińska, B., Jarek, M., Nowaczyk, G., Wiesner, M., Załęski, K., Babutina, T., Warowicka, A., & Jurga, S. (2015). Synthesis and characterization of magnetite/silver/antibiotic nanocomposites for targeted antimicrobial therapy. *Materials Science and Engineering C, 55*, 343–359. doi:10.1016/j.msec.2015.05.023 PMID:26117765

Iyer, A. K., Khaled, G., Fang, J., & Maeda, H. (2006). Exploiting the enhanced permeability and retention effect for tumor targeting. *Drug Discovery Today, 11*(17-18), 812–818. doi:10.1016/j.drudis.2006.07.005 PMID:16935749

Jackson, J. M., Witek, M. A., Kamande, J. W., & Soper, S. A. (2017). Materials and microfluidics: Enabling the efficient isolation and analysis of circulating tumour cells. *Chemical Society Reviews, 46*(14), 4245–4280. doi:10.1039/C7CS00016B PMID:28632258

Jaidev, L. R., Bhavsar, D. V., Sharma, U., Jagannathan, N. R., Krishnan, U. M., & Sethuraman, S. (2014). Engineered multifunctional nanomaterials for multimodal imaging of retinoblastoma cellsin vitro. *Journal of Biomaterials Science. Polymer Edition, 25*(11), 1093–1109. doi:10.1080/09205063.2014.917040 PMID:24911385

Jain, M. D., Bachmeier, C. A., Phuoc, V. H., & Chavez, J. C. (2018). Axicabtagene ciloleucel (KTE-C19), an anti-CD19 CAR T therapy for the treatment of relapsed/refractory aggressive B-cell non-Hodgkin's lymphoma. *Ther Clin Risk Manag., 14*, 1007-1017. doi:10.2147/TCRM.S145039 PMID:29910620

Jain, P. K., Lee, K. S., El-sayed, I. H., & El-sayed, M. A. (2006). Calculated Absorption and Scattering Properties of Gold Nanoparticles of Different Size, Shape, and Composition: Applications in Biological Imaging and Biomedicine. *Journal of Photochemistry and Photobiology. B, Biology, 110*, 7238–7248. PMID:16599493

Jain, T. K., Richey, J., Strand, M., Leslie-Pelecky, D. L., Flask, C. A., & Labhasetwar, V. (2008). Magnetic nanoparticles with dual functional properties: Drug delivery and magnetic resonance imaging. *Biomaterials, 29*(29), 4012–4021. doi:10.1016/j.biomaterials.2008.07.004 PMID:18649936

Jaiswal, S., Bhattacharya, K., McHale, P., & Duffy, B. (2015). Dual effects of beta-cyclodextrin-stabilised silver nanoparticles: Enhanced biofilm inhibition and reduced cytotoxicity. *Journal of Materials Science. Materials in Medicine, 26*(1), 5367. doi:10.100710856-014-5367-1 PMID:25596861

Jalalian, S. H., Taghdisi, S. M., Hamedani, N. S., Kalat, S. A. M., Lavaee, P., ZandKarimi, M., Ghows, N., Jaafari, M.R., Naghibi, S., & Danesh, N.M. (2013). *EpirubicinLoaded Super Paramagnetic Iron Oxide Nanoparticle? Aptamer Bioconjugate for Combined Colon Cancer*. Academic Press.

Jansook, P., Ogawa, N., & Loftsson, T. (2018). Cyclodextrins: Structure, physicochemical properties and pharmaceutical applications. *International Journal of Pharmaceutics*, *535*(1-2), 272–284. doi:10.1016/j.ijpharm.2017.11.018 PMID:29138045

Janßen, H. C., Angrisani, N., Kalies, S., Hansmann, F., Kietzmann, M., Warwas, D. P., Behrens, P., & Reifenrath, J. (2020). Biodistribution, biocompatibility and targeted accumulation of magnetic nanoporous silica nanoparticles as drug carrier in orthopedics. *Journal of Nanobiotechnology*, *18*(1), 14. doi:10.118612951-020-0578-8 PMID:31941495

Janvier, C., Villeneuve, F., Alabouvette, C., Edel-Hermann, V., Mateille, T., & Steinberg, C. (2007). Soil health through soil disease suppression: Which strategy from descriptors to indicators? *Soil Biology & Biochemistry*, *39*(1), 1–23. doi:10.1016/j.soilbio.2006.07.001

Jan, Y. J., Chen, J. F., Zhu, Y., Lu, Y. T., Chen, S. H., Chung, H., Smalley, M., Huang, Y.-W., Dong, J., Chen, L.-C., Yu, H.-H., Tomlinson, J. S., Hou, S., Agopian, V. G., Posadas, E. M., & Tseng, H.-R. (2018). NanoVelcro rare-cell assays for detection and characterization of circulating tumor cells. *Advanced Drug Delivery Reviews*, *125*, 78–93. doi:10.1016/j.addr.2018.03.006 PMID:29551650

Javani, S., Lorca, R., Latorre, A., Flors, C., Cortajarena, A. L., & Somoza, Á. (2016). Antibacterial Activity of DNA-Stabilized Silver Nanoclusters Tuned by Oligonucleotide Sequence. *ACS Applied Materials & Interfaces*, *8*(16), 10147–10154. doi:10.1021/acsami.6b00670 PMID:27058628

Jayachandran, R., Gatfield, J., Massner, J., Albrecht, I., Zanolari, B., & Pieters, J. (2008). RNA interference in J774 macrophages reveals a role for coronin 1 in mycobacterial trafficking but not in actin-dependent processes. *Molecular Biology of the Cell*, *19*(3), 1241–1251. doi:10.1091/mbc.e07-07-0640 PMID:18162581

Jayaprakash, R., Sah, S. k., Hemalatha, S., & Easwaramoorthy, D. (2016a). Synthesis, Characterization, QSAR, Docking, Antibacterial Activity, and Brine Shrimp Lethal Studies on L-Phenylalanine Schiff Bases. Asian J. Pharma. clin. *Res.*, *9*, 3.

Jayaprakash, R., Sah, S. k., Hemalatha, S., & Easwaramoorthy, D. (2016b). QSAR, Brine Shrimp Lethal Assay and antimicrobial studies on Synthesized L-Tryptophan-2,4-dihydroxy benzaldehyde Schiff Base. *Int. J. Chem. Tech*, *9*, 48–54.

Jayaraman, R. (2009). Antibiotic resistance: An overview of mechanisms and a paradigm shift. *Current Science*, *96*(11), 1475–1484.

Jazayeri, M. H., Amani, H., Pourfatollah, A. A., Pazoki-Toroudi, H., & Sedighimoghaddam, B. (2016). Various methods of gold nanoparticles (GNPs) conjugation to antibodies. *Sensing and Bio-Sensing Research*, *9*, 17–22. doi:10.1016/j.sbsr.2016.04.002

Jean-François, F., Elezgaray, J., Berson, P., Vacher, P., & Dufourc, E. J. (2008). Pore formation induced by an antimicrobial peptide: Electrostatic effects. *Biophysical Journal*, *95*(12), 5748–5756. doi:10.1529/biophysj.108.136655 PMID:18820233

Jeevanandham, J., Barhoum, A., Chan, Y., Dufresne, A., & Danquah, M. (2018). Review on nanoparticles and nanostructured materials: History, sources, toxicity and regulations. *Beilstein Journal of Nanotechnology*, *9*, 1050–1074. doi:10.3762/bjnano.9.98 PMID:29719757

Jemima, S. A., & Verghese, S. (2008). Multiplex PCR for blaCTX-M & blaSHV in the extended spectrum beta lactamase (ESBL) producing Gram-negative isolates. *The Indian Journal of Medical Research*, *128*, 313–317. PubMed

Jena, M., Mishra, S., Jena, S., & Mishra, S. S. (2013). Nanotechnology-future prospect in recent medicine: A review. *International Journal of Basic and Clinical Pharmacology*, *2*, 353–359. doi:10.5455/2319-2003.ijbcp20130802

Jensen, M., & Mehlhorn, H. (2009). Seventy-five years of Resochin in the fight against malaria. *Parasitology Research*, *105*(3), 609–627. doi:10.100700436-009-1524-8 PMID:19593586

Jeong, W. J., Kwon, S. H., & Lim, Y. B. (2018). Modular self-assembling peptide platform with a tunable thermo responsiveness via a single amino acid substitution. *Advanced Functional Materials*, *28*, 35.

Jeong, W. J., Lee, M. S., & Lim, Y. B. (2013). Helix stabilised, thermostable, and protease-resistant self-assembled peptide nanostructures as potential inhibitors of protein–protein interactions. *Biomacromol*, *14*(8), 2684–2689. doi:10.1021/bm400532y PMID:23844716

Jeyarani, S., Vinita, N. M., Puja, P., Senthamilselvi, S., Devan, U., Velangani, A. J., Biruntha, M., Pugazhendhi, A., & Kumar, P. (2020). Biomimetic gold nanoparticles for its cytotoxicity and biocompatibility evidenced by fluorescence-based assays in cancer (MDA-MB-231) and non-cancerous (HEK-293) cells. *Journal of Photochemistry and Photobiology. B, Biology*, *202*, 111715. doi:10.1016/j.jphotobiol.2019.111715 PMID:31790882

Jiang, H. L., Wang, L. Q., & Zhu, K. J. (2014). Coaxial electrospinning for encapsulation and controlled release of fragile water-soluble bioactive agents. *Journal of Controlled Release*, *193*, 296–303. doi:10.1016/j.jconrel.2014.04.025 PMID:24780265

Jiang, H., Fang, D., Hsiao, B., & Chen, W. (2004). Preparation and characterization of ibuprofen loaded poly(lactide-co-glycolide)/poly(ethylene glycol)-g-chitosan electrospun memberanes. *Journal of Biomaterials Science. Polymer Edition*, *15*(3), 279–296. doi:10.1163/156856204322977184 PMID:15147162

Jiang, W., Mashayekhi, H., & Xing, B. (2009). Bacterial toxicity comparison between nano- and micro-scaled oxide particles. *Environmental Pollution*, *157*(5), 1619–1625. doi:10.1016/j.envpol.2008.12.025 PMID:19185963

Jiang, W., Rutherford, D., Vuong, T., & Liu, H. (2017). Nanomaterials for treating cardiovascular diseases: A review. *Bioactive Materials*, *2*(4), 185–198. doi:10.1016/j.bioactmat.2017.11.002 PMID:29744429

Ji, F., Zhang, K., Li, J., Gu, Y., Zhao, J., & Zhang, J. (2018). A dual pH/magnetic responsive nanocarrier based on PEGylated Fe3 O4 nanoparticles for doxorubicin delivery. *Journal of Nanoscience and Nanotechnology*, *18*(7), 4464–4470. doi:10.1166/jnn.2018.15275 PMID:29442620

Jijie, R., Barras, A., Teodorescu, F., Boukherroub, R., & Szunerits, S. (2017). Advancements on the molecular design of nanoantibiotics: Current level of development and future challenges. *Molecular Systems Design & Engineering*, *2*(4), 349–369. doi:10.1039/C7ME00048K

Jiling, C., Youzhi, F., Xiangui, L., & Junhua, W. (2016). Arbuscular mycorrhizal fungi alleviate the negative effects of iron oxide nanoparticles on bacterial community in rhizospheric soils. *Frontiers in Environmental Science*, *4*, 10.

Jin, Y., Kannan, S., Wu, M., & Zhao, J.X. (2007). Toxicity of luminescent silica nanoparticles to living cells. *Chemical Research in Toxicology*, *20*, 1126–1133.

Jin, T., & He, Y. (2011). Antibacterial activities of magnesium oxide (MgO) nanoparticles against foodborne pathogens. *Journal of Nanoparticle Research*, *13*(12), 6877–6885. doi:10.100711051-011-0595-5

Jnawali, H. N., & Ryoo, S. (2013). First-and second-line drugs and drug resistance. *Tuberculosis-Current Issues in Diagnosis and Management*, 163-180.

Jo, A., Ding, T., & Ahn, J. (2016). Synergistic antimicrobial activity of bacteriophages and antibiotics against Staphylococcus aureus. *Food Science and Biotechnology*, *25*(3), 935–940. doi:10.100710068-016-0153-0 PMID:30263357

Joerger, R., Klaus, T., & Granqvist, C. G. (2000). Biologically produced silver–carbon composite materials for optically functional thin-film coatings. *Advanced Materials*, *12*(6), 407–409. doi:10.1002/(SICI)1521-4095(200003)12:6<407::AID-ADMA407>3.0.CO;2-O

Joginder, S. D., Ravinder, K., Naresh, K., Pawan, K., Kiran, N., & Surekha, D. (2017). Nanotechnology: The new perspective in precision agriculture. *Biotechnology Reports (Amsterdam, Netherlands)*, *15*, 11–13. doi:10.1016/j.btre.2017.03.002 PMID:28603692

Johanesen, P. A., Mackin, K. E., Hutton, M. L., Awad, M. M., Larcombe, S., Amy, J. M., & Lyras, D. (2015). Disruption of the Gut Microbiome: Clostridium difficile Infection and the Threat of Antibiotic Resistance. *Genes*, *6*(4), 1347–1360. doi:10.3390/genes6041347 PMID:26703737

Johansen, S. K., Maus, C. E., Plikaytis, B. B., & Douthwaite, S. (2006). Capreomycin binds across the ribosomal subunit interface using tlyA-encoded 2′-O-methylations in 16S and 23S rRNAs. *Molecular Cell*, *23*(2), 173–182. doi:10.1016/j.molcel.2006.05.044 PMID:16857584

Johnson, J. R., Johnston, B., Clabots, C., Kuskowski, M. A., & Castanheira, M. (2010). Escherichia coli sequence type ST131 as the major cause of serious multidrug-resistant E. coli infections in the United States. *Clinical Infectious Diseases*, *51*(3), 286–294. doi:10.1086/653932 PMID:20572763

Johnston, R. L. (2012). Metal nanoparticles and nanoalloys. *Frontiers of Nanoscience*, *3*, 1–42. doi:10.1016/B978-0-08-096357-0.00006-6

Jokerst, J. V., Lobovkina, T., Zare, R. N., & Gambhir, S. S. (2011). Nanoparticle PEGylation for imaging and therapy. *Nanomedicine (London)*, *6*(4), 15–28. doi:10.2217/nnm.11.19 PMID:21718180

Joliot, A., Pernelle, C., Deagostini-Bazin, H., & Prochiantz, A. (1991). Antennapedia homeobox peptide regulates neural morphogenesis. *Proceedings of the National Academy of Sciences of the United States of America*, *88*(5), 1864–1868. doi:10.1073/pnas.88.5.1864 PMID:1672046

Jordan, A., Scholz, R., Wust, P., Fahling, H., Krause, J., Wlodarczyk, W., Sander, B., Vogl, T., & Felix, R. (1997). Effects of Magnetic Fluid Hyperthermia (MFH) on C3H Mammary Carcinoma in vivo. *International Journal of Hyperthermia*, *13*(6), 587–605. doi:10.3109/02656739709023559 PMID:9421741

Jores, K., Mehnert, W., Drechsler, M., Bunjes, H., Johann, C., & Mader, K. (2004). Investigations on the structure of solid lipid ¨ nanoparticles (SLN) and oil-loaded solid lipid nanoparticles by photon correlation spectroscopy, field-flow fractionation and transmission electron microscopy. *Journal of Controlled Release*, *95*(2), 217–227. doi:10.1016/j.jconrel.2003.11.012 PMID:14980770

Jorfi, M., & Foster, E. J. (2015). Recent advances in nanocellulose for biomedical applications. *Journal of Applied Polymer Science*, *132*(14), 1–19. doi:10.1002/app.41719

Jotai, A. W., Kim, I. S., & Ni, Q. Q. (2019). Cellulose acetate nanofibers embedded with AgNPs anchored TiO2 nanoparticles for long term excellent antibacterial applications. *Carbohydrate Polymers*, *207*, 640–649. doi:10.1016/j.carbpol.2018.12.029 PMID:30600049

Jo, Y.-K., Jung, G., & Kim, B. H. (2009). Antifungal Activity of Silver Ions and Nanoparticles on Phytopathogenic Fungi. *Plant Disease*, *93*(10), 1037–1042. doi:10.1094/PDIS-93-10-1037 PubMed

Juréen, P., Werngren, J., Toro, J. C., & Hoffner, S. (2008). Pyrazinamide resistance and pncA gene mutations in Mycobacterium tuberculosis. *Antimicrobial Agents and Chemotherapy*, *52*(5), 1852–1854. doi:10.1128/AAC.00110-08 PMID:18316515

Kaakoush, N. O., Castano-Rodriguez, N., Mitchell, H. M., & Man, S. M. (2015). Global Epidemiology of Campylobacter Infection. *Clinical Microbiology Reviews*, *28*(3), 687–720. doi:10.1128/CMR.00006-15 PMID:26062576

Kajani, A. A., Bordbar, A. K., Zarkesh Esfahani, S. H., & Razmjou, A. (2016). Gold nanoparticles as potent anticancer agent: Green synthesis, characterization, and: In vitro study. *RSC Advances*, *6*(68), 63973–63983. doi:10.1039/C6RA09050H

Kalaitzidou, K., Fukushima, H., & Drzal, L. T. (2007). A new compounding method for exfoliated graphite- polypropylene nanocomposites with enhanced flexural properties and lower percolation threshold. *Composites Science and Technology*, *67*(10), 2045–2051. doi:10.1016/j.compscitech.2006.11.014

Kalan, L., & Wright, G. D. (2011). Antibiotic adjuvants: Multicomponent anti-infective strategies. *Expert Reviews in Molecular Medicine*, *13*, e5. doi:10.1017/S1462399410001766 PMID:21342612

Kalhapure, R. S., Sonawane, S. J., Sikwal, D. R., Jadhav, M., Rambharose, S., Mocktar, C., & Govender, T. (2015). Solid lipid nanoparticles of clotrimazole silver complex: An efficient nano antibacterial against *Staphylococcus aureus* and MRSA. *Colloids and Surfaces. B, Biointerfaces*, *136*, 651–658. doi:10.1016/j.colsurfb.2015.10.003 PMID:26492156

Kalimuthu, K., Lin, S., Tseng, L., Murugan, K., & Hwang, J.-S. (2013). Bio-efficacy potential of seaweed Gracilaria firma with copepod, Megacyclops formosanus for the control larvae of dengue vector Aedes aegypti. *Hydrobiologia*, *741*(1), 113–123. doi:10.100710750-013-1745-9

Kalimuthu, K., Panneerselvam, C., Chou, C., Tseng, L.-C., Murugan, K., Tsai, K.-H., Alarfaj, A. A., Higuchi, A., Canale, A., Hwang, J.-S., & Benelli, G. (2017). Control of dengue and Zika virus vector Aedes aegypti using the predatory copepod Megacyclops formosanus: Synergy with Hedychium coronarium-synthesized silver nanoparticles and related histological changes in targeted mosquitoes. *Process Safety and Environmental Protection*, *109*, 82–96. doi:10.1016/j.psep.2017.03.027

Kalishwaralal, K., Banumathi, E., Pandian, S. R. K., Deepak, V., Muniyandi, J., Eom, S. H., & Gurunathan, S. (2009). Silver nanoparticles inhibit VEGF induced cell proliferation and migration in bovine retinal endothelial cells. *Colloids and Surfaces. B, Biointerfaces*, *73*(1), 51–57. doi:10.1016/j.colsurfb.2009.04.025 PubMed

Kalita, S., Kandimalla, R., Bhowal, A. C., Kotoky, J., & Kundu, S. (2018). Functionalization of β-lactam antibiotic on lysozyme capped gold nanoclusters retrogress MRSA and its persisters following awakening. *Scientific Reports*, *8*(1), 1–13. doi:10.103841598-018-22736-5 PMID:29636496

Kalombo, L., Lemmer, Y., Semete-Makokotlela, B., Ramalapa, B., Nkuna, P., Booysen, L. L., Naidoo, S., Hayeshi, R., Verschoor, J. A., & Swai, H. S. (2019). Spray-Dried, Nanoencapsulated, Multi-Drug Anti-Tuberculosis Therapy Aimed at Once Weekly Administration for the Duration of Treatment. *Nanomaterials (Basel, Switzerland)*, *9*(8), 1167. doi:10.3390/nano9081167 PMID:31443150

Kamal, F., & Dennis, J. J. (2015). Burkholderia cepacia complex Phage-Antibiotic Synergy (PAS): Antibiotics stimulate lytic phage activity. *Applied and Environmental Microbiology*, *81*(3), 1132–1138. doi:10.1128/AEM.02850-14 PMID:25452284

Kamaruzzaman, N. F., Tan, L. P., Hamdan, R. H., Choong, S. S., Wong, W. K., Gibson, A. J., Chivu, A., & Pina, M. D. F. (2019). Antimicrobial polymers: The potential replacement of existing antibiotics? *International Journal of Molecular Sciences*, *20*(11), 2747. doi:10.3390/ijms20112747 PMID:31167476

Kamat, M., El-Boubbou, K., Zhu, D. C., Lansdell, T., Lu, X., Li, W., & Huang, X. (2010). Hyaluronic Acid Immobilized Magnetic Nanoparticles for Active Targeting and Imaging of Macrophages. *Bioconjugate Chemistry*, *21*(11), 2128–2135. doi:10.1021/bc100354m PMID:20977242

Kang, S. J., Park, S. J., Mishig-Ochir, T., & Lee, B. J. (2014). Antimicrobial peptides: Therapeutic potentials. *Expert Review of Anti-Infective Therapy*, *12*(12), 1477–1486. doi:10.1586/14787210.2014.976613 PMID:25371141

Kang, S., Pinault, M., Pfefferle, L. D., & Elimelech, M. (2007). Single-walled carbon nanotubes exhibit strong antimicrobial activity. *Langmuir, 23*(17), 8670–8673. doi:10.1021/la701067r PMID:17658863

Kao, J. C., HuangFu, W.-C., Tsai, T.-T., Ho, M.-R., Jhan, M.-K., Shen, T.-J., Tseng, P.-C., Wang, Y.-T., & Lin, C.-F. (2018). The antiparasitic drug niclosamide inhibits dengue virus infection by interfering with endosomal acidification independent of mTOR. *PLoS Neglected Tropical Diseases, 12*(8), e0006715. doi:10.1371/journal.pntd.0006715 PMID:30125275

Kappen, M., & Budke-Gieseking, A. (2018). MagnetogeneticActivationof Small GTPaseswithMagIcSNanoparticlesInside Living Cells. *12th International conference on Scientific and clinical applications of Magnetic carriers.*

Karagiota, A., Kourti, M., Simos, G., & Mylonis, I. (2019). HIF-1-derived cell-penetrating peptides inhibit ERK-dependent activation of HIF-1 and trigger apoptosis of cancer cells under hypoxia. *Cellular and Molecular Life Sciences, 76*(4), 809–825. doi:10.100700018-018-2985-7 PMID:30535970

Karmali, M. A., Gannon, V., & Sargeant, J. M. (2010). Verocytotoxin-producing Escherichia coli (VTEC). *Veterinary Microbiology, 140*(3-4), 360–370. doi:10.1016/j.vetmic.2009.04.011 PMID:19410388

Kaspar, A. A., & Reichert, J. M. (2013). Future directions for peptide therapeutics development. *Drug Discovery Today, 18*(17-18), 807–817. doi:10.1016/j.drudis.2013.05.011 PMID:23726889

Kasparis, G., Wang, L., Plan, A., LaGrow, A., Sergides, A., Wilhelm, C., & Nguyen, T. K. T. (2018). Synthesis and Application of Zinc Ferrite Nanoparticles as Agents for Cancer Thermotherapy. *12th International conference on Scientific and clinical applications of Magnetic carriers.*

Katayama, Y., Ito, T., & Hiramatsu, K. (2010). A new class of genetic element, staphylococcus cassette chromosome mec, encodes methicillin resistance in Staphylococcus aureus. *Antimicrobial Agents and Chemotherapy, 44*(6), 1549–1555. doi:10.1128/AAC.44.6.1549-1555.2000 PMID:10817707

Katz, E., & Willner, I. (2004). Integrated nanoparticle-biomolecule hybrid systems: Synthesis, properties, and applications. *Angewandte Chemie International Edition in English, 43*(45), 6042–6108. doi:10.1002/anie.200400651 PMID:15538757

Kaur, I. P., & Singh, H. (2014). Nanostructured drug delivery for better management of tuberculosis. *Journal of Controlled Release, 184*, 36–50. doi:10.1016/j.jconrel.2014.04.009 PMID:24732260

Kayal, & Ramanujan. (2010). Anti-Cancer Drug Loaded Iron–Gold Core–Shell Nanoparticles (Fe@Au) for Magnetic Drug Targeting. *Journal of Nanoscience and Nanotechnology*. PMID:21133071

Kearns, H., Goodacre, R., Jamieson, L. E., Graham, D., & Faulds, K. (2017). SERS detection of multiple antimicrobial-resistant pathogens using nanosensors. *Analytical Chemistry, 89*(23), 12666–12673. doi:10.1021/acs.analchem.7b02653 PMID:28985467

Kelkar, S. S., & Reineke, T. M. (2011). Theranostics: Combining imaging and therapy. *Bioconjugate Chemistry, 22*(10), 1879–1903. doi:10.1021/bc200151q PMID:21830812

Kenawy, E. R., Abdel-Hay, F. I., El-Newehy, M. H., & Wnek, G. E. (2009). Processing of polymer nanofibers through electrospinning as drug delivery systems. *Materials Chemistry and Physics, 113*(1), 296–302. doi:10.1016/j.matchemphys.2008.07.081

Kenawy, E. R., Worley, S. D., & Broughton, R. (2017). The chemistry and applications of antimicrobial polymers: A state-of-the-art review. *Biomacromolecules, 8*(5), 1359–1384. doi:10.1021/bm061150q PMID:17425365

Keren, I., Kaldalu, N., Spoering, A., Wang, Y., & Lewis, K. (2004). Persister cells and tolerance to antimicrobials. *FEMS Microbiology Letters, 230*(1), 13–18. doi:10.1016/S0378-1097(03)00856-5 PMID:14734160

Kettler, K., Veltman, K., van de Meent, D., van Wezel, A., & Hendriks, A. J. (2014). Cellular uptake of nanoparticles as determined by particle properties, experimental conditions, and cell type. *Environmental Toxicology and Chemistry*, *33*(3), 481–492. doi:10.1002/etc.2470 PMID:24273100

Khan, M. G., Yezdani, U., Chakravorty, A., &Shukla, T. (2020). Efforts and Challenges paved by India to confront of Corona Virus (COVID-19). *Bangladesh Journal of Medical Science*, 88-S.

Khandelia, H., & Kaznessis, Y. N. (2007). Cation-pi interactions stabilize the structure of the antimicrobial peptide indolicidin near membranes: Molecular dynamics simulations. *The Journal of Physical Chemistry B*, *111*(1), 242–250. doi:10.1021/jp064776j PMID:17201448

Khan, I., Abdalla, A., & Qurashi, A. (2017). Synthesis of hierarchical WO3 and Bi2O3/WO3 nanocomposite for solar-driven water splitting applications. *International Journal of Hydrogen Energy*, *42*(5), 3431–3439. doi:10.1016/j.ijhydene.2016.11.105

Khan, I., Yamani, Z. H., & Qurashi, A. (2017). Sonochemical-driven ultrafast facile synthesis of SnO2 nanoparticles: Growth mechanism structural electrical and hydrogen gas sensing properties. *Ultrasonics Sonochemistry*, *34*, 484–490. doi:10.1016/j.ultsonch.2016.06.025 PMID:27773272

Kharissova, O. V., Dias, H. V. R., Kharisov, B. I., Perez, B. O., & Perez, V. M. J. (2013). The greener synthesis of nanoparticles. *Trends in Biotechnology*, *31*(4), 240–248. doi:10.1016/j.tibtech.2013.01.003 PMID:23434153

Khashan, K. S., Sulaiman, G. M., & Abdulameer, F. A. (2016). Synthesis and Antibacterial Activity of CuO Nanoparticles Suspension Induced by Laser Ablation in Liquid. *Arabian Journal for Science and Engineering*, *41*(1), 301–310. doi:10.100713369-015-1733-7

Khatak, S., Semwal, U. P., Pandey, M. K., & Dureja, H. (2018). *Investigation of Antimycobacterial potential of solid lipid nanoparticles against M. smegmatis*. Academic Press.

Khil, M.S., Cha, D.I., Kim, H.Y., Kim, I.S., & Bhattarai, N. (2003). Electrospun nanofibrous polyurethane membrane as wound dressing. *Journal* of *Biomedical* Materials *Research Part B*: *Applied Biomaterial*, *67B*(2), 675–679.

Khodashenas, B., & Ghorbani, H. R. (2015). Synthesis of silver nanoparticles with different shapes. *Arabian Journal of Chemistry*, 1–16.

Khoder, R., & Korri-Youssoufi, H. (2020). E-DNA biosensors of M. tuberculosis based on nanostructured polypyrrole. *Materials Science and Engineering C*, *108*, 110371. doi:10.1016/j.msec.2019.110371 PMID:31924004

Khosa, A., Reddi, S., & Saha, R. N. (2018). Nanostructured lipid carriers for site-specific drug delivery. *Biomedicine and Pharmacotherapy*, *103*, 598–613. doi:10.1016/j.biopha.2018.04.055 PMID:29677547

Khoshnevisan, K., Maleki, H., Samadian, H., Shahsavari, S., Sarrafzadeh, M. H., Larijani, B., Dorkoosh, F. A., Haghpanah, V., & Khorramizadeh, M. R. (2018). Cellulose Acetate ElectrospunNanofibers for Drug Delivery Systems: Applications and Recent Advances. *Carbohydrate Polymers*, *198*, 131–141. doi:10.1016/j.carbpol.2018.06.072 PMID:30092983

Kianfara, P., Vitalea, A., Dalle Vacchea, S., & Bongiovanni, R. (2019). Photo-crosslinking of chitosan/poly(ethylene oxide) electrospun nanofibers. *Carbohydrate Polymers*, *217*, 144–151. doi:10.1016/j.carbpol.2019.04.062 PMID:31079670

Kidd, S. E., Goeman, E., Meis, J. F., Slavin, M. A., & Verweij, P. E. (2015). Multi-triazole-resistant Aspergillus fumigatus infections in Australia. *Mycoses*, *58*(6), 350–355. doi:10.1111/myc.12324 PMID:25885568

Kim, Lee, Park, HeeMun, Park, & Yoo. (2015). Drug-loaded gold/iron/gold plasmonic nanoparticles for magnetic targeted chemo photothermal treatment of rheumatoid arthritis. *J. Biomaterials*.

Kim, D., Kim, J., Park, Y. I., Lee, N., & Hyeon, T. (2018). Recent development of inorganic nanoparticles for biomedical imaging. *ACS Central Science*, *4*(3), 324–336. doi:10.1021/acscentsci.7b00574 PMID:29632878

Kim, E. J., Choi, M., Park, H. Y., Hwang, J. Y., Kim, H. E., Hong, S. W., Lee, J., Yong, K., & Kim, W. (2019). Thorn like TiO2 nanoarrays with broad spectrum antimicrobial activity through physical puncture and photocatalytic action. *Scientific Reports*, *9*(1), 13697. doi:10.103841598-019-50116-0 PMID:31548584

Kim, H., Jo, A., Baek, S., Lim, D., Park, S.-Y., Cho, S. K., Chung, J. W., & Yoon, J. (2017). Synergistically enhanced selective intracellular uptake of anticancer drug carrier comprising folic acidconjugated hydrogels containing magnetite nanoparticles. *Scientific Reports*, *7*(1), 41090. doi:10.1038rep41090 PMID:28106163

Kim, H., Kitamatsu, M., & Ohtsuki, T. (2018). Enhanced intracellular peptide delivery by multivalent cell-penetrating peptide with bioreducible linkage. *Bioorganic & Medicinal Chemistry Letters*, *28*(3), 378–381. doi:10.1016/j.bmcl.2017.12.035 PMID:29275934

Kim, K. J., Sung, W. S., Suh, B. K., Moon, S. K., Choi, J. S., Kim, J. G., & Lee, D. G. (2009). Antifungal activity and mode of action of silver nano-particles on Candida albicans. *Biometals*, *22*(2), 235–242. doi:10.100710534-008-9159-2 PMID:18769871

Kim, M. H., Jeong, H., Sim, Y. M., Lee, S., Yong, D., Ryu, C. M., & Choi, J. Y. (2020). Using comparative genomics to understand molecular features of carbapenem-resistant Acinetobacter baumannii from South Korea causing invasive infections and their clinical implications. *PLoS One*, *15*(2), e0229416. doi:10.1371/journal.pone.0229416 PMID:32084241

Kim, M. R., & Park, T. G. (2002). Temperature-responsive and degradable hyaluronic acid/Pluronic composite hydrogels for controlled release of human growth hormone. *Journal of Controlled Release*, *80*(1-3), 69–77. doi:10.1016/S0168-3659(01)00557-0 PMID:11943388

Kim, M. S., Kim, H. J., Jang, J. Y., & Shin, H. S. (2018). Development of Coaxial Alginate-PCL Nanofibrous Dressing for Controlled Release of Spirulina Extract. *Journal of Biomaterials Science*, *29*(12), 1–0. doi:10.1080/09205063.2018.1462931 PMID:29667511

Kim, M., Park, J. H., Jeong, H., Hong, J., Choi, W. S., Lee, B. H., & Park, C. Y. (2017). An evaluation of the in vivo safety of nonporous silica nanoparticles: Ocular topical administration versus oral administration. *Scientific Reports*, *7*(1), 8238. doi:10.103841598-017-08843-9 PMID:28811672

Kim, S., Do, I., & Drzal, L. T. (2009). Thermal stability and dynamic mechanical behavior of exfoliated graphite nanoplatelets-LLDPE nanocomposites. *Polymer Composites*, *31*(5), 755–761. doi:10.1002/pc.20781

Kim, S., Lim, Y. T., Soltesz, E. G., De Grand, A. M., Lee, J., Nakayama, A., Parker, J. A., Mihalijevic, T., Laurence, R. G., Dor, D. M., Cohn, L. H., Bawendi, M. G., & Frangioni, J. V. (2004). Near-infrared fluorescent type II quantum dots for sentinel lymph node mapping. *Nature Biotechnology*, *22*(1), 93–97. doi:10.1038/nbt920 PMID:14661026

Kinaret, P., Ilves, M., Fortino, V., Rydman, E., Karisola, P., Lähde, A., …Alenius, H. (2017). *Inhalation and oropharyngeal aspiration exposure to rod-like carbon.* Academic Press.

Kinaret, P., Ilves, M., Fortino, V., Rydman, E., Karisola, P., Lähde, A., Koivisto, J., Jokiniemi, J., Wolff, H., Savolainen, K., Greco, D., & Alenius, H. (2017, January 24). nanotubes induce similar airway inflammation and biological responses in mouse lungs. *ACS Nano*, *11*(1), 291–303. doi:10.1021/acsnano.6b05652 PMID:28045493

Kiristi, M., Singh, V. V., Esteban-Fernández de Ávila, B., Uygun, M., Soto, F., Aktaş Uygun, D., & Wang, J. (2015). Lysozyme-Based Antibacterial Nanomotors. *ACS Nano*, *9*(9), 9252–9259. doi:10.1021/acsnano.5b04142 PMID:26308491

Kitching, M., Choudhary, P., Inguva, S., Guo, Y., Ramani, M., Das, S. K., & Marsili, E. (2016). Fungal surface protein mediated one-pot synthesis of stable and hemocompatible gold nanoparticles. *Enzyme and Microbial Technology*, *95*, 76–84. doi:10.1016/j.enzmictec.2016.08.007 PMID:27866629

Kitov, P. I., & Bundle, D. R. (2003). on the nature of the multivalency effect: A thermodynamic model. *Journal of the American Chemical Society*, *125*(52), 16271–16284. doi:10.1021/ja038223n PMID:14692768

Klaine, S. J., Alvarez, P. J., Batley, G. E., Fernandes, T. F., Handy, R. D., Lyon, D. Y., Mahendra, S., McLaughlin, M. J., & Lead, J. R. (2008). Nanomaterials in the environment: Behavior, fate, bioavailability, and effects. *Environmental Toxicology and Chemistry*, *27*(9), 1825–1851. doi:10.1897/08-090.1 PMID:19086204

Klębowski, B., Depciuch, J., Parlińska-Wojtan, M., & Baran, J. (2018). Applications of Noble Metal-Based Nanoparticles in Medicine. *International Journal of Molecular Sciences*, *19*(12), 4031. doi:10.3390/ijms19124031 PMID:30551592

Klingelfus, T., Lirola, J. R., Oya Silva, L. F., Disner, G. R., Vicentini, M., Nadaline, M. J. B., Robles, J. C. Z., Trein, L. M., Voigt, C. L., Silva de Assis, H. C., Mela, M., Leme, D. M., & Cestari, M. M. (2017). Acute and longterm effects of trophic exposure to silver nanospheres in the central nervous system of a neotropical fish Hoplias intermedius. *Neurotoxicology*, *63*, 146–154. doi:10.1016/j.neuro.2017.10.003 PMID:29031576

Knetsch, M. L., & Koole, L. H. (2011). New strategies in the development of antimicrobial coatings: The example of increasing usage of silver and silver nanoparticles. *Polymers*, *3*(1), 340–366. doi:10.3390/polym3010340

Kobayashi, H., Hama, Y., Koyama, Y., Barrett, T., Regino, C. A. S., Urano, Y., & Choyke, P. L. (2007). Simultaneous multicolor imaging of five different lymphatic basins using quantum dots. *Nano Letters*, *7*(6), 1711–1716. doi:10.1021/nl0707003 PMID:17530812

Koch, K., Bhushan, B., Jung, Y. C., & Barthlott, W. (2009). Fabrication of artificial lotus leaves and significance of hierarchical structure for super hydrophobicity and low adhesion. *Soft Matter*, *5*(7), 1386–1393. doi:10.1039/b818940d

Koh, C. L., Sam, C. K., Yin, W. F., Tan, L. Y., Krishnan, T., Chong, Y. M., & Chan, K. G. (2013). Plant-derived natural products as sources of anti-quorum sensing compounds. *Sensors (Basel)*, *13*(5), 6217–6228. doi:10.3390/s130506217 PubMed

Kohler, N., Sun, C., Wang, J., & Zhang, M. (2005). Methotrexate- modified superparamagnetic nanopar- ticles and their intracellular uptake into human cancer cells. *Langmuir*, *21*(19), 8858–8864. doi:10.1021/la0503451 PMID:16142971

Kohl, S., & Bock, R. (2009). Transposition of a bacterial insertion sequence in chloroplasts. *The Plant Journal*, *58*(3), 423–436. doi:10.1111/j.1365-313X.2009.03787.x PMID:19144000

Kokkinos, J., Ignacio, R. M. C., Sharbeen, G., Boyer, C., Gonzales-Aloy, E., Goldstein, D., Mccaroll, J. A., & Philips, P. A. (2020). Targeting the Undruggable in Pancreatic Cancer Using Nano-Based Gene Silencing Drugs. *Biomaterials*, *240*, 119742. doi:10.1016/j.biomaterials.2019.119742 PMID:32088410

Kolosnjaj-Tabi, J., Di Corato, R., Lartigue, L., Marangon, I., Guardia, P., Silva, A. K. A., Luciani, N., Clement, O., Flaud, P., Singh, J. V., Decuzzi, P., Pellegrino, T., Wilhelm, C., & Gazeau, F. (2014). Heat-generating iron oxide nanocubes: Subtle "Destructurators" of the tumoral microenvironment. *ACS Nano*, *8*(5), 4268–4283. doi:10.1021/nn405356r PMID:24738788

Komarala, E. P., Nigam, S., Aslam, M., & Bahadur, D. (2016). In-vitro evaluation of layered double hydroxide–Fe3 O4 magnetic nanohybrids for thermochemotherapy. *New Journal of Chemistry*, *40*(1), 423–433. doi:10.1039/C5NJ01701G

Kong, C., Neoh, H. M., & Nathan, S. (2016). Targeting *Staphylococcus aureus* toxins: A potential form of anti-virulence therapy. *Toxins*, *8*(3), 72. doi:10.3390/toxins8030072 PMID:26999200

Kong, F.-Y., Zhang, J.-W., Li, R.-F., Wang, Z.-X., Wang, W.-J., & Wang, W. (2017). Unique Roles of Gold Nanoparticles in Drug Delivery, Targeting and Imaging Applications. *Molecules (Basel, Switzerland)*, *22*(9), 1445. doi:10.3390/molecules22091445 PMID:28858253

Kong, L., & Ziegler, G. R. (2012). Patents on fiber spinning from starches. *Recent Patents on Food, Nutrition & Agriculture*, *4*, 210–2019. doi:10.2174/2212798411204030210 PMID:23013410

Kong, M., Chen, X. G., Xing, K., & Park, H. J. (2010). Antimicrobial properties of chitosan and mode of action: A state of the art review. *International Journal of Food Microbiology*, *144*(1), 51–63. doi:10.1016/j.ijfoodmicro.2010.09.012 PMID:20951455

Koper, O. B., Klabunde, J. S., Marchin, G. L., Klabunde, K. J., Stoimenov, P., & Bohra, L. (2002). Nanoscale powders and formulations with biocidal activity toward spores and vegetative cells of bacillus species, viruses, and toxins. *Current Microbiology*, *44*(1), 49–55. doi:10.100700284-001-0073-x PMID:11727041

Kostevšek, N., Monaco, I., & Al-Jamalc, W. (2018). Magneto-liposomes as Thernostic agents. *12th International conference on Scientific and clinical applications of Magnetic carriers.*

Kotek, R. (2008). Recent Advances in Polymer Fibers. *Polymer Reviews (Philadelphia, Pa.)*, *48*(2), 221–229. doi:10.1080/15583720802020038

Kreuter, J. (1991). Peroral administration of nanoparticles. *Advanced Drug Delivery Reviews*, *7*(1), 71–86. doi:10.1016/0169-409X(91)90048-H

Kreyling, W. G., Semmler-Behnke, M., Seitz, J., Scymczak, W., Wenk, A., Mayer, P., Takenaka, S., & Oberdörster, G. (2009). Size dependence of the translocation of inhaled iridium and carbon nanoparticle aggregates from the lung of rats to the blood and secondary target organs. *Inhalation Toxicology*, *21*(sup1, Suppl 1), 55–60. doi:10.1080/08958370902942517 PMID:19558234

Krishnamoorthy, K., Manivannan, G., Kim, S. J., Jeyasubramanian, K., & Premanathan, M. (2012). Antibacterial activity of MgO nanoparticles based on lipid peroxidation by oxygen vacancy. *Journal of Nano Research*, 14.

Krishnamoorthy, P., & Jayalakshmi, T. (2012a). Preparation, characterization and synthesis of silver nanoparticles by using phyllanthusniruri for the antimicrobial activity and cytotoxic effects. *Journal of Chemical and Pharmaceutical Research*, *4*(11), 4783–4794.

Kruk, T., Szczepanowicz, K., Stefańska, J., Socha, R. P., & Warszyński, P. (2015). Synthesis and antimicrobial activity of monodisperse copper nanoparticles. *Colloids and Surfaces. B, Biointerfaces*, *128*, 17–22. doi:10.1016/j.colsurfb.2015.02.009 PubMed

Krzyściak, W., Pluskwa, K. K., Jurczak, A., & Kościelniak, D. (2013). The pathogenicity of the Streptococcus genus. *European Journal of Clinical Microbiology & Infectious Diseases*, *32*(11), 1361–1376. doi:10.1007/s10096-013-1914-9 PubMed

Kulkarni, P., Rawtani, D., & Barot, T. (2019). Formulation and optimization of long acting dual niosomes using box-Behnken experimental design method for combinative delivery of ethionamide and D-cycloserine in tuberculosis treatment. *Colloids and Surfaces. A, Physicochemical and Engineering Aspects*, *565*, 131–142. doi:10.1016/j.colsurfa.2019.01.004

Kulshrestha, S., Qayyum, S., & Khan, A. U. (2017). Antibiofilm efficacy of green synthesized graphene oxide-silver nanocomposite using *Lagerstroemia speciosa* floral extract: A comparative study on inhibition of gram-positive and gram-negative biofilms. *Microbial Pathogenesis*, *103*, 167–177. doi:10.1016/j.micpath.2016.12.022 PMID:28027944

Kumar, A., Bansal, A., Behera, B., Jain, S. L., & Ray, S. S. (2016). Ternary hybrid polymeric nanocomposites through grafting of polystyrene on graphene oxide-TiO2 by surface initiated atom transfer radical polymerization (SI-ATRP). *Materials Chemistry and Physics*, *172*, 189–196. doi:10.1016/j.matchemphys.2016.01.064

Kumar, A., Sahoo, B., Montpetit, A., Behera, S., Lockey, R. F., & Mohapatra, S. S. (2007). Development of hyaluronic acid–Fe2O3 hybrid magnetic nanoparticles for targeted delivery of peptides. *Nanomedicine; Nanotechnology, Biology, and Medicine*, *3*(2), 132–137. doi:10.1016/j.nano.2007.03.001 PMID:17572355

Kumari, M., Mukherjee, A., & Chandrasekaran, N. (2009). Genotoxicity of silver nanoparticles in *Allium cepa*. *The Science of the Total Environment*, *407*(19), 5243–5246. doi:10.1016/j.scitotenv.2009.06.024 PMID:19616276

Kumari, P., Panda, P. K., Ealisha, J., Khushboo, K., & Mallick, A. M. (2017). Mechanistic insight to ROS and apoptosis regulated cytotoxicity inferred by green synthesized CuO nanoparticles from *Calotropis gigantea* to embryonic Zebraish. *Scientific Reports*, *7*(1), 16284. doi:10.103841598-017-16581-1 PMID:29176605

Kumar, P. M., Murugan, K., Madhiyazhagan, P., Kovendan, K., Amerasan, D., Chandramohan, B., Dinesh, D., Suresh, U., Nicoletti, M., Alsalhi, M. S., Devanesan, S., Wei, H., Kalimuthu, K., Hwang, J. S., Lo Iacono, A., & Benelli, G. (2016). Biosynthesis, characterization, and acute toxicity of Berberis tinctoria-fabricated silver nanoparticles against the Asian tiger mosquito, Aedes albopictus, and the mosquito predators Toxorhynchites splendens and Mesocyclops thermocyclopoides. *Parasitology Research*, *115*(2), 751–759. doi:10.100700436-015-4799-y PMID:26499804

Kumar, R., Roopan, S. M., Prabhakar, A., Khanna, V. G., & Chakroborty, S. (2012b). Agricultural waste *Annona squamosa* peel extract: Biosynthesis of silver nanoparticles. *Spectrochimica Acta Part A*, *90*, 173–176. doi:10.1016/j.saa.2012.01.029 PMID:22336049

Kumar, S., Meena, V. K., Hazari, P. P., & Sharma, R. K. (2017). PEG coated and doxorubicin loaded multimodal Gadolinium oxide nanoparticles for simultaneous drug delivery and imaging applications. *International Journal of Pharmaceutics*, *527*(1-2), 142–150. doi:10.1016/j.ijpharm.2017.05.027 PMID:28506803

Kunjachan, S., Pola, R., Gremse, F., Theek, B., Ehling, J., Moeckel, D., ... Storm, G. (2014). Passive versus active tumor targeting using RGD-and NGR-modified polymeric nanomedicines. *Nano Letters*, *14*(2), 972–981. doi:10.1021/nl404391r PMID:24422585

Kuntzler, S. G., Almedia, A. C. A., Costa, J. A. V., & Morais, M. G. (2018). Polyhydroxybutyrate and phenolic compounds microalgae electrospun nanofibers: A novel nanomaterial with antibacterial activity. *International Journal of Biological Macromolecules*, *1*(113), 1008–1014. doi:10.1016/j.ijbiomac.2018.03.002 PMID:29505877

Kuo, C.-Y., Liu, T.-Y., Hardiansyah, A., & Chiu, W.-Y. (2016). Magnetically polymeric nanocarriers for targeting delivery of curcumin and hyperthermia treatments toward cancer cells. *Journal of Polymer Science. Part A, Polymer Chemistry*, *54*(17), 2706–2713. doi:10.1002/pola.28150

Kuppan, P., Vasanthan, K. S., Sundaramurthi, D. M., Krishnan, U., & Sethuraman, S. (2011). Development of poly (3-hydroxybutyrate-co-3-hydroxyvalerate) fibers for skin tissue engineering: Effects of topography, mechanical, and chemical stimuli. *Biomacromolecules*, *12*(9), 3156–3165. doi:10.1021/bm200618w PMID:21800891

Kuriyama, R., & Sakai, H. (1974). Role of tubulin-SH group in polymerization to microtubules. *Journal of Biochemistry*, *76*(3), 651–654. doi:10.1093/oxfordjournals.jbchem.a130609 PMID:4474165

Kutateladze, M., & Adamia, R. (2010). Bacteriophages as potential new therapeutics to replace or supplement antibiotics. *Trends in Biotechnology*, *28*(12), 591–595. doi:10.1016/j.tibtech.2010.08.001 PMID:20810181

Kuznetsov, A. A., Filippov, V. I., Alyautdin, R. N., Torshina, N. L., & Kuznetsov, O. A. (2001). Application of magnetic liposomes for magnetically guided transport of muscle relaxants and anti-cancer photodynamic drugs. *Journal of Magnetism and Magnetic Materials*, *225*(1-2), 95–100. doi:10.1016/S0304-8853(00)01235-X

Kwon, H., & Lok, A. S. (2011). Hepatitis B therapy. *Nature Reviews. Gastroenterology & Hepatology*, *8*(5), 275–284. doi:10.1038/nrgastro.2011.33 PMID:21423260

Kwon, S., Singh, R., Perez, R., Abou Neel, E., Kim, H., & Chrzanowski, W. (2013). Silica-based mesoporous nanoparticles for controlled drug delivery. *Journal of Tissue Engineering*, *4*, 4. doi:10.1177/2041731413503357 PMID:24020012

Kyziol, A., Michna, J., Morneo, I., Gamez, E., & Irusta, S. (2017). Preparation and characterization of electrospun alginate nanofibers loaded with ciprofloxacin hydrochloride. *European Polymer, 96*, 350–360.

L¨ubbe, A. S., Alexiou, C., & Bergemann, C. (2001). Clinical applications of magnetic drug targeting. *The Journal of Surgical Research*, *95*(2), 200–206. doi:10.1006/jsre.2000.6030 PMID:11162046

Labbate, M., Case, R. J., & Stokes, H. W. (2009). The integrin /gene cassette system: An active player in bacterial adaptation. *Methods in Molecular Biology (Clifton, N.J.)*, *532*, 103–125. doi:10.1007/978-1-60327-853-9_6 PMID:19271181

Labrie, S. J., Samson, J. E., & Moineau, S. (2010). Bacteriophage resistance mechanisms. *Nature Reviews. Microbiology*, *8*(5), 317–327. doi:10.1038/nrmicro2315 PMID:20348932

Lahive, E., Jurkschat, K., Shaw, B. J., Handy, R. D., Spurgeon, D. J., & Svendsen, C. (2014). Toxicity of cerium oxide nanoparticles to the earthworm Eisenia fetida: Subtle effects. *Environmental Chemistry*, *11*(3), 268. doi:10.1071/EN14028

Lancuski, A., Abu Ammar, A., Avrahami, R., Vilensky, R., Vasilyev, G., & Zussman, E. (2017). Design of starch formate compound fibers as encapsulation platform for biotherapeutics. *Carbohydrate Polymers*, *20*(158), 68–76. doi:10.1016/j.carbpol.2016.12.003 PMID:28024544

Lancuski, A., Vasilyev, G., Luc Putaux, J., & Zussman, E. (2015). Rheological properties and electrospinnability of high amylose starch in formic acid. *Biomacromolecules*, *16*(8), 2529–2536. doi:10.1021/acs.biomac.5b00817 PMID:26192477

Landsiedel, R., Sauer, U. G., Ma-Hock, L., Schnekenburger, J., & Wiemann, M. (2014). Pulmonary toxicity of nanomaterials: A critical comparison of published *in vitro* assays and *in vivo* inhalation or instillatiogn studies. *Nanomedicine (London)*, *9*(16), 2557–2585. doi:10.2217/nnm.14.149 PMID:25490426

Langer, R. (1990). New methods of drug delivery. *Science*, *249*(4976), 1527–1533. doi:10.1126cience.2218494 PMID:2218494

Lara, H. H., Ayala-Nunez, N. V., Turrent, L. C. I., & Padilla, C. R. (2010). Bactericidal effect of silver nanoparticles against multi-drug resistant bacteria. *World Journal of Microbiology & Biotechnology*, *26*(4), 615–621. doi:10.100711274-009-0211-3

Lasic, D. D. (1998). Novel applications of liposomes. *Trends in Biotechnology*, *16*(7), 307–321. doi:10.1016/S0167-7799(98)01220-7 PMID:9675915

Lau, N. S., Tsuge, T., & Sudesh, K. (2011). Formation of new polyhydroxyalkanoate containing 3-hydroxy-4-methylvalerate monomer in *Burkholderia* sp. *Applied Microbiology and Biotechnology*, *89*(5), 1599–1609. doi:10.100700253-011-3097-6 PMID:21279348

Laurent, S., Forge, D., Port, M., Roch, A., Robic, C., Vander Elst, L., & Muller, R. N. (2008). Magnetic iron oxide nanoparticles synthesis, stabilization, vectorization, physicochemical characterizations, and biological applications. *Chemical Reviews*, *108*(6), 2064–2110. doi:10.1021/cr068445e PMID:18543879

Laurent, S., & Mahmoudi, M. (2011). Superparamagnetic iron oxide nanoparticles: Promises for diagnosis and treatment of cancer. *International Journal of Molecular Epidemiology and Genetics*, *2*, 367–390. PMID:22199999

LaVan, D. A., McGuire, T., & Langer, R. (2003). Small-scale systems for in vivo drug delivery. *Nature Biotechnology*, *21*(10), 1184–1191. doi:10.1038/nbt876 PMID:14520404

Law, J. W., Ab Mutalib, N. S., Chan, K. G., & Lee, L. H. (2014). Rapid methods for the detection of foodborne bacterial pathogens: Principles, applications, advantages and limitations. *Frontiers in Microbiology*, *5*, 770. doi:10.3389/fmicb.2014.00770 PMID:25628612

Lawrie, C., Gal, S., Dunlop, H., Pushkaran, B., Liggins, A., Pulford, K., Banham, A. H., Pezzella, F., Boultwood, J., Wainscoat, J. S., Hatton, C. S., & Harris, A. L. (2008). Detection of elevated levels of tumour-associated microRNAs in serum of patients with diffuse large B-cell lymphoma. *British Journal of Haematology*, *141*, 672–675. PubMed

Laxminarayan, R., & Chaudhury, R. R. (2016). Antibiotic Resistance in India: Drivers and Opportunities for Action. *PLoS Medicine*, *13*(3), e1001974. doi:10.1371/journal.pmed.1001974 PMID:26934098

Lebeaux, D., Ghigo, J. M., & Beloin, C. (2014). Biofilm-related infections: Bridging the gap between clinical management and fundamental aspects of recalcitrance toward antibiotics. *Microbiology and Molecular Biology Reviews*, *78*(3), 510–543. doi:10.1128/MMBR.00013-14 PMID:25184564

Lebreton, F., Willems, R. J. L., & Gilmore, M. S. (2014). Enterococcus Diversity, Origins in Nature, and Gut Colonization. In M. S. Gilmore, D. B. Clewell, Y. Ike, & N. Shankar (Eds.), *Enterococci: From Commensals to Leading Causes of Drug Resistant Infection* (pp. 5–65). Massachusetts Eye and Ear Infirmary.

Lee, C.-S., Gong, J., Oh, D.-S., Jeon, J.-R., & Chang, Y.-S. (2018). Zerovalent-iron/platinum Janus micromotors with spatially separated functionalities for efficient water decontamination. *ACS Applied Nano Materials*, *1*(2), 768–776. doi:10.1021/acsanm.7b00223

Lee, C.-S., Kim, H., Yu, J., Su, H. Y., Ban, S., Oh, S., Jeong, D., Im, J., Baek, M. J., & Kim, T. H. (2017). Doxorubicin-loaded oligonucleotide conjugated gold nanoparticles: A promising in vivo drug delivery system for colorectal cancer therapy. *European Journal of Medicinal Chemistry*, *142*, 416–423. Advance online publication. doi:10.1016/j.ejmech.2017.08.063 PMID:28870452

Lee, D. E., Koo, H., Sun, I. C., Ryu, J. H., Kim, K., & Kwon, I. C. (2012). Multifunctional nanoparticles for multimodal imaging and theragnosis. *Chemical Society Reviews*, *41*(7), 2656–2672. doi:10.1039/C2CS15261D PMID:22189429

Lee, H. J., Lee, G., Jang, N. R., Yun, J. H., Song, J. Y., & Kim, B. S. (2011). Biological synthesis of copper nanoparticles using plants extract. *Nanotechnology*, *1*, 371–374.

Lee, J. E., Lee, N., Kim, T., Kim, J., & Hyeon, T. (2011). Multifunctional mesoporous silica nanocomposite nanoparticles for thernaoustic applications. *Accounts of Chemical Research*, *44*(10), 893–902. doi:10.1021/ar2000259 PMID:21848274

Lee, J. H., Chen, K. J., Noh, S. H., Garcia, M. A., Wang, H., Lin, W. Y., Jeong, H., Kong, B. J., Stout, D. B., Cheon, J., & Tseng, H. R. (2013). On-demand drug release system for in vivo cancer treatment through self- assembled magnetic nanoparticles. *Angewandte Chemie International Edition in English*, *52*(16), 4384–4388. doi:10.1002/anie.201207721 PMID:23519915

Lee, J., & Zhang, L. (2015). The hierarchy quorum sensing network in Pseudomonas aeruginosa. *Protein & Cell*, *6*(1), 26–41. doi:10.1007/s13238-014-0100-x PubMed

Lee, K. (2003). *Health impacts of globalization: towards global governance.* Palgrave Macmillan. doi:10.1057/9781403943828

Lee, K. Y., & Mooney, D. J. (2012). Alginate: Properties and biomedical applications. *Progress in Polymer Science, 37*(1), 106–126. doi:10.1016/j.progpolymsci.2011.06.003 PMID:22125349

Lee, K., & Yu, Y. (2017). Janus nanoparticles for T cell activation: Clustering ligands to enhance stimulation. *Journal of Materials Chemistry. B, Materials for Biology and Medicine, 5*(23), 4410–4415. doi:10.1039/C7TB00150A PMID:28966791

Lee, S., Bi, X., Reed, R. B., Ranville, J. F., Herckes, P., & Westerhoff, P. (2014). Nanoparticle size detection limits by single particle ICPMS for 40 elements. *Environmental Science & Technology, 48*(17), 10291–10300. doi:10.1021/es502422v PMID:25122540

Lei, J., Sun, L., Huang, S., Zhu, C., Li, P., He, J., ... He, Q. (2019). The antimicrobial peptides and their potential clinical applications. *American Journal of Translational Research, 11*(7), 3919. PMID:31396309

Lei, R., Wu, C., Yang, B., Ma, H., Shi, C., Wang, Q., Wang, Q., Yuan, Y., & Liao, M. (2008). Integrated metabolomic analysis of the nano-sized copper particle-induced hepatotoxicity and nephrotoxicity in rats: A rapid in vivo screening method for nanotoxicity. *Toxicology and Applied Pharmacology, 232*(2), 292–301. doi:10.1016/j.taap.2008.06.026 PMID:18706438

Leroux, S., & Ullmann, A. J. (2013). Management and diagnostic guidelines for fungal diseases in infectious diseases and clinical microbiology: Critical appraisal. *Clinical Microbiology and Infection, 19*(12), 1115–1121. doi:10.1111/1469-0691.12426 PMID:24118188

Lesniak, A., Salvati, A., Santos-Martinez, M. J., Radomski, M. W., Dawson, K. A., & Aberg, C. (2013). Nanoparticle adhesion to the cell membrane and its effect on nanoparticle uptake efficiency. *Journal of the American Chemical Society, 135*(4), 1438–1444. doi:10.1021/ja309812z PMID:23301582

Levy, S. B. (2007). *Antibiotic Resistance: An Ecological Imbalance, in Ciba Foundation Symposium 207 - Antibiotic Resistance: Origins, Evolution, Selection and Spread.* John Wiley & Sons, Ltd.

Lewis, K. (2005). Persister cells and the riddle of biofilm survival. *Biochemistry (Moscow), 70*(2), 267–274. doi:10.100710541-005-0111-6 PMID:15807669

Li, X., Xu, H., Chen, Z.-S., & Chen, G. (2011). Biosynthesis of Nanoparticles by Microorganisms and Their Applications. Journal of Nanomaterials, 2011, 270974. doi:10.1155/2011/270974

Liang, M., Chen, X., Xu, Y., Zhu, L., Jin, X., & Huang, C. (2017). Double-grooved nanofibre surfaces with enhanced anisotropic hydrophobicity. *Nanoscale, 9*(42), 16214–16222. doi:10.1039/C7NR05188C PMID:29043355

Li, C., Liu, H., Sun, Y., Wang, H., Guo, F., Rao, S., Deng, J., Zhang, Y., Miao, Y., Guo, C., Meng, J., Chen, X., Li, L., Li, D., Xu, H., Wang, H., Li, B., & Jiang, C. (2009). PAMAM nanoparticles promote acute lung injury by inducing autophagic cell death through the Akt- TSC2-mTOR signaling pathway. *Journal of Molecular Cell Biology, 1*(1), 37–45. doi:10.1093/jmcb/mjp002 PMID:19516051

Liibbe, A. S., & Bergemann, C. (1997). Scientific and Clinical Applications of Magnetic Carriers. Plenum Press.

Li, J. (2019). A new tool to attack biofilms: Driving magnetic iron-oxide nanoparticles to disrupt the atrix. *Nanoscale, 11*, 6905–6915. doi:10.1039/C8NR09802F PMID:30912773

Li, L. L., & Wang, H. (2018). Infection-targeted bactericidal nanoparticles. *Nature Biomedical Engineering, 2*(2), 56–57. doi:10.103841551-018-0199-9 PMID:31015624

Li, L., Gao, F., Jiang, W., Wu, X., Cai, Y., Tang, J., Gao, X., & Gao, F. (2016). Folic acid-conjugated superparamagnetic iron oxide nanoparticles for tumor-targeting MR imaging. *Drug Delivery, 23*(5), 1726–1733. doi:10.3109/10717544.2015.1006404 PMID:25715808

Li, L., Rashidi, L. H., Yao, M., Ma, L., Chen, L., Zhang, J., Zhang, Y., & Chen, W. (2017). CuS nanoagents for photodynamic and photothermal therapies: Phenomena and possible mechanisms. *Photodiagnosis and Photodynamic Therapy, 19*, 5–14. doi:10.1016/j.pdpdt.2017.04.001 PMID:28389371

Li, L., Wang, Y., Pan, L., Shi, Y., Cheng, W., Shi, Y., & Yu, G. A. (2015). Nanostructured conductive hydrogels-based biosensor platform for human metabolite detection. *Nano Letters, 15*(2), 1146–1151. doi:10.1021/nl504217p PMID:25569673

Li, M., Zhu, L., & Lin, D. (2011). Toxicity of ZnO nanoparticles to *Escherichia coli*: Mechanism and the influence of medium components. *Environmental Science & Technology, 45*(5), 1977–1983. doi:10.1021/es102624t PMID:21280647

Lima, E., Guerra, R., Lara, V., & Guzmán, A. (2013). Gold nanoparticles as efficient antimicrobial agents for Escherichia coli and Salmonella typhi. *Chemistry Central Journal, 7*(1), 11. doi:10.1186/1752-153X-7-11 PMID:23331621

Limbach, L. K., Li, Y., Grass, R. N., Brunner, T. J., Hintermann, M. A., Muller, M., Gunther, D., & Stark, W. J. (2005). Oxide nanoparticle uptake in human lung fibroblasts: Effects of particle size, agglomeration, and diffusion at low concentrations. *Environmental Science & Technology, 39*(23), 9370–9376. doi:10.1021/es051043o PMID:16382966

Lim, C. T. (2017). Nanofiber technology: Current status and emerging developments. *Progress in Polymer Science, 70*, 1–17. doi:10.1016/j.progpolymsci.2017.03.002

Lim, E. K., Chung, B. H., & Chung, S. J. (2018). Recent advances in pH-sensitive polymeric nanoparticles for smart drug delivery in cancer therapy. *Current Drug Targets, 19*(4), 300–317. doi:10.2174/1389450117666160602202339 PMID:27262486

Lim, J. A., Shin, H., Heu, S., & Ryu, S. (2014). Exogenous lytic activity of SPN9CC endolysin against gram-negative bacteria. *Journal of Microbiology and Biotechnology, 24*(6), 803–811. doi:10.4014/jmb.1403.03035 PMID:24690638

Lim, K. J., Sung, B. H., Shin, J. R., Lee, Y. W., Kim, D. J., Yang, K. S., & Kim, S. C. (2013). A cancer specific cell-penetrating peptide, BR2, for the efficient delivery of an scFv into cancer cells. *PLoS One, 8*(6), e66084. doi:10.1371/journal.pone.0066084 PMID:23776609

Li, N., Zhao, P., & Astruc, D. (2014). Anisotropic Gold Nanoparticles: Synthesis, Properties, Applications, and Toxicity. *Angewandte Chemie International Edition, 53*(7), 1756–1789. doi:10.1002/anie.201300441 PMID:24421264

Lipovsky, A., Nitzan, Y., Gedanken, A., & Lubart, R. (2011). Antifungal activity of ZnO nanoparticles - The role of ROS mediated cell injury. *Nanotechnology, 22*(10), 105101. doi:10.1088/0957-4484/22/10/105101 PMID:21289395

Li, Q., Chen, L., Gadinski, M. R., Zhang, S., Zhang, G., Li, H. U., Iagodkine, E., Haque, A., Chen, L.-Q., Jackson, T. N., & Wang, Q. (2015). Flexible high-temperature dielectric materials from polymer nanocomposites. *Nature, 523*(7562), 576–579. doi:10.1038/nature14647 PMID:26223625

Li, Q., Mahendra, S., Lyon, D. Y., Brunet, L., Liga, M. V., Li, D., & Alvarez, P. J. (2018). Antimicrobial nanomaterials for water disinfection and microbial control: Potential applications and implications. *Water Research, 42*(18), 4591–4602. doi:10.1016/j.watres.2008.08.015 PMID:18804836

Li, S., Gao, Y., Ding, Y., Xu, A., & Tan, H. (2020). Supramolecular nano drug delivery systems mediated via host-guest chemistry of cucurbit[n]uril (n = 6 and 7). *Chinese Chemical Letters*. Advance online publication. doi:10.1016/j.cclet.2020.04.049

Li, S., Lu, X., Li, X., Xue, Y., Zhang, C., Lei, J., & Wang, C. (2012). Preparation of bamboo-like ppy nanotubes and their application for removal of Cr(VI) ions in aqueous solution. *Journal of Colloid and Interface Science, 378*(1), 30–35. doi:10.1016/j.jcis.2012.03.065 PMID:22551477

Liu, Men, Perry, & Smith. (1998). *Nanoparticle and Iron Chelators as a Potential Novel Alzheimer Therapy*. Humana Press.

Liu, C.-H., Hsu, H.-L., Chen, J.-P., & Ma, Y.-H. (2018). Intravenous Delivery of Tissue Plasminogen Activator by Thermosensitive Magneto-liposomes for Target Thrombolysis. *12th International conference on Scientific and clinical applications of Magnetic carriers.*

Liu, G., Men, P., Harris, P. L. R., Rolston, R. K., Perry, G., & Smith, M. A. (2006). Nanoparticle iron chelators: A new therapeutic approach in Alzheimer disease and other neurologic disorders associated with trace metal imbalance. *Neuroscience Letters, 406*(3), 189–193. doi:10.1016/j.neulet.2006.07.020 PMID:16919875

Liu, H., Tu, Z., Feng, F., Shi, H., Chen, K., & Xu, X. (2015). Virosome, a hybrid vehicle for efficient and safe drug delivery and its emerging application in cancer treatment. *Acta Pharmaceutica (Zagreb, Croatia), 65*(2), 105–116. doi:10.1515/acph-2015-0019 PMID:26011928

Liu, J. L., Zhang, W. J., Li, X. D., Yang, N., Pan, W. S., Kong, J., & ... . (2016). Sustained-release genistein from nanostructured lipid carrier suppresses human lens epithelial cell growth. *International Journal of Ophthalmology, 9*(5), 643–649. PMID:27275415

Liu, M., Duan, X. P., Li, Y. M., Yang, D. P., & Long, Y. Z. (2017). Electrospunnanofibers for wound healing. *Materials Science and Engineering C, 76*, 1413–1423. doi:10.1016/j.msec.2017.03.034 PMID:28482508

Liu, X., Wu, L., Wang, L., & Jiang, W. (2018). A dual-targeting DNA tetrahedron nanocarrier for breast cancer cell imaging and drug delivery. *Talanta, 179*, 356–363. doi:10.1016/j.talanta.2017.11.034 PMID:29310244

Liu, Y., Busscher, H. J., Zhao, B., Li, Y., Zhang, Z., van der Mei, H. C., Ren, Y., & Shi, L. (2016). Surface-adaptive, antimicrobially loaded, micellar nanocarriers with enhanced penetration and killing efficiency in staphylococcal biofilms. *ACS Nano, 10*(4), 4779–4789. doi:10.1021/acsnano.6b01370 PMID:26998731

Liu, Y., Chen, X., & Xin, J. H. (2008). Hydrophobic duck feathers and their simulation on textile substrates for water repellent treatment. *Bioinspiration & Biomimetics, 3*(4), 046007. doi:10.1088/1748-3182/3/4/046007 PMID:18997276

Liu, Y., & He, J. H. (2010). Bubble electrospinning for mass production of nanofibers. *International Journal of Nonlinear Sciences and Numerical Simulation, 8*(3), 393–396.

Liu, Y., Ren, Y., Li, Y., Su, L., Zhang, Y., Huang, F., ... Shi, L. (2018). Nanocarriers with conjugated antimicrobials to eradicate pathogenic biofilms evaluated in murine in vivo and human ex vivo infection models. *Acta Biomaterialia, 79*, 331–343. doi:10.1016/j.actbio.2018.08.038 PMID:30172935

Liu, Y., Zhang, X., Xia, Y., & Yang, H. (2010). Magnetic-Field-Assisted Electrospinning of Aligned Straight and Wavy Polymeric Nanofibers. *Advanced Materials, 22*(22), 2454–2457. doi:10.1002/adma.200903870 PMID:20376855

Liu, Z., Jiang, W., Nam, J., Moon, J. J., & Kim, B. Y. S. (2018). Immunomodulating Nanomedicine for Cancer Therapy. *Nano Letters, 18*(11), 6655–6659. doi:10.1021/acs.nanolett.8b02340 PMID:30185039

Li, W. R., Xie, X. B., Shi, Q. S., Duan, S. S., Ouyang, Y. S., & Chen, Y. B. (2011). Antibacterial effect of silver nanoparticles on *Staphylococcus aureus. Biometals, 24*(1), 135–141. doi:10.100710534-010-9381-6 PMID:20938718

Li, W., Wu, X. R., Qin, H., Zhao, Z. Q., & Liu, H. W. (2016). Light-driven and lightguided microswimmers. *Advanced Functional Materials, 26*(18), 3164–3171. doi:10.1002/adfm.201505378

Li, X. Z., Plésiat, P., & Nikaido, H. (2015). The challenge of efflux-mediated antibiotic resistance in Gram-negative bacteria. *Clinical Microbiology Reviews, 28*(2), 337–418. doi:10.1128/CMR.00117-14 PMID:25788514

Li, X., Chen, H., & Yang, B. (2016). Centrifugally spun starch-based fibers from amylopectin rich starches. *Carbohydrate Polymers, 137*, 459–465. doi:10.1016/j.carbpol.2015.10.079 PMID:26686151

Li, X., Robinson, S. M., Gupta, A., Saha, K., Jiang, Z., Moyano, D. F., Sahar, A., Riley, M. A., & Rotello, V. M. (2014). Functional gold nanoparticles as potent antimicrobial agents against multi-drug-resistant bacteria. *ACS Nano*, *8*(10), 10682–10686. doi:10.1021/nn5042625 PMID:25232643

Li, X., Yang, Y., Jia, Y., Pu, X., Yang, T., Wang, Y., Ma, X., Chen, Q., Sun, M., Wei, D., Kuang, Y., Li, Y., & Liu, Y. (2017). Enhanced tumor targeting effects of a novel paclitaxel-loaded polymer: PEG-PCCL-modified magnetic iron oxide nanoparticles. *Drug Delivery*, *24*(1), 1284–1294. doi:10.1080/10717544.2017.1373167 PMID:28891337

Li, Y., Chang, Y., Lian, X., Zhou, L., Yu, Z., Wang, H., & An, F. (2018). Silver nanoparticles for enhanced cancer theranostics: In vitro and in vivo perspectives. *Journal of Biomedical Nanotechnology*, *14*(9), 1515–1542. doi:10.1166/jbn.2018.2614 PMID:29958548

Li, Y., Mou, F. Z., Chen, C. R., You, M., Yin, Y. X., Xu, L. L., & Guan, J. G. (2016). Light controlled bubble propulsion of amorphous TiO2/Au Janus micromotors. *RSC Advances*, *6*(13), 10697–10703. doi:10.1039/C5RA26798F

Li, Y., & Peng, N. (2019). Endogenous CRISPR-Cas System-Based Genome Editing and Antimicrobials: Review and Prospects. *Frontiers in Microbiology*, *10*, 2471. doi:10.3389/fmicb.2019.02471 PMID:31708910

Li, Y., Zhang, W., Niu, J., & Chen, Y. (2012). Mechanism of photogenerated reactive oxygen species and correlation with the antibacterial properties of engineered metal oxide nanoparticles. *ACS Nano*, *6*(6), 1–22. doi:10.1021/nn300934k PMID:22587225

Li, Z., Brecher, M., Deng, Y. Q., Zhang, J., Sakamuru, S., Liu, B., Huang, R., Koetzner, C. A., Allen, C. A., Jones, S. A., Chen, H., Zhang, N. N., Tian, M., Gao, F., Lin, Q., Banavali, N., Zhou, J., Boles, N., Xia, M., ... Li, H. (2017). Existing drugs as broad-spectrum and potent inhibitors for Zika virus by targeting NS2B-NS3 interaction. *Cell Research*, *27*(8), 1046–1064. doi:10.1038/cr.2017.88 PMID:28685770

Locatelli, E., & Franchini, M. C. (2012). Biodegradable PLGA-b-PEG polymeric nanoparticles: Synthesis, properties, and nanomedical applications as drug delivery system. *Journal of Nanoparticle Research*, *14*(12), 1. doi:10.100711051-012-1316-4

Lockman, P. R., Koziara, J. M., Mumper, R. J., & Allen, D. D. (2004). Nanoparticle surface charges alter blood–brain barrier integrity and permeability. *Journal of Drug Targeting*, *12*(9-10), 635–641. doi:10.1080/10611860400015936 PMID:15621689

Lodish, H., Berk, A., Zipursky, S. L., Matsudaira, P., Baltimore, D., & Darnell, J. (2000). *"Tumor cells and the onset of cancer," in Molecular Cell Biology* (4th ed.). W. H. Freeman.

Lohr, J. G., Adalsteinsson, V. A., Cibulskis, K., Choudhury, A. D., Rosenberg, M., Cruz-Gordillo, P., Francis, J. M., Zhang, C.-Z., Shalek, A. K., Satija, R., Trombetta, J. J., Lu, D., Tallapragada, N., Tahirova, N., Kim, S., Blumenstiel, B., Sougnez, C., Lowe, A., Wong, B., ... Love, J. C. (2014). Whole-exome sequencing of circulating tumor cells provides a window into metastatic prostate cancer. *Nature Biotechnology*, *32*(5), 479–484. doi:10.1038/nbt.2892 PMID:24752078

Lok, A. S., & McMahon, B. J. (2009). Chronic hepatitis B: Update. *Hepatology (Baltimore, Md.)*, *50*(3), 661–662. doi:10.1002/hep.23190 PMID:19714720

Long, R., Liu, Y., Wu, S., Wang, S., & Wang, P. (2017). Co-delivery of doxorubicin and small interfering RNA with genipin-cross-linked iron (III) oxide-polyetherimide nanoparticles for combined cancer therapy. *Nanomaterials Nanotechnology.*, *7*, 184798041770409. doi:10.1177/1847980417704094

Look, M., Bandyopadhyay, A., Blum, J. S., & Fahmy, T. M. (2010). Application of nanotechnologies for improved immune response against infectious diseases in the developing world. *Advanced Drug Delivery Reviews*, *62*(4-5), 378–393. doi:10.1016/j.addr.2009.11.011 PMID:19922750

Lopez-Abarrategui, C., Figueroa-Espi, V., Lugo-Alvarez, M. B., Pereira, C. D., Garay, H., Barbosa, J. A., ... Franco, O. L. (2016). The intrinsic antimicrobial activity of citric acid-coated manganese ferrite nanoparticles is enhanced after conjugation with the antifungal peptide Cm-p5. *International Journal of Nanomedicine*, *11*, 3849–3857. doi:10.2147/IJN.S107561 PMID:27563243

Lopez-Carrizales, M., Velasco, K. I., Castillo, C., Flores, A., Magana, M., Martinez-Castanon, G. A., & Martinez-Gutierrez, F. (2018). In Vitro Synergism of Silver Nanoparticles with Antibiotics as an Alternative Treatment in Multiresistant Uropathogens. *Antibiotics (Basel, Switzerland)*, *7*(2), 50. Advance online publication. doi:10.3390/antibiotics7020050 PMID:29921822

Love, M. J., Bhandari, D., Dobson, R. C. J., & Billington, C. (2018). Potential for Bacteriophage Endolysins to Supplement or Replace Antibiotics in Food Production and Clinical Care. *Antibiotics (Basel, Switzerland)*, *7*(1), 17. Advance online publication. doi:10.3390/antibiotics7010017 PMID:29495476

Lovleva, A., & Bonomo, R. A. (2017). The Ecology of Extended-Spectrum β-lactamases (ESBLs) in the Developed World. *Journal of Travel Medicine*, *24*(suppl_1), S44–S51. doi:10.1093/jtm/taw102 PMID:28521000

Lowy, F. D. (2003). Antimicrobial resistance: The example of *Staphylococcus aureus*. *The Journal of Clinical Investigation*, *111*(9), 1265–1273. doi:10.1172/JCI18535 PMID:12727914

Lu, B., Huang, X., Mo, J., & Zhao, W. (2016). Drug delivery using nanoparticles for cancer stem-like cell targeting. *Frontiers in Pharmacology*, *7*(065103), 84. doi:10.3389/fphar.2016.00084 PMID:27148051

Lubbe, A. S., Bergemann, C., Huhnt, W., Fricke, T., & Riess, H. (1996). Preclinical experiences with magnetic drug targeting: Tolerance and efficacy. *Cancer Research*, *56*, 4694–4701. PMID:8840986

Lü, J. M., Wang, X., Marin-Muller, C., Wang, H., Lin, P. H., Yao, Q., & Chen, C. (2009). Current advances in research and clinical applications of PLGA-based nanotechnology. *Expert Review of Molecular Diagnostics*, *9*(4), 325–341. doi:10.1586/erm.09.15 PMID:19435455

Lu, J., Struewing, I., Buse, H. Y., Kou, J., Shuman, H. A., Faucher, S. P., & Ashbolt, N. J. (2013). Legionella pneumophila transcriptional response following exposure to CuO nanoparticles. *Applied and Environmental Microbiology*, *79*(8), 2713–2720. doi:10.1128/AEM.03462-12 PubMed

Luo, L., Wu, Y., Liu, C., Huang, L., Zou, Y., Shen, Y., & Lin, Q. (2019). Designing soluble soybean polysaccharides-based nanoparticles to improve sustained antimicrobial activity of nisin. *Carbohydrate Polymers*, *225*, 115251. doi:10.1016/j.carbpol.2019.115251 PMID:31521298

Luo, X., Peng, X., Hou, J., Wu, S., Shen, J., & Wang, L. (2017). Folic acid-functionalized polyethylenimine superparamagnetic iron oxide nanoparticles as theranostic agents for magnetic resonance imaging and PD-L1 siRNA delivery for gastric cancer. *International Journal of Nanomedicine*, *12*, 5331–5343. doi:10.2147/IJN.S137245 PMID:28794626

Luo, X., Wang, Y., Lin, H., & Qu, F. (2016). DOX-Fe3 O4 @mSiO2 -PO-FA nanocomposite for synergistic chemo- and photothermal therapy. *RSC Advances*, *6*(113), 112232–112240. doi:10.1039/C6RA23292B

Lu, P., & Ding, B. (2008). Applications of electrospun fibers. *Recent Patents on Nanotechnology*, *2*(3), 169–182. doi:10.2174/187221008786369688 PMID:19076051

Lu, T. K., & Collins, J. J. (2007). Dispersing biofilms with engineered enzymatic bacteriophage. *Proceedings of the National Academy of Sciences of the United States of America*, *104*(27), 11197–11202. doi:10.1073/pnas.0704624104 PMID:17592147

Lu, T. K., & Collins, J. J. (2009). Engineered bacteriophage targeting gene networks as adjuvants for antibiotic therapy. *Proceedings of the National Academy of Sciences of the United States of America*, *106*(12), 4629–4634. doi:10.1073/pnas.0800442106 PMID:19255432

Luten, J., van Nostrum, C. F., De Smedt, S. C., & Hennink, W. E. (2008). Biodegradable polymers as non-viral carriers for plasmid DNA delivery. *Journal of Controlled Release*, *126*(2), 97–110. doi:10.1016/j.jconrel.2007.10.028 PMID:18201788

Lux Research. (2016). Revenue from nanotechnology (Figure 11). In Nanotechnology Update: U.S. Leads in Government Spending Amidst Increased Spending Across Asia. Lux Research, Inc.

Lu, X., Wang, L., Yang, Z., & Lu, H. (2013). Strategies of polyhydroxyalkanoates modification for the medical application in neural regeneration/nerve tissue engineering. *Advances in Bioscience and Biotechnology*, *4*(6), 731–740. doi:10.4236/abb.2013.46097

Lu, Y. T., Zhao, L., Shen, Q., Garcia, M. A., Wu, D., Hou, S., Song, M., Xu, X., OuYang, W.-H., OuYang, W. W.-L., Lichterman, J., Luo, Z., Xuan, X., Huang, J., Chung, L. W. K., Rettig, M., Tseng, H.-R., Shao, C., & Posadas, E. M. (2013). NanoVelcro Chip for CTC enumeration in prostate cancer patients. *Methods (San Diego, Calif.)*, *64*(2), 144–152. doi:10.1016/j.ymeth.2013.06.019 PMID:23816790

Lynch, I., Cedervall, T., Lundqvist, M., Cabaleiro-Lago, C., Linse, S., & Dawson, K. A. (2007). The nanoparticle-protein complex as a biological entity; a complex fluids and surface science challenge for the 21st century. *Advances in Colloid and Interface Science*, *135*, 167–174. doi:10.1016/j.cis.2007.04.021 PMID:17574200

Maarseveen, N., & Boucher, C. (2006). *Resistance to protease inhibitors*. Antiretroviral Resistance in Clinical Practice.

Mabena, L. F., Ray, S. S., Mhlanga, S. D., & Coville, N. J. (2011). Nitrogen-doped carbon nanotubes as a metal catalyst support. *Applied Nanoscience*, *1*(2), 67–77. doi:10.100713204-011-0013-4

Ma, C., White, J. C., Dhankher, O. P., & Xing, B. (2015). Metal-based nanotoxicity and detoxification pathways in higher plants. *Environmental Science & Technology*, *49*(12), 7109–7122. doi:10.1021/acs.est.5b00685 PMID:25974388

Machelart, A., Salzano, G., Li, X., Demars, A., Debrie, A. S., Menendez-Miranda, M., ... Belhaouane, I. (2019). Intrinsic antibacterial activity of nanoparticles made of β-cyclodextrins potentiates their effect as drug nanocarriers against tuberculosis. *ACS Nano*, *13*(4), 3992–4007. doi:10.1021/acsnano.8b07902 PMID:30822386

Mackevica, A., & Foss Hansen, S. (2016). Release of nanomaterials from solid nano-composites and consumer exposure assessment—A forward-looking review. *Nanotoxicology*, *10*(6), 641–653. doi:10.3109/17435390.2015.1132346 PMID:26667577

MacLean, E., Broger, T., Yerlikaya, S., Fernandez-Carballo, B. L., Pai, M., & Denkinger, C. M. (2019). A systematic review of biomarkers to detect active tuberculosis. *Nature Microbiology*, *4*(5), 748–758. doi:10.103841564-019-0380-2 PMID:30804546

Madaan, K., Kumar, S., Poonia, N., Lather, V., & Pandita, D. (2014). Dendrimers in drug delivery and targeting: Drug-dendrimer interactions and toxicity issues. *Journal of Pharmacy & Bioallied Sciences*, *6*(3), 139–150. doi:10.4103/0975-7406.130965 PMID:25035633

Madhiyazhagan, P., Murugan, K., Kumar, A. N., Nataraj, T., Dinesh, D., Panneerselvam, C., Subramaniam, J., Mahesh Kumar, P., Suresh, U., Roni, M., Nicoletti, M., Alarfaj, A. A., Higuchi, A., Munusamy, M. A., & Benelli, G. (2015). S argassum muticum-synthesized silver nanoparticles: An effective control tool against mosquito vectors and bacterial pathogens. *Parasitology Research*, *114*(11), 4305–4317. doi:10.100700436-015-4671-0 PMID:26281786

Madhumathi, K., Sudheesh Kumar, P. T., Abhilash, S., Sreeja, V., Tamura, H., Manzoor, K., Nair, S. V., & Jayakumar, R. (2010). Development of novel chitin/nanosilver composite scaffolds for wound dressing applications. *Journal of Materials Science. Materials in Medicine*, *21*(2), 807–813. doi:10.100710856-009-3877-z PMID:19802687

Magdanz, V., Sanchez, S., & Schmidt, O. G. (2013). Development of a sperm-flagella driven micro-bio-robot. *Advanced materials (Deerfield Beach, Fla.)*, *25*(45), 6581–6588. doi:10.1002/adma.201302544 PMID:23996782

Magnusson, K.-E., & Bayer, M. E. (1982). Anionic sites on the envelope of Salmonella typhimurium mapped with cationized ferritin. *Cell Biophysics*, *4*(2-3), 163–175. doi:10.1007/BF02918311 PubMed

Mahlapuu, M., Hakansson, J., Ringstad, L., & Bjorn, C. (2016). Antimicrobial Peptides: An Emerging Category of Therapeutic Agents. *Frontiers in Cellular and Infection Microbiology*, *6*, 194. doi:10.3389/fcimb.2016.00194 PMID:28083516

Mahmoudi, M., Shokrgozar, M. A., Simchi, A., Imani, M., Milani, A. S., Stroeve, P., Vali, H., Hafeli, U. O., & Bonakdar, S. (2009). Multiphysics flow modeling and in vitro toxicity of iron oxide nanoparticles coated with poly(vinyl alcohol). *The Journal of Physical Chemistry C*, *113*(6), 2322–2331. doi:10.1021/jp809453v

Mahmoudi, M., Yu, M., Serpooshan, V., Wu, J. C., Langer, R., Lee, R. T., Karp, J. M., & Farokhzad, O. C. (2017). Multiscale technologies for treatment of ischemic cardiomyopathy. *Nature Nanotechnology*, *12*(9), 845–855. doi:10.1038/nnano.2017.167 PMID:28875984

Maiti, D., Saha, A., & Devi, P. S. (2016). Surface modified multifunctional ZnFe2O4 nanoparticles for hydrophobic and hydrophilic anti-cancer drug molecule loading. *Physical Chemistry Chemical Physics*, *18*(3), 1439–1450. doi:10.1039/C5CP05840F PMID:26524183

Ma, J., Mercer, R. R., Barger, M., Schwegler-Berry, D., Cohen, J. M., Demokritou, P., & Castranova, V. (2015). Effects of amorphous silica coating on cerium oxide nanoparticles induced pulmonaryresponses. *Toxicology and Applied Pharmacology*, *288*(1), 63–73. doi:10.1016/j.taap.2015.07.012 PMID:26210349

Ma, J., Zhang, Y., Chen, X., Jin, Y., Chen, D., Wu, Y., Cui, J., Wang, H., Liu, J., Li, N., & Gao, F. (2013). Association of Preexisting Drug-Resistance Mutations and Treatment Failure in Hepatitis B Patients. *PLoS One*, *8*(7), e67606. doi:10.1371/journal.pone.0067606 PMID:23935839

Majidi, S. S., Slemming-Adamsen, P., Hanif, M., Zhang, Z., Wang, Z., & Chen, M. (2018). Wet electrospun alginate/gelatin hydrogel nanofibers for 3D cell culture. *International Journal of Biological Macromolecules*, *118*, 1648–1654. doi:10.1016/j.ijbiomac.2018.07.005 PMID:29981331

Ma, K., Dong, P., Liang, M., Yu, S., Chen, Y., & Wang, F. (2020). Facile Assembly of Multifunctional Antibacterial Nanoplatform Leveraging Synergistic Sensitization between Silver Nanostructure and Vancomycin. *ACS Applied Materials & Interfaces*, *12*(6), 6955–6965. doi:10.1021/acsami.9b22043 PMID:31977179

Manchanda, V., Sanchaita, S., & Singh, N. P. (2010). Multidrug resistant acinetobacter. *Journal of Global Infectious Diseases*, *2*(3), 291. doi:10.4103/0974-777X.68538 PMID:20927292

Mandal, A. K. (2017). Copper Nanomaterials as Drug Delivery System against Infectious Agents and Cancerous Cells. JALSI, 15(4), 1–8. doi:10.9734/JALSI/2017/38444

Mandali, P. K., & Chand, D. K. (2011). Palladium nanoparticles catalysed Suzuki cross-coupling reactions in ambient conditions. *Catalysis Communications*, *31*, 16–20. doi:10.1016/j.catcom.2012.10.020

Mandli, J., Mohammadi, H., & Amine, A. (2017). Electrochemical DNA sandwich biosensor based on enzyme amplified microRNA-21 detection and gold nanoparticles. *Bioelectrochemistry (Amsterdam, Netherlands)*, *116*, 17–23. PubMed

Mane, P. C., Chaudhari, R. D., Shinde, M. D., Kadam, D. D., Song, C. K., Amalnerkar, D. P., & Lee, H. (2017). Designing Ecofriendly Bionanocomposite Assembly with Improved Antimicrobial and Potent on-site Zika Virus Vector Larvicidal Activities with its Mode of Action. *Scientific Reports*, *7*(1), 15531. doi:10.103841598-017-15537-9 PMID:29138496

Mann, A. P., Scodeller, P., Hussain, S., Joo, J., Kwon, E., Braun, G. B., ... Krajewski, S. (2016). A peptide for targeted, systemic delivery of imaging and therapeutic compounds into acute brain injuries. *Nature Communications*, *7*(1), 1–11. doi:10.1038/ncomms11980 PMID:27351915

Mann, J. (1999). Fighting bacteria: the mould that grew on sewage: evolution of the cephalosporins. In *The Elusive Magic Bullet: The Search for the Perfect Drug* (pp. 57–58). Oxford University Press.

Mano, N., & Heller, A. (2005). Bioelectrochemical Propulsion. *Journal of the American Chemical Society*, *127*(33), 11574–11575. doi:10.1021/ja053937e PMID:16104713

Man, S. M. (2011). The clinical importance of emerging Campylobacter species. *Nature Reviews. Gastroenterology & Hepatology*, *8*(12), 669–685. doi:10.1038/nrgastro.2011.191 PMID:22025030

Mansfield, C. M., Alloy, M. M., Hamilton, J., Verbeck, G. F., Newton, K., Klaine, S. J., & Roberts, A. P. (2015). Photo-induced toxicity of titanium dioxide nanoparticles to Daphnia magna under natural sunlight. *Chemosphere*, *120*, 206–210. doi:10.1016/j.chemosphere.2014.06.075 PMID:25062026

Manshadi, M. K. D., Saadat, M., Mohammadi, M., Kamali, R., Shamsi, M., Naseh, M., & Sanati-Nezhad, A. (2019). Magnetic aerosol drug targeting in lung cancer therapy using permanent magnet. *Drug Delivery*, *26*(1), 120–128. doi:10.1080/10717544.2018.1561765 PMID:30798633

Mansha, M., Qurashi, A., Ullah, N., Bakare, F. O., Khan, I., & Yamani, Z. H. (2016). Synthesis of In2O3/graphene heterostructure and their hydrogen gas sensing properties. *Ceramics International*, *42*(9), 11490–11495. doi:10.1016/j.ceramint.2016.04.035

Mansha, M., Qurashi, A., Ullah, N., Bakare, F. O., Khan, I., & Yamani, Z. H. (2017). Synthesis, characterization and visible-light-driven photoelectrochemical hydrogen evolution reaction of carbazole-containing conjugated polymers. *International Journal of Hydrogen Energy*, *42*(16), 10952–10961. doi:10.1016/j.ijhydene.2017.02.053

Mantravadi, P. K., Kalesh, K. A., Dobson, R. C., Hudson, A. O., & Parthasarathy, A. (2019). The quest for novel antimicrobial compounds: Emerging trends in research, development, and technologies. *Antibiotics (Basel, Switzerland)*, *8*(1), 8. doi:10.3390/antibiotics8010008 PMID:30682820

Manzoor-ul-Haq, Rathod, V., Singh, D., Singh, A. K., Ninganagouda, S., & Hiremath, J. (2015). Dried Mushroom *Agaricus bisporus* mediated synthesis of silver nanoparticles from Bandipora District (Jammu and Kashmir) and their efficacy against Methicillin Resistant *Staphylococcus aureus* (MRSA) strains. *Nanoscience and Nanotechnology*, *5*(1), 1–8. PMID:26328301

Ma, P., & Zhang, R. (1999). Synthetic nano-scale fibrous extracellular matrix. *Journal of Biomedical Materials Research*, *46*(1), 60–72. doi:10.1002/(SICI)1097-4636(199907)46:1<60::AID-JBM7>3.0.CO;2-H PMID:10357136

Marasco, D., Perretta, G., Sabatella, M., & Ruvo, M. (2008). Past and future perspectives of synthetic peptide libraries. *Current Protein & Peptide Science*, *9*(5), 447–467. doi:10.2174/138920308785915209 PMID:18855697

Marchant, G. E., Sylvester, D. J., & Abbott, K. W. (2008). Risk management principles for nanotechnology. *NanoEthics*, *2*(1), 43–60. doi:10.100711569-008-0028-9

Maretti, E., Costantino, L., Rustichelli, C., Leo, E., Croce, M. A., Buttini, F., Truzzi, E., & Iannuccelli, V. (2017). Surface engineering of Solid Lipid Nanoparticle assemblies by methyl α-d-mannopyranoside for the active targeting to macrophages in anti-tuberculosis inhalation therapy. *International Journal of Pharmaceutics*, *528*(1-2), 440–451. doi:10.1016/j.ijpharm.2017.06.045 PMID:28624659

Mariana, C., Alina, M. H., Carmen, C., Grigore, M., & Veronica, L. (2016). *Antibiotic Drug Delivery Systems for the Intracellular Targeting of Bacterial Pathogens*. Smart Drug Delivery System INTECH. Retrieved from https://www.intechopen.com/books/smart-drug-delivery-system

Marshall, B. M., & Levy, S. B. (2011). Food animals and antimicrobials: Impacts on human health. *Clinical Microbiology Reviews*, *24*(4), 718–733. doi:10.1128/CMR.00002-11 PMID:21976606

Marta, B., Jakab, E., Potara, M., Simon, T., Imre-Lucaci, F., Barbu-Tudoran, L., Popescu, O., & Astilean, S. (2014). Pluronic-coated silver nanoprisms: Synthesis, characterization and their antibacterial activity. *Colloids and Surfaces. A, Physicochemical and Engineering Aspects*, *441*, 77–83. doi:10.1016/j.colsurfa.2013.08.076

Martel, S., Mohammadi, M., Felfoul, O., Lu, Z., & Pouponneau, P. (2009). Flagellated Magnetotactic Bacteria as Controlled MRI-trackable Propulsion and Steering Systems for Medical Nanorobots Operating in the Human Microvasculature. *The International Journal of Robotics Research*, *28*(4), 571–582. doi:10.1177/0278364908100924 PMID:19890435

Martin, C. (1996). Membrane-based of nanomaterials. *Chemistry of Materials*, *8*(43), 1739–1746. doi:10.1021/cm960166s

Martínez-Carmona, M., Gun'ko, Y. K., & Vallet-Regí, M. (2018). Mesoporous silica materials as drug delivery: "The Nightmare" of bacterial infection. *Pharmaceutics*, *10*(4), 279. doi:10.3390/pharmaceutics10040279 PMID:30558308

Martinez, J. L., & Baquero, F. (2018). Mutation frequencies and antibiotic resistance. *Antimicrobial Agents and Chemotherapy*, *44*(7), 1771–1777. doi:10.1128/AAC.44.7.1771-1777.2000 PMID:10858329

Martins, A., Chung, S., Pedro, A. J., Sousa, R. A., Marques, A. P., Reis, R. L., & Neves, N. M. (2009). Hierarchical starch-based fibrous scaffold for bone tissue engineering applications. *Journal of Tissue Engineering and Regenerative Medicine*, *3*(1), 37–42. PMID:19021239

Martins, R., Baptista, P., Raniero, L., Doria, G., Silva, L., Franco, R., & Fortunato, E. (2007). Amorphous/nanocrystalline silicon biosensor for the specific identification of unamplified nucleic acid sequences using gold nanoparticle probes. *Applied Physics Letters*, *90*(2), 023903. doi:10.1063/1.2431449

Maruyama, C. R., Guilger, M., Pascoli, M., Bileshy-José, N., Abhilash, P., Fraceto, L. F., & de Lima, R. (2016). Nanoparticles based on chitosan as carriers for the combined herbicides imazapic and imazapyr. *Scientific Reports*, *6*(1), 19768. doi:10.1038rep19768 PMID:26813942

Mashaghi, S., Jadidi, T., Koenderink, G., & Mashaghi, A. (2013). Lipid Nanotechnology. *International Journal of Molecular Sciences*, *14*(2), 4242–4282. doi:10.3390/ijms14024242 PMID:23429269

Masri, A., Anwar, A., Ahmed, D., Siddiqui, R. B., Raza Shah, M., & Khan, N. A. (2018). Silver nanoparticle conjugation-enhanced antibacterial efficacy of clinically approved drugs cephradine and vildagliptin. *Antibiotics (Basel, Switzerland)*, *7*(4), 100. doi:10.3390/antibiotics7040100 PMID:30445704

Mastrangelo, E., Pezzullo, M., De Burghgraeve, T., Kaptein, S., Pastorino, B., Dallmeier, K., de Lamballerie, X., Neyts, J., Hanson, A. M., Frick, D. N., Bolognesi, M., & Milani, M. (2012). Ivermectin is a potent inhibitor of flavivirus replication specifically targeting NS3 helicase activity: New prospects for an old drug. *The Journal of Antimicrobial Chemotherapy*, *67*(8), 1884–1894. doi:10.1093/jac/dks147 PMID:22535622

Matthew, D. (2018). *Externally Induced Drug Release Systems with Magnetic Nanoparticle Carriers: An Emerging Field in Nanomedicine*. Adv. Therap.

Matyjiaszewski, K. (2012). Atom transfer radial polymerization (ATRP): Current status and future perspectives. *Macromolecules*, *45*(10), 4015–4039. doi:10.1021/ma3001719

Maurer-Jones, M. A., Ian, L., Gunsolus, I. L., Murphy, C. J., & Haynes, C. L. (2013). Toxicity of engineered nanoparticles in the environment. *Analytical Chemistry*, *85*(6), 3036–3049. doi:10.1021/ac303636s PMID:23427995

Maurer, L., & Meyer, J. (2016). A Systematic Review of Evidence for Silver Nanoparticle-Induced Mitochondrial Toxicity. *Environmental Science. Nano*, *3*(2), 311–322. Advance online publication. doi:10.1039/C5EN00187K

Mavon, A., Miquel, C., Lejeune, O., Payre, B., & Moretto, P. (2007). In vitro percutaneous absorption and in vivo stratum corneum distribution of an organic and a mineral sunscreen. *Skin Pharmacology and Physiology*, *20*(1), 10–20. doi:10.1159/000096167 PMID:17035717

Mayes, J. M. (2011). Gum tragacanth and karaya. In A. Imeson (Ed.), *Food stabilisers, thickeners and gelling agents Singapore* (pp. 167–179). John Wiley & Sons.

Mazel, D. (2006). Integrons: Agents of bacterial evolution. *Nature Reviews. Microbiology*, *4*(8), 608–620. doi:10.1038/nrmicro1462 PMID:16845431

McBain, A. J., & Gilbert, P. (2001). Biocide tolerance and the harbingers of doom. *International Biodeterioration & Biodegradation*, *47*(2), 55–61. doi:10.1016/S0964-8305(01)00037-3

McClements, D., & Xiao, H. (2017). Is nano safe in foods? Establishing the factors impacting the gastrointestinal fate and toxicity of organic and inorganic food-grade nanoparticles. *NPJ Science of Food, 1*. doi:10.103841538-017-0005-1

McDonnell, G. E. (2007). Chemical Disinfection. In *Antisepsis, disinfection and sterilization* (pp. 111–115). STERIS Limited.

McEwen, S. A. (2006). Antibiotic use in animal agriculture: What have we learned and where are we going? *Animal Biotechnology*, *17*(2), 239–250. doi:10.1080/10495390600957233 PMID:17127534

McKee, M. S., & Filser, J. (2016). Impacts of metal-based engineered nanomaterials on soil communities. *Environmental Science: Nano, 3*(3), 506-533. doi:10.1039/C6EN00007J

McMahon, H., & Boucrot, E. (2011). Molecular mechanism and physiological functions of clathrin-mediated endocytosis. *Nature Reviews. Molecular Cell Biology*, *12*(8), 517–533. doi:10.1038/nrm3151 PMID:21779028

McNamara, K., & Tofail, S. A. (2017). Nanoparticles in biomedical applications. *Advances in Physics: X*, *2*(1), 54–88. doi:10.1080/23746149.2016.1254570

McNeill, W. H. (1976). Plagues and peoples. Bantam, Doubleday Dell Publishing Group, Inc.

McNeil, S. E. (2009). Nanoparticle therapeutics: A personal perspective. *Wiley Interdisciplinary Reviews*, *1*(3), 264–271. PMID:20049796

Medeiros, E. S., Glenn, G. M., Klamczynski, A., Orts, W. J., & Mattoso, L. H. C. (2009). Solution blow spinning: A new method to produce micro-and nanofibers from polymer solutions. *Journal of Applied Polymer Science*, *113*(4), 2322–2330. doi:10.1002/app.30275

Medina, P.P., & Slack, F.J. (2008). MicroRNAs and cancer: an overview. *Cell Cycle, 7*, 2485-2492.

Medina, E., Goldmann, O., Toppel, A. W., & Chhatwal, G. S. (2003). Survival of Streptococcus pyogenes within host phagocytic cells: A pathogenic mechanism for persistence and systemic invasion. *The Journal of Infectious Diseases*, *187*(4), 597–603. doi:10.1086/373998 PubMed

Mehanna, M. M., Mohyeldin, S. M., & Elgindy, N. A. (2019). Rifampicin-Carbohydrate Spray-Dried Nanocomposite: A Futuristic Multiparticulate Platform For Pulmonary Delivery. *International Journal of Nanomedicine*, *14*, 9089–9112. doi:10.2147/IJN.S211182 PMID:31819421

Mehmood, A., Ghafar, H., Yaqoob, S., Gohar, U. F., & Ahmad, B. (2017). Mesoporous silica nanoparticles: A review. *Journal of Developing Drugs*, *6*(02). Advance online publication. doi:10.4172/2329-6631.1000174

Mehrad, B., Clark, N. M., Zhanel, G. G., & Lynch, J. P. III. (2015). Antimicrobial Resistance in Hospital-Acquired Gram-Negative Bacterial Infections. *Chest*, *147*(5), 1413–1421. doi:10.1378/chest.14-2171 PMID:25940252

Mehta, M., Sharma, N., Vyas, M., Khurana, N., Maurya, P. K., Singh, H., ... Wadhwa, R. (2019). Interactions with the macrophages: An emerging targeted approach using novel drug delivery systems in respiratory diseases. *Chemico-Biological Interactions*, *304*, 10–19. doi:10.1016/j.cbi.2019.02.021 PMID:30849336

Mehta, S. K., & Jindal, N. (2013). Formulation of Tyloxapol niosomes for encapsulation, stabilization and dissolution of anti-tubercular drugs. *Colloids and Surfaces. B, Biointerfaces*, *101*, 434–441. doi:10.1016/j.colsurfb.2012.07.006 PMID:23010052

Mei, Y., Solovev, A. A., Sanchez, S., & Schmidt, O. G. (2011). Rolled-up nanotech on polymers: From basic perception to self-propelled catalytic microengines. *Chemical Society Reviews*, *40*(5), 2109. doi:10.1039/c0cs00078g PMID:21340080

Meng, C., Tian, J., Li, Y., & Zheng, S. (2010). Inñuence of native bacterial magnetic particles on mouse immune response. Wei Sheng Wu Xue Bao, 50, 817–821. PubMed

Menon, J.U., Kuriakose, A., & Iyer, R. (2017). Dual-Drug Containing Core-Shell Nanoparticles for Lung Cancer Therapy. *Sci Rep.*, *7*(1), 13249. . doi:10.103841598-017-13320-4

Menon, M. P., & Selvakumar, R. (2017). Extraction and modification of cellulose nanofibers derived from biomass for environmental application. *RSC Advances*, *7*(68), 42750–42773. doi:10.1039/C7RA06713E

Mérida, F., Juan, E., & Torres-Lugo, M. (2018). Ultrasonic Potentiation of Magnetic Fluid Hyperthermia/Pifithrin-μ Combined Therapies in Ovarian Cancer Cells. *12th International conference on Scientific and clinical applications of Magnetic carriers.*

Merino, S., Camprubi, S., Alberti, S., Benedi, V. J., & Tomás, J. M. (1992). Mechanisms of *Klebsiella pneumoniae* resistance to complement-mediated killing. *Infection and Immunity*, *60*(6), 2529–2535. doi:10.1128/IAI.60.6.2529-2535.1992 PMID:1587619

Metz, S. W., Thomas, A., Brackbill, A., Xianwen, Y., Stone, M., Horvath, K., Miley, M. J., Luft, C., DeSimone, J. M., Tian, S., & de Silva, A. M. (2018). Nanoparticle delivery of a tetravalent E protein subunit vaccine induces balanced, type-specific neutralizing antibodies to each dengue virus serotype. *PLoS Neglected Tropical Diseases*, *12*(9), e0006793. doi:10.1371/journal.pntd.0006793 PMID:30248097

Miclăuş, T., Beer, C., Chevallier, J., Scavenius, C., Bochenkov, V. E., Enghild, J. J., & Sutherland, D. S. (2016). Dynamic protein coronas revealed as a modulator of silver nanoparticle sulphidation in vitro. *Nature Communications*, *7*(1), 1–10. doi:10.1038/ncomms11770

Miethling-graff, R., Rumpker, R., Richter, M., Verano-Braga, T., Kjeldsen, F., Brewer, J., Hoyland, J., Rubahn, H.-G., & Erdmann, H. (2014). Toxicology in vitro exposure to silver nanoparticles induces size- and dose-dependent oxidative stress and cytotoxicity in human colon carcinoma cells. *Toxicology In Vitro*, *28*(7), 1280–1289. doi:10.1016/j.tiv.2014.06.005 PMID:24997297

Mirkin, C. A., Letsinger, R. L., Mucic, R. C., & Storhoff, J. J. (1996). A DNA-based method for rationally assembling nanoparticles into macroscopic materials. *Nature*, *382*(6592), 607–609. doi:10.1038/382607a0 PMID:8757129

Miro, E., Verges, C., Garcia, I., Mirelis, B., Navarro, F., Coll, P., Prats, G., & Martinez-Martinez, L. (2004). Resistencia a quinolonas y betalactámicos en Salmonella enterica, y su relación con mutaciones en las topoisomerasas, alteraciones en la permeabilidad celular y expresión de un mecanismo de expulsión activa. *Enfermedades Infecciosas y Microbiologia Clinica*, *22*(4), 204–211. doi:10.1016/S0213-005X(04)73067-0 PMID:15056435

Mishra, A., Kumari, M., Pandey, S., Chaudhry, V., Gupta, K. C., & Nautiyal, C. S. (2014). Biocatalytic and antimicrobial activities of gold nanoparticles synthesized by *Trichoderma* sp. *Bioresource Technology*, *166*, 235–242. doi:10.1016/j.biortech.2014.04.085 PMID:24914997

Missoum, K., Sadocco, P., Causio, J., Belgacem, M. N., & Bras, J. (2014). Antibacterial activity and biodegradability assessment of chemically grafted nanofibrillated cellulose. *Materials Science and Engineering C*, *45*, 477–483. doi:10.1016/j.msec.2014.09.037 PMID:25491853

Mo, H. Z., Jia, Q. C., Hui, M. G., Li, R. T., Yi, Q. G., Yuan, T., Zun, J. Z., & Yin, H. (2017). Combination of LCMS and GCMS based metabolomics to study the hepatotoxic effect of regular nanoparticles in rats. *Chinese Journal of Natural Medicines*, *15*(9), 684–694. doi:10.1016/S1875-5364(17)30098-5 PMID:28991530

Mohamed Hamouda, I. (2012). Current perspectives of nanoparticles in medical and dental biomaterials. *Journal of Biomedical Research*, *26*(3), 143–151. doi:10.7555/JBR.26.20120027 PMID:23554743

Mohamed, M. M., Fouad, S. A., Elshoky, H. A., Mohammed, G. M., & Salaheldin, T. A. (2017). Antibacterial effect of gold nanoparticles against *Corynebacterium pseudotuberculosis*. *International Journal of Veterinary Science and Medicine*, *5*(1), 23–29. doi:10.1016/j.ijvsm.2017.02.003 PMID:30255044

Mohammadalizadeh, Z., Karbasi, S., & Arasteh, S. (2019). (in press). Physical, mechanical and biological evaluation of poly (3-hydroxybutyrate)-chitosan/MWNTs as a novel electrospun scaffold for cartilage tissue engineering applications. *Polymer-Plastics Technology and Materials*.

Mohammadi Gazestani, A., Khoei, S., Khoee, S., Emamgholizadeh Minaei, S., & Motevalian, M. (2018). In vivo evaluation of the combination effect of near-infrared laser and 5-fluorouracil-loaded PLGA-coated magnetite nanographene oxide. *Artificial Cells, Nanomedicine, and Biotechnology*, *10*(sup2), 1–9. doi:10.1080/21691401.2018.1450265 PMID:29543075

Mohammadi, G., Nokhodchi, A., Barzegar-Jalali, M., Lotfipour, F., Adibkia, K., Ehyaei, N., & Valizadeh, H. (2011). Physicochemical and anti-bacterial performance characterization of clarithromycin nanoparticles as colloidal drug delivery system. *Colloids and Surfaces. B, Biointerfaces*, *88*(1), 39–44. doi:10.1016/j.colsurfb.2011.05.050 PMID:21752610

Mohanpuria, P., Rana, N. K., & Yadav, S. K. (2008). Biosynthesis of nanoparticles: Technological concepts and future applications. *Journal of Nanoparticle Research*, *10*(3), 507–517. doi:10.100711051-007-9275-x

Mohanta, D., Patnaik, S., Sood, S., & Das, N. (2019). Carbon nanotubes: Evaluation of toxicity at biointerfaces. *Journal of Pharmaceutical Analysis*, *9*(5), 293–300. doi:10.1016/j.jpha.2019.04.003 PMID:31929938

Moken, M. C., Mc Murry, L. M., & Levy, S. B. (1997). Selection of multiple-antibiotic-resistant (mar) mutants of Escherichia coli by using the disinfectant pine oil: Roles of the mar and acrAB loci. *Antimicrobial Agents and Chemotherapy*, *41*(12), 2270–2772. doi:10.1128/AAC.41.12.2770 PMID:9420057

Monopoli, M. P., Aberg, C., Salvati, A., & Dawson, K. (2012). A biomolecular coronas provide the biological identity of nanosized materials. *Nature Nanotechnology*, *7*(12), 779–786. doi:10.1038/nnano.2012.207 PMID:23212421

Monteiro-riviere, N. A., & Inman, A. O. (2006). Challenges for assessing carbon nanomaterial toxicity to the skin. *Carbon N Y*, *44*(6), 1070–1078. doi:10.1016/j.carbon.2005.11.004

Monteiro-Riviere, N. A., Nemanich, R. J., Inman, A. O., Wang, Y. Y., & Riviere, J. E. (2005). Multi-walled carbon nanotube interactions with human epidermal keratinocytes. *Toxicology Letters*, *155*(3), 377–384. doi:10.1016/j.toxlet.2004.11.004 PMID:15649621

Moo, J. G. S., Presolski, S., & Pumera, M. (2016). Photochromic spatiotemporal control of bubble-propelled micromotors by a spiropyran molecular switch. *ACS Nano*, *10*(3), 3543–3552. doi:10.1021/acsnano.5b07847 PMID:26919161

Moore, A. M., Patel, S., Forsberg, K. J., Wang, B., Bentley, G., Razia, Y., Qin, X., Tarr, P. I., & Dantas, G. (2013). Pediatric fecal microbiota harbor diverse and novel antibiotic resistance genes. *PLoS One*, *8*(11), e78822. doi:10.1371/journal.pone.0078822 PMID:24236055

Moore, T. L., Rodriguez-Lorenzo, L., Hirsch, V., Balog, S., Urban, D., Jud, C., Rothen-Rutishauser, B., Lattuada, M., & Petri-Fink, A. (2015). Nanoparticle colloidal stability in cell culture media and impact on cellular interactions. *Chemical Society Reviews*, *44*(17), 6287–6305. doi:10.1039/C4CS00487F PMID:26056687

Morad, F. N., Rashid, A., Khavi, R., Rahimi, M. K., & Bahador, A. (2019). Producing of loxacin Loaded Gum Tragacanth/Bassorin Electrospun Nano Fibers and Evaluation Its Antibacterial Activity. *Ekoloji*, *28*(107), 391–400.

Morejón, Pilaquinga, Domenech, Ganchala, & & Debut. (2018). *Larvicidal Activity of Silver Nanoparticles Synthesized Using Extracts of Ambrosia arborescens (Asteraceae) to Control Aedes aegypti L. (Diptera: Culicidae).* Article 6917938.

Morejon, B., Pilaquinga, F., Domenech, F., Ganchala, D., Debut, A., & Neira, M. (2019). Larvicidal Activity of Silver Nanoparticles SynthesizedUsing Extracts ofAmbrosia arborescens(Asteraceae) toControlAedes aegyptiL. (Diptera: Culicidae). *Journal of Nanotechnology.*

Moreno-Vega, A. I., Gomez-Quintero, T., Nunez-Anita, R. E., Acosta-Torres, L. S., & Castaño, V. (2012). Polymeric and ceramic nanoparticles in biomedical applications. *Journal of Nanotechnology*, *2012*, 2012. doi:10.1155/2012/936041

Moretton, M. A., Cagel, M., Bernabeu, E., Gonzalez, L., & Chiappetta, D. A. (2015). Nanopolymersomes as potential carriers for rifampicin pulmonary delivery. *Colloids and Surfaces. B, Biointerfaces*, *136*, 1017–1025. doi:10.1016/j.colsurfb.2015.10.049 PMID:26590894

Mostaghasi, E., Zarepour, A., & Zarrabi, A. (2018). Folic acid armed Fe3O4-HPG nanoparticles as a safe nano vehicle for biomedical theranostics. *Journal of the Taiwan Institute of Chemical Engineers*, *82*, 33–41. doi:10.1016/j.jtice.2017.11.004

Mou, F. Z., Li, Y., Chen, C. R., Li, W., Yin, Y. X., Ma, H. R., & Guan, J. G. (2015). Single component TiO2 tubular micro engines with motion controlled by light-induced bubbles. *Small*, *11*(21), 2564–2570. doi:10.1002mll.201403372 PMID:25627213

Moulton, S. E., & Wallace, G. G. (2014). 3-dimensional (3D) fabricated polymer based drug delivery systems. *Journal of Controlled Release*, *193*, 27–34. doi:10.1016/j.jconrel.2014.07.005 PMID:25020039

Mourdikoudis, S., Pallares, R. M., & Thanh, N. T. K. (2018). Characterizationtechniques for nanoparticles: Comparison and complementarity upon studying nanoparticle properties. *Nanoscale*, *10*(27), 12871–12934. doi:10.1039/C8NR02278J PMID:29926865

Moutevelis, E., & Woolfson, D. N. (2009). A Periodic Table of Coiled-Coil Protein Structures. *Journal of Molecular Biology*, *385*(3), 726–732. doi:10.1016/j.jmb.2008.11.028 PMID:19059267

Moye, Z. D., Woolston, J., & Sulakvelidze, A. (2018). Bacteriophage Applications for Food Production and Processing. *Viruses*, *10*(4), 205. Advance online publication. doi:10.3390/v10040205 PMID:29671810

Mozafari, M. R. (2010). Nanoliposomes: preparation and analysis. In Liposomes (pp. 29-50). Humana Press. doi:10.1007/978-1-60327-360-2_2

Mozafari, M. R., Khosravi-Darani, K., Borazan, G. G., Cui, J., Pardakhty, A., & Yurdugul, S. (2008). Encapsulation of food ingredients using nanoliposome technology. *International Journal of Food Properties*, *11*(4), 833–844. doi:10.1080/10942910701648115

Mudshinge, S. R., Deore, A. B., Patil, S., & Bhalgat, C. M. (2011). Nanoparticles: Emerging carriers for drug delivery. *Saudi Pharmaceutical Journal*, *19*(3), 129–141. doi:10.1016/j.jsps.2011.04.001 PMID:23960751

Mudunkotuwa, I. A., Pettibone, J. M., & Grassian, V. H. (2012). Environmental implications of nanoparticle aging in the processing and fate of copper-based nanomaterials. *Environmental Science & Technology*, *46*(13), 7001–7010. doi:10.1021/es203851d PMID:22280489

Mueller, P., Massner, J., Jayachandran, R., Combaluzier, B., Albrecht, I., Gatfield, J., Blum, C., Ceredig, R., Rodewald, H.-R., Rolink, A. G., & Pieters, J. (2008). Regulation of T cell survival through coronin-1–mediated generation of inositol-1, 4, 5-trisphosphate and calcium mobilization after T cell receptor triggering. *Nature Immunology*, *9*(4), 424–431. doi:10.1038/ni1570 PMID:18345003

Muetgreet, J. & Hiemstra. (1959). *Process for the production of amylose articles by extrusion of aqueous sodium hydroxide solution thereof into concentrated aqueous ammonium sulphate solution.* US Patent, 2,902,336.

Muhlen, A. Z., Uhlen, E. Z. M., Niehus, H., & Mehnert, W. (1996). Atomic force microscopy studies of solid lipid nanoparticles. *Pharmaceutical Research*, *13*(9), 1411–1416. doi:10.1023/A:1016042504830 PMID:8893284

Mühling, M., Bradford, A., Readman, J. W., Somerfield, P. J., & Handy, R. D. (2019). An investigation into the effects of silver nanoparticles on antibiotic resistance of naturally occurring bacteria. *Marine Environmental Research*, *68*(5), 278–283. doi:10.1016/j.marenvres.2009.07.001 PMID:19665221

Mukherjee, A., Mohammed Sadiq, I., Prathna, T. C., & Chandrasekaran, N. (2011). Antimicrobial activity of aluminium oxide nanoparticles for potential clinical applications. *Sci. Microb. Pathog. Commun. Curr. Res. Technol. Adv.*, *1*, 245–251.

Mukherjee, S., Chowdhury, D., Kotcherlakota, R., Patra, S., Vinothkumar, B., Bhadra, M. P., ... Patra, C. R. (2014). Potential theranostics application of bio-synthesized silver nanoparticles (4-in-1 system). *Theranostics*, *4*(3), 316–335. doi:10.7150/thno.7819 PMID:24505239

Müller, R. H., Gohla, S., & Keck, C. M. (2011). State of the art of nanocrystals–special features, production, nanotoxicology aspects and intracellular delivery. *European Journal of Pharmaceutics and Biopharmaceutics*, *78*(1), 1–9. doi:10.1016/j.ejpb.2011.01.007 PMID:21266197

Munawar, A., Mohammed, J. T. M., Syeda, K. M. W., & Ellen, K. W. (2017). An Overview of Chitosan Nanoparticles and Its Application in Non-Parenteral Drug Delivery. *Pharmaceutics*, *9*(4), 53. doi:10.3390/pharmaceutics9040053 PMID:29156634

Mun, E. J., Babiker, H. M., Weinberg, U., Kirson, E. D., & Von Hoff, D. D. (2018). Tumor-treating fields: A fourth modality in cancer treatment. *Clinical Cancer Research*, *24*(2), 266–275. doi:10.1158/1078-0432.CCR-17-1117 PMID:28765323

Mu, Q., Kievit, F. M., Kant, R. J., Lin, G., Jeon, M., & Zhang, M. (2015). Anti-HER2/neu peptide-conjugated iron oxide nanoparticles for targeted delivery of paclitaxel to breast cancer cells. *Nanoscale*, *7*(43), 18010–18014. doi:10.1039/C5NR04867B PMID:26469772

Mura, S., Nicolas, J., & Couvreur, P. (2013). Stimuli-responsivenanocarriers for drug delivery. *Nature Materials*, *12*(11), 991–1003. doi:10.1038/nmat3776 PMID:24150417

Murugan, K., Aruna, P., Panneerselvam, C., Madhiyazhagan, P., Paulpandi, M., Subramaniam, J., Rajaganesh, R., Wei, H., Alsalhi, M. S., Devanesan, S., Nicoletti, M., Syuhei, B., Canale, A., & Benelli, G. (2016b). Fighting arboviral diseases: Low toxicity on mammalian cells, dengue growth inhibition (in vitro), and mosquitocidal activity of Centroceras clavulatum-synthesized silver nanoparticles. *Parasitology Research*, *115*(2), 651–662. doi:10.100700436-015-4783-6 PMID:26462804

Murugan, K., Priyanka, V., Dinesh, D., Madhiyazhagan, P., Panneerselvam, C., Subramaniam, J., Suresh, U., Chandramohan, B., Roni, M., Nicoletti, M., Alarfaj, A. A., Higuchi, A., Munusamy, M. A., Khater, H. F., Messing, R. H., & Benelli, G. (2015c). Predation by Asian bullfrog tadpoles, Hoplobatrachus tigerinus, against the dengue vector, Aedes aegypti, in an aquatic environment treated with mosquitocidal nanoparticles. *Parasitology Research*, *114*(10), 3601–3610. doi:10.100700436-015-4582-0 PMID:26091763

Muthamil Selvan, S., Vijai Anand, K., Govindaraju, K., Tamilselvan, S., Kumar, V. G., Subramanian, K. S., Kannan, M., & Raja, K. (2018). Green synthesis of copper oxide nanoparticles and mosquito larvicidal activity against dengue, zika and chikungunya causing vector Aedes aegypti. *IET Nanobiotechnology / IET*, *12*(8), 1042–1046. doi:10.1049/iet-nbt.2018.5083 PMID:30964011

Muthu, K., & Priya, S. (2017). Green synthesis, characterization and catalytic activity of silver nanoparticles using Cassia auriculata flower extract separated fraction. *Spectrochimica Acta. Part A: Molecular and Biomolecular Spectroscopy*, *179*, 66–72. doi:10.1016/j.saa.2017.02.024 PMID:28219038

Muzammil, S., & Hayat, S. (2018). Nanoantibiotics: Future nanotechnologies to combat antibiotic resistance. *Frontiers in Bioscience (Elite Edition)*, *10*(2), 352–374. doi:10.2741/e827 PMID:29293463

Nabeshi, H., Yoshikawa, T., Matsuyama, K., Nakazato, Y., Arimori, A., Isobe, M., ... Yamashita, T. (2010). Size-dependent cytotoxic effects of amorphous silica nanoparticles on Langerhans cells. *Die Pharmazie-An International Journal of Pharmaceutical Sciences*, *65*(3), 199–201. PMID:20383940

Nabil, G., Bhise, K., Sau, S., Atef, M., El-Banna, H. A., & Iyer, A. K. (2019). Nanoengineered delivery systems for cancer imaging and therapy: Recent advances, future directions and patent evaluation. *Drug Discovery Today*, *24*(2), 462–491. doi:10.1016/j.drudis.2018.08.009 PMID:30121330

Nagamune, T. (2017). Biomolecular engineering for nanobio/bionanotechnology. *Nano Convergence*, *4*(1), 9. doi:10.118640580-017-0103-4 PMID:28491487

Nagarajan, K., Marimuthu, S. K., Palanisamy, S., & Subbiah, L. (2018). Peptide therapeutics versus superbugs: Highlight on current research and advancements. *International Journal of Peptide Research and Therapeutics*, *24*(1), 19–33. doi:10.100710989-017-9650-0

Naghizadeh, S., Hassanzadeh Nemati, N., Hassani Najafabadi, A., Niknejad, H., & Khani, M.-M. (2018). Controlled release of fluorouracil (5-FU) from chitosan-co-poly(ethylene glycol)/poly(glycerol sebacate)-copoly(ethylene glycol)-coated iron oxide. *Int J Polymer Mater Polymer Biomater.*, *67*(4), 212–220. doi:10.1080/00914037.2017.1320657

Nagiah, N., Madhavi, L., Anitha, R., Anandan, C. Srinivasan, N.T., & Sivagnanam, U.T. (2013). Development and characterization of coaxially electrospun gelatin coated poly (3-hydroxybutyric acid) thin films as potential scaffolds for skin regeneration. *Materials Science and Engineering C*, 33-4444-4452.

Nagrath, S., Sequist, L. V., Maheswaran, S., Bell, D. W., Irimia, D., Ulkus, L., Smith, M. R., Kwak, E. L., Digumarthy, S., Muzikansky, A., Ryan, P., Balis, U. J., Tompkins, R. G., Haber, D. A., & Toner, M. (2007). Isolation of rare circulating tumour cells in cancer patients by microchip technology. *Nature*, *450*(7173), 1235–1239. doi:10.1038/nature06385 PMID:18097410

Naika, H. R., Lingaraju, K., Manjunath, K., Kumar, D., Nagaraju, G., Suresh, D., & Nagabhushana, H. (2015). Green synthesis of CuO nanoparticles using Gloriosa superba L. extract and their antibacterial activity. *Journal of Taibah University for Science: JTUSCI*, *9*(1), 7–12. doi:10.1016/j.jtusci.2014.04.006

Nanda, A., & Saravanan, M. (2009). Biosynthesis of silver nanoparticles from *Staphylococcus aureus* and its antimicrobial activity against MRSA and MRSE. *Nanomedicine (London)*, *5*(4), 452–456. doi:10.1016/j.nano.2009.01.012 PMID:19523420

Nandagopal, G. D., Periyathambi, P., & Sastry, T. P. (2016). Fabrication of a multifunctional nanocomposite containing INP coated with chitosan coupled with folic acid and loaded with quercetin in diagnosis and its possible treatment of cancer. *RSC Advances*, *6*(101), 99514–99523. doi:10.1039/C6RA19901A

Nanoantibiotics: Future nanotechnologies to combat antibiotic resistance. (n.d.). Available at: http://www.bioscience.org/2018/v10e/af/827/2.htm#a249

Nano-Liposomes Formulation Production - Creative Biostructure. (n.d.). Available at: https://www.creative-biostructure.com/nano-liposomes-formulation-production-617.htm

Naqvi, S. Z. H., Kiran, U., Ali, M. I., Jamal, A., Hameed, A., Ahmed, S., & Ali. (2013). Combined eficacy of biologically synthesized silver nanoparticles and different antibiotics against multi-drug-resistant bacteria. *International Journal of Nanomedicine*, *8*, 3187–3195. doi:10.2147/IJN.S49284 PMID:23986635

Narayanan, K. B., & Sakthivel, N. (2010). Biological synthesis of metal nanoparticles by microbes. *Advances in Colloid and Interface Science*, *156*(1–2), 1–13. doi:10.1016/j.cis.2010.02.001 PMID:20181326

Narayanan, K. B., & Sakthivel, N. (2011). Green synthesis of biogenic metal nanoparticles by terrestrial and aquatic phototrophic and heterotrophic eukaryotes and biocompatible agents. *Advances in Colloid and Interface Science*, *169*(2), 59–79. doi:10.1016/j.cis.2011.08.004 PMID:21981929

Nasiri, M., & Hassanzadeh-Tabrizi, S. A. (2018). Synthesis and characterization of folate-decorated cobalt ferrite nanoparticles coated with poly(ethylene glycol) for biomedical applications. *Journal of the Chinese Chemical Society (Taipei)*, *65*(2), 231–242. doi:10.1002/jccs.201700271

Naskar, A., Bera, S., Bhattacharya, R., Roy, S. S., & Jana, S. (2016). Synthesis, characterization and cytotoxicity of polyethylene glycol coupled zinc oxide–chemically converted graphene nanocomposite on human OAW42 ovarian cancer cells. *Polymers for Advanced Technologies*, *27*(4), 436–443. doi:10.1002/pat.3689

National Science and Technology Council Committee on Technology. (2005). *The National Nanotechnology Initiative: Research and Development Leading to a Revolution in Technology and Industry*. Office of Sciences and Technology Policy.

Navale, G. R., Thripuranthaka, M., Late, D. J., & Shinde, S. S. (2015). Antimicrobial activity of ZnO nanoparticles against pathogenic bacteria and fungi. *JSM Nanotechnology and Nanomedicine*, *3*, 1033.

Nayak, R., Padhye, R., Kyratzis, I. L., Truong, Y. B., & Arnold, L. (2012). Recent advances in nanofibre fabrication techniques. *Textile Research Journal*, *82*(2), 129–147. doi:10.1177/0040517511424524

Ndieyira, J. W., Watari, M., Barrera, A. D., Zhou, D., Vögtli, M., Batchelor, M., ... Rayment, T. (2008). Nanomechanical detection of antibiotic–mucopeptide binding in a model for superbug drug resistance. *Nature Nanotechnology*, *3*(11), 691–696. doi:10.1038/nnano.2008.275 PMID:18989336

Neethu, S., Midhun, S. J., Radhakrishnan, E. K., & Jyothis, M. (2018). Green synthesized silver nanoparticles by marine endophytic fungus Penicillium polonicum and its antibacterial efficacy against biofilm forming, multidrug-resistant Acinetobacter baumanii. *Microbial Pathogenesis*, *116*, 263–272. doi:10.1016/j.micpath.2018.01.033 PMID:29366864

Nehra, P., Chauhan, R. P., Garg, N., & Verma, K. (2018). Antibacterial and antifungal activity of chitosan coated iron oxide nanoparticles. *British Journal of Biomedical Science*, *75*(1), 13–18. doi:10.1080/09674845.2017.1347362 PMID:28945174

Nelson, D. C., Schmelcher, M., Rodriguez-Rubio, L., Klumpp, J., Pritchard, D. G., Dong, S., & Donovan, D. M. (2012). Endolysins as antimicrobials. *Advances in Virus Research*, *83*, 299–365. doi:10.1016/B978-0-12-394438-2.00007-4 PMID:22748813

Nemati, E., Mokhtarzadeh, A., Panahi-Azar, V., Mohammadi, A., Hamishehkar, H., Mesgari-Abbasi, M., Ezzati Nazhad Dolatabadi, J., & de la Guardia, M. (2019). Ethambutol-loaded solid lipid nanoparticles as dry powder inhalable formulation for tuberculosis therapy. *AAPS PharmSciTech*, *20*(3), 120. doi:10.120812249-019-1334-y PMID:30796625

Neu, H. (1992). The Crisis in Antibiotic Resistance. *Science*, *257*(5073), 1064–1073. doi:10.1126cience.257.5073.1064 PMID:1509257

Neupane, S., Pan, Y., Takalkar, S., Bentz, K., Farmakes, J., Xu, Y., Chen, B., Liu, G., Qian, S. Y., & Yang, Z. (2017). Probing the aggregation mechanism of gold nanoparticles triggered by a globular protein. *The Journal of Physical Chemistry C*, *121*(2), 1377–1386. doi:10.1021/acs.jpcc.6b11963

Newman, L., Rowley, J., Vander Hoorn, S., Wijesooriya, N. S., Unemo, M., Low, N., Stevens, G., Gottlieb, S., Kiarie, J., & Temmerman, M. (2015). Global Estimates of the Prevalence and Incidence of Four Curable Sexually Transmitted Infections in 2012 Based on Systematic Review and Global Reporting. *PLoS One*, *10*(12), e0143304. doi:10.1371/journal.pone.0143304 PMID:26646541

Ngoy, J. M., Wagner, N., Riboldi, L., & Bolland, O. (2014). A CO2 capture technology using multi-walled carbon nanotubes with polyaspartamide surfactant. *Energy Procedia*, *63*, 2230–2248. doi:10.1016/j.egypro.2014.11.242

Nguyen, H.N., Ha,, P.T., Sao Nguyen,, A., Nguyen,, D.T., Do,, H.D., Thi,, Q.N., & Thi,, M.N.H. (2016). Curcumin as fluorescent probe for directly monitoring in vitro uptake of curcumin combined paclitaxel loaded PLA-TPGS nanoparticles. *Advances in Natural Sciences: Nanoscience and Nanotechnology*, *7*(2), 02500.

Nguyen, N. T., Grelling, N., Wetteland, C. L., Rosario, R., & Liu, H. (2018). Antimicrobial Activities and Mechanisms of Magnesium Oxide Nanoparticles (nMgO) against Pathogenic Bacteria, Yeasts, and Biofilms. *Scientific Reports*, *8*(1), 16260. doi:10.103841598-018-34567-5 PMID:30389984

Nichols, J. W., & Bae, Y. H. (2014). EPR: Evidence and fallacy. *Journal of Controlled Release*, *190*, 451–464. doi:10.1016/j.jconrel.2014.03.057 PMID:24794900

Nichols, R. J., Sen, S., Choo, Y. J., Beltrao, P., Zietek, M., Chaba, R., Lee, S., Kazmierczak, K. M., Lee, K. J., Wong, A., Shales, M., Lovett, S., Winkler, M. E., Krogan, N. J., Typas, A., & Gross, C. A. (2011). Phenotypic landscape of a bacterial cell. *Cell*, *144*(1), 143–156. doi:10.1016/j.cell.2010.11.052 PMID:21185072

Nicolosi, D., Cupri, S., Genovese, C., Tempera, G., Mattina, R., & Pignatello, R. (2015). Nanotechnology approaches for antibacterial drug delivery: Preparation and microbiological evaluation of fusogenic liposomes carrying fusidic acid. *International Journal of Antimicrobial Agents*, *45*(6), 622–626. doi:10.1016/j.ijantimicag.2015.01.016 PMID:25816979

Niemirowicz, K., & Bucki, R. (2017). Enhancing the fungicidal activity of antibiotics: Are magnetic nanoparticles the key? *Nanomedicine (London)*, *12*(15), 1747–1749. doi:10.2217/nnm-2017-0051 PMID:28703013

Niemirowicz, K., Car, H., Sadowska, A., Wątek, M., Kręowski, R., Cechowska-Pasko, M., Wilczewska, A. Z., Mystkowska, J., Kasacka, I., Torres, A., & Bucki, R. (2017). Pharmacokinetics and anticancer activity of folic acid-functionalized magnetic nanoparticles. *Journal of Biomedical Nanotechnology*, *13*(6), 665–677. doi:10.1166/jbn.2017.2363

Niemirowicz, K., Piktel, E., Wilczewska, A. Z., Markiewicz, K. H., Durnaś, B., Wątek, M., Puszkarz, I., Wróblewska, M., Niklińska, W., Savage, P., & Bucki, R. (2016). Core–shell magnetic nanoparticles display synergistic antibacterial effects against Pseudomonas aeruginosa and Staphylococcus aureus when combined with cathelicidin LL-37 or selected ceragenins. *International Journal of Nanomedicine*, *11*, 5443–5455. doi:10.2147/IJN.S113706 PMID:27799768

Nigam, S., & Bahadur, D. (2018). Doxorubicin-loaded dendritic-Fe3 O4 supramolecular nanoparticles for magnetic drug targeting and tumor regression in spheroid murine melanoma model. *Nanomedicine; Nanotechnology, Biology, and Medicine*, *14*(3), 759–768. doi:10.1016/j.nano.2018.01.005 PMID:29339187

Nikaido, H. (2009). Multidrug resistance in bacteria. *Annual Review of Biochemistry*, *78*(1), 119–146. doi:10.1146/annurev.biochem.78.082907.145923 PMID:19231985

Nikjoo, D., & Aroguz, A. Z. (2016). Magnetic field responsive methylcellulosepolycaprolactone nanocomposite gels for targeted and controlled release of 5-fluorouracil. *International Journal of Polymeric Materials*, *65*(8), 421–432. doi:10.1080/00914037.2015.1129954

Nikolova, M. P., & Chavali, M. S. (2020). Metal Oxide Nanoparticles as Biomedical Materials. *Biomimetics*, *5*(2), 27. doi:10.3390/biomimetics5020027 PMID:32521669

Ninganagouda, S., Rathod, V., Jyoti, H., Singh, D., & Prema, K., & Manzoor-Ul-Haq. (2013). Extracellular biosynthesis of silver nanoparticles using *Aspergillus flavus* and their antimicrobial activity against gram negative MDR strains. *International Journal of Pharma and Bio Sciences*, *4*(2), 222–229.

Nino-Martinez, N., Salas Orozco, M. F., Martinez-Castanon, G. A., Torres Mendez, F., & Ruiz, F. (2019). Molecular Mechanisms of Bacterial Resistance to Metal and Metal Oxide Nanoparticles. *International Journal of Molecular Sciences*, *20*(11), 2808. Advance online publication. doi:10.3390/ijms20112808 PMID:31181755

Nirmala, M., Nair, M. G., Rekha, K., Anukaliani, A., Samdarshi, S., & Nair, R. G. (2010). Photocatalytic activity of ZnO nanopowders synthesized by DC thermal plasma. *African Journal of Basic & Applied Sciences*, *2*, 161–166.

Nista, S. V., Bettini, J., & Mei, L. H. (2015). Coaxial nanofibers of chitosan-alginate-PEO polycomplex obtained by electrospinning. *Carbohydrate Polymers*, *127*, 222–228. doi:10.1016/j.carbpol.2015.03.063 PMID:25965478

Nizamov, T. R., Garanina, A. S., Grebennikov, I. S., Zhironkina, O. A., Strelkova, O. S., Alieva, I. B., Kireev, I. I., Abakumov, M. A., Savchenko, A. G., & Majouga, A. G. (2018). Effect of iron oxide nanoparticle shape on doxorubicin drug delivery toward LNCaP and PC-3 cell lines. *BioNanoScience*, *8*(1), 394–406. doi:10.100712668-018-0502-y

Njagi, E. C., Huang, H., Stafford, L., Genuino, H., Galindo, H. M., Collins, J. B., Hoag, G. E., & Suib, S. L. (2011). Biosynthesis of iron and silver nanoparticles at room temperature using aqueous sorghum bran extracts. *Langmuir*, *27*(1), 264–271. doi:10.1021/la103190n PMID:21133391

Nkanga, C. I., & Krause, R. W. M. (2019). encapsulation of isoniazid-conjugated phthalocyanine-in-cyclodextrin-in-Liposomes Using Heating Method. *Scientific Reports*, *9*(1), 1–16. doi:10.103841598-019-47991-y PMID:31391517

Noorani, B., Yazdian, F., Khoshraftar, A., Ghaemi, R. V., Alihemmati, Z., Rashedi, H., & Shahmoradi, S. (2018). Fabrication and evaluation of nanofibrouspolyhydroxybutyratevalerate scaffolds containing hydroxyapatite particles for bone tissue engineering. *International Journal of Polymeric Materials and Polymeric Biomaterials*, *67*(17), 987–995. doi:10.1080/00914037.2017.1417283

Nordmann, P., & Poirel, L. (2014). The difficult-to-control spread of carbapenemase producers among Enterobacteriaceae worldwide. *Clinical Microbiology and Infection*, *20*(9), 821–830. doi:10.1111/1469-0691.12719 PMID:24930781

Nowack, B., & Bucheli, T. D. (2007). Occurrence, behavior and effects of nanoparticles in the environment. *Environmental Pollution*, *150*(1), 5–22. doi:10.1016/j.envpol.2007.06.006 PMID:17658673

NRC. (2009). *Science and Decisions: Advancing Risk Assessment. Committee on Improving Risk Analysis Approaches Used by the U.S. EPA, Board on Environmental Studies and Toxicology, Division on Earth and Life Studies, National Research Council of the National Academies*. The National Academies Press.

Nyang'wa, B. T., Berry, C., Fielding, K., & Nunn, A. J. (2019). Multidrug-resistant tuberculosis. *Lancet*, *394*(10195), 298–299. doi:10.1016/S0140-6736(19)30691-9 PMID:31354136

O'Neill, J., Davies, S., Rex, J., White, L. J., & Murray, R. (2016). *Tackling Drug-Resistant Infections Globally: Final Report and Recommendations*. Wellcome Trust and UK Government.

O'Sullivan, L., Buttimer, C., McAuliffe, O., Bolton, D., & Coffey, A. (2016). Bacteriophage-based tools: Recent advances and novel applications. *F1000 Research*, *5*, 2782. doi:10.12688/f1000research.9705.1 PMID:27990274

Oakley, A. J., Barrett, S., Peat, T. S., Newman, J., Streltsov, V. A., Waddington, L., Saito, T., Tashiro, M., & McKimm-Breschkin, J. L. (2010). Structural and Functional Basis of Resistance to Neuraminidase Inhibitors of Influenza B Viruses. *Journal of Medicinal Chemistry*, *53*(17), 6421–6431. doi:10.1021/jm100621s PMID:20695427

Oberdörster, G., Maynard, A., Donaldson, K., Castranova, V., Fitzpatrick, J., Ausman, K., & Olin, S. (2005). Principles for characterizing the potential human health effectsfrom exposure to nanomaterials: Elements of a screening strategy. *Particle and Fibre Toxicology*, *2*(1), 1.

Oberdorster, E. (2004). Manufactured nanomaterials (fullerenes, C60) induce oxidative stress in the brain of juvenile largemouth bass. *Environmental Health Perspectives*, *112*(10), 1058–1062. doi:10.1289/ehp.7021 PMID:15238277

Oberdörster, G., Ferin, J., & Lehnert, B. E. (1994). Correlation between particle size, in vivo particle persistence, and lung injury. *Environmental Health Perspectives*, *102*(suppl 5), 173–179. PMID:7882925

Oda, K., Arakawa, H., Tanaka, T., Matsuda, K., Tanikawa, C., Mori, T., Nishimori, H., Tamai, K., Tokino, T., Nakamura, Y., & Taya, Y. (2000). p53AIP1, a potential mediator of p53-dependent apoptosis, and its regulation by Ser-46- phosphorylated p53. *Cell*, *15*(102(6)), 849–862. doi:10.1016/S0092-8674(00)00073-8 PMID:11030628

Ogunyemi, S. O., Abdallah, Y., Zhang, M., Fouad, H., Hong, X., Ibrahim, E., Masum, M. M. I., Hossain, A., Mo, J., & Li, B. (2019). Green synthesis of zinc oxide nanoparticles using different plant extracts and their antibacterial activity against Xanthomonas oryzae pv. oryzae. *Artificial Cells, Nanomedicine, and Biotechnology*, *47*(1), 341–352. doi:10.1080/21691401.2018.1557671 PMID:30691311

Ohlsen, K., Ternes, T., Werner, G., Wallner, U., Löffler, D., Ziebuhr, W., Witte, W., & Hacker, J. (2013). Impact of antibiotics on conjugational resistance gene transfer in Staphylococcus aureus in sewage. *Environmental Microbiology*, *5*(8), 711–716. doi:10.1046/j.1462-2920.2003.00459.x PMID:12871238

Okamoto, M., Morita, S., & Kotaka, T. (2001). Dispersed structure and ionic conductivity of smectic clay/polymer nanocomposites. *Polymer*, *42*(6), 2685–2688. doi:10.1016/S0032-3861(00)00642-X

Okamoto, S., Tamaru, A., Nakajima, C., Nishimura, K., Tanaka, Y., Tokuyama, S., Suzuki, Y., & Ochi, K. (2007). Loss of a conserved 7-methylguanosine modification in 16S rRNA confers low-level streptomycin resistance in bacteria. *Molecular Microbiology*, *63*(4), 1096–1106. doi:10.1111/j.1365-2958.2006.05585.x PMID:17238915

Oktay, B., Basturk, E., Kayaman Apohan, N., & Vezir Kahraman, M. (2013). Highly Porous starch/ Poly (ethylene-alt-maleic anhyride) composite nanofiber mesh. *Polymer Composites*, *34*(8), 1321–1324. doi:10.1002/pc.22545

Olson, E. S., Orozco, J., Wu, Z., Malone, C. D., Yi, B., Gao, W., Eghtedari, M., Wang, J., & Mattrey, R. F. (2013). Toward *in vivo* detection of hydrogen peroxide with ultrasound molecular imaging. *Biomaterials*, *34*(35), 8918–8924. doi:10.1016/j.biomaterials.2013.06.055 PMID:23958028

Omoike, A., Chorover, J., Kwon, K. D., & Kubicki, J. D. (2004). Adhesion of bacterial exopolymers to α-FeOOH: Inner-sphere complexation of phosphodiester groups. *Langmuir*, *20*(25), 11108–11114. doi:10.1021/la048597+ PMID:15568864

Ondarçuhu, T., & Joachim, C. (1998). Drawing a single nanofibre over hundreds of microns. *Europhysics Letters*, *42*(2), 215–220. doi:10.1209/epl/i1998-00233-9

Onyebujoh, P., Zumla, A., Ribeiro, I., Rustomjee, R., Mwaba, P., Gomes, M., & Grange, J. M. (2005). Treatment of tuberculosis: Present status and future prospects. *Bulletin of the World Health Organization*, *83*, 857–865. PMID:16302043

Oppenheim, J. J. (2003). Roles of antimicrobial peptides such as defensins in innate and adaptive immunity. *Annals of the Rheumatic Diseases*, *62*(90002), 17–21. doi:10.1136/ard.62.suppl_2.ii17 PMID:14532141

Ossendorp, F., Toes, R. E., Offringa, R., van der Burg, S. H., & Melief, C. J. M. (2000). Importance of CD4(+) T helper cell responses in tumor immunity. *Immunology Letters*, *74*(1), 75–79. doi:10.1016/S0165-2478(00)00252-2 PMID:10996631

Otsuka, H., Nagasaki, Y., & Kataoka, K. (2003). PEGylated nanoparticles for biological and pharmaceutical applications. *Advanced Drug Delivery Reviews*, *55*(3), 403–419. doi:10.1016/S0169-409X(02)00226-0 PMID:12628324

Otter, J. A., Vickery, K., Walker, J. D., Pulcini, E. D., Stoodley, P., Goldenberg, S. D., ... Edgeworth, J. D. (2015). Surface-attached cells, biofilms and biocide susceptibility: Implications for hospital cleaning and disinfection. *The Journal of Hospital Infection*, *89*(1), 16–27. doi:10.1016/j.jhin.2014.09.008 PMID:25447198

Owens, E. A., Henary, M., El Fakhri, G., & Choi, H. S. (2016). Tissue-specific near-infrared fluorescence imaging. *Accounts of Chemical Research*, *49*(9), 1731–1740. doi:10.1021/acs.accounts.6b00239 PMID:27564418

Ozcelikkale, A., Ghosh, S., & Han, B. (2013). Multifaceted transport characteristics of nanomedicine: Needs for characterization in dynamic environment. *Molecular Pharmaceutics*, *10*(6), 2111–2126. doi:10.1021/mp3005947 PMID:23517188

Paavola, A., Yliruusi, J., Kajimoto, Y., Kalso, E., Wahlström, T., & Rosenberg, P. (1995). Controlled release of lidocaine from injectable gels and efficacy in rat sciatic nerve block. *Pharmaceutical Research*, *12*(12), 1997–2002. doi:10.1023/A:1016264527738 PMID:8786979

Pacheco-Blandino, I., Vanner, R., & Buzea, C. (2012). Toxicity of nanoparticles. In *Toxicity of building materials* (pp. 427–475). Woodhead Publishing. doi:10.1533/9780857096357.427

Padil, V. V. T., & Černík, M. (2013). Green synthesis of copper oxide nanoparticles using gum karaya as a biotemplate and their antibacterial application. *International Journal of Nanomedicine*, *8*, 889. PMID:23467397

Pageni, P., Yang, P., Bam, M., Zhu, T., Chen, Y. P., Decho, A. W., Nagarkatti, M., & Tang, C. (2018). Recyclable magnetic nanoparticles grafted with antimicrobial metallopolymer-antibiotic bioconjugates. *Biomaterials*, *178*(September), 363–372. doi:10.1016/j.biomaterials.2018.05.007 PMID:29759729

Pai, M., & Memish, Z. A. (2016). Antimicrobial resistance and the growing threat of drug-resistant tuberculosis. *Journal of Epidemiology and Global Health*, *6*(2), 45–47. doi:10.1016/j.jegh.2016.02.001 PMID:26896704

Palabiyikoglu, I. (2003). Pathogenesis of intensive care infections. *Turkish Journal of Intensive Care Medicine*, *3*, 81–101.

Palacci, J., Sacanna, S., Vatchinsky, A., Chaikin, P. M., & Pine, D. J. (2013). Photoactivated colloidal dockers for cargo transportation. *Journal of the American Chemical Society*, *135*(43), 15978–159781. doi:10.1021/ja406090s PMID:24131488

Pal, D., Sahu, C. K., & Haldar, A. (2014). Bhasma: The ancient Indian nanomedicine. *Journal of Advanced Pharmaceutical Technology & Research*, *5*(1), 4. doi:10.4103/2231-4040.126980 PMID:24696811

Palmer, K. L., Kos, V. N., & Gilmore, M. S. (2010). Horizontal gene transfers and the genomics of enterococcal antibiotic resistance. *Current Opinion in Microbiology*, *13*(5), 632–639. doi:10.1016/j.mib.2010.08.004 PMID:20837397

Palmerston Mendes, L., Pan, J., & Torchilin, V. P. (2017). Dendrimers as nanocarriers for nucleic acid and drug delivery in cancer therapy. *Molecules (Basel, Switzerland)*, *22*(9), 1401. doi:10.3390/molecules22091401 PMID:28832535

Palomäki, J., Välimäki, E., Sund, J., Vippola, M., Clausen, P. A., Jensen, K. A., Savolainen, K., Matikainen, S., & Alenius, H. (2011, September 27). NLRP3 inflammasome through a similar mechanism. *ACS Nano*, *5*(9), 6861–6870. doi:10.1021/nn200595c PMID:21800904

Palomino, J. C., & Martin, A. (2014). Drug resistance mechanisms in Mycobacterium tuberculosis. *Antibiotics (Basel, Switzerland)*, *3*(3), 317–340. doi:10.3390/antibiotics3030317 PMID:27025748

Pal, S., Tak, Y. K., & Song, J. M. (2007). Does the antibacterial activity of silver nanoparticles depend on the shape of the nanoparticle? A study of the gram-negative bacterium *Escherichia coli*. *Applied and Environmental Microbiology*, *73*(6), 1712–1720. doi:10.1128/AEM.02218-06 PMID:17261510

Pamies, R., Cifre, J. G. H., Espín, V. F., Collado-González, M., Baños, F. G. D., & de la Torre, J. G. (2014). Aggregation behaviour of gold nanoparticles in saline aqueous media. *Journal of Nanoparticle Research*, *16*(4), 2376. doi:10.100711051-014-2376-4

Panacek, A., Kvítek, L., Prucek, R., Kolar, M., Vecerova, R., Pizúrova, N., Sharma, V. K., Nevecna, T., & Zboril, R. (2006). Silver colloid nanoparticles: Synthesis, characterization and their antibacterial activity. *The Journal of Physical Chemistry B*, *110*(33), 16248–16253. doi:10.1021/jp063826h PMID:16913750

Panacek, A., Smekalova, M., Kilianova, M., Prucek, R., & Bogdanova, K.,V ec erova, R., et al. (2016). Strong and non-specific synergistic antibacterial efficiency of antibiotics combined with silver nanoparticles at very low concentrations showing no cytotoxic effect. *Molecules (Basel, Switzerland)*, *21*, 1–17. PMID:26729075

Panagi, Z., Beletsi, A., Evangelatos, G., Livaniou, E., Ithakissios, D. S., & Avgoustakis, K. (2001). Effect of dose on the biodistribution and pharmacokinetics of PLGA and PLGA–mPEG nanoparticles. *International Journal of Pharmaceutics*, *221*(1-2), 143–152. doi:10.1016/S0378-5173(01)00676-7 PMID:11397575

Panda, T., & Deepa, K. (2011). Biosynthesis of gold nanoparticles. *Journal of Nanoscience and Nanotechnology*, *11*(12), 10279–10294. doi:10.1166/jnn.2011.5021 PMID:22408900

Panigrahi, S., Kundu, S., Ghosh, S., Nath, S., & Pal, T. (2004). General method of synthesis for metal nanoparticles. *Journal of Nanoparticle Research*, *6*(4), 411–414. doi:10.100711051-004-6575-2

Pankhurst, Q. A., Connolly, J., Jones, S. K., & Dobson, J. (2003). Applications of magnetic nanoparticles in biomedicine. *Journal of Physics. D, Applied Physics*, *36*(13), 167–181. doi:10.1088/0022-3727/36/13/201

Panneerselvam, C., Murugan, K., Roni, M., Aziz, A. T., Suresh, U., Rajaganesh, R., Madhiyazhagan, P., Subramaniam, J., Dinesh, D., Nicoletti, M., Higuchi, A., Alarfaj, A. A., Munusamy, M. A., Kumar, S., Desneux, N., & Benelli, G. (2016). Fern-synthesized nanoparticles in the fight against malaria: LC/MS analysis of Pteridium aquilinum leaf extract and biosynthesis of silver nanoparticles with high mosquitocidal and antiplasmodial activity. *Parasitology Research*, *115*(3), 997–1013. doi:10.100700436-015-4828-x PMID:26612497

Papageorgiou, I., Brown, C., Schins, R., Singh, S., Newson, R., Davis, S., Fisher, J., Ingham, E., & Case, C. P. (2007). The effect of nano- and micron-sized particles of cobalt-chromium alloy on human fibroblasts in vitro. *Biomaterials*, *28*(19), 2946–2958. doi:10.1016/j.biomaterials.2007.02.034 PMID:17379299

Paques, J. P., van der Linden, E., van Rijn, C. J. M., & Sagis, L. M. C. (2014). Preparation methods of alginate nanoparticles. *Advances in Colloid and Interface Science*, *209*, 163–171. doi:10.1016/j.cis.2014.03.009 PMID:24745976

Parikh, P., Zala, D., & Makwana, B. A. (2014). Biosynthesis of Copper Nanoparticles and Their Antimicrobial Activity. OAlib, 01(01), 1–15. doi:10.4236/oalib.preprints.1200067

Park, K. (2013). Facing the truth about nanotechnology in drug delivery. *ACS Nano*, *7*(9), 7442–7447. doi:10.1021/nn404501g PMID:24490875

Park, K. H., Chhowalla, M., Iqbal, Z., & Sesti, F. (2003). Single-walled carbon nanotubes are a new class of ion channel blockers. *The Journal of Biological Chemistry*, *278*(50), 50212–50216. doi:10.1074/jbc.M310216200 PMID:14522977

Park, M. Z., Neigh, A. M., Vermeulen, J. P., de la Fonteyne, L. J., Verharen, H. W., Briedé, J. J., van Loveren, H., & de Jong, W. H. (2011). The effect of particle size on the cytotoxicity, inflammation, developmental toxicity and genotoxicity of silver nanoparticles. *Biomaterials*, *32*(36), 9810–9817. doi:10.1016/j.biomaterials.2011.08.085 PMID:21944826

Park, S. S., Dong, H., Liu, X., Harrington, S. M., Krco, C. J., Grams, M. P., & Kwon, E. D. (2015). PD-1 restrains radiotherapy-induced abscopal effect. *Cancer Immunology Research*, *3*(6), 610–619. doi:10.1158/2326-6066.CIR-14-0138 PMID:25701325

Parolo, C., de la Escosura-Muñiz, A., Polo, E., Grazú, V., De La Fuente, J. M., & Merkoçi, A. (2013). Design, preparation, and evaluation of a fixed-orientation antibody/gold-nanoparticle conjugate as an immunosensing label. *ACS Applied Materials & Interfaces*, *5*(21), 10753–10759. doi:10.1021/am4029153 PMID:24095174

Partridge, S., Tsafnat, G., Coiera, E., & Iredell, J. (2009). Gene cassettes and cassette arrays in mobile resistance integrons. *FEMS Microbiology Reviews*, *33*(4), 757–784. doi:10.1111/j.1574-6976.2009.00175.x PMID:19416365

Parveen, S., & Sahoo, S. K. (2006). Nanomedicine. *Clinical Pharmacokinetics*, *45*(10), 965–988. doi:10.2165/00003088-200645100-00002 PMID:16984211

Parveen, S., & Sahoo, S. K. (2008). Polymeric nanoparticles for cancer therapy. *Journal of Drug Targeting*, *16*(2), 108–123. doi:10.1080/10611860701794353 PMID:18274932

Patel, K., Raj, B. S., Chen, Y., & Lou, X. (2017). Novel folic acid conjugated Fe3 O4 - ZnO hybrid nanoparticles for targeted photodynamic therapy. *Colloids and Surfaces. B, Biointerfaces*, *150*, 317–325. doi:10.1016/j.colsurfb.2016.10.045 PMID:27810128

Patel, S., Singh, D., Srivastava, S., Singh, M., Shah, K., Chauahn, D. N., & Chauhan, N. S. (2017). Nanoparticles as a platform for antimicrobial drug delivery. *Adv Pharma Pharmacy*, *5*(3), 31–43. doi:10.13189/app.2017.050301

Pati, R., Mehta, R. K., Mohanty, S., Padhi, A., Sengupta, M., Vaseeharan, B., Goswami, C., & Sonawane, A. (2014). Topical application of zinc oxide nanoparticles reduces bacterial skin infection in mice and exhibits antibacterial activity by inducing oxidative stress response and cell membrane disintegration in macrophages. *Nanomedicine (London)*, *10*(6), 1195–1208. doi:10.1016/j.nano.2014.02.012 PMID:24607937

Patlolla, A. K., Berry, A., May, L., & Tchounwou, P. B. (2012). Genotoxicity of silver nanoparticles in Viciafaba: A pilot study on the environmental monitoring of nanoparticles. *International Journal of Environmental Research and Public Health*, *9*(5), 1649–1662. doi:10.3390/ijerph9051649 PMID:22754463

Patra, J.K., & Baek, K.H. (2014). Green nanobiotechnology: factor affecting synthesis and characterization techniques. *Journal of Nanomaterials,* 1-12. doi:10.1155/2014/417305

Patra, B., Gautam, R., Priyadarsini, E., Rajamani, P., Pradhan, S. N., Saravanan, M., & Meena, R. (2019). Piper betle : Augmented synthesis of gold nanoparticles and its invitro cytotoxicity assessment on hela and hek 293 cells. *Journal of Cluster Science*, 1–13.

Patra, J. K., Das, G., Fraceto, L. F., Campos, E. V. R., del Pilar Rodriguez-Torres, M., Acosta-Torres, L. S., ... Habtemariam, S. (2018). Nano based drug delivery systems: Recent developments and future prospects. *Journal of Nanobiotechnology*, *16*(1), 71. doi:10.118612951-018-0392-8 PMID:30231877

Paul Das, M., Rebecca Livingstone, J., Veluswamy, P., & Das, J. (2018). Exploration of Wedelia chinensis leaf-assisted silver nanoparticles for antioxidant, antibacterial and in vitro cytotoxic applications. *Yao Wu Shi Pin Fen Xi*, *26*(2), 917–925. doi:10.1016/j.jfda.2017.07.014 PMID:29567263

Paul, A. M., Sajeev, A., Nivetha, R., Gothandapani, K., Bhardwaj, P., Govardhan, K., ... Grace, A. N. (2020). Cuprous oxide (Cu2O)/graphitic carbon nitride (g-C3N4) nanocomposites for electrocatalytic hydrogen evolution reaction. *Diamond and Related Materials*, *107*, 107899. doi:10.1016/j.diamond.2020.107899

Pavela, R. (2015a). Essential oils for the development of eco-friendly mosquito larvicides: A review. *Industrial Crops and Products*, *76*, 174–187. doi:10.1016/j.indcrop.2015.06.050

Paxton, W. F., Kistler, K. C., Olmeda, C. C., Sen, A., St Angelo, S. K., Cao, Y., Mallouk, T. E., Lammert, P. E., & Crespi, V. H. (2004). Catalytic nanomotors: Autonomous movement of striped nanorods. *Journal of the American Chemical Society*, *126*(41), 13424–13431. doi:10.1021/ja047697z PMID:15479099

Payne, D. J., Gwynn, M. N., Holmes, D. J., & Pompliano, D. L. (2007). Drugs for bad bugs: Confronting the challenges of antibacterial discovery. *Nature Reviews. Drug Discovery*, *6*(1), 29–40. doi:10.1038/nrd2201 PMID:17159923

Pecot, C. V., Bischoff, F. Z., Mayer, J. A., Wong, K. L., Pham, T., Bottsford-Miller, J., Stone, R. L., Lin, Y. G., Jaladurgam, P., Roh, J. W., Goodman, B. W., Merritt, W. M., Pircher, T. J., Mikolajczyk, S. D., Nick, A. M., Celestino, J., Eng, C., Ellis, L. M., Deavers, M. T., & Sood, A. K. (2011). A novel platform for detection of CK+ and CK- CTCs. *Cancer Discovery*, *1*(7), 580–586. doi:10.1158/2159-8290.CD-11-0215 PMID:22180853

Peer, D., Karp, J. M., Hong, S., Farokhzad, O. C., Margalit, R., & Langer, R. (2007). Nanocarriers as an emerging platform for cancer therapy. *Nature Nanotechnology*, *2*(12), 751–760. doi:10.1038/nnano.2007.387 PMID:18654426

Peleg, A. Y., Seifert, H., & Paterson, D. L. (2008). Acinetobacter baumannii: Emergence of a successful pathogen. *Clinical Microbiology Reviews*, *21*(3), 538–582. doi:10.1128/CMR.00058-07 PMID:18625687

Pelgrift, R. Y., & Friedman, A. J. (2013). Nanotechnology as a therapeutic tool to combat microbial resistance. *Advanced Drug Delivery Reviews*, *65*(13-14), 1803–1815. doi:10.1016/j.addr.2013.07.011 PMID:23892192

Peña, J., Bargar, J. R., & Sposito, G. (2011). Role of bacterial biomass in the sorption of Ni by biomass-birnessite assemblages. *Environmental Science & Technology*, *45*(17), 7338–7344. doi:10.1021/es201446r PMID:21780745

Peng, C., Zhang, T., Ortiz-Ortiz, D. N., Vishwakarma, A., Barton, H. A., & Joy, A. (2019). Modification of narrow-spectrum peptidomimetic polyurethanes with fatty acid chains confers broad-spectrum antibacterial activity. *Polymer International*, *68*(7), 1255–1262. doi:10.1002/pi.5773

Peng, F., Setyawati, M. I., Tee, J. K., Ding, X., Wang, J., Nga, M. E., Ho, H. K., & Leong, D. T. (2019). Nanoparticles promote in vivo breast cancer cell intravasation and extravasation by inducing endothelial leakiness. *Nature Nanotechnology*, *14*(3), 279–286. doi:10.103841565-018-0356-z PMID:30692675

Peng, N., Wu, B., Wang, L., He, W., Ai, Z., Zhang, X., Wang, Y., Fan, L., & Ye, Q. (2016). High drug loading and pH-responsive targeted nanocarriers from alginate-modified SPIONs for anti-tumor chemotherapy. *Biomaterials Science*, *4*(12), 1802–1813. doi:10.1039/C6BM00504G PMID:27792228

Peng, Q., Zhang, S., Yang, Q., Zhang, T., Wei, X. Q., Jiang, L., Zhang, C.-L., Chen, Q.-M., Zhang, Z.-R., & Lin, Y. F. (2013). Preformed albumin corona, a protective coating for nanoparticles based drug delivery system. *Biomaterials*, *34*(33), 8521–8530. doi:10.1016/j.biomaterials.2013.07.102 PMID:23932500

Pennycook, T. J., McBride, J. R., Rosenthal, S. J., Pennycook, S. J., & Pantelides, S. T. (2012). Dynamic fluctuations in ultrasmall nanocrystals induce white light emission. *Nano Letters*, *12*(6), 3038–3042. doi:10.1021/nl3008727 PMID:22568665

Pereira, R., Rocha-Santos, T. A. P., Antunes, F. E., Rasteiro, M. G., Ribeiro, R., Goncalves, F., Soares, A. M. V. M., & Lopes, I. (2011). Screening evaluation of the ecotoxicity and genotoxicity of soils contaminated with organic and inorganic nanoparticles: The role of ageing. *Journal of Hazardous Materials*, *194*, 345–354. doi:10.1016/j.jhazmat.2011.07.112 PMID:21871729

Perelson, A. S., & Ribeiro, R. M. (2013). Modeling the within-host dynamics of HIV infection. *BMC Biology*, *11*(1), 96. doi:10.1186/1741-7007-11-96 PMID:24020860

Petersen, E. J., Huang, Q., & Weber, W. J. Jr. (2008). Bioaccumulation of radio-labeled carbon nanotubes by Eisenia foetida. *Environmental Science & Technology*, *42*(8), 3090–3095. doi:10.1021/es071366f PMID:18497171

Petersen, J., Dandri, M., Mier, W., Lutgehetmann, M., Volz, T., von Weizsacker, F., Haberkorn, U., Fischer, L., Pollok, J.-M., Erbes, B., Seitz, S., & Urban, S. (2008). Prevention of hepatitis B virus infection in vivo by entry inhibitors derived from the large envelope protein. *Nature Biotechnology*, *26*(3), 335–341. doi:10.1038/nbt1389 PMID:18297057

Peyser, L. A., Vinson, A. E., Bartko, A. P., & Dickson, R. M. (2001). Photoactivated fluorescence from individual silver nanoclusters. *Science*, *291*(5501), 103–106. doi:10.1126cience.291.5501.103 PMID:11141556

Pfaller, M. A., & Diekema, D. J. (2012). Progress in antifungal susceptibility testing of Candida spp. by use of Clinical and Laboratory Standards Institute broth microdilution methods, 2010 to 2012. *Journal of Clinical Microbiology*, *50*(9), 2846–2856. doi:10.1128/JCM.00937-12 PMID:22740712

Pfaller, M. A., & Wenzel, R. (1992). Impact of the changing epidemiology of fungal infections in the 1990s. *European Journal of Clinical Microbiology & Infectious Diseases*, *11*(4), 287–291. doi:10.1007/BF01962067 PubMed

Phillips, A. N., Cambiano, V., Miners, A., Revill, P., Pillay, D., Lundgren, J. D., Bennett, D., Raizes, E., Nakagawa, F., De Luca, A., Vitoria, M., Barcarolo, J., Perriens, J., Jordan, M. R., & Bertagnolio, S. (2014). Effectiveness and cost-effectiveness of potential responses to future high levels of transmitted HIV drug resistance in antiretroviral drug-naive populations beginning treatment: Modelling study and economic analysis. *The Lancet. HIV*, *1*(2), 85–93. doi:10.1016/S2352-3018(14)70021-9 PMID:26423990

Philosof-Mazor, L., Dakwar, G. R., Popov, M., Kolusheva, S., Shames, A., Linder, C., Greenberg, S., Heldman, E., Stepensky, D., & Jelinek, R. (2013). Bolaamphiphilic vesicles encapsulating iron oxide nanoparticles: New vehicles for magnetically targeted drug delivery. *International Journal of Pharmaceutics*, *450*(1-2), 241–249. doi:10.1016/j.ijpharm.2013.04.017 PMID:23623794

Pichardo, S., Gutiérrez-Praena, D., Puerto, M., Sánchez, E., Grilo, A., Cameán, A. M., & Jos, Á. (2012). Oxidative stressresponses to carboxylic acid functionalized single wall carbon nano-tubes on the human intestinal cell line Caco-2. *Toxicology In Vitro*, *26*(5), 672–677. doi:10.1016/j.tiv.2012.03.007 PMID:22449549

Picheth, G. F., Pirich, C. L., Sierakowski, M. R., Woehl, M. A., Sakakibara, C. N., de Souza, C. F., Martin, A. A., da Silva, R., & de Freitas, R. A. (2017). Bacterial cellulose in biomedical applications: A review. *International Journal of Biological Macromolecules*, *104*, 97–106. doi:10.1016/j.ijbiomac.2017.05.171 PMID:28587970

Pieters, J. (2008). Mycobacterium tuberculosis and the macrophage: Maintaining a balance. *Cell Host & Microbe*, *3*(6), 399–407. doi:10.1016/j.chom.2008.05.006 PMID:18541216

Pietroiusti, A., Stockmann-Juvala, H., Lucaroni, F., & Savolainen, K. (2018). Nanomaterial exposure, toxicity, and impact on human health. *Wiley Interdisciplinary Reviews. Nanomedicine and Nanobiotechnology*, *10*(5), e1513. doi:10.1002/wnan.1513 PMID:29473695

Piktel, E., Wnorowska, U., Cieśluk, M., Deptula, P., Pogoda, K., Misztalewska-Turkowicz, I., Paprocka, P., Niemirowicz-Laskowska, K., Wilczewska, A. Z., Janmey, P. A., & Bucki, R. (2019). Inhibition of inflammatory response in human keratinocytes by magnetic nanoparticles functionalized with PBP10 peptide derived from the PIP2-binding site of human plasma gelsolin. *Journal of Nanobiotechnology*, *17*(1), 22. doi:10.118612951-019-0455-5 PMID:30711007

Pileni, M. P. (1997). Nanosized particles made in colloidal assemblies. *Langmuir*, *13*(13), 3266–3276. doi:10.1021/la960319q

Pinheiro, M., Lúcio, M., Lima, J. L., & Reis, S. (2011). Liposomes as drug delivery systems for the treatment of TB. *Nanomedicine (London)*, *6*(8), 1413–1428. doi:10.2217/nnm.11.122 PMID:22026379

Pinheiro, M., Ribeiro, R., Vieira, A., Andrade, F., & Reis, S. (2016). Design of a nanostructured lipid carrier intended to improve the treatment of tuberculosis. *Drug Design, Development and Therapy*, *10*, 2467–2475. doi:10.2147/DDDT.S104395 PMID:27536067

Pinho, S. S., & Reis, C. A. (2015). Glycosylation in cancer: Mechanisms and clinical implications. *Nature Reviews. Cancer*, *15*(9), 540–555. doi:10.1038/nrc3982 PMID:26289314

Pinnell, S. R., Fairhurst, D., Gillies, R., Mitchnick, M. A., & Kollias, N. (2000). Microfine zinc oxide is a superior sunscreen ingredient to microfine titanium dioxide. *Dermatologic Surgery*, *26*(4), 309–314. doi:10.1046/j.1524-4725.2000.99237.x PMID:10759815

Pinto, R. J. B., Diana, S., Sadocco, P., Pascol Neto, C., & Trindade, T. (2013). Antibacterial Activity of Nanocomposites of Copper and Cellulose. *BioMed Research International*, *2013*, 1–6. doi:10.1155/2013/280512 PMID:24455681

Pires, D. P., Cleto, S., Sillankorva, S., Azeredo, J., & Lu, T. K. (2016). Genetically engineered phages: A review of advances over the last decade. *Microbiology and Molecular Biology Reviews*, *80*(3), 523–543. doi:10.1128/MMBR.00069-15 PMID:27250768

Pirmardvand Chegini, S., Varshosaz, J., & Taymouri, S. (2018). Recent approaches for targeted drug delivery in rheumatoid arthritis diagnosis and treatment. *Artificial Cells, Nanomedicine, and Biotechnology*, *46*(sup2), 502–514. doi:10.1080/21691401.2018.1460373 PMID:29661045

Pisani, E. (2015). Antimicrobial resistance: What does medicine quality have to do with it? *Antimicrobial Resistance Review,* 1-40. Retrieved August 14, 2020, from https://amr-review.org/sites/default/files/ElizabethPisaniMedicinesQualitypaper.pdf

Pissuwan, D., Niidome, T., & Cortie, M. B. (2011). The forthcoming applications of gold nanoparticles in drug and gene delivery systems. *Journal of Controlled Release, 149*(1), 65–71. doi:10.1016/j.jconrel.2009.12.006 PMID:20004222

Plant, A. L., Benson, D. M., & Smith, L. C. (1985). Cellular uptake and intercellular localization of benzo (a) pyrene by digital fluorescence imaging microscopy. *The Journal of Cell Biology, 100*(4), 1295–1308. doi:10.1083/jcb.100.4.1295 PMID:3980583

Plassat, V., Wilhelm, C., Marsaud, V., Ménager, C., Gazeau, F., Renoir, J.-M., & Lesieur, S. (2011). Anti-estrogen- loaded superparamagnetic liposomes for intracellular magnetic targeting and treatment of breast cancer tumors. *Advanced Functional Materials, 21*(1), 83–92. doi:10.1002/adfm.201001450

Pokhrel, L. R., & Dubey, B. (2013). Evaluation of developmental responses of two crop plants exposed to silver and zinc oxide nanoparticles. *The Science of the Total Environment, 452*, 321–332. doi:10.1016/j.scitotenv.2013.02.059 PMID:23532040

Politano, A. D., Campbell, K. T., Rosenberger, L. H., & Sawyer, R. G. (2013). Use of silver in the prevention and treatment of infections. *Silver review. Surgical Infections, 14*(1), 8–20. doi:10.1089ur.2011.097 PMID:23448590

Polo, E., del Pino, P., Pardo, A., Taboada, P., & Pelaz, B. (2018). Magnetic Nanoparticles for Cancer Therapy and Bioimaging. In G. Gonçalves & G. Tobias (Eds.), Nanooncology, Nanomedicine and Nanotoxicology. Springer International Publishing AG. doi:10.1007/978-3-319-89878-0_7

Polymers to Combat Antibiotic-Resistant Bacterial Infections - Advanced Science News. (n.d.). Available at: https://www.advancedsciencenews.com/polymers-to-combat-antibiotic-resistant-bacterial-infections/

Pommier, Y., Johnson, A. A., & Marchand, C. (2005). Integrase inhibitors to treat HIV/AIDS. *Nature Reviews. Drug Discovery, 4*(3), 236–248. doi:10.1038/nrd1660 PMID:15729361

Poole, K. (2002). Mechanisms of bacterial biocide and antibiotic resistance. *Journal of Applied Microbiology, 92*(S1), 55S–64S. doi:10.1046/j.1365-2672.92.5s1.8.x PMID:12000613

Potekhin, S. A., Melnik, T. N., Popov, V., Lanina, N. F., Vazina, A. A., Rigler, P., Verdini, A. S., Corradin, G., & Kajava, A. V. (2001). De novo design of fibrils made of short helical coiled coil peptides. *Chemistry & Biology, 8*(11), 1025–1032. doi:10.1016/S1074-5521(01)00073-4 PMID:11731294

Potter, T. M., Neun, B. W., & Dobrovolskaia, M. A. (2018). Invitro and invivo methods for analysis of nanoparticle potential to induce delayed type hypersensitivity reactions. *Methods in Molecular Biology (Clifton, N.J.), 1682*, 197–210. doi:10.1007/978-1-4939-7352-1_17 PMID:29039104

Poudel, K., Gautam, M., Jin, S. G., Choi, H. G., Yong, C. S., & Kim, J. O. (2019). Copper sulfide: An emerging adaptable nanoplatform in cancer theranostics. *International Journal of Pharmaceutics, 562*(March), 135–150. doi:10.1016/j.ijpharm.2019.03.043 PMID:30904728

Poulsen, L. K., Ballard, G., & Stahl, D. A. (1993). Use of rRNA fluorescence in situ hybridization for measuring the activity of single cells in young and established biofilms. *Applied and Environmental Microbiology, 59*(5), 1354–1360. doi:10.1128/AEM.59.5.1354-1360.1993 PMID:7685999

Prabaharan, M. (2015). Chitosan-based nanoparticles for tumor-targeted drug delivery. *International Journal of Biological Macromolecules, 72*, 1313–1322. doi:10.1016/j.ijbiomac.2014.10.052 PMID:25450550

Prabha, G., & Raj, V. (2016). Formation and characterization of β-cyclodextrin (β-CD)–polyethyleneglycol (PEG)–polyethyleneimine (PEI) coated Fe3O4 nanoparticles for loading and releasing 5-Fluorouracil drug. *Biomedicine and Pharmacotherapy*, *80*, 173–182. doi:10.1016/j.biopha.2016.03.015 PMID:27133054

Prabha, G., & Raj, V. (2017). Sodium alginate–polyvinyl alcohol–bovin serum albumin coated Fe3 O4 nanoparticles as anticancer drug delivery vehicle: Doxorubicin loading and in vitro release study and cytotoxicity to HepG2 and L02 cells. *Materials Science and Engineering*, *79*, 410–422. doi:10.1016/j.msec.2017.04.075 PMID:28629035

Prabhakar, U., Maeda, H., Jain, R. K., Sevick-Muraca, E. M., Zamboni, W., Farokhzad, O. C., ... Blakey, D. C. (2013). *Challenges and key considerations of the enhanced permeability and retention effect for nanomedicine drug delivery in oncology*. Academic Press.

Prabhu, S., & Poulose, E. K. (2012). Silver nanoparticles: Mechanism of antimicrobial action, synthesis, medical applications, and toxicity effects. *International Nano Letters*, *32*(1), 2–10. doi:10.1186/2228-5326-2-32

Prabhu, Y. T., Rao, K. V., Kumari, B. S., Kumar, V. S., & Pavani, T. (2015). Synthesis of Fe3O4 nanoparticles and its antibacterial application. *International Nano Letters*, *5*(2), 85–92. doi:10.100740089-015-0141-z

Prasad Rao, J., & Geckeler, K. E. (2011). Polymer nanoparticles: Preparation techniques and size-control parameters. *Progress in Polymer Science*, *36*(7), 887–913. doi:10.1016/j.progpolymsci.2011.01.001

Prasad, K. S., Patel, H., Patel, T., Patel, K., & Selvaraj, K. (2012). Biosynthesis of Se nanoparticles and its effect on UV-induced DNA damage. *Colloids and Surfaces. B, Biointerfaces*, *103*, 261–266. doi:10.1016/j.colsurfb.2012.10.029 PMID:23201746

Prasanna, A., Pooja, R., Suchithra, V., Ravikumar, A., Kumar Gupta, P., & Niranjan, V. (2018). Akhil Ravikumar, Praveen Kumar Gupta, Vidya Niranjan. Smart Drug Delivery Systems for Cancer Treatment Using Nanomaterials. *Materials Today: Proceedings*, *5*(10), 21047–21054. doi:10.1016/j.matpr.2018.06.498

Prasanna, V. L., & Vijayaraghavan, R. (2015). Insight into the mechanism of antibacterial activity of ZnO: Surface defects mediated reactive oxygen species even in the dark. *Langmuir*, *31*(33), 9155–9162. doi:10.1021/acs.langmuir.5b02266 PMID:26222950

Prashanthini, L. K., & Kalyani, D. (2015). Preparation of scaffold using rice husk and starch for bone regeneration. *International Journal for Research in Applied Science and Engineering Technology*, *3*(6), 598–603.

Preethi, B., Sha, S. K., Jayaprakash, R., Kutti Rani, S., & Hemalatha, S. (2019). Synthesis, characterization and biological studies on 4-bromo-2-{(z)-[(furan-2-ylmethyl) imino]methyl}phenol praseodymium complex. *Rasayan Journal of Chemistry*, *12*(3), 1455–1462. doi:10.31788/RJC.2019.1235296

Prestinaci, F., Patrizio, P., & Annalisa, P. (2015). Antimicrobial Resistance: A Global Multifaceted Phenomenon. *Pathogens and Global Health*, *109*(7), 309–318. doi:10.1179/2047773215Y.0000000030 PubMed

Prestinaci, F., Pezzotti, P., & Pantosti, A. (2015). Antimicrobial resistance: A global multifaceted phenomenon. *Pathogens and Global Health*, *109*(7), 309–318. doi:10.1179/2047773215Y.0000000030 PMID:26343252

Primeau, A. J., Rendon, A., Hedley, D., Lilge, L., & Tannock, I. F. (2005). The distribution of the anticancer drug Doxorubicin in relation to blood vessels in solid tumors. *Clinical Cancer Research*, *11*(24), 8782–8788. doi:10.1158/1078-0432.CCR-05-1664 PMID:16361566

Pujals, S., & Giralt, E. (2008). Proline-rich, amphipathic cell-penetrating peptides. *Advanced Drug Delivery Reviews*, *60*(4-5), 473–484. doi:10.1016/j.addr.2007.09.012 PMID:18187229

Pursey, E., Sunderhauf, D., Gaze, W. H., Westra, E. R., & van Houte, S. (2018). CRISPR-Cas antimicrobials: Challenges and future prospects. *PLoS Pathogens*, *14*(6), e1006990. doi:10.1371/journal.ppat.1006990 PMID:29902258

Pushpanathan, M., Gunasekaran, P., & Rajendhran, J. (2013). Antimicrobial Peptides: Versatile Biological Properties. *International Journal of Peptides*, *12*, 92–119. PMID:23935642

Pushpanathan, M., Gunasekaran, P., & Rajendhran, J. (2013). Antimicrobial peptides: Versatile biological properties. *International Journal of Peptides*. PMID:23935642

Puvanakrishnan, P., Park, J., Chatterjee, D., Krishnan, S., & Tunnel, J. W. (2012). In vivo tumor targeting of gold nanoparticles: Effect of particle type and dosing strategy. *International Journal of Nanomedicine*, *7*, 1251–1258. doi:10.2147/IJN.S29147 PMID:22419872

Pyle, M. M. (1947, October). Relative numbers of resistant tubercle bacilli in sputa of patients before and during treatment with streptomycin. *Proceedings of the Staff Meetings. Mayo Clinic*, *22*(21), 465. PMID:20267460

Qadir, M. I., & Chauhdary, Z. (2018). Antibacterial Activity of Novel Strains of Bacteriophages: An Experimental Approach. *Critical Reviews in Eukaryotic Gene Expression*, *28*(1), 1–12. doi:10.1615/CritRevEukaryotGeneExpr.2017019608 PMID:29773011

Qian, L. P., Zhou, L. H., Too, H. P., & Chow, G. M. (2010). Gold decorated NaYF4:Yb,Er/NaYF4/silica (core/shell/shell) upconversion nanoparticles for photothermal destruction of BE(2)-C neuroblastoma cells. *Journal of Nanoparticle Research*, *13*(2), 499–510. doi:10.100711051-010-0080-6

Qi, L., Xu, Z., Jiang, X., Hu, C., & Zou, X. (2014). Preparation and antibacterial activity of chitosan nanoparticles. *Carbohydrate Research*, *339*(16), 2693–2700. doi:10.1016/j.carres.2004.09.007 PMID:15519328

Qiu, Y., & Park, K. (2001). Environment-sensitive hydrogels for drug delivery. *Advanced Drug Delivery Reviews*, *53*(3), 321–339. doi:10.1016/S0169-409X(01)00203-4 PMID:11744175

Qi, W., Zhang, X., & Wang, H. (2018). Self-assembled polymer nanocomposites for biomedical application. *Current Opinion in Colloid & Interface Science*, *35*, 36–41. doi:10.1016/j.cocis.2018.01.003

Quirós, J., Boltes, K., & Rosal, R. (2016). Bioactive Applications for Electrospun Fibers. *Polymer Reviews (Philadelphia, Pa.)*, *56*(4), 631–667. doi:10.1080/15583724.2015.1136641

Raavé, R., van Kuppevelt, T. H., & Daamen, W. F. (2018). Chemotherapeutic drug delivery by tumoral extracellular matrix targeting. *Journal of Controlled Release*, *274*, 1–8. doi:10.1016/j.jconrel.2018.01.029 PMID:29382546

Radovic-Moreno, A. F., Lu, T. K., Puscasu, V. A., Yoon, C. J., Langer, R., & Farokhzad, O. C. (2012). Surface charge-switching polymeric nanoparticles for bacterial cell wall-targeted delivery of antibiotics. *ACS Nano*, *6*(5), 4279–4287. doi:10.1021/nn3008383 PMID:22471841

Raffi, M., Mehrwan, S., Bhatti, T. M., Akhter, J. I., Hameed, A., Yawar, W., & Hasan, M. M. (2010). Investigations into the antibacterial behavior of copper nanoparticles against Escherichia coli. *Annals of Microbiology*, *60*(1), 75–80. doi:10.1007/s13213-010-0015-6

Raghavan, V., Deb, A., & Grace, A. N. (2020). *Honokiol-Camptothecin loaded graphene oxide nanoparticle towards combinatorial anti-cancer drug delivery*. IET Nanobiotechnology.

Raghunath, A., & Perumal, E. (2017). Metal oxide nanoparticles as antimicrobial agents: A promise for the future. *International Journal of Antimicrobial Agents*, *49*(2), 137–152. doi:10.1016/j.ijantimicag.2016.11.011 PMID:28089172

Raghupathi, K. R., Koodali, R. T., & Manna, A. C. (2011). Size-dependent bacterial growth inhibition and mechanism of antibacterial activity of zinc oxide nanoparticles. *Langmuir*, *27*(7), 4020–4028. doi:10.1021/la104825u PMID:21401066

Rahman, A. E., Iqbal, A., Hoque, D. E., Moinuddin, M., Zaman, S. B., Rahman, Q., Begum, T., Chowdhury, A. I., Haider, R., Arifeen, S. E., Kissoon, N., & Larson, C. P. (2017). Managing neonatal and early childhood syndromic sepsis in sub-district hospitals in resource poor settings: Improvement in quality of care through introduction of a package of interventions in rural Bangladesh. *PLoS One*, *12*(1), e0170267. doi:10.1371/journal.pone.0170267 PMID:28114415

Rahman, K., Khan, S. U., Fahad, S., Chang, M. X., Abbas, A., Khan, W. U., Rahman, L., Haq, Z. U., Nabi, G., & Khan, D. (2019). Nano-biotechnology: A new approach to treat and prevent malaria. *International Journal of Nanomedicine*, *14*, 1401–1410. doi:10.2147/IJN.S190692 PMID:30863068

Rai, A., Prabhune, A., & Perry, C. C. (2010). Antibiotic mediated synthesis of gold nanoparticles with potent antimicrobial activity and their application in antimicrobial coatings. *Journal of Materials Chemistry*, *20*(32), 6789–6798. doi:10.1039/c0jm00817f

Rai, A., Yadav, A. G., & Gade, A. (2009). Silver nanoparticles as a new generation of antimicrobials. *Biotechnology Advances*, *27*(1), 76–83. doi:10.1016/j.biotechadv.2008.09.002 PMID:18854209

Rai, M., Ingle, A. P., Gaikwad, S., Gupta, I., Gade, A., & Silvério da Silva, S. (2016). Nanotechnology based anti-infectives to fight microbial intrusions. *Journal of Applied Microbiology*, *120*(3), 527–542. doi:10.1111/jam.13010 PMID:26599354

Rai, V. K., Mishra, N., Yadav, K. S., & Yadav, N. P. (2018). Nanoemulsion as pharmaceutical carrier for dermal and transdermal drug delivery: Formulation development, stability issues, basic considerations and applications. *Journal of Controlled Release*, *270*, 203–225. doi:10.1016/j.jconrel.2017.11.049 PMID:29199062

Raja, M., Shanmugharaj, A. M., & Hun, R. S. (2010). An efficient growth of silver and copper nanoparticles on multiwalled carbon nanotube with enhanced antimicrobial activity. *Journal of Biomedical Materials Research*, *96B*, 119–126. PubMed

Rajan, R., Chandran, K., Harper, S. L., Yun, S.-I., & Kalaichelvan, P. T. (2015). Plant extract synthesized silver nanoparticles: An ongoing source of novel biocompatible materials. *Industrial Crops and Products*, *70*, 356–373. doi:10.1016/j.indcrop.2015.03.015

Rajender, N., & Suresh, K. I. (2016). Surface-initiated atomtransfer radical polymerization (SI-ATRP) from graphene oxide: Effect of functionalized graphene sheet (FGS) on the synthesis and material properties of PMMA nanocomposites. *Macromolecular Materials and Engineering*, *301*(1), 81–92. doi:10.1002/mame.201500256

Rajeshkumar, S. (2016). Anticancer activity of eco-friendly gold nanoparticles against lung and liver cancer cells. *Journal of Genetic Engineering and Biotechnology*, *14*(1), 195–202. doi:10.1016/j.jgeb.2016.05.007 PMID:30647615

Rajeshkumar, S., & Bharath, L. V. (2017). Mechanism of plant-mediated synthesis of silver nanoparticles – A review on biomolecules involved, characterisation and antibacterial activity. *Chemico-Biological Interactions*, *273*, 219–227. doi:10.1016/j.cbi.2017.06.019 PMID:28647323

Ramacharyulu, P. V., Muhammad, R., Kumar, J. P., & Prasad, G. K. (2015). Iron phthalocyanine modified mesoporous titania nanoparticles for photocatalytic activity and CO 2 capture applications. *Physical Chemistry Chemical Physics*, *17*(39), 26456–26462. doi:10.1039/C5CP03576G PMID:26393761

Ramakrishna, S., Fujihara, K., Teo, W. E., Yong, T., Ma, Z., & Ramaseshan, R. (2006). Electrospun nanofibers: Solving global issues. *Materials Today*, *9*(3), 40–50. doi:10.1016/S1369-7021(06)71389-X

Raman, S. S., Hecht, J. R., & Chan, E. (2019). Talimogene laherparepvec: Review of its mechanism of action and clinical efficacy and safety. *Immunotherapy*, *11*(8), 705–723. doi:10.2217/imt-2019-0033 PMID:31045464

Rana, S., & Kalaichelvan, P. (2013). Ecotoxicity of Nanoparticles. *ISRN Toxicology, 574648*, 1–11. Advance online publication. doi:10.1155/2013/574648 PMID:23724300

Ranghar, S., Sirohi, P., Verma, P., & Agarwal, V. (2012). Nanoparticle-based drug delivery systems: Promising approaches against infections. *Brazilian Archives of Biology and Technology, 57*(2), 209–222. doi:10.1590/S1516-89132013005000011

Ranjani, S., Shariq Ahmed, M., Ruckmani, K., & Hemalatha, S. (2019b). Green nanocolloids control Multi drug resistant pathogenic bacteria. Journal of Cluster Science. Advance online publication. doi:10.1007/s10876-019-01694-6

Ranjani, S., Tamanna, K., & Hemalatha, S. (2019a) Triphala green nano colloids: Synthesis, characterization and screening biomarkers. J. App. Nano Sci.

Ranjbar-Mohammadi, M., (2018). Characteristics of aloe vera incorporated poly (ε-caprolactone)/gum tragacanthnanofibers as dressings for wound care. *Journal of Industrial Textiles, 47*(7), 144–1477.

Ranjbar-Mohammadi, M., & Bahrami, S. H. (2016). Electrospun curcumin loaded poly (ε-caprolactone)/gum tragacanth nanofibers for biomedical application. *International Journal of Biological Macromolecules, 84*, 448–456. doi:10.1016/j.ijbiomac.2015.12.024 PMID:26706845

Ranjbar-Mohammadi, M., Bahrami, S. H., & Joghataei, M. T. (2013). Fabrication of novel nanofiber scaffolds from gum tragacanth/poly (vinyl alcohol) for wound dressing application: In vitro evaluation and antibacterial properties. *Materials Science and Engineering C, 33*(8), 4935–4943. doi:10.1016/j.msec.2013.08.016 PMID:24094207

Rasheed, T., Bilal, M., Li, C., Nabeel, F., Khalid, M., & Iqbal, H. M. N. (2018). Catalytic potential of bio-synthesized silver nanoparticles using Convolvulus arvensis extract for the degradation of environmental pollutants. *Journal of Photochemistry and Photobiology B: Biology, 181*(2017), 44–52. doi:10.1016/j.jphotobiol.2018.02.024

Rasheed, T., Bilal, M., Iqbal, H. M. N., & Li, C. (2017). Green biosynthesis of silver nanoparticles using leaves extract of Artemisia vulgaris and their potential biomedical applications. *Colloids and Surfaces. B, Biointerfaces, 158*, 408–415. doi:10.1016/j.colsurfb.2017.07.020 PMID:28719862

Rashida, M. M. O., Islam, M. S., Haque, M. A., Rahman, M. A., Hossain, M. T., & Hamid, M. A. (2016). Antibacterial activity of polyaniline coated silver nanoparticles synthesized from Piper betle leaves extract. *IJPR, 15*(2), 591. PMID:27642330

Rastogi, A., Zivcak, M., Sytar, O., Kalaji, H. M., He, X., Mbarki, S., & Brestic, M. (2017). Impact of Metal and Metal Oxide Nanoparticles on Plant: A Critical Review. *Frontiers in Chemistry, 5*, 78. doi:10.3389/fchem.2017.00078 PMID:29075626

Rau, R. B., de Lima-Morales, D., Wink, P. L., Ribeiro, A. R., Martins, A. F., & Barth, A. L. (2018). Emergence of mcr-1 Producing Salmonella enterica serovar Typhimurium from Retail Meat: First Detection in Brazil. *Foodborne Pathogens and Disease, 15*(1), 58–59. doi:10.1089/fpd.2017.2346 PMID:29048947

Raveendran, S., Dhandayuthapani, B., Nagaoka, Y., Yoshida, Y., Maekawa, T., & Sakthi Kumar, D. (2013). Biocompatible nanofibers based on extremophilic bacterial polysaccharide, Mauran from HalomonasMaura. *Carbohydrate Polymers, 92*(2), 1225–1233. doi:10.1016/j.carbpol.2012.10.033 PMID:23399150

Ravindran, A., Chandran, P., & Khan, S. S. (2013). Biofunctionalized silver nanoparticles: Advances and prospects. *Colloids and Surfaces. B, Biointerfaces, 105*, 342–352. doi:10.1016/j.colsurfb.2012.07.036 PMID:23411404

Ravishankar, R. V., & Jamuna, B. A. (2011). Nanoparticles and their potential application as antimicrobials. Science against microbial pathogens: communicating current research and technological advances, 197-209.

Ravishankar, R. V., & Jamuna, B. A. (2011). Nanoparticles and their potential application as antimicrobials. In A. Méndez-Vilas (Ed.), *Science against Microbial Pathogen: Communicating Current Research and Technological Advances* (Vol. 2, pp. 197–209). Formatex Research Center.

Rawal, I., & Kaur, A. (2013). Synthesis of mesoporous polypyrrole nanowires/nanoparticles for ammonia gas sensing application. *Sensors and Actuators. A, Physical, 203*, 92–102. doi:10.1016/j.sna.2013.08.023

Rawat, R., Whitty, A., & Tonge, P. J. (2003). The isoniazid-NAD adduct is a slow, tight-binding inhibitor of InhA, the Mycobacterium tuberculosis enoyl reductase: Adduct affinity and drug resistance. *Proceedings of the National Academy of Sciences of the United States of America, 100*(24), 13881–13886. doi:10.1073/pnas.2235848100 PMID:14623976

Ray Chowdhuri, A., Bhattacharya, D., & Sahu, S. K. (2016). Magnetic nanoscale metal organic frameworks for potential targeted anticancer drug delivery, imaging and as an MRI contrast agent. *Dalton Transactions (Cambridge, England), 45*(7), 2963–2973. doi:10.1039/C5DT03736K PMID:26754449

Raygada, J. L., & Levine, D. P. (2009). Methicillin-Resistant Staphylococcus aureus: A Growing Risk in the Hospital and in the Community. *American Health & Drug Benefits, 2*, 86–95. PubMed

Ray, S. S., Chen, S. S., Li, C. W., Nguyen, N. C., & Nguyen, H. T. (2016). A comprehensive review: Electrospinning technique for fabrication and surface modification of membranes for water treatment application. *RSC Advances, 6*(88), 85495–85514. doi:10.1039/C6RA14952A

Reches, M., & Gazit, E. (2003). Casting metal nanowires within discrete self-assembled peptide nanotubes. *Science, 300*(5619), 625–627. doi:10.1126cience.1082387 PMID:12714741

Reddy, K. M., Feris, K., Bell, J., Wingett, D. G., Hanley, C., & Punnoose, A. (2007). Selective toxicity of zinc oxide nanoparticles to prokaryotic and eukaryotic systems. *Applied Physics Letters, 90*(213902), 2139021–2139023. doi:10.1063/1.2742324 PMID:18160973

Reddy, L. H., Arias, J. L., Nicolas, J., & Couvreur, P. (2012). Magnetic Nanoparticles: Design and Characterization, Toxicity and Biocompatibility, Pharmaceutical and Biomedical Applications. *Chemical Reviews, 112*(11), 5818–5878. doi:10.1021/cr300068p PMID:23043508

Reddy, P. M., Chang, K. C., Liu, Z. J., Chen, C. T., & Ho, Y. P. (2014). Functionalized magnetic iron oxide (Fe_3O_4) nanoparticles for capturing gram-positive and gram-negative bacteria. *Journal of Biomedical Nanotechnology, 10*(8), 1429–1439. doi:10.1166/jbn.2014.1848 PMID:25016643

Reijnders, L. (2006). Cleaner nanotechnology and hazard reduction of manufactured nanoparticles. *Journal of Cleaner Production, 14*(2), 124–133. doi:10.1016/j.jclepro.2005.03.018

Rengarajan, J., Sassetti, C. M., Naroditskaya, V., Sloutsky, A., Bloom, B. R., & Rubin, E. J. (2004). The folate pathway is a target for resistance to the drug para-aminosalicylic acid (PAS) in mycobacteria. *Molecular Microbiology, 53*(1), 275–282. doi:10.1111/j.1365-2958.2004.04120.x PMID:15225321

Resnik, D., & Tinkle, S. (2007). Ethics in Nanomedicine. *Nanomedicine (London), 2*(3), 345–350. doi:10.2217/17435889.2.3.345 PMID:17716179

Reza Mozafari, M., Johnson, C., Hatziantoniou, S., & Demetzos, C. (2008). Nanoliposomes and their applications in food nanotechnology. *Journal of Liposome Research, 18*(4), 309–327. doi:10.1080/08982100802465941 PMID:18951288

Rhee, S. Y., Blanco, J. L., Jordan, M. R., Taylor, J., Lemey, P., Varghese, V., Hamers, R. L., Bertagnolio, S., de Wit, T. F. R., Aghokeng, A. F., Albert, J., Avi, R., Avila-Rios, S., Bessong, P. O., Brooks, J. I., Boucher, C. A. B., Brumme, Z. L., Busch, M. P., Bussmann, H., ... Shafer, R. W. (2015). Geographic and Temporal Trends in the Molecular Epidemiology and Genetic Mechanisms of Transmitted HIV-1 Drug Resistance: An Individual-Patient- and SequenceLevel Meta-Analysis. *PLoS Medicine*, *12*(4), e1001810. doi:10.1371/journal.pmed.1001810 PMID:25849352

Ribeiro, S. M., Felício, M. R., Boas, E. V., Goncalves, S., Costa, F. F., Samy, R. P., Santos, N. C., & Franco, O. L. (2016). New frontiers for anti-biofilm drug development. *Pharmacology & Therapeutics*, *160*, 133–144. doi:10.1016/j.pharmthera.2016.02.006 PMID:26896562

Ribeiro, T. P., Monteiro, F. J., & Laranjeira, M. S. (2020). Duality of iron (III) doped nano hydroxyapatite in triple negative breast cancer monitoring and as a drug-free therapeutic agent. *Ceramics International*, *46*(10), 16590–16597. doi:10.1016/j.ceramint.2020.03.231

Richard, J. P., Melikov, K., Brooks, H., Prevot, P., Lebleu, B., & Chernomordik, L. V. (2005). Cellular uptake of unconjugated TAT peptide involves clathrin-dependent endocytosis and heparan sulfate receptors. *The Journal of Biological Chemistry*, *280*(15), 15300–15306. doi:10.1074/jbc.M401604200 PMID:15687490

Richard, J. W., Spencer, B. A., McCoy, L. F., Carina, E., Washington, J., Edgar, P., Rosenblatt, J., & Goodheart, R. (2002). Acticoat versus silverlon: The truth. *Journal of Burns and Surgical Wound Care*, *1*, 11–20.

Richard, P. F. (1959). There's plenty of room at the bottom: An invitation to enter a new field of physics. *Annual Meeting of the American Physical Society*.

Rico, C. M., Hong, J., Morales, M. I., Zhao, L., Barrios, A. C., Zhang, J.-Y., Peralta-Videa, J. R., & Gardea-Torresdey, J. L. (2013). Effect of cerium oxide nanoparticles on rice: A study involving the antioxidant defense system and in vivo fluorescence imaging. *Environmental Science & Technology*, *47*(11), 5635–5642. doi:10.1021/es401032m PMID:23662857

Rinaudo, M. (2006). Chitin and chitosan: Properties and applications. *Progress in Polymer Science*, *31*(7), 603–632. doi:10.1016/j.progpolymsci.2006.06.001

Rivera-Rodriguez, A., Chiu-Lam, A., Morozov, V.M., Ishov, A.M., & Rinaldi, C. (2018). Magnetic nanoparticle hyperthermia potentiates paclitaxel activity in sensitive and resistant breast cancer cells. *Int J Nanomedicine*, *13*, 4771-4779. . doi:10.2147/IJN.S171130

Rizvi, S. A. A., & Saleh, A. M. (2018). Applications of nanoparticle systems in drug delivery technology. *Saudi Pharmaceutical Journal*, *26*(1), 64–70. doi:10.1016/j.jsps.2017.10.012 PMID:29379334

Roberts, A. P., Chandler, M., Courvalin, P., Guédon, G., Mullany, P., Pembroke, T., Rood, J. I., Jeffery Smith, C., Summers, A. O., Tsuda, M., & Berg, D. E. (2008). Revised nomenclature for transposable genetic elements. *Plasmid*, *60*(3), 167–173. doi:10.1016/j.plasmid.2008.08.001 PMID:18778731

Roberts, A. P., & Mullany, P. (2009). A modular master on the move: The Tn916 family of mobile genetic elements. *Trends in Microbiology*, *6*(6), 251–258. doi:10.1016/j.tim.2009.03.002 PMID:19464182

Robles-Gracia, M. A., Del-Toro-Sanchez, C. L., Marquez-Rios, E., Barrera-Rodriguez, A., Aguilar, J., Aguilar, J. A., Reynoso-Marin, F. J., Ceja, I., Dorame-Miranda, R., & Rodriguez-Felix, F. (2018). Nanofibers of cellulose bagasse from Agave tequilana Weber var. azul by electrospinnimg: Preparation and Characterization. *Carbohydrate Polymers*, *15*(192), 69–74. doi:10.1016/j.carbpol.2018.03.058

Rocha, T. L., Gomes, T., Sousa, V. S., Mestre, N. C., & Bebianno, M. J. (2015). Ecotoxicological impact of engineered nanomaterials in bivalve molluscs: An overview. *Marine Environmental Research*, *111*, 74–88. doi:10.1016/j.marenvres.2015.06.013 PMID:26152602

Roco, M. C. (2011). The long view of nanotechnology development: The National Nanotechnology Initiative at 10 years. *Journal of Nanoparticle Research*, 13.

Roduner, E. (2006). Size matters: Why nanomaterials are different. *Chemical Society Reviews*, *35*(7), 583–592. doi:10.1039/b502142c PMID:16791330

Roe, D., Karandikar, B., Bonn-Savage, N., Gibbins, B., & Roullet, J. B. (2008). Antimicrobial surface functionalization of plastic catheters by silver nanoparticles. *The Journal of Antimicrobial Chemotherapy*, *61*(4), 869–876. doi:10.1093/jac/dkn034 PMID:18305203

Rohini, B., Tahira, A., Waseem, M., Khan, J., Kashif, M., & Hemalatha, S. (2019). AgNPs from Nigella sativa control breast cancer: An In vitro study. *Journal of Environmental Pathology, Toxicology and Oncology*, *38*(2), 185–194. doi:10.1615/JEnvironPatholToxicolOncol.2019027318 PubMed

Roh, J. Y., Sim, S. J., Yi, J., Park, K., Chung, K. H., Ryu, D. Y., & Choi, J. (2009). Ecotoxicity of silver nanoparticles on the soil nematode Caenorhabditis elegans using functional ecotoxicogenomics. *Environmental Science & Technology*, *43*(10), 3933–3940. doi:10.1021/es803477u PMID:19544910

Roopan, S., Mathew, R., Mahesh, S., Titus, D., Aggarwal, K., Bhatia, N., Kasinathan, D., Elumalai, K., & Samuel, J. (2018). Environmental friendly synthesis of zinc oxide nanoparticles and estimation of its larvicidal activity against Aedes aegypti. *International Journal of Environmental Science and Technology*, *16*(12), 8053–8060. Advance online publication. doi:10.100713762-018-2175-z

Rosemary, M. J., MacLaren, I., & Pradeep, T. (2006). Investigations of the antibacterial properties of ciprofloxacin@SiO2. *Langmuir*, *22*(24), 10125–10129. doi:10.1021/la061411h PMID:17107009

Roshmi, T., Soumya, K. R., Jyothis, M., & Radhakrishnan, E. K. (2015). Effect of biofabricated gold nanoparticle-based antibiotic conjugates on minimum inhibitory concentration of bacterial isolates of clinical origin. *Gold Bulletin*, *48*(1-2), 63–71. doi:10.100713404-015-0162-4

Rossi, E. D., Aínsa, J. A., & Riccardi, G. (2006). Role of mycobacterial efflux transporters in drug resistance: An unresolved question. *FEMS Microbiology Reviews*, *30*(1), 36–52. doi:10.1111/j.1574-6976.2005.00002.x PMID:16438679

Rossi, E. M., Pylkkänen, L., Koivisto, A. J., Nykäsenoja, H., Wolff, H., Savolainen, K., & Alenius, H. (2010). Inhalation exposure to nanosized and fine TiO2 particles inhibits features of allergic asthma in a murine model. *Particle and Fibre Toxicology*, *7*(1), 35. doi:10.1186/1743-8977-7-35 PMID:21108815

Ruber Perez, C. M., Stephanopoulos, N., Sur, S., Lee, S. S., Newcomb, C., & Stupp, S. I. (2015). The Powerful Functions of Peptide-Based Bioactive Matrices for Regenerative Medicine. *Annals of Biomedical Engineering*, *43*(3), 501–514. doi:10.100710439-014-1166-6 PMID:25366903

Ruck, T., Bittner, S., Wiendl, H., & Meuth, S. G. (2015). Alemtuzumab in Multiple Sclerosis: Mechanism of Action and Beyond. *International Journal of Molecular Sciences, 16*(7), 16414-16439. doi:10.3390/ijms160716414

Ruiz-Hernández, E., Baeza, A., & Vallet-Regí, M. (2011). Smart Drug Delivery through DNA/Magnetic Nanoparticle Gates. *ACS Nano*, *5*(2), 1259–1266. doi:10.1021/nn1029229 PMID:21250653

Ruparelia, J. P., Chatterjee, A. K., Duttagupta, S. P., & Mukherji, S. (2008). Strain specificity in antimicrobial activity of silver and copper nanoparticles. *Acta Biomaterialia*, *4*(3), 707–716. doi:10.1016/j.actbio.2007.11.006 PubMed

Ruppé, É., Woerther, P.-L., & Barbier, F. (2015). Mechanisms of Antimicrobial Resistance in Gram-Negative Bacilli. *Annals of Intensive Care*, *5*(1), 21. doi:10.118613613-015-0061-0 PMID:26261001

Russell, A. D., Tattawasart, U., Maillard, J. Y., & Furr, J. R. (1998). Possible link between bacterial resistance and use of antibiotics and biocides. *Antimicrobial Agents and Chemotherapy*, *42*(8), 2151. doi:10.1128/AAC.42.8.2151 PMID:9722471

Rutherford, S. T., & Bassler, B. L. (2012). Bacterial quorum sensing: Its role in virulence and possibilities for its control. *Cold Spring Harbor Perspectives in Medicine*, *2*(11), 12427. doi:10.1101/cshperspect.a012427 PMID:23125205

Ryvkin, H., Ashkenazy, H., Weiss-Ottolenghi, Y., Piller, C., Pupko, T., & Gershoni, J. M. (2018). Phage display peptide libraries: Deviations from randomness and correctives. *Nucleic Acids Research*, *46*(9), e52. doi:10.1093/nar/gky077 PMID:29420788

Sabahi, A., Salahandish, R., Ghaffarinejad, A., & Omidinia, E. (2019). Electrochemical nano-genosensor for highly sensitive detection of miR-21 biomarker based on SWCNT-grafted dendritic Au nanostructure for early detection of prostate cancer. *Talanta*. PMID:31892044

Sabella, S., Carney, R. P., Brunetti, V., Malvindi, M. A., Al-Juffali, N., Vecchio, G., Janes, S. M., Bakr, O. M., Cingolani, R., Stellacci, F., & Pompa, P. P. (2014). A general mechanism for intracellular toxicity of metal-containing nanoparticles. *Nanoscale*, *6*(12), 7052–7061. doi:10.1039/c4nr01234h PMID:24842463

Sadhasivam, L., Dey, N., Francis, A. P., & Devasena, T. (2015). Innovare Academic Sciences. *Transdermal Patches of Chitosan Nanoparticles for Insulin Delivery*, *7*(5).

Sadhu, P. K., Saisivam, S., & Debnath, S. K. (2019). *Design and characterization of niosomes of ethionamide for multi drug resistance tuberculosis*. Academic Press.

Saeb, A. T. M., Alshammari, A. S., Al-Brahim, H., & Al-Rubeaan, K. A. (2014). Production of silver nanoparticles with strong and stable antimicrobial activity against highly pathogenic and multidrug resistant bacteria. *TheScientificWorldJournal*, *2014*, 1–9. doi:10.1155/2014/704708 PMID:25093206

Saeed, K., & Khan, I. (2016). Preparation and characterization of single-walled carbon nanotube/nylon 6, 6 nanocomposites. *Instrumentation Science & Technology*, *44*(4), 435–444. doi:10.1080/10739149.2015.1127256

Sager, T. M., Porter, D. W., Robinson, V. A., Lindsley, W. G., Schwegler-Berry, D. E., & Castranova, V. (2007). Improved method to disperse nanoparticles for in vitro and in vivo investigation of toxicity. *Nanotoxicology*, *1*(2), 118–129. doi:10.1080/17435390701381596

Saggar, J. K., Yu, M., Tan, Q., & Tannock, I. F. (2013). The tumor microenvironment and strategies to improve drug distribution. *Frontiers in Oncology*, *3*, 154. doi:10.3389/fonc.2013.00154 PMID:23772420

Sah, S. K., Rasool, U., & Hemalatha, S. (2019a) Andrographis paniculata extract inhibit growth, biofilm formation in multidrug resistant strains of Klebsiella pneumonia. J. Trad. Compl. Med.

Saha, B., Bhattacharya, J., Mukherjee, A., Ghosh, A., Santra, C., Dasgupta, A. K., & Karmakar, P. (2007). In vitro structural and functional evaluation of gold nanoparticles conjugated antibiotics. *Nanoscale Research Letters*, *2*(12), 614–622. doi:10.100711671-007-9104-2

Sahoo, S. K., & Labhasetwar, V. (2003). Nanotech approaches to drug delivery and imaging. *Drug Discovery Today*, *8*(24), 1112–1120. doi:10.1016/S1359-6446(03)02903-9 PMID:14678737

Sah, S. K., & Hemalatha, S. (2015). Extended spectrum Beta lactamase (ESBL) Mechanism of antibiotic resistance and Epidemiology. *International Journal of Pharm Tech Research*, *7*, 303–309.

Sah, S. K., Rasool, U., & Hemalatha, S. (2019b). Efficacy of Andrographis paniculata against AmpC producing multi drug resistant E. coli Biocat. *Agricultural Biotechnology*, *21*, 101139. doi:10.1016/j.bcab.2019.101139

Sahu, S. C., & Hayes, A. W. (2017). *Toxicity of nanomaterials found in human environment: A literature review.* doi:10.1177/2397847317726352

Saifi, M. A., Khurana, A., & Godugu, C. (2018). Nanotoxicology: toxicity and risk assessment of nanomaterials. In Nanomaterials in chromatography (pp. 437-465). Elsevier.

Saikia, C., Das, M. K., Ramteke, A., & Maji, T. K. (2017). Evaluation of folic acid tagged aminated starch/ZnO coated iron oxide nanoparticles as targeted curcumin delivery system. *Carbohydrate Polymers*, *157*, 391–399. doi:10.1016/j.carbpol.2016.09.087 PMID:27987943

Saksena, R., Matlani, M., Singh, V., Kumar, A., Anveshi, A., Kumar, D., & Gaind, R. (2017). Early treatment failure in concurrent dengue and mixed malaria species infection with suspected resistance to artemisinin combination therapy from a tertiary care center in Delhi: A case report. *International Medical Case Reports Journal*, *10*, 289–294. doi:10.2147/IMCRJ.S139729 PMID:28860870

Sakthieaswari, P. (2016). *Biosynthesis and characterization of iron oxide nanoparticles produced from a few selected seaweeds collected from Mandapam Coast, Tamil Nadu* (Thesis). Madurai Kamaraj University.

Salavati, M., Davar, F., & Mir, N. (2008). Synthesis and characterization of metallic copper nanoparticles via thermal decomposition. *Polyhedron*, *27*(17), 3514–3518. doi:10.1016/j.poly.2008.08.020

Saleh, M. M., & Petri, W. A. Jr. (2019). Type 3 Immunity during Clostridioides difficile Infection: Too Much of a Good Thing? *Infection and Immunity*, *88*(1), e00306-19. Advance online publication. doi:10.1128/IAI.00306-19 PMID:31570564

Salouti, M. (2014). *Nanoparticle based Drug Delivery Systems for Treatment of Infectious Diseases in book Application of Nanotechnology in Drug Delivery*. Academic Press.

Sampath, M., Vijayan, R., Tamilarasu, E., Tamilselvan, A., & Sengottuvelan, B. (2014). Green synthesis of novel jasmine bud-shaped copper nanoparticles. *Journal of Nanotechnology*, *2014*, 1–7. Advance online publication. doi:10.1155/2014/626523

Sandhiya, S., Dkhar, S. A., & Surendiran, A. (2009). Emerging trends of nanomedicine - an overview. *Fundamental & Clinical Pharmacology*, *23*(3), 263–269. doi:10.1111/j.1472-8206.2009.00692.x PMID:19527298

Sandhu, B. K., & McBride, S. M. (2018). Clostridioides difficile. *Trends in Microbiology*, *26*(12), 1049–1050. doi:10.1016/j.tim.2018.09.004 PMID:30297117

Sanpui, P., Chattopadhyay, A., & Ghosh, S. S. (2011). Induction of apoptosis in cancer cells at low silver nanoparticle concentrations using chitosan nanocarrier. *ACS Applied Materials & Interfaces*, *3*(2), 218–228. doi:10.1021/am100840c PMID:21280584

Santajit, S., & Indrawattana, N. (2016). Mechanisms of Antimicrobial Resistance in ESKAPE Pathogens. *BioMed Research International*, *2016*, 1–8. doi:10.1155/2016/2475067 PMID:27274985

Santhoshkumar, J., Agarwal, H., Menon, S., Rajeshkumar, S., & Venkat Kumar, S. (2019). A biological synthesis of copper nanoparticles and its potential applications. In *Green Synthesis* (pp. 199–221). Characterization and Applications of Nanoparticles. doi:10.1016/B978-0-08-102579-6.00009-5

Sarabandi, K., Rafiee, Z., Khodaei, D., & Jafari, S. M. (2019). Encapsulation of food ingredients by nanoliposomes. In *Lipid-Based Nanostructures for Food Encapsulation Purposes* (pp. 347–404). Academic Press. doi:10.1016/B978-0-12-815673-5.00009-X

Saravanan, M., Duche, K., Asmelash, T., Gebreyesus, A., Negash, L., Tesfay, A., Hailekiros, H., Niguse, S., Gopinath, V., & Barik, S. (2015). "Nano-Biomaterials"–A New Approach Concerning Multi-Drug Resistant Tuberculosis (MDR-TB). *Pharmaceutical Nanotechnology*, *3*(1), 5–18. doi:10.2174/2211738503666150415230743

Saroja, Ch., Lakshmi, P. K., & Bhaskaran, S. (2011). Recent trends in vaccine delivery systems: A review. *International Journal of Pharmaceutical Investigation*, *1*(2), 64–74. doi:10.4103/2230-973X.82384 PMID:23071924

Sasmal, P., & Datta, P. (2019). Tranexamic acid-loaded chitosan electrospunnanofibers as drug delivery system for hemorrhage control applications. *Journal of Drug Delivery Science and Technology*, *52*, 559–567. doi:10.1016/j.jddst.2019.05.018

Sathish Kumar, S., Fazeela, S. M., Sah, S. K., Kulandaisamy, S., Mahasampath Gowri, S., Hemalatha, S., & Balasubramanian, K. K. (2018). Synthesis, characterization, in vitro, and in silico studies of 4-(20-hydroxybenzoyl) and 4-(20-hydroxynaphthoyl)- thiabenzene-1-methyl-1-oxides. *Synthetic Communications*, *48*(5), 553–560. doi:10.1080/00397911.2017.1411950

Sathishkumar, G., Logeshwaran, V., Sarathbabu, S., Jha, P. K., Jeyaraj, M., Rajkuberan, C., Senthilkumar, N., & Sivaramakrishnan, S. (2018). Green synthesis of magnetite Fe_3O_4 nanoparticles using *Couroupita guianensis* Aubl. fruit extract for their antibacterial and cytotoxicity activities. *Artificial Cells, Nanomedicine, and Biotechnology*, *46*(3), 589–598. doi:10.1080/21691401.2017.1332635 PMID:28554257

Sathishkumar, M., Sneha, K., Won, S. W., Cho, C. W., Kim, S., & Yun, Y. S. (2009). Cinnamon zeylanicum bark extract and powder mediated green synthesis of nano-crystalline silver particles and its bactericidal activity. *Colloids and Surfaces*, *73*(2), 332–338. doi:10.1016/j.colsurfb.2009.06.005 PMID:19576733

Savjani, J. K., Gajjar, A. K., & Savjani, K. T. (2009). Mechanisms of resistance: Useful tool to design antibacterial agents for drug-resistant bacteria. *Mini-Reviews in Medicinal Chemistry*, *9*(2), 194–205. doi:10.2174/138955709787316038 PMID:19200024

Savolainen, K., Alenius, H., Norppa, H., Pylkkänen, L., Tuomi, T., & Kasper, G. (2010). Risk assessment of engineered nanomaterials and nanotechnologies—A review. *Toxicology, 269*, 92–104.

Savolainen, K., & Alenius, H. (2013). Disseminating widely. *Nature Nanotechnology*, *8*(2), 72. doi:10.1038/nnano.2013.3 PMID:23380925

Savolainen, K., & Norppa, H. (2016). In vitro and in vivo genotoxic effects of straight versus tangled multi-walled carbon nanotubes. *Nanotoxicology*, *10*(6). PMID:26674712

Sayyed-Alangi, S. Z., & Nematzadeh, M. (2019). Formulation, development and evaluation of bifunctionalized nanoliposomes containing Trifolium resupinatum sprout methanolic extract: As effective natural antioxidants on the oxidative stability of soybean oil. *BMC Chemistry*, *13*(1), 77. doi:10.118613065-019-0594-7 PMID:31384824

Scallan, E., Hoekstra, R. M., Angulo, F. J., Tauxe, R. V., Widdowson, M. A., Roy, S. L., Jones, J. L., & Griffin, P. M. (2011). Foodborne illness acquired in the United States—Major pathogens. *Emerging Infectious Diseases*, *17*(1), 7–15. doi:10.3201/eid1701.P11101 PMID:21192848

Schaeublin, N. M., Braydich-Stolle, L. K., Schrand, A. M., Miller, J. M., Hutchison, J., Schlager, J. J., & Hussain, S. M. (2011). Surface charge of gold nanoparticles mediates mechanism of toxicity. *Nanoscale*, *3*(2), 410–420. doi:10.1039/c0nr00478b PMID:21229159

Schaffer, J. N., & Pearson, M. M. (2015). Proteus mirabilis and Urinary Tract Infections. *Microbiology Spectrum*, *3*(5), UTI-0017–UTI-2013. doi:10.1128/microbiolspec.UTI-0017-2013 PubMed

Schaufler, K., Nowak, K., Düx, A., Semmler, T., Villa, L., Kourouma, L., Bangoura, K., Wieler, L. H., Leendertz, F. H., & Guenther, S. (2018). Clinically Relevant ESBL-Producing K. pneumoniae ST307 and E. coli ST38 in an Urban West African Rat Population. Frontiers in Microbiology, 9, 150. doi:10.3389/fmicb.2018.00150 PubMed

Schiro, P. G., Zhao, M., Kuo, J. S., Koehler, K. M., Sabath, D. E., & Chiu, D. T. (2012). Sensitive and High-Throughput Isolation of Rare Cells from Peripheral Blood with Ensemble-Decision Aliquot Ranking. *Angewandte Chemie International Edition*, *51*(19), 4618–4622. doi:10.1002/anie.201108695 PMID:22359315

Schmelcher, M., Donovan, D. M., & Loessner, M. J. (2012). Bacteriophage endolysins as novel antimicrobials. *Future Microbiology*, *7*(10), 1147–1171. doi:10.2217/fmb.12.97 PMID:23030422

Schmelcher, M., Powell, A. M., Camp, M. J., Pohl, C. S., & Donovan, D. M. (2015). Synergistic streptococcal phage lambdaSA2 and B30 endolysins kill streptococci in cow milk and in a mouse model of mastitis. *Applied Microbiology and Biotechnology*, *99*(20), 8475–8486. doi:10.100700253-015-6579-0 PMID:25895090

Schmidt, A., Molano, V., Hollering, M., Pöthig, A., Casini, A., & Kühn, F. (2016). Evaluation of New Palladium Cages as Potential Delivery Systems for the Anticancer Drug Cisplatin. *Chemistry (Weinheim an der Bergstrasse, Germany)*, *22*(7), 2253–2256. doi:10.1002/chem.201504930 PMID:26756963

Schmieder, R., & Edwards, R. (2012). Insights into antibiotic resistance through metagenomic approaches. *Future Microbiology*, *7*(1), 73–89. doi:10.2217/fmb.11.135 PMID:22191448

Schneider, G. (1984). Silver nitrate prophylaxis. *Canadian Medical Association Journal*, *131*, 193–196. PMID:6378348

Schroeder, A., Heller, D. A., Winslow, M. M., Dahlman, J. E., Pratt, G. W., Langer, R., & Anderson, D. G. (2012). Treating metastatic cancer with nanotechnology. *Nature Reviews. Cancer*, *12*(1), 39–50. doi:10.1038/nrc3180 PMID:22193407

Schuler, M., Bossy-Wetzel, E., Goldstein, J. C., Fitzgerald, P., & Green, D. R. (2000). p53 induces apoptosis by caspase activation through mitochondrial cytochrome c release. *The Journal of Biological Chemistry*, *275*(10), 7337–7342. doi:10.1074/jbc.275.10.7337 PMID:10702305

Schwendener, R. A. (2014). Liposomes as vaccine delivery systems: A review of the recent advances. *Therapeutic Advances in Vaccines*, *2*(6), 159–182. doi:10.1177/2051013614541440 PMID:25364509

Scriboni, A. B., Couto, V. M., de Morais Ribeiro, L. N., Freires, I. A., Groppo, F. C., De Paula, E., ... Cogo-Müller, K. (2019). Fusogenic liposomes increase the antimicrobial activity of vancomycin against staphylococcus aureus biofilm. *Frontiers in Pharmacology*, *10*, 10. doi:10.3389/fphar.2019.01401 PMID:31849660

Seiffert, S. N., Hilty, M., Perreten, V., & Endimiani, A. (2013). Extended spectrum cephalosporin-resistant gram-negative organisms in livestock: An emerging problem for human health? *Drug Resistance Updates*, *16*(1), 22–45. doi:10.1016/j.drup.2012.12.001 PMID:23395305

Sekar, A. D., Kumar, V., Muthukumar, H., Gopinath, P., & Matheswaran, M. (2019). Electrospinning of Fe-doped ZnO nanoparticles incorporated polyvinyl alcohol nanofibers for its antibacterial treatment and cytotoxic studies. *European Polymer Journal*, *118*, 27–35. doi:10.1016/j.eurpolymj.2019.05.038

Selck, H., Handy, R. D., Fernandes, T. F., Klaine, S. J., & Petersen, E. J. (2016). Nanomaterials in the aquatic environment: A European Union-United States perspective on the status of ecotoxicity testing, research priorities, and challenges ahead. *Environmental Toxicology and Chemistry*, *35*(5), 1055–1067. doi:10.1002/etc.3385 PMID:27089437

Selvaraj, V., & Alagar, M. (2007). Analytical detection and biological assay of antileukemic drug 5-fluorouracil using gold nanoparticles as probe. *International Journal of Pharmaceutics*, *337*(1-2), 275–281. doi:10.1016/j.ijpharm.2006.12.027 PubMed

Seoane, I. T., Cerrutti, P., Vazquez, A., Manfredi, B., & Cyras, V. P. (2017). Polyhydroxybutyrate-Based Nanocomposites with Cellulose Nanocrystals and Bacterial Cellulose. *Journal of Polymers and the Environment*, *25*(3), 586–598. doi:10.100710924-016-0838-8

Sergio, N., & García, H. (2016). Nanoparticles for Catalysis. *Ed. Thomas Nann. Nanomaterials*, *6*, 123.

Serrano, A. L., Olivas, R. M., Landaluze, J. S., & Camara, C. (2014). Nanoparticles: A global vision. Characterization, separation and quantification methods. Potential environmental and health impact. *Analytical Methods*, *6*(1), 38–56. doi:10.1039/C3AY40517F

Seung, K. J., Keshavjee, S., & Rich, M. L. (2015). Multidrug-resistant tuberculosis and extensively drug-resistant tuberculosis. *Cold Spring Harbor Perspectives in Medicine*, *5*(9), a017863. doi:10.1101/cshperspect.a017863 PMID:25918181

Shaalan, M., Saleh, M., El-Mahdy, M., & El-Matbouli, M. (2016). Recent progress in applications of nanoparticles in fish medicine: A review. *Nanomedicine; Nanotechnology, Biology, and Medicine*, *12*(3), 701–710. doi:10.1016/j.nano.2015.11.005 PMID:26656532

Shabbir, M. A. B., Shabbir, M. Z., Wu, Q., Mahmood, S., Sajid, A., Maan, M. K., Ahmed, S., Naveed, U., Hao, H., & Yuan, Z. (2019). CRISPR-cas system: Biological function in microbes and its use to treat antimicrobial resistant pathogens. *Annals of Clinical Microbiology and Antimicrobials*, *18*(1), 21. doi:10.118612941-019-0317-x PMID:31277669

Shah, K., Chan, L. W., & Wong, T. W. (2017). Critical physicochemical and biological attributes of nanoemulsions for pulmonary delivery of rifampicin by nebulization technique in tuberculosis treatment. *Drug Delivery*, *24*(1), 1631–1647. doi:10.1080/10717544.2017.1384298 PMID:29063794

Shahverdi, A. R., Fakhimi, A., Shahverdi, H. R., & Minaian, S. (2007). Synthesis and effect of silver nanoparticles on the antibacterial activity of different antibiotics against *Staphylococcus aureus* and *Escherichia coli*. *Nanomedicine (London)*, *3*(2), 168–171. doi:10.1016/j.nano.2007.02.001 PMID:17468052

Shaikh, S., Fatima, J., Shakil, S., Rizvi, S. M. D., & Kamal, M. A. (2015). Antibiotic resistance and extended spectrum beta-lactamases: Types, epidemiology and treatment. *Saudi Journal of Biological Sciences*, *22*(1), 90–101. doi:10.1016/j.sjbs.2014.08.002 PubMed

Shaker, M. A., & Shaaban, M. I. (2017). Formulation of carbapenems loaded gold nanoparticles to combat multi-antibiotic bacterial resistance: In vitro antibacterial study. *International Journal of Pharmaceutics*, *525*(1), 71–84. doi:10.1016/j.ijpharm.2017.04.019 PMID:28411141

Shakil, M. S., Hasan, M. A., & Sarker, S. R. (2019). Iron Oxide Nanoparticles for Breast Cancer Theranostics. *Current Drug Metabolism*, *20*(6), 446–456. doi:10.2174/1389200220666181122105043 PMID:30465497

Shamaila, S., Noshin, Z., Saira, R., Rehana, S., Jawad, N., & Shahzad, N. (2016). Gold Nanoparticles: An Efficient Antimicrobial Agent against Enteric Bacterial Human Pathogen. *Nanomaterials (Basel, Switzerland)*, *6*(4), 71. doi:10.3390/nano6040071 PMID:28335198

Shandilya, N., Le, B. O., Bressot, C., & Morgeneyer, M. (2015). Emission of titanium dioxide nanoparticles from building materials to the environment by wear and weather. *Environmental Science & Technology*, *49*(4), 2163–2170. doi:10.1021/es504710p PMID:25590625

Shan, G., Xing, J., Zhang, H., & Liu, H. (2005). Biodesulfurization of Dibenzothiophene by Microbial Cells Coated with Magnetite Nanoparticles. *Applied and Environmental Microbiology*, *71*(8), 4497–4502. doi:10.1128/AEM.71.8.4497-4502.2005 PubMed

Shang, D., Liu, Y., Jiang, F., Ji, F., Wang, H., & Han, X. (2019). Synergistic Antibacterial Activity of Designed Trp-Containing Antibacterial Peptides in Combination With Antibiotics Against Multidrug-Resistant Staphylococcus epidermidis. *Frontiers in Microbiology, 10*, 2719. doi:10.3389/fmicb.2019.02719 PMID:31824473

Shanmugasundaram, P., Saroj, S. K., Ubaid, R., Mahasampath Gowri, S., Easwaramoorthy, D., & Hemalatha, S. (2019). *Novel curcumin analogs act as antagonists to control nosocomial infection causing Pseudomonas aeruginosa*. Biocat. Agri. Biotech., doi:10.1016/j.bcab.2019

Sharifabad, M. E., Mercer, T., & Sen, T. (2016). Drug-loaded liposome-capped mesoporous core–shell magnetic nanoparticles for cellular toxicity study. *Nanomedicine (London), 11*, 2757–2767. doi:10.2217/nnm-2016-0248 PMID:27759497

Sharifi, S., Behzadi, S., Laurent, S., Forrest, M. L., Stroeve, P., & Mahmoudi, M. (2012). Toxicity of nanomaterials. *Chemical Society Reviews, 41*(6), 2323–2343. doi:10.1039/C1CS15188F PMID:22170510

Shariq, A. M., Ranjani, S., Tahira, A., Waseem, M., Khan, J., Kashif, M., & Hemalatha, S. (2019). Biogenic AgNps synthesized via endophytic bacteria and its biological applications. *Environmental Science and Pollution Research International, 26*(26), 26939–26946. Advance online publication. doi:10.1007/s11356-019-05869-6 PubMed

Sharma, C., Dhiman, R., Rokana, N., & Panwar, H. (2017). Nanotechnology: An Untapped Resource for Food Packaging. *Frontiers in Microbiology, 8*, 1735. doi:10.3389/fmicb.2017.01735 PMID:28955314

Sharma, S. K., & Mohan, A. (2019). Extrapulmonary Tuberculosis. In *Mycobacterium Tuberculosis: Molecular Infection Biology, Pathogenesis, Diagnostics and New Interventions* (pp. 37–53). Springer. doi:10.1007/978-981-32-9413-4_4

Sharma, S., Kelly, T. K., & Jones, P. A. (2009). Epigenetics in cancer. *Carcinogenesis, 31*(1), 27–36. doi:10.1093/carcin/bgp220 PMID:19752007

Sharma, S., Umar, A., Sood, S., Mehta, S. K., & Kansal, S. K. (2018). Photoluminescent C-dots: An overview on the recent development in the synthesis, physiochemical properties and potential applications. *Journal of Alloys and Compounds, 748*, 818–853. doi:10.1016/j.jallcom.2018.03.001

Sharma, V. K., Johnson, N., Cizmas, L., McDonald, T. J., & Kim, H. (2016). A review of the influence of treatment strategies on antibiotic resistant bacteria and antibiotic resistance genes. *Chemosphere, 150*, 702–714. doi:10.1016/j.chemosphere.2015.12.084 PMID:26775188

Shaw, B. J., & Handy, R. D. (2011). Physiological effects of nanoparticles on fish: A comparison of nano - metals versus metal ions. *Environment International, 37*(6), 1083–1097. doi:10.1016/j.envint.2011.03.009 PMID:21474182

Shaw, D., Clamp, A., & Jayson, G. C. (2013). Angiogenesis as a target for the treatment of ovarian cancer. *Current Opinion in Oncology, 25*(5), 558–565. doi:10.1097/CCO.0b013e328363e0da PMID:23942301

Shen, B. B., Gao, X. C., Yu, S. Y., Ma, Y., & Ji, C. H. (2016). Fabrication and potential application of a di-functional magnetic system: Magnetic hyperthermia therapy and drug delivery. *CrystEngComm, 18*(7), 1133–1138. doi:10.1039/C5CE02267C

Shen, B., Ma, Y., Yu, S., & Ji, C. (2016). Smart multifunctional magnetic nanoparticle-based drug delivery system for cancer thermochemotherapy and intracellular imaging. *ACS Applied Materials & Interfaces, 8*(37), 24502–24508. doi:10.1021/acsami.6b09772 PMID:27573061

Shen, W., Cetinel, S., & Montemagno, C. (2018). Application of biomolecular recognition via magnetic nanoparticle in nanobiotechnology. *Journal of Nanoparticle Research, 20*(5), 130. doi:10.100711051-018-4232-4

Shen, Z., Wu, A., & Chen, X. (2017). Iron oxide nanoparticle based contrast agents for magnetic resonance imaging. *Molecular Pharmaceutics, 14*(5), 1352–1364. doi:10.1021/acs.molpharmaceut.6b00839 PMID:27776215

Shiehzadeh, F., Hadizadeh, F., Mohammadpour, A., Aryan, E., Gholami, L., & Tafaghodi, M. (2019). Streptomycin sulfate dry powder inhalers for the new tuberculosis treatment schedule. *Journal of Drug Delivery Science and Technology*, *52*, 957–967. doi:10.1016/j.jddst.2019.05.052

Shi, H. G., Farber, L., Michaels, J. N., Dickey, A., Thompson, K. C., Shelukar, S. D., Hurter, P. N., Reynolds, S. D., & Kaufman, M. J. (2003). Characterization of crystalline drug nanoparticles using atomic force microscopy and complementary techniques. *Pharmaceutical Research*, *20*(3), 479–484. doi:10.1023/A:1022676709565 PMID:12669972

Shi, J., Xiao, Z., Kamaly, N., & Farokhzad, O. C. (2011). Self-assembled targeted nanoparticles: Evolution of technologies and bench to bedside translation. *Accounts of Chemical Research*, *44*(10), 1123–1134. doi:10.1021/ar200054n PMID:21692448

Shimizu, A., Takada, H., Koike, T., Takeshita, A., Saha, M., Rinawati, Nakada, N., Murata, A., Suzuki, T., Suzuki, S., Chiem, N. H., Tuyen, B. C., Viet, P. H., Siringan, M. A., Kwan, C., Zakaria, M. P., & Reungsang, A. (2013). Ubiquitous occurrence of sulfonamides in tropical Asian waters. *The Science of the Total Environment*, *452-453*, 108–115. doi:10.1016/j.scitotenv.2013.02.027 PMID:23500404

Shin, W. K., Cho, J., Kannan, A. G., Lee, Y. S., & Kim, D. W. (2016). Cross-linked composite gel polymer electrolyte using mesoporous methacrylate-functionalized SiO2 nanoparticles for lithium-ion polymer batteries. *Scientific Reports*, *18*(6), 26332. doi:10.1038rep26332 PMID:27189842

Shipunoval, Stremovskiy, Nikitin, Care, Sunna, Zvyagin, & Deyev. (2018). A novel method of nanoparticle surface biomodification for the development of effective agents for cancer theranostics. *12th International conference on Scientific and clinical applications of Magnetic carriers.*

Shirin, M., Asghar, E., & Somayeh, H.-N. (2018). Copper Nanoparticles as Antibacterial Agents. *Journal of Molecular Pharmaceutics & Organic Process Research*, *6*, 140.

Shor, E., & Perlin, D. S. (2015). Coping with stress and the emergence of multidrug resistance in fungi. *PLoS Pathogens*, *11*(3), e1004668. doi:10.1371/journal.ppat.1004668 PMID:25790300

Shrivastava, P., Vyas, S., Sharma, R., Mody, N., Gautam, L., Jain, A., Vishwakarma, N., & Vyas, S. P. (2020). Nanotechnology for oral drug delivery and targeting. In *Nanoengineered Biomaterials for Advanced Drug Delivery* (pp. 473–497). Elsevier., doi:10.1016/B978-0-08-102985-5.00020-6

Shukla, G. S., & Krag, D. N. (2005). Phage display selection for cell-specific ligands: Development of a screening procedure suitable for small tumor specimens. *Journal of Drug Targeting*, *13*(1), 7–18. doi:10.1080/10611860400020464 PMID:15848950

Siddiqi, K. S., Husen, A., & Rao, R. A. K. (2018). A review on biosynthesis of silver nanoparticles and their biocidal properties. *Journal of Nanobiotechnology*, *16*(1), 14. Advance online publication. doi:10.118612951-018-0334-5 PMID:29452593

Siegel, R. L., Miller, K. D., Fedewa, S. A., Ahnen, D. J., Meester, R. G., Barzi, A., & Jemal, A. (2017). Colorectal cancer statistics. *CA: a Cancer Journal for Clinicians*, *67*(3), 177–193. doi:10.3322/caac.21395 PMID:28248415

Sietze, R. (2007). The endothelial glycocalyx: Composition, functions, and visualization. *European Journal of Phycology*, *454*(3), 345–359. PMID:17256154

Si, H. Y., Li, D. P., Wang, T. M., Zhang, H. L., Ren, F. Y., Xu, Z. G., & Zhao, Y. Y. (2010). Improving the anti-tumor effect of genistein with a biocompatible superpara- magnetic drug delivery system. *Journal of Nanoscience and Nanotechnology*, *10*(4), 2325–2331. doi:10.1166/jnn.2010.1913 PMID:20355429

Silbergeld, E. K., Graham, J., & Price, L. B. (2008). Industrial food animal production, antimicrobial resistance, and human health. *Annual Review of Public Health, 29*(1), 151–169. doi:10.1146/annurev.publhealth.29.020907.090904 PMID:18348709

Silhavy, T. J., Kahne, D., & Walker, S. (2010). The bacterial cell envelope. *Cold Spring Harbor Perspectives in Biology, 5*, 1–16. PMID:20452953

Silva, S., Negri, M., Henriques, M., Oliveira, R., Williams, D. W., & Azeredo, J. (2012). Candida glabrata, Candida parapsilosis and Candida tropicalis: Biology, epidemiology, pathogenicity and antifungal resistance. *FEMS Microbiology Reviews, 36*(2), 288–305. doi:10.1111/j.1574-6976.2011.00278.x PubMed

Simmchen, J., Baeza, A., Miguel-Lopez, A., Stanton, M. M., Vallet-Regi, M., Ruiz-Molina, D., & Sánchez, S. (2017). Dynamics of Novel Photoactive AgCl Microstars and Their Environmental Applications. *ChemNanoMat: Chemistry of Nanomaterials for Energy, Biology and More, 3*(1), 65–71. doi:10.1002/cnma.201600300

Simonin, M., & Richaume, A. (2015). Impact of engineered nanoparticles on the activity, abundance, and diversity of soil microbial communities: A review. *Environmental Science and Pollution Research International, 22*(18), 13710–13723. doi:10.100711356-015-4171-x PMID:25647498

Sims, M. J. (1884). *The Story of My Life Marion-Sims* (Vol. H). D. Appleton & Co.

Singh, R., Cheng, S., & Singh, S. (2020). Oxidative stress-mediated genotoxic effect of zinc oxide nanoparticles on Deinococcus radiodurans. *3 Biotech, 10*(2), 66.

Singh, A., Dar, M., Joshi, B., Sharma, B., Shrivastava, S., & Shukla, S. (2018). Phytofabrication of Silver nanoparticles: Novel Drug to overcome hepatocellular ailments. *Toxicology Reports, 5*(5), 333–342. doi:10.1016/j.toxrep.2018.02.013 PMID:29854602

Singh, B. N., Prateeksha, U. D. K., Singh, B. R., Defoirdt, T., Gupta, V. K., De Souza, A. O., & … . (2017). Bactericidal, quorum quenching and anti-biofilm nanofactories: A new niche for nanotechnologists. *Critical Reviews in Biotechnology, 37*(4), 525–540. doi:10.1080/07388551.2016.1199010 PMID:27684212

Singh, G., Dwivedi, H., Saraf, S. K., & Saraf, S. A. (2011). Niosomal delivery of isoniazid-development and characterization. *Tropical Journal of Pharmaceutical Research, 10*(2). Advance online publication. doi:10.4314/tjpr.v10i2.66564

Singh, H., Bhandari, R., & Kaur, I. P. (2013). Encapsulation of Rifampicin in a solid lipid nanoparticulate system to limit its degradation and interaction with Isoniazid at acidic pH. *International Journal of Pharmaceutics, 446*(1-2), 106–111. doi:10.1016/j.ijpharm.2013.02.012 PMID:23410991

Singh, K., Abha, M., Deepankar, S., & Kavita, S. (2019). *Antiviral and antimicrobial potentiality of Nano drugs*. Indian Institute of Technology, Banaras Hindu University. doi:10.1016/B978-0-12-814029-1.00013-2

Singh, P., Garg, A., Pandit, S., Mokkapati, V., & Mijakovic, I. (2018). Antimicrobial Effects of Biogenic Nanoparticles. *Nanomaterials (Basel, Switzerland), 8*(12), 1009. Advance online publication. doi:10.3390/nano8121009 PMID:30563095

Singh, P., Kim, Y. J., Zhang, D., & Yang, D. C. (2016). Biological Synthesis of Nanoparticles from Plants and Microorganisms. *Trends in Biotechnology, 34*(7), 588–599. doi:10.1016/j.tibtech.2016.02.006 PMID:26944794

Singh, R., Wagh, P., Wadhwani, S., Gaidhani, S., Kumbhar, A., Bellare, J., & Chopade, B. A. (2013). Synthesis, optimization, and characterization of silver nanoparticles from Acinetobacter calcoaceticus and their enhanced antibacterial activity when combined with antibiotics. *International Journal of Nanomedicine, 8*, 4277. PMID:24235826

Singh, S. K., Chidrawar, V. R., Ushir, Y. V., Vadalia, K. R., Sheth, N. R., & Singh, S. (2010). Pharmaceutical characterization of amoxicillin trihydrate as mucoadhesive microspheres in management of *H. pylori. International Journal of Pharm Tech Research*, *2*(1), 348–358.

Singh, V. K., Tuladhar, R., & Chaudhary, M. K. (2015). Beta Lactamase Producing Escherichia coli, Klebsiella pneumoniae and Methicillin Resistant Staphylococcus aureus among Uropathogens. *Nepal Journal of Science and Technology*, *16*(1), 105–112. doi:10.3126/njst.v16i1.14364

Singh, V., & Srivastava, V. C. (2020). Self-engineered iron oxide nanoparticle incorporated on mesoporousbiochar derived from textile mill sludge for the removal of an emerging pharmaceutical pollutant. *Environmental Pollution*, *259*, 113822. doi:10.1016/j.envpol.2019.113822 PMID:31887588

Singh, Y., Meher, J. G., Raval, K., Khan, F. A., Chaurasia, M., Jain, N. K., & Chourasia, M. K. (2017). Nanoemulsion: Concepts, development and applications in drug delivery. *Journal of Controlled Release*, *252*, 28–49. doi:10.1016/j.jconrel.2017.03.008 PMID:28279798

Sirelkhatim, A., Mahmud, S., Seeni, A., Kaus, N. H. M., Ann, L. C., Bakhori, S. K. M., Hasan, H., & Mohamad, D. (2015). Review on zinc oxide nanoparticles: Antibacterial activity and toxicity mechanism. *Nano-Micro Letters*, *7*(3), 219–242. doi:10.100740820-015-0040-x PMID:30464967

Sivaraj, R., Rahman, P. K., Rajiv, P., Narendhran, S., & Venckatesh, R. (2014). Biosynthesis and characterization of Acalypha indica mediated copper oxide nanoparticles and evaluation of its antimicrobial and anticancer activity. *Spectrochimica Acta. Part A: Molecular and Biomolecular Spectroscopy*, *129*, 255–258. doi:10.1016/j.saa.2014.03.027 PubMed

Siwinska-Stefanska, K., Kubiaka, A., Piasecki, A., Goscianska, J., Nowaczyk, G., Jurga, S., & Jesionowski, T. (2018). TiO_2-ZnO binary oxide systems: Comprehensive characterization and tests of photocatalytic activity. *Materials (Basel)*, *11*(5), 841. doi:10.3390/ma11050841 PMID:29783700

Si, Y., Fu, Q., Wang, X., Zhu, J., Yu, J., Sun, G., & Ding, B. (2015). Superelastic and super- hydrophobic nanofiber-assembled cellular aerogels for effective separation of oil/ water emulsions. *ACS Nano*, *9*(4), 3791–3799. doi:10.1021/nn506633b PMID:25853279

Si, Y., Wang, L., Wang, X., Tang, N., Yu, J., & Ding, B. (2017). Ultrahigh-water-content, super- elastic, and shape-memory nanofiber-assembled hydrogels exhibiting pressure-re- sponsive conductivity. *Advanced Materials*, *29*(24), 1700339. doi:10.1002/adma.201700339 PMID:28417597

Slavin, Y. N., Asnis, J., Hafeli, U. O., & Bach, H. (2017). Metal nanoparticles: Understanding the mechanisms behind antibacterial activity. *Journal of Nanobiotechnology*, *15*(1), 65. doi:10.118612951-017-0308-z PMID:28974225

Smartt, A. E., Xu, T., Jegier, P., Carswell, J. J., Blount, S. A., Sayler, G. S., & Ripp, S. (2012). Pathogen detection using engineered bacteriophages. *Analytical and Bioanalytical Chemistry*, *402*(10), 3127–3146. doi:10.100700216-011-5555-5 PMID:22101465

Smetana, A., Klabunde, K., Marchin, G., & Sorenson, C. (2008). Biocidal activity of nanocrystalline silver powders and particles. *Langmuir*, *24*(14), 7457–7464. doi:10.1021/la800091y PMID:18543995

Smillie, C., Garcillán-Barcia, M. P., Francia, M. V., Rocha, E. P. C., & de la Cruz, F. (2010). Mobility of plasmids. *Microbiology and Molecular Biology Reviews*, *74*(3), 434–452. doi:10.1128/MMBR.00020-10 PMID:20805406

Smith, H. O., Gwinn, M. L., & Salzberg, S. L. (1999). DNA uptake signal sequences in naturally transformable bacteria. *Research in Microbiology*, *150*(9-10), 603–616. doi:10.1016/S0923-2508(99)00130-8 PMID:10673000

Smyth, M., Fournier, C., Driemeier, C., Picart, C., Foster, E. J., & Bras, J. (2017). Tunable Structural and Mechanical Properties of Cellulose Nanofiber Substrates in Aqueous Conditions for Stem Cell Culture. *Biomacromolecules*, *18*(7), 2034–2044. doi:10.1021/acs.biomac.7b00209 PMID:28485582

Solairaj, D., & Rameshthangam, P. (2017). Silver Nanoparticle Embedded α-Chitin Nanocomposite for Enhanced Antimicrobial and Mosquito Larvicidal Activity. *Journal of Polymers and the Environment*, *25*(2), 435–452. doi:10.100710924-016-0822-3

Somorjai, G. A., & Park, J. Y. (2008). Colloid science of metal nanoparticle catalysts in 2D and 3D structures. Challenges of nucleation, growth, composition, particle shape, size control and their influence on activity and selectivity. *Topics in Catalysis*, *49*(3-4), 126–135. doi:10.100711244-008-9077-0

Sondi, I., & Salopek-Sondi, B. (2004). Silver nanoparticles as antimicrobial agent: A case study on *E. coli* as a model for Gram-negative bacteria. *Journal of Colloid and Interface Science*, *275*(1), 177–182. doi:10.1016/j.jcis.2004.02.012 PMID:15158396

Song, J., Kong, H., & Jang, J. (2011). Bacterial adhesion inhibition of the quaternary ammonium functionalized silica nanoparticles. *Coll. Surf.*, *82*(2), 651–656. doi:10.1016/j.colsurfb.2010.10.027 PMID:21115282

Song, S., Zhao, Y., Yuan, X., & Zhang, J. (2019). β-Chitin nanofiber hydrogel as a scaffold to in situ fabricate monodispersed ultra-small silver nanoparticles. *Colloids and Surfaces. A, Physicochemical and Engineering Aspects*, *574*, 36–43. doi:10.1016/j.colsurfa.2019.04.047

Song, X., Lin, Q., Guo, L., Fu, Y., Han, J., Ke, H., Sun, X., Gong, T., & Zhang, Z. (2015). Rifampicin loaded mannosylated cationic nanostructured lipid carriers for alveolar macrophage-specific delivery. *Pharmaceutical Research*, *32*(5), 1741–1751. doi:10.100711095-014-1572-3 PMID:25407545

Sonohara, R., Muramatsu, N., Ohshima, H., & Kondo, T. (1995). Difference in surface properties between Escherichia coli and Staphylococcus aureus as revealed by electrophoretic mobility measurements. *Biophysical Chemistry*, *55*(3), 273–277. doi:10.1016/0301-4622(95)00004-H PubMed

Sood, K., Kaur, J., Singh, H., Arya, S. K., & Khatri, M. (2019). Comparative toxicity evaluation of graphene oxide (GO) and zinc oxide (ZnO) nanoparticles on Drosophila melanogaster. *Toxicology Reports*, *6*, 768–781. doi:10.1016/j.toxrep.2019.07.009 PMID:31428565

Sood, S., Kapil, A., Das, B., Jain, Y., & Kabra, S. K. (1999). Re-emergence of chloramphenicol-sensitive Salmonella typhi. *Lancet*, *353*(9160), 1241–1242. doi:10.1016/S0140-6736(99)00637-6 PubMed

Soothill, G., Hu, Y., & Coates, A. (2013). Can we prevent antimicrobial resistance by using antimicrobials better? *Pathogens (Basel, Switzerland)*, *2*(2), 422–435. doi:10.3390/pathogens2020422 PMID:25437042

Souza, J. P., Venturini, F. P., Santos, F., & Zucolotto, V. (2018). Chronic toxicity in Ceriodaphniadubia induced by graphene oxide. *Chemosphere*, *190*, 218–224. doi:10.1016/j.chemosphere.2017.10.018 PMID:28992473

Souza, S. F., Mariano, M., Reis, D., Lombello, C. B., Ferreira, M., & Sain, M. (2018). Cell interactions and cytotoxic studies of cellulose nanofibers from curaua natural fibers. *Carbohydrate Polymers*, *201*, 87–95. doi:10.1016/j.carbpol.2018.08.056 PMID:30241866

Sperling, R. A., Gil, P. R., Zhang, F., Zanella, M., & Parak, W. J. (2008). Biological applications of gold nanoparticles. *Chemical Society Reviews*, *37*(9), 1896–1908. doi:10.1039/b712170a PMID:18762838

Spicer, C. D., Jumeaux, C., Gupta, B., & Stevens, M. M. (2018). Peptide and protein nanoparticle conjugates: Versatile platforms for biomedical applications. *Chemical Society Reviews*, *47*(10), 3574–3620. doi:10.1039/C7CS00877E PMID:29479622

Srinath, B. S., Namratha, K., & Byrappa, K. (2018). Eco-Friendly Synthesis of Gold Nanoparticles by *Bacillus subtilis* and Their Environmental Applications. *Advanced Science Letters*, *24*(8), 5942–5946. doi:10.1166/asl.2018.12224

Stanković, A., Dimitrijević, S., & Uskoković, D. (2013). Influence of size scale and morphology on antibacterial properties of ZnO powders hydrothemally synthesized using different surface stabilizing agents. *Colloids and Surfaces. B, Biointerfaces*, *102*, 21–28. doi:10.1016/j.colsurfb.2012.07.033 PMID:23010107

Starodub, M. E., & Trevors, J. T. (1989). Silver resistance in *Escherichia coli* R1. *Journal of Medical Microbiology*, *29*(2), 101–110. doi:10.1099/00222615-29-2-101 PMID:2659795

Stauber, D. J., Debler, E. W., Horton, P. A., Smith, K. A., & Wilson, I. A. (2006). Crystal structure of the IL-2 signaling complex: Paradigm for a heterotrimeric cytokine receptor. *Proceedings of the National Academy of Sciences of the United States of America*, *103*(8), 2788–2793. doi:10.1073/pnas.0511161103 PMID:16477002

Stewart, M. E., Anderton, C. R., Thompson, L. B., Maria, J., Gray, S. K., Rogers, J. A., & Nuzzo, R. G. (2008). Nanostructured Plasmonic Sensors. *Chemical Reviews*, *108*(2), 494–521. doi:10.1021/cr068126n PMID:18229956

Stoimenov, P. K., Klinger, R. L., Marchin, G. L., & Klabunde, K. J. (2002). Metal oxide nanoparticles as bactericidal agents. *Langmuir*, *18*(17), 6679–6686. doi:10.1021/la0202374

Stoumpos, S., Hennessy, M., Vesey, A. T., Radjenovic, A., Kasthuri, R., Kingsmore, D. B., Mark, P. B., & Roditi, G. (2019). Ferumoxytol magnetic resonance angiography: A dose-finding study in patients with chronic kidney disease. *European Radiology*, *29*(7), 3543–3552. doi:10.100700330-019-06137-4 PMID:30919067

Stout, R. D., Ferguson, K. P., Li, Y. N., & Lambe, D. W. Jr. (1992). Staphylococcal exopolysaccharides inhibit lymphocyte proliferative responses by activation of monocyte prostaglandin production. *Infection and Immunity*, *60*(3), 922–927. doi:10.1128/IAI.60.3.922-927.1992 PMID:1541565

Strasfeld, L., & Chou, S. (2010, June). Antiviral Drug Resistance: Mechanisms and Clinical Implications. *Infectious Disease Clinics of North America*, *24*(2), 413–437. doi:10.1016/j.idc.2010.01.001 PMID:20466277

Strawn, L. K., Fortes, E. D., Bihn, E. A., Nightingale, K. K., Grohn, Y. T., Worobo, R. W., Wiedmann, M., & Bergholz, P. W. (2013). Landscape and meteorological factors affecting prevalence of three food-borne pathogens in fruit and vegetable farms. *Applied and Environmental Microbiology*, *79*(2), 588–600. doi:10.1128/AEM.02491-12 PMID:23144137

Stylianopoulos, T., Poh, M.-Z., Insin, N., Bawendi, M. G., Fukumura, D., Munn, L. L., & Jain, R. K. (2010). Diffusion of particles in the extracellular matrix: The effect of repulsive electrostatic interactions. *Biophysical Journal*, *99*(5), 1342–1349. doi:10.1016/j.bpj.2010.06.016 PMID:20816045

Subhankari, I., & Nayak, P. (2013). Antimicrobial Activity of Copper Nanoparticles Synthesised by Ginger (Zingiber officinale) Extract. *World J Nano Sci Technol*, *2*, 10–13.

Subramaniam, J., Murugan, K., Jebanesan, A., Pontheckan, P., Dinesh, D., Nicoletti, M., Wei, H., Higuchi, A., Kumar, S., Canale, A., & Benelli, G. (2016). Do Chenopodium ambrosioides-Synthesized Silver Nanoparticles Impact Oryzias melastigma Predation Against Aedes albopictus Larvae? *Journal of Cluster Science*, *28*(1), 413–436. doi:10.100710876-016-1113-9

Sugano, M., Morisaki, H., Negishi, Y., Endo-Takahashi, Y., Kuwata, H., Miyazaki, T., & Yamamoto, M. (2016). Potential effect of cationic liposomes on interactions with oral bacterial cells and biofilms. *Journal of Liposome Research*, *26*(2), 156–162. PMID:26152278

Sukhanova, A., Bozrova, S., Sokolov, P., Berestovoy, M., Karaulov, A., & Nabiev, I. (2018). Dependence of nanoparticle toxicity on their physical and chemical properties. *Nanoscale Research Letters*, *13*(1), 44. doi:10.118611671-018-2457-x PMID:29417375

Suman, T. Y., Ravindranath, R. R. S., Elumalai, D., Kaleena, P. K., Ramkumar, R., Perumal, P., Aranganathan, L., & Chitrarasu, P. S. (2015). Larvicidal activity of titanium dioxide nanoparticles synthesized using Morinda citrifolia root extract against Anopheles stephensi, Aedes aegypti and Culex quinquefasciatus and its other effect on non-target fish. *Asian Pacific Journal of Tropical Disease*, *5*(3), 224–230. doi:10.1016/S2222-1808(14)60658-7

Sun, C., Lee, J., & Zhang, M. (2008). Magnetic nanoparticles in MR imaging and drug delivery. *Advanced Drug Delivery Reviews*, *60*(11), 1252–1265. doi:10.1016/j.addr.2008.03.018 PMID:18558452

Sun, H., Zhou, J., Huang, Z., Qu, L., Lin, N., Liang, C., Dai, R., Tang, L., & Tian, F. (2017). Carbon nanotube-incorporated collagen hydrogels improve cell alignment and the performance of cardiac constructs. *International Journal of Nanomedicine*, *12*, 3109–3120. doi:10.2147/IJN.S128030 PMID:28450785

Sun, J. B., Duan, J. H., Dai, S. L., Ren, J., Zhang, Y. D., Tian, J. S., & Li, Y. (2007). In vitro and in vivo antitumor effects of doxorubicin loaded with bacterial magnetosomes (DBMs) on H22 cells: The magnetic bionanoparticles as drug carriers. *Cancer Letters*, *258*(1), 109–117. doi:10.1016/j.canlet.2007.08.018 PubMed

Sun, S., & Murray, C. B. (1999). Synthesis of monodisperse cobalt nanocrystals and their assembly into magnetic superlattices. *Journal of Applied Physics*, *85*(8), 4325–4330. doi:10.1063/1.370357

Sun, T., Zhang, Y. S., Pang, B., Hyun, D. C., Yang, M., & Xia, Y. (2014). Engineered nanoparticles for drug delivery in cancer therapy. *Angewandte Chemie International Edition*, *53*(46), 12320–12364. doi:10.1002/anie.201403036 PMID:25294565

Sun, Z., Zhang, J., Zhang, X., Wang, S., Zhang, Y., & Li, C. (2008). Comparison of gyrA gene mutations between laboratory-selected ofloxacin-resistant Mycobacterium tuberculosis strains and clinical isolates. *International Journal of Antimicrobial Agents*, *31*(2), 115–121. doi:10.1016/j.ijantimicag.2007.10.014 PMID:18164184

Suresh, A. K., Pelletier, D. A., & Doktycz, M. J. (2013). Relating nanomaterial properties and microbial toxicity. *Nano*, *5*, 463–474. PMID:23203029

Suresh, U., Murugan, K., Benelli, G., Nicoletti, M., Barnard, D. R., Panneerselvam, C., Kumar, P. M., Subramaniam, J., Dinesh, D., & Chandramohan, B. (2015). Tackling the growing threat of dengue: Phyllanthus niruri-mediated synthesis of silver nanoparticles and their mosquitocidal properties against the dengue vector Aedes aegypti (Diptera: Culicidae). *Parasitology Research*, *114*(4), 1551–1562. doi:10.100700436-015-4339-9 PMID:25669140

Sutthanont, N., Attrapadung, S., & Nuchprayoon, S. (2019). Larvicidal Activity of Synthesized Silver Nanoparticles from Curcuma zedoaria Essential Oil against Culex quinquefasciatus. *Insects*, *10*(1), 27. doi:10.3390/insects10010027 PMID:30641859

Su, Y., Zhao, L., Meng, F., Qiao, Z., Yao, Y., & Luo, J. (2018). Triclosan loaded polyurethane micelles with pH and lipase sensitive properties for antibacterial applications and treatment of biofilms. *Materials Science and Engineering C*, *93*, 921–930. doi:10.1016/j.msec.2018.08.063 PMID:30274129

Su, Y., Zheng, X., Chen, Y., Li, M., & Liu, K. (2015). Alteration of intracellular protein expressions as a key mechanism of the deterioration of bacterial denitrification caused by copper oxide nanoparticles. *Scientific Reports*, *5*(1), 15824. doi:10.1038rep15824 PMID:26508362

Suzuki, E. (2002). High-resolution scanning electron microscopy of immunogold-labelled cells by the use of thin plasma coating of osmium. *Journal of Microscopy*, *208*(3), 153–157. doi:10.1046/j.1365-2818.2002.01082.x PMID:12460446

Svobodova, Z., Kucerova, J., Autebert, J., Horak, D., Bruckova, L., Viovy, J.-L., & Bilkova, Z. (2014). Application of an improved magnetic immunosorbent in an Ephesia chip designed for circulating tumor cell capture. *Electrophoresis*, *35*(2-3), 323–329. doi:10.1002/elps.201300196 PMID:23868447

Sykora, D., Kasicka, V., Miksik, I., Rezanka, P., Zaruba, K., Matejka, P., & Kral, V. (2010). Applications of gold nanoparticles in separation sciences. *Journal of Separation Science*, *30*(3), 372–387. doi:10.1002/jssc.200900677 PMID:20099261

Tabatabaei Mirakabad, F., Nejati-Koshki, K., Akbarzadeh, A., Yamchi, M., Milani, M., Zarghami, N., Zeighamian, V., Rahimzadeh, A., Alimohammadi, S., Hanifehpour, Y., & Joo, S. W. (2014). PLGA-Based Nanoparticles as Cancer Drug Delivery Systems. *Asian Pacific Journal of Cancer Prevention*, *15*(2), 517–535. doi:10.7314/APJCP.2014.15.2.517 PMID:24568455

Taghavi, S. M., Momenpour, M., Azarian, M., Ahmadian, M., Souri, F., Taghavi, S. A., Sadeghain, M., & Karchani, M. (2013). Effects of Nanoparticles on the Environment and Outdoor Workplaces. *Electronic Physician*, *5*(4), 706–712. PMID:26120406

Tahira, A., Priya, S., Sah, S. K., Khan, M. S., & Hemalatha, S. (2019c). Ta-AgNps are potential antimicrobial resistance breakers. J. Nano. Str.

Tahira, A., Vabeiryureilai, M., & Senthil Kumar, N. (2019a). Fungal-mediated synthesis of pharmaceutically active silver nanoparticles and anticancer property against A549 cells through apoptosis Environ. Sci. Pol. Res., 26(13), 13649–13657. doi:10.1007/s11356-019-04718-w

Tahira, A., & Hemalatha, S. (2019b). Mycosilver Nanoparticles: Synthesis, characterization and screening the efficacy against plant pathogenic fungi. *BioNanoScience*.

Tahira, A., Khan, M. S., & Hemalatha, S. (2017). Extraction of Flavonoids from various Parts of Couroupita guianensis and Its Efficacy Against Pathogenic Bacteria. *Asian Journal of Pharmaceutical and Clinical Research*, *10*, 4.

Tahira, A., Khan, M. S., & Hemalatha, S. (2018a). Novel Silver Nanoparticles Synthesized from Anthers of Couroupita guianensis Abul. Control Growth and Biofilm Formation in Human Pathogenic Bacteria. *Nano Biomedicine and Engineering*, *10*, 3.

Tahira, A., Khan, M. S., & Hemalatha, S. (2018b). A facile and rapid method for green synthesis of Silver Myco nanoparticles from endophytic fungi. *International Journal of Nanodimension*, *9*, 435–441.

Taju, G., Majeed, S. A., Nambi, K. S. N., & Hameed, A. S. S. (2014). In vitro assay for the toxicity of silver nanoparticles using heart and gill cell lines of Catlacatla and gill cell line of Labeorohita. *Comparative Biochemistry and Physiology. Toxicology & Pharmacology: CBP*, *161*, 41–52. doi:10.1016/j.cbpc.2014.01.007 PMID:24524868

Takenaga, M., Ohta, Y., Tokura, Y., Hamaguchi, A., Igarashi, R., Disratthakit, A., & Doi, N. (2008). Lipid microsphere formulation containing rifampicin targets alveolar macrophages. *Drug Delivery*, *15*(3), 169–175. doi:10.1080/10717540801952530 PMID:18379929

Takenaka, S., Karg, E., Roth, C., Schulz, H., Ziesenis, A., Heinzmann, U., ... Heyder, J. (2001). Pulmonary and systemic distribution of inhaled ultrafine silver particles in rats. *Environmental Health Perspectives*, *109*(suppl 4), 547–551. PMID:11544161

Talamini, L., Violatto, M. B., Cai, Q., Monopoli, M. P., Kantner, K., Krpetić, Z., ... Boselli, L. (2017). Influence of size and shape on the anatomical distribution of endotoxin-free gold nanoparticles. *ACS Nano*, *11*(6), 5519–5529. doi:10.1021/acsnano.7b00497 PMID:28558193

Tamayo, L. A., Zapata, P. A., Vejar, N. D., Azocar, M. I., Gulppi, M. A., Zhou, X., Thompson, G. E., Rabagliati, F. M., & Páez, M. A. (2014). Release of silver and copper nanoparticles from polyethylene nanocomposites and their penetration into *Listeria monocytogenes*. *Materials Science and Engineering*, *40*, 24–31. doi:10.1016/j.msec.2014.03.037 PMID:24857461

Tang, J., Shalabi, A., & Hubbard-Lucey, V. M. (2018). Comprehensive analysis of the clinical immuno-oncology landscape. *Annals of Oncology: Official Journal of the European Society for Medical Oncology*, *29*(1), 84–91. doi:10.1093/annonc/mdx755 PMID:29228097

Tang, S., Zhao, Z., Chen, G., Su, Y., Lu, L., Li, B., Liang, D., & Jin, R. (2016). Fabrication of ampicillin/starch/polymer composite nanofibers with controlled drug release properties by electrospinning. *Journal of Sol-Gel Science and Technology*, *77*(3), 594–603. doi:10.100710971-015-3887-x

Tang, Y., Lana, X., Lianga, C., Zhonga, Z., Xiea, R., Zhoua, Y., Miaoa, X., Wangc, H., & Wang, W. (2019). Honey loaded alginate/PVA nanofibrous membrane as potential bioactive wound dressing. *Carbohydrate Polymers*, *219*, 113–120. doi:10.1016/j.carbpol.2019.05.004 PMID:31151507

Tang, Z. X., & Feng, B. (2014). MgO nanoparticles as antibacterial agent: Preparation and activity. *Brazilian Journal of Chemical Engineering*, *31*(3), 591–601. doi:10.1590/0104-6632.20140313s00002813

Tanser, F., Bärnighausen, T., Grapsa, E., Zaidi, J., & Newell, M. L. (2013). High coverage of ART associated with decline in risk of HIV acquisition in rural KwaZulu-Natal, South Africa. *Science*, *339*(6122), 966–971. doi:10.1126cience.1228160 PMID:23430656

Taran, M., Rad, M., & Alavi, M. (2016). Characterization of Ag nanoparticles biosynthesized by *Bacillus* sp. HAI4 in different conditions and their antibacterial effects. *Journal of Applied Pharmaceutical Science, 6*(11), 94-99.

Taviad, K. K., Vekariya, S., Bedarkar, P., Galib, R., & Patgiri, B. J. (2018). Process standardization of Swarna Makshika Shodhana (purification) in Triphala Kwatha (decoction). *Ayu*, *39*(3), 187. doi:10.4103/ayu.AYU_26_18 PMID:31000997

Tayel, A. A., El-Tras, W. F., Moussa, S., El-Baz, A. F., Mahrous, H., Salem, M. F., & Brimer, L. (2011). Antibacterial action of zinc oxide nanoparticles against foodborne pathogens. *Journal of Food Safety*, *31*(2), 211–218. doi:10.1111/j.1745-4565.2010.00287.x

Teixeira, V., Feio, M. J., & Bastos, M. (2012). Role of lipids in the interaction of antimicrobial peptides with membranes. *Progress in Lipid Research*, *51*(2), 149–177. doi:10.1016/j.plipres.2011.12.005 PMID:22245454

Telenti, A., Imboden, P., Marchesi, F., Matter, L., Schopfer, K., Bodmer, T., Lowrie, D., Colston, M. J., & Cole, S. (1993). Detection of rifampicin-resistance mutations in Mycobacterium tuberculosis. *Lancet*, *341*(8846), 647–651. doi:10.1016/0140-6736(93)90417-F PMID:8095569

Temsamani & Vidal. (2004). The use of cell penetrating peptides for drug delivery. *Research Focus DDT, 9*, 1012-1019.

Tenover, F. C. (2006). Mechanisms of Antimicrobial resistance in bacteria. *The American Journal of Medicine*, *119*(6, 6A), 3–10. doi:10.1016/j.amjmed.2006.03.011 PMID:16735149

Tenover, F. C., & Hughes, J. M. (1996). Development and spread of multiply-resistant bacterial pathogens. *Journal of the American Medical Association*, *275*, 300–304. doi:10.1001/jama.1996.03530280052036 PMID:8544270

Teo, B. K., & Sun, X. H. (2007). Silicon-based low-dimensional nanomaterials and nanodevices. *Chemical Reviews*, *107*(5), 1454–1532. doi:10.1021/cr030187n PMID:17488056

Teow, Y., Asharani, P. V., Hande, M. P., & Valiyaveettil, S. (2011). Health impact and safety of engineered nanomaterials. *Chemical Communications (Cambridge)*, *47*(25), 7025–7038. doi:10.1039/c0cc05271j PMID:21479319

Ter-Avetisyan, G., Tünnemann, G., Nowak, D., Nitschke, M., Herrmann, A., Drab, M., & Cardoso, M. C. (2009). Cell entry of arginine-rich peptides is independent of endocytosis. *The Journal of Biological Chemistry*, *284*(6), 3370–3378. doi:10.1074/jbc.M805550200 PMID:19047062

Tesauro, D., Accardo, A., Diaferia, C., Milano, V., Guillon, J., Ronga, L., & Rossi, F. (2019). Peptide-Based Drug-Delivery Systems in Biotechnological Applications: Recent Advances and Perspectives. *Molecules (Basel, Switzerland)*, *24*(2), 351. doi:10.3390/molecules24020351 PMID:30669445

Thadavirul, N., Pavasant, P., & Supaphol, P. (2017). Fabrication and evaluation of poly- caprolactone–poly (hydroxy-butyrate) or poly (3-hydroxybutyrate-co-3-hydro- xyvalerate) dual-leached porous scaffolds for bone tissue engineering applications. *Macromolecular Materials and Engineering*, *302*(3), 1600289. doi:10.1002/mame.201600289

Thai, P. V. K., Ha, D. T. M., Hanh, N. T., Day, J., Dunstan, S., Nhu, N. T. Q., ... Thuong, N. T. (2018). Bacterial risk factors for treatment failure and relapse among patients with isoniazid resistant tuberculosis. *BMC Infectious Diseases*, *18*(1), 112. doi:10.118612879-018-3033-9 PMID:29510687

Thakor, A. S., Jokerst, J. V., Ghanouni, P., Campbell, J. L., Mittra, E., & Gambhir, S. S. (2016). Clinically approved nanoparticle imaging agents. *Journal of Nuclear Medicine*, *57*(12), 1833–1837. doi:10.2967/jnumed.116.181362 PMID:27738007

Thavasi, V., Singh, G., & Ramakrishna, S. (2008). Electrospun nanofibers in energy and environmental applications. *Energy & Environmental Science*, *1*(2), 205–221. doi:10.1039/b809074m

Thaxton, C.S., Rink, J.S., Naha, P.C., & Cormode, D.P. (2016). Lipoproteins and lipoprotein mimetics for imaging and drug delivery. *Adv Drug Deliv Rev.*, *15*(106), 116-131.

Theivasanthi, T., & Alagar, M. (2011). Studies of Copper Nanoparticles Effects on Micro-organisms. *Annals of Biological Research*, *2*, 368–373.

Thomas, M., & Klibanov, A. M. (2003). Conjugation to gold nanoparticles enhances polyethylenimine's transfer of plasmid DNA into mammalian cells. *Proceedings of the National Academy of Sciences of the United States of America*, *100*(16), 9138–9143. doi:10.1073/pnas.1233634100 PMID:12886020

Thomas, S., Kumar Mishra, P., & Talegaonkar, S. (2015). Ceramic nanoparticles: Fabrication methods and applications in drug delivery. *Current Pharmaceutical Design*, *21*(42), 6165–6188. doi:10.2174/1381612821666151027153246 PMID:26503144

Tian, J., Yan, C., Liu, K., Tao, J., Guo, Z., Liu, J., Zhang, Y., Xiong, F., & Gu, N. (2017). Paclitaxel-loaded magnetic nanoparticles: Synthesis, characterization, and application in targeting. *Journal of Pharmaceutical Sciences*, *106*(8), 2115–2122. doi:10.1016/j.xphs.2017.04.023 PMID:28456728

Tian, Z. M., He, J., Chen, X., Wen, T., Zhai, C., Zhang, Z. X., Cho, J. D., Chou, X. J., & Xue, C. Y. (2018). Core-shell coaxially structured triboelectric nanogenerator for energy harvesting and motion sensing. *Rsc Adv.*, *8*(6), 2950–2957. doi:10.1039/C7RA12739A

Tinkle, S., McNeil, S. E., Mühlebach, S., Bawa, R., Borchard, G., Barenholz, Y., Tamarkin, L., & Desai, N. (2014). Nanomedicines: Addressing the scientific and regulatory gap. *Annals of the New York Academy of Sciences*, *1313*(1), 35–56. doi:10.1111/nyas.12403 PMID:24673240

Tippayawat, P., Phromviyo, N., Boueroy, P., & Chompoosor, A. (2016). Green synthesis of silver nanoparticles in aloe vera plant extract prepared by a hydrothermal method and their synergistic antibacterial activity. *PeerJ*, *4*, 2589. doi:10.7717/peerj.2589 PMID:27781173

Tiwari, D. K., Behari, J., & Sen, P. (2008). Time and dose-dependent antimicrobial potential of Ag nanoparticles synthesized by top down approach. *Current Science*, *95*, 647–655.

Tiwari, G., Tiwari, R., Sriwastawa, B., Bhati, L., Pandey, S., Pandey, P., & Bannerjee, S. K. (2012). Drug delivery systems: An updated review. *International Journal of Pharmaceutical Investigation*, *2*(1), 2–11. doi:10.4103/2230-973X.96920 PMID:23071954

Tiwari, J. N., Tiwari, R. N., & Kim, K. S. (2012). Progress in materials science three-dimensional nanostructured materials for advanced electrochemical energy devices. *Progress in Materials Science*, *57*(4), 724–803. doi:10.1016/j.pmatsci.2011.08.003

Tock, M. R., & Dryden, D. T. (2005). The biology of restriction and anti-restriction. *Current Opinion in Microbiology*, *8*(4), 466–472. doi:10.1016/j.mib.2005.06.003 PMID:15979932

Tokajuk, G., Niemirowicz, K., Deptuła, P., Piktel, E., Cie'sluk, M., Wilczewska, A. Z., Dabrowski, J. R., & Bucki, R. (2017). Use of magnetic nanoparticles as a drug delivery system to improve chlorhexidine antimicrobial activity. *International Journal of Nanomedicine*, *12*, 7833–7846. doi:10.2147/IJN.S140661 PMID:29123396

Toloue, E. B., Karbasi, S., Salehi, H., & Rafienia, M. (2019). Potential of an electrospun composite scaffold of poly (3-hydroxybutyrate)- chitosan/alumina nanowires in bone tissue engineering applications. *Materials Science and Engineering C*, *99*, 1075–1091. doi:10.1016/j.msec.2019.02.062 PMID:30889640

Torchilin, V. P. (2005). Recent advances with liposomes as pharmaceutical carriers. *Nature Reviews. Drug Discovery*, *4*(2), 145–160. doi:10.1038/nrd1632 PMID:15688077

Tourinho, P. S., van Gestel, C. A., Lofts, S., Svendsen, C., Soares, A. M., & Loureiro, S. (2012). Metal-based nanoparticles in soil: Fate, behavior, and effects on soil invertebrates. *Environmental Toxicology and Chemistry*, *31*(8), 1679–1692. doi:10.1002/etc.1880 PMID:22573562

Tozzi, V. (2010). Pharmacogenetics of antiretrovirals. *Antiviral Research*, *85*(1), 190–200. doi:10.1016/j.antiviral.2009.09.001 PMID:19744523

Treguer, M., Rocco, F., Lelong, G., Le Nestour, A., Cardinal, T., Maali, A., & Lounis, B. (2005). Fluorescent silver oligomeric clusters and colloidal particles. *Solid State Sciences*, *7*(7), 812–818. doi:10.1016/j.solidstatesciences.2005.01.017

Trouiller, B., Reliene, R., Westbrook, A., Solaimani, P., & Schiestl, R. H. (2009). Titanium dioxide nanoparticles induce DNA damage and genetic instability *in vivo* in mice. *Cancer Research*, *69*(22), 8784–8789. doi:10.1158/0008-5472.CAN-09-2496 PMID:19887611

Tshokey, T., Tshering, T., Pradhan, A. R., Adhikari, D., Sharma, R., Gurung, K., Dorji, T., Wangmo, S., Dorji, U., & Wangdi, K. (2018). Antibiotic resistance in Neisseria gonorrhoea and treatment outcomes of gonococcal urethritis suspected patients in two large hospitals in Bhutan, 2015. *PLoS One*, *13*(8), e0201721. doi:10.1371/journal.pone.0201721 PMID:30067836

Tunçsoy, B. S. (2018). Toxicity of nanoparticles on insects: A Review. *Adana Science and Technology University Journal of Science, 49*.

Türk, S., Altınsoy, I., Efe, G. Ç., İpek, M., Özacar, M., & Bindal, C. (2018). 3D porous collagen/functionalized multi-walled carbon nanotube/chitosan/hydroxyapatite composite scaffolds for bone tissue engineering. *Materials Science and Engineering C, 92*, 757–768. doi:10.1016/j.msec.2018.07.020 PMID:30184804

Turner, M., Golovko, V. B., Vaughan, O. P., Abdulkin, P., Berenguer-Murcia, A., Tikhov, M. S., Johnson, B. F. G., & Lambert, R. M. (2008). Selective oxidation with dioxygen by gold nanoparticle catalysts derived from 55-atom clusters. *Nature, 454*(7207), 981–983. doi:10.1038/nature07194 PMID:18719586

Turos, E., Reddy, G. S., Greenhalgh, K., Ramaraju, P., Abeylath, S. C., Jang, S., Dickey, S., & Lim, D. V. (2017). Penicillin-bound polyacrylate nanoparticles: Restoring the activity of β- lactam antibiotics against MRSA. *Bioorganic & Medicinal Chemistry Letters, 17*(12), 3468–3472. doi:10.1016/j.bmcl.2007.03.077 PMID:17420125

Tyagi, P. K., Tyagi, S., Verma, C., & Rajpal, A. (2013). Estimation of toxic effects of chemically and biologically synthesized silver nanoparticles on human gut microflora containing *Bacillus subtilis. J. Toxicology and Environmental Health Sciences, 5*(9), 172–177. doi:10.5897/JTEHS2013.0271

Tyers, M., & Wright, G. D. (2019). Drug combinations: A strategy to extend the life of antibiotics in the 21st century. *Nature Reviews. Microbiology, 17*(3), 141–155. doi:10.103841579-018-0141-x PMID:30683887

Ubaid, R., & Hemalatha, S. (2017). Marine endophytic actinomycetes assisted synthesis of copper nanoparticles (CuNPs): Characterization and antibacterial efficacy against human pathogens. *Materials Letters, 194*, 176–180. doi:10.1016/j.matlet.2017.02.055

Ubaid, R., & Hemalatha, S. (2019). Effect of biosynthesized copper nanoparticles (Cunps) on growth and biofilm formation in fluconazole resistant Candida albicans. *Journal of Microbiology, Biotechnology and Food Sciences, 9*, 21–24.

Ubaid, R., Priya, S., Afsana, P., Sah, S. K., & Hemalatha, S. (2018b). Efficacy of Andrographis paniculata against extended spectrum β-lactamase (ESBL) producing E. coli BMC comp. *Alt. Med., 18*, 244.

Ubaid, R., Sah, S. K., & Hemalatha, S. (2018a). Growth inhibitory effect of oven dried copper nanoparticles (cunps) on drug resistant clinical isolates Ira. *Journal of Marine Science and Engineering, 15*, 12–20.

Ubeyitogullari, A., & Ciftci, O. N. (2019). A novel and green nanoparticle formation approach to forming low-crystallinity curcumin nanoparticles to improve curcumin's bioaccessibility. *Scientific Reports, 9*(1), 1–11. doi:10.103841598-019-55619-4 PMID:31836788

Uchegbu, I. F., & Vyas, S. P. (1998). Non-ionic surfactant based vesicles (niosomes) in drug delivery. *International Journal of Pharmaceutics, 172*(1-2), 33–70. doi:10.1016/S0378-5173(98)00169-0

Uetrecht, A. C., & Bear, J. E. (2006). Coronins: The return of the crown. *Trends in Cell Biology, 16*(8), 421–426. doi:10.1016/j.tcb.2006.06.002 PMID:16806932

Uğur, S. S., Sarıışık, M., Aktaş, A. H., Uçar, M. C., & Erden, E. (2010). Modifying of cotton fabric surface with nano-ZnO multilayer films by layer-by-layer deposition method. *Nanoscale Research Letters, 5*(7), 1204–1210. doi:10.100711671-010-9627-9 PMID:20596450

Ulbrich, K., Hola, K., Subr, V., Bakandritsos, A., Tucek, J., & Zboril, R. (2016). Targeted drug delivery with polymers and magnetic nanoparticles: Covalent and noncovalent approaches, release control, and clinical studies. *Chemical Reviews, 116*(9), 5338–5431. doi:10.1021/acs.chemrev.5b00589 PMID:27109701

Ullah Khan, S., Saleh, T. A., Wahab, A., Khan, M. H. U., Khan, D., Ullah Khan, W., Rahim, A., Kamal, S., Ullah Khan, F., & Fahad, S. (2018). Nanosilver: New ageless and versatile biomedical therapeutic scaffold. *International Journal of Nanomedicine*, *13*, 733–762. doi:10.2147/IJN.S153167 PMID:29440898

Ullah, H., Wahid, F., Santos, H. A., & Khan, T. (2016). Advances in biomedical and pharmaceutical applications of functional bacterial cellulose-based nanocomposites. *Carbohydrate Polymers*, *150*, 330–352. doi:10.1016/j.carbpol.2016.05.029 PMID:27312644

Ulloa-Ogaz, A. L., Pinon-Castillo, H. A., Munoz-Castellanos, L. N., Athie-Garcia, M. S., Ballinas-Casarrubias, M. D. L., Murillo-Ramirez, J. G., Flores-Ongay, L. Á., Duran, R., & Orrantia-Borunda, E. (2017). Oxidative damage to *Pseudomonas aeruginosa* ATCC 27833 and *Staphylococcusaureus* ATCC 24213 induced by CuO-NPs. *Environmental Science and Pollution Research International*, *24*(27), 22048–22060. doi:10.100711356-017-9718-6 PMID:28791555

Ungureanu, B. S., Teodorescu, C. M., & Săftoiu, A. (2016, September). Magnetic Nanoparticles for Hepatocellular Carcinoma Diagnosis and Therapy. *Journal of Gastrointestinal and Liver Diseases; JGLD*, *25*(3), 375–383. doi:10.15403/jgld.2014.1121.253.nan PMID:27689203

Upadhyay, S., Khan, I., Gothwal, A., Pachouri, P. K., Bhaskar, N., Gupta, U. D., Chauhan, D. S., & Gupta, U. (2017). Conjugated and entrapped HPMA-PLA Nano-polymeric micelles based dual delivery of first line anti TB drugs: Improved and safe drug delivery against sensitive and resistant Mycobacterium tuberculosis. *Pharmaceutical Research*, *34*(9), 1944–1955. doi:10.100711095-017-2206-3 PMID:28685299

Uppuluri, P., Pierce, C. G., & López-Ribot, J. L. (2009). Candida albicans Biofilm Formation and its Clinical Consequences. *Future Microbiology*, *4*(10), 1235–1237. doi:10.2217/fmb.09.85 PubMed

US National Library of Medicine. (2012). https://clinicaltrials.gov/ct2/show/NCT00356980?term

US National Library of Medicine. (2016). https://clinicaltrials.gov/ct2/show/NCT02191878?term

Usha, C., & Rachel, G. A. (2014). Biogenic synthesis of silver nanoparticles by *Acacia nilotica* and their Antibacterial activity. *International Journal of Scientific Research*, *3*(6), 27–29. doi:10.15373/22778179/June2014/11

Usman, M. S., El Zowalaty, M. E., Shameli, K., Zainuddin, N., Salama, M., & Ibrahim, N. A. (2013). Synthesis, characterization, and antimicrobial properties of copper nanoparticles. *International Journal of Nanomedicine*, *8*, 4467–4479. PubMed

Valant, J., Drobne, D., & Novak, S. (2012). Effect of ingested titanium dioxidenanoparticles on the digestive gland cell membrane of terrestrialisopods. *Chemosphere*, *87*(1), 19–25. doi:10.1016/j.chemosphere.2011.11.047 PMID:22189379

Vale, G., Mehennaoui, K., Cambier, S., Libralato, G., Jomini, S., & Domingos, R. F. (2016). Manufactured nanoparticles in the aquatic environment-biochemical responses on freshwater organisms: A critical overview. *Aquatic Toxicology (Amsterdam, Netherlands)*, *170*, 162–174. doi:10.1016/j.aquatox.2015.11.019 PMID:26655660

Valsami-Jones, E., & Lynch, I. (2015). NANOSAFETY. How safe are nanomaterials? *Science*, *350*(6259), 388–389. doi:10.1126cience.aad0768 PMID:26494749

Valsesia, A., Desmet, C., Jimenez, I. O., Oddo, A., Capomaccio, R., Rossi, F., & Colpo, P. (2018). Direct quantification of nanoparticle surface hydrophobicity. *Communications Chemistry*, *1*(1), 53. doi:10.103842004-018-0054-7

Van der Linden, J. W., Arendrup, M. C., Warris, A., Lagrou, K., Pelloux, H., Hauser, P. M., Chryssanthou, E., Mellado, E., Kidd, S. E., Tortorano, A. M., Dannaoui, E., Gaustad, P., Baddley, J. W., Uekötter, A., Lass-Flörl, C., Klimko, N., Moore, C. B., Denning, D. W., Pasqualotto, A. C., ... Verweij, P. E. (2015). Prospective multicenter international surveillance of azole resistance in Aspergillus fumigatus. [*Emerging Infectious Diseases*, *21*(6), 1041–1044. doi:10.3201/eid2106.140717 PMID:25988348

Van der Zee, J. (2002). Heating the patient: A promising approach? *Annals of Oncology: Official Journal of the European Society for Medical Oncology*, *13*(8), 1173–1184. doi:10.1093/annonc/mdf280 PMID:12181239

van Duin, D., & Doi, Y. (2017). The global epidemiology of carbapenemase-producing Enterobacteriaceae. *Virulence*, *8*(4), 460–469. doi:10.1080/21505594.2016.1222343 PMID:27593176

Van Giau, V., An, S. S. A., & Hulme, J. (2019). Recent advances in the treatment of pathogenic infections using antibiotics and nano-drug delivery vehicles. *Drug Design, Development and Therapy*, *13*, 327–343. doi:10.2147/DDDT.S190577 PMID:30705582

Van Hoek, A., Mevius, D., Guerra, B., Mullany, P., Roberts, A., & Aarts, H. (2011). Acquired antibiotic resistance genes: An overview. *Frontiers in Microbiology*, *2*, 203. doi:10.3389/fmicb.2011.00203 PMID:22046172

Vance, M. E., Kuiken, T., Vejerano, E. P., McGinnis, S. P., Hochella, M. F. Jr, Rejeski, D., & Hull, M. S. (2015). Nanotechnology in the real world: Redeveloping the nanomaterial consumer products inventory. *Beilstein Journal of Nanotechnology*, *6*, 1769–1780. doi:10.3762/bjnano.6.181 PMID:26425429

Vanhee, van der Sloot, Verschueren, Serrano, Rousseau, & Schymkowitz. (2011). Computational design of peptide ligands. *Trends Biotechnol.*, *29*(5), 231-239.

Varma, J. R., Kumar, T. S., Prasanthi, B., & Ratna, J. V. (2015). Formulation and characterization of pyrazinamide polymeric nanoparticles for pulmonary tuberculosis: Efficiency for alveolar macrophage targeting. *Indian Journal of Pharmaceutical Sciences*, *77*(3), 258. doi:10.4103/0250-474X.159602 PMID:26180270

Varshosaz, J., Hassanzadeh, F., Aliabadi, H.S., Khoraskani, F.R., Mirian, M., & Behdadfar, B. (2016). Targeted delivery of doxorubicin to breast cancer cells by magnetic LHRH chitosan bioconjugated nanoparticles. *Int J Biol Macromol.*, *93*(Pt A), 1192–1205.

Vasanth, S. B., & Kurian, G. A. (2017). Toxicity evaluation of silver nanoparticles synthesized by chemical and green route in different experimental models. *Artificial Cells, Nanomedicine, and Biotechnology*, *45*(8), 1721–1727. doi:10.1080/21691401.2017.1282500 PMID:28278585

Vasilev, S., Zelenovskiy, P., Vasileva, D., Nuraeva, A., Shur, V. Y., & Kholkin, A. L. (2016). ShurYa, V.; Kholkin, A.L. Piezoelectric properties of diphenylalanine microtubes prepared from the solution. *Journal of Physics and Chemistry of Solids*, *93*, 68–72. doi:10.1016/j.jpcs.2016.02.002

Veiga, A. S., Sinthuvanich, C., Gaspar, D., Franquelim, H. G., Castanho, M. A. R. B., & Schneider, J. P. (2012). Arginine-rich self-assembling peptides as potent antibacterial gels. *Biomaterials*, *33*(35), 8907–8916. doi:10.1016/j.biomaterials.2012.08.046 PMID:22995710

Velappan, S., Nivedhita, P., Vimala, R., & Raja, S. (2020). Role of nanotitania on the thermomechanical properties of silicon carbide refractories. *Ceramics International*, *46*(16), 25921–25926. doi:10.1016/j.ceramint.2020.07.077

Velusamy, V., Arshak, K., Korostynska, O., Oliwa, K., & Adley, C. (2010). An overview of foodborne pathogen detection: In the perspective of biosensors. *Biotechnology Advances*, *28*(2), 232–254. doi:10.1016/j.biotechadv.2009.12.004 PMID:20006978

Ventola, C. L. (2017). Cancer immunotherapy, part 3: Challenges and future trends. *P&T*, *42*(8), 514. PMID:28781505

Veres, T., Molnár, K., Nesztor, D., Tombácz, E., Voniatis, C., Fehér, D., Weber, Gy., Zrinyi, M., & Jedlovszky-Hajdu, A. (2018). Magnetic gelfiber based scaffoldfortheranosticapplication. *12th International conference on Scientific and clinical applications of Magnetic carriers.*

Verma, A., & Mehata, M. S. (2016). Controllable synthesis of silver nanoparticles using Neem leaves and their antimicrobial activity. *Journal of Radiation Research and Applied Sciences*, *9*(1), 109–115. doi:10.1016/j.jrras.2015.11.001

Verma, A., Uzun, O., Hu, Y., Hu, Y., Han, H. S., Watson, N., Chen, S., Irvine, D. J., & Stellacci, F. (2008). Surface-structure-regulated cell-membrane penetration by monolayer-protected nanoparticles. *Nature Materials*, *7*(7), 588–595. doi:10.1038/nmat2202 PMID:18500347

Verma, Crosbie-Staunton, Amrosatti, Gallaagher, Rayn, Doody, McAtamney, MacLoughlin, Galvin, Burke, Volkov, & Gun'ko. (2013). Magnetic core shell nanoparticle for Drug delivery by nebulization. *Journal of Nanobiotechnology.*

Versiani, A. F., Andrade, L. M., Estefania, M. N., Sergio, S., Geraldo, J. M., Claudilne, R. C., ... Fonseca, F. G. (2016). Gold nanoparticles and their applications in biomedicine. *Future Virology*, *11*(4), 293–309. doi:10.2217/fvl-2015-0010

Versluis, A., van Geel, P., Oppelaar, H., van Berkel, T. J. C., & Bijsterbosch, M. K. (1996). Receptor-mediated uptake of low-density lipoprotein by B16 melanoma cells *in vitro* and *in vivo* in mice. *British Journal of Cancer*, *74*(4), 525–532. doi:10.1038/bjc.1996.396 PMID:8761365

Versluis, F., Marsden, H. R., & Kros, A. (2010). Power struggles in peptide-amphiphile nanostructures. *Chemical Society Reviews*, *39*(9), 3434–3444. doi:10.1039/b919446k PMID:20644886

Vester, B., & Douthwaite, S. (2001). Macrolide resistance conferred by base substitutions in 23S rRNA. *Antimicrobial Agents and Chemotherapy*, *45*(1), 1–12. doi:10.1128/AAC.45.1.1-12.2001 PMID:11120937

Viale, P., Giannella, M., Tedeschi, S., & Lewis, R. (2015). Treatment of MDR-Gram negative infections in the 21st century: A never ending threat for clinicians. *Current Opinion in Pharmacology*, *24*, 30–37. doi:10.1016/j.coph.2015.07.001 PMID:26210268

Vieira, A. C., Chaves, L. L., Pinheiro, M., Lima, S. A. C., Ferreira, D., Sarmento, B., & Reis, S. (2018). Mannosylated solid lipid nanoparticles for the selective delivery of rifampicin to macrophages. *Artificial cells, Nanomedicine, and Biotechnology, 46*(sup1), 653-663.

Vieira, A. C., Chaves, L. L., Pinheiro, S., Pinto, S., Pinheiro, M., Lima, S. C., Ferreira, D., Sarmento, B., & Reis, S. (2018). Mucoadhesive chitosan-coated solid lipid nanoparticles for better management of tuberculosis. *International Journal of Pharmaceutics*, *536*(1), 478–485. doi:10.1016/j.ijpharm.2017.11.071 PMID:29203137

Vieira, A. C., Magalhães, J., Rocha, S., Cardoso, M. S., Santos, S. G., Borges, M., Pinheiro, M., & Reis, S. (2017). Targeted macrophages delivery of rifampicin-loaded lipid nanoparticles to improve tuberculosis treatment. *Nanomedicine (London)*, *12*(24), 2721–2736. doi:10.2217/nnm-2017-0248 PMID:29119867

Vieira, A., Ramesh, A., Seddon, A. M., & Karlyshev, A. V. (2017). CmeABC Multidrug Efflux Pump Contributes to Antibiotic Resistance and Promotes Campylobacter jejuni Survival and Multiplication in Acanthamoeba polyphaga. *Applied and Environmental Microbiology*, *83*(22), e01600-17. Advance online publication. doi:10.1128/AEM.01600-17 PMID:28916560

Viertel, T. M., Ritter, K., & Horz, H. P. (2014). Viruses versus bacteria—Novel approaches to phage therapy as a tool against multidrug-resistant pathogens. *The Journal of Antimicrobial Chemotherapy*, *69*(9), 2326–2336. doi:10.1093/jac/dku173 PMID:24872344

Vilchèze, C., & Jacobs, W. R. Jr. (2019). The isoniazid paradigm of killing, resistance, and persistence in Mycobacterium tuberculosis. *Journal of Molecular Biology*, *431*(18), 3450–3461. doi:10.1016/j.jmb.2019.02.016 PMID:30797860

Vinothini, K., Rajendran, N. K., Ramu, A., Elumalai, N., & Rajan, M. (2019). Folate receptor targeted delivery of paclitaxel to breast cancer cells via folic acid conjugated graphene oxide grafted methyl acrylate nanocarrier. *Biomedicine and Pharmacotherapy*, *110*, 906–917. doi:10.1016/j.biopha.2018.12.008 PMID:30572195

Virmani, I., Sasi, C., Priyadarshini, E., Kumar, R., Sharma, S.K., Singh, G.P., Pachwarya, R.B., Paulraj, R., Barabadi, H., Saravanan, M., &Meena, R. (2019). Comparative anticancer potential of biologically and chemically synthesized gold nanoparticles. *Journal of Cluster Science*, 1-10.

Vismara, E., Bongio, C., Coletti, A., Edelman, R., Serafini, A., Mauri, M., Simonutti, R., Bertini, S., Urso, E., Assaraf, Y., & Livney, Y. (2017). Albumin and hyaluronic acid-coated superparamagnetic iron oxide nanoparticles loaded with paclitaxel for biomedical applications. *Molecules (Basel, Switzerland)*, *22*(7), 1030. doi:10.3390/molecules22071030 PMID:28640222

Vlierberghe, S. V., Dubruel, P., & Schacht, E. (2011). Biopolymer-based hydrogels as scaffolds for tissue engineering applications: A review. *Biomacromolecules*, *12*(2), 1387–1408. doi:10.1021/bm200083n PMID:21388145

Volkmar, W., Pettinger, T. K., & Murdock, N. (2014). Nanopharmaceuticals (part 1): Products on the Market. *International Journal of Nanomedicine*, *9*, 4357–4373. PubMed

Von Nussbaum, F., Brands, M., Hinzen, B., Weigand, S., & Habich, D. (2006). Antibacterial natural products in medicinal chemistry—Exodus or revival? *Angewandte Chemie International Edition*, *45*(31), 5072–5129. doi:10.1002/anie.200600350 PMID:16881035

Vu, B. T., Shahin, S. A., Croissant, J., Fatieiev, Y., Matsumoto, K., Doan, T. L., Yik, T., Simargi, S., Conteras, A., Ratliff, L., Jimenez, C. M., Raehm, L., Khashab, N., Durand, J. O., Glackin, C., & Tamanoi, F. (2018). Chick chorioallantoic membrane assay as an in vivo model to study the effect of nanoparticle based anticancer drugs in ovarian cancer. *Scientific Reports*, *8*(1), 8524. doi:10.103841598-018-25573-8 PMID:29867159

Vyas, S. P., Kannan, M. E., Jain, S., Mishra, V., & Singh, P. (2004). Design of liposomal aerosols for improved delivery of rifampicin to alveolar macrophages. *International Journal of Pharmaceutics*, *269*(1), 37–49. doi:10.1016/j.ijpharm.2003.08.017 PMID:14698575

Wagner, D. E., Philips, C. L., Ali, W. M., Nybakken, G. E., Crawford, E. D., Schwab, A. D., Smith, W. F., & Fairman, R. (2005). Toward the development of peptide nanofilaments and nanopores as smart materials. *Proceedings of the National Academy of Sciences of the United States of America*, *102*(36), 12656–12661. doi:10.1073/pnas.0505871102 PMID:16129839

Wagner, S. C., Roskamp, M., Pallerla, M., Araghi, R. R., Schlecht, S., & Koksch, B. (2010). pallerla M, Araghi RR, Schlecht S, koksch B. Nanoparticle-induced folding and fibril formation of coiled coil-based model peptides. *Small*, *6*(12), 1321–1328. doi:10.1002mll.200902067 PMID:20517875

Wahaba, J. A., Kimc, I. S., & Ni, Q. Q. (2019). A comparative study on synthesis of AgNPs on cellulose nanofibers by thermal treatment and DMF for antibacterial activities. *Materials Science and Engineering C*, *98*, 1179–1195. doi:10.1016/j.msec.2019.01.017 PMID:30813001

Waite, C.L., & Roth, C.M. (2012). Nanoscale drug delivery systems for enhanced drug penetration into solid tumors: current progress and opportunities. *Critical Reviews™ in Biomedical Engineering, 40*(1), 21-41.

Waldhauer, I., & Steinle, A. (2008). NK cells and cancer immunosurveillance. *Oncogene*, *27*(45), 5932–5943. doi:10.1038/onc.2008.267 PMID:18836474

Walkey, C. D., Olsen, J. B., Song, F., Liu, R., Guo, H., Olsen, D. W. H., Cohen, Y., Emili, A., & Chan, W. C. (2014). Protein corona fingerprinting predicts the cellular interaction of gold and silver nanoparticles. *ACS Nano*, *8*(3), 2439–2455. doi:10.1021/nn406018q PMID:24517450

Wallwiener, M., Riethdorf, S., Hartkopf, A. D., Modugno, C., Nees, J., Madhavan, D., Sprick, M. R., Schott, S., Domschke, C., Baccelli, I., Schönfisch, B., Burwinkel, B., Marmé, F., Heil, J., Sohn, C., Pantel, K., Trumpp, A., & Schneeweiss, A. (2014). Serial enumeration of circulating tumor cells predicts treatment response and prognosis in metastatic breast cancer: A prospective study in 393 patients. *BMC Cancer*, *14*(1), 512. doi:10.1186/1471-2407-14-512 PMID:25015676

Wanga, W. P., & Pana, C. Y. (2004). Preparation and characterization of polystyrene/graphite composite prepared by cationic grafting polymerization. *Polymer*, *45*(12), 3987–3995. doi:10.1016/j.polymer.2004.04.023

Wang, D. Y., Van Der Mei, H. C., Ren, Y., Busscher, H. J., & Shi, L. (2019). Lipid-based antimicrobial delivery-systems for the treatment of bacterial infections. *Frontiers in Chemistry*, 7. PMID:31998680

Wang, G., Li, X., & Wang, Z. (2015). The antimicrobial peptide database as a tool for research and education. *Nucleic Acids Research*, *44*(D1), 1087–1093. doi:10.1093/nar/gkv1278 PMID:26602694

Wang, H., Boerman, O., Sariibrahimoglu, K., Li, Y., Jansen, J., & Leeuwenburgh, S. (2012). Comparison of micro- vs. nanostructured colloidal gelatin gels for sustained delivery of osteogenic proteins: Bone morphogenetic protein-2 and alkaline phosphatase. *Biomaterials*, *33*(33), 8695–8703. doi:10.1016/j.biomaterials.2012.08.024 PMID:22922022

Wang, J., Chen, B., & Cheng, J. (2011b). Apoptotic mechanism of human leukemia K562/A02 cells induced by magnetic iron oxide nanoparticles coloaded with daunorubicin and 5-bromotetrandrin. *International Journal of Nanomedicine*, *6*, 1027. PMID:21720514

Wang, J., Chen, B., & Chen, J. (2011a). Synthesis and antitumor efficacy of daunorubicin-loaded magnetic nanoparticles. *International Journal of Nanomedicine*, *6*, 203. PMID:21445276

Wang, J., Du, L., Fu, Y., Jiang, P., & Wang, X. (2019). ZnO nanoparticles inhibit the activity of Porphyromonas gingivalis and Actinomyces naeslundii and promote the mineralization of the cementum. *BMC Oral Health*, *19*(1), 1–11. doi:10.118612903-019-0780-y PMID:31088450

Wang, J., Hsu, S.-W., Tao, A. R., & Jokerst, J. V. (2017). Copper sulfide nanodisk as photoacoustic contrast agent for ovarian tumor detection. *Photons Plus Ultrasound: Imaging and Sensing*, *2017*(10064), 100642D. doi:10.1117/12.2251184

Wang, K., Ma, Q., Pang, K., Ding, B., Zhang, J., & Duan, Y. (2018). One-pot Synthesis of Graphene/Chitin Nanofibers Hybrids and Their Remarkable Reinforcement on Poly (vinyl alcohol). *Carbohydrate Polymers*, *194*, 146–153. doi:10.1016/j.carbpol.2018.04.036 PMID:29801822

Wang, L. (2016). Synthetic methods of CuS nanoparticles and their applications for imaging and cancer therapy. *RSC Advances*, *6*(86), 82596–82615. doi:10.1039/C6RA18355G

Wang, L., Du, J., Zhou, Y., & Wang, Y. (2017). Safety of nanosuspensions in drug delivery. *Nanomedicine; Nanotechnology, Biology, and Medicine*, *13*(2), 455–469. doi:10.1016/j.nano.2016.08.007 PMID:27558350

Wang, L., Hu, C., & Shao, L. (2017). The antimicrobial activity of nanoparticles: Present situation and prospects for the future. *International Journal of Nanomedicine*, *12*, 1227–1249. doi:10.2147/IJN.S121956 PMID:28243086

Wang, M., Du, L., Wu, X., Xiong, S., & Chu, P. K. (2011). Charged Diphenylalanine Nanotubes and Controlled Hierarchical Self-Assembly. *ACS Nano*, *5*(6), 4448–4454. doi:10.1021/nn2016524 PMID:21591732

Wang, Q., Alshaker, H., Böhler, T., Srivats, S., Chao, Y., Cooper, C., & Pchejetski, D. (2017). Core shell lipid-polymer hybrid nanoparticles with combined docetaxel and molecular targeted therapy for the treatment of metastatic prostate cancer. *Scientific Reports*, *7*(7), 5901. doi:10.103841598-017-06142-x PMID:28724986

Wang, R., Billone, P. S., & Mullett, W. M. (2013). Nanomedicine in action: An overview of cancer nanomedicine on the market and in clinical trials. *Journal of Nanomaterials*, *2013*, 2013. doi:10.1155/2013/629681

Wang, R., Schmidt, J. W., Harhay, D. M., Bosilevac, J. M., King, D. A., & Arthur, T. M. (2017). Biofilm Formation, Antimicrobial Resistance, and Sanitizer Tolerance of Salmonella enterica Strains Isolated from Beef Trim. *Foodborne Pathogens and Disease*, *14*(12), 687–695. doi:10.1089/fpd.2017.2319 PMID:29035101

Wang, S. H., & Yu, J. (2018). Structure-based design for binding peptides in anticancer therapy. *Biomaterials*, *156*, 1–15. doi:10.1016/j.biomaterials.2017.11.024 PMID:29182932

Wang, S., Lu, W., Tovmachenko, O., Rai, U. S., Yu, H., & Ray, P. C. (2008). Challenge in understanding size and shape dependent toxicity of gold nanomaterials in humanskin keratinocytes. *Chemical Physics Letters*, *463*(1-3), 145–149. doi:10.1016/j.cplett.2008.08.039 PMID:24068836

Wang, S., Wang, H., Jiao, J., Chen, K. J., Owens, G. E., Kamei, K., Sun, J., Sherman, D. J., Behrenbruch, C. P., Wu, H., & Tseng, H.-R. (2009). Three-dimensional nanostructured substrates toward efficient capture of circulating tumor cells. *Angewandte Chemie International Edition in English*, *48*(47), 8970–8973. doi:10.1002/anie.200901668 PMID:19847834

Wang, S., Yang, Z., Gong, G., Wang, J., Wu, J., Yang, S., & Jiang, L. (2016). Icephobicity of penguins spheniscushum-boldti and an artificial replica of penguin feather with air infused hierarchical rough structures. *The Journal of Physical Chemistry C*, *120*(29), 15923–15929. doi:10.1021/acs.jpcc.5b12298

Wang, S., Zeng, X., Yang, Q., & Qiao, S. (2016). Antimicrobial peptides as potential alternatives to antibiotics in food animal industry. *International Journal of Molecular Sciences*, *17*(5), 603. doi:10.3390/ijms17050603 PMID:27153059

Wang, W., Castro, L. A., Hoyos, M., & Mallouk, T. E. (2012). Autonomous Motion of Metallic Microrods Propelled by Ultrasound. *ACS Nano*, *6*(7), 6122–6132. doi:10.1021/nn301312z PMID:22631222

Wang, W., Itoh, S., Konno, K., Kikkawa, T., Ichinose, S., Sakai, K., Ohkuma, T., & Watabe, K. (2009). Effects of Schwann cell alignment along the oriented electrospun chitosan nanofibers on nerve regeneration. *Journal of Biomedical Materials Research. Part A*, *91*(4), 994–1005. doi:10.1002/jbm.a.32329 PMID:19097155

Wang, X., Chang, Y., Zhang, D., Tian, B., Yang, Y., & Wei, F. (2016). Transferrinconjugated drug/dye-co-encapsulated magnetic nanocarriers for active-targeting fluorescent/magnetic resonance imaging and antitumor effects in human brain tumor cells. *RSC Advances*, *6*(107), 105661–105675. doi:10.1039/C6RA20903C

Wang, X., Jing, X., Zhang, X., Liu, Q., Liu, J., Song, D., Wang, J., & Liu, L. (2016). A versatile platform of magnetic microspheres loaded with dual-anticancer drugs for drug release. *Materials Chemistry and Physics*, *177*, 213–219. doi:10.1016/j.matchemphys.2016.04.021

Wang, X., Tu, M., Tian, B., Yi, Y., Wei, Z., & Wei, F. (2016). Synthesis of tumortargeted folate conjugated fluorescent magnetic albumin nanoparticles for enhanced intracellular dual-modal imaging into human brain tumor cells. *Analytical Biochemistry*, *512*, 8–17. doi:10.1016/j.ab.2016.08.010 PMID:27523645

Wang, Y. F., Liu, L., Xue, X., & Liang, X. J. (2017). Nanoparticle-based drug delivery systems: What can they really do in vivo? *F1000 Research*, *6*, 681. doi:10.12688/f1000research.9690.1 PMID:28620465

Wang, Y., Hernandez, R. M., Bartlett, D. J., Bingham, J. M., Kline, T. R., Sen, A., & Mallouk, T. E. (2006). Bipolar Electrochemical Mechanism for the Propulsion of Catalytic Nanomotors in Hydrogen Peroxide Solutions. *Langmuir, 22*(25), 10451–10456. doi:10.1021/la0615950 PMID:17129015

Wang, Y., Li, P., Truong-Dinh Tran, T., Zhang, J., & Kong, L. (2016). Manufacturing Techniques and Surface Engineering of Polymer Based Nanoparticles for Targeted Drug Delivery to Cancer. *Nanomaterials (Basel, Switzerland), 6*(2), 26. doi:10.3390/nano6020026 PMID:28344283

Wang, Y.-P., Liao, Y.-T., Liu, C.-H., Yu, J., Alamri, H. R., Alothman, Z. A., Hossain, M. S. A., Yamauchi, Y., & Wu, K. C.-W. (2017). Trifunctional Fe3 O4 /CaP/ Alginate core–shell–corona nanoparticles for magnetically guided, pH-responsive, and chemically targeted chemotherapy. *ACS Biomaterials Science & Engineering, 3*(10), 2366–2374. doi:10.1021/acsbiomaterials.7b00230 PMID:30854421

Wang, Y., Yi, S., Sun, L., Huang, Y., Lenaghan, S. C., & Zhang, M. (2014). Doxorubicin-loaded cyclic peptide nanotube bundles overcome chemoresistance in breast cancer cells. *Journal of Biomedical Nanotechnology, 10*(3), 445–45. doi:10.1166/jbn.2014.1724 PMID:24730240

Wang, Z., & Rothberg, L. J. (2005). Origins of blinking in single-molecule raman spectroscopy. *The Journal of Physical Chemistry B, 109*(8), 3387–3391. doi:10.1021/jp0460947 PMID:16851369

Warheit, D. B. (2018). Hazard and risk assessment strategies for nanoparticle exposures: How far have we come in the past 10 years? *F1000 Research, 7*, 376. doi:10.12688/f1000research.12691.1 PMID:29636906

Weber, C., & Falkenhagen, D. (1997). Specific blood purification by means of antibodyconjugated magnetic microspheres. In *Scientific and Clinical Applications of Magnetic Carriers* (pp. 371–378). Plenum Press. doi:10.1007/978-1-4757-6482-6_28

Wei, D., Sun, W., Qian, W., Ye, Y., & Ma, X. (2009). The synthesis of chitosan-based silver nanoparticles and their antibacterial activity. *Carbohydrate Research, 344*(17), 2375–2382. doi:10.1016/j.carres.2009.09.001 PMID:19800053

Weiner, L. M., Dhodapkar, M. V., & Ferrone, S. (2009). Monoclonal antibodies for cancer immunotherapy. *Lancet, 373*(9668), 1033–1040. doi:10.1016/S0140-6736(09)60251-8 PMID:19304016

Weir, E., Lawlor, A., Whelan, A., & Regan, F. (2008). The use of nanoparticles in anti-microbial materials and their characterization. *Analyst (London), 133*(7), 835–845. doi:10.1039/b715532h PMID:18575632

Weitz, I. S., Maoz, M., Panitz, D., Eichler, S., & Segal, E. (2015). Combination of CuO nanoparticles and fluconazole: Preparation, characterization, and antifungal activity against Candida albicans. *Journal of Nanoparticle Research, 17*(8), 342. doi:10.100711051-015-3149-4

Weitz, R., Harnau, L., Rauschenbach, S., Burghard, M., & Kern, K. (2008). Polymer nanofibers via nozzle-free centrifugal spinning. *Nano Letters, 8*(4), 1187–1191. doi:10.1021/nl080124q PMID:18307320

Wei, X., Zhou, W., Sanjay, S. T., Zhang, J., Jin, Q., Xu, F., Dominguez, D. C., & Li, X. J. (2018). Multiplexed instrument-free bar-chart spin chip integrated with nanoparticle-mediated magnetic apt sensors for visual quantitative detection of multiple pathogens. *Analytical Chemistry, 90*(16), 9888–9896. doi:10.1021/acs.analchem.8b02055 PMID:30028601

Weng, W., Chen, G., & Wu, D. (2005). Transport properties of electrically conducting nylon 6/foliated graphite nanocomposites. *Polymer, 46*(16), 6250–6257. doi:10.1016/j.polymer.2005.05.071

Whaley, S. R., English, D. S., Hu, E. L., Barbara, P. F., & Belcher, A. M. (2000). Selection of peptides with semiconductor binding specificity for directed nanocrystal assembly. *Nature, 405*(6787), 665–668. doi:10.1038/35015043 PMID:10864319

White, C. J., & Yudin, A. K. (2011). Contemporary strategies for peptide macrocyclization. *Nature Chemistry*, *3*(7), 509. doi:10.1038/nchem.1062 PMID:21697871

White, N. J. (2004). Antimalarial drug resistance. *The Journal of Clinical Investigation*, *113*(8), 1084–1092. doi:10.1172/JCI21682 PMID:15085184

WHO. (2015). *Guidelines for the Treatment of Malaria*. WHO.

WHO. (2019). *Outbreak update – Dengue in Pakistan*. WHO.

Wicki, A., Witzigmann, D., Balasubramanian, V., & Huwyler, J. (2015). Nanomedicine in cancer therapy: Challenges, opportunities, and clinical applications. *Journal of Controlled Release*, *200*, 138–157. doi:10.1016/j.jconrel.2014.12.030 PMID:25545217

Wick, P., Manser, P., Limbach, L. K., Dettlaff-Weglikowska, U., Krumeich, F., Roth, S., Stark, W., & Bruinink, A. (2007). The degree and kind of agglomeration affect carbon nanotube cytotoxicity. *Toxicology Letters*, *168*(2), 121–131. doi:10.1016/j.toxlet.2006.08.019 PMID:17169512

Widder, K. J., Morris, R. M., Howard, D. P., & Senyei, A. E. (1981). Tumor remission in Yoshida sarcoma-bearing rats by selective targeting of magnetic albu- min microspheres containing doxorubicin. *Proceedings of the National Academy of Sciences of the United States of America*, *78*(1), 579–581. doi:10.1073/pnas.78.1.579 PMID:6941258

Widder, K. J., Senyei, A. E., & Ranney, D. F. (1979). Magneti- cally responsive microspheres and other carri- ers for the biophysical targeting of antitumor agents. *Advances in Pharmacology and Chemotherapy*, *16*, 213–271. doi:10.1016/S1054-3589(08)60246-X PMID:382799

Willi, K., Sandmeier, H., Kulik, E. M., & Meyer, J. (1997). Transduction of antibiotic resistance markers among *Actinobacillus actinomycetemcomitans* strains by temperate bacteriophages Aaϕ23. *Cellular and Molecular Life Sciences*, *53*(11-12), 904–910. doi:10.1007000180050109 PMID:9447241

Wnorowska, U., Fiedoruk, K., Piktel, E., Prasad, S. V., Sulik, M., Janion, M., Daniluk, T., Savage, P. B., & Bucki, R. (2020). Nanoantibiotics containing membrane-active human cathelicidin LL-37 or synthetic ceragenins attached to the surface of magnetic nanoparticles as novel and innovative therapeutic tools: Current status and potential future applications. *Journal of Nanobiotechnology*, *18*(1), 3. doi:10.118612951-019-0566-z PMID:31898542

Wong, D., Nielsen, T. B., Bonomo, R. A., Pantapalangkoor, P., Luna, B., & Spellberg, B. (2017). Clinical and Pathophysiological Overview of Acinetobacter Infections: A Century of Challenges. *Clinical Microbiology Reviews*, *30*(1), 409–447. doi:10.1128/CMR.00058-16 PMID:27974412

Woodard, L. E., Dennis, C. L., Borchers, J. A., Attaluri, A., Velarde, E., Dawidczyk, C., Searson, P. C., Pomper, M. G., & Ivkov, R. (2018). Nanoparticle architecture preserves magnetic properties during coating to enable robust multi-model functionality. *Scientific Reports*, *8*(12706), 1–13. doi:10.103841598-018-29711-0 PMID:30139940

Woolhouse, M. E., & Ward, M. J. (2013). Microbiology. Sources of antimicrobial resistance. *Science*, *341*(6153), 1460–1461. doi:10.1126cience.1243444 PMID:24030495

World Health Organization. (2013). *Global tuberculosis report 2013*. World Health Organization.

World Health Organization. (2019). The WHO global task force on TB impact measurement (No. WHO/CDS/TB/2019.2). World Health Organization.

World Health Organization. (n.d.). *Antibacterial Agents in Clinical Development: An Analysis of the Antibacterial Clinical Development Pipeline, Including Tuberculosis*. World Health Organization. https://apps.who.int/iris/handle/10665/258965

World Health Organization. (n.d.). *Hepatitis B, Fact sheet N° 204* [updated July 2016]. Available from: http: // www. who. int/mediacentre /factsheets/fs204/en/

Wozniak, R. A. F., & Waldor, M. K. (2010). Integrative and conjugative elements: Mosaic mobile genetic elements enabling dynamic lateral gene flow. *Nature Reviews. Microbiology*, *8*(8), 552–563. doi:10.1038/nrmicro2382 PMID:20601965

Wu, D., Wang, L., Li, W., Xu, X., & Jiang, W. (2017, November). DNA nanostructure-based drug delivery nanosystems in cancer therapy. (2017). *International Journal of Pharmaceutics*, *533*(1), 169–178. Advance online publication. doi:10.1016/j.ijpharm.2017.09.032 PMID:28923770

Wu, L., Zhang, Z., Li, H., Wei, W., & Liu, S. (2016). Fe3 O4 -capped mesoporous silica foam for pH-responsive drug delivery. *Journal of Nanoscience and Nanotechnology*, *16*(7), 6781–6787. doi:10.1166/jnn.2016.11385

Wu, M., & Huang, S. (2017). Magnetic nanoparticles in cancer diagnosis, drug delivery and treatment. *Molecular and Clinical Oncology*, *7*(5), 738–746. doi:10.3892/mco.2017.1399 PMID:29075487

Wu, R. H., Nguyen, T. P., Marquart, G. W., Miesen, T. J., Mau, T., & Mackiewicz, M. R. (2014). A facile route to tailoring peptide-stabilized gold nanoparticles using glutathione as a synthon. *Molecules (Basel, Switzerland)*, *19*(5), 6754–6775. doi:10.3390/molecules19056754 PMID:24858266

Wust, P., Hildebrandt, B., Sreenivasa, G., Rau, B., Gellermann, J., Riess, H., Felix, R., & Schlag, P. M. (2002). Hyperthermia in combined treatment of cancer. *The Lancet. Oncology*, *3*(8), 487–497. doi:10.1016/S1470-2045(02)00818-5 PMID:12147435

Wu, X., Li, Z., Li, X., Tian, Y., Fan, Y., Yu, C., Zhou, B., Liu, Y., Xiang, R., & Yang, L. (2017). Synergistic effects of antimicrobial peptide DP7 combined with antibiotics against multidrug-resistant bacteria. *Drug Design, Development and Therapy*, *11*, 939–946. doi:10.2147/DDDT.S107195 PMID:28356719

Wu, Y. J., Si, T. Y., Shao, J. X., Wu, Z. G., & He, Q. (2016). Near-infrared light-driven Janus capsule motors: Fabrication, propulsion, and simulation. *Nano Research*, *9*(12), 3747–3756. doi:10.100712274-016-1245-0

Wu, Y., Wu, Z., Lin, X., He, Q., & Li, J. (2012). Autonomous Movement of Controllable Assembled Janus Capsule Motors. *ACS Nano*, *6*(12), 10910–10916. doi:10.1021/nn304335x PMID:23153409

Wu, Z., Lin, X., Zou, X., Sun, J., & He, Q. (2014). Biodegradable Protein-Based Rockets for Drug Transportation and Light-Triggered Release. *ACS Applied Materials & Interfaces*, *7*(1), 250–255. doi:10.1021/am507680u PMID:25496011

Wu, Z., Si, T., Gao, W., Lin, X., Wang, J., & He, Q. (2015). Superfast Near-Infrared Light-Driven Polymer Multilayer Rockets. *Small*, *12*(5), 577–582. doi:10.1002mll.201502605 PMID:26690728

Wu, Z., Wu, Y., He, W., Lin, X., Sun, J., & He, Q. (2013). Self-Propelled Polymer-Based Multilayer Nanorockets for Transportation and Drug Release. *Angewandte Chemie International Edition*, *52*(27), 7000–7003. doi:10.1002/anie.201301643 PMID:23703837

Wyllie, A. H. (1980). Glucocorticoid-induced thymocyte apoptosis is associated with endogenous endonuclease activation. *Nature*, *284*(5756), 555–556. doi:10.1038/284555a0 PMID:6245367

Xenariou, S., Griesenbach, U., Ferrari, S., Dean, P., Scheule, R. K., Cheng, S. H., Geddes, D. M., Plank, C., & Alton, E. W. F. W. (2006). Using mag- netic forces to enhance non-viral gene transfer to airway epi- thelium in vivo. *Gene Therapy*, *13*(21), 1545–1552. doi:10.1038j.gt.3302803 PMID:16738690

Xi, A., Solovev, A., Ananth, N., Gracias, D. H., Sanchez, S., & Schmidt, O. G. (2013). Rolled-up magnetic microdrillers: Towards remotely controlled minimally invasive surgery. *Nanoscale*, *5*(4), 1294–1297. doi:10.1039/C2NR32798H PMID:23154823

Xiang, L., Wei, J., Jianbo, S., Guili, W., Feng, G., & Ying, L. (2007). Puriÿed and sterilized magnetosomes from Magnetospirillum gryphiswaldense MSR-1 were not toxic to mouse ÿbroblasts in vitro. *Letters in Applied Microbiology*, *45*(1), 75–81. doi:10.1111/j.1472-765X.2007.02143.x PubMed

Xiao, Y., Ju, H.-X., & Chen, H.-Y. (1999). Hydrogen peroxide sensor based on horseradish peroxidase-labeled Au colloids immobilized on gold electrode surface by cysteamine monolayer. *Analytica Chimica Acta*, *391*(1), 73–82. doi:10.1016/S0003-2670(99)00196-8

Xia, T., Zhao, Y., Sager, T., George, S., Pokhrel, S., Li, N., Schoenfeld, D., Meng, H., Lin, S., Wang, X., Wang, M., Ji, Z., Zink, J. I., Mädler, L., Castranova, V., Lin, S., & Nel, A. E. (2011). Decreased dissolution of ZnO by iron doping yields nanoparticles with reduced toxicity in the rodent lung and zebrafish embryos. *ACS Nano*, *5*(2), 1223–1235. doi:10.1021/nn1028482 PMID:21250651

Xie, J., Chen, K., & Chen, X. (2009). Production, modiÿcation and bio-applications of magnetic nanoparticles gestated by magnetotactic bacteria. *Nano Research*, *2*(4), 261–278. doi:10.1007/s12274-009-9025-8 PubMed

Xie, W., Gao, Q., Wang, D., Guo, Z., Gao, F., Wang, X., Cai, Q., Feng, S., Fan, H., Sun, X., & Zhao, L. (2018). Doxorubicin-loaded Fe3O4@MoS2- PEG-2DG nanocubes as a theranostic platform for magnetic resonance imaging-guided chemo-photothermal therapy of breast cancer. *Nano Research*, *11*(5), 2470–2487. doi:10.100712274-017-1871-1

Xie, Y., He, Y., Irwin, P. L., Jin, T., & Shi, X. (2011). Antibacterial activity and mechanism of action of zinc oxide nanoparticles against Campylobacter jejuni. *Applied and Environmental Microbiology*, *77*(7), 2325–2331. doi:10.1128/AEM.02149-10 PMID:21296935

Xie, Y., Zhao, J., Huang, R., Qi, W., Wang, Y., Su, R., & He, Z. (2016). Calcium-Ion-Triggered Co-assembly of Peptide and Polysaccharide into a Hybrid Hydrogel for Drug Delivery. *Nanoscale Research Letters*, *11*(1), 184. doi:10.118611671-016-1415-8 PMID:27067732

Xiliang, L., Aoife, M., Anthony, J. K., & Malcolm, S. (2011). Nanoparticles in Electrochemical Biosensors. In H. S. Nalwa (Ed.), *Encyclopedia of Nanoscience and Nanotechnology* (pp. 473–501). Academic Press.

Xuan, M. J., Shao, J. X., Lin, X. K., Dai, L. R., & He, Q. (2015). Light-activated Janus self-assembled capsule micromotors. *Colloids and Surfaces. A, Physicochemical and Engineering Aspects*, *482*, 92–99. doi:10.1016/j.colsurfa.2015.04.032

Xu, C., Peng, C., Sun, L., Zhang, S., Huang, H., Chen, Y., & Shi, J. (2015). Distinctive effects of TiO 2 and CuO nanoparticles on soil microbes and their community structures in flooded paddy soil. *Soil Biology & Biochemistry*, *86*, 24–33. doi:10.1016/j.soilbio.2015.03.011

Xu, D., Xu, T., Guo, X., Liu, Q., Liu, J., Lv, W., Jing, X., Zhang, H., & Wang, J. (2017). Effect of the synthesis method on the performance of Fe3 O4 inositol hexaphosphate as a drug delivery vehicle for combination therapeutics with doxorubicin. *New Journal of Chemistry*, *41*(13), 5305–5312. doi:10.1039/C7NJ00599G

Xue, X., & Liang, X. J. (2012). Overcoming drug efflux-based multidrug resistance in cancer with nanotechnology. *Chinese Journal of Cancer*, *31*(2), 100–109. doi:10.5732/cjc.011.10326 PMID:22237039

Xue, Y., Xiao, H., & Zhang, Y. (2015). Antimicrobial Polymeric Materials with Quaternary Ammonium and Phosphonium Salts. *International Journal of Molecular Sciences*, *16*(2), 3626–3655. doi:10.3390/ijms16023626 PMID:25667977

Xu, S., & Liu, G. (1997). Nanometer-scale fabrication by simultaneous nanoshaving and molecular self assembly. *Langmuir*, *13*(2), 127–129. doi:10.1021/la962029f

Xu, T. L., Han, Y., Liu, W., Pang, X. Y., Zheng, B., Zhang, Y., & Zhou, X. N. (2018). Antivirus effectiveness of ivermectin on dengue virus type 2 in Aedes albopictus. *PLoS Neglected Tropical Diseases*, *12*(11), e0006934. doi:10.1371/journal.pntd.0006934 PMID:30452439

Xu, Y., Huang, Y., Zhang, X., Lu, W., Yu, J., & Liu, S. (2018). Carrier-free Janus nano-prodrug based on camptothecin and gemcitabine: Reduction-triggered drug release and synergistic in vitro antiproliferative effect in multiple cancer cells. *International Journal of Pharmaceutics*, *550*(1-2), 45–56. doi:10.1016/j.ijpharm.2018.08.041 PMID:30138703

Xu, Y., Lu, H., & Kang, T. (2017). Effect of different solvent systems on PHBV/PEO electrospun fibers. *RSC Advances*, *7*(7), 4000–4010. doi:10.1039/C6RA26783A

Xu, Y., Wei, M. T., Ou-Yang, H. D., Walker, S. G., Wang, H. Z., Gordon, C. R., Guterman, S., Zawacki, E., Applebaum, E., Brink, P. R., Rafailovich, M., & Mironava, T. (2016). Exposure to TiO_2 nanoparticles increases *Staphylococcus aureus* infection of HeLa cells. *Journal of Nanobiotechnology*, *14*(1), 34. doi:10.118612951-016-0184-y PMID:27102228

Yacaman, M. J., Ascencio, J. A., Liu, H. B., & Torresdey, G. J. (2001). Structure shape and stability of nanometric sized particles. *Journal of Vacuum Science and Technology*, *19*(4), 1091–1103. doi:10.1116/1.1387089

Yadavalli, T., Ramasamy, S., Chandrasekaran, G., Michael, I., Therese, H. A., & Chennakesavulu, R. (2015). Dual responsive PNIPAM–chitosan targeted magnetic nanopolymers for targeted drug delivery. *Journal of Magnetism and Magnetic Materials*, *380*, 315–320. doi:10.1016/j.jmmm.2014.09.035

Yadav, H. K., Almokdad, A. A., Sumia, I. M., & Debe, M. S. (2019). Polymer-based nanomaterials for drug-delivery carriers. In *Nanocarriers for Drug Delivery* (pp. 531–556). Elsevier. doi:10.1016/B978-0-12-814033-8.00017-5

Yadav, V., Paniliatis, B. J., Shi, H., Lee, K., Cebe, P., & Kaplan, D. L. (2010). Novel in vivo- degradable cellulose chitin copolymer from metabolically engineered *Gluconacetobacterxylinus*. *Applied and Environmental Microbiology*, *76*(18), 6257–6265. doi:10.1128/AEM.00698-10 PMID:20656868

Yaghini, E., Turner, H. D., LeMarois, A. M., Suhling, K., Naasani, I., & MacRobert, A. J. (2016). *In vivo* bio-distribution studies and ex vivo lymph node imaging using heavy metal-free quantum dots. *Biomaterials*, *104*, 182–191. doi:10.1016/j.biomaterials.2016.07.014 PMID:27454064

Yamamoto, O. (2011). Influence of particle size on the antibacterial activity of zinc oxide. *International Journal of Inorganic Materials*, *3*(7), 643–646. doi:10.1016/S1466-6049(01)00197-0

Yamauchi, K. A., & Herr, A. E. (2017). Subcellular western blotting of single cells. *Microsystems & Nanoengineering*, *3*(1), 16079. doi:10.1038/micronano.2016.79 PMID:29333327

Yamauchi, Y., & Kuroda, K. (2008). Rational design of mesoporous metals and related nanomaterials by a soft-template approach. *Chemistry, an Asian Journal*, *3*(4), 664–676. doi:10.1002/asia.200700350 PMID:18327878

Yang, H., Liu, C., Yang, D., Zhang, H., & Xi, Z. (2009). Comparative study of cytotoxicity, oxidative stress and genotoxicity induced by four typical nanomaterials: The role of particle size, shape and composition. *Journal of Applied Toxicology*, *29*(1), 69–78. doi:10.1002/jat.1385 PMID:18756589

Yang, J., Lee, C.-H., Ko, H.-J., Suh, J.-S., Yoon, H.-G., Lee, K., Huh, Y.-M., & Haam, S. (2007). Multifunctional magneto-polymeric nanohybrids for targeted detection and synergistic therapeutic effects on breast cancer. *Angewandte Chemie International Edition in English*, *46*(46), 8836–8839. doi:10.1002/anie.200703554 PMID:17943947

Yang, L., Haagensen, J. A., Jelsbak, L., Johansen, H. K., Sternberg, C., Høiby, N., & Molin, S. (2008). In situ growth rates and biofilm development of *Pseudomonas aeruginosa* populations in chronic lung infections. *Journal of Bacteriology*, *190*(8), 2767–2776. doi:10.1128/JB.01581-07 PMID:18156255

Yang, S., Damiano, M. G., Zhang, H., Tripathy, S., Luthi, A. J., Rink, J. S., ... Thaxton, C. S. (2013). Biomimetic, synthetic HDL nanostructures for lymphoma. *Proceedings of the National Academy of Sciences of the United States of America*, *110*(7), 2511–2516. doi:10.1073/pnas.1213657110 PMID:23345442

Yang, X., Dang, Y., Lou, J., Shao, H., & Jiang, X. (2018). D-alanyl-D-alanine- modified gold nanoparticles form a broad-spectrum sensor for bacteria. *Theranostics*, *8*(5), 1449–1457. doi:10.7150/thno.22540 PMID:29507633

Yao, Q., Gao, Y., Gao, T., Zhang, Y., Harnoode, C., Dong, A., Liu, Y., & Xiao, L. (2016). Surface arming magnetic nanoparticles with amine N-halamines as recyclable antibacterial agents: Construction and evaluation. *Colloids and Surfaces. B, Biointerfaces*, *144*, 319–326. doi:10.1016/j.colsurfb.2016.04.024 PMID:27108209

Yao, V. J., D'Angelo, S., Butler, K. S., Theron, C., Smith, T. L., Marchiò, S., ... Bradbury, A. R. (2016). Ligand-targeted theranostic nanomedicines against cancer. *Journal of Controlled Release*, *240*, 267–286. doi:10.1016/j.jconrel.2016.01.002 PMID:26772878

Yar, Y., Khodadust, R., Akkoc, Y., Utkur, M., Saritas, E. U., Gozuacik, D., & Yagci Acar, H. (2018). Development of tailored SPIONPNIPAM nanoparticles by ATRP for dually responsive doxorubicin delivery and MR imaging. *Journal of Materials Chemistry*, *6*(2), 289–300. doi:10.1039/C7TB00646B PMID:32254171

Yashavantha Rao, H. C., Nagendra-Prasad, M. N., Prasad, A., Harini, B. P., Azmath, P., Rakshith, D., & Satish, S. (2016). Biomimetic synthesis of silver nanoparticles using endosymbiotic bacterium inhabiting Euphorbia hirta L. and their bactericidal potential. *Scientifica*. PMID:27403378

Yatvin, M. B., Weinstein, J. N., Dennis, W. H., & Blumenthal, R. (1978). Design of liposomes for enhanced local release of drugs by hyperthermia. *Science*, *202*(4374), 1290–1293. doi:10.1126cience.364652 PMID:364652

Yeh, C. Y., Hsiao, J. K., Wang, Y. P., Lan, C. H., & Wu, H. C. (2016). Peptide-conjugated nanoparticles for targeted imaging and therapy of prostate cancer. *Biomaterials*, *99*, 1–15. doi:10.1016/j.biomaterials.2016.05.015 PMID:27209258

Yehl, K., Lemire, S., Yang, A. C., Ando, H., Mimee, M., Torres, M. T., de la Fuente-Nunez, C., & Lu, T. K. (2019). Engineering Phage Host-Range and Suppressing Bacterial Resistance through Phage Tail Fiber Mutagenesis. *Cell*, *179*(2), 459–469 e459. doi:10.1016/j.cell.2019.09.015 PMID:31585083

Yien, L., Zin, N.M., Sarwar, A., & Katas, H. (2012). Antifungal activity of chitosan nanoparticles and correlation with their physical properties. *Int. J. Biomater.*. doi:10.1155/2012/632698

Yonezawa, R., Kuwana, T., Kawamura, K., & Inamo, Y. (2015). Invasive Community-Acquired Methicillin-Resistant Staphylococcus aureus in a Japanese Girl with Disseminating Multiple Organ Infection: A Case Report and Review of Japanese Pediatric Cases. *Case Reports in Pediatrics*, *291025*, 1–6. Advance online publication. doi:10.1155/2015/291025 PMID:26819794

Yoon, H. J., Kozminsky, M., & Nagrath, S. (2014). Emerging role of nanomaterials in circulating tumor cell isolation and analysis. *ACS Nano*, *8*(3), 1995–2017. doi:10.1021/nn5004277 PMID:24601556

Yoo, S. Y., Jin, H. E., Choi, D. S., Kobayashi, M., Farouz, Y., Wang, S., & Lee, S. W. (2016). M13 Bacteriophage and Adeno-Associated Virus Hybrid for Novel Tissue Engineering Material with Gene Delivery Functions. *Advanced Healthcare Materials*, *5*(1), 88–93. doi:10.1002/adhm.201500179 PMID:26010471

Yosef, I., Manor, M., Kiro, R., & Qimron, U. (2015). Temperate and lytic bacteriophages programmed to sensitize and kill antibiotic-resistant bacteria. *Proceedings of the National Academy of Sciences of the United States of America*, *112*(23), 7267–7272. doi:10.1073/pnas.1500107112 PMID:26060300

Yoshihiko, T., Iichi, Z., Miki, Y., Sumio, C., Juichi, S., Keiji, I., & Kazuhiro, M. (2000). Inhibitory Effect of Dexamethasone Palmitate-Low Density Lipoprotein Complex on Low Density Lipoprotein-Induced Macrophage Foam Cell Formation. *Biological & Pharmaceutical Bulletin*, *23*(4), 466–471. doi:10.1248/bpb.23.466 PMID:10784429

Young, L. S., & Hindler, J. (1986). Aminoglycoside resistance: A worldwide perspective. *The American Journal of Medicine*, *80*(6, Suppl 5C), 35–39. doi:10.1016/0002-9343(86)90474-2 PMID:3717187

Yu, C., Xuan, T., Chen, Y., Zhao, Z., Liu, X., Lian, G., & Li, H. (2016). Gadolinium-doped carbon dots with high quantum yield as an effective fluorescence and magnetic resonance bimodal imaging probe. *Journal of Alloys and Compounds*, *688*, 611–619. doi:10.1016/j.jallcom.2016.07.226

Yue, Y., Li, X., Sigg, L., Suter, M. J., Pillai, S., Behra, R., & Schirmer, K. (2017). Interaction of silver nanoparticles with algae and fish cells: A side by side comparison. *Journal of Nanobiotechnology*, *15*(1), 16. doi:10.118612951-017-0254-9 PMID:28245850

Yu, K., Liu, M., Dai, H., & Huang, X. (2020). Targeted drug delivery systems for bladder cancer therapy. *Journal of Drug Delivery Science and Technology*, *56*, 101535. Advance online publication. doi:10.1016/j.jddst.2020.101535

Yu, M., Bardia, A., Aceto, N., Bersani, F., Madden, M. W., Donaldson, M. C., Desai, R., Zhu, H., Comaills, V., Zheng, Z., Wittner, B. S., Stojanov, P., Brachtel, E., Sgroi, D., Kapur, R., Shioda, T., Ting, D. T., Ramaswamy, S., Getz, G., ... Haber, D. A. (2014). Cancer therapy. Ex vivo culture of circulating breast tumor cells for individualized testing of drug susceptibility. *Science*, *345*(6193), 216–220. doi:10.1126cience.1253533 PMID:25013076

Yu, M., Ting, D. T., Stott, S. L., Wittner, B. S., Ozsolak, F., Paul, S., Ciciliano, J. C., Smas, M. E., Winokur, D., Gilman, A. J., Ulman, M. J., Xega, K., Contino, G., Alagesan, B., Brannigan, B. W., Milos, P. M., Ryan, D. P., Sequist, L. V., Bardeesy, N., ... Haber, D. A. (2012). RNA sequencing of pancreatic circulating tumour cells implicates WNT signalling in metastasis. *Nature*, *487*(7408), 510–513. doi:10.1038/nature11217 PMID:22763454

Yun, S. H., & Kwok, S. J. (2017). Light in diagnosis, therapy and surgery. *Nature Biomedical Engineering*, *1*(1), 1–16. doi:10.103841551-016-0008 PMID:28649464

Zaidi, S., Misba, L., & Khan, A. U. (2017). Nano-therapeutics : A revolution in infection control in post antibiotic era. *Nanomedicine; Nanotechnology, Biology, and Medicine*, *13*(7), 2281–2301. doi:10.1016/j.nano.2017.06.015 PMID:28673854

Zainab, G., Iqbal, N., Babar, A., Huang, C., Wang, X., Yu, J., & Ding, B. (2017). Free-standing,spider-web-like polyamide/carbon nanotube composite nanofibrous membrane impregnated with polyethyleneimine for CO2 capture. *Composites Communications*, *6*, 41–47. doi:10.1016/j.coco.2017.09.001

Zaloga, J., Janko, C., Nowak, J., Matuszak, J., Knaup, S., Eberbeck, D., Tietze, R., Unterweger, H., Friedrich, R. P., Duerr, S., Heimke-Brinck, R., Baum, E., Cicha, I., D¨orje, F., Odenbach, S., Lyer, S., Lee, G., & Alexiou, C. (2014). Development of a lauric acid/albumin hybrid iron oxide nanoparticle system with improved biocompatibility. *International Journal of Nanomedicine*, *9*, 4847. doi:10.2147/IJN.S68539 PMID:25364244

Zaman, S., Hussain, M., Nye, R., Mehta, V., Mamun, K. T., & Hossain, N. (2017). A Review on Antibiotic Resistance: Alarm Bells are Ringing. *Cureus Journal of Medical Science*, *9*(6), e1403. doi:10.7759/cureus.1403 PMID:28852600

Zamora-Mora, V., Fernández-Gutiérrez, M., González-Gómez, Á., Sanz, B., Román, J. S., Goya, G. F., & Mijangos, C. (2017). Chitosan nanoparticles for combined drug delivery and magnetic hyperthermia: From preparation to in vitro studies. *Carbohydrate Polymers*, *157*, 361–370. doi:10.1016/j.carbpol.2016.09.084 PMID:27987939

Zare, E. N., Makvandi, P., & Tay, F. R. (2019). Recent progress in the industrial and biomedical applications of tragacanth gum: A review. *Carbohydrate Polymers*, *212*, 450–467. doi:10.1016/j.carbpol.2019.02.076 PMID:30832879

Zarei, M., Jamnejad, A., & Elahe, K. (2014). Antibacterial Effect of Silver Nanoparticles Against Four Foodborne Pathogens. *Jundishapur Journal of Microbiology*, *7*(1), e8720. doi:10.5812/jjm.8720 PubMed

Zarei, M., & Karbasi, S. (2018). Evaluation of the effects of multiwalled carbon nanotubes on electrospun poly (3-hydroxybutirate) scaffold for tissue engineering applications. *Journal of Porous Materials*, *25*(1), 259–272. doi:10.100710934-017-0439-5

Zarekhalili, S.Z., Bahrami, H.M., Mohammadi, R., & Milan, P.B. (2016).Fabrication and characterization of PVA/Gum tragacanth/PCL hybrid nanofibrous scaffolds for skin substitutes. *International Journal of Biological Macromolecules*, *94*(Pt A), 679-690.

Zarrin, A., Sadighian, S., Rostamizadeh, K., Firuzi, O., Hamidi, M., Mohammadi-Samani, S., & Miri, R. (2016). Design, preparation, and in vitro characterization of a trimodally-targeted nanomagnetic onco-theranostic system for cancer diagnosis and therapy. *International Journal of Pharmaceutics*, *500*(1–2), 62–76. doi:10.1016/j.ijpharm.2015.12.051 PMID:26721723

Zaunbrecher, M. A., Sikes, R. D., Metchock, B., Shinnick, T. M., & Posey, J. E. (2009). Overexpression of the chromosomally encoded aminoglycoside acetyltransferase eis confers kanamycin resistance in Mycobacterium tuberculosis. *Proceedings of the National Academy of Sciences of the United States of America*, *106*(47), 20004–20009. doi:10.1073/pnas.0907925106 PMID:19906990

Zawrah, M. F., & Sherein, I. A. E.-M. (2011). Antimicrobial Activities of Gold Nanoparticles against Major Foodborne Pathogens. *Life Science Journal*, *8*, 37–44.

Zazo, H., Colino, C. I., & Lanao, J. M. (2016). Current applications of nanoparticles in infectious diseases. *Journal of Controlled Release*, *224*, 86–102. doi:10.1016/j.jconrel.2016.01.008 PMID:26772877

Zelga, P. J., Gornicz, M. M., Gluszkiewicz, J. M., & Piasecka-Zelga, J. (2016). Outcomes of acute dermal irritation and sensitisation tests on active dressings for chronic wounds: A comparative study. *Journal of Wound Care*, *25*(12), 722–729. doi:10.12968/jowc.2016.25.12.722 PMID:27974010

Zeng, H., Wasylczyk, P., Parmeggiani, C., Martella, D., Burresi, M., & Wiersma, D. S. (2015). Light-fueled microscopic walkers. *Advanced Materials*, *27*(26), 3883–3887. doi:10.1002/adma.201501446 PMID:26033690

Zhang, B., Yan, X., He, H.W., Yu, M., Ning, X., & Long, Y.Z. (2017). Solvent-free electrospinning: opportunities and challenges. *Polymer Chemistry, 8*(2) 333–352.

Zhang, H., Li, J., Hu, Y., Shen, M., Shi, X., & Zhang, G. (2016). Folic acid-targeted iron oxide nanoparticles as contrast agents for magnetic resonance imaging of human ovarian cancer. *Journal of Ovarian Research*, *9*(1), 19. doi:10.118613048-016-0230-2 PMID:27025582

Zhang, L. H., Duan, X. P., Yan, X., Yu, M., Ning, X., Zhao, Y., & Long, Y.-Z. (2016). Recent advances in melt electrospinning. *RSC Advances*, *6*(58), 53400–53414. doi:10.1039/C6RA09558E

Zhang, L., Abbott, J. J., Dong, L., Kratochvil, B. E., Bell, D., & Nelson, B. J. (2009). Artificial bacterial flagella: Fabrication and magnetic control. *Applied Physics Letters*, *94*(6), 064107. doi:10.1063/1.3079655

Zhang, L., & Gallo, R. L. (2016). a Gallo, RL: Antimicrobial peptides. *Current Biology*, *26*(1), R14–R19. doi:10.1016/j.cub.2015.11.017 PMID:26766224

Zhang, Q., Zhang, K., Xu, D., Yang, G., Huang, H., Nie, F., Liu, C., & Yang, S. (2014). CuO nanostructures: Synthesis, characterization, growth mechanisms, fundamental properties, and applications. *Progress in Materials Science*, *4*(60), 208–337. doi:10.1016/j.pmatsci.2013.09.003

Zhang, S., Wu, L., Cao, J., Wang, K., Ge, Y., Ma, W., & Shen, S. (2018). Effect of magnetic nanoparticles size on rheumatoid arthritis targeting and photothermal therapy. *Colloids and Surfaces. B, Biointerfaces*, *170*, 224–232. doi:10.1016/j.colsurfb.2018.06.016 PMID:29929166

Zhang, W., Li, Y., Niu, J., & Chen, Y. (2013). Photogeneration of reactive oxygen species on uncoated silver, gold, nickel, and silicon nanoparticles and their antibacterial effects. *Langmuir*, *29*(15), 4647–4651. doi:10.1021/la400500t PMID:23544954

Zhang, W., Qiao, X., & Chen, J. (2007). Synthesis of silver nanoparticles-Effects of concerned parameters in water/oil microemulsion. *Materials Science and Engineering B. Solid-State Materials for Advanced Technology*, *142*(1), 1–15. doi:10.1016/j.mseb.2007.06.014

Zhang, X. Y., & Zhang, P. Y. (2017). Polymersomes in nanomedicine-A review. *Current Nanoscience*, *13*(2), 124–129. doi:10.2174/1573413712666161018144519

Zhang, X., He, S., Ding, B., Qu, C., Zhang, Q., Chen, H., Sun, Y., Fang, H., Long, Y., Zhang, R., Lan, X., & Cheng, Z. (2019). Cheng, Z. Cancer Cell Membrane-Coated Rare Earth Doped Nanoparticles for Tumor Surgery Navigation in NIR-II Imaging Window. *Chemical Engineering Journal*, *385*, 123959. doi:10.1016/j.cej.2019.123959

Zhang, X., Hu, S., Wang, M., Yu, J., Khan, Q., Shang, J., & Ba, L. (2015). Continuous graphene and carbon nanotube based high flexible and transparent pressure sensor arrays. *Nanotechnology*, *26*(11), 115501. doi:10.1088/0957-4484/26/11/115501 PMID:25712583

Zhang, X., Young, M. A., Lyandres, O., & Van Duyne, R. P. (2005). Rapid detection of an anthrax biomarker by surface-enhanced Raman spectroscopy. *Journal of the American Chemical Society*, *127*(12), 4484–4489. doi:10.1021/ja043623b PMID:15783231

Zhang, X., Zhi-Guo, L., Wei, S., & Sangiliyandi, G. (2016). Silver Nanoparticles: Synthesis, Characterization, Properties, Applications, and Therapeutic Approaches. *International Journal of Molecular Sciences*, *7*(9), 1534. doi:10.3390/ijms17091534

Zhang, Y., Kohler, N., & Zhang, M. (2002). Surface modification of superparamagnetic magnetite nanoparticles and their intracellular uptake. *Biomaterials*, *23*(7), 1553–1561. doi:10.1016/S0142-9612(01)00267-8 PMID:11922461

Zhang, Y., & Mitchison, D. (2003). The curious characteristics of pyrazinamide: A review. *The International Journal of Tuberculosis and Lung Disease*, *7*(1), 6–21. PMID:12701830

Zhang, Z., Kleinstreuer, C., Donohue, J. F., & Kim, C. S. (2005). Comparison of micro- and nano-size particledepositions in a human upper airway model. *Journal of Aerosol Science*, *36*(2), 123–129. doi:10.1016/j.jaerosci.2004.08.006

Zhan, Z., Wei, F., Zheng, J., Yang, W., Luo, J., & Yao, L. (2018). Recent advances of light-driven micro/nanomotors: Toward powerful thrust and precise. control. *Nanotechnology Reviews*, *7*(6), 555–581. doi:10.1515/ntrev-2018-0106

Zhao, G., & Pumera, M. (2012). Magnetotactic Artificial Self-Propelled Nanojets. *Langmuir*, *29*(24), 7411–7415. doi:10.1021/la303762a PMID:23078532

Zhao, L., Hu, J., Wang, H., Adeleye, A., Ortiz, C., & Keller, A. A. (2017). 1H NMR and GCMS based metabolomics reveal nano-cu altered cucumber fruit nutritional supply. *Plant Physiology and Biochemistry*, *110*, 138–146. doi:10.1016/j.plaphy.2016.02.010 PMID:26922143

Zhao, L., Sun, Y., Hernandez-Viezcas, J. A., Servin, A. D., Hong, J., Niu, G., & Gardea-Torresdey, J. L. (2013). Influence of CeO2 andZnO nanoparticles on cucumber phys-iological markers and bioaccumulation of Ce and Zn: A life cycle study. *Journal of Agricultural and Food Chemistry*, *61*(49), 11945–11951. doi:10.1021/jf404328e PMID:24245665

Zhao, P., Xu, Q., Tao, J., Jin, Z., Pan, Y., Yu, C., & Yu, Z. (2018). Near infrared quantum dots in biomedical applications: Current status and future perspective. *Wiley Interdisciplinary Reviews. Nanomedicine and Nanobiotechnology*, *10*(3), e1483. doi:10.1002/wnan.1483 PMID:28719080

Zhao, X., Hilliard, L. R., Mechery, S. J., Wang, Y., Bagwe, R. P., Jin, S., & Tan, W. (2004). A rapid bioassay for single bacterial cell quantitation using bioconjugated nanoparticles. *Proceedings of the National Academy of Sciences of the United States of America*, *101*(42), 15027–15032. doi:10.1073/pnas.0404806101 PMID:15477593

Zhao, Y., Cao, X., & Jiang, L. (2007). Bio-mimic multichannel microtubes by a facile method. *Journal of the American Chemical Society*, *129*(4), 764–765. doi:10.1021/ja068165g PMID:17243804

Zhao, Y., Tian, Y., Cui, Y., Liu, W., Ma, W., & Jiang, X. (2010). Small molecule-capped gold nanoparticles as potent antibacterial agents that target gram-negative bacteria. *Journal of the American Chemical Society*, *132*(35), 12349–12356. doi:10.1021/ja1028843 PMID:20707350

Zharkova, M. S., Orlov, D. S., Golubeva, O. Y., Chakchir, O. B., Eliseev, I. E., Grinchuk, T. M., & Shamova, O. V. (2019). Application of Antimicrobial Peptides of the Innate Immune System in Combination With Conventional Antibiotics-A Novel Way to Combat Antibiotic Resistance? *Frontiers in Cellular and Infection Microbiology*, *9*, 128. doi:10.3389/fcimb.2019.00128 PMID:31114762

Zha, Z., Zhang, S., Deng, Z., Li, Y., Li, C., & Dai, Z. (2013). Enzyme-responsive copper sulphide nanoparticles for combined photoacoustic imaging, tumor-selective chemotherapy and photothermal therapy. *Chemical Communications*, *49*(33), 3455–3457. doi:10.1039/c3cc40608c PMID:23507786

Zheng, K., Setyawati, M. I., Leong, D. T., & Xie, J. (2017). Antimicrobial gold nanoclusters. *ACS Nano*, *11*(7), 6904–6910. doi:10.1021/acsnano.7b02035 PMID:28595000

Zheng, X. C., Ren, W., Zhang, S., Zhong, T., Duan, X., Yin, Y.-F., Xu, M.-Q., Hao, Y.-L., Li, Z.-T., Li, H., Liu, M., Li, Z.-Y., & Zhang, X. (2018). The theranostic efficiency of tumor-specific, pH-responsive, peptide-modified, liposome-containing paclitaxel and superparamagnetic iron oxide nanoparticles. *International Journal of Nanomedicine*, *13*, 1495–1504. doi:10.2147/IJN.S157082 PMID:29559778

Zhijiang, C., Cong, Z., Jie, G., Qing, Z., & Kongyin, Z. (2018). Electrospun carboxy multiwalled carbon nanotubes grafted polyhydroxybutyrate composite nanofibers membrane scaffolds: Preparation, characterization and cytocompatibility. *Materials Science and Engineering C*, *82*(1), 29–30. doi:10.1016/j.msec.2017.08.005 PMID:29025660

Zhijiang, C., Yi, X., Haizheng, Y., Jia, J., & Liu, Y. (2016). Poly(hydroxybutyrate)/cellulose Acetate Blend Nanofiber Scaffolds: Preparation, Characterization and Cytocompatibility. *Materials Science and Engineering C*, *58*, 757–767. doi:10.1016/j.msec.2015.09.048 PMID:26478369

Zhou, C., Qi, X., Li, P., Chen, W. N., Mouad, L., Chang, M. W., Leong, S. S. J., & Chan-Park, M. B. (2010). High potency and broad-spectrum antimicrobial peptides synthesized via ring-opening polymerization of α-aminoacid-N-carboxyanhydrides. *Biomacromolecules*, *11*(1), 60–67. doi:10.1021/bm900896h PMID:19957992

Zhou, D. K., Li, Y. C. G., Xu, P. T., Ren, L. Q., Zhang, G. Y., Mallouk, T. E., & Li, L. Q. (2017). Visible-light driven Si-Au micromotors in water and organic solvents. *Nanoscale*, *9*(32), 11434–11438. doi:10.1039/C7NR04161F PMID:28786464

Zhou, D. K., Ren, L. Q., Li, Y. G. C., Xu, P. T., Gao, Y., Zhang, G. Y., Wang, W., Mallouk, T. E., & Li, L. Q. (2017). Visible light-driven, magnetically Steerable gold/iron oxide nanomotors. *Chemical Communications*, *53*(83), 11465–11468. doi:10.1039/C7CC06327J PMID:28983536

Zhou, J., Leuschner, C., Kumar, C., Hormes, J. F., & Soboyejo, W. O. (2006). Sub-cellular accumulation of magnetic nanoparticles in breast tumors and metastases. *Biomaterials*, *27*(9), 2001–2008. doi:10.1016/j.biomaterials.2005.10.013 PMID:16280161

Zhu, C., & Xia, Y. (2017). Chunlei zhu and Younan Xia, "Biomimetics: Reconstitution of low-density lipoprotein for targeted drug delivery and related theranostic applications. *Chemical Society Reviews*, *46*(24), 7668–7682. doi:10.1039/C7CS00492C PMID:29104991

Zhu, M., Nie, G., Meng, H., Xia, T., Nel, A., & Zhao, Y. (2013). Physicochemical properties determine nanomaterial cellular uptake, transport, and fate. *Accounts of Chemical Research*, *46*(3), 622–631. doi:10.1021/ar300031y PMID:22891796

Zhuo, Y., Yuan, R., Chai, Y., Tang, D., Zhang, Y., Wang, N., Li, X., & Zhu, Q. (2005). A reagentless amperometric immunosensor based on gold nanoparticles/thionine/Nafion-membrane-modified gold electrode for determination of α-1-fetoprotein. *Electrochemistry Communications*, *7*(4), 355–360. doi:10.1016/j.elecom.2005.02.001

Zhu, W., Murray, P. R., Huskins, W. C., Jernigan, J. A., McDonald, L. C., Clark, N. C., Anderson, K. F., McDougal, L. K., Hageman, J. C., Olsen-Rasmussen, M., Frace, M., Alangaden, G. J., Chenoweth, C., Zervos, M. J., Robinson-Dunn, B., Schreckenberger, P. C., Reller, L. B., Rudrik, J. T., & Patel, J. B. (2010). Dissemination of an Enterococcus Inc18-Like vanA plasmid associated with vancomycin-resistant Staphylococcus aureus. *Antimicrobial Agents and Chemotherapy*, *54*(10), 4314–4320. doi:10.1128/AAC.00185-10 PMID:20660665

Zimmer, A. T., Baron, P. A., & Biswas, P. (2002). The influence of operating parameters on number-weighted aerosol size distribution generated from a gas metal arc welding process. *Journal of Aerosol Science*, *33*(3), 519–531. doi:10.1016/S0021-8502(01)00189-6

Zorova, L. D., Popkov, V. A., Plotnikov, E. J., Silachev, D. N., Pevzner, I. B., Jankauskas, S. S., Zorov, S. D., Babenko, V. A., & Zorov, D. B. (2018). Mitochondrial membrane potential. *AnnalBiochem*, *552*, 50–59. PMID:28711444

About the Contributors

Muthupandian Saravanan Graduated in Microbiology from Madurai Kamaraj University, India and Doctorate with Specialization in Medical Microbiology and Nanomedicine from Sathyabama University, India. Thereafter, He worked as a Post Doctoral Researcher through Israel Government postdoctoral research fellowship in 2011-2012 in Institute of Drug Research, Faculty of Medicine, The Hebrew University of Jerusalem, Israel, focusing his research on Nano-biomaterials & their Biomedical Applications. Prior to postdoc, he worked as an Assistant Professor (SG), SRM University, and Department of Biotechnology for six Years, from 2005 -2011. Presently, Dr. Saravanan working as a Professor, Department of Pharmacology Saveetha Dental college, SIMATS, Chennai, India and worked as Associate Professor Department of Medical Microbiology and Immunology, Division of Bio-medical Sciences, School of Medicine, College of Health Science, Mekelle University, Ethiopia (2012-2021). His Research Specialization: Development of Novel biomaterials Emerging and Re-emerging Infectious disease and Cancer. He has published more than 150 research paper in Peer-reviewed Journal including high impact Journal with more than 7500 citations and h-index of 42 and i10 index of 80 and also three edited books. He has participated more than 75 National and International conferences, Reviewers of more than 100 international peer-reviewed journals and Guest editor/ Editors for various reputed PubMed and Scopus indexed journals. He has received many fellowships and awards notably, IET- Nanobiotechnology premium Awards in two times continuously in the year 2019 and 2020, DST-SERC Young Scientist fellowship in 2011, International Fellowship "Advanced Course on Diagnostics" Sponsored by LSH&TM & Fondation Mérieux, in France 2013, International Fellowship "Pertussis: biology, epidemiology and prevention" meeting Sponsored by Fondation Mérieux & WHO in France 2014, International Union of Microbiological Societies (IUMS) travel grant in 2015 to Canada, International Fellowship "Advanced Course on Antibiotics" (AdCAb) Sponsored by Institute of Pasteur and Fondation Mérieux France, 2016. His area of research includes Medical Microbiology, Antimicrobial Resistance, Nanomedicine, Nanomaterials to combat AMR and Cancer.

* * *

Toqeer Ahmed is currently working as Assistant Professor, Centre for Climate Research and Development (CCRD), COMSATS University Islamabad. Since his education, he has worked at responsible positions in different Government departments and R & D organizations. He has completed several national as well as international projects in collaboration with the University of Edinburgh, UK and International NGOs. He has 34 research publications (32.153 IF) in peer reviewed research journals and 4 book chapters in the field of food and water quality assessment, wastewater treatment, climate impacts on health and Nano-biotechnology. Dr. Ahmed has lectured and presented Research Papers in both national and international conferences. He is currently working in the field of Environmental Microbiology, Climate Change, Water Management and Environmental Nanotechnology. He is on the Editorial Boards of few International impact factor Journals and active member of scientific societies.

Juhee Ahn received his Ph.D. degree majoring in Food Microbiology at the University of Missouri, Columbia, USA and continued his work as a post-doctoral research assistant in Food Microbiology Lab at the University of Missouri (2003-2004) and High pressure processing and Food Safety Engineering Lab at the Ohio State University (2004-2006). He was a visiting scholar at the University of Maryland (2012-2013) as well as Zhejiang University (2016-2017). Currently, he is a professor at the department of Medical Biomaterials Engineering, Kangwon National University, South Korea. His primary research field is the microbial pathogenesis, including the mechanistic studies of antibiotic resistance, bacterial infection, and bacteriophage control. Research Interest: Microbial Pathogenesis: - Antibiotic resistance mechanisms - Biofilm formation - Bacteriophage control.

Palanisamy Babu has completed his Ph.D. in Biotechnology from Anna University Chennai, Tamil Nadu, India. He is currently working as Senior Lecturer at the Department of Biotechnology, School of Mathematics and Natural sciences, The Copperbelt University, Kitwe, Zambia. And also as Adjunct Lecturer at Faculty of Pharmaceutical Sciences, UCSI University, Kuala Lumpur Malaysia. He also worked as Associate Professor at the Department of Biotechnology, Addis Ababa Science and Technology University, Addis Ababa, Ethiopia. His main area of research is Pharmaceutical Biotechnology and Environmental Biotechnology. He has published several scientific research articles in international peer-reviewed journals and has authored various book chapters and review papers. He has bagged the Young Scientist Award from the University of Cagliari, Italy for best presentation, and various recognition from different national and international organizations for his academic and scientific contributions. He served as an Executive Committee Member (an honorary position) in Indian Education Forum (IEF) organized by the Embassy of India, Addis Ababa, Ethiopia for 3 years since 2017. Dr. Suresh is an associate member of many scientific societies and professional bodies.

R. Balachandar is working as an assistant professor in Aarupadai Veedu Institute of Technology, Chennai.

Mythili Gnanamangai Balasubramanian finished her academic excellence in biotechnology and continue to do research in nanotechnology, biotechnology, biomedical and environmental biotechnology. She has developed many biomedical products in her research career.

Malathi Balasubramaniyan received her PhD in Nanobiotechnology from Anna University, Chennai. She is currently working as Assistant Professor in Bannari Amman Institute of Technology, Sathyamangalam. Her main research interest is synthesising nanocarriers for efficacious drug & vaccine delivery against infectious diseases. She has received TNSCST-SPSR grant for the year 2019-2020.

Melaku Belete is a lecturer, researcher and community service provider in Department of Medical Laboratory Science, College of Medicine and Health Sciences, Wollo University, Ethiopia.

Mohd Yasin Bhat has done his PhD from Department from Department of Plant sciences, School of life sciences, University of Hyderabad. During his PhD, he has worked on bioactive constituents of medicinal plant for cancer and inflammation. He has several national and international publications to his name apart from attending several scientific workshops and conferences.

Arghya Chakravorty is 25 years old young, passionate & enthusiastic researcher, as well as a registered PhD scholar, enrolled at Centre for Nanotechnology Research, Vellore Institute of Technology - Vellore; holding four Indian patents, while his research interest is interdisciplinary material science & nanobiotechnology along with to deals with human resources and professional relationships. He contributed to a total number of ten book chapters to different renowned international publishers like Elsevier, Springer, CRC Press, and IGI Global - USA; three research papers, and three review papers. In this year he have been added as a reviewer of one international standard book and six reputed journals, which are indexed in SCOPUS, WOS, SCIE, ESCI etc.; and as an Editorial Board Member at "Algerian Journal of Renewable Energy and Sustainable Development". Recently he has worked with the world's pioneer nanotechnologist Prof. Seeram Ramakrishna & the manuscript has been published by a renowned journal of Elsevier. Before joining Ph.D., he has earned first-class M.Sc in Microbiology from HNB Garhwal (Central) University, Uttrakhand as well as PG Diploma in Aviation & Hospitality Services, and served himself as PGT - Biology at Brilliant Stars School (Affiliated by CBSE – New Delhi), Udaipur, Tripura. As Principal Investigator, he has also guided two non-profit projects in the field of public health & community medicine in collaboration with North University of China & HUST Wuhan as an Adjacent Research Associate of SANCHARAN, (A socio-cultural organization) 4289/2003, Agartala 799001, India; those projects were successfully presented at RECYCLE 2020, hosted by IIT- Guwahati.

Sudharshan Reddy Dachani is working as Associate Professor of Pharmacy Practice & Pharmacology, College of Pharmacy at Shaqra University, Kingdom of Saudi Arabia. He has 14 years of Academic, Research & Clinical Training Experience in Reputed Universities and Organizations of India as well as Abroad. He specialized in Pharmacology and offers courses to U.G, P.G and Ph.D. scholars, guided 30 PharmD, M. Pharm 18 students. He has published more than 26 papers in International and National Research Peer Reviewed Journals, Attended Invited Oral/Poster Presentations, Conferences, Seminars and workshops. Research interests focus on Diabetes and Metabolic disorders In-Vivo Preclinical Screening Methods and Toxicological Studies. Current Research Focuses on the Medication Related Problems/ Drug safety, Drug Interaction, Adverse Reporting and Medication Adherence at Clinical Studies. He has been serving as a Director of Pranathi Herbals PVT Ltd, Editorial board member of Pharmaceutical Innovations. Life member of various Professional Affiliations and Bodies/organizations

Maheswaran Easwaran received his doctoral degree in veterinary medicine in 2016 under the supervision of Prof. Dr. Hyun-Jin Shin at Chungnam National University, South Korea. He continued his research as a postdoctoral researcher in the same lab for 1 year. Then, he moved to Kangwon National University, South Korea, to continue his research as a senior researcher with Prof. Dr. Juhee Ahn. He is actively involved in various infectious diseases such as coronavirus and multidrug resistant bacterial pathogens.

Laxmikant Gautam is currently pursuing his PhD under the supervision of Prof. Suresh P. Vyas at Drug Delivery Research Laboratory, Department of Pharmaceutical Sciences, Dr. Harisingh Gour University, Sagar, M.P., India. He is awarded with ICMR-SRF Fellow in 2018 for the cancer project. He is currently working on site-directed drug delivery system(s), and Nanotechnology.

Mir Zahoor Gul is currently working as Post-Doctoral Fellow at the Department of Biochemistry, University College of Science, Osmania University, Hyderabad, India. He has earned his Ph.D. degree at School of Life Sciences, University of Hyderabad, India in the year 2015. His research interests lie in the area of Nanophytomedicine, Drug delivery, Nanobiocomposites, Nanobioremdiation of synthetic wastes, to name a few. Dr. Gul has over 30 research publications in peer reviewed international journals, including 10 book chapters in edited volumes with international publishers. Dr. Gul is a recipient of several national fellowships for his research. endeavours.

Muhammad Zeeshan Hyder is working as Associate Professor at Department of Biosciences, CUI, Islamabad.

Anamika Jain is pursuing her Ph.D. under the supervision of Prof. S. P. Vyas at Drug Delivery Research Laboratory, Department of Pharmaceutical Sciences, Dr. H.S. Gour University, Sagar (MP), India. She is currently working on the targeted drug delivery system(s).

Yasodha Krishna Janapati is working as Associate Professors at the School of Pharmacy & Health Sciences, USIU – AFRICA. He has 15 years of unblemished service in Teaching, Research, and extension in reputed Universities of India as well as abroad. He is specialized in Pharmaceutical Chemistry and offers courses to U.G, P.G and Ph.D. scholars. He established U.G and P.G laboratories with modern teaching aids to maximize the learning ability of the students. In addition to the above, he published more than 30 articles in reputed journals, guided 17 M. Pharm and 2 Ph.D. scholars in their research Programmes, life member for various professional organization, editor and reviewer for various journals.

Sunil Junapudi is working as Senior Faculty member in the Department of Pharmaceutical chemistry at Geethanjali college of Pharmacy, Hyderabad, India. He completed PhD in Pharmaceutical Sciences from JNTUK Kakinada, India. He served as editorial board and reviewer for a number of national and international journals. He has more than 12 years of teaching and research experience at the university level. International Publications - 22: National Publications-11, one patient.

Muddukrishnaiah K. is an Indian Council of Medical Research Senior Research Fellow.

Hemalatha Karuppiah is currently working as Professor in Sethu Institute of Technology, Tamilnadu, India. She completed her M.Tech and PhD in Indian Institute of Technology Madras in the field of Biomedical Engineering. She has nearly 25 years of experience in teaching and research. His areas of research are Computational physiology, sensor design, signal and image processing and Biomedical nanotechnology.

Mohd Hashim Khan completed his B. Sc.(Hons), M. Sc. (Microbiology) PDTC (Tissue Culture) from Aligarh Muslim University, Aligarh- UP, India. Ph. D. (Microbiology) from B.S. Abdur Rahman Crescent Institute of Science and Technology, Vandalur, Chennai- Tamil Nadu- India. Currently working on formulations of Dual nanomaterials to combat Durg resistance in human ESKAPE pathogens.

Parameswaran Kiruthika Lakshmi has been working as Assistant Professor in the Department of Microbiology, The Madura College, Madurai. She graduated her Ph.D degree in Microbiology from Madurai Kamaraj University and received CSIR-Senior Research Fellowship for her research work. Previously, she worked as Research Associate in an Indo-UK (DBT-BBSRC, UK sponsored) Major Research Project under the guidance of Dr. S. Srisudha, Head, Department of Botany and Microbiology, Lady Doak College, Madurai and has got four years of Post-Doctoral research experience. Totally, she has got 9 years of research experience in the field of Microbiology and is a Lifetime member in Indian Association of Medical Microbiologists (IAMM). She possessed a major passion towards molecular diagnosis and development of therapeutic molecules for treating high-risk infectious agents.

Subbiah Latha completed a bachelor's degree in pharmacy at the Dr.M.G.R. Medical University, Chennai, a master's degree in pharmacy and a doctor of philosophy in pharmacy at Jadavpur University, Kolkata, India. Currently, Dr. Latha is an Associate Professor in the Department of Pharmaceutical Technology at Anna University. She has 20 years of experience in higher education and research gained at Jadavpur University, Bharathidasan University, and Anna University. Dr. Latha has broad knowledge in Pharmaceutics / Pharmaceutical engineering and Pharmacology. Her research interests include the development of novel drug carriers using magnetic nanoparticles for targeted drug delivery. Dr. Latha's lab has received competitive research funds of ₹75.52 Lakhs. She is a visiting scientist to Technische Universität, Ilmenau, Germany, Yokohama National University, Japan and Shizuoka University, Japan. She is the author or co-author of over 63 research publications and 12 book chapters and edited a book titled 'Nanobiopharmaceutical Technology–Applications and Perspectives' published by Elsevier'. She has supervised over 1 Post-doc, 3 PhD and 42 Master's students. She has been the recipient of the 2008 Tamil Nadu Young Scientist Fellowship for Young Scientists; the 2014 Tamil Nadu Young Women Scientist Award in Medical Sciences Discipline in recognition of outstanding research in Pharmacy; the 2014 Late Smt. Shyama Dahiya Memorial Award for the Outstanding Pharmacy Teacher and 2018 Young Pharmacist Award from the Association of Pharmacy Professionals. She is a life member of many professional bodies.

Irfan Liaqat did his MS in Biochemistry & Molecular Biology from COMSATS University Islamabad and worked on the Impacts of nanostructures against vectors causing dengue fever under the changing climate.

Karthikeyan Mahendhran is working as Assistant Professor in Department of Zoology and Microbiology, Thiagarajar College, Affiliated to Madurai Kamaraj University, Madurai, Tamil Nadu, India. He obtained his MSc., PhD., from Madurai Kamaraj University, Madurai, Tamil Nadu, India in the field of Microbiology. He has 8 years of teaching and research experience. He worked as a Visiting faculty for M.Tech Environmental Engineering Programme for period of 2 years at Thiagarajar College of Engineering, Madurai. He has awarded Teachers Associate ship for Research Excellence (TARE Award) from Department of Science and Technology, Science and Engineering Research Board (DST-SERB)

Govt of India. He is a recipient of Indian National Science Academy (INSA) Visiting Scientist Fellowship-2018 and worked as a visiting scientist at National Institute of technology, Karnataka, India. He worked as Junior Research fellow in Ministry of Environment and Forest funded Major research project on Biopolymer production. He is a recipient of Summer Research Fellowship Award sponsored by Indian Academy of Sciences and Tamil Nadu State Council for Science and Technology and Department of Biotechnology, Govt of India sponsored Student project Fellowship award. He has received different research projects sponsored by State and National funding agencies. His main research interest focuses on the commercial applications of microbial metabolites in medical and environmental applications, computational studies of Indian traditional medicines for therapeutic applications. He has published many research articles and book chapters in international reputed journals with good impact factors.

Sathish Kumar Marimuthu graduated from Bharathidasan University with a Master's degree in Bioinformatics, now pursuing PhD at Anna University.

Lakshmipathy Muthukrishnan has done his Master's degree in Applied Microbiology from the University of Madras in 2008 and completed his Doctorate degree in the field of Life Sciences from Sathyabama University in 2016. The author has an overall research experience of 12 years at various research institutes. The author has started his research career in the Department of Virology, King Institute of Preventive Medicine and Research as a Research Assistant in National Polio Surveillance Project (NPSP) funded by World Health Organization (WHO) from 2008 – 2010. Then he was offered JRF in Sathyabama University and worked for a DBT project on encountering multidrug resistant strains of bacteria using silver nanoparticles. In the meantime the author pursued Ph.D under the guideship of Prof. Dr. Anima Nanda and awarded with the same during 2016. The author has worked in ICMR funded Viral research and Diagnostic laboratory (VRDL) project in the diagnosis of Swine flu. The author was conferred with CSIR – Research Associateship (RA) at CSIR – CLRI under the mentorship of Chief Scientists, Dr. C. Rose and Dr. C. Muralidharan from 2017 to 2020. Currently, the author is working as an Assistant Professor (Research) in the Department of Conservative Dentistry and Endodontics, Saveetha Dental College & Hospitals, Saveetha Institute of Medical and Technical Sciences, Chennai, Tamil Nadu, India till date. The author has published 25 research articles and 3 book chapters in various peer reviewer international journals and books. The author has a life-membership in various bodies such as Indian Association of Applied Microbiology, Indian Association of Nanoscience and Nanotechnology and International Society for Fluoride Research Inc. Passion toward interdisciplinary research to tackle societal issues related to emergence of drug resistant strains by converging Microbiology, Microbial Nanotechnology and Nanobiotechnology remains his vision.

Ramovatar Meena is presently working as an Assistant Professor at the School of Environmental Sciences, Jawaharlal Nehru University, New Delhi, India.

Abdul Nazeer has the baccalaureate degree on Biotechnology and pursued his Masters in Nanotechnology from Anna University, Chennai. The gradient blend of biotechnology and nanotechnology has helped him understand and explore the nanoscience of the biotech concepts. He did his PhD thesis work on the influence of biosurfactant on the nanocarriers to improve the pharmacokinetics and bioavailability of the drugs. Having worked as the Research Fellow in Nano-Bio Translational Research Laboratory, Erode, India, he has gained the expertise in fabricating nanoparticles, nanoemulsions, and vesicular drug delivery systems for drug delivery, larvicidal and repellent actions, and treatment of chlorosis in plants. He has publications on all these topics in peer-reviewed journals apart from 2 patents on herbal phytosome and herbal mosquito repellent naphthalene balls.

Selvamani Palanisamy completed a bachelor's degree in pharmacy at the Dr.M.G.R. Medical University, Chennai, a master's degree in pharmacy and a doctor of philosophy in pharmacy at Jadavpur University, Kolkata, India. Currently, Dr. Selvamani is an Associate Professor in the Department of Pharmaceutical Technology at Anna University. He has 18 years of experience in higher education and research gained at Jadavpur University, Bharathidasan University and Anna University. Dr. Selvamani has broad knowledge in Medicinal Chemistry, Pharmaceutical Analysis and Pharmacology. His research interests include the discovery and development of natural antimicrobials, novel drug carriers using magnetic nanoparticles for targeted drug delivery. Dr. Selvamani's lab has received competitive research funds of ₹761.58 Lakhs. He is a visiting scientist to Technische Universität, Ilmenau, Germany, Yokohama National University, Japan and Shizuoka University, Japan. He is the author or co-author of over 80 research publications and 12 book chapters and edited a book titled 'Nanobiopharmaceutical Technology–Applications and Perspectives' published by Elsevier'. He has supervised over 2 Post-doc, 3 PhD and 42 Master's students. He has been the recipient of the 2015 Tamil Nadu Young Scientist Fellowship for Young Scientists; the 2015 Young Investigator Award from the Association of Pharmacy Professionals and 2013 Active Researcher Award of Anna University, Chennai. He is a life member of many professional bodies.

Sathish Kumar Perumal, Post-Doctoral fellow, Department of Pharmaceutical Technology, Bharathidasan Institute of Technology Campus, Anna University, Tiruchirappalli, Tamil Nadu, India. Ph.D. in Plant Biotechnology, Department of Plant Science, Bharathidasan University, Tiruchirappalli, Tamil Nadu, India. Published More than 8 Research articles and 1 Book chapter in national and international journals.

Shanmugam Poornima specialized in chemical engineering and keen interest in nanotechnology. She has published more than ten research papers in peer reviewed journals and submitted ten nucleotide sequences to NCBI, Maryland, USA.

Surya Narayan Pradhan is presently working at School of Life Sciences, Sambalpur University, Odisha, India.

Paulraj R. currently works at the School of Environmental Sciences, Jawaharlal Nehru University. Paulraj does research in Cancer Research, Molecular Biology and Cell Biology. Their current project is 'In vitro and in vivo study' on nanotechnology.

Nageshwari Raja graduated in Biotechnology from St.Peter's Engineering College and Master Degree with specialization in Biotechnology from Anna University. Thereafter she worked as Research Fellow for 2 years (2013 - 2015) in the Department of Immunology, Madurai Kamaraj University, India, focusing her research on HLA immunogenetics and continued as a Research Coordinator in Alpha Hospital & Research Center for 1 year (2015 - 2016) and formulated a molecular diagnostic kit for congenital adrenal hyperplasia. For the past two years (2016 - 2018), she worked as a Lecturer in Biochemistry, Meenakshi Mission Hospital & Research Center, Madurai, in actively all areas of academics, engaging and challenging paramedical students to realize their full potential and develop their skills. Presently, Mrs. Nageshwari Raja working as an Assistant Professor for the Department of Biomedical Engineering, Sethu Institute of Technology, India. Her Research Specialization: Development of molecular diagnostic kit for endocrinology related disorders and design of biomedical devices and machines for diagnosing medical problems and solving the societal issues.

Murugappan Ramanathan is an Associate Professor in Zoology and Microbiology, Thiagarajar College, Madurai, Tamil Nadu India with a rich teaching and research experience of 14 years with 25 publications. Received several awards and accolades to list a few, • Overseas Associateship Award for Young Scientists (Young Scientist Fellowship Award) in Niche areas of Biotechnology; 2006-2007, sponsored by DBT. Underwent training on Molecular Techniques, Stem Cell Research at National University of Singapore. • Indian National Science Academy (INSA) Fellowship to visit The James Hutton Institute, Scotland UK. International Collaboration/Exchange Programme 2016. • Young Scientist Fellowship from Tamil Nadu State Council for Science and Technology (TNSCST). Received four major research projects sanctioned by Department of Science and Technology (DST), Ministry of Environment and Forest (MoEF), Ministry of Earth Science (MoES) and University Grants Commission (UGC), India respectively and four minor research projects. Seven students received doctoral studies under the guidance of Dr. Murugappan Ramanathan. Apart from research projects and publications Dr. Murugappan Ramanathan is a member in Science forums viz., Association of Microbiologists of India (AMI), Asian PGPR society of Sustainable Agriculture etc., Project Review Committee (Task force) Member-DSIR-DST, Ministry of Science and Technology, New Delhi for the TDUPW project. Organised two National level workshops and twenty conference. Presented research findings at International and National level conferences.

Subbaiya Ramasamy completed his UG, PG in Microbiology and Ph.D in Bioengineering and published more than 40 Research and Review articles.

Gulzar Ahmed Rather works in the department of biomedical engineering Sathyabama University Rajiv Gandhi Salai, Chennai, Tamil Nadu, 600119 India as a full-time Ph.D. Scholar. He has registered the department in the year 2017. He is working on nanobiotechnology under the guidance of Prof. Anima Nanda, Dean, and Prof. in the Department of Biomedical Engineering. He has completed his M.Sc in Biochemistry in 2014, from the Department of Biochemistry, Kashmir University. He has done the project on the level of evaluation of knowledge, attitude, and behavior of people about Alternative Medicine. He has got more than two-years of teaching experience as a biology lecturer, from 2014-2016. He has attended the 8th JK Science Congress at the University of Kashmir. International seminar on Accu-Pressure at the University of Kashmir. Expert Lecture on Stem Cell Therapy for Brain Tumors by Dr. Khalid Shah, Faculty at Harvard Medical School, Boston. Attend Fivedays First National workshop on techniques in biotechnology (NWTBT-17) Which was organized by the Department of Biotechnology, School of Bio and chemical engineering Sathyabama University (18th to 22nd, 2017). Attended three days National workshop on Animal cell line techniques (NACL-2017) 22nd to 24th August 2017. He has received the best paper award in a 'International conference on Bioengineering on health and environment-2018. He has attended and published papers in the 6thInternational Conference on Nanotechnology for the better living 7-11April 2019 jointly organized by the National Institute of Technology (NIT) Srinagar and the Indian Institute of Technology (IIT) Kharagpur. Besides doing a Ph.D. he is also teaching students' biochemistry at B.Tech. level in the Department of Biomedical Engineering, from time to time he also attended ISRO exam duties as an invigilator.

Pandi Sakthieaswari is working as an Assistant Professor in the Department of Botany and Microbiology, Lady Doak College, Madurai since 2009. She had completed her Ph.D in the field of Biosynthesis of Nanoparticles. She is passionate towards doing research with the experience of guiding nearly 20 PG projects and 10 UG projects in various fields like Nanoscience, Microbiology, Phytochemistry and Environmental Science. She also received an opportunity to write a chapter under the title 'Natural Resources' in the book titled 'Textbook of Environmental Studies' published and released by Centre for Environmental Studies, Lady Doak College.

Palanisamy Selvamani completed a Bachelor's degree in Pharmacy at the Dr.M.G.R. Medical University, Chennai, a Master's degree in Pharmaceutical Engineering and a Doctor of Philosophy in Pharmacy at Jadavpur University, Kolkata, India. Currently, Dr.Selvamani is an Associate Professor in the Department of Pharmaceutical Technology, at Anna University with 20 years of experience in academic research involved in the identification and evaluation of Phyto-drugs for Rheumatoid arthritis, Alzheimer's disease and antimicrobial peptides from marine sources. He is a visiting scientist to Technische Universität, Ilmenau, Germany, Yokohama National University, Japan and Shizuoka University, Japan and the recipient of the 2014 Active Researcher Award of Anna University.

Hyun-Jin Shin has completed his PhD from University of Minnesota and Post-doctoral studies from Northwestern Medical School. He is the member of current research workers in animal diseases, Korean Society of Viruses, Korean Society of Veterinary Medicine. He has published more than 40 papers in reputed journals and has been serving as an Editorial Board Member of Journal of Veterinary Science.

Priya Shrivastava is currently pursuing her PhD under the supervision of Prof. Suresh P. Vyas at Drug Delivery Research Laboratory, Department of Pharmaceutical Sciences, Dr. Harisingh Gour University, Sagar (M.P.), India. She is Gold Medallist in B. Pharm and M. Pharm from Dr. Harisingh Gour University, Sagar and has been awarded with Pharmaceutical Science Alumni Award, Prof. C. S. Chauhan gold medal and Nagarjuna Award from Dr. Harisingh Gour University Sagar (M.P.). She has received prestigious 'M. P. Young Scientist Award 2019' in the field of pharmaceutical and medical sciences. She is currently working on targeted drug delivery and combination drug therapy for the treatment of tuberculosis.

Chockaiyan Usha has been working as an Assistant Professor of Biotechnology at Lady Doak College, Tamil Nadu. Her Doctoral research work was on phytofabrication of silver nanoparticles from medicinal plants and its therapeutic applications. Her research arena is to fabricate and evaluate the application of wound dressing materials impregnated with nanomaterials. Dr. Usha, has published many articles in Peer reviewed and International journals. She is the recipient of Summer research fellowship award sponsored the Indian Academy of Sciences and worked under the supervision of Dr. Kasbekar, Centre for DNA Fingerprinting & Diagnostics, Hyderabad. She is serving as Reviewer in International journal of Science & Research, International journal of Life Science and Pharma Research. She also holds as Life Member in Asian PGPR Society of Sustainable Agriculture.

A. Vigneshwaran is currently pursuing Master of Technology in Biotechnology in School of Biosciences and Technology, Vellore Institute of Technology, Katpadi, Vellore district, Tamil Nadu, India. He has published a research article in a peer reviewed journal and submitted three nucleotide sequences to NCBI, Mary Land, USA. He served as a member of IEEE professional society.

Sudarshana Vijaykumar is a dynamic researcher with 21 valuable publications accumulating 267 citations along with two patents on on herbal phytosome and herbal mosquito repellent naphthalene balls. She was the Head of the Nano-Bio Translational Research Laboratory and the recipient of the grants from DST-SERB (YSS/2014/000637), and TNSCST. She has supervised two PhD candidates on Nanopharmaceutical stream. Her research expertise has an interesting spectrum of plant tissue culture in PhD and Nano-Biotechnology array in post PhD research career. Her science communication skill has been widely appreciated and she has been invited for the talks in well reputed Institutes in and outside India. Her contribution to the science forum is growing to be prominent as she is producing a lot of scientific students delivering the ethical research towards the wellbeing of the society.

Vimalraj Vinayagam received his PhD in Organic chemistry from Annamalai University, Chidambaram. He is currently working as Assistant Professor in Bannari Amman Institute of Technology, Sathyamangalam. He has a background of strong research and publications in the field of organic chemistry. His current interest and research are oriented towards the application of nanochemistry for medical applications.

Nikhar Vishwakarma is currently pursuing his PhD under the supervision of Prof. Suresh P. Vyas at Drug Delivery Research Laboratory, Department of Pharmaceutical Sciences, Dr. Harisingh Gour University, Sagar, M.P., India.

Suresh P. Vyas is a professor in the Department of Pharmaceutical Sciences, Dr. Hari Singh Gour Vishwavidyalaya, Sagar, India. He has about 36 years of teaching and research experience at postgraduate level. He is pioneer scientist in the field of nano-biotechnology and immunology. He has judiously exploited bioligands for targeting of bioactives and antigens. He has supervised 53 PhDs and more than 147 M. Pharm students. He has over 338 research publications to his credit published in journals of high scientific impact. He authored 19 reference books including Pharmaceutical Biotechnology, Liposomal Therapeutics, Novel Controlled Drug Delivery Systems, Targeted Drug Delivery Systems, Theory and Practices in Novel Drug Delivery Systems, contributed many chapters in various renowned books and also granted Indian patent on "Bile salt stabilized non-ionic surfactant vesicles". He is a commonwealth postdoctoral fellowship awardee and has worked under fellowship at School of Pharmacy University of London (U.K.). Prof. Vyas is a recipient of Best Scientist award by Association of Pharmacy Professionals (APP) 2014, National Teacher Excellence Award by IES group of Institution, 2014, Pandit Lajja Shankar Jha award, Best Pharmacy Teacher of the year 2001 by Association of Pharmaceutical Teachers of India and many more. He has been on various administrative positions of the university including Dean, Faculty of technology; Head, Department of Pharmaceutical Sciences and Department of Biotechnology, Chairman, board of studies etc. He serves on the editorial board of International Journal of Pharmaceutics (Elsevier), Recent Patents on Drug Delivery and Formulation (Bentham Science Publishers), Therapeutic Delivery (Future Science Groups), Journal of Drug Delivery (Hindawi) and Journal of Nanoscience Letters (Simplex publishers). Prof. Vyas is widely visited scientist and delivered invited/

popular/keynote addresses in national and international conferences in India and abroad and ranked first in India for research impact, citations (16,927), h-index (69) and i-10 index (238) evaluated and published by Department of Science and Technology, Government of India, 2016. His area of research includes developing nanocarriers for site specific drug delivery, oncology and Immunology.

Index

D

E

G

H

I

M

N

P

Q

R

S

T

www.ingramcontent.com/pod-product-compliance
Lightning Source LLC
LaVergne TN
LVHW080530300926
849162LV00007B/1576